Title VI

The Encyclopedic Dictionary of

Physical Geography

The Encyclopedic Dictionary of
Physical Geography

Second Edition

Edited by

Andrew Goudie

B W Atkinson
K J Gregory
I G Simmons
D R Stoddart
David Sugden

Copyright © Basil Blackwell Ltd 1985, 1994
© Editorial organization Andrew Goudie 1985, 1994

First edition published 1985

Published (with corrections) in paperback 1988
Reprinted 1990

Second edition, revised and updated, first published 1994

Basil Blackwell Ltd
108 Cowley Road, Oxford OX4 1JF, UK

Basil Blackwell Inc.
238 Main Street
Cambridge, Massachusetts 02142, USA

British Library Cataloguing in Publication Data

The Encyclopedic dictionary of physical geography.
(Blackwell Reference)
1. Physical geography — Dictionaries
I. Goudie, Andrew
910'.02'0321 GB10
ISBN 0-631-13292-9
ISBN 0-631-15581-3 Pbk

Library of Congress Cataloging-in-Publication Data

The Encyclopedic dictionary of physical geography/
edited by Andrew Goudie . . . [et al.]. — 2nd ed.
p.   cm.
Rev. ed. of: The Encyclopedic dictionary of physical geography. 1985.
Includes bibliographical references and index.
ISBN 0-631-18607-7
ISBN 0-631-18608-5 (pbk.)
1. Geography — Dictionaries. I. Goudie, Andrew.
II. Encyclopaedic dictionary of physical geography.
GB10.E53   1994
910'.02'03–dc20   93-29556
                                                    CIP
Typeset in 9½ on 10pt Plantin by TecSet Ltd, Wallington, Surrey.
Printed in Great Britain by
T.J. Press (Padstow) Ltd, Padstow, Cornwall.
This book is printed on acid-free paper

Contents

Preface to the First Edition   vi

Preface to the Second Edition   vii

Acknowledgements   viii

Editors' Introduction   ix

Contributors   x

Abbreviations in Physical Geography   xii

THE ENCYCLOPEDIC DICTIONARY OF PHYSICAL
GEOGRAPHY   1

Index   569

Preface to the First Edition

The preparation of a dictionary of this complexity has involved many people, and all deserve thanks for the efficiency with which they have prepared their material on time and in the format required. We have been fortunate in having as a model our companion volume, *The dictionary of human geography*, which was so expertly edited by R.J. Johnston and his team. I would like to express particular thanks to Janet Godden for having taken over so much of the organizational burden, and to Andrew Watson for being willing to prepare many of the short entries.

ASG

Preface to the Second Edition

In this second edition we have taken the opportunity to update many of the entries and their illustrations, and have added a substantial number of new entries. These new entries include some that should doubtless have been in the first edition, but most are entries that relate to new developments that have taken place in the discipline, especially with respect to increasing concerns over major environmental issues. We have also made substantial additions to the list of acronyms and abbreviations, and have updated many of the references and guides to further reading.

ASG

Acknowledgements

The author and publishers wish to thank the following for permission to use copyright material.

American Meteorological Society for figures in **urban meteorology** from Matson, McClain, McGinnis and Pritchard, 'Satellite detection of urban heat islands', *Monthly Weather Review*, Vol.106, No.12, pp.1725–34;

Blackwell Publishers for figures in **floristic realms** and **thermocline** from Andrew S. Goudie, *The Nature of the Environment*, 1993;

Blackwell Scientific Publishers for figure in **carrying capacity** from Begon et al., *Ecology, Individuals, Population and Communities*, 1986;

Cambridge University Press for table in **geological time-scale** from W.B. Harland, R.L. Armstrong, A.V. Cole, L. E. Craig, A.G. Smith and D.G. Smith, *A Geological Timescale*, 1989;

Chapman & Hall for table in **Beaufort scale** from J.E. Oliver and R.W. Fairbridge, *The Encyclopedia of Climatology*, 1987; and figures in **global ocean circulation** and **water mass** from D. Tolmazin, *Elements of Dynamic Oceanography*, 1985;

Gebrueder Borntraeger Verlagsbuchhandlung for table in **rock mass strength** modified from M.J. Selby, *Zeitschrift fur Geomorphologie*, 24, table 6, 1980;

Hodder & Stoughton Ltd. for table in **periglacial** based on A.R. Washburn, *Cryology*, pp. 7–8, 1979, Edward Arnold;

Longman Group Ltd. for figures and tables in **plate tectonics, rift valley, stress, mass movement types, dune** and **karren** from M.A. Summerfield, *Global Geomorphology*, 1991; and for table in **glaciofluvial** from R.J. Price, *Glacial and Fluvialglacial Landforms*, 1973, table 3, Oliver & Boyd;

McGraw-Hill, Inc. for table in **Manning equation** based on. V.T. Chow, ed., *Handbook of Applied Hydrology*, 1964;

Oxford University Press for table in **sea level** from A.S. Goudie, *Environmental Change*, table 6.1, 1992;

Pergamon Press Ltd. for figures in **water mass** from Open University Oceanography Course Team, *Ocean Circulation*, figs. 6.12, 6.13, 6.17, 1989;

The University of Chicago Press for table in **albedo** from W.D. Sellers, *Physical Climatology*, table 1, 1965.

Every effort has been made to trace all the copyright holders, but if any have been inadvertently overlooked the publishers will be pleased to make the necessary arrangement at the first opportunity.

Editors' Introduction

The prime virtue of our companion volume, *The dictionary of human geography*, is that it provides digestible short discussions on many of the new, and often complex, concepts that have arisen in that field in the past few decades. We have emulated this approach so far as we can, but because of the large array of technical terms with which the physical geographer has to contend we have also tried to provide a comprehensive range of short definitions of these terms to complement our conceptual reviews.

We have designed this dictionary for professional geographers and for earth, environmental and life scientists who work on the boundaries of our discipline. It is also intended for use by tertiary-level students, and secondary school teachers, all of whom need up-to-date definitions of words and terms in current usage. Furthermore, we hope that it will provide comprehensive but select guidance to the literature.

As in the companion volume, two systems are used to facilitate navigation through our complex sea of entries. The first is *cross-referencing*. Within an entry, certain other entries are referred to in small capital letters. Reading of these entries will expand the understanding of the term originally referred to and will also place it in a broader context. Secondly, there is an *index*, from which the reader will be able to find other entries in which a term is used and thereby obtain a wider sense of its usage. Most entries are followed by references or by suggestions for further reading as appropriate. References which are also suitable for use as further reading are indicated with a dagger.

ASG
BWA
KJG
IGS
DRS
DS

Contributors

Clive T Agnew **CTA**
University College London

JE Allen **JEA**
*Queen Mary and Westfield College,
London*

BW Atkinson **BWA**
*Queen Mary and Westfield College
London*

Keith Barber **KEB**
University of Southampton

Eric C Barrett **ECB**
University of Bristol

Roger G Barry **RGB**
University of Colorado, Boulder

Denys Brunsden **DB**
King's College London

Peter A Bull **PAB**
University of Oxford

Ian Burton **IB**
Lasalle Academy, Ottawa

Stanley A Changnon **SAC**
*Illinois Department of Energy,
Champaign Ill.*

Paul J Curran **PJC**
University of Southampton

Hugh M French **HMF**
University of Ottawa

Peter A Furley **PAF**
University of Edinburgh

Andrew S Goudie **ASG**
University of Oxford

William L Graf **WLG**
Arizona State University

John SA Green **JSAG**
University of East Anglia

Kenneth J Gregory **KJG**
Goldsmiths College, London

Angela M Gurnell **AMG**
University of Southampton

A Henderson-Sellers **AH-S**
MacQuarie University

Alan R Hill **ARH**
York University, Ontario

Robert L Jones **RLJ**
Coventry University

Barbara A Kennedy **BAK**
University of Oxford

Cuchlain A M King **CAMK**
formerly of University of Nottingham

M J Kirkby **MJK**
University of Leeds

John Lewin **JL**
University College of Wales, Aberystwyth

John G Lockwood **JGL**
University of Leeds

Harry van Loon **HvL**
*National Center for Atmospheric
Research, Boulder, Colo.*

Judith Maizels **JM**
University of Aberdeen

John A Matthews **JAM**
University of Cardiff

TR Oke **TRO**
*University of British Columbia,
Vancouver*

J Orford **JO**
Queen's University of Belfast

Adrian Parker **AP**
University of Oxford

Susan M Parker **SMP**
London

Allen H Perry **AHP**
University College, Swansea

David T Pugh **DTP**
IOS Deacon Laboratory, Godalming

Ross Reynolds **RR**
University of Reading

Keith S Richards **KSR**
University of Cambridge

MJ Selby **MJS**
University of Waikato

William D Sellers **WDS**
University of Arizona, Tucson

IG Simmons **IGS**
University of Durham

BJ Smith **BJS**
Queen's University of Belfast

Keith Smith **KS**
University of Stirling

Peter Smithson **PS**
University of Sheffield

Rodney H Squires **RHS**
*University of Minnesota,
 Minneapolis*

Philip A Stott **PAS**
School of Oriental and African Studies

David Sugden **DES**
University of Edinburgh

MA Summerfield **MAS**
University of Edinburgh

Bruce G Thom **BGT**
University of Sydney

John E Thornes **JET**
University of Birmingham

David G Tout **DGT**
University of Manchester

Michael H Unsworth **MHU**
*Institute of Terrestrial Ecology,
Midlothian*

Heather Viles **HV**
St Catherine's College, Oxford

DE Walling **DEW**
University of Exeter

Andrew Watson **AW**
University of Oxford

David Watts **DW**
University of Hull

Keith J Weston **KJW**
University of Edinburgh

W Brian Whalley **WBW**
Queen's University of Belfast

Paul W Williams **PWW**
University of Auckland

Abbreviations in Physical Geography

One of the horrors of the second half of the twentieth century has been the proliferation of abbreviations and acronyms. In physical geography a prime cause of this has been the growth of world organizations, such as the United Nations, and the tendency for much research to be carried out by multi-disciplinary and multi-institutional research teams. It is a matter for regret that so many books and papers fail to record the full meaning of abbreviations used. We have therefore listed below the abbreviations most frequently encountered in the current literature of physical geography.

AAAS	American Association for the Advancement of Science
AAG	Association of American Geographers
AAR	Accumulation area ratio
AAS	Atomic absorption spectrophotometer
ACCAD	WMO Advisory Committee on the World Climate
ACMAD	African Centre of Meteorological Applications for Development
AE	Actual evapotranspiration
AEM	Auger electron microscopy *see* also SAM/SAEM
AES	Auger electron spectroscopy
AFOS	Agriculture Forestry and Other Human Activities
AGFG	American Geomorphology Field Group
AGGG	Advisory Group on Greenhouse Gases
AGRHYMET	Centre d'Agrométéorologie et d'Hydrologie Operationnelle, Niamey
AIRE	Association Internationale des Resources en Eau
AIS	Airborne imaging spectrometer
AMCEN	African Ministerial Conference on the Environment
AMDASS	Agrometeorological Data System
AMISU	Advanced microwave sounding unit
AMQUA	American Quaternary Association
AMRA	American Water Resources Association
AMRT	Apparent mean residence time
AMSR	Advanced mechanically scanned radiometer

AOGCM	Atmosphere ocean general circulation model
AOSIS	Association of Small Island States
APT	Automatic picture transmission
ARMA	Autoregressive-moving average
ASAE	American Society of Agricultural Engineers
ASCA	Agricultural Stabilization and Conservation Service
ASCE	American Society of Civil Engineering
ASEAN	Association of South East Asian Nations
ASTEX	Azores Stratocumulus Transition Experiment
ASTM	American Society for Testing Materials
ASV	Anode-stripping voltometry
ATS	Antarctic Treaty System
ATS	Applications technology satellite
AVHRR	Advanced very high resolution radiometry
AVIRIS	Airborne visible and infrared imaging spectrometer
AWS	Automatic weather station
BAHC	Biological Aspects of the Hydrological Cycle (IGBP, WCRP)
BAPMoN	Background Air Pollution Monitoring Network
BAS	British Antarctic Survey
BE	Backscattered electrons
BES	British Ecological Society
BESI	Backscattered electron scanning images
BGRG	British Geomorphological Research Group
BGS	British Geological Survey
BHS	British Hydrological Society
BLE	Bombardment-induced light emission
BOD	Biochemical oxygen demand
BOFS	Biogeochemical Ocean Flux Study
BP	Before present
BRDF	Biodirectional reflectance-distribution function
BRGM	Bureau de Recherches Géologiques et Minières (Fr)
BRW	Barrow Observatory, Barrow, Alaska
BS	British standard
BSE	Backscattered electrons
BSI	Backscattered electron imaging
CAC	Climate Analysis Center, Washington
CACGP	Commission for Atmospheric Chemistry and Global Pollution (IAMAP)
CAeM	Commission for Aeronautical Meteorology of WMO
CAgM	Commission for Agricultural Meterology of WMO
CAS	Committee on Atmospheric Sciences
CASAFA	Inter-Union Commission on the Application of Science to Agriculture, Forestry and Aquaculture
CAZRI	Central Arid Zone Research Institute, India
CBR	California bearing ratio
CBS	Commission for Basic Systems of WMO
CCAMLR	Commission for the Conservation of Antarctic Marine Living Resources

CCC	Canadian Climate Centre, Downsview, Ontario, Canada
CCCO	Committee on Climate Changes and the Ocean
CCD	Charge-coupled device
CCDP	Climate Change Detection Project
CCI	Commission for Climatology of WMO
CCN	Cloud condensation nucleus
CCT	Computer compatible tape
CDIAC	Carbon Dioxide Information Analysis Center
CEA	Commissariat à l'Energie Atomique (Fr)
CEC	Commission of the European Communities
CERC	Coastal Engineering Research Station Washington
CFC	Chlorofluorocarbon
CGIAR	Consultative Group on International Agricultural Research
CHy	Commission for Hydrology of WMO
CIAP	Climate Impact Assessment Program
CIDIE	Committee of International Development Institutions on the Environment
CILSS	Permanent Inter-State Committee for Drought Control in the Sahel
CIMO	Commission for Instruments and Methods of Observations of WMO
CL	Cathodoluminescence
CLICOM	Climate-Computer System
CLIMAP	Climate Long-Range Investigation Mapping and Prediction
CMEA	Council for Mutual Economic Assistance
CMM	Commission for Marine Meteorology of WMO
CNES	Centre National d'Etudes Spatiales (Fr)
CNRM	Centre National de Recherches Météorologiques, France
CNRS	Centre Nationale de la Recherche Scientifique
COADS	Comprehensive Ocean Air Data Set
COBIOTECH	Scientific Committee for Biotechnology
CODATA	Committee on Data for Science and Technology
COLE	Coefficient of linear extensibility
COSPAR	Committee on Space Research (of ICSU)
COWAR	Committee on Water Research (of ICSU)
CPD	Critical point drying
CRREL	(US Army) Cold Regions Research and Engineering Laboratory, Hanover (New Hampshire)
CRU	Climatic Research Unit (University of East Anglia)
CSE	Centre du Suivi Ecologique (Sénégal)
CSERGE	Centre for Social and Economic Research on the Global Environment (University of East Anglia and University College London, England)
CSIRO	Commonwealth Scientific and Industrial Research Organization (Australia)
CSM	Climate System Monitoring
CSSA	Crop Science Society of America
CTEM	Conventional transmission electron microscopy

CZCS	Coastal Zone Colour Scanner
DALR	Dry adiabatic lapse rate
DARE	Data Rescue Programme
DBCP	Drifting Buoy Co-operation Panel
DC/PAC	Desertification Control Programme Activity Centre (UNEP)
DCP	Data collection platform
DCS	Data collection service
DDT	Dichloro-diphenyl-trichloroethane
DECARP	Desertification Encroachment and Rehabilitation Programme (Sudan)
DEFORPA	Dépérissement des Forêts et Pollution Atmosphérique (SRETIE, MAF, MRT)
DESCON	Consultative Group for Desertification Control
DIESA	United Nations Department for International Economic and Social Affairs
DIS	Data Information Systems (IGBP)
DMN	Direction de la Météorologie Nationale (Fr)
DMS	Dimethylsulphide
DMSP	Defense Meteorological Satellite Program
DOE	Department of Energy
DOGGIE	Deep Ocean Geological and Geophysical Instrument Explorer
DOLPHIN	Deep Ocean Long Path Hydrographic Instrument
DRED	Direction de la Recherche et des Etudes Doctorales, Ministère de l'Education Nationale de la Jeunesse et des Sports (Fr)
DRET	Direction des Recherches, Etudes et Techniques, Délégation Générale pour l'Armement, Ministère de la Défense (Fr)
DSC	Differential scanning colourimetry
DTA	Differential thermal analysis
DTG	Differential thermo-gravimetry
DVI	H.H. Lamb's Dust Veil Index
EAP	Environmental Action Plan
ECA	Economic Commission for Africa
ECB	United Nations Environment Co-ordination Board
ECE	Economic Commission for Europe
ECMWF	European Centre for Medium-range Weather Forecasts
ECOSOC	United Nations Economic and Social Council
ECP	Electron channelling patterns
EDAX	*See* EDXRA
EDF	Environmental Defense Fund, USA
EDS	Energy-dispersive spectroscopy = EDXRA
EDX	*See* EDXRA
EDXRA	Energy dispersive X-ray analysis
EEC	European Economic Community
EELS	Electron energy-loss spectrometry
EEZ	Exclusive Economic Zone
EFL	Effective Focal Length
EFTA	European Free Trade Association
EIA	Environmental impact assessment

EIS	Environmental impact statement
EIS	Environmental Information System
ELA	Equilibrium line altitude
ELC	Environment Liaison Centre
ELR	Environmental lapse rate
ELS	*See* EELS
EMA	Electron microprobe analysis = EPMA
EMEP	Monitoring and Evaluation of Pollution in Europe
EMP	Electron microprobe analysis, i.e. EDXRA (EDX) and WDXRA (WDX)
ENSO	El Niño-southern oscillation
EOS	Earth Observing System
EOSDIS	EOS Data and Information System
EPA	Environmental Protection Agency (USA)
EPMA	Electron probe microanalysis
EPOCH	European Programme on Climatology and Natural Hazards (EEC)
ERB	Earth Radiation Budget
ERBE	Earth Radiation Budget Experiment
ERL	Environmental Research Laboratory, Boulder
EROS	Earth Resources Observation Service
ERS	ESA Remote Sensing Satellite
ERTS	Earth Resources Technology Satellite
ESA	European Space Agency
ESCA	Electron spectroscopy for chemical analysis
ESMR	Electrically scanning microwave radiometer
ESR	Electron spin resonance (spectroscopy)
ESTAR	Electronically Scanned Thinned Array Radiometer
EUCREX	European Cloud and Radiation Experiment
FAGS	Federation of Astronomical and Geophysical Data Sources
FAO	Food and Agriculture Organization of the United Nations
FCCC	Framework Convention on Climate Change
FDA	Food and Drug Administration (USA)
FET	Field effect transistor
FEWS	Famine Early Warning System
FGGE	First GARP Global Experiment
FID	International Federation for Information and Documentation
FIRE	First ISCCP Regional Experiment
FOV	Field-of-view
FRAM	Fine Resolution Antarctic Model
FWC	Full Well Capacity
GA	General Assembly of the United Nations
GACC	General Agreement on Climate Change
GADS	Global Aerosol Data System
GAIM	Global Analysis, Integration and Modelling (IGBP)
GAMETAG	Global Atmospheric Measurements Experiment of Tropospheric Aerosols and Gases
GARP	Global Atmospheric Research Programme

GATE	GARP Atlantic Tropical Experiment
GATT	General Agreement on Tariffs and Trade
GAW	Global Atmospheric Watch
GCIP	GEWEX Continental-Scale International Project
GCM	General Circulation Model
GCOS	Global Climate Observing System
GCTE	Global Change and Terrestrial Ecosystems
GDP	Gross domestic product
GDPS	Global Data Processing System
GEDEX	Greenhouse Effect Detection Experiment
GEMS	Global Environment Monitoring System
GEOSECS	Geochemical Ocean Sections Study
GESAMP	Group of Experts on Scientific Aspects of Marine Pollution
GEWEX	Global Energy and Water Cycle Experiment
GFDL	Geophysical Fluid Dynamics Laboratory, Princeton, USA
GHOST	Global Horizontal Sounding Technique
GIEWS	Global Information and Early Warning System on Food and Agriculture
GIFOV	Ground instantaneous field-of-view
GIS	Geographic Information System
GISP	Greenland Ice Sheet Program
GISS	Goddard Institute for Space Sciences
GLASOD	Global Assessment of Soil Degradation
GLOSS	Global Sea-Level Observing System
GLRS	Geodynamics Laser Ranging System
GLU	Grazing Livestock Unit
GMCC	Geophysical Monitoring for Climatic Change
GMS	Geostationary meteorological satellite
GNP	Gross national product
GOES	Geostationary operational environmental satellite
GOFS	Global Ocean Flux Study
GOOS	Global Ocean Observing System
GO_3OS	Global Ozone Observing System
GOS	Global Observing System
GPCP	Global Precipitation Climatology Project
GRDC	Global Run-off Data Centre
GRID	Global Resources Information Database of GEMS
GSA	Geological Society of America
GSFC	Goddard Space Flight Center
GTCE	Global Change and Terrestrial Ecosystems (IGBP)
GTCP	Global Tropospheric Chemistry Programme
GTS	Global Telecommunication System
GTSPP	Global Temperature–Salinity Pilot Project
HABITAT	United Nations Conference on Human Settlements
HAPEX	Humidity and Precipitation Experiment
HCMM	Heat Capacity Mapping Mission
HDGEC	Human Dimensions of Global Environmental Change (ISSC)
HEIS	High energy ion scattering = RBS

HIRIS	High-resolution imaging spectrometer
HMMR	High-resolution multifrequency microwave radiometer
HOMS	Hydrological Operational Multipurpose Subprogramme (of WMO)
HREM	High-resolution electron microscopy
HRGC	Human Response to Global Change
HRIR	High-resolution infrared radiometer
HRS	Hydraulics Research Station
HVEM	High-voltage electron microscopy
HWP	Hydrology and Water Resources Programme of WMO
IAEA	International Atomic Energy Agency
IAEG	International Association of Engineering Geology
IAH	International Association of Hydrogeologists
IAHR	International Association of Hydraulic Research
IAHS	International Association of Hydrological Sciences
IAMAP	International Association of Meteorology and Atmospheric Physics
IAPSO	International Association of Physical Sciences of the Oceans
IASH	International Association for Scientific Hydrology
IAU	International Astronomical Union
IAWGD	Inter-Agency Working Group on Desertification
IBG	Institute of British Geographers
IBN	International Biosciences Networks
IBP	International Biological Programme
ICCE	International Commission on Continental Erosion (of IAHS)
ICE	Institution of Civil Engineers (London)
ICES	International Council for Exploration of the Sea
ICID	International Commission on Irrigation and Drainage
ICIHI	Independent Commission on International Humanitarian Issues
ICL	Inter-Union Commission on the Lithosphere
ICOLP	Industry Co-operative Programme for Ozone Layer Protection
ICP	Inductively coupled plasma (spectrometry)
ICRAF	International Centre for Research in Agro-Forestry
ICRCCM	Intercomparison of Radiation Codes in Climate Models
ICRISAT	International Crops Research Institute for the Semi-Arid Tropics
ICRP	International Commission on Radiological Protection
ICSI	International Commission on Snow and Ice (of IAHS)
ICSTI	International Council for Scientific and Technical Information
ICSU	International Council of Scientific Unions
IDA	International Development Association
IDNDR	International Decade for Natural Disaster Reduction
IEA	International Energy Agency
IEMVT	Institut d'Elevage et de Médecine Vétérinaire des Pays Tropicaux
IFAD	International Fund for Agricultural Development
IFOV	Instantaneous field-of-view

IFREMER	Institut Français de Recherche pour l'Exploitation de la Mer (Fr)
IFREPOLE	Institut Français pour la Recherche et Technologie Polaires – Expéditions Paul-Emile Victor (Fr)
IGAC	International Global Atmospheric Chemistry Study
IGADD	Inter-Governmental Authority on Drought and Development
IGAP	International Global Aerosol Programme
IGBP	International Geosphere Biosphere Project (of ICSU)
IGCP	International Geological Correlation Programme
IGFA	International Group of Funding Agencies for Global Change Research
IGN	Institut Géographique National (France)
IGOSS	Integrated Global Ocean Services System
IGS	Institute of Geological Sciences
IGS	International Glaciological Society
IGU	International Geographical Union
IGY	International Geophysical Year
IH	Institute of Hydrology
IHD	International Hydrological Decade
IHP	International Hydrological Programme (of UNESCO)
IIASA	International Institute for Applied Systems Analysis
IIED	International Institute for Environment and Development
IIT	Indian Institute of Technology
IITA	International Institute of Tropical Agriculture
IJC	International Joint Commission
ILCA	International Livestock Centre for Africa
ILO	International Labour Organization
IMC	Image motion compensation
IMCO	Intergovernmental Maritime Consultative Organization
IMF	International Monetary Fund
IMMA	Ion microprobe mass analysis
IMO	International Meteorological Organization
INC	Intergovernmental Negotiating Committee on a Framework Convention on Climate Change
INC/FCCC	International Negotiating Committee, FCCC
INFOCLIMA	Climate Data Information Referral System
INQUA	International Quaternary Association
INRA	Institut National de la Recherche Agronomique (Fr)
INRAN	Institut National pour la Recherche Agronomique du Niger
INSAH	Institut du Sahel
INSU	Institut National des Sciences de l'Univers (Fr)
INSULA	International Scientific Council for Island Development
IOC	Intergovernmental Oceanographic Commission
IOC	International Ozone Commission
IODE	International Oceanographic Data Exchange
IOSDL	Institute of Oceanographic Sciences Deacon Laboratory
IPCC	Intergovernmental Panel on Climate Change
IPS	International Peat Society

IQSY	International Years of the Quiet Sun
IR	Infrared radiation
IRA	Infrared spectrometry
ISCCP	International Satellite Cloud Climatology Program
ISLSCP	International Satellite Land Surface Climatology Project
ISRM	International Society for Rock Mechanics
ISS	Ion scattering spectrometry = LEIS
ISSAG	Imaging Spectrometer Science Advisory Group
ISSC	International Social Sciences Council
ISSMFE	International Society of Soil Mechanics and Foundation Engineering
ISSS	International Society of Soil Science
ISY	International space year
ITC	International Institute for Aerial Surveys and Earth Sciences (Dutch)
ITCB	Intertropical cloud band
ITCZ	Intertropical convergence zone
ITE	Institute of Terrestrial Ecology
ITTA	International Tropical Timber Agreement
ITTO	International Tropical Timber Organization
ITU	International Telecommunications Union
IUB	International Union of Biochemistry
IUBS	International Union of Biological Science (ICSU)
IUCN	International Union for the Conservation of Nature and Natural Resources
IUFRO	International Union of Forest Research Organizations
IUGG	International Union of Geodesy and Geophysics
IUH	Instantaneous unit hydrograph
IUHPS	International Union of the History and Philosophy of Science
IUMS	International Union of Microbiological Sciences
IUPAC	International Union of Pure and Applied Chemistry
IUPAP	International Union for Pure and Applied Physics
IUPESM	International Union for Physical and Engineering Sciences in Medicine
IWC	International Whaling Commission
IWRA	International Water Resources Association
JGOFS	Joint Global Ocean Flux Study
JPL	Jet Propulsion Laboratory
JSC	Joint Steering Committee for the WCRP
JSTC	Joint Scientific and Technical Committee for the Global Climate Observing System (GCOS)
KREMU	Kenya Rangeland Ecological Monitoring Unit
LAI	Leaf area index
LAMMA	Laser microprobe mass analysis = LMP
LASA	Lidar Atmospheric Sounder and Alimeter
Lat.	Latitude
LDC	Less-developed country
LEIS	Low-energy ion scattering = ISS

LFC	Large format camera
LMP	Laser microprobe analysis = LAMMA
LMT	Local meridian time
LoA	Law of the atmosphere
LOES	Laser optical emission spectrometry
LOICZ	Land Ocean Interactions in the Coastal Zone (IGBP)
Long.	Longitude
LRDC	Land Resources Development Centre
LTA	Low temperature ashing
M-region	Maunder Region
MAB	Man and the Biosphere Programme (UNESCO)
MAF	Ministère de l'Agriculture et des Forêts (Fr)
MARS	Monitoring agro-ecological resources by means of remote sensing and simulation
MAST	Marine Science and Technology (EEC)
MCA	Multi-channel analyser
MDD	METEOSAT Data Dissemination
MECCA	Model Evaluation Consortium for Climate Assessment
MEDI	Marine Environmental Data Referral System
MENJS	Ministère de l'Education Nationale, de la Jeunesse et des Sports (Fr)
MFD	Multi-function detector
MGO	Main Geophysical Laboratory, St Petersburg, Russia
MIT	Massachusetts Institute of Technology, Boston
MITRE	Meteorological Office Institute of Hydrology Terrestrial Project
MLO	Mauna Loa Observatory, Hawaii
MNR	Marine Nature Reserve (Nature Conservancy Council)
MODIS	Moderate-resolution imaging spectrometer
MODIS-N	MODIS Nadir
MODIS-T	MODIS Tilt
MOMS	Modular opto-electronic multispectral scanner
MOS	Marine observation satellite
MPI	Max Planck Institut, Germany
MRCS	Multi-Region Cloud Study (of GEWEX)
MRI	Meteorological Research Institute, Japan
MRIR	Medium resolution infrared radiometer
MRT	Ministère de la Recherche et de la Technologie (Fr)
MSL	Mean sea level
MSS	Multispectral scanning system
MSU	Microwave sounding unit
MTF	Modulation transfer function
MVA	Manufacturing value added
NAS	National Academy of Sciences (USA)
NASA	National Aeronautics and Space Administration (USA)
NATO	North Atlantic Treaty Organization
NBS	National Bureau of Standards
NCAR	National Center for Atmospheric Research (USA)
NCC	Nature Conservancy Council (UK)

NCDC	National Climate Data Center, Asheville
NCS	National Conservation Strategy
NDVI	Normalized difference vegetation index
NEPA	National Environmental Policy Act, 1970 (UK)
NERC	Natural Environment Research Council (UK)
NGO	Non-governmental organization
NH	Northern hemisphere
NICs	Newly industrialized countries
NIST	National Institute of Standards and Technology
NMR	Nuclear magnetic resonance (spectroscopy)
NNR	National Nature Reserve (UK)
NOAA	National Oceanic and Atmospheric Administration (USA)
NOAA/ERL	National Oceanic and Atmospheric Administration/Environmental Research Laboratory
NODC	National Oceanographic Data Center
NORAGRIC	Norwegian Centre for International Agricultural Development, Agricultural University of Norway
NPACD	National Plan of Action to Combat Desertification
NPP	Net primary productivity
NPTEC	National Power Technical and Environmental Centre
NRC	National Research Council (Canada)
NRI	Natural Resources Institute (UK)
NRL	Naval Research Laboratory
NSF	National Science Foundation (USA)
NTIS	National Technical Information Service
NUSS	Nuclear Safety Standards
NWP	Numerical Weather Prediction
NZARP	New Zealand Antarctic Research Programme
NZGS	New Zealand Geological Society
OAU	Organization of African Unity
ODA	Official development assistance
OECD	Organization for Economic Cooperation and Development
OES	Optical emission spectrometry
OHP	Operational Hydrology Programme (of WMO)
OM	Optical microscopy
ONR	Office of Naval Research
OPEC	Organization of Petroleum Exporting Countries
ORSTOM	Office de la Recherche Scientific et Technique Outre-Mer
OSS	Observatoire du Sahara et du Sahel
OSU	Oregon State University, USA
OTM	Optical transmission microscopy
OZONET	Ozone network
PACD	(World) Plan of Action to Combat Desertification
PAGES	Past global changes
PAMOY	Programme Atmosphère Moyenne (Fr)
PBL	Planetry boundary layer
PCSP	Polar Continental Shelf Project (Ottawa)
PDSI	Palmer Drought Severity Index

PE	Potential evapotranspiration
P-E	Precipitation-effectiveness
PFA	Principal factor analysis
PFO	Programme Flux Océaniques (France-JGOFS)
PIE	Polar Ice Extent Project
PIGB	Programme International Géosphère Biosphère (ICSU)
PIPOR	Programme for International Polar Ocean Research
PIXE	Particle-induced X-ray emission
pixel	Picture element
PML	Plymouth Marine Laboratory
PNEDC	Programme National d'Etude de la Dynamique du Climat (Fr)
PNOC	Programme National d'Océanographie Citière (Fr)
ppm	Parts per million
PPP	Polluter pays principle
ppt	Parts per thousand
PRs	Permanent Representatives of Members with WMO
PSA	Pacific Science Association
PSO	Polar Stratospheric Ozone Project
PVC	Potential volume change
PWP	Pore water pressure
QBO	Quasi-biennial oscillation
QRA	Quaternary Research Association (UK)
R	Wolf sunspot number
R_2	Zürich relative sunspot number
RADAR	Radio detection and ranging
RAL	Rutherford Appleton Laboratory
RBS	Rutherford backscattering spectrometry = HEIS
RBV	Return beam vidicon camera system
REE	Rare earth elements
RES	Radio-echo sounding
RGS	Royal Geographical Society
RISP	Ross Ice Shelf Project
RMS	Rock mass strength
rms	Root mean square
RY	Recurrence surface (Swedish: *rekurrensytor*)
S/N	Signal-to-noise ratio
SAC	Scientific Advisory Committee for the World Climate Impact Assessment and Response Strategies Programme (WCIRP)
SADCC	Southern African Development Coordination Conference
SAEM	Scanning auger (electron) microscopy
SAGE	Stratospheric Aerosol and Gas Experiment
SALR	Saturated adiabatic lapse rate
SAM	Scanning auger microprobe
SAR	Synthetic aperture radar
SCAR	Scientific Committee on Antarctic Research
SCAT	Scatterometer
SCOPE	Scientific Committee on Problems of the Environment of the ICSU

SCOR	Scientific Committee on Ocean Research
SCOSTEP	Scientific Committee on Solar-Terrestrial Physics
SCP	Single cell protein
SCS	Soil Conservation Service
SCSA	Soil Conservation Society of America
SE	Secondary electrons
SEAREX	Sea-Air Exchange Program
SEI	Secondary electron images
SEI	Stockholm Environment Institute
SEM	Scanning electron microscope
SH	Southern hemisphere
SHOM	Service Hydrographique de la Marine (Fr)
SI	System International d'Unités
SIMS	Secondary ion mass spectrometry
SIO	Scripps Institution of Oceanography
SIPRE	(former name of CRREL)
SIRS	Satellite infrared spectrometer
SISEX	Shuttle Imaging Spectrometer Experiment
SLAM	Scanning laser acoustic microscope
SLAR	Sideways looking airborne radar
SMD	Soil moisture deficit
SMIRR	Shuttle Multispectral Infrared Radiometer
SMM	Solar maximum mission
SMMR	Scanning multifrequency microwave radiometer
SMO	Samoa Observatory, American Samoa
SPAM	Spectral Analysis Manager
SPO	South Pole Observatory, Antarctica
SPONG	Société Permanent des Organisations Non-Gouvernementaux (Burkina Faso)
SPOT	Satellite Probatoire d'Observation de la Terre
SPREP	South Pacific Regional Environment Programme
SPRI	Scott Polar Research Institute
SRETIE	Service de la Recherche, des Etudes et du Traitement de l'Information sur l'Environment, Ministère de l'Environment (Fr)
SSMS	Spark source mass spectrometry
SSSA	Soil Science Society of America
SSSI	Site of special scientific interest
SST	Sea surface temperature
START	System for Analysis, Research and Training (IGBP)
STEM	Scanning transmission electron microscope
STEP	Science and Technology for Environmental Protection (EEC) Subgroup of IPCC WG III (Response Studies).
SWCC	Second World Climate Conference
SWIR	Short-wave infrared wavelengths (1.1 to 2.5 μm)
TAAF	Terres Australes et Antarctiques Françaises (Fr)
TCP	Tropical Cyclone Programme
TDCN	Topologically distinct channel network

TEELS	Transmission electron energy loss spectrometry
TEM	Transmission electron microscopy (*see* also CTEM)
TFAP	Tropical Forestry Action Plan
TG	Thermo-gravimetry
THI	Temperature humidity index
TIGER	Terrestrial Initiative in Global Environmental Research
TIMS	Thermal infrared multispectral scanner
TIROS	Television and infrared observation satellite
TL	Thermoluminescence
TM	Thematic mapper
TNCs	Transnational corporations
TOGA	Tropical Ocean Global Atmosphere
TOVS	TIROS Operational Vertical Sounder
TRUCE	Tropical Urban Climate Experiment
TTO	Transient Tracers in the Ocean
TWAS	Third World Academy of Sciences
UARS	Upper Atmosphere Research Satellite
UCAR	University Corporation for Atmospheric Research
UEA	University of East Anglia
UGAMP	Universities Global Atmospheric Modelling Project
UKMO	Meteorological Office, Bracknell, UK
UMIST	University of Manchester Institute of Science and Technology
UNCED	United Nations Conference on Environment and Development
UNCHS	United Nations Centre for Human Settlements (HABITAT)
UNCOD	United Nations Conference on Desertification
UNCTAD	United Nations Conference on Trade and Development
UNCTC	United Nations Centre on Transnational Corporations
UNDP	United Nations Development Programme
UNDRC	United Nations Disaster Relief Coordinator
UNDTCD	United Nations Division of Technical Cooperation for Development
UNECE	United Nations Economic Commission for Europe
UNEP	United Nations Environment Programme
UNESCO	United Nations Education, Scientific and Cultural Organization
UNFPA	United Nations Population Fund
UNIDO	United Nations Industrial Development Organization
UNITAR	United Nations Institute for Training and Research
UNOEOA	United Nations Office for Emergency Operations in Africa
UNPAAERD	United Nations Programme of Action for African Economic Recovery and Development
UNSO	United Nations Sudano-Sahelian Office
UNU	United Nations University (Tokyo)
USARP	United States Antarctic Research Program
USDA	United States Department of Agriculture
USDOE	United States Department of Energy
USGS	United States Geological Survey
USLE	Universal soil loss equation

UVS	Ultraviolet spectrometry
VCP	Voluntary Cooperation Programme (WMO)
VEI	Volcano explosivity index
VNIR	Visible and near-infrared wavelengths (0.4 to 1.1 μm)
WB	World Bank
WCAP	World Climate Applications Programme
WCASP	World Climate Applications and Services Programme
WCDMP	World Climate Data and Monitoring Programme
WCDP	World Climate Data Programme
WCIP	World Climate Impact Studies Programme
WCIRP	World Climate Impact Assessment and Response Strategies Programme
WCP	World Climate Programme
WCRP	World Climate Research Programme
WCS	World Conservation Strategy
WDC	World Data Centre
WDS	Wavelength dispersive spectroscopy = WDXRA
WDX	*See* WDXRA
WDXRA	Wavelength dispersive X-ray analysis
WHO	World Health Organization
WIPO	World Intellectual Properties Organization
WMO	World Meteorological Organization
WOCE	World Ocean Circulation Experiment
WRI	World Resources Institute
WWF	Worldwide Fund for Nature
WWW	World Weather Watch
WWWDM	World Weather Watch Data Management
XPS	X-ray photoelectron spectroscopy
XRD	X-ray diffraction
XRF	X-ray fluorescence spectrometry

A

abîme A vertical shaft in karstic limestone terrain.

abiotic The abiotic components of an ECOSYSTEM are those which are not living. These include mineral soil particles, water, atmospheric gases and inorganic salts; sometimes simple organic substances that have resulted from excretion or decomposition may be included. The term abiotic is also used for physical and chemical influences upon organisms, for example humidity, temperature, pH and salinity. An abiotic environment is one which is devoid of life.

PHA

ablation The process by which snow or ice is lost from a GLACIER, floating ice or snow. Examples are melting and run-off, calving of icebergs, evaporation, sublimation and removal of snow by wind. Melting followed by refreezing at another part of a glacier is not regarded as ablation because the glacier does not lose mass. Melting is the most important process in temperate and subpolar regions and accounts for seasonal and diurnal meltwater floods. Most such ablation occurs at the glacier surface, and at the snouts of glaciers in many mid-latitude areas it lowers the ice surface by the order of 10m each year. A small amount of melting occurs within and beneath glaciers whose ice is at the pressure melting point. In the Antarctic the most important ablation process is the calving of ice shelves, though considerable losses may also occur through bottom melting of ice shelves and the removal of snow by offshore katabatic winds. DES

Reading
Paterson, W.S.B. 1981: *The physics of glaciers.* 2nd edn. Oxford: Pergamon.

abrasion The process of wearing down or wearing away by friction as by windborne sand or material frozen into glacial ice.

absolute age The age of an event or rock, mineral or fossil, measured in years.

absolute humidity See HUMIDITY.

abundance The total number of individuals of a particular species present in an area. Various methods are used to measure the abundance of organisms but in view of the time and effort involved it is usually impractical to count all individuals within an area. Instead, population size is often estimated by collecting data from small plots (quadrats) selected by a random sampling procedure. Population size is influenced by a complex array of factors which include, for example, the physical environment, weather conditions, available resources (food, nesting sites, etc.), competition both within and between species, and predation. ARH

Reading
Mueller-Dombois, D. and Ellenberg, H. 1974: *Aims and methods of vegetation ecology.* New York and London: Wiley. Chapter 6, pp. 67–92.
Watts, D. 1971: *Principles of biogeography.* New York and London: McGraw-Hill. Chapter 5, pp. 197–241.

abyss *a.* A deep part of the ocean, especially one more than about 3000 m below sea level.
b. A ravine or deep gorge.

abyssobenthic zone The bottom of a deep lake, sea or ocean inhabited by characteristic organisms.

abyssopelagic zone The portion of deep lakes, seas and oceans in which specific forms of plankton and nekton are found.

accelerated erosion See SOIL EROSION.

accessory mineral The mineral components of a rock which do not occur in sufficient quantities to merit their inclusion in the definition or classification of the rock, that is, not an essential mineral.

accommodation A term used in soil science referring to the extent to which faces of adjacent aggregates are moulds one of another. Where adjacent faces meet and leave virtually no void (such as in the regular packing of cuboids) there is said to be good accommodation. On the other hand, a packing of spheres displays no accommodation.

accordant junctions, law of The law which states that tributary rivers join main rivers at the same level, that is there is usually no sudden drop (Playfair's law).

accordant summits The phenomenon of hill crests and mountain peaks in a region being within a similar plane, horizontal or inclined, attesting that they are remnants of a former plain or plateau.

accretion *a.* The gradual increase in the area of land as a result of sedimentation. *b.* The process by which inorganic objects increase in size through the attachment of additional material to their surface as with the growth of hailstones.

accumulated departure The amount, which may be positive or negative, by which, over a period of time, the value of a meteorological element, such as mean annual temperature, departs from the long-term mean value.

accumulated temperature Normally the total number of days (or hours) since a given date, during which the mean temperature has been above or below a given threshold. The threshold value for agriculture is usually 6°C and accumulated mean temperatures above this value can be correlated with the growth of vegetation. For heating purposes the threshold is usually 15.5°C and accumulated mean temperatures below this value can be correlated with energy use. Generally

accumulated temperature is used in agriculture and DEGREE DAYS are used in energy management.   JET

acid precipitation Rain and snow with a pH of less than 5.6. The latter is the hydrogen ion concentration of natural precipitation subject to normal concentrations and pressures of atmospheric carbon dioxide. As the pH scale is logarithmic, a one-point change on it represents a tenfold increase or decrease in acidity (Kemp 1990). The slight natural acidity of precipitation is largely due to weak carbonic acid formed by dissolved atmospheric carbon dioxide, and to sulphur compounds from volcanic eruptions which are converted to sulphuric acid in the atmosphere. The chemical analysis and dating of fossil ice has revealed that some two centuries ago, precipitation possessed a pH that was

Acid precipitation is one of the most serious and contentious environmental issues at the present time. Emissions of sulphate and nitrate rich pollutants from power stations, smelters, and other sources can increase the natural acidity of rainfall, with unfortunate ecological consequences, including possible damage to the foliage of fir trees (Abies alba) in the Black Forest, Germany.

generally in excess of 5. Since that time, industrial–urban development, particularly in northern hemisphere mid-latitudes, has resulted in the release of increasing quantities of sulphur and nitrogen oxides into the atmosphere. These emissions are caused by fossil fuel burning and sulphide ore smelting, the oxides being transformed into sulphuric and nitric acids in the atmosphere. These relatively strong acids undergo ionic separation in weakly acidic natural precipitation, with the dissociated hydrogen ions causing its pH to fall below 5.6. (Likens *et al.* 1979). *Sensu stricto*, acid precipitation is thus wet deposition. However, a related process, dry deposition, whereby oxides of sulphur and nitrogen fall out from the atmosphere either as dry gases or adsorbed on other AEROSOLS such as soot, is also operational. These particles become acidic when they join with moisture; fog or surface water, for example (Park 1987; Kemp 1990).

Atmospheric circulation patterns mean that pollutants (POLLUTION) can travel substantial distances before being deposited as acid precipitation. As the loci of acidic pollution are within the westerly wind-belt, their discharges are usually routed eastward. The rate and distance of movement are associated with the height of pollutant emission (Kemp 1990). Tall stacks enhance long-distance transfer (Elsom 1987), with upper westerly winds or jet streams more effective than boundary-layer circulation, both in this respect and in increasing the residence time of pollutants in the atmosphere (Kemp 1990).

The phenomenon of acid precipitation was first recognized in England during the mid-nineteenth century, but has been studied in detail only during the past three decades (Park 1987). Its role is still imperfectly understood. Complex interactions between environmental factors (geology, hydrology and land use, for example) mean that acid precipitation is probably one of numerous components operative in a particular locality (Kemp 1990). Aquatic ecosystems appear to respond more rapidly to acidification than terrestrial ones. Acid water is thought to diminish biological productivity in lakes and rivers developed on siliceous substrates, with the reproductive capacity of fish being impaired (Pearce 1982). The interception of acid precipitation by trees is considered to increase the chance of tissue death, nutrient leaching

and chlorophyll degradation in leaves (Shriner and Johnston 1985). Increased acidity of soil water seems to check bacterial activity, results in the replacement of nutrient cations by hydrogen ions and the displacement of the former in solution, and stimulates the mobilization of toxic heavy metals such as aluminium and lead (Kemp 1990). The liberation of heavy metal cations from soils and sites of toxic waste disposal can contaminate drinking water, while supply pipes of the latter may be leached of copper and lead by acidic water (Elsom 1987). Building stone containing calcium and magnesium carbonates could be subject to the reaction of these with the sulphuric component of acid precipitation and to the production of soluble sulphates. In urban areas, dry deposition is frequently of most importance, with chemical reactions initiated by the addition of moisture. In towns and cities, sulphuric acid in the atmosphere, inhaled during episodes of smog, can lead to respiratory difficulties (Kemp 1990).

In a global climatic context, nitrogen oxides are part of the process whereby tropospheric ozone (a greenhouse gas) is produced. Thus more or less of these oxides could contribute to higher or lower ozone concentrations and increase or reduce climatic warming (Martin 1989). Between 1970 and 1984 there was a 40 per cent decrease in the emission of SO_2 in Britain (Caulfield and Pearce 1984). Chemical and biological evidence of a slight reduction in the acidity of water in Galloway since 1980 may reflect this trend (Battarbee *et al.* 1988).   RLJ

Reading and References

Battarbee, R.W., Flower, R.J., Stevenson, A.C., Jones, V.J., Harriman, R. and Appleby, P.G. 1988: Diatom and chemical evidence for reversibility of acidification of Scottish lochs. *Nature* 332, pp. 530–2.

Caulfield, C. and Pearce, F. 1984: Ministers reject clean-up of acid rain. *New Scientist* 104, pp. 1433–6.

†Elsom, D. 1987: *Atmospheric pollution: causes, effects and control policies.* Oxford and New York: Basil Blackwell.

†Kemp, D.D. 1990: *Global environmental issues: a climatological approach.* London and New York: Routledge.

Likens, G.E., Wright, R.F., Galloway, J.N. and Butler, T.J. 1979: Acid rain. *Scientific American* 241, pp. 39–47.

Martin, H.C. 1989: The linkages between climate change and acid rain. In J.C. White ed., *Global climate change linkages: acid rain, air quality and stratospheric ozone.* New York: Elsevier. Pp 59–66.

†Park, C.C. 1987: *Acid rain: rhetoric and reality.* London: Methuen.

Pearce, F. 1982: The menace of acid rain. *New Scientist* 95, pp. 419–23.

Shriner, D.S. and Johnston, J.W. 1985: Acid rain interactions with leaf surfaces: a review. In D.D. Adams and W.P. Page eds, *Acid deposition: environmental, economic and policy issues*. New York and London: Plenum Press. Pp 241–53.

Wellburn, A. 1988: *Air pollution and acid rain: the biological impact*. Harlow: Longman Scientific and Technical.

acid rocks Commonly used term for igneous rocks which contain more than 66 per cent silica, free or combined, or any igneous rock composed predominantly of highly siliceous minerals (see BASIC ROCKS).

acidity profile The acid concentration in ice core layers as a function of depth as determined from electrical measurements. The magnitudes of some volcanic eruptions in the northern hemisphere have been estimated from the acidity of annual layers in ice cores taken in Greenland. This methodology is sometimes referred to as 'acidity signal' or 'acidity record'.   ASG

aclinic line The magnetic equator, an irregularly curved line near the equator along which the compass needle does not dip from the horizontal.

actinometer An instrument used for measuring the chemical and heating influences of the sun's radiation.

active layer The top layer of ground above the permafrost table which thaws each summer and refreezes each autumn. In temperature terms, it is the layer which fluctuates above and below 0°C during the year. In permafrost areas seasonally frozen and thawed ground can be equated with the active layer. Other synonyms include 'depth of thaw', 'depth to permafrost', and 'annually thawed layer'. These terms are acceptable in areas where the active layer extends downwards to the permafrost table, but they are misleading where the active layer is separated from the permafrost by a layer of ground which remains in an unfrozen state throughout the year. The thickness of the active layer varies from as little as 15–30 cm in high latitudes to over 1.5 m in subarctic continental regions. Thickness depends on many factors, including the degree and orientation of the slope, vegetation, drainage, snow cover, soil and rock type, and ground moisture conditions.

Processes operating in the active layer include FROST CREEP and FROST HEAVE or cryoturbation, the lateral and vertical displacement of soil which accompanies seasonal and/or diurnal freezing and thawing. During thaw, water movement through the active layer assists various mass wasting processes, especially gelifluction. Most patterned ground phenomena form in the active layer.   HMF

Reading
Brown, R.J.E. and Kupsch, W.O. 1974: *Permafrost terminology*, Publication 14274. Ottawa: National Research Council of Canada.

French, H.M. 1976: *The periglacial environment*. London and New York: Longman.

French, H.M. 1988: Active layer processes. In M.J. Clark ed., *Advances in periglacial geomorphology*. Chichester: Wiley. Pp. 151–79.

activity (ratio) An empirical relationship of a soil, defined by Skempton as the plasticity index divided by percentage weight less than 2 μm in size.

As the plasticity index often varies according to cations in the clay mineral structure, the activity is a useful measure of the swelling potential of a soil (its ability to take up moisture). Skempton suggested three main classes of activity: active, normal, and inactive. Most British soils tend to be in the normal to inactive ranges. Active soils tend to have high CATION EXCHANGE capacities. There are sometimes problems with using the activity values, especially if particles aggregate, so the AGGREGATION RATIO has also been used to express percentage of clay mineral relationships.   WBW

Reading
Bell, F.G. 1992: *Engineering properties of soils and rocks*. 3rd edn. Oxford: Butterworth Heinemann

Skempton, A.W. 1953: The colloidal activity of clays. *Proceedings of the Third International Conference of Soil Mechanics*. Zurich.

actualism See UNIFORMITARIANISM.

adaptive radiation The evolutionary diversification of a group of organisms in response to the ecological pressures of different habitats. When new groups of organisms evolve or occupy a newly accessible environment they will tend to fill, over time, all the available niches; they will thus 'radiate' along evolutionary lines in a genetical response to the stimulus of environmental diversity. The sum of the various lines leading away from the ancestral stock comprises an adaptive radiation.

Among mammals, for example, there have arisen specialist grazers and carnivores, burrowers, fliers and aquatic species, with different species filling the equivalent NICHE in various BIOMES, e.g. the bison in North America and the kangaroo in Australia and many others. Adaptive radiation is an important process in the diversification of island floras and faunas (see ISLAND BIO-GEOGRAPHY). PAS

Reading
Stebbins, G.L. 1977: *Processes of organic evolution.* 3rd edn. Englewood Cliffs, NJ: Prentice-Hall.

adhesion ripple An irregular sand ridge transverse to wind direction, formed when dry sand is blown across a smooth moist surface. It may be 30–40 cm long and a few centimetres high. The crest is symmetrical and migrates upwind. The stoss (windward) side is steeper than the lee side. ASG

adiabatic An adiabatic process is a thermodynamic change of state of a system in which there is no transfer of heat or mass across the boundaries of the system. In the atmosphere the most commonly related variables in the adiabatic process are temperature and pressure. If a mass of air experiences lower pressure than in its initial condition it will expand and do mechanical work on the surrounding air. The energy required to do this work is taken from the heat energy of the air mass and consequently the temperature of the air falls. Conversely, when pressure increases, work is done on the mass of air and the temperature rises. A diabatic process is a thermodynamic change of state of a system in which there is transfer of heat across the boundaries of the system. BWA

adobe *a.* Sun-dried mud bricks, the material from which they are made or the buildings made from them, especially in the south-western USA.
b. Loessial deposits of the south and western USA.

adret The side of a hill or valley that receives the most sunlight (see UBAC), and which may therefore in high-altitude areas have the most intensive land use and settlement.

adsorption The physical or chemical bonding of molecules, gaseous, liquid or dissolved, to the surfaces of solid bodies or their interiors, if the surfaces are porous or permeable.

advection The movement of a property in the fluid natural environment (air and water) due solely to the velocity field of the fluid. Thus heat is transferred through the atmosphere by winds. (We should note that heat is also transferred by radiation – a non-advective process.) Advection may be resolved into two components: horizontal and vertical. In meteorology, advection refers frequently only to the horizontal motion whereas vertical advection, particularly on the scale of the individual cloud, is often called CONVECTION. BWA

advection fog See FOG.

adventitious The term applied to the roots and buds of plants that grow from unusual portions of the plant, e.g. roots growing from tree trunks and branches.

adventitive cone A small volcanic crater or cone which develops on the flanks of a larger volcano.

aeolianite Cemented dune sand, calcium carbonate being the most frequent cement. The degree of cementation is very variable, the end product being a hardened dune rock with total occlusion of pore space. Aeolianite of Quaternary age is generally found in coastal areas within 40° of the equator, especially those that experience at least one dry season. The balance between leaching and lime production is the prime control of this overall distribution. Most examples contain between 30 and 60 per cent calcium carbonate, although not all of this may occur as cement. According to Yaalon (1967) a minimum of 8 per cent calcium carbonate is required for cementation of dune sands under semi-arid conditions. Sources of calcium carbonate include biogenic skeletal fragments, dust, spray, and groundwater. ASG

Reading and Reference
†Gardner, R.A.M. 1983: Aeolianite. In A.S. Goudie and K. Pye eds, *Chemical sediments and geomorphology.* London: Academic Press.
Yaalon, D. 1967: Factors affecting the lithification of aeolianite and interpretation of its environmental significance in the coastline plain of Israel. *Journal of sedimentary petrology* 37, pp. 1189–99.

aeration zone In the context of the HYDROLOGICAL CYCLE, the zone between the soil moisture zone and the capillary zone immediately above the water table. In this zone vadose water is moving downwards under the influence of gravity and the zone may vary in thickness from 0 to several hundred metres in arid regions.   KJG

aerial camera A camera designed specially to hold AERIAL FILM for the taking of AERIAL PHOTOGRAPHY. The five main types of aerial camera are the mapping camera, reconnaissance camera, strip camera, panoramic camera and multiband camera.

The majority of aerial photographs are taken with high quality *mapping cameras*. These are relatively simple in design and comprise a low distortion lens and a very large film magazine. The *reconnaissance camera* is cheaper than the mapping camera both to buy and to operate. Their disadvantages are their relatively large levels of geometric distortion and their unsuitability for colour aerial film.

The *strip camera* focuses light onto an adjustable slit, under which the film moves at a speed that is proportional to the ground speed of the aircraft. The resulting long strips of photography are not usually appropriate for use in physical geography. The lens in the *panoramic camera* oscillates, scanning from horizon to horizon while focusing the light onto a cylinder, upon which a photographic film is held. This results in distorted photographs which are rarely used by physical geographers.

The *multiband camera* makes use of the fact that objects on the earth's surface vary in the way in which they reflect ELECTROMAGNETIC RADIATION. To use this phenomenon to differentiate between objects, a scene is photographed through different filters using either a multi-lens camera or a multi-camera array. A multi-lens camera uses one film which is exposed by radiation passing through several filters and lenses whereas a multi-camera array comprises a number of small format cameras, each camera having its own film and filter.   PJC

Reading
Paine, D.P. 1981: *Aerial photography and image interpretation for resource management*. New York: Wiley.
Wolf, P.F. 1974: *Elements of photogrammetry*. New York: McGraw-Hill.

aerial film A specially designed roll film supplied in many widths and lengths to fit AERIAL CAMERAS for the taking of AERIAL PHOTOGRAPHY. The four types of aerial film are black and white, black and white/near infrared, colour and false colour/near infrared.

Black and white film is sensitive to a broad waveband of visible light and is often called panchromatic film. The two types of this film are mapping film which has equal sensitivity to all visible wavelengths and reconnaissance film which has reduced sensitivity to blue wavelengths.

Black and white/near infrared film has characteristics similar to black and white film. The main difference is its spectral sensitivity which extends beyond visible wavelengths to the region of near infrared ELECTROMAGNETIC RADIATION.

Colour film comprises three separate layers. Each layer is sensitive to a particular waveband of light and is coloured with the complementary colour to that waveband of light. A yellow layer therefore records blue radiation, a magenta layer records green radiation and a cyan layer records red radiation. In the positive product, e.g. a paper print, the colours are similar to those in the original scene.

False colour/near infrared film has a structure similar to colour film. However, its relationship between radiation sensitivity and colour is different as it uses a yellow layer to record green radiation, a magenta layer to record red radiation and a cyan layer to record near infrared radiation. In the positive product, e.g. a transparency, the colours are dissimilar to those in the original scene as primarily blue reflective objects appear black, primarily green reflective objects appear blue, primarily red reflective objects appear green, and primarily near infrared reflective objects (e.g. vegetation) appear red.   PJC

Reading
Lillesand, T.M. and Kiefer, R.W. 1987: *Remote sensing and image interpretation*. 2nd edn. New York: Wiley.
Slater, P.N. 1980: *Remote sensing: optics and optical systems*. Reading, Mass. and London: Addison-Wesley.

aerial photography is taken using AERIAL FILM in an AERIAL CAMERA that is usually mounted in an aircraft. It is the most widely used type of REMOTE SENSING. The characteristics of aerial photography that make it so popular are:

1 Availability: aerial photographs are readily available at a range of scales for much of the world.

2 Economy: aerial photographs are cheaper than field surveys and are

often cheaper and more accurate than maps for many countries of the world.

3 Synoptic viewpoint: aerial photographs make possible the detection of both small features and spatial relationships that would not be evident on the ground.

4 Time freezing ability: an aerial photograph is a record of the earth's surface at a particular moment and can therefore be used as a historical record.

5 Spectral and spatial resolution: aerial photographs are sensitive to ELECTROMAGNETIC RADIATION in wavelengths and for areas that are outside the spectral sensitivity range of the human eye.

6 Three-dimensional perspective: a stereoscopic view of the earth's surface can be created and measured both horizontally and vertically.

The angle from which the aerial photography is taken determines whether it is vertical, high oblique or low oblique. Vertical aerial photography results when the camera axis is pointing vertically downwards and oblique aerial photography results when the camera axis is pointing obliquely downwards. Low oblique aerial photography incorporates the horizon into the photograph, while high oblique aerial photography does not. Vertical aerial photographs are the most widely used type as they have an approximately constant scale over the whole photograph and can be used for mapping and measurement. Oblique aerial photographs have their advantages, as they cover many times the area of a vertical aerial photograph taken from the same height using the same focal length lens and in addition present a view that is more natural to the interpreter.

Aerial photographs are taken at a wide range of scales. A small-scale aerial photograph at 1-50,000 will provide a synoptic, low spatial resolution overview of a large area, while a large-scale aerial photograph at 1-2000 will provide a detailed and high spatial resolution view of a small area.

Once obtained, aerial photographs are interpreted for the identification of objects and assessment of their significance. During this process the interpreters usually undertake several tasks of detection, recognition and identification, analysis, deduction, classification, idealization and accuracy determination. Detection involves selec-

tively picking out visible objects. Recognition and identification involve naming the objects or areas, and analysis involves trying to detect their spatial order. Deduction involves the principle of convergence of evidence in order to predict the occurrence of certain relationships on the aerial photographs. Classification is used to arrange the objects and elements identified into an orderly system before the photographic interpretation is idealized using lines which are drawn to summarize the spatial distribution of objects or areas. The final stage is accuracy determination in which random points are visited in the field to confirm or correct the interpretation.

Recognition and identification of objects or areas comprise the most important link in this chain of events. An interpreter uses seven characteristics of the aerial photography to help with this stage: tone, texture, pattern, place, shape, shadow and size. Tone is the single most important characteristic of the aerial photograph as it represents a record of the ELECTROMAGNETIC RADIATION that has been reflected from the earth's surface onto the aerial film. Texture is the frequency of tonal changes which arise within an aerial photograph when several features are viewed together. Pattern is the spatial arrangement of objects on the aerial photograph. Place is a statement of an object's position on the aerial photograph in relation to others in its vicinity. Shape is a qualitative statement of the general form, configuration or outline of an object on an aerial photograph. Shadows of objects on an aerial photograph are used to help in identifying them, e.g. by enhancing geological boundaries. Size of an object is a function of the scale of the aerial photograph. The sizes of objects can be estimated by comparing them with objects for which the size is known.   PJC

Reading
Lo, C.P. 1976: *Geographical applications of aerial photography*. Newton Abbot: David & Charles; New York: Crane, Russak.
Ritchie, W., Wood, M., Wright, R. and Tait, D. 1988: *Surveying and mapping for field scientists*. Harlow: Longman Scientific and Technical.

aerobic Living or active exclusively in the presence of air or oxygen.

aerobiology The study of the recognition, conveyance and behaviour of passive airborne organic particles (Gregory 1973). These particles, or aerosols, may be viable or non-viable and occur both in and out of doors. They are dispersed in air and move, dependent upon atmospheric properties and conditions. Most have a diameter of approximately 0.5–100 μm and may occur singly or in groups. The viability of airborne microorganisms can be affected by other atmospheric particulates (such as dusts), and by gaseous pollutants (ACID PRECIPITATION). Both living and non-living organisms are considered in an aerobiological context (Edmonds 1979). (See also POLLEN ANALYSIS.)   RLJ

Reading and References
Edmonds, R.L. ed. 1979: *Aerobiology: the ecological systems approach*. Stroudsburg, Pa: Dowden, Hutchinson & Ross.
†Gregory, P.H. 1973: *The microbiology of the atmosphere*. 2nd edn. Aylesbury: Leonard Hill.
†Knox, R.B. 1979: *Pollen and allergy*. London: Edward Arnold.

aerography Describes studies of the geographical ranges of species, genera, families, etc.

aerology The study of the atmosphere, especially the study of the atmosphere above the surface layers.

aeronomy The branch of atmospheric physics which is concerned with those regions, generally above 50 km, where ionization and dissociation are fundamental properties.

Reading
Brasseur, G. and Solomon, S. 1986: *Aeronomy of the middle atmosphere: chemistry and physics of the stratosphere and mesosphere*. Dordrecht: Reidel.

aerosol An intimate mixture of two substances, one of which is in the liquid or solid state dispersed uniformly within a gas. The term is normally used to describe smoke, condensation nuclei, freezing nuclei or fog contained within the atmosphere, or other pollutants such as droplets containing sulphur dioxide or nitrogen dioxide. Aerosols tend to obscure visibility by scattering light. They tend to vary in size between about a milli-micron (10^{-9} m) and one micron (10^{-6} m). Clouds are not normally considered to be aerosols because the droplets are too large and tend to fall due to gravity. Aerosols can remain in the atmosphere for long periods, collisions with air molecules keeping them aloft. JET

aesthetic degradation The deterioration in quality of an environmental re-

source. Aesthetic quality results from personal perception of an environmental phenomenon, such as a range of mountains, or the central business district of a city. The quantification and assessment of aesthetic degradation is complex. Perhaps the most widespread example of its occurrence is in the deterioration of the quality of life as a consequence of damage to, or the destruction of, rural landscape by urban industrial development. As Tivy and O'Hare (1981) stress, aesthetic values are bound up with cultural ones. The most forceful opponents of aesthetic degradation are frequently, as Popper (1981) points out, upper-middle-class urban and surburban dwellers who wish to preserve the landscapes in which they live, and to which they periodically go for recreation. The more vociferous of these campaigners seem to have scant regard for opposing economic and/or social objectives in the localities in question. (See also POLLUTION.)   RLJ

Reading and References
†Ehrlich, P.R., Ehrlich, A.H. and Holdren, J.P. 1977: *Ecoscience: population, resources, environment*. 2nd edn. San Francisco and Oxford: W.H. Freeman.
Popper, F.J. 1981: *The politics of land use reform*. Madison: University of Wisconsin Press.
Tivy, J. and O'Hare, G. 1981: *Human impact on the ecosystem*. Edinburgh and New York: Oliver & Boyd.

aestivation The dormancy of certain animals during the summer season, the dry season or prolonged droughts. It is an important means of adaptation for desert animals, and can be contrasted with hibernation – a state of dormancy in the winter months.

affluent A stream or river flowing into another, a tributary.

after-glow A faint glow in the western sky seen occasionally just after sunset when the sun is 3° to 4° below the horizon. Probably the result of scattering of white light by dust particles in the atmosphere.

aftershocks A series of small earthquakes following a major tremor and originating at or near its focus. Aftershocks generally decrease in frequency over time but may occur over a period of several days or months.

ageostrophic A type of atmospheric motion in the troposphere in which the horizontal pressure gradient is not balanced with the deviating force (CORIOLIS FORCE) owing to the wind velocity. It is associated with vertical motion and hence the formation of cloud and weather.

agglomerate A rock composed of angular fragments of lava, generally more than 20 mm in diameter, which have been fused by heat.

aggradation The building upwards of the landsurface by accumulation of material deposited by various geomorphological agencies (e.g. by wind, wave or water).

aggregation ratio of a soil is the ratio of the percentage weight of clay minerals, determined by mineralogical analysis, to the percentage weight of clay particles, determined by sedimentation methods. The ratio is meant to account for problems with the ACTIVITY of a soil in which clay mineral particles aggregate and act as clays, but have a size corresponding to a value in the silt size range.   WBW

aggressivity In the context of limestone solution, the propensity of water to dissolve calcium carbonate. When water comes in contact with air it dissolves an amount of carbon dioxide into the water. The resultant carbonic acid can dissolve calcium carbonate (via the theoretical compound $CaHCO_3$) until the aggressiveness of the water diminishes (when the CO_2 is used up). It is then said to be saturated with respect to calcium carbonate. When two saturated solutions of calcium carbonate are mixed it is possible for the resulting water to be aggressive. This phenomenon, first identified by Bögli (1971, English version) is called mixing corrosion or, in German, *Mischungskorrosion*.   PAB

Reference
Bögli, A. 1971: Corrosion by mixing of karst water. *Transactions of the Cave Research Group of Great Britain* 13, pp. 109–14.

agonic line A shifting, irregular imaginary line running through the earth's north and south magnetic poles along which the compass needle points to true north, hence the line of no magnetic variation.

agroclimatology The study of the interaction between climatological and hydrological factors and agriculture, including animal husbandry and forestry. Its aim is to apply climatological information for the purpose of improving farming practices and increasing agricultural productivity in quantity and in quality. Agroclimatology and AGROMETEOROLOGY share nearly the same aims, scope and methodology. However, in their application the latter tends to emphasize weather forecasting in dealing with daily problems, whereas the former is more concerned with the use of mean data as a guide to long-range planning.   ASG

Reading
Chang, J.H. 1968: *Climate and agriculture*. Chicago: Aldine.

agroforestry Any system where trees are deliberately left, planted or encouraged on land where crops are grown or animals grazed. It includes practices as diverse as slash-and-burn agriculture, the growth of shade trees and the use of living fences either to contain or to exclude animals. Deep-rooted trees tap nutrient sources that are out of reach of most crops; these nutrients become readily available when the leaves fall. Leguminous trees improve soil fertility directly through nitrogen fixation. Tree roots help to bind the soil and increase aeration. Mixtures of trees and crops provide more complete ground cover which helps to prevent soil erosion and weed invasion, while making full and productive use of available solar radiation. Leaf litter from the trees adds organic matter to the soil and acts as a mulch. Tree cover helps to regulate temperatures, reducing extremes. Farming communities benefit from a regular supply of wood and other tree products. Multi-purpose trees can provide fodder for livestock, edible fruits and nuts, fuel, timber, supports for climbing vegetables and medicinal products.   ASG

agrometeorology The science concerned with the application of meteorology to the measurement and analysis of the physical environment in agricultural systems. The influence of the weather on agriculture can be on a wide range of scales in space and time, and this is reflected in the scope of agrometeorology. At the smallest scale the subject involves the study of microscale processes taking place within the layers of air adjacent to leaves of crops, soil surfaces or animals' coats (see MICROCLIMATE; MICROMETEOROLOGY). These processes determine rates of exchange of energy, and mass between the surface and the surrounding air. Such exchange rates are the essential link between the biological response and the physical environment. For example, the capture of radiant energy and its use to convert carbon dioxide and water into carbohydrates are essential elements in crop growth. Agrometeorologists have studied how the structure of leaf canopies affects the capture of light and how measurements of the atmospheric carbon dioxide concentration may be used to determine rates of crop growth.

On a broader scale agrometeorologists attempt to use standard weather records to analyse and predict responses of plants and animals. An area of particular interest concerns the estimation of water use by crops as a basis for planning irrigation requirements. Methods based on empirical correlations with windspeed, sunshine, temperature and humidity have been superseded by methods with a sounder physical basis, often using new measuring techniques (see EVAPOTRANSPIRATION). Other examples of this scale of agrometeorology include procedures for forecasting the occurrence of damaging frosts, or plant or animal diseases. Although the subject implies a primary concern with atmospheric processes, the agrometeorologist is also commonly interested in the soil environment because of the large influence which the weather can have on soil temperature and on the availability of water and nutrients to plant roots. Agrometeorology is also increasingly concerned with the study of the environment of glasshouses, animal houses and other protected environments designed for improving agricultural production.   MHU

Reading
Campbell, G.S. 1977: *An introduction to environmental biophysics*. New York: Springer-Verlag.

Monteith, J.L. ed. 1975: *Vegetation and the atmosphere*. Vols. I and II. London: Academic Press.

Smith, L.P. 1975: *Methods in agricultural meteorology*. Amsterdam: Elsevier.

aiguille A sharply pointed rock outcrop or mountain peak. Often applied to pinnacles which are the products of frost action.

air mass A body of air which is quasi-homogeneous in terms of TEMPERATURE and HUMIDITY characteristics in the horizontal plane and has similar LAPSE RATE features. An ideal air mass is a BAROTROPIC fluid in which isobaric and isosteric (constant specific volume) surfaces do not intersect. Air mass classification (see diagram) is based on the nature of the source

The 'Chamonix Aiguilles' in the Mont Blanc Range of the Alps show the denudational effects of high altitude weathering associated with nivation and frost wedging.

	Tropical	Polar	Arctic/Antarctic
Maritime	Maritime tropical (mT) warm and very moist; near Azores in N. Atlantic	Maritime polar (mP) cool and fairly moist; Atlantic south of Greenland	Arctic or Antarctic (A) (AA) very cold and dry;
Continental	Continental tropical (cT) hot and dry; Sahara desert	Continental polar (Cp) cold and dry; Siberia in winter	frozen Arctic Ocean central Antarctica

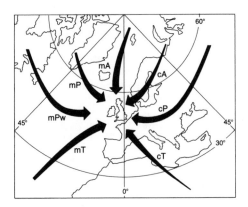

Air mass classification and properties with examples of source regions.

area (e.g. polar, tropical) and the characteristics of the surface during the outward trajectory (e.g. maritime, continental). Thermodynamic and dynamic factors will modify the properties of air masses in their transit from source areas.   AHP

Reading
Balasco, J.E. 1952: *Characteristics of air masses over the British Isles.* Meteorological Office, Geophysical Memoir 11. London: HMSO.
Harvey, J.G. 1976: *Atmosphere and ocean.* Sussex: Artemis.
Miller, A.A. 1953: Air mass climatology. *Geography* 38, pp. 55–67.

air parcel An imaginary body of air to which may be assigned any or all of the dynamic and thermodynamic properties of the atmosphere. Investigation of the STABILITY of the atmosphere is made most simply by the 'parcel method' in which it is hypothesized that a test parcel of air moves vertically with respect to its environment as represented by an ascent curve on a TEPHIGRAM.   AHP

Reading
Barry, R.G. and Chorley, R.J. 1992: *Atmosphere, weather and climate.* 6th edn. London: Routledge.

air pollution If man had never evolved on this planet the composition of the earth's ATMOSPHERE would be different from what

it is today. Air pollution can therefore be defined as the presence in the atmosphere of natural or man-made or man-caused contaminants in a given quantity, and for a given time, that are damaging to human, plant or animal life; or to property; or sensually interfere with the comfortable enjoyment of life. Common air pollutants include smoke, sulphur dioxide, carbon monoxide, carbon dioxide, nitric oxide, nitrogen dioxide, ozone and lead.   JET

Reading
Lyons, T.J. and Scott, W.D. 1990: *Principles of air pollution meteorology.* London: Belhaven.
Seinfeld, J.H. 1986: *Atmospheric chemistry and physics of air pollution.* New York and Chichester: Wiley.

air–sea interaction Occurs because the atmosphere and the oceans constitute a single mechanical and thermodynamic system of two coupled fluids, and transfers of heat, momentum, solids and gases take place across the air–sea boundary. Because the air–sea boundary is a moving surface of great complexity, observation of these fluxes is frequently difficult; but we know that interaction occurs in a manner which is so complex that cause and effect cannot always be distinguished. Such small-scale processes as wind stress, heat transfer and EVAPORATION are of fundamental importance in the generation and maintenance of all atmospheric and oceanic circulations (Perry and Walker 1977). Wind stress is responsible for a remarkable variety of phenomena. In the air it causes a reduction of windspeed near the surface, while on the sea the stress generates surface waves and leads to a stirring of the water in its surface layers, or if the degree of stability of the water permits, throughout its depth.

Because the area of the ocean surface is more than twice that of the land, energy exchange at that surface dominates world climate, and the oceans drive the atmosphere above them through the transfer of heat energy. The ocean's physical characteristics serve to make it the climatic system's largest reservoir of heat and momentum. The mass of the ocean and its high specific heat combine to make the ocean's heat capacity approximately four thousand times that of the atmosphere, while its total momentum is several times that of the atmosphere in spite of the fact that the ocean moves much more slowly than the air. These facts help to explain why the ocean is relatively sluggish in its reaction to outside perturbations. The ocean acts as a low-pass filter on the high-frequency perturbations of the atmospheric field. The oceans, because of their great heat capacity, can act as a stabilizing influence on an otherwise frenetic atmosphere, and the role of the oceans in global climate processes has become a major priority of scientific investigation.

The climate system has variability on many time-scales, some of which result from internal instabilities in the ocean–atmosphere system. Because changes of TEMPERATURE at the sea surface affect the fluxes of sensible and LATENT HEAT and are believed to have profound climatological consequences, SSTAs (sea surface temperature anomalies) analysis has become a focal point in climatology since the 1960s.

Namias (1975) has given many examples of the effects of anomalous sea temperature patterns in both the North Atlantic and the North Pacific on the overlying atmosphere, and the subsequent creation of large-scale climatic anomalies in those areas. A series of such relationships has been established based on these case studies and those of Bjerknes (1969) who found very large-scale air–sea systems in tropical latitudes. Verification of these observations and hypotheses have come from computer simulations using hemispheric and global numerical models. Such experiments may be regarded as sensitivity tests but they provide no information on the possible origin and behaviour of sea surface temperature anomalies. The complete physical consequences of interactions between the sea and the air can ultimately be calculated only through the use of coupled air–sea models in which each fluid is free to respond to the influence of the other, and such models are still in their infancy. Recent empirical evidence suggests that it is the atmosphere that drives the ocean rather than vice versa.

Several large-scale international observation programmes were initiated under the Global Atmospheric Research Programme (GARP), including the Atlantic Tropical Experiment (GATE) to increase understanding of air–sea interaction processes. Since 1985 the Tropical Ocean Global Atmosphere (TOGA) programme has sought to predict climatic variability as a result of better understanding of many aspects of ENSO (El Niño Southern Oscillation) events (Anderson and Davey 1992). Over periods from years to decades the ocean at great depths and in higher latitudes is thought to be involved in the regulation of climate and its changes and the World Ocean Circulation Experiment (WOCE) is designed to improve understanding, using satellite-borne sensors to monitor the global ocean.   AHP

References and Reading
Anderson, D. and Davey, M.K. 1992: *The Tropical Ocean Global Atmosphere Programme*. World Meteorological Organization, Bulletin 41, pp. 402–13.
Bjerknes, J. 1969: Atmospheric teleconnections from the equatorial Pacific. *Monthly weather review* 97, pp. 163–72.
Namias, J. 1975: *Short period climatic variations*. 2 vols. San Diego, Calif.: University of California Press.
Perry, A.H. and Walker, J.M. 1977: *The ocean–atmosphere system*. London: Longman.

air shed   The conceptual boundary of an atmospheric 'catchment' used in crude modelling of air pollution around known sources of pollution under expected conditions of air movement and rates of chemical change. A term analogous with 'watershed'.

aklé   A type of parallel wavy dune pattern in which the interdune areas are enclosed by crescentic elements of the dune ridges. It is sometimes called a fishscale dune pattern.

alas   A steep-sided, flat-floored depression, sometimes containing a lake, found in areas where local melting of permafrost has taken place. It is one manifestation of thermokarst.

albedo   A measure of the reflectivity of a body or surface derived from the Latin *albus* white. The albedo is defined as the total RADIATION reflected by the body divided by the total incident radiation. Numerical values are expressed between the ranges of either 0–1 or 0–100 per cent. It is therefore wavelength integrated across the full solar spectrum, while the term reflectivity is generally associated with a single wave-length or narrow waveband, i.e. a spectral reflectivity. The term albedo was already in common use by astronomers at

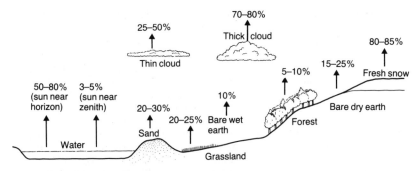

The albedo of the earth's surface. *The fraction of the total radiation from the sun that is reflected by a surface is called its albedo. The albedo for the earth as a whole, called the* planetary albedo, *is about 35 per cent. The albedo varies for different surface types. Note also that the angle at which the sun's rays strike a water surface greatly affects the albedo value.*
Source: *A.S. Goudie 1984: The nature of the environment. Oxford and New York: Basil Blackwell. Figure 2.2.*

Albedos for selected objects

Water surfaces		
Winter:	0° latitude	6
	30° latitude	9
	60° latitude	21
Summer:	0° latitude	6
	30° latitude	6
	60° latitude	7
Bare areas and soils		
Snow, fresh-fallen		75–95
Snow, several days old		40–70
Ice, sea		30–40
Sand dune, dry		35–45
Sand dune, wet		20–30
Soil, dark		5–15
Soil, moist grey		10–20
Soil, dry clay or grey		20–35
Soil, dry light sand		25–45
Concrete, dry		17–27
Road, black top		5–10
Natural surfaces		
Desert		25–30
Savannah, dry season		25–30
Savannah, wet season		15–20
Chaparral		15–20
Meadows, green		10–20
Forest, deciduous		10–20

Forest, coniferous	5–15
Tundra	15–20
Crops	15–25
Cloud overcast	
Cumuliform	70–90
Stratus (500–1000 ft thick)	59–84
Altostratus	39–59
Cirrostratus	44–50
Planets	
Earth	34–42
Jupiter	73
Mars	16
Mercury	5.6
Moon	6.7
Neptune	84
Pluto	14
Saturn	76
Uranus	93
Venus	76
Human skin	
Blond	43–45
Brunette	35
Dark	16–22

Source: Sellers, W.D. 1965: *Physical climatology.* Chicago: University of Chicago Press. P. 21.

the beginning of the twentieth century (e.g. Russell 1916) when it referred to the whole-planet value. The earth's albedo is close to 0.3, contrasting strongly with its highly reflective neighbour, Venus, which has an albedo of 0.7. The advent of orbiting satellites permitted measurement of reflected radiation at the top of the atmosphere above specific geographical locations. Maps of albedo are now produced illustrating the variation of reflectivity over the globe. This top-of-the-atmosphere albedo is termed the *planetary* or *system albedo*. The satellite-sensed radiation is composed of the reflected beams from the surface, the atmosphere and clouds. Strictly, surface albedos can be measured only with albedo-meters mounted close to the surface but they may also be calculated from clear-sky albedos measured by satellites. Surface albedos range in value: oceans: 0.07, dense forests: 0.10, grass and farmlands: 0.16–0.20, bright deserts: 0.25–0.40, and highly reflective ice: 0.40–0.60 and snow surfaces: 0.50–1.0. CLOUDS generally have high albedos though they too exhibit a considerable range from *cumuliform* clouds (0.80) to some *cirriform* clouds which barely reflect solar radiation making their detection from space very difficult. AH-S

Reference
Russell, H.N. 1916: On the albedo of the planets and their satellites. *Astrophysical journal* 43.3, pp. 173–96.

alcove An arcuate, steep-sided cavity on the side of a rock outcrop which has been produced by erosion by water, especially spring sapping or solutional processes.

alcrete Aluminium-rich duricrusts, often in the form of indurated bauxites. Generally the products of the accumulation of aluminium sesquioxides within the zone of weathering.

alfisol Relatively young, acid soils characterized by a clay-enriched B horizon, commonly occurring beneath deciduous forest in humid, subhumid, temperate and subtropical climates. A soil order of the US SEVENTH APPROXIMATION.

algae A large group of simple plants containing chlorophyll. They are either aquatic or live in damp places and range in form from minute plankton to enormous seaweeds.

algal bloom A spontaneous proliferation of microscopic algae in water bodies as a result of changes in water temperature or chemistry. Algal blooms may be characteristic of lakes where eutrophication has been caused by the addition of pollutants.

alidade *a.* The sighting device, index and reading device of a surveying instrument.
b. A straight-edged rule with a sighting device mounted parallel to the ruler used to plot the direction of objects.

aliens Organisms deliberately or accidentally introduced by man into regions outside the range of their natural distribution. The term is normally applied to species which are able to establish 'wild' populations, however small and temporary, in their new location. Some aliens are able to spread very successfully through wide areas and may become serious and invasive pests; Elton (1958) regards the more aggressive alien arrivals as 'invasions'.

Many alien species have been deliberately introduced into zoos and gardens, from which they have later escaped to form naturalized populations. In the plant kingdom, a classic case is that of the Oxford ragwort (*Senecio squalidus*), which spread from the Oxford Botanic Garden along the railway network of Britain. Many introductions are totally accidental, arriving on ships, in packing or even on aircraft tyre-treads. Unfortunately certain aliens have created serious ecological problems in their new homes, such as the European rabbit in Australia, and island ecosystems are especially susceptible to disruption by alien invaders: the rare organisms of Galápagos Islands made famous by Darwin have been put under pressure by the arrival of a range of alien mammals, including dogs, cats, rats, pigs and goats. Finally, many pets escape to pose threats to both agriculture and native ecosystems, such as the gerbil (*Meriones unguiculatus*) in Florida. (See also EXOTIC and ISLAND BIOGEOGRAPHY.) PAS

Reading and Reference
Elton, C. 1958: *The ecology of invasions by animals and plants.* London and New York: Methuen.
†Lever, C. 1977: *The naturalized animals of the British Isles.* London: Hutchinson.
†Salisbury, E. 1961: *Weeds and aliens.* London: Collins.
†Simmons, I.G. 1979: *Biogeography: natural and cultural.* London: Edward Arnold.

alimentation The accumulation in quantity of ice, through snowfall or avalanching, in a firn field contributing to a glacier.

allelopathy The production of chemicals by plants in order to inhibit or depress the growth of competing plants. The phenomenon is probably far more widespread than generally thought and is known to occur in plant communities as diverse as desert shrubs, tropical and temperate forests and heathlands. Examples include the checking of spruce growth on heather moorland by the production of a chemical by the heather roots which inhibits the growth of the mycorrhizal fungi essential for good growth of the trees; the inhibition of herbaceous species by the shrubs of the Californian chaparral, and the suppression of their own seedlings (autotoxicity) by a number of forest trees, such as black walnut (*Juglans nigra*) and the silky oak (*Grevillea robusta*). Several plants also use chemical defences against herbivores, e.g. oak leaves with a high tannin content are less palatable to certain defoliating caterpillars. KEB

Reading
Ashton, D.H. and Willis, E.J. (1982) *Antagonisms in the regeneration of Eucalyptus regnans in the mature forest.* In E.I. Newman, ed., *The plant community as a working mechanism.* British Ecological Society: special publication no. 1. Oxford: Blackwell Scientific. Pp. 113–28.
Krebs, C.J. 1985: *Ecology: the experimental analysis of distribution and abundance.* 3rd edn. New York: Harper & Row.

Allen's rule The rule which states that the relative size of the limbs and other appendages of warm-blooded animals tends to decrease away from the equator. This correlates with the increased need to conserve body heat.

Allerød The name given to an INTER-STADIAL of the *Late Glacial* of the last glaciation of the Pleistocene in Europe. The classic threefold division into two cold zones (I and III) separated by a milder interstadial (zone II) emanates from a type section at Allerød, north of Copenhagen, where an organic lake mud was exposed between an upper and lower clay, both of which contained pollen of *Dryas octopetula*, a plant tolerant of severely cold climates. The lake muds contained a cool temperature flora, and the milder stage which they represented was called the Allerød Interstadial. The interstadial itself and the following Younger Dryas temperature reversal are sometimes called the Allerød Oscillation. The classic date for the interstadial is 11,350–12,000 BP, but its exact date and status are in dispute.   ASG

Reading
Lowe, J.J. and Gray, M.J. 1980: The stratigraphic subdivision of the Lateglacial of NW Europe: a discussion. In J.J. Lowe, M.J. Gray and J.E. Robinson eds, *Studies in the Lateglacial of north-west Europe.* Oxford: Pergamon. Pp. 157–75.
Mercer, J.H. 1969: The Allerød oscillation: a European climatic anomaly. *Arctic and alpine research* 1, pp. 227–34.

allochthonous Refers to the material forming rocks which have been transported to the site of deposition, whereas an autochthonous sediment is one in which the main constituents have been formed *in situ* (e.g. evaporites, coal etc.).

allogenic stream A stream which derives its discharge from outside the local area. The term is particularly used where local conditions do not generate much streamflow, for example in arid areas or ones with permeable rocks. Here stream-flow may be derived from distant parts of the topographic catchment where precipita-

tion and run-off are effective. Appearances can sometimes be misleading because streamflow may be augmented by contributions from local groundwater that are not readily appreciated.   JL

allogenic succession This process is caused by an external environmental factor rather than by the organisms themselves. Instances are the change in vegetation induced by the inflow and accumulation of sediment in a pond (a geomorphological process), or a change in regional climate. (See also AUTOGENIC SUCCESSION; CLISERE.)   JAM

allometric growth A biological concept which derives from 'the study of proportional changes correlated with variation in size of either the total organism or the part under consideration. The variates may be morphological, physiological or chemical' (Gould 1966, p. 629). Allometric growth therefore defines a condition in which a change in size of the whole is accompanied by scale-related changes in the proportions of aspects of the object under study. In terms of physical geography, investigation of such scale-related changes has generally concentrated on morphological variables so that, as Church and Mark (1980) point out in their major review of the concept, one is dealing with scale distortions of geometric relationships (compare D'Arcy Thompson 1961). If no such distortions occur, *isometric* growth has taken place. To give two examples: on the one hand, it is widely observed that the gradient of the principal stream channel in a drainage basin is reduced at an ever-decreasing rate as the drainage area enlarges; on the other hand, the relationship between channel width and the wavelength of meanders appears to be roughly constant, regardless of actual channel dimensions.

Following Church and Mark (1980), we can define some basic concepts. First, allometry refers to a proportional relationship of the form: $A_1/A_2 = b$. If the resulting ratio is constant for all values of A_2, isometry exists. If A_1 increases at a faster rate than A_2, there is positive allometric growth; if the reverse, negative. It is often the case that b represents the exponent in the general form of the power equation $y = ax^b$ (where a is a constant dependent upon the units of measurement): if y and x have the same scale dimensions, b will be

1.0. If the relationship is positively allometric b will be >1.0, and negatively allometric if <1.0. If y and x have different scale dimensions, then the value of b indicative of isometry will vary accordingly. (For example, if y is a length, L^1 and x is an area, L^2, a value of b of 0.5 would indicate isometry.) A clear departure from isometry is also indicated if the relationship between x and y, plotted as a power function, is curved rather than linear.

Church and Mark further indicate (1980, p. 345) that *dynamic* and *static* allometry should be distinguished. In the former case one is dealing with the changing proportions of an individual landform over time; in the latter data are taken from a number of individuals of different sizes at one moment. Strictly, a study of static allometry should include only individuals of equivalent age. In practice, this requirement may be difficult to meet and has frequently been ignored. It should be clear, however, that the interpretation of the results of studies which aim to investigate static allometry can be very readily complicated by extraneous sources of inter-individual variation, particularly those due to differences in materials and detailed history.

The concept of allometric growth was explicitly introduced into geomorphological literature by Woldenberg (1966) in an investigation of HORTON'S LAWS of drainage basin composition. The widest application, however, has been suggested by Bull (1975), who suggested that *all* proportional relationships of the form $y = ax^b$ are essentially allometric. The preceding definitions should have made it clear that this is far from necessarily the case and Bull's suggestion seems likely simply to obscure the true nature of the underlying concept.

Given the persistent concern of geomorphologists with the size and shape of landforms the idea of allometry is of obvious potential interest. Which relationships are scale-dependent? Which are isometric? Why? All these would seem to be valid and valuable questions to ask. Moreover, as landforms *do* alter in size over time, the adoption of this particular concept from biology appears to be quite permissible, especially since it involves very little modification of the underlying biological ideas. Nevertheless, not all geomorphological use seems to have been governed by a clear grasp of the basic principles and, in particular, workers have sometimes failed to appreciate the need for equations to be properly balanced in dimensional terms.

Church and Mark (1980), in their painstaking discussion of cases of allometric and isometric relationships in geomorphology, conclude with a very important theoretical proposition: that isometry indicates relationships which are, physically, completely defined. Allometric situations suggest the intrusion of as yet unidentified, scale-dependent controls. In brief, they consider that a reduction of allometric equations to properly dimensioned, isometric forms should produce valuable insights into the manner in which 'growth' influences 'form'.　　　　BAK

References
Bull, W.B. 1975: Allometric change of landforms. *Geological Society of America bulletin* 86, pp. 1489–98.
Church, M.A. and Mark, D.M. 1980: On size and scale in geomorphology. *Progress in physical geography* 4, pp. 342–90.
D'Arcy Thompson, A.W. 1961: *On growth and form*, ed. J.T. Bonner. Cambridge: Cambridge University Press.
Gould, S.J. 1966: Allometry and size in ontogeny and phylogeny. *Biological review* 41, pp. 587–640.
Woldenberg, M.J. 1966: Horton's laws justified in terms of allometric growth and steady state in open systems. *Geological Society of America bulletin* 77, pp. 431–4.

allopatric Adjective applied to the condition of two species or subspecies being distributed in areas which do not overlap.

allophane An amorphous hydrated aluminosilicate gel. The chemical composition is highly variable. The name is applied to any amorphous substance in clays.

alluvial channel A river channel that is cut in ALLUVIUM. This applies to most larger natural rivers; ones that are entirely developed in bedrock may be found in high-relief areas, but even there incising streams transporting the material they have eroded may have a discontinuous veneer of alluvial material. More generally, even those channels that have bedrock on the floors of their deepest scour pools have banks in alluvial materials which have been deposited during floods or during the lateral movement of such rivers.

Non-alluvial channels may be different in form and development from alluvial rivers. Bedrock channels may be constrained by rock outcrop, while meltwater streams flowing on glacier ice create channels by removing ice rather than interacting with the bed materials. This is an essential characteristic of alluvial channels: they are self-formed in their own transportable sediments and can adjust their morphology according to discharges and the sediment sizes and loads present.

Alluvial channels can be characterized in terms of their cross-section, planform and long profile properties. These are inter-related. For example, a BRAIDED RIVER planform pattern is usually associated with a shallow cross-section and with relatively steep gradients. But these form elements are also dependent on river discharge and stream power and on sediment properties. Alluvial channel systems are therefore complex ones to study. In the short term stable or equilibrium forms may be developed and these may alter when controlling conditions vary. Unfortunately, equilibrium states are not easy to define, nor are changes precisely predictable. The study of alluvial channels is a large area of scientific enquiry. (See also CHANNELS, TYPES OF RIVER/ STREAM.)   JL

Reading
Richards, K. 1982: *Rivers*. London and New York: Methuen.
Schumm, S.A. 1977: *The fluvial system*. New York: Wiley.

alluvial fans Depositional landforms whose surface forms a segment of a cone that radiates downslope from the point where the stream leaves the source area. The coalescing of many fans form a depositional piedmont that is commonly called a *bajada*. Each fan is derived from a source area with a drainage net that transports the erosional products of the source area to the fan apex in a single trunk stream. The plan view of the cone-shaped deposit is broadly fan-shaped with the contours bowing downslope. Overall radial profiles are concave and cross-fan profiles are convex. They vary greatly in size from less than 10 m in length to more than 20 km, and many large fans are thicker than 300 m. The debris that makes up fans decreases in size downfan, but is frequently coarse, and much of it has been transported by mudflow activity. Deposition is caused by decreases in depth and velocity where streamflow spreads out on a fan, and by infiltration of water into permeable super-ficial deposits.

Alluvial fans are widespread, especially in arid, mountainous and periglacial areas, but are especially notable in particular tectonic environments, where there is a marked contrast between mountain front and depositional area. Uplift creates moun-tainous areas that provide debris and increased stream competence.   ASG

Reading
Bull, W.B. 1977: The alluvial-fan environment. *Progress in physical geography* 1.2, pp. 222–70.
Rachocki, A. and Church, M.A. 1990: *Alluvial fans: a field approach*. Chichester: Wiley.

alluvial fill Sedimentary material depos-ited by water flowing in stream channels. During the seventeenth century the term included all water-laid deposits (including

Alluvial fans or cones in West Greenland. Note the raised margins to debris flow channels. Such landforms are especially significant in many arid and periglacial areas where coarse debris is produced and run-off is sporadic.

marine sediments), but in 1830 Lyell restricted its use to materials deposited by rivers (Stamp 1961). Particle sizes range from fine clays deposited by overbank waters to boulders deposited in the channel bed by large floods. Materials may be massively bedded if deposited by a single event, or they may occur in a variety of bed forms related to flow variation or location related to the active channel.

Alluvial fill may assume a variety of surface forms depending on the location of the deposition. Valley fills occur when stream deposits fill a rock-bound valley, resulting in a surface that is relatively flat in cross-section and that assumes a relatively gentle down-valley slope which is related to the channel slope during deposition. If these valley fill materials are subsequently eroded by the stream, remnants may be left in the form of terraces that appear as flat-topped elevated deposits along the valley side. When the gradient of the depositing stream rapidly declines deposition occurs in a cone-like arrangement with the apex of the cone at the upstream point where the stream first encounters the decline in gradient. When this cone-shaped deposit occurs in the water of a lake or ocean, a delta results, with only its uppermost portions exposed above water level. When the cone is deposited in a subaerial situation, such as when the stream debouches from a mountain front onto a large valley floor, an alluvial fan is formed.

The process of alluvial filling is usually initiated in one of three ways (Schumm 1977, pp. 203–34). First, a change in base level causes a reduced channel gradient resulting in progressive backfilling from lower to upper reaches of the valley. A change in base level might result from a relative rise in sea level or from tectonic activity that elevates the lower portions of the river course in relation to the upper portions. Secondly, the water-sediment relationships in the stream system may change to initiate sedimentation along the channel. Such changes could be the product of land use or climatic changes that increase sediment production or that decrease run-off. Thirdly, changes in channel configuration may trigger deposition, such as an engineered adjustment in the width/depth ratio so that the channel becomes relatively more wide and shallow.

Alluvial fill is economically significant in that it is frequently a reservoir of ground-water, especially when it is in the form of valley fill and alluvial fans. Sand and gravel in economically viable concentrations may also occur depending on the mode of deposition (Happ 1971). Soils developed on the surface of alluvial fill are frequently productive for agriculture.　　　WLG

Reading and References
Happ, S.C. 1971: Genetic classification of valley sediment deposits. *American Society of Civil Engineers: journal of the hydraulics division* 97, pp. 43–53.
†Schumm, S.A. 1977: *The fluvial system*. New York: Wiley.
Stamp, L.D. 1961: *A glossary of geographical terms*. London: Longman.

alluvial terrace See TERRACE.

alluvium Material deposited by running water. The term is not usually applied to lake or marine sediments and may be restricted to unlithified, size-sorted fine sediments (silt and clay). Fine material of marine and fluvial origin may not in practice be easy to distinguish; on geological maps this is often not attempted and both are classed together. Coarser sediment may not by convention be included, but there is no good reason for this if no particular grain size is intended. The term can be prefixed by 'fine-grained' or 'coarse-grained'. Studies of alluvial channels do not imply any particular grain sizes.

Distinguishing characteristics of alluvium are its stratification, sorting and structure. Coarser sediment deposited on the channel bed or in BARS is overlain by finer materials deposited from suspension, either in channel slackwater areas or as an OVERBANK DEPOSIT following floods. In detail, sedimentary structures may be complex and dependent on the type of river activity. Large-scale contrasts are often drawn, for example between the deposits of braided and meandering rivers, while different scales and types of CURRENT BEDDING may be present. The size of sediments involved may depend on that supplied to streams from slopes or BANK EROSION and on the distance from such sources that a particular reach may be, because rivers sort and modify alluvial materials as they are transported. Thus fine-grained alluvium can be dominant in the lower courses of present rivers; coarser alluvium can be found close to the supply points for such material (e.g. in alluvial cones in high-relief semi-arid areas or at glacier margins) and in earlier Pleistocene deposits in mid-latitudes.

Here the slope- and glacier-derived coarser materials produced under former cold-climate conditions contrast with the finer alluvium coming from more recent slope inputs.   JL

Reading
Allen, J.R.L. 1970: *Physical processes of sedimentation*. London: Allen & Unwin.
Reading, H.G. ed. 1986: *Sedimentary environments and facies*. 2nd edn. Oxford: Blackwell Scientific.

alp *a.* A shoulder high on the side of a glacial trough.
b. (In Switzerland) a summer pasture below the snow level.

alpha diversity The diversity that results when competition between species produces greater specialization within the different niches that they occupy and reduces the variation within specific species. Also known as niche diversification.

alpine The zone of a mountain above the tree line and below the level of permanent snow.

alpine orogeny The period of mountain-building during the Tertiary era, ending during the Miocene, that produced the Alpine-Himalayan belt.

altimetric frequency curve A frequency curve constructed by dividing an area into squares, determining the maximum altitude in each square and plotting the frequency of these determinations. A useful technique for rapid determination of generalized altitudes in an area, and much used by denudation chronologists for identifying erosion surface remnants.

altiplanation A form of solifluction, i.e. earth movement in cold regions, that produces terraces and flat summits that consist of accumulations of loose rock. An alternative term is cryoplanation.

altithermal During the Holocene there was a phase, of varying date, when conditions were warmer (perhaps by 1–3°C) than at present. In the Camp Century Ice Core (Greenland) a warm phase lasted from 4100 to 8000 BP, whereas in the Dome Ice Core (Antarctica) the warmest phase was between 11,000 and 8000 BP (Dansgaard *et al.* 1970; Lorius *et al.* 1979). Rainfall conditions may also have changed, aridity having triggered off both renewed sand dune activity and a decline in human population levels in areas such as the High Plains from Texas to Nebraska, USA. ASG

References
Dansgaard, W., Johnsen, S.J. and Clausen, H.B. 1970: Ice cores and paleoclimatology. In I.U. Ollson ed., *Radiocarbon variation and absolute chronology*. New York: Wiley. Pp. 337–51.
Lorius, C., Merlivat, L. Jouzel, J. and Pourclet, M. 1979: A 30,000-year isotope climatic record from Antarctic ice. *Nature* 280, pp. 644–8.

altocumulus See CLOUDS.

altostratus See CLOUDS.

alveolar Composed of, or having, small cellular structures akin to a honeycomb.

ambient Preceding or surrounding a phenomenon, e.g. ambient temperature refers to the temperature of the surrounding atmosphere, water or soil.

amensalism A type of interspecific interaction. Amensal species are inhibited by an inhibitor, but the inhibitor for its part is not particularly affected by the amensal. A common type of amensalism involves antibiotics secreted by plants which effectively keep other plants from growing near or under them.

amino acid racemization A dating technique based on the fact that protein preserved in the skeletal remains of animals undergoes a series of chemical reactions, many of which are time dependent. After the death of the organism the protein slowly degrades, as does the nature of the amino acids which form the basis of the protein. Thus an examination of the amino acid composition of bone and of carbonate fossils has potential for dating purposes (Miller and Mangerud 1985) of Quaternary materials. The degree of change in amino acid composition, however, depends on factors other than time (e.g. temperature), and therefore involves certain assumptions, particularly about temperature conditions, that may create substantial errors.   ASG

Reference
Miller, G.H., and Mangerud J., 1985: Aminostratigraphy of European marine interglacials *Quaternary Science Reviews* 4, 215–78.

amphidromic point See TIDES.

anabatic flows Upslope winds usually produced by local heating of the ground during the day. The most common type is the VALLEY WIND. Anabatic flows develop best on east or west facing slopes on days with clear skies. The air along the slope is heated by contact with the warm surface much more rapidly than air at the same elevation away from the slope. The resulting temperature difference sets up a thermal circulation with the air ascending along the slope and descending over the adjoining plain or valley. Under ideal conditions anabatic winds can reach speeds of 10–15 m s^{-1} and can be a factor in the spreading of forest fires in dry weather. If the air is moist the anabatic flow may produce anabatic clouds above the crest of the slope.   WDS

Reading
Atkinson, B.W. 1981: *Meso-scale atmospheric circulations.* New York and London: Academic Press.
Geiger, R. 1965: *The climate near the ground.* Cambridge, Mass.: Harvard University Press.
Oke, T.R. 1987: *Boundary layer climates.* 2nd edn. London: Routledge.

anabranching A type of channel pattern which resembles braiding but in which the islands are wide relative to the width of the river channel at average discharge. Brice (see CHANNELS, TYPES OF RIVER/STREAM) defines anabranching as the pattern where the width of the islands is more than three times the river width at average discharge. Anastomosing channels are those which have major distributaries that branch and then rejoin the main channel and these distributaries can be meandering or braided.   KJG

anaclinal Refers to a feature, especially a river or valley, which is transverse to strike and against the dip of strata.

andromy The migration of some fish species from salt to fresh water for breeding.

anaerobic Living or active in the absence of air or oxygen, e.g. some bacteria which obtain oxygen from sulphates and nitrates in aqueous environments.

ana-front A front which has ascending air at one side, particularly one experiencing the unusual phenomenon of rising cold air.

analemna A scale drawn on a globe to show the daily declination of the sun, enabling the determination of those parallels where the sun is directly overhead at any specific time of year.

anamolistic cycle The tidal cycle, normally taken as lasting 27.5 days, which is related to the varying distance between the earth and the moon.

anaseism The vertical component of the waves moving upwards from the focus of an earthquake.

anastomosis Braiding of rivers, i.e. the tendency of some streams to divide and reunite producing a complex pattern of channels.

anchor ice Submerged ice which is attached to the bottom. It occurs in polar seas where the upper water layer some metres in thickness is overturned by convection during freezing (see SEA ICE). Under these circumstances ice may attach itself to the bottom or around seaweed, if the water depth is sufficiently shallow. With continued growth the anchor ice may become buoyant and float to the surface, sometimes complete with seaweed. Anchor ice can also form on the beds of river channels in PERMAFROST areas.   DES

andosols Dark soils developed on volcanic rock and ash.

anemograph A self-recording instrument for measuring the speed and sometimes the direction of the wind.

anemometer A device for measuring the speed and direction of the wind.

angiosperms Plants which possess flowers and specialized seeds. The class is divided into Monocotyledoneae and Dicotyledoneae.

angle of dilation (θ) The angle by which the grains of a granular material are displaced and reorientated on a shearing surface (SHEAR STRENGTH) cutting through the mass of particles. The reorientation movement is a response to the interlocking of particles which provide the frictional resistance or shear strength. The angle (θ) is related to the ANGLE OF INTERNAL

SHEARING RESISTANCE (ϕ) and the static ANGLE OF PLANE SLIDING FRICTION (ϕ_{us}) by $\theta = \phi - \phi_{us}$. WBW

Reading
Statham, I. 1977: *Earth surface sediment transport*. Oxford: Clarendon Press.

angle of initial yield (φ_i) The angle of a slope of granular material at which movement, often as 'avalanching', is seen to start. The angle depends upon the type of material, particularly its packing and bulk density, which together create interlocking particles. It has a higher value than the ANGLE OF RESIDUAL SHEAR (φ_r). WBW

angle of internal shearing resistance (ϕ) The angle, usually measured in a TRIAXIAL or SHEAR BOX apparatus to give the friction angle φ, for granular materials, of the MOHR–COULOMB equation. It is not a constant for any material but depends upon the VOID RATIO or POROSITY, as well as other frictional properties which relate to the interlocking of particles. WBW

Reading
Statham, I. 1977: *Earth surface sediment transport*. Oxford: Clarendon Press.

angle of plane sliding friction (ϕ_{us}) The angle at which non-cohesive (granular) particles just begin to slide down a surface. Strictly, this is the static angle; if it is the angle at which particles just stop moving it is the dynamic angle. It can apply to individual granular particles, a mass of such particles, or a slab of rock. In the latter case it is related to joint friction. For any of these, the static angle ϕ_{us} and the dynamic angle ϕ_{ud} are approximately constant but ϕ_{us} is greater than ϕ_{ud}. WBW

Reading
Statham, I. 1977: *Earth surface sediment transport*. Oxford: Clarendon Press.

angle of repose A term which has been applied to the angle at which granular material comes to rest (also called the angle of rest) and approximates to the angle of scree slopes. Strictly, natural slopes for one material may have a variable angle of repose according to whether the material has just come to rest or is about to move, hence it is also related to the ANGLE OF INITIAL YIELD or the ANGLE OF RESIDUAL SHEAR. WBW

angle of residual shear (ϕ_r) The angle at which granular material comes to rest after movement. The angle is less than the ANGLE OF INITIAL YIELD (ϕ_i) and is comparable to the ANGLE OF INTERNAL SHEARING RESISTANCE (ϕ) for cohesionless material in its most loosely packed state; thus the difference between ϕ_r and ϕ_i represents a loss of strength of the material to define a residual shear strength. WBW

Reading
Statham, I. 1977: *Earth surface sediment transport*. Oxford: Clarendon Press.

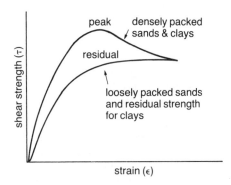

angular momentum A rather abstract quantity of great generality, essentially the product of the mass of a particle, the distance to an axis, and tangential components of velocity. It can be shown that forces directed towards (or away from) the axis cannot change the angular momentum. Many natural systems, including hurricanes and tornadoes, are dominated by forces acting towards a centre of low pressure. The (large) tangential components of velocity V at distance r are constrained by conservation of angular momentum to have V nearly proportional to l/r. This gives such systems a characteristic shape, flow field, and qualities of persistence. Angular momentum on the global scale is similarly constrained. (See also MOMENTUM BUDGET.) JSAG

angular unconformity A stratigraphic unconformity represented by younger strata overlying older strata which dip at a different angle, usually a steeper one.

anisotrophy The condition of a mineral or geological stratum having different optical or physical properties in different directions.

annual exceedance series The series obtained by selecting a number of discharges equal to the number of years of record but not using only one from each hydrological or water year as in the case of the ANNUAL SERIES.

annual series A term used in flood frequency analysis for the series of discharges obtained by selecting the maximum instantaneous discharge from each year of the period of record. The annual series is therefore equal to the number of years of hydrological record analysed. (See also FLOOD FREQUENCY.)   KJG

annular drainage A circular or ring-like drainage pattern produced when streams and rivers drain a dissected dome or basin.

Antarctic meteorology Deals with the circulation of the atmosphere and the *climate* south of the Antarctic Circle. Within this boundary the mean monthly temperature reaches a peak soon after the summer solstice, after which it drops quickly to its winter level (figure 1). The mean *temperature* of the warmest month in Antarctica is below freezing, except on the north and west coasts of the Antarctic Peninsula which are under the influence of the sea in addition to lying north of the Antarctic Circle. The fact that the annual temperature curve, both in single years and in the mean, usually has two or more minima is characteristic of most places in Antarctica. This coreless winter is due partly to the absence of INSOLATION for several months, and partly to the frequent exchange of air with lower latitudes. A secondary maximum in the mean temperature of May/June at many stations (Vostok and McMurdo Sound in figure 1) is the result of the usual enhancement of the long waves in the WESTERLIES from March to June, which frequently enables air from middle latitudes to reach Antarctica. Places on the east side of the preferred position of a long wave (see ROSSBY WAVE) then have not uncommon invasions of maritime air, while those on the west side of the wave (d'Urville in figure 1) lie in a flow of colder air. The northward extent of the long waves in the westerlies diminishes on average from June to September and at the same time the pack

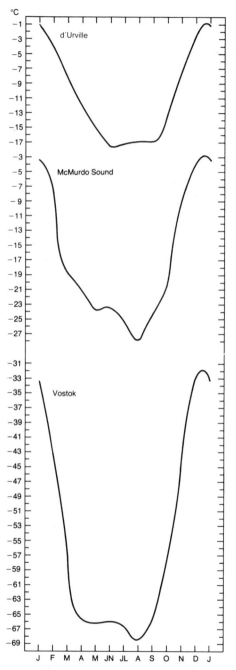

1. *Mean monthly surface air temperature at d'Urville (66°42'S, 140°00' E, 41 m); McMurdo Sound Station (77°53'S, 166°44'E, 24 m); and Vostok (78°28'S, 106°48'E, 3488 m).*

ice reaches its northernmost limit. The maritime air then has a shorter fetch, is colder, and travels much of the way to Antarctica over ice. This circulation change and the absence of the sun are responsible for the late minimum in August and September (van Loon 1967). When the sun reappears the temperature rises quickly.

The lowest temperature recorded in Antarctica is −88.3°C at Vostok; at the South Pole the temperature has fallen as low as −82.8°C. The cold pole is thus not at the geographic pole but over the highest parts of the plateau. An INVERSION OF TEMPERA-TURE in the lowest levels of the TROPO-SPHERE is a common feature of all months except December and January. The average strength of the inversion in winter varies

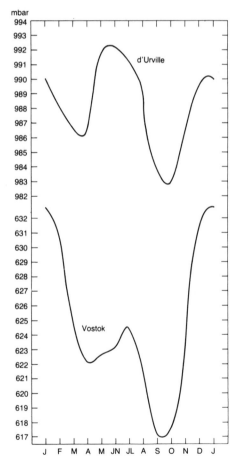

2. Mean monthly sea-level pressure at d'Urville, and Vostok's pressure at the level of the station.

from 5°C on the coast to 25°C on the plateau of East Antarctica. Under extreme conditions on a winter day the temperature on the plateau may rise by as much as 30°C with increasing height in the first kiometre above the surface. Sometimes, when an EXTRA-TROPICAL CYCLONE approaches, the combination of strong wind, advection of warmer air, and extensive cloudiness can destroy the inversion with the result that the temperature of the air at the surface rises drastically.

Antarctica is on average surrounded by a belt of low air pressure (see PRESSURE, AIR) which contains three to four minima with a pressure of 975–985 mb, depending on the season. This subantarctic TROUGH is the statistical result of a succession of lows from the north which reach the waters off the coast where they usually stagnate and decay. The pressure rises from the trough towards the continent, but as the surface mean pressure is a function of height it may be less than 600 mb on the highest parts of the continent. The subantarctic trough moves a few degrees of latitude towards the north from spring to summer and from autumn to winter, and towards the continent from summer to autumn and from winter to spring. While it moves south it also becomes deeper. A marked half-yearly wave in the pressure over the Antarctic with maxima in summer and winter is associated with this seasonal movement of the trough. The wave's amplitude is biggest near the coast where a second harmonic in the annual pressure curve explains as much as 80 per cent of the mean annual variance in several places. The decreasing amplitude of the half-yearly wave as one moves inland is illustrated by d'Urville and Vostok in figure 2. Note also that at Vostok, as over most of Antarctica, the pressure reaches its max-imum in summer.

The average winds in the interior blow downslope. As this KATABATIC FLOW approaches the coast it acquires an easterly component and a meridional component that follows the direction in which the coast trends. When a low from the north comes near the continent the wind associated with the pressure gradient in the low often reinforces the katabatic wind which results in violent gales and BLIZZARDS. The zonal wind along the coast and over the Antarctic Ocean has an appreciable half-yearly com-ponent with the weakest easterlies/strongest westerlies in the transitional seasons. The mean winds over the interior are light in comparison with the coastal winds, they change little with season, but are lightest in summer. The constancy of Antarctic winds is on the whole high.

Almost all PRECIPITATION in the Antarctic falls as snow whose measurement is complicated by frequent drifting and blowing. The precipitation decreases inland from the coast and, judged by the annual accumulation, the amount may be as high as 40–60 cm of water on the coasts. The comparatively large amounts on the coasts and the escarpment are owing to the combination of cyclonic and OROGRAPHIC PRECIPITATION. On the high plateau the precipitation is below a water equivalent of 5 cm a year. A modest amount of the precipitation in the interior consists of deposition of ice particles.

Cloudiness decreases from the coast toward the interior. Along the coast the average total cloudiness is 70 to 80 per cent in summer and 50 to 70 per cent in winter. On the plateau the lowest averages are 40 to 50 per cent in summer and less than 20 per cent in winter.   HVL

Reading and References
†Hisdal, V. 1960: *Norwegian–British–Swedish Antarctic Expedition 1949–1952. Scientific results: temperature.* Norsk Polarinstitutt., Oslo. Vol. 1, pp. 125–81.

†Mawson, D. 1915: *The home of the blizzard.* Reprinted 1969, New York: Greenwood Press.

†Meinardus, W. 1938: Klimakunde der Antarktis. In W. Koppen and R. Geiger eds, *Handbuch der Klimatologie,* Band IV (U). Berlin: Bornträger.

†Schwerdfeger, W. 1970: The climate of the Antarctic. In S. Orvig ed., *World survey of climatology,* vol. 14. Amsterdam: Elsevier.

†South African Weather Bureau 1957: *Meteorology of the Antarctic,* ed. M.P. van Rooy. Pretoria: Weather Bureau.

van Loon, H. 1967: The half-yearly oscillations in middle and high southern latitudes and the coreless winter. *Journal of the atmospheric sciences* 24, pp. 472–86.

†van Loon, H., Taljaard, J.J., Sasamori, T., London, J., Hoyt, D.V., Labitzke, K. and Newton, C.W. 1972: *Meteorology of the southern hemisphere.* Meteorological Monographs 13.35. Boston: American Meteorological Society.

antecedent drainage A drainage system which has maintained its general direction across an area of localized uplift.

antecedent moisture The soil moisture condition in an area before a rainfall, which moderates the area's run-off response to the rainfall. Antecedent moisture condition is usually expressed as an index and may be estimated by weighting past rainfall events to derive an antecedent precipitation index

(API). The API is a weighted sum of preceding rainfall within given time units and if a daily time base is used the API is often calculated as:

$$API_t = API_t^{-1}.k + P_t$$

where API_t is the antecedent precipitation index for day t, P_t is the precipitation on day t, and k is a decay factor ($k < 1.0$ and usually $0.85 < k < 0.98$).

Alternatively, a water budget based upon the preceding rainfall and evapotranspiration rates can be used to provide an estimate of soil moisture storage (see WATER BALANCE) or base flow in the area may be used as an index of soil moisture or antecedent moisture condition.   AMG

Reading
Chow, V.T. 1964: *Handbook of applied hydrology.* New York: McGraw-Hill.

antecedent platform theory The theory, propounded by Sir John Murray, that coral reefs and atolls developed through upward growth from submarine platforms.

antecedent precipitation index An index of moisture conditions in a catchment area used to assess the amount of effective rainfall that will form direct surface run-off. If there has been no rain for several weeks less rainfall will get into the streams than if the ground surface is already saturated. The index is calculated on a daily basis and assumes that soil moisture declines exponentially when there is no rainfall. Thus we have:

$$API_t = k. API_{t-1}$$

where API_t is the index t days after the starting point. The value of k will depend upon the potential loss of moisture so has a seasonal variation between 0.85 and 0.98. An allowance is made for any precipitation input during the period.   PS

anteconsequent stream A stream which flows consequent on an early uplift but antecedent to later stages of the same tectonic uplift.

Reading
Shelley, D. 1989: Anteconsequent drainage: an unusual example formed during constructive volcanism. *Geomorphology* 2, pp. 363–7.

anthropochore A plant which has been introduced to a specific area by man.

anthropogene A primarily Russian term for the period during which man has been an inhabitant of the earth (i.e. the past two to three million years).

anthropogeomorphology The study of the role of man as a geomorphological agent. There are very few spheres of human activity which do not create landforms (table 1). There are those landforms produced by direct anthropogenic processes. These are relatively obvious in form and origin and are frequently created deliberately and knowingly (table 2). Landforms produced by indirect anthropogenic processes are often less easy to recognize, not least because they do not so much involve the operation of a new process or processes as the acceleration of natural processes. They are the result of environmental changes brought about inadvertently by man's technology.   ASG

Reading
Brown, E.H. 1970: Man shapes the earth. *Geographical journal* 136, pp. 74–85.

Goudie, A.S. 1993: *The human impact*. 4th edn. Oxford: Basil Blackwell; Cambridge, Mass.: MIT Press.

Haigh, M.J. 1978: Evolution of slopes on artificial landforms, Blaenavon, UK. *University of Chicago, Department of Geography research paper* 183.

Jennings, J.N. 1966: Man as a geological agent. *Australian journal of science* 28, pp. 150–6.

Sherlock, R.L. 1922: *Man as a geological agent*. London: Witherby.

Table 1   *Some anthropogenic landforms*

Feature	Cause
Pits and ponds	Mining, marling
Broads	Peat extraction
Spoil heaps	Mining
Terracing, lynchets	Agriculture
Ridge and furrow	Agriculture
Cuttings	Transport
Embankments	Transport; river and coast management
Dikes	River and coast management
Mounds	Defence, memorials
Craters	War; *qanat* construction
City mounds (*tells*)	Human occupation
Canals	Transport, irrigation
Reservoirs	Water management
Subsidence depressions	Mineral and water extraction
Moats	Defence

Table 2   *Classification of anthropogenic landforming processes*

1   *Direct anthropogenic processes*
　1.1   Constructional
　　　tipping: loose, compacted, molten
　　　graded: moulded, ploughed, terraced
　1.2   Excavational
　　　digging, cutting, mining, blasting of cohesive or non-cohesive materials
　　　cratered
　　　trampled, churned
　1.3   Hydrological interference
　　　flooding, damming, canal construction
　　　dredging, channel modification
　　　draining
　　　coastal protection

2   *Indirect anthropogenic processes*
　2.1   Acceleration of erosion and sedimentation
　　　agricultural activity and clearances of vegetation
　　　engineering, especially road construction and urbanization
　　　incidental modifications of hydrological regime
　2.2   Subsidence: collapse, settling
　　　mining
　　　hydraulic
　　　thermokarst
　2.3   Slope failure: landslide, flow, accelerated creep
　　　loading
　　　undercutting
　　　shaking
　　　lubrication
　2.4   Earthquake generation
　　　loading (reservoirs)
　　　lubrication (fault plane)

Source: Haigh 1978.

antibiosis A specific form of antagonism which involves the formation by one organism of a substance which is harmful to another organism.

anticentre The point opposite the epicentre, above the focus, of an earthquake.

anticline A fold in geological strata that is convex upwards, i.e. forming an arch. The residual portion of such a fold.

anti-cyclone An extensive region of relatively high atmospheric pressure, typically a few thousand kilometres across, in which the low level winds spiral out clockwise in the northern hemisphere and counter-clockwise in the southern. Anticyclones are common features of surface

Small anticline in Ordovician mudstone at Penrhyn Castle, Cardigan, Wales. Such features are the result of folding of strata due to tectonic activity.

weather maps and are generally associated with calm, dry weather.

They originate either from strong radiative cooling at the earth's surface or from extensive subsidence through the depth of the troposphere. The first kind, known as cold anti-cyclones or highs, form across the wintertime continents and are shallow features produced by cold, dense air which is confined to the lower troposphere. The mobile ridges which occur in the polar air between frontal systems are also cold highs. The second kind, called warm anti-cyclones, are semipermanent features of the subtropical regions of the world. Here the descending branch of the HADLEY CELL ensures the persistence of a large downward flow of mass to supply the outflowing surface winds. The compression of the subsiding air leads to a deep, anomalously warm troposphere within which the anti-cyclonic circulation persists with height. Fine summers in Britain are normally associated with an unusual north-eastward excursion of the warm Azores anti-cyclone.

Although the subsidence and dry air in the highs tend to dampen convective activity, extensive and sometimes persistent low layer cloud can occur in some regions.   RR

Reading
Palmer, E. and Newton, C.W. 1969: *Atmospheric circulation systems*. New York and London: Academic Press.

antidune A ripple on the bed of a stream or river similar in form to a sand dune but which migrates against the direction of flow (i.e. upstream).

antiforms Upfolds of strata in the earth's crust; *synforms* are downfolds. In both cases the precise stratigraphic relationships of the rocks are not known, whereas in the case of anticlines and synclines they would be.

antipleion An area or a specific meteorological station where the mean annual temperature is lower than the average for the region.

antipodal bulge The tidal effect occurring at the point on the earth's surface opposite that where the pull of the moon's gravity is strongest. Hence it is the tidal effect at the point where lunar attraction is weakest.

antipodes Any two points on the earth's surface which are directly opposite each other so that a straight line joining them passes through the centre of the earth.

antitrades A deep layer of westerly winds in the TROPOSPHERE above the surface TRADE WINDS. In a simple way they represent the upper limits of the HADLEY CELL within which occurs the poleward transfer of heat, and momentum and water vapour.   BWA

antitriptic wind See WIND.

aphanitic Microcrystalline and cryptocrystalline rock textures. Pertaining to a texture of which the crystalline components are not visible with the naked eye.

aphelion The point of the orbit of a planet, or other solar satellite, which is farthest from the sun.

aphotic zone The portion of lakes, seas and oceans at a depth to which sunlight does not penetrate.

aphytic zone The portion of the floor of lakes, seas and oceans which, owing to their depth, are not colonized by plants.

apogee The point of the orbit of the moon or a planet that is farthest from the earth.

aposematic coloration The conspicuous and distinctive markings on a plant or animal which communicate that it is poisonous or distasteful to potential predators.

applied geomorphology The application of geomorphology to the solution of miscellaneous problems, especially to the development of resources and the diminution of hazards (Hails 1977). The great American geomorphologists of the second half of the nineteenth century – Powell, Dutton, McGee and Gilbert – were employed by the US Government to undertake surveys to enable the development of the West, and R.E. Horton, one of the founders of modern quantitative geomorphology, was active in the soil conservation movement generated by the 'Dust Bowl' conditions of the 1930s. It is notable that all these workers made fundamental contributions to theory, and thus to 'pure' geomorphology, even though much of their work was concerned with the solution of immediate environmental problems. More recently there has been much interest in the USA in engineering geomorphology and environmental impact assessment (Coates 1976): in the UK in the geomorphological problems of arid environments (Cooke et al. 1983), in Australia in the preparation of inventories of environmental conditions in different parts of that island, in Canada in the development of permafrost areas (Williams 1979), and among French geomorphologists working for ORSTOM and other bodies, etc. In the past two decades the role of the geomorphologist has developed, partly because with increasing population pressures and technological developments the impact of man on geomorphological processes has increased (Goudie 1993) and partly because the increasing competence of geomorphologists in the study of materials and processes proved to be more valuable than denudation chronology to the solution of environmental problems.

The role of the geomorphologist in environmental management (Cooke and Doornkamp 1990) can be subdivided into six main categories. First of these is the mapping of geomorphological phenomena. Landforms, especially depositional ones, may be important resources of useful materials for construction, while maps of slope angle categories may help in the planning of land use, and maps of hazardous ground may facilitate the optimal location of engineering structures. Secondly, because landforms are relatively easily recognized on air photographs and remote sensing imagery, they can be used as the basis for mapping other aspects of the environment, the distribution of which is related to their position on different landforms. An important example of this is the use of landform mapping to provide the basis of a soil map (Gerrard 1992; see also CATENA). The third category is the recognition and measurement of the speed at which geomorphological change is taking place. Such changes may be hazardous to man. By using sequential maps and air photographs, or archival information, and by monitoring processes with appropriate instrumentation (Goudie 1990), areas at potential risk can be identified, and predictions can be made as to the amount and direction of change. For example, by calculating rates of soil erosion in different parts of a river catchment an estimate can be made of the likely life of a dam before it is silted up, and measures can be taken to reduce the rates of erosion in the areas where the erosion is highest. Indeed, the fourth category of applied geomorphology is to assess the causes of the observed changes and hazards, for without a knowledge of cause, attempts at amelioration may have limited success. Fifthly, having decided on the speed, location and causes of change, appropriate solutions can be made by employing engineering and other means. Sixthly, because such means may themselves create a series of sequential changes in geomorphological systems, the applied geomorphologist may make certain recommendations as to the likely consequences of building, for example, a groyne to reduce coastal erosion. Examples of engineering solutions having unforeseen environmental consequences, sometimes to the extent that the original problem is heightened and intensified rather than reduced, are all too common, especially in many coastal situations (Bird 1979). ASG

Reading and References
Bird, E.C.F. 1979: Coastal processes. In K.J. Gregory and D.E. Walling eds, *Man and environmental processes*. Folkestone: Dawson. Pp. 81–101.

Coates, D.R. ed. 1976: *Geomorphology and engineering*. Stroudsburg, Pa: Dowden, Hutchinson & Ross.

Cooke, R.U., Brunsden, D., Doornkamp, J.C. and Jones, D.K.C. 1983: *Urban geomorphology in drylands*. Oxford: Oxford University Press.

Cooke, R.U. and Doornkamp, J.C. 1990: *Geomorphology in environmental management*. 2nd edn. Oxford: Oxford University Press.

Gerrard, J. 1992: *Soil geomorphology*. London: Chapman and Hall.

Goudie, A.S. ed. 1990. *Geormorphological techniques*. 2nd edn. London: Unwin Hyman.

—1993: *The human impact*. 4th edn. Oxford: Basil Blackwell; Cambridge, Mass.: MIT Press.

Hails, J.R. ed. 1977: *Applied geomorphology*. Amsterdam: Elsevier.

†Verstappen, H.Th. 1983: *Applied geomorphology*. Amsterdam: Elsevier.

Williams, P.J. 1979: *Pipelines and permafrost; physical geography and development in the circumpolar North*. London: Longman.

applied meteorology The use of archived and real-time atmospheric data to solve practical problems in a wide range of economic, social and environmental fields. The need to analyse and apply atmospheric information for such purposes arises because weather and climate impinge directly on vital human concerns such as agriculture, water resources, energy, health and transport. Specific operational problems are usually raised by clients or managers involved with weather-sensitive activities but a continuing dialogue between meteorologist and customer may well be necessary to ensure the optimum application of atmospheric knowledge in any particular field or industry. For example, atmospheric data can be usefully applied throughout the construction industry from the initial planning of location, through the design of buildings and the on-site construction phase to the control of energy and other

Applied meteorology: sectors and activities where climate has significant social, economic and environmental significance

Primary sectors	General activities	Specific activities
Food	Agriculture	Land use, crop scheduling and operations, hazard control, productivity, livestock and irrigation, pests and diseases, soil tractionability
	Fisheries	Management, operations, yield
Water	Water disasters	Flood/droughts/pollution abatement
	Water resources	Engineering design, supply, operations
Health and community	Human biometeorology	Health, disease, morbidity and mortality
	Human comfort	Settlement design, heating and ventilation, clothing, acclimatization
	Air pollution	Potential, dispersion, control
	Tourism and recreation	Sites, facilities, equipment, marketing, sports activities
Energy	Fossil fuels	Distribution, utilization, conservation
	Renewable resources	Solar/wind/water power development
Industry and trade	Building and construction	Sites, design, performance, operations, safety
	Communications	Engineering design, construction
	Forestry	Regeneration, productivity, biological hazards, fire
	Transportation	Air, water and land facilities, scheduling, operations, safety
	Commerce	Plant operations, product design, storage of materials, sales planning, absenteeism, accidents
	Services	Finance, law, insurance, sales

Source: Thomas, M.K. 1981: *The nature and scope of climate applications*. Canadian Climate Centre (unpublished).

running costs when the building is complete.

The comprehensive scope of applied meteorology is illustrated in the table, which lists the activities which can benefit from the application of atmospheric knowledge. Such applications cover all time-scales of atmospheric behaviour. For example, the type of agricultural production (such as the crop range) is dependent on the climate, including knowledge not only of average conditions but also of the variation around the mean and the frequency distribution of extreme events. On the other hand the annual yields and the profitability of farming are likely to be determined by the weather. For long-term strategic decisions a farmer may need to know the percentage probability of a frost occurring on any date near the beginning and the end of the growing season, as well as the average dates for the last spring frost and the first frost of autumn, while day-to-day activities will fall within the synoptic scale of atmospheric events and may well be influenced by forecasts of low night minimum temperature. Viewed in this sense, the basic purpose of applied meteorology is to help all sections of society to achieve a better adjustment to their atmospheric environment. Man exists in an ecological relationship with the atmosphere but the evidence of crop failure, pollution episodes and storm damage reveals the lack of an ideal relationship. The atmosphere is neutral – offering both resources and risks – and in order to attain improved adjustment it is necessary for us to understand, and then respond to, the processes of weather and climate both natural and man-modified. In particular a better physical knowledge of the atmosphere and its variations is required together with a greater understanding of the effects of weather and climate on other aspects of the natural environment as well as on man's activities. Given these requirements, applied meteorology can achieve its basic aims in three different, but related, ways:

1 the alleviation of atmospheric hazards, e.g. hurricanes, floods or urban snowfalls;
2 the improved economic efficiency of atmospherically sensitive enterprises, e.g. agriculture, transport; and
3 the satisfaction of social needs, e.g. planning of leisure time, participation in weather-sensitive recreations.

The evolution of applied meteorology has had three recognizable stages. The first stage, which in a heavily modified form is still with us, originated in the nineteenth and early twentieth centuries with the rise of scientific weather forecasting. Rapid developments in synoptic meteorology and the associated technology ensured that, almost until 1950, the chief practical benefits associated with atmospheric understanding were perceived to relate to the short-term forecast. With the more recent advent of computers and REMOTE SENSING the reliability of such forecasts has greatly increased. Indeed, the dissemination of weather forecasts to both specialist users and to the general public is the most visible activity currently undertaken by virtually all national agencies, such as the British Meteorological Office. The second stage of evolution dates from around 1940 when the term 'applied climatology' began to be used as the realization grew that meteorological agencies could provide additional valuable services to the community beside weather forecasts. According to Jacobs (1947), applied climatology was born in the USA out of the military requirements for planning months ahead on a 'probability risk' basis during the Second World War. At about the same time the concept of climate as a natural resource was being formulated by Landsberg (1946) and others so that the wartime experience was built upon by growing appointment of industrial and agricultural climatologists in the USA during the immediate postwar years. As the problems themselves became better defined the need for climate data became more explicit. Thus fundamental work on potential evapotranspiration soon led to the use of the climatic water budget as an aid to precise irrigation scheduling.

From these beginnings, applied meteorology (embracing both the climatological and the meteorological time-scales) entered the modern phase during the 1970s. First, the climatic time-scale was given new attention by a series of extreme and unexpected atmospheric fluctuations which confirmed the continuing vulnerability of the world to climatic variability. Many of these fluctuations resulted in major natural disasters, of which the Sahelian drought was the most prominent and persistent. At the same time, emerging concern about the enhanced GREENHOUSE EFFECT, mainly brought about by growing atmospheric concentrations of carbon dioxide and other radiatively active gases arising from the consumption of fossil fuels, provided further uncertainties related to longer-term climate change. For example, it became apparent that the stationarity of climate could no longer be taken for granted by planners or engineers when designing or operating long-term strategic facilities, such as water resource systems. The greater awareness of the need for better applications of climatic information prompted both academic and political responses. Within the discipline, a new wave of textbooks attempted to assess the existing state of knowledge in applied climatology (Maunder 1970, 1986; Mather 1974). Outside the discipline, the practical significance of applied climatology was rediscovered by policy-makers at national and international levels. Several countries (e.g. Canada and Australia) established a national climate programme which typically comprised a mix of database development, data applications, climate impact studies and applied research. In the USA, for example, the National Climate Program Act of 1978 (PL95-367) followed the declaration by Congress that atmospheric fluctuations can have major impacts on the supply of critical national resources, such as food, energy and water. This emphasis on climate as an agent of global change has led to greater support for research programmes into relevant climatology, an overview of which is maintained through the World Climate Applications Programme, approved by the World Meteorological Organization in 1979.

Secondly, attention has been directed to the shorter time-scale of weather events by increasing political pressures for meteorological agencies to become more cost-effective. The existing infrastructure for sensing, recording and reporting weather conditions has been developed largely in response to long-standing national needs, such as airline navigation, agriculture or severe storm warning. During the 1980s, the advent of new technologies, together with the changing needs of existing clients and the emergence of a new breed of weather sensitive customers, encouraged many governments to examine their expenditure on meteorology, estimated in 1990 to be in excess of $US 12,000 million globally (Maunder, 1990). Some governments then began to question their return on investment in such public facilities and suggested

that some of the costs of meteorology, traditionally wholly borne by the taxpayer, should be reduced (or at lest redistributed) by applying a more market-led approach to weather services. As a result, some state meteorological agencies (e.g. those of the United Kingdom and New Zealand) have been semi-privatized.

These changes have prompted a more competitive and commercial attitude towards applied meteorology based on better marketing skills and more targeted services, especially those directed towards new business clients, e.g. transport companies and supermarket chains. In turn, this has led to the search by meteorologists for better assessments of the economic benefits to be derived from weather services in order to:

1 Provide governments with a justification to spend some taxpayers' money to support the meteorological infrastructure.
2 Provide meteorologists with in-house guidance in allocating scarce resources to specific services and research and development programmes.
3 Provide customers with an incentive to take a revenue-earning service, thus increasing income and demand.

The valuation of weather information and services is notoriously difficult. Cost-benefit analysis is increasingly giving way to more sophisticated decision-making techniques which show how the economic value of (often) imperfect information changes as its quality increases, as shown by Katz and Murphy (1990). In addition, there has been some redefinition of the term 'meteorological services'. While the supply of timely sensed raw, or synthesized, data and general, public-service, weather forecasts can be regarded as meteorological *information*, meteorological *services* are now best thought of as client-tailored, value-added products which aid operational decision-making. These products can rarely be delivered without close attention to the client's needs. Therefore, most meteorological agencies now offer their clients a specialized consultancy service which aims to define the weather sensitivity of the client's activities and to demonstrate how weather services can be cost-effectively employed to achieve a more competitive edge.

Future progress in applied meteorology depends not only on an improved knowledge and prediction of the atmospheric fluctuations themselves but also on a better understanding of the linkage between atmospheric events and human activities. The provision of reliable seasonal climatic forecasts is an obvious goal but, as Lamb (1981) has pointed out, expensive research into climate system dynamics is predicated on the belief that responses and strategies for dealing with such climatic fluctuations can be attained. At present, the economic and social management responses necessary to achieve such optimal adjustment to climatic variation are still largely unknown. Technical improvements in weather forecasting, based on an assumed relationship between forecast skill (as measured by the meteorologists) and forecast value (as assessed by the end-user), need to be complemented with applied studies which validate, or modify, such relationships.

Most major research programmes in applied meteorology now include an element of weather and climate impact assessment. The challenge of such work, based on inter-disciplinary collaboration and synthesis, has been well stated by Kates (1980). Although growing attention is being given to the needs of the end-user of weather and climate information, with a particular stress on identifying weather sensitivity and improving decision-making, major questions remain. These surround the effective application of weather and climate forecasts both for existing levels of predictive skill and for future potential improvements in skill (Livezey, 1990). In turn, the development of appropriate human responses will depend on improvements in the supply of information and services. Given an increased awareness and understanding of atmospheric information by the user, combined with improved data delivery, new techniques of analysis will also be necessary before the promise of the future can be achieved. For example, more detailed, site-specific weather parameters need to be included in various numerical models (e.g. for crop yield or water use), while economic decision-making models capable of handling probabilistic meteorological information also require more development.   KS

Reading and References
Jacobs, W.C. 1947: Wartime developments in applied climatology. *Meteorological monographs* 1.1.

Kates, R.W. 1980: Climate and society; lessons from recent events. *Weather* 35, pp. 17–25.
Katz, R.W. and Murphy, A.H. 1990: Quality/value relationships for imperfect weather forecasts in a prototype multistage decision making model. *Journal of forecasting* 9, pp. 75–86.
Lamb, P.J. 1981: Do we know what we should be trying to forecast – climatically? *Bulletin of the American Meteorological Society* 62, pp. 1000–1.
Landsberg, H. 1946: Climate as a natural resource. *Scientific monthly* 63, pp. 293–8.
— and Jacobs, W.C. 1951: Applied climatology. In T.F. Malone ed., *Compendium of meteorology*. Boston: American Meteorological Society. Pp. 976–92.
Livezey, R.E. 1990: Variability of skill of long-range forecasts and implications for their use and value. *Bulletin of the American Meteorological Society* 71, pp. 300–9.
Mather, J.R. 1974: *Climatology: fundamentals and applications*. New York: McGraw-Hill.
Maunder, W.J. 1970: *The value of the weather*. London: Methuen.
— 1986: *The uncertainty business*. London: Methuen.
— 1990: Economic and social benefits of meteorological and hydrological services. In *Economic benefits of meteorological and hydrological services*. Proceedings of the Technical Conference 26–30 March 1990 (WMO No. 733). Geneva: World Meteorological Organization. Pp. 1–11.
WMO 1979: *Proceedings of the World Climate Conference*. Publication No. 537. Geneva: World Meteorological Organization.

aquaculture Fish farming. The breeding and rearing of marine and freshwater fish in captivity.

aquatic macrophyte A water plant large enough to be seen without a microscope. In some cases these plants, nourished by nutrients washed into streams and lakes from agricultural fertilizers, grow to such density that they exclude desirable fish species and clog waterways originally intended for navigation.

aquiclude See AQUIFUGE; GROUNDWATER.

aquifer See GROUNDWATER.

aquifuge An impermeable rock incapable of absorbing or transmitting significant amounts of water. Unfissured, unweathered granite is an example. Certain rocks such as clay and mudstone are very porous and absorb water, but when saturated are unable to transmit it in significant amounts under natural conditions. Such formations are known as *aquicludes*. The term *aquitard* is also sometimes used to describe the hydrological characteristics of the less permeable bed in a strategic sequence, that may be capable of transmitting some water, but not in economically significant quantities.   PWW

Reading
Freeze, R.A. and Cherry, J.A. 1979: *Groundwater.* Englewood Cliffs, NJ: Prentice-Hall.
Ward, R.C. 1975: *Principles of hydrology.* 2nd edn. Maidenhead, Berks.: McGraw-Hill.

aquitard See AQUIFUGE; GROUNDWATER.

arboreal Pertaining to trees.

arches, natural A bridge or arch of rock joining two rock outcrops which has been produced by natural processes of weathering and erosion.

archipelago A sea or lake containing numerous islands or a chain or cluster of islands.

arctic Variously defined depending on one's interest. The popular definition is to include all areas north of the Arctic Circle (Lat. $66\frac{1}{2}°N$) which is the latitude at which the sun does not rise in mid-winter or set in mid-summer. A more useful natural definition includes land areas north of the tree line and oceans normally affected by Arctic water masses.   DES

Reading
Sugden, D.E. 1982: *Arctic and Antarctic.* Oxford: Basil Blackwell; Totowa, NJ: Barnes & Noble.

arctic–alpine flora A group of plants displaying a disjunct geographical distribution that embraces both the lowland regions of the arctic and the high-altitude mountain areas of the temperate and even the tropical zones. The main mountain systems involved are the Rockies, the Alps, the Himalayas and high intertropical mountains, such as those of East Africa. Classic examples of the flora include *Anemone alpina* (Alps/Arctic), *Polygonum viviparum* (Alps, Altai, Himalayas/Arctic), *Ranunculus pygmaeus* (Alps, Rockies/Arctic), *Salix herbacea* (Alps, Urals, Rockies/Arctic) and *Saxifraga oppositifolia* (Alps/Arctic).   PAS

Reading
Löve, A. and Löve, D. 1974: Origin and evolution of the arctic and alpine floras. In J.D. Ives and R.G. Barry eds, *Arctic and alpine environments.* London: Methuen. Pp. 571–603.

arctic haze A reddish-brown atmospheric haze, which is often observed in the Arctic, especially in the winter and spring, when atmospheric conditions are calm. It consists primarily of atmospheric pollutants derived from industrial sources in Europe and northern Asia. These include sooty and acidic particles.   ASG

arctic meteorology Meteorological conditions in the Arctic reflect the combined influence of the earth's orbital geometry on the daylength and angle of the sun, the surface properties, especially the seasonal snow and sea ice cover, and the global atmospheric circulation. The Arctic Circle delimits the polar cap which experiences an annual cycle of winter 'night' and summer 'day', but the timing of the actual seasons is substantially out of phase.

A natural arch formed by marine erosion of limestone at Durdle Door, Dorset, England. Steeply dipping beds of highly variable lithology contribute to the morphological variety of this stretch of coastline.

February is typically the coldest month and the coastal zones are usually ice-free only in August and early September. In a climatic sense, the Arctic can be approximately defined with respect to the treeless zone of TUNDRA vegetation and permanently frozen ground (PERMAFROST) on land and the ocean areas with sea ice cover for most of the year.

There are no distinctive weather systems in high northern latitudes. In the middle TROPOSPHERE there is a polar vortex with a central Arctic closed low in summer and deep troughs over the eastern Canadian Arctic and eastern Siberia in winter. At the surface the mean pressure patterns are weak. In winter there is a ridge of high pressure across the Beaufort–East Siberian seas with deep Icelandic and Aleutian lows. Frontal systems enter the Arctic basins mainly via the Norwegian Sea and then move slowly around the central Arctic. Both highs and lows are persistent features in winter months, but they have little effect on the surface weather as a result of an intense ground-based radiatively controlled temperature inversion which persists throughout the polar night. The snow-covered pack ice essentially isolates the atmosphere from the Arctic Ocean so that temperatures average −30° to 35°C at the surface, but are 10°C higher at 1000 m altitude. Air from lower latitudes may move above this temperature inversion without disrupting it unless there are strong winds and a thick cloud cover which can radiate infrared radiation back to the snow-covered surface, causing the temperature to rise. This occurs rather infrequently when a deep depression system moves northward into the Arctic basin.

Winter cloud amounts are generally small, except near the open oceans or local open water areas, and snowfall is modest with average depths in spring of the order of 40 cm (12 cm water equivalent). Winds give rise to frequent drifting or blowing snow, causing drifts to accumulate wherever there are surface irregularities on land or on the ice. When temperatures are close to −40°C or below ice crystals commonly form in suspension in the air. This phenomenon, known as 'diamond dust', produces optical phenomena such as haloes and sun pillars when the sun is above the horizon. Any domestic or industrial emission at arctic settlements produces ice fog,

through crystallization, and in valley locations this will become persistent and dense.

An arctic high pressure is present in spring when the best weather for field operations is usually experienced. Despite clear skies and long days, however, the snow cover causes most of the incoming solar radiation to be reflected back to space instead of being absorbed and warming the surface. In summer, pressure systems are relatively weak although frontal disturbances tend to occur along the Siberian–Alaskan coastline where the land/sea ice thermal contrast gives rise to a frontal zone. These sytems bring thicker clouds and light precipitation, mainly falling as rain.

Snow cover on land melts rapidly in early June leaving a swampy tundra as a result of the underlying permafrost only about a metre below the surface. On the High Arctic islands, however, there is vegetation in only a few locally favoured spots and most of the surface is a barren polar desert. The Arctic sea ice loses its snow cover more gradually during June and July, but eventually 20–30 per cent of the bare ice surface is puddlecovered and ice motion assists in the mechanical breakup of the floes.

The snow and ice melt creates extensive wet surfaces in June, July and August and keeps air temperatures over the pack ice close to 0°C. Skies are commonly overcast with stratus or stratocumulus cloud over the Arctic Ocean and coastal areas. Inland, daytime heating will often give broken cloud cover and afternoon temperatures may rise to 15–20°C. Consequently, there are steep local temperature gradients in summer near the northern Arctic coasts of Alaska and Siberia.   RGB

Reading
Barry, R.G. 1983: Arctic ocean ice and climate: perspectives on a century of polar research. *Annals of the Association of American Geographers*, 73, pp. 485–501.

— 1989: The present climate of the Arctic Ocean and possible past and future states. In Y. Herman ed., *The Arctic Seas: climatology, oceanography, biology and geology*. New York: Van Nostrand Reinhold. Pp. 1–46.

— , Courtin, G.M. and Labine, C. 1981: Tundra climates. In L.C. Bliss, J.B. Cragg, D.W. Heal and J.J. Moore eds, *Tundra ecosystems: a comparative analysis*. Cambridge: Cambridge University Press. Pp. 81–114.

— and Hare, F.K. 1974: Arctic climate. In J.D. Ives and R.G. Barry eds, *Arctic and alpine environments*. London: Methuen. Pp. 17–54.

Ohmura, A. 1982: Climate and energy balance on the Arctic tundra. *Journal of climatology*, 2, pp. 65–84.

Orvig, S. ed. 1970: *The climates of polar regions*. World Survey of Climatology, ed. H.E. Landsberg, vol. 14. Amsterdam: Elsevier.

The Polar Group (eds D.J. Baker, U. Radok and G. Weller) 1980: Polar atmosphere-ice-ocean processes: a review of polar problems in climate research. *Reviews of geophysics and space physics* 18, pp. 525–43.

arctic smoke See FROST SMOKE.

areic Without streams or rivers.

arena A shallow, broadly circular basin hemmed in by a rim of higher land.

arenaceous Pertaining to, containing or composed of sand. Applied to sedimentary rocks composed of cemented sand, usually quartz sand.

areography The study of the geographical ranges of plant and animal taxa. It focuses on the form and size of taxonomic ranges, and differs in emphasis from BIOGEOGRAPHY (concerned with the delimitation of floral and faunal sets and the origins of their constituent elements) and ecogeography (concerned with the reasons for the form and size of taxonomic ranges).   ASG

Reading
Rapoport, E.H. 1982: *Areography: geographical strategies of species*. Oxford: Pergamon.

arête A fretted, steep-sided rock ridge separating valley or cirque glaciers. The basic form is the result of undercutting or BASAL SAPPING by glaciers which evacuate any rock debris and thus maintain steep rock slopes. Arêtes are common whenever mountains rise above glaciers, for example in mountain chains and as nunataks protruding above ice sheets.   DES

argillaceous Pertaining to, containing or composed of clay. Applied to rocks which contain clay-sized material and clay minerals.

arid zone The portion of the earth between latitude 15° and 30° north of the equator and the same latitudes south which contains most of the world's warm deserts. Any area of the earth that receives less than a specific amount of rainfall annually, usually about 250 mm.

aridisol Desert soil with predominantly mineral profiles and frequently containing water-soluble salt accumulations. A soil order of the US SEVENTH APPROXIMATION.

arkose A sandstone containing more than 25 per cent feldspar. Any feldspar-rich sandstone.

armoured mud balls (also called clay balls, pudding balls, mud pebbles and mud balls). Roughly spherical lumps of cohesive sediment, which generally have diameters of a few centimetres, though much larger examples have been reported. Many examples are lumps of clay or cohesive mud that have been gouged from stream beds or banks by vigorous currents. They often occur in badlands and along ephemeral streams, but can also be found on beaches, in tidal channels, etc.   ASG

Reading
Bell, H.S. 1940: Armored mud balls: their origin, properties and role in sedimentation. *Journal of geology* 48, pp. 1–31.
Picard, M.D. and High, L.R. 1973: *Sedimentary structures of ephemeral streams*. Amsterdam: Elsevier Scientific. Pp. 49–51.

armouring A term used of heterogeneous river bed material when coarse grains are concentrated sufficiently at the bed surface to stabilize the bed and inhibit transportation of underlying finer material. An armour layer is coarser and better sorted than the substrate, and is typically only 1–2 grains thick. Genetically, a distinction exists between armoured and paved beds. An armoured bed is mobile during floods, but the coarser veneer reforms on or after the falling limb of a flood capable of disrupting and transporting the armouring grains as finer grains are winnowed and the protective layer is recreated. The substrate remains protected until the next event capable of entraining the armour grains. A paved surface, however, is more stable and is markedly coarser than the substrate. Whereas an armour layer is characteristic of an equilibrium channel in heterogeneous sediment, a paved layer often occurs in a channel experiencing degradation, and arises because of scour and removal of a substantial thickness of bed material such that the coarsest component is left as a lag deposit. The immobility of a paved surface is often indicated by discoloration and staining of the component grains; the occasional transport of armour grains keeps them clean.   KSR

Reading
Gomez, B. 1984: Typology of segregated (armoured paved) surfaces: some comments. *Earth surface processes and landforms* 9, pp. 19–24.

arroyo A trench with a roughly rectangular cross-section excavated in valley-bottom alluvium with a through-flowing stream channel on the floor of the trench (Graf 1983). Although the term gully is used for similar features, a gully is V-shaped in cross-section instead of rectangular and is excavated in colluvium instead of alluvial fill. The term arroyo in the sense of a stream bed has been used in Spanish since at least the year 775, but its modern use in English physical geography dates from the exploration and survey of the American West in the 1860s. Dodge (1902) first used the term in geomorphological research, indentifying arroyos as the product of stream-channel entrenchment resulting from changes in climate and land use that caused increased run-off.   WLG

Reading and References
†Cooke, R.U. and Reeves, R.W. 1976: *Arroyos and environmental change in the American South-West.* Oxford: Clarendon Press.

Dodge, R.E. 1902: Arroyo formation. *Science* 15, p. 746.

Graf, W.L. 1983: The arroyo problem – paleohydrology and paleohydraulics in the short term. In K.J. Gregory ed., *Background to palaeohydrology.* Chichester: Wiley.

artesian A term referring to water existing under hydrostatic pressure in a confined aquifer. The water level in a borehole penetrating an artesian aquifer will usually rise well above the upper boundary of the water-bearing rocks and may even flow out of the borehole at the surface, in which case it is known as a flowing artesian well.

Water moves in artesian aquifers as a result of differences in fluid potential (see EQUIPOTENTIALS), water moving towards areas of lower HYDRAULIC HEAD. Natural outflow points are artesian springs, where water boils up under pressure. As a consequence, the surface of an artesian spring is usually domed upwards. Artesian waters are often highly mineralized as a result of a long residence time underground. Hence artesian springs may sometimes build mounds of chemical precipitates deposited from emerging GROUNDWATER, especially if they are located in a tropical arid environment where evaporation is great.   PWW

artificial recharge See RECHARGE.

aspect The direction towards which a slope faces.

association, plant A unit for describing vegetation. It has been given many different interpretations (see Carpenter 1962), both as an abstract community type and as a concrete piece of vegetation (an individual stand) on the ground. In America it has been widely used to describe large-scale units of CLIMAX VEGETATION. In Europe, where it has been applied to smaller-scale units, its usage ranges from a synonym for plant community to a unit of specific rank. In PHYTOSOCIOLOGY the plant association is the fundamental unit, characterized by a particular combination of plant species. In the landscape, plant associations may be clearly defined units with sharp boundaries, they may merge with a transition zone (ECOTONE) or they may form a continuum of variation.   JAM

Reading and References
Carpenter, J.R. 1962: *An ecological glossary.* New York and London: Hafner.

†Langford, A.N. and Buell, M.F. 1969: Integration, identity and stability in the plant association. *Advances in ecological research* 6, pp. 84–135.

†Mueller-Dombois, D. and Ellenberg, H. 1974: *Aims and methods of vegetation ecology.* New York and London: Wiley.

asthenosphere A zone within the earth's upper MANTLE, extending from 50 to 300 km from the surface to a depth of around 700 km, characterized by a lower mechanical strength and lower resistance to deformation than the regions above and below it. It is approximately, though not exactly, equivalent to the zone in the mantle which transmits seismic waves at a low velocity, due to its partially melted state. In the PLATE TECTONICS model the asthenosphere is regarded as the deformable zone over which the relatively rigid LITHOSPHERE moves.   MAS

Reading
Davies, P.A. and Runcorn, S.K. eds 1980: *Mechanisms of continental drift and plate tectonics.* London: Academic Press.

Mörner, N.-A. ed. 1980: *Earth rheology, isostasy and eustasy.* Chichester: Wiley.

Wyllie, P.J. 1976: The earth's mantle. In J.T. Wilson ed., *Continents adrift and continents aground.* San Francisco: W.H. Freeman. Pp. 46–57.

astrobleme A term put forward by Dietz (1961) meaning 'star wound' and referring to the erosional remnant or scar of a structure of extra-terrestrial origin produced before the Pliocene by the impact of a meteorite on the earth's surface.   ASG

Reading
Dietz, R.S. 1961: Astroblemes. *Scientific American* 205, pp. 51–8.

asymmetric valley A river valley or glacial valley of which one side is inclined in a different angle to the other. Such valleys are a feature of periglacial areas where differences in aspect cause considerable differences in the strength of frost weathering and solifluction, but they can also be caused by structural circumstances (see UNICLINAL SHIFTING). ASG

Reading
Churchill, R.R. 1982: Aspect-induced differences in hillslope processes. *Earth surface processes and landforms* 7, pp. 171–82.

asymmetrical fold A fold in geological strata which has one side dipping more steeply than the other.

Atlantic coastlines Coastlines where the trend of the mountain ranges is at right angles or oblique to the coastline (e.g. south-west Ireland), whereas the Pacific (or concordant) type of coastline is parallel to the general trend-lines of relief (e.g. the Adriatic coast of former Yugoslavia). ASG

atmometer An instrument for measuring the rate of evaporation.

atmosphere The gaseous envelope of air surrounding the earth and bound to it by gravitational attraction. Up to a height of about 80 km the relative proportions of the major constituent gases (apart from water vapour) are more or less constant, as given in the table. The percentage volume of water vapour varies between less than 1 per cent and more than 3 per cent. Only carbon dioxide, ozone and water vapour vary locally in concentration. The atmosphere has been divided into vertical regions based on temperature as shown in the diagram. The atmosphere also contains natural pollutants in the form of AEROSOLS, dust and smoke from volcanoes, forest fires, soil

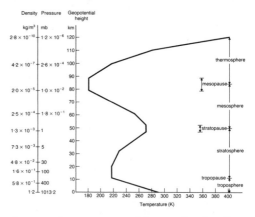

Average temperature structure of the atmosphere from 0 to 120 km.
Source: *Meteorological Glossary. 1972.*

erosion etc., and man-made pollutants such as sulphur dioxide, smoke, nitrogen dioxide, nitric oxide, ozone, lead and carbon monoxide. JET

atmospheric composition A unit mass of dry air is made up of 75.5 per cent nitrogen (N_2), 23.2 per cent oxygen (O_2), 1.3 per cent argon (A), 0.01 per cent carbon dioxide (CO_2) and smaller proportions of gases such as neon and helium. At a height of about 100 km molecular diffusion becomes comparable with mixing due to air motion and the light gases tend to float upwards. Also at these heights diatomic molecules (specially O_2) become split into

Composition of dry air

	Molecular weight (12C=12.000)	Volume (%)	Weight (%)
Dry air	28.966	100.0	100.0
Nitrogen	28.013	78.09	75.54
Oxygen	31.999	20.95	23.14
Argon	39.948	0.93	1.27
Carbon dioxide	44.010	0.03	0.05
Neon	20.183	0.0018	0.0012
Helium	4.003	5.2×10^{-4}	7.2×10^{-5}
Krypton	83.800	1.0×10^{-4}	3.0×10^{-4}
Hydrogen	2.016	5.0×10^{-5}	4.0×10^{-6}
Xenon	131.300	8.0×10^{-6}	3.6×10^{-5}
Ozone	47.998	1.0×10^{-6}	1.7×10^{-6}

Source: *Meteorological Glossary* 1972. London: HMSO

their atomic components. Atmospheric air contains water vapour from about 4 per cent by mass at temperatures of 30°C such as found in the tropics to 1 per cent at temperatures of 0°C such as found in subpolar latitudes and in the middle TRO-POSPHERE. Many molecules are spectacularly peculiar. Water has a large latent heat and changes state readily, consequently it plays a major role as a store of latent energy. Its high solvency encourages chemical reactions in the sea and in living matter. Carbon dioxide dissolves readily in sea water where it can be utilized by living matter ultimately to form carboniferous rocks. Otherwise it would clog up wavelengths through which terrestrial RADIA-TION escapes to space and lead to a hot 'run away GREENHOUSE' atmosphere like that of Venus. Ozone (at 0.0001 per cent by mass) removes virtually all solar radiation with wavelength less than 0.3 μm which would otherwise destroy living matter.

JSAG

Reading
Goody, R.M. and Walker, J.C.G. 1972: *Atmospheres.* Englewood Cliffs, NJ: Prentice Hall.

atmospheric energetics Concerns the energy content of the atmosphere and how it is changed from one form to another. Almost all energy is heat, latent heat and POTENTIAL ENERGY. THERMODYNAMIC DIAGRAMS relate these. A small fraction of the total of sensible, latent and potential energy (known as available potential energy) can be converted into the KINETIC ENERGY of the WIND.

JSAG

Reading
Lorenz, E.N. 1967: *The nature and theory of the general circulation of the atmosphere.* Geneva: World Meteorological Organization.

atmospheric instability In general a system is unstable if an introduced disturbance increases in magnitude through time. Conversely the system is stable if the introduced disturbance is damped out through time. In the atmosphere this idea is applied at two main scales: cyclone-scale and cloud-scale.

Cyclone-scale instability is exemplified by the growth of cyclones within the extratropical westerlies. They are a manifestation of baroclinic instability. Cloud-scale instability results in vertical displacement of air parcels in an atmosphere initially in hydrostatic equilibrium. Cumulus clouds

frequently result from this type of instability. (See also VERTICAL STABILITY/INSTABIL-ITY.)

BWA

Reading
Atkinson, B.W. 1972: The atmosphere. In D.Q. Bowen ed., *A concise physical geography.* Amersham: Hulton Educational. Esp. pp. 45–9.
— 1981: Atmospheric waves. In B.W. Atkinson ed., *Dynamical meteorology: an introductory selection.* London and New York: Methuen.
Barry, R.G. and Chorley, R.J. 1992: *Atmosphere, weather and climate.* 6th edn. London and New York: Routledge.

atmospheric layers These are principally the TROPOSPHERE (or overturning layer) 0–10 km above the earth's surface, the STRATOSPHERE (or layer of constant temperature) 10–25 km, the ozonosphere (warmed through photochemistry involving oxygen) 25–60 km, the MESOSPHERE (some similarity with troposphere) 60–100 km and between 100 and 500 km the thermosphere (molecular conductivity balancing energy input), the IONOSPHERE (electrical charge on particles significant) and the exosphere (molecules liable to escape into orbit). Similar layers occur at similar values of the pressure (which is related to height) in atmospheres of other planets. Near the ground, there is a hierarchy of boundary layers: convective (containing the active regions of cumulus-scale motion) about 1 km deep, mechanical or Ekman (mixing due to mechanical stirring) about 300 m deep, logarithmic or constant flux about 10 m deep, and finally an unnamed layer penetrated by material objects such as trees, grass and waves, that interfere with the flow of air and transfer momentum, heat, moisture, salt, pollen, etc. into the atmosphere.

JSAG

Reading
Goody, R.M. and Walker, J.C.G. 1972: *Atmospheres.* Englewood Cliffs, NJ: Prentice-Hall.
McIntosh, D.H. and Thom, A.S. 1969: *Essentials of meteorology.* London: Wykeham Publications.

atmospheric predictability A property of the type of information predicted as well as a certain length of time ahead. Weather systems of middle latitudes take only a few days to grow from unobservable beginnings: clearly this means that explicit weather prediction must in turn be limited to a few days. The behaviour of a system over its lifetime depends little on exactly when it was initiated, so there is a time-scale of about ten days over which the average weather may be predictable. Over longer time-scales (weeks, months, seasons) inter-action with the sea and land surface begin to be important, and current (1994) atmospheric models do not predict conditions well. Evaluation of predictions is arguable and politically loaded. The need to justify large investment of government money, the effects on insurance of over-prediction of catastrophes and complacency over public appreciation are obvious hazards. Some economically important weather is incredibly difficult to forecast. For example, the first time in winter that snow reaches the ground depends on whether it will melt below cloud base, so that forecasting of snow needs accurate temperatures up to a height of 1 km, and this, in turn, may depend on whether some apparently insignificant cloud sheets disperse overnight. Similarly, forecasting of severe convection is difficult. Disastrous thunderstorms may be set off by a temperature anomaly of only $0.2\ K$. JSAG

Reading
McIntosh, D.H. and Thom, A.S. 1969: *Essentials of meteorology*. London: Wykeham Publications.

atmospheric waves An abstraction convenient for describing some phenomena closely related to the physical processes responsible for propagating characteristics through the atmosphere.

Elastic waves
Propagate at the speed of sound (330 ms^{-1}) and transmit pressure pulses. They are usually of tiny amplitude but represent the practical limiting signal velocity (playing a role like that of the speed of light in classical physics).

Gravity (-inertial) waves
Represent the action of restoring forces (gravity and Coriolis) acting because the atmosphere is stably stratified. They are analogous to the waves on the interface between two immiscible liquids when the lower is slightly denser than the upper. Short waves (gravity waves) have a characteristic period of about 600s which is evident in many natural phenomena such as LEE WAVES and overshoot of cumulus tops. Very long waves (gravity-inertia waves) are dominated by Coriolis accelerations (GEOSTROPHIC WIND) with an upper bound to the duration of the period of 2π/Coriolis parameter ($\simeq$ 12h); they are exemplified by cloud bands caused by mountain chains.

Cyclone waves
Eddies in the western circumpolar flow around a hemisphere with horizontal dimensions of at most a few thousand kilometres. Although frequently quasi-circular in plan view, vertical cross-sections reveal wave forms in the temperature and pressure distribution. These waves, known also as baroclinic waves, lie within and are inextricably linked to ROSSBY WAVES.

Rossby waves
These are very large (both wavelength and amplitude of several thousand kilometres) perturbations in the extra-tropical high atmosphere. They are usually two to six in number, encircle the globe from west to east in each hemisphere, contain the jet streams and are vital to the formation of extra-tropical cyclones and anti-cyclones and hence extra-tropical climate. JSAG

Reading
Atkinson, B.W. 1981: Atmospheric waves. In B.W. Atkinson ed., *Dynamical meteorology: an introductory selection*. London and New York: Methuen. Pp. 110–15.
Eliassen, A. and Kleinschmidt, E. 1957: Dynamic meteorology. In S. Flügge ed., *Encyclopedia of physics*. Berlin: Springer-Verlag. Vol. 48, pp. 1–154.

atoll An annular form of CORAL ALGAL REEF consisting of an irregular elliptical reef, often breached by channels, around a central lagoon. There are over 400 atolls recorded in the world, most of which are found in tropical waters of the Indo-Pacific. Atolls vary greatly in size and shape, as well as in the depth of the central lagoon. The largest atoll is Kwajalein in the Marshall Islands (120 × 32 km). As with all coral algal reefs, atolls are sensitive to fluctuations in relative sea level. Some atolls, for example Aldabra Atoll in the Seychelles archipelago, are now elevated by a few metres above present sea level; others have become drowned as their growth has failed to keep pace with changing sea level (e.g. Saya de Malha, Indian Ocean). Low, sandy islands (called cays) may form on the reef rim of atolls. Micro-atolls are rounded forms found often on reef flats, usually single colonies of massive corals less than 6 m in diameter with a flat or concave upper surface devoid of living coral. They grow preferentially where water is ponded at low tide. HAV

Reading
Guilcher, A. 1988: *Coral reef geomorphology*. Chichester: Wiley.

Woodroffe, C.D. and McLean, R. 1990: Microatolls and recent sea level change on coral reefs. *Nature* 244, pp. 531–4.

atollon A small atoll which lies on the flank of a larger one.

atterberg limits The results of tests (Index Tests) which, arbitrarily defined, show the properties of soils which have COHESION in that they represent changes, in state or water content, from solid to plastic to liquid materials. The Plastic Limit (PL) is the minimum moisture content at which the soil can be rolled into a thread 3 mm in diameter without breaking. The Liquid Limit (LL) is the minimum moisture content at which the soil can flow under its own weight. These are the ones most commonly used, but the Shrinkage Limit (SL) is the moisture content at which further loss of moisture does not further decrease the volume of the sample. The PL and LL are often combined to give the Plasticity Index (PI) from $PI = LL-PL$ and the Liquidity Index (LI) from $LI = (100m-PL)/(LL-PL)$, where m is the natural moisture content of the soil. A chart of PI as ordinate, plotted against LL, is often used for comparing different types of soil and for classifying them.   WBW

Reading
Mitchell, J.K. 1976: *Fundamentals of soil behaviour.* New York: Wiley.

Whalley, W.B. 1976: *Properties of materials and geomorphological explanation.* Oxford: Oxford University Press.

aufeis See ICING.

auge A hot, dry wind which blows from the south of France to the Bay of Biscay.

aulacogens can be thought of as a continental rifting system in which seafloor spreading proceeded for a while and then ceased. Such 'aborted pull-aparts' most commonly occur at places in continental lithosphere where three directions of sea-floor spreading are tending to occur at the same time. The common occurrence is for two of these directions to become predominant and for the third axis to become an aulacogen.   ASG

Reading
Dewey, J.F. and Burke, K. 1974: Hot spots and continental break-up: implications for collisional orogeny. *Geology* 2, pp. 57–60.

aureole, metamorphic The zone of metamorphosed rock adjacent to an intrusion of igneous rock.

aurora borealis The 'Northern Lights'. Flashing white and coloured luminescence in the ionized layers of the earth's atmosphere about 400 km above the poles. The result of solar particles being trapped in the earth's magnetic field. The term 'aurora australis' has been applied to the phenomenon in the southern hemisphere.

autecology The ECOLOGY of individual organisms and of particular species. Originally used for the study of relationships between a single organism and its environment, the term is now equally widely used for the study of relationships between plants or animals of the same species, particularly species populations, and their environments (population ecology). Autecology provides the fundamental basis for understanding the distribution of organisms and their behaviour in communities, and involves the ecology of organisms at different stages of their life histories together with the environmental controls on germination, establishment, growth, reproduction, dispersal and survival. Modern autecology is an experimental science with important branches developed to physiological processes (physiological ecology) and genetics (evolutionary ecology). (See also SYNECOLOGY.)

JAM

Reading
Bannister, P. 1976: *Introduction to physiological plant ecology.* Oxford and New York: Blackwell Scientific and Halsted-Wiley.

Daubenmire, R.F. 1974: *Plants and environment: a textbook of plant autecology.* New York and London: Wiley.

Macfadyen, A. 1963: *Animal ecology: aims and methods.* London and New York: Pitman.

Pianka, E.R. 1974: *Evolutionary ecology.* New York and London: Harper & Row.

Vernberg, F.J. and Vernberg, W. 1970: *The animal and the environment.* New York and London: Holt Rinehart & Winston.

autochthonous A feature which has formed *in situ* and is not the product of transport processes. The term is often applied to such sediments as evaporites which have developed in place. In tectonics the term is often used to describe rock formations in Alpine structures which have not been displaced by major thrusting, although they have been folded and faulted.

autocorrelation The property of persistence in sequences of values measured over time or space. It is usually measured by comparing each value in the sequence either with its immediately previous (lag = 1) value or with the value at a fixed previous time or distance (lag > 1). For a sequence $x_1...x_n...x_N$ the comparisons for lag $r \, (< = N)$ are between the $(N - r)$ pairs x_n and $x_{(n-r)}$ as n ranges from $r + 1$ to N. The pairs of values are then used to calculate a correlation coefficient which measures the degree of dependence for the given lag. The correlation coefficient is calculated as for a normal least-squares correlation, although in this context is called the coefficient of autocorrelation. A correlogram may then be constructed in which the successive values of the coefficient for different lags is plotted against the lag. Comparison between correlograms for different types of sequence gives some idea of the type of persistence present, if any (for lag zero the coefficient is necessarily 1.0). A sequence of independent values clearly shows zero coefficients – that is no autocorrelation. A simple Markov chain of values shows an exponential decline in the coefficients. For a sequence with a linear trend, the coefficient varies about a constant value; while for a regular wave form the coefficient also varies regularly over a range of positive and negative values. Linear or harmonic trends should be removed from a sequence to see whether other sources of persistence are present. Many if not most time and space sequences show some degree of autocorrelation, so care must be exercised in obtaining valid independent samples or applying parametric statistical tests.   MJK

autogenic succession The process of community change (SUCCESSION) caused by the reaction of organisms, particularly plants, on their own environment. By the reaction mechanisms, organisms may so change their environment that other species are given a competitive advantage. The original organisms are eventually replaced, having brought about their own destruction. A.G. Tansley first used the term in the context of vegetation to distinguish this classic mechanism of succession from ALLOGENIC SUCCESSION, which results from the action of external environmental factors independent of the organisms themselves. Consider, for example, a vegetation change in response to a change in soil pH. If this was caused by the *in situ* accumulation of acidic plant litter the succession would be autogenic, but if the cause was prolonged leaching due to heavy rainfall the succession would be allogenic. In reality the two concepts may be difficult, if not impossible, to separate. (See also COMPETITION; ECOSYSTEM.)

JAM

Reading and References
†Botkin, D.B. 1981: Causality and succession. In D.C. West, H.H. Shugart and D.B. Botkin eds, *Forest succession: concepts and applications*. New York and Heidelberg: Springer. Pp. 36–55.
Connell, J.H. and Slatyer, R.O. 1977: Mechanisms of succession in natural communities and their role in community stability and organisation. *American naturalist* 111, pp. 1119–44.
Tansley, A.G. 1935: The use and abuse of vegetational concepts and terms. *Ecology* 16, pp. 284–307.

autotrophic Pertaining to organisms which produce organic substances from inorganic compounds. Included in these are chlorophyll-containing plants which produce organic materials from water and carbon dioxide.

avalanche The sudden and rapid movement of ice, snow, earth or rock down a

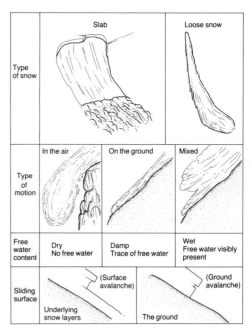

Avalanches: *classification*
Source: *C. Embleton and J. Thornes eds 1979:* Progress in geomorphology. *London: Edward Arnold.*

slope. Avalanches are an obvious and important mechanism of mass wasting in mountainous parts of the earth; they are also highly significant on subaqueous continental margins and deltas as well as in extra-terrestrial environments, for example on Mars. Avalanches occur when the shear stresses on a potential surface of sliding exceed the shear strength on the same plane. Failure is sometimes associated with increased shear stress in response to slope steepening or loading (for example, slope undercutting or snow or deltaic sediment accumulation), to reduced shear strength within the material (for example, increased PORE WATER PRESSURE or the growth of weak snow crystals), and sometimes to a combination of the two, especially when associated with an external trigger such as an earthquake.

Avalanches are commonly subdivided according to the material involved. *Snow avalanches* occur in predictable locations in snowy mountains and create distinctive ground features as they plunge down the mountain side (Rapp 1960). *Debris avalanches* involve the rapid downslope movement of sediment. On land they are commonly associated with saturated ground conditions. In subaqueous environments they reflect sediment overloading. One large example off the Spanish Sahara involved 18,000 km² of disturbance (Embley and Jacobi 1977). *Rock avalanches* are very rapid downslope movements of bedrock which become shattered during movement. These avalanches sometimes achieve velocities as high as 400 km per hour owing to the presence of trapped interstitial air; they can travel tens of kilometres from their source, sometimes with devastating effects on human life.   DES

Reading and References
†Clapperton, C.M. and Hamilton, P. 1971: Peru beneath its eternal threat. Analysis of a major catastrophe. *Geographical magazine* 43.9, pp. 632–9. (Huascaran rock avalanche.)
Embley, R.W. and Jacobi, R.D. 1977: Distribution and morphology of large submarine sediment slides and slumps on Atlantic continental margins. *Marine geotechnology* 2, pp. 205–28.
Nicoletti, P.G. and Sorriso-Valvo, M. 1991: Geomorphic controls on the shape and mobility of rock avalanches. *Bulletin of the Geological Society of America* 103, pp. 1365–73.
†Perla, R.I. and Martinelli, M. 1976: *Avalanche handbook*. Agriculture handbook 489, (*Snow*). US Department of Agriculture, Forest Service.
Rapp, A. 1960: Recent development of mountain slopes in Karkevägge and surroundings, northern Scandinavia. *Geografiska annaler* 42A, pp. 71–200.
†Williams, G.P. and Guy, H.P. 1971: Debris avalanches – a geomorphic hazard. In D.R. Coates ed., *Environmental geomorphology*. Binghamton: State University of New York.

avalanche tarns Small water-filled depressions produced by repeated avalanche impact.

Reading
Fitzharris, B.B. and Owens, I.F. 1984: Avalanche tarns. *Journal of glaciology* 30, pp. 308–12.

aven A vertical passage or shaft which connects a cave with the surface or overlying chambers and passages.

avulsion The diversion of a river channel to a new course at a lower elevation on its floodplain as a result of floodplain aggradation. It causes established meander belts to become abandoned and new ones to form.

Reading
Smith, N.D., Cross, T.A., Dufficy, J.P. and Clough, S.R. 1989: Anatomy of an avulsion. *Sedimentology* 36, pp. 1–23.

azimuth The arc of the sky extending from the zenith to the point of the horizon where it intersects at 90°.

azoic Without life. Pertaining to the period of earth history before organic life evolved or to portions of the seas and oceans where organisms cannot exist.

azonal soils Soils which have not developed marked horizons owing to their youth.

azotobacter The principal nitrogen-fixing bacteria. An aerobic bacteria which obtains energy from carbohydrates in the soil zone.k

B

backing wind See WIND.

backshore The backshore of the coastal zone lies between the highest point reached by marine action and the normal high-tide level. On a low coast, the backshore zone is often in the form of a berm, above the normal reach of the tide. The berm often slopes gently landwards. On shingle BEACHES, the berm crest can attain 13 m above normal high-tide level, as on Chesil Beach, Dorset. On sandy coasts, foredunes may form in the backshore zone and washover fans are associated with barrier island backshore zones. On a steep coast, the backshore is that part of the platform and cliff foot affected by waves under storm conditions. A wave-cut notch is a common feature.   CAMK

Reading
Davies, J.L. 1980: *Geographical variation in coastal development*. 2nd edn. London: Longman.

backswamp Low-lying marshy or swampy area on a FLOODPLAIN where overbank flood or tributary drainage water may become ponded between river levées and valley sides or other relatively elevated alluvial sediments. Such areas can be extensive where levées are well-developed and where near-channel deposits are aggrading. In settled areas such environments may be artificially drained leaving fine-grained soils which can be rich in organic materials and which may have characteristics developed under water-logged conditions.   JL

backwall The arcuate cliffed head of a cirque basin or a landslide.

backwash The return flow of water down a BEACH after a wave has broken. It is the return to the sea of the swash or uprush of the wave. It plays an important part in determining the gradient of the swash slope in association with the size of the beach material. On a coarse pebble beach the backwash is reduced in volume through percolation, so that a steeper slope is necessary to maintain equilibrium between swash and backwash. On a fine sand or wet beach the backwash is a large proportion of the swash, so a flat beach can remain in equilibrium. Rhomboid ripple marks are sometimes formed by backwash. Long waves and steep waves enhance the back-wash and are associated with flatter swash slope gradients.   CAMK

Reading
Demarest, D.E. 1947: Rhomboid ripples marks and their relationship to beach slope. *Journal of sedimentary petrology* 17, pp. 18–22.

backwearing The parallel retreat of a slope without a change in overall form or inclination. The term may be applied to escarpments and to side slopes: it is commonly used to contrast with down-wearing of a slope in which material is lost from the upper segments of the slope with consequent decreases in inclination. Parallel retreat implies that the resistance of rock and soil of the slope is constant into the hill which is being eroded, or that resistance does not control the slope form.   MJS

badlands A term originally used to describe intensely dissected natural land-scapes where vegetation is sparse or absent and which are useless for agriculture (Bryan and Yair 1982). Badlands were thought to be areas where intensive fluvial activity had produced high drainage densities, though it is now recognized that non-fluvial processes such as piping, tunnel erosion, and mass wasting also play an important role. Man may have contributed to their development

Badland scenery created by erosion of colluvium derived from granite in Swaziland. The colluvium is of late Pleistocene age and is rendered very erodible as a result of its high exchangeable sodium content. This promotes slaking.

by removing vegetation cover. Extensive badland development is usually associated with unconsolidated or poorly cemented materials, such as shales, clays, loess and colluvium (see DONGA) and with materials that slake easily through their high exchangeable sodium percentage. It is also favoured by a relatively dry climate, though striking examples are known from Hong Kong.   ASG

Reference
Bryan, R. and Yair, A. 1982: *Badland geomorphology and piping*. Norwich: GeoBooks.

bajada Confluent alluvial and pediment fans at the foot of mountain and hill slopes encircling desert basins.

ball lightning See LIGHTNING.

bank erosion Removal of material from the side of a river channel. This may be accomplished by several processes: particle by particle removal following surface wash, frost heave, groundwater sapping and the dislodgement and fall of material and the subsequent entrainment of particles by flowing water; abrasion by transported ice; and mass failure of the bank. Bank erosion is usually associated with high flows, but may be at a maximum as water levels fall after a high-flow period and the lateral

support provided by the deep water and the apron of previously slumped material are removed. Rates of erosion depend on prior conditions affecting bank strength, and in particular water content, which may vary seasonally. Bankside vegetation and root systems are also an important control on bank stability.

In detail bank erosion forms also reflect bank composition. In coarser materials erosion may be by particle entrainment with steep banks that have talus slopes at the base as long as such material is not picked up by the flowing water. In cohesive sediment bank failure may be in the form of shallow slips. In composite banks, with an upper cohesive fine layer above coarser materials, the process may be by removal of the underlying material accompanied by periodic collapse of overhanging 'cantilevers' of fine bank-top material. Until they are broken up and disaggregated, collapsed blocks may temporarily protect the bank base especially where given additional strength by vegetation and a root mat.

Bank erosion rates are highest where flow in river channels is asymmetric, as on the outer bank of meander bends. The exact location of such points may be a determinant in evolving channel patterns. For example, if located downstream of the bend apex, a meander loop may tend to translate downvalley rather than expand laterally. High rates of bank erosion may also relate to flow diverted by BARS in the channel, and the fact that non-cohesive banks may erode relatively rapidly may, among other factors, promote the formation of braided channel patterns (see BRAIDED RIVER).   JL

Reading
Richards, K. 1982: *Rivers*. London and New York: Methuen.

Schumm, S.A. 1977: *The fluvial system*. New York: Wiley.

bank storage Water retained in permeable channel-side deposits as soil and ground water. Drainage of such water by effluent seepage may contribute to the maintenance of river flows at low water, while under dry conditions there may be a downstream decrease in river discharge because of percolation into bank storage. This is particularly important in semi-arid environments, in allogenic streams, and on regulated rivers where low flow discharges are augmented by reservoir releases. A loss in the volume of water transmitted downstream has to be allowed for because of the bank storage factor. JL

bankfull discharge The river discharge which exactly fills the river channel to the bankfull level without spilling on to the floodplain, and which depends upon the definition of CHANNEL CAPACITY. Bankfull discharge has often been assumed to be a significant or critical channel-forming flow which is important in determining the size and shape of the river channel and it has therefore sometimes been regarded as equivalent to DOMINANT DISCHARGE. The frequency of occurrence of bankfull discharge varies considerably and has been quoted as ranging from 4 months to 5 years in Scotland, 1.3 to 14 years in lowland England and 1 to 30 years in twenty-eight basins in the western USA but care should be taken in quoting such recurrence intervals because at least sixteen different methods are available (Williams 1978) to calculate bankfull discharge. KJG

Reading and References
†Nixon, M. 1959: A study of bankfull discharges of the rivers of England and Wales. *Proceedings of the Institute of Civil Engineers* 12, pp. 157–74.
Williams, G.P. 1978: Bankfull discharge of rivers. *Water resources research* 14, pp. 1141–54.

banner cloud Often seen to extend downward from isolated peaks rather like a flag. The classic example is that on the Matterhorn. The mechanism of the cloud is not yet understood. BWA

bar, coastal Coastal bars are mainly submarine, sandy features. They may be straight and parallel to the coast, less often oblique and occasionally crescentic, with a convex seaward form. They may be single, double or multiple. The most common submarine bars form at the break-point of steep, destructive waves, which move landward outside the break-point and seawards inside it. Crescentic submarine bars may be associated with edge waves. Transverse and multiple bars are formed in low-energy environments and are usually low features requiring special conditions, and are, therefore, less common than the break-point bars. The latter are found on many exposed coasts with low to moderate tidal range. CAMK

Reading
Bowen, A.J. and Inman, D.L. 1971: Edge waves and crescentic bars. *Journal of geophysical research* 76, pp. 8662–71.

Carter, R.W.G. 1988: *Coastal environments*. London: Academic Press.
Davies, R.A. and Fox, W.T. 1972: Coastal processes and nearshore sand bars. *Journal of sedimentary petrology* 42, pp. 401–12.
Greenwood, B. and Davidson-Arnott, R.G.D. 1979: A tentative classification of bars. *Canadian journal of earth science* 16, pp. 312–32.
King, C.A.M. 1972: *Beaches and coasts*. 2nd edn. London: Edward Arnold.
Schwartz, M.L. ed. 1972: *Spits and bars*. Stroudsburg, Penn.: Dowden, Hutchinson and Ross.

barchan A crescentic sand dune whose horns point in the direction of dune movement. The windward slope is gentle and the lee slope or slipface is at the angle of repose of dry sand. Barchans tend to develop in areas of limited sand supply and with directional wind regimes. They range in size from features just a few metres high to megabarchans that may be several hundred metres high. They may coalesce to form transverse barchanoid ridges. ASG

Reading
McKee, E.D. 1979: A study of global sand seas. *US Geological Survey professional paper* 1052.

baroclinity A measure of the fractional rate of change of density in the horizontal. It is also frequently defined as a state where surfaces of equal pressure and surfaces of equal density intersect. In zones of large baroclinity spontaneous generation of new motion systems is likely: primary depressions (see CYCLONE) in the baroclinic zone of middle (30–60°) latitudes, secondary depressions in the frontal zones of the primary depressions, and thunderstorms in the fronts of the secondary depressions. Large baroclinity also defines the regions of transition between air masses, a concept which has lost favour in recent years with the recognition that these regions are often the result, rather than a cause, of the air motion in the air masses. JSAG

barometer An instrument for measuring atmospheric pressure. There are two types: mercury and aneroid. A mercury barometer works on the principle that atmospheric pressure is sufficient to support a column of mercury in a glass tube. As the pressure varies, so does the height of the column. An aneroid barometer comprises a series of vacuum chambers with springs inside to prevent total collapse. Atmospheric pressure changes are recorded by the compression or expansion of the chambers. This mechanism is frequently used in the barograph, which is a barometer that gives a graph of pressure changes through time.

BWA

Reading
Meteorological Office 1956: *Handbook of meteorological instruments. Part I Instruments for surface observations.* London: HMSO.

barotropic Technically the opposite of baroclinic, but in practice represents a special case of motion in which all processes connected with horizontal variations of DENSITY are excluded. Confusion can arise because the working substance may be barotropic (as with a fluid of constant density) or the motion may be barotropic (as with purely horizontal motion of a fluid in which density varies with height).

JSAG

Reading
Eliassen, A. and Kleinschmidt, E. 1967: Dynamic meteorology. In S. Flügge ed., *Encyclopaedia of physics*. Berlin: Springer-Verlag. Vol. 48, pp. 1–154.

barranca A steep-sided gully or ravine.

barrier island Barrier islands are elongated, mainly sandy features parallel to the coast and separated by a lagoon. They are not attached at either end, and are often separated by tidal inlets. Barrier islands make up 10–15 per cent of all the world coastlines. They normally consist of a berm in front of dune ridges, and are backed by washover fans and tidal marshes in the lagoon. They occur worldwide, but are particularly common in low to middle latitudes with low to moderate tidal range and swell wave regimes. They tend to migrate landwards on coasts where sea level is rising slowly, by washover under storm conditions. They are common on gently sloping coasts with wide shelves and coastal plains. The east coast of the USA provides good examples.

CAMK

Reading
Hoyt, J.H. 1967: Barrier island formation. *Geological Society of America bulletin* 78, pp. 1125–36.

Oertel, G.F. and Leatherman, S.P. eds 1985: Barrier islands. *Marine geology* 63, pp. 1–396.

Schwartz, M.J. ed. 1973: *Barrier islands*. Stroudsburg, Penn.: Dowden, Hutchinson and Ross.

Swift, D.J.P. 1975: Barrier island genesis: evidence from the central Atlantic shelf, eastern U.S.A. *Sedimentary geology* 14, pp. 1–13.

barrier reef One of the whole suite of coral reef forms. It is characterized by the presence of a lagoon, or body of water, in between the reef and its associated coastline. This feature differentiates a barrier from a fringing reef. Charles Darwin identified barrier reefs as occupying the mid position of a developmental sequence of reefs caused by the submergence of the central land mass, or island. Fringing reefs form the start of this sequence and atolls the end. Like other coral reefs, barrier reefs are generally limited in occurrence to the tropical latitudes and are found on sediment-free coasts. Good examples of barrier reefs are found off the coast of Belize, and around the island of Mayotte, Comores Islands. Most barrier reefs are up to a kilometre in width and their lagoons are often up to a few kilometres wide. The Great Barrier Reef forms the largest stretch of reefs in the world, being almost 2000 km

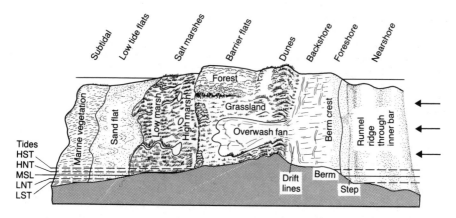

Barrier beach *idealized section.*

long, and is actually a compound form of many different reefs. It is backed by a lagoonal sea tens of kilometres wide.   HV

barrier spit The part of a barrier island system that is attached to an eroding headland. It is generally parallel to the coast and separated from it by a lagoon. An inlet, which is often tidal, separates a barrier split from a detached barrier island. One theory, first put forward by G.K. Gilbert, suggests that barrier spits are the first stage of barrier island formation. They are fed by longshore drift of material derived from the eroding headland, and are generally formed mainly of sand. Some barrier spits may be attached to older beach ridges, but one end of the elongated feature is always attached to the mainland, while the other is free but associated with a barrier island.   CAMK

Reading
Carter, R.W.G. 1988: *Coastal environments*. London: Academic Press.

Gilbert, G.K. 1880: *Lake Bonneville*. US Geological Survey monograph 1, pp. 23–65.

Schwartz, M.L. 1971: The multiple causality of barrier islands. *Journal of geology* 79, pp. 91–4.

bars A generic term for ridge-like accumulations of sediment in coastal and fluvial environments. These are developed as subaqueous BEDFORMS, but may be exposed at low tide and low river flows. Bars are larger features than ripples, with dimensions in metres and comparable in height to the depth of generating flows. They can be single features or in a repeating pattern, as at the bends of a river channel. Indeed, some are characteristically associated with particular positions in river channels or wave environments, but in the case of BRAIDED RIVERS the whole channel floor may consist of bar forms among which the channel divides at low flow.

Because of the physical similarity of some aeolian and subaqueous features, large-scale submerged forms may also be known as dunes or sand waves, like their aeolian equivalents. Further confusion in terminology also arises because of early colloquial use of 'bar' for any emergent or near-surface barrier to water transport and this use might even include permanent islands in coastal environments. It is best to avoid this particular usage (using 'barrier' instead) and to treat bars as larger subaqueous bedforms.

Bars may be subcategorized and identified in terms of their geometric shape (cuspate, linguoid, lunate, scroll, spool), orientation with respect to current direction (transverse, longitudinal, diagonal), or position in channel or shore environment (braid, medial, side, attached, point, riffle, delta-mouth, break-point, offshore). The result is a full terminology which is sometimes confusing because alternative names can be used for similar features. Thus a linguoid (tongue-shaped) bar may also be a transverse, braided and medial one. Bars are also commonly complex sedimentary bodies and the product of more than one phase or even style of sedimentation.

Bar dimensions may be given in terms of height (amplitude), cross-current (transverse) and down-current length, and the spacing (wavelength) of repeated forms. The morphology of river bars may be internally complex, consisting of more than one accretionary unit, and modified by phases of erosion. It may be possible to separate a coarser upstream 'bar head' from a finer lee downstream 'bar tail', while there may be size and shape gradients in the sediment present even when sharp discontinuities are absent.

It is possible to observe bar development under laboratory experimental conditions or in the field, both as bars develop in response to hydraulic conditions at high flows and as bar topography evolves over a matter of years in response to repeated high-flow events. Bar sedimentation may be an important component of the development of floodplains in general, as well as being a significant changing element of channel form requiring, on larger navigable rivers, both careful pilotage and possible engineering control by dredging.   JL

Reading
Collinson, J.D. and Lewin, J. eds 1983: *Modern and ancient fluvial systems*. Special publication no. 6, International Association of Sedimentologists. Oxford: Blackwell Scientific.

barysphere The part of the earth's interior (i.e. core, mantle, and asthenosphere) which lies beneath the lithosphere.

basal complex Describes the rocks which lie beneath the Pre-Cambrian shields that make up large portions of the continents.

basal ice A relatively thin layer (with a vertical extent of up to tens of metres) of debris-rich ice present at the base of glaciers, which is produced at and interacts with the glacier bed. As a result of its different mode and environment of formation it may differ from 'normal' glacier ice in terms of its overall extent and properties (including its rheology), debris, solutes and gases.   ASG

Reading
Hubbard, B. and Sharp, M. 1989: Basal ice formation and deformation: a review. *Progress in physical geography* 13, pp. 529–58.

basal sapping The process whereby a slope is undercut at the base. It can describe the retreat of a hillslope or escarpment by erosion along a spring line (mid-latitudes) by salt weathering (arid environments) or by glacial erosion (CIRQUE).

basal sliding The process by which a glacier slides over bedrock. It is distinguished from processes associated with the internal deformation of ice within the body of a glacier. Two important processes of sliding are *enhanced basal creep*, where pressures caused by irregularities on the bed increase stresses within the ice and allow it to deform round the obstacles and *regelation*, where the pressures induced by the bed obstacles cause ice to melt on the upstream sides, flow round the obstacles as water, and refreeze on to the glacier on the downstream sides. Basal sliding is closely related to the presence and nature of water at the ice–rock interface. If no water is present basal sliding is restricted to enhanced basal creep and the ice–rock interface remains immobile. Water allows movement at the ice–rock interface and the higher the water pressure the quicker the rate of sliding. Glacier sliding takes place in a stick-slip fashion with periodic movements of a few centimetres, and probably relates to the impeding effect and sudden release of small frozen patches at the ice-rock interface. A recently recognized additional form of basal sliding incorporates the deformation of weak sediments beneath the weight of a glacier.   DES

Reading
Paterson, W.S.B. 1981: *The physics of glaciers*. Oxford: Pergamon. Chapter 7.

basal slip See BASAL SLIDING.

basalt A fine-grained and dark-coloured igneous rock. The lava extruded from volcanic and fissure eruptions. Rocks composed primarily of calcic plagioclase and pyroxene with or without olivine.

base exchange (or CATION EXCHANGE) A vital soil reaction whereby bases or cations such as calcium, magnesium, sodium and potassium are made available as plant nutrients. These cations are usually loosely bonded on the surface of clay and organic colloidal particles in the soil complex, and cation exchange takes place when hydrogen cations, derived from organic decomposition and atmospheric sources, replace the metal cations, releasing the latter into the soil water around the root hairs, where the plant may absorb them.   KEB

Reading
Pears, N. 1985: *Basic biogeography*. 2nd edn. London and New York: Longman.

Trudgill, S.T. 1988: *Soil and vegetation systems*. 2nd edn. Oxford: Oxford University Press.

base flow A term often used to describe the reliable background river flow component from a drainage basin. Base flow, or delayed flow, was originally considered to be the result of effluent seepage from groundwater storage but in some drainage basins this sustained component of flow is also produced by interflow from zones above the water table and particularly by throughflow or lateral drainage through the soil. The DEPLETION CURVE or base flow recession curve describes the way in which base flow discharge recedes at a particular site during a period without rainfall.   AMG

base level The lower limit to the operation of subaerial erosion processes, usually defined with reference to the role of running water. The level of the sea surface at any moment acts as a general base level for the continents, although there can be a wide range of local base levels, some above and some below sea level. The term was first used by J.W. Powell (1875) who defined it in a very broad fashion to include: first, the concept of the sea as 'a grand base-level, below which the dry lands cannot be eroded'; secondly, the existence of local or temporary base levels 'which are the levels of the beds of the principal streams which carry away the products of erosion'; and, finally, as 'an imaginary surface, inclining slightly in all its parts towards the lower end of the principal stream draining the area'. W.M. Davis (1902) considered that the breadth of Powell's definition had led to variety of practice and confusion. He therefore proposed that base level be restricted to 'simply . . . the level base with respect to which normal sub-aerial erosion proceeds'. This suggestion has largely been adopted.

Base levels can be identified on a number of different scales. A stream channel acts as the base level for processes on the adjacent slope. A major stream is the base level for the courses of its tributaries. A lake or a reservoir or a waterfall is the base level for the entire basin upstream. Clearly, these base levels can and will alter over time, with implications for the operation of subaerial erosional processes. By and large a fall in base level creates an increase in potential energy, by increasing the total relief, and may result in an acceleration of rates of erosion and down cutting. A rise in base level, on the other hand, reduces relief and potential energy and is frequently marked by aggradation.

One major feature of the Pleistocene has been the frequency with which the 'grand base-level' of the sea has fluctuated under the combined influence of EUSTASY and ISOSTASY. These fluctuations have had very complex repercussions in many drainage basins, even those where the effects of direct glaciation or tectonism are absent. An experimental study of the COMPLEX RESPONSE to a single change in base level is provided by Schumm and Parker (1973), who demonstrate that *two* sets of paired terraces may be produced in the lower reaches of a drainage basin. Such experiments must cast real doubt on many geomorphological studies which have considered that a separate base level change is required to account for each set of paired terraces observed.   BAK

Reading and References
†Davis, W.M. 1902: Base-level, grade and peneplain. *Journal of geology* 1, pp. 77–111.
Powell, J.W. 1875: *Exploration of the Colorado River of the West and its tributaries*. Washington: Government Printing Office.
†Schumm, S.A. and Parker, R.S. 1973: Implications of complex response of drainage systems for Quaternary alluvial stratigraphy. *Nature (physical science)* 243, pp. 99–100.

base saturation The condition arising when the cation exchange capacity of a soil

is saturated with exchangeable bases – calcium, magnesium, sodium and potassium – when expressed as a percentage of the total cation exchange capacity.

basement complex A broad term for the older rocks, usually Archean igneous and metamorphic rocks, which underlie more recent sedimentary rocks in any region. Such rocks are a feature of the ancient shield areas of the earth's surface.

basic rocks Igneous rocks containing less than about 55 per cent silica, basalt a typical example (see ACID ROCKS).

basin-and-range A type of terrain, as found in Utah and Nevada, where there are fault block mountains interspersed with basins. The basins often contain lakes, some of which may only fill up in pluvials.

basin discharge The total flow of water through a river cross-section as it represents drainage from the catchment area or drainage basin above that point.

Total basin discharge depends upon the water balance of the drainage basin and this is climatically controlled, but the temporal distribution of the discharge is a result of the ways in which water is routed through the basin hydrological cycle and this reflects both the climate and the catchment characteristics.

The water balance of the catchment determines the proportion of the precipitation which will be either lost through evapotranspiration processes or available to contribute to storage and run-off processes in the catchment area. The distribution of the water balance through the year will control the timing and magnitude of both quick flow and delayed flow because it will affect the state of the various water stores in the drainage basin and thus their response to inputs of water (see HYDROGRAPHS). However, the magnitude of the water stores and the ease with which water may enter or leave them is mainly a function of drainage basin characteristics, notably the size, shape, slope, soil, rock and vegetation characteristics of the catchment. The size of the catchment, its rock types and soil types and their extent will place an upper limit on the capacity of the soil moisture and groundwater stores, whereas the slope of the catchment and the combined effect of its vegetation and soils will control the degree to which precipitation may be able to infiltrate and reach subsurface stores. As a result the water balance will control the supply of water for storage and run-off processes but the drainage basin characteristics and antecedent conditions will govern the rate at which this water may enter or leave storage and thus the type of discharge regime experienced at the catchment outlet.   AMG

Reading
Chorley, R.J. 1978: The hillslope hydrological cycle. In M.J. Kirkby ed., *Hillslope hydrology*. Chichester: Wiley.
Petts, G.E. and Foster, I.D.L. 1985: *Rivers and landscape*. London: Edward Arnold.

batholith A large mass of intrusive igneous rock which extends to great depth and can cover or underlie very large areas.

bathymetry The measurement of water depth.

bauxite The main ore of aluminium. An impure aluminium hydroxide associated with the clay deposits of weathering zones, especially in tropical regions.

baydzharakh A type of THERMOKARST formed when the ACTIVE LAYER is becoming deeper so that ice wedges begin to thaw and high-centred polygons with conical surface expression form. They may be 3–4 m high and 3–20 m wide. Collapse of decaying mounds leads to the development of more continous steep-sided depressions called duyodas.   ASG

beach A coastal accumulation of varied types of sediment, usually of sand size or above. The sediment is derived from rivers and other sources and is moved by tides and waves to form a beach. Beaches have characteristic profile forms, which are determined by the steepness of the waves and the size of the sediments. These profiles may be highly changeable and many beaches display either swell or storm profiles at different times of year. Within the beach profile several different zones exist. These zones are called the breaker, surf and swash zones. These zones are related to the changes occurring to waves as they hit the shoreline. Within each zone are found different features, such as BERMS, BEACH RIDGES, and BARS (see Komar 1976 for a detailed description of different coastal profiles). Beaches also possess long profile shapes which are influenced by the nature of the coastline. As sea level changes, beaches may become elevated above sea level, producing RAISED BEACHES.   HV

Reading and Reference
Carter, R.W.G. 1988: *Coastal environments*. London: Academic Press.
Hardisty, J. 1990: *Beaches form and process*. London: Unwin Hyman.
Komar, P.D. 1976: *Beach processes and sedimentation*. Englewood Cliffs, NJ: Prentice Hall.

beach ridge Accumulations of sediment forming a prominent feature on many beaches. Beach ridges can take a wide variety of forms. Ridges may be formed near the top of shingle or sandy beaches (in

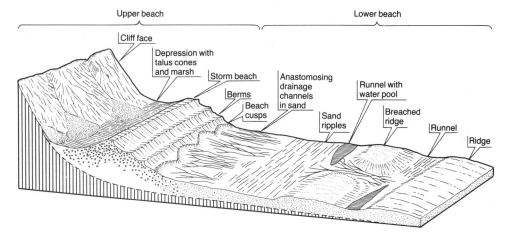

***Beach 1**. The idealized features of a sand and shingle beach.*
Source: *A.S. Goudie 1984:* The nature of the environment. *Oxford and New York: Basil Blackwell. Figure 8.4.*

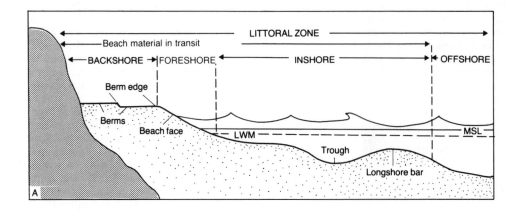

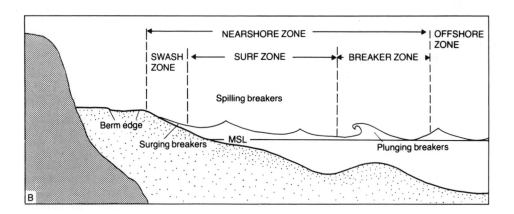

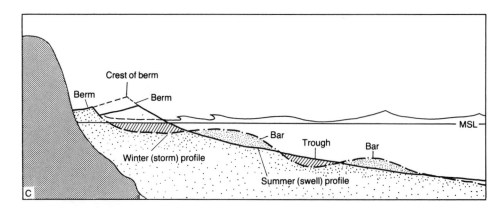

Beach 2. *Schematic diagrams showing zones used to describe* (a) *beach profiles;* (b) *wave and current action;* (c) *profile changes due to storm and swell waves.*
Source: E. Derbyshire, K.J. Gregory and J.R. Hails eds 1979: Geomorphological processes. *Folkestone: Dawson. Figure 3.9.*

Beach rock at Arsuz, near Iskenderun, south-east Turkey. The example is of relatively recent origin as it cements pottery shards.

sand these are called BERMS). Ridge features may also be formed in association with runnels near low-tide level on shallow gradient beaches. Multiple ridge forms often occur when sediment has been left by successive storm events. HV

Reading
Komar, P.D. 1976: *Beach processes and sedimentation.* Englewood Cliffs, NJ: Prentice Hall.

beach rock Beach rock is a sedimentary rock, or consolidated chemical sediment, which forms in the intertidal zone on beaches, most notably in the tropics. Beach rock may also develop along extra-tropical coastlines, such as around the Mediterranean and the Red Sea. It forms where a layer of beach sand and other material becomes consolidated by the secondary deposition of calcium carbonate at about the level of the water table. The cementing material may be aragonite or calcite and may come from groundwater or sea water (Stoddart and Cann 1965). Biochemical precipitation of calcium carbonate by micro-organisms may also be involved. Beach rock often forms relatively quickly, and is forming today in some areas as evidenced by the inclusion of recent hu-

man artefacts in the consolidated layer. Once formed, beach rock is relatively persistent, but often contains a suite of erosional features zoned according to their position relative to sea level. HV

Reference
Stoddart, D.R. and Cann, J.R. 1965: Nature and origin of beachrock. *Journal of sedimentary petrology* 35, pp. 243–7.

beaded drainage A series of small pools connected by streams. The pools result from the thawing of ground ice and may be 1–3 m deep and up to 30 m in diameter (see THERMOKARST).

Reading
Hopkins, D.M., Karlstrom, T.N.V., Black, R.F., Williams, J.R., Péwé, T.F., Fernald, A.T. and Muller, E.H. 1955: Permafrost and ground water in Alaska. *US Geological Survey professional paper* 264–F, pp. 113–47.

beaded esker See ESKER.

bearing capacity of a soil is the value of the average contact pressure between a foundation and the soil below it which will produce a (shear) failure in the soil. Strictly, this is the 'ultimate' bearing capacity; in soil mechanics a 'maximum safe' value is used, which is the ultimate value divided by a factor of safety. Various

Reasonable current specifications for the Beaufort wind force scale

Beaufort no.	Descriptive term	Windspeed (knots)		Effect on sea surface
		Mean	Limits	
0	Calm	0	<1	Sea like a mirror.
1	Light air	3	1–4	Ripples with the appearance of scales, no foam crests.
2	Light breeze	7	5–8	Small wavelets, crests have glassy appearance and do not break.
3	Gentle breeze	11	9–12	Large wavelets, crests begin to break, perhaps scattered white horses.
4	Moderate breeze	15	13–16	Small waves becoming longer, fairly frequent white horses.
5	Fresh breeze	19	17–21	Moderate waves, many white horses, chance of some spray.
6	Strong breeze	24	22–26	Large waves form, white foam crests extensive, probably some spray.
7	Near gale	29	27–31	Sea heaps up and white foam from breaking waves blown in streaks.
8	Gale	34	32–36	Moderately high waves of great length, edges of crests begin to break into the spin-drift, foam blown in streaks.
9	Strong gale	39	37–42	High waves, dense streaks of foam, crests of waves begin to topple, tumble and roll over.
10	Storm	45	43–48	Very high waves with long overhanging crests, sea surface takes white appearance, visibility affected by spray.
11	Violent storm	52	49–55	Exceptionally high waves, sea completely covered by long white patches of foam, everywhere edges of wave crests blown into froth.
12	Hurricane		>55	Air filled with foam and spray, sea completely white with driving spray, visibility seriously affected.

Source: Mather, J.R. 1987: Beaufort wind scale. In J.E. Oliver and R.W. Fairbridge eds, *The encylopedia of climatology*. New York: Van Nostrand Reinhold. Table 2, p. 162.

methods are available to enable calculation of appropriate values, according to the soil properties and the likely mode of failure.

WBW

Reading
Smith, G.N. 1974: *Elements of soil mechanics for civil mining engineers.* London: Crosby Lockwood Staples.

Beaufort scale Admiral Sir Francis Beaufort formalized a scale of WIND based on its effect on a man-of-war. The scale was later adapted for use on land. It was still internationally useful in 1946, when it was argued that the anemometers in common use were unable to register the shorter-term gust capable of destroying a ship.　JSAG

Reading
List, R.J. ed. 1951: *Smithsonian meteorological tables.* 6th revised edn. Washington Smithsonian Institution. P. 119. (Contains full listing of the scale.)

Beaumont period One of 48 consecutive hours during which the screen dry bulb temperature has been 10°C or above and the relative humidity has been at or above 75 per cent on at least 46 of the 48 hourly observations. It is used as a criterion for the issue of a potato blight warning.

bed load, bed load equation Fluid-transported sediment that moves along or in close proximity to the bed of the flow. This movement of the heavier and larger particles may be by rolling, sliding or saltation (the last being movement in a series of hops resulting from grains impacting on one another). For a given transporting flow the particles that move in this way are ones that are too heavy to be kept suspended in the flow itself since their fall velocity is greater than the upward velocity component of the turbulent fluid.

For rivers, bed load usually amounts to less than 10 per cent of total sediment transport, though higher proportions have been reported for mountain streams. Measurements available are, however, relatively few and possibly unreliable. The problem is that any device inserted in flowing water to trap just that sediment moving at or near the bed itself interferes with the pattern of fluid flow and sediment movement. Techniques include permanent traps or slots in the bed, both ones which are periodically emptied and therefore assess just the total amount of sediment transport in the sampling period, and ones which include a conveyor system or weighing device which samples sediment continuously (e.g. Leopold and Emmett 1976). Other portable basket-, bag- or pan-type samplers may be lowered onto the bed and then retrieved. These measurements are liable to varying and often unknown error (especially through faulty positioning on the bed) and trapping efficiency, but some long-term series of observations on European and North American rivers are now available. These show that even for a given discharge, bed load transport rates are unsteady and uneven across streams; it is also well known that bed material (which may include an additional component of suspension load) may accrete and move discontinuously in the form of migratory BARS. When bed load movement is in this form, it has been suggested that transport rates may be approximated by the volumetric transfer rate of such BEDFORMS over time.

Bed load movement has also been studied using tracers and the acoustic monitoring of inter-particle collisions; a particular aim may be to ascertain the hydraulic conditions at which bed material starts to move. Difficulties in measuring bed load transport, both practical and conceptual, have led to the development of a series of empirically calibrated bed load equations. Given certain information (for example, stream velocity, discharge or stream power and measures of bed material size and sorting), bed load transport rates may be calculated rather than directly measured. Some equations involve prediction of transport rates in terms of 'excess' shear or power above the threshold value at which transport starts. Some involve computation

of total sediment transport rates rather than the bed load alone. From a practical point of view the problem is often that such equations may have to be used in conditions beyond those for which they have been designed and calibrated and, in practice, equations may give very different estimates from one another. None the less, properly used, such estimates have proved very useful as an aid to the design of reservoirs and other engineering works on rivers where bed load movement may be a practical problem. JL

Reading and Reference
†Graf, W.H. 1971: *Hydraulics of sediment transport*. New York: McGraw-Hill.
Leopold, L.B. and Emmett, W.W. 1976: Bed load measurements, East Fork River, Wyoming. *Proceedings of the National Academy of Sciences* 73, pp. 1000–4.
†Richards, K. 1982: *Rivers*. London and New York: Methuen.

bed roughness The surface relief at the base of a fluid flow, as on the bed of a river channel. This may consist of several elements: particle roughness, commonly defined with reference to size of larger particles in relation to the depth of fluid flow, form roughness produced by bedforms and the distorting effects of channel bends and 'spill' resistance, as where rapid changes occur where flow spills around protruding boulders. Velocity formulae (see FLOW EQUATIONS), relating flow velocity to the hydraulic radius and slope of river channels for UNIFORM STEADY FLOWS, incorporate a roughness coefficient (Manning's n), or related Darcy–Wiesbach friction coefficient or CHÉZY EQUATION. Sometimes roughness coefficients are estimated simply from the grain size on the bed, but it has to be remembered that this size may be variably effective at different depths and discharges, as may other forms of roughness or resistance. JL

Reading
Graf, W.H. 1971: *Hydraulics of sediment transport*. New York: McGraw-Hill.
Richards, K. 1982: *Rivers*. London and New York: Methuen.

bedding plane The interface between two strata of sedimentary rocks, often a plane of weakness between two such strata.

bedforms Features developed by fluid flow over a deformable bed, as developed by wind on a bed of sand or streamflow over alluvial sediments. These forms may vary in size from small-scale ripples to larger BARS or dunes. A hierarchy of forms may be present at any one place superimposed on one another and possibly related to different formative flows. The dimensions of bedforms may relate to flow magnitudes, as is particularly evident when comparing the giant 'ripples' tens of metres in length in gravel-sized material produced by the catastrophic draining of the Pleistocene ice-dammed Lake Missoula in North America with the ripples of dimensions in centimetres developed in sand in many shallow flows.

In flume experiments with sands, sequences of bedforms have been shown to develop with increasing stream power, bed shear stress and sediment transport rate. An initial PLANE BED with little sediment movement develops into one with ripples and then dunes of increasing dimension; there then follows re-establishment of a transitional plane bed phase under conditions of high sediment transport, finally with standing waves and antidunes developing into a chute and pool morphology. In detail the changes observed may be complex. Overlap of bedform types under given conditions of flow and sediment transport occurs. In part this may depend on the existing degree of development of bedforms of particular types. Under field conditions, while similar features and developments may be observed, things become even more varied, because of discharge variability and lag effects and variable flow conditions across channels, among other factors.

The existence of bedforms may be important in generating flow resistance and variability of channel flow conditions. This can lead, for example, to flow separation and the creation of localized areas of slackwater or upstream flow. These may be particularly important for flora and fauna, so that it may be desirable in channelized rivers to allow for or mimic such environments where bedforms are absent.   JL

Reading
Allen, J.R.L. 1970: *Physical processes of sedimentation*. London: Allen & Unwin.

— 1984: *Sedimentary structures*. Amsterdam: Elsevier.

Collinson, J.D. and Lewin, J. eds 1983: *Modern and ancient fluvial systems*. Special publication no. 6, International Association of Sedimentologists. Oxford: Blackwell Scientific.

bedrock The consolidated, unweathered rock exposed at the landsurface or underlying the soil zone and unconsolidated surficial deposits..

benefit–cost ratio See COST–BENEFIT RATIO.

benioff zone The zone or plane of earthquake foci beneath some continental margins. The inclined plane dips deeper on the island of the continental margin.

benthic Pertaining to plants, animals and other organisms that inhabit the floors of lakes, seas and oceans.

berg wind A type of FÖHN wind blowing, mainly in winter, off the interior plateau of South Africa, roughly at right angles to the coast.

Bergeron–Findeisen mechanism See CLOUD MICROPHYSICS.

berghlaup An Icelandic term (literally, rock-leaping) which is sometimes used to refer to a large fallen earth or rock mass. No genetic interpretation is usually intended and it may be considered as an equivalent of BERGSTURZ.   WBW

Bergmann's rule The rule which states that the size of the bodies of warm-blooded animal species and subspecies tends to increase in environments with lower mean annual temperatures.

bergschrund The crevasse occurring at the head of a cirque or valley glacier because of the movement of the glacier ice away from the rock wall.

bergsturz A German word which has been used to denote large-scale falls ($>10^6 m^3$), usually of rock from mountain sides. There is usually no implied mechanism and the released rock debris may or may not travel very long distances on an air cushion.   WBW

berm A ridge of sand parallel to the coastline, commonly found on the landward side of steeply sloping beaches. It is a nearly horizontal feature formed by deposition at the upper limit of the swash zone. When steep beaches are transformed into more shallow gradient beaches, due to a change in wave regime, the berm is removed and a long-shore bar deposited just below low-tide level.   HV

Bernouilli's theorem and effect Describe the relationship between the pressure in a fluid flow in the direction of flow (dynamic pressure) and at right angles to it (static pressure).

The theorem states that the sum of these two pressures is equal to the local HYDRO-STATIC PRESSURE. The dynamic pressure exceeds the hydrostatic pressure by an amount $\rho u^2/2$ where u is the local flow velocity and ρ is the density of the fluid. The static pressure is thus less than the hydrostatic pressure by the same amount. The Bernouilli effect is produced where flow velocities differ laterally. The difference in static pressures tends to push objects towards the streams of more rapid flow, e.g. blowing between two oranges which are hanging so that they almost touch, causes them first to swing in to hit each other.   MJK

best units analysis The method used for slope profile examination whereby the profile is divided into segments and elements such that the respective coefficients of variation of angle or curvature do not exceed specified values. Where two or more segments overlap the overlapping portion is attributed to the longest unit. Allied to this concept is that of 'best segments analysis' and 'best elements analysis' where the slope is divided into segments or elements alone.   WBW

Reading
Young, A. 1972: *Slopes*. Edinburgh: Oliver & Boyd.

beta diversity A measure of the degree of change in species composition of biological communities in relation to variations in local environments within a landscape. Beta or between-habitat diversity can be distinguished from alpha or within-habitat diversity, which measures the number of species within a single community. Landscapes in which the species composition changes rapidly in relation to variations in factors such as elevation, soil moisture, and degree of disturbance will have higher beta diversities than landscapes which have an absence of species changes along such habitat gradients.   ARH

Reading
Whittaker, R.H. 1975: *Communities and ecosystems*. 2nd edn. London: Collier Macmillan. Chapter 4, pp. 111–91.

biennial oscillation See MACRO-METEOROLOGY.

bifurcation ratio The ratio $(B_b = \Sigma n)/\Sigma(n+1)$ of number of streams of a paritcular order (Σn) to number of streams of the next highest order $\Sigma(n+1)$. It is therefore dependent upon techniques of stream ordering and gives an expression of the rate at which a stream network bifurcates. It has been correlated with hydrograph parameters and sometimes with sediment delivery factors. In a fifth Strahler order drainage basin, four values of R_b could be calculated and so a weighted mean bifurcation ratio (WR_b) was recommended by Schumm (1956) as:

$$WR_b = \frac{\Sigma[R_{b_{n:n+1}} \times (N_n + N_{n+1})]}{N}$$

(See also ORDER, STREAM.)   KJG

Reading and Reference
†Gregory, K.J. and Walling, D.E. 1976: *Drainage basin form and process*. London: Edward Arnold.
Schumm, S.A. 1956: The evolution of drainage systems and slopes in badlands at Perth Amboy, New Jersey. *Bulletin of the Geological Society of America* 67, pp. 597–646.

biochemical oxygen demand (BOD) A useful measure of organic pollution (e.g. by sewage) whereby the amount of dissolved oxygen (in milligrams per litre of water) used up by micro-organisms feeding on the organic matter is measured over five days at a temperature of $20°C$. Water cleanliness may therefore be expressed in BOD_5 terms: clean river water will have a BOD_5 of less than 4, effluent a BOD_5 of 20 or 80.   KEB

Reading
Simmons, I.G. 1979: *Biogeography: natural and cultural*. London: Edward Arnold.

biocides See PESTICIDES.

bioclastic Pertaining to rock composed of fragmented organic remains.

biocoenosis (biocenose) The mixed community of plant and animals (a biotic community) in particular a HABITAT. It may be artificially partitioned into three components: a plant community (phyto-coenosis), an animal community (zoocoenosis) and a community of micro-organisms (micro-biocoenosis). First introduced by K. Möbius in 1877, the word is normally used in the sense of an actual community in the landscape rather than in the sense of an abstract concept or community type. The physical environment of a biocoenosis is its ECOTOPE; the biocoenosis with its ecotope make up a BIOGEOCOENOSIS.   JAM

Reference
Möbius, K. 1877: *Die Auster und die Austernwirtschaft.* Berlin: Wiegundt, Hempel & Parey.

biodegradation The decomposition of organic substances by micro-organisms. The metabolism of aerobic (oxygen-utilizing) bacteria is primarily responsible for this breakdown.

The death and decay of organic matter is essential for the replenishment of raw materials in the biosphere. Reactions involved in the degradation of organic compounds are frequently those of their synthesis in reverse (Horne 1978). Catalysts for the degradation reactions are produced by micro-organisms, and are often identical or similar enzymes to those employed in synthesis (Bailey *et al.* 1978). Natural toxins and some complex structures are resistant to breakdown. For example, parts of the intricate chlorophyll molecule can be found fossilized in ancient sedimentary rocks. Also, organic compounds synthesized by industrial chemistry (plastics, and chlorinated hydrocarbons, for example) are not readily biodegradable, and may persist in the environment for a considerable time (see BIOLOGICAL MAGNIFICATION). The presence of biodegradable material in bodies of water can lead to a significant reduction in the dissolved oxygen content of the water, as oxygen is needed for micro-organic activity to take place (BIOCHEMICAL OXYGEN DEMAND).   RLJ

References
Bailey, R.A., Clark, H.M., Ferris, J.P., Krause, S. and Strong, R.L. 1978: *Chemistry of the environment.* New York and London: Academic Press.
Horne, R.A. 1978: *The chemistry of our environment.* New York and Chichester: Wiley.

biodiversity Biological diversity – 'an enormous cornucopia of wild and cultivated species, diverse in form and function, with beauty and usefulness beyond the wildest imagination' (Iltis, 1988). This has recently become a major environmental issue because environments are being degraded at an accelerating rate, much diversity is being irreversibly lost through the destruction of natural habitats, and science is discovering new uses for biological diversity (Wilson, 1988). The number of species of organism on earth is still imperfectly understood as is the rate at which they are being lost, but particular fears are expressed about the loss of species caused by rainforest exploitation.

Biodiversity has five main aspects:

1 The distribution of different kinds of ecosystems, which comprise communities of plant and animal species and the surrounding environment and which are valuable not only for the species they contain, but also in their own right.
2 The total number of species in a region or area.
3 The number of endemic species in an area.
4 The genetic diversity of an individual species.
5 The subpopulations of an individual species which embrace the genetic diversity.   ASG

Reading and References
Iltis, H.H. 1988: Serendipity in the exploration of biodiversity – what good are weedy tomatoes? In E.O. Wilson ed., *Biodiversity.* Washington DC: National Academy Press. Pp. 99–105.
Wilson, E.O. ed. 1988: *Biodiversity.* Washington DC: National Academy Press.

biogeochemical cycles The cycling, at various scales, of minerals and compounds through the ecosystem. The cycles (see CARBON CYCLE and NITROGEN CYCLE) involve phases of weathering of inorganic material, uptake and storage by organisms, and return to the pool of the soil, the atmosphere or ocean sediments. An increasing amount of research has been focused on the working out of the details of such cycles during the past decade as a result of concern over global environmental change and nutrient budgets. Classic studies, such as that of the Hubbard Brook Experimental Forest by Likens and associates, are often cited (Begon *et al.* 1990, pp. 688–91) as examples of well-documented small-scale studies. Diagrammatic representations of major element cycles are common in textbooks but there are fewer *quantified* studies on a global scale (Mannion, 1991).

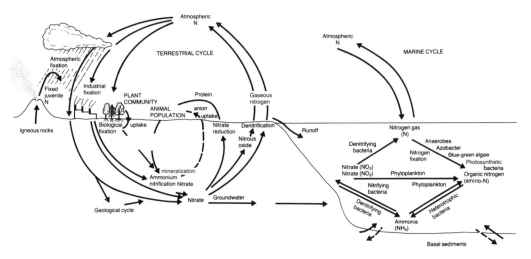

Biogeochemical cycles: *generalized diagrams of the terrestrial and marine nitrogen cycles.*
Source: *P. Furley and W. Newey 1983:* Geography of the biosphere. *London: Butterworths. Figure 3.7*

The biogeochemistry of carbon has attracted particular attention because of the concern over GLOBAL WARMING and the GREENHOUSE EFFECT.   KEB

Reading and References
Begon, M., Harper, J.L. and Townsend, C.R. 1990: *Ecology.* 2nd edn. Oxford: Blackwell Scientific Publications.

Degens, E.T., Kempe, S. and Richey, J.E. eds 1991: *Biogeochemistry of major world rivers.* Chichester: Wiley.

Mannion, A.M. 1991: *Global environmental change.* Harlow, Essex: Longman Scientific and Technical.

biogeocoenosis (biogeocenose) A combination on a specific area of the earth's surface of a particular BIOCOENOSIS (biotic community) and its ECOTOPE (physical environment), e.g. a forest, a peat bog or an oyster bank. Introduced by V.N. Sukachev in 1944, the term was widely used in the former USSR as an equivalent to the western term ECOSYSTEM; a biogeocoenosis type being equivalent to an abstract ecosystem type. A biogeocoenosis is generally considered to possess a degree of homogeneity in its structure and a certain coherence in its functioning.   JAM

Reading and Reference
Sukachev, V.N. and Dylis, N. 1964: *Fundamentals of forest biogeocoenology.* Edinburgh and London: Oliver & Boyd.

†Troll, C. 1971: Landscape ecology (geoecology) and biogeocenology – a terminological study. *Geoforum* 8, pp. 43–6.

biogeography A difficult term used rather differently by two groups of scholars. To biologists it is the study of the distribution, past and present, of various taxa of plants and animals at various scales (usually global, continental or regional) as a prelude to explaining how such distributions came about and how these have played a part in the evolutionary history of the taxon. Large-scale world phenomena such as continental drift, orogeny, the Pleistocene glaciations, and global climatic belts contribute to explanations at this scale. Smaller-scale explanations such as local variations in topography and soils (for plants) and vegetation type (for animals) are commonly thought of as the province of ECOLOGY. To geographers the term biogeography has meant the study of the biosphere and of human effects on plants and animals. Some parts of the study have been spatial-synoptic (so overlapping with PHYTO-GEOGRAPHY and ZOOGEOGRAPHY), while other parts have been orientated towards processes such as Quaternary ecology and human impact, hence overlapping with ecology. There seems to be a good case for a different term, but possible alternatives such as ecological geography, geographical ecology, and geoecology have gained no general acceptance, so biogeography continues to be used. If we accept that in geography viewed as a whole the human species is central, then presumably a definitely geographical biogeography must include the human species, along with the uniquely manipulative capabilities of our taxon. We might, with Pierre Dansereau, think of man as creating new genotypes and creating new ECOSYSTEMS. We may study the progress of the creation of new plants and animals by human selection, from the dog to recombinant DNA in genetic engineering, and put this into the context of changing ecosystems which are manipulated by man, either deliberately, as in the burning of forests to provide open ground for cereal agriculture, or accidentally, as in the gradual change in vegetation communities brought about by nomadic pastoralists. Modern changes in ecosystems from contamination by urban-industrial wastes may equally be part of such a study. In general, such a view is orientated towards processes and functions, rather than being spatial and synoptic: this follows the trend in ecology. It is fair to say that changes in thinking and methods in biogeography (as understood by geographers) are heavily influenced by events in ecological thinking, a set of processes not confined, however, to ecological science since it interacts with many disciplines in which a holistic perspective has relevance as well as a reductionist one. Of all the branches of physical geography, biogeography is perhaps the one where linkages with human geography are most easily formed, possibly because they deal with the same organism. It is commonplace to point out that *Homo sapiens* is an animal that lives in two worlds: the metabolic world of needs for food, water, clothing, shelter and reproductive success, and the cultural world of *Homard à l'Americaine*, single malt whisky, culottes by Yves St Laurent, Castle Howard and blue videos. In the 1970s and 1980s biogeographers were among the leaders in suggesting that a resurgence of the man–environment relations tradition of thinking in geography was now called for as a corrective to the excesses of quantitative social science exhibited during the 1960s.   IGS

Reading and References
†Dansereau, P. 1957: *Biogeography: an ecological perspective.* New York: Ronald Press.
†Seddon, B. 1971: *Introduction to biogeography.* London: Duckworth.
†Simmons, I.G. 1979: *Biogeography: natural and cultural.* London: Edward Arnold.
— 1980: Biogeography. In E.H. Brown ed., *Geography yesterday and tomorrow.* Oxford: Oxford University Press. Pp. 146–66.
†Stott, P. 1981: *Historical plant geography.* London: Allen & Unwin.
†Watts, D. 1971: *Principles of biogeography: an introduction to the functional mechanisms of ecosystems.* London: McGraw-Hill.

biogeomorphology Encapsulates concisely a developing and previously much neglected approach to geomorphology which explicitly considers the role of organisms. Two main foci are (*a*) the influence of landforms/geomorphology on the distribution and development of plants, animals and micro-organisms; and (*b*) the influence of plants, animals and micro-organisms on earth surface processes and the development of landforms.   ASG

Reading
Viles, H.A. ed. 1988: *Biogeomorphology.* Oxford: Basil Blackwell.

bioherm *a.* An ancient coral reef.
b. An organism which plays a role in reef formation.

biokarst A KARST landform, usually small in scale, produced mainly by organic action. Strictly speaking, the term PHYTOKARST should be restricted to phenomena produced by plants alone. Biokarst features can either be erosional (as where organisms bore into or abrade carbonate rock surfaces) or constructional (as in the case of certain tufas and reef forms).   ASG

Reading
Viles, H.A. 1984: Biokarst: review and prospect. *Progress in physical geography* 8, pp. 523–43.

biological control (biocontrol) The control of pestilential organisms such as insects and fungi through biological means rather than the application of man-made chemicals. This can include breeding resistant crop strains, inducing fertility in the pest species, disruption of breeding patterns through the release of sterilized animals or spraying juvenile hormones to interrupt life cycles, breeding viruses that attack the pests, or the introduction or encouragement of natural or exotic predators to control pest outbreaks.   ASG

biological magnification The increased concentration of toxic material at consecutive TROPHIC LEVELS in an ecosystem. Toxins, such as persistent pesticides and heavy metals, are incorporated into living tissue from the physical environment. Physical, chemical and biological processes operate to amplify the harmful substances in food chains by concentrating the quantities in individual organisms. The effect of these substances on organisms varies. In general, reproductive capacity is impaired at low concentrations and death occurs at high ones, sometimes via disease. Approximately 10 per cent of food at one trophic level are transferred to the next, the remainder being removed by respiratory and executive activity, but as toxic materials are not so readily broken down as other components of organic tissue (BIODEGRADATION), their transfer efficiency is higher. Thus a build-up of them occurs at successive trophic levels (Woodwell 1967).

DDT (PESTICIDES) residues are an example of accumulation in the physical environment. They have a low biodegradability, and are excreted slowly from organisms because they become dissolved in fatty tissues. Woodwell reports (in a study undertaken with Wurster and Isaacson) DDT concentrations of up to 36 kg ha^{-1} after two decades of application of the pesticide to a New York marsh ecosystem. Marsh plankton contained 0.04 ppm, minnows 1 ppm and a carnivorous gull 75 ppm of DDT in their tissues.

RLJ

Reading and Reference
†Jorgensen, S.E. and Johnsen, I. 1981: *Principles of environmental science and technology.* Amsterdam, Oxford and New York: Elsevier.
†Odum, E.P. 1975: *Ecology: The link between the natural and social sciences.* 2nd edn. London and New York: Holt, Reinhart & Winston.
Woodwell, G.M. 1967: Toxic substances and ecological cycles. *Scientific American* 216, pp. 24–31.

biological productivity The growth of organic material per unit area per unit time. It is usually represented as the dry weight of the material, expressed as g m^{-2} day^{-1} or year, or t ha^{-1} year^{-1}; but it could also be given in energy equivalents. Primary productivity is that of plants, secondary that of animals.

Concepts of biological productivity differentiate between gross and net productivity. The latter is, strictly, as defined above; the former is net productivity plus a respiration factor, normally taken to be equivalent to the energy used in RESPIRATION by the organisms which gave rise to the biological increment. For primary producers gross productivity may also be related to the proportion of incoming radiant energy which is directly utilized in PHOTO-SYNTHESIS, but even now there are relatively few measurements of this. Odum (1971) suggests that c.5 per cent of incident visible light is used year-round in gross production in the restricted freshwater environment of Silver Springs, Florida, but for most communities the figure is likely to be much less, at c.0.5–2.5 per cent. Despite these low values, enough energy is fixed in biological systems to ensure their growth to maturity, and their efficient functioning. In mid and high latitudes, there may be considerable seasonal variation in such figures, and they can also change substantially over the daytime period.

The amount of gross primary productivity taken up by respiration is determined by the type and age of the plant community, and also by certain climatic factors, notably temperature. But the mean world figure might be 35–55 per cent. However, young communities tend to respire less than the old ones; and, in contrast, rates of respiration might be expected to increase in mature woods and forests, in which the ratio of respiring and non-photosynthetic woody tissue to photosynthetic leaf tissue is greater than elsewhere, and to be augmented still further in zones of high temperature, where metabolic activity rises. Respiration in mature tropical forest communities might be of the order of 70–75 per cent, whereas it would only be c.30–40 per cent in temperate latitude forests year-round (Whittaker 1975). For agricultural crops, alfalfa has respiration rates of only 12 per cent in its phase of rapid early growth, but this reaches 38 per cent as a six month average. Those crops with the lowest respiration rates of all will probably have the highest leaf–stalk ratios, as sugar cane. Relatively little is known about respiration rates in marine communities, though for plankton, these are high, at 30–40 per cent; and in the Sargasso Sea, they move up to 53 per cent of gross primary productivity. In the subtropical Silver Springs, Florida, respiration rates have been measured at 57.5 per cent (figures from Odum 1971).

The differences in community respiration clearly will have their effect on NET PRIMARY PRODUCTIVITY.

For some of the larger animals details of secondary productivity at an individual level are available. For instance, it is the case that once birds and mammals reach maturity, their productivity effectively is zero, except when they are child-rearing. But at the community level there will be a definable secondary production over time, though this is extremely difficult to measure. The terms 'net' and 'gross' are rarely applied to secondary production at a community level, but when they are, the latter refers to the total assimilated food energy. This in turn may be defined as food ingestion less food egestion, for some food which is taken in will be egested unused, and so plays no part in animal productivity. In the case of certain animal groups, e.g. caterpillars, egested food may form up to 80 per cent of the total food intake. Secondary productivities are usually compared to the net primary productivities of the plant communities in which they live, and inevitably they are much lower than these. For example, in an arthropod community of an old-field grassland in Tennessee, above-ground net primary productivity was 270 g m^{-2} year^{-1}, the ingestion of this by herbivores and carnivores was 10.6 per cent of this (29 g m^{-2} year^{-1}), and the secondary productivity of both groups combined, after respiration and egestion, was only 6.4 g m^{-2} year^{-1} (Whittaker 1975).

From all the above, it will be deduced that rates of biological productivity should vary on a regional scale, and this they do. Enough is not yet known about secondary productivity to generalize; and the details of net primary productivity are given elsewhere. For gross primary productivity Odum (1971) has summarized the available evidence. On land, deserts, semi-arid grasslands and tundra have low values, at less than 0.5 g m^{-2} day^{-1}. The open oceans are slightly more productive, at up to 1.0 g m^{-2} day^{-1}: in contrast, continental shelf values lie between 0.5 and 3 g m^{-2} day^{-1}. Tropical reefs and some estuaries are among the most productive of world systems, with values of 10–25 g m^{-2} day^{-1}, though some marine turtle-grass flats reach 34 g m^{-2} day^{-1}. Tropical forest systems have gross productivity rates equivalent to those of reefs, as also do certain alluvial plains. Among other terrestrial communities moist temperate woodlands, moist grasslands and most agricultural land (not intensive agriculture) are reasonably highly productive (3–10 g m^{-2} day^{-1}); and somewhat drier (not semi-arid) grasslands, mountain forests and less productive agricultural land are less so, at 0.3–3 g m^{-2} day^{-1}.   DW

Reading and References
†Odum, E.P. 1971: *Fundamentals of ecology*. 3rd edn. Philadelphia: W.B. Saunders.
†Whittaker, R.H. 1975: *Communities and ecosystems*. 2nd edn. New York: Macmillan.

bioluminescence The light produced by some living organisms or the process of producing that light, characteristic of glow-worms and some marine fish.

biomass The mass of biological material present per plant or animal, per community, or per unit area. On a world scale, and largely because of the efficiency with which vegetation colonizes land, most of the biomass is found in terrestrial rather than oceanic environments, in a ratio of several hundred to one. This is particularly the case for phytomass (the mass of growing and dead plant material). Whittaker and Likens (1975) placed the total continental phytomass at 1837×10^9 tonnes dry weight and that in oceans and estuaries at only 3.9 × 10^9 tonnes dry weight. On the other hand, the balance between terrestrial and oceanic animal biomass, which is at much lower levels all round, is much more even: similar estimates (Whittaker and Likens 1973) have set the former at 1005 × 10^6 tonnes, and the latter at 997 × 10^6 tonnes, both dry weights.

On land, about 90 per cent of phytomass is located in the world's forests. Tropical rain forests produce a mean phytomass of 45 kg m^{-2}, other tropical and temperate forests one of 30–35 kg m^{-2}, and boreal forests one of 20 kg m^{-2}. Natural grasslands and tundra have phytomasses of 0.5 to 5 kg m^{-2}; and desert phytomasses range from 0.7 kg m^{-2} to much less. Within all vegetation communities a good deal of phytomass is accounted for by root growth: thus, roots give rise to over 50 per cent of the phytomass in many prairie grasslands of North America, and to an even higher percentage in most deserts. Animal biomasses are greatest per unit area on land in tropical forests, and secondarily in the savannah grasslands of Africa; and in oceanic environments in tropical reef zones, estuaries and on continental shelves.   DW

References
Whittaker, R.H. and Likens, G.E. 1973: The primary production of the biosphere. *Human ecology* 1, pp. 299–369.
— 1975: Net primary production and plant biomass for the earth. In H. Reith and R.H. Whittaker eds, *The primary production of the biosphere*. New York: Springer-Verlag, Pp. 305–28.

biome A mixed community of plants and animals (a biotic community) occupying a major geographical area on a continental scale. Usually applied to terrestrial environments, each biome is characterized by

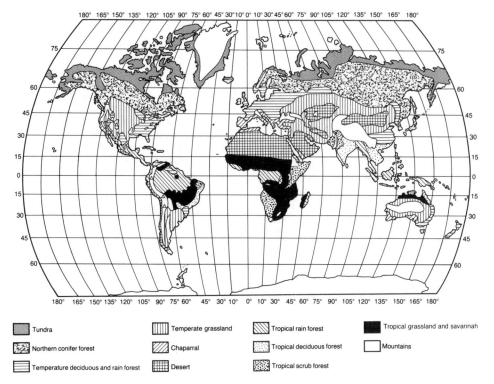

Major biomes of the world as if unaffected by human activity.
Source: *I. Simmons 1979:* Biogeography, natural and cultural. *London: Edward Arnold. Figure 3.3.*

similarity of vegetation structure or phy-siognomy rather than by similarity of species composition, and is usually related to climate. Within a particular biome the plants and animals are regarded as being well adapted to each other and to broadly similar environmental conditions, especially climate. Both CLIMAX VEGETATION and SERAL COMMUNITIES are represented. A group of biomes in which the plant and animal communities exhibit similar adapta-tions form a biome type. Thus the tropical rain forests of the Congo Basin and Amazonia are two biomes within the tropical rain forest biome type. Other biome types respectively include tundra, taiga, savannah and hot desert biomes (see LIFE FORMS). Needless to say, large areas of many biomes have undergone transform-ation by human societies and small-scale maps that appear in texts and atlases are misleading (Holzner *et al.* 1983).   JAM

Reading and Reference
†Brewer, R. 1979: *Principles of ecology.* Philadelphia and London: W.B. Saunders.

†Carpenter, J.R. 1939: The biome. *American midland naturalist* 21, pp. 75–91.

†Furley, P.A. and Newey, W.W. 1983: *Geography of the biosphere.* London: Butterworths.

Holzner, W., Ikusima, I. and Werger, M.J.A. eds 1983: *Man's impact on vegetation.* The Hague, Boston and London: W. Junk.

†Kendeigh, S.C. 1974: *Ecology with specific reference to animals and man.* Englewood Cliffs, NJ: Prentice Hall.

†Shelford, V.E. 1963: *The ecology of North America.* Urbana: University of Illinois Press.

†Walter, H. 1979: *Vegetation of the earth.* 2nd edn. Berlin: Springer-Verlag.

biometeorology The study of the effects of weather and climate on plants, animals and man. The International Society of Biometeorology, founded in 1956, has classified the subject into six main groups: phytological, zoological, human, cosmic, space and palaeo. Human biometeorology includes the study of the influence of weather and climate on healthy man and on his diseases and the effect of micro-climates in houses and cities on health. Although Hippocrates discussed some of these topics over 2000 years ago it is only in the second half of the twentieth century that the main developments in this science have taken place. In some countries the old name of bioclimatology is still used.   DGT

Reading
Tromp, S.W. 1980: *Biometeorology.* London: Heyden.

biosphere The zone at the interface of the earth's crust, ocean, and atmosphere where life is found. It is in general placed at the junction of the lithosphere and atmosphere, though the situation is complicated in the oceans, where life is found throughout their depths, but rather sparsely away from the zone near the surface. The depth of the biosphere on land is commonly from about 3 m below ground to 40 m above it; in the seas the zone of photosynthesis is less than 200 m deep. The separation of spheres by this terminology belies the fact that they are all interactive and indeed one hypothesis avers, for example, that the gaseous composition of the atmosphere is a result of life, not that life adapted to given geophysical conditions. Vernadsky suggested that the 'envelope' of human thought and action should be added: this he termed the noosphere, a label often attributed to Teilhard de Chardin.   IGS

Reading
Furley, P.A. and Newey, W.W. 1983: *Geography of the biosphere.* London: Butterworths.
Scientific American 1970: *The biosphere.* San Francisco and Reading: W.H. Freeman.

biostasy A term that was applied by Erhart (1956) to periods of soil formation, with rhexistasy referring to phases of denudation. In periods of biostasy there is normal vegetation, while in phases of rhexistasy there is dying out or lack of vegetation as a result of soil erosion resulting from climatic changes, tectonic displacement, etc. The period of rhexistasy is characterized by mechanical reworking, whereas biostasy is characterized by chemical decomposition.   ASG

Reference
Erhart, H. 1956: *La genèse des sols en tant que phénomene géologique.* Paris: Masson.

biota The entire complement of species of organisms, plants and animals, found within a given region.

biotechnology The use of microbial, animal or plant cells or enzymes to synthesize, break down or transform materials (Smith 1988). However, as Bull et al. (1982) and Bu'Lock (1987) note, alternative definitions of the term have arisen as a result of the philosophies and interests of the practitioners of a rapidly developing and sometimes controversial subject. It involves the interaction of biology and engineering to their mutual benefit, and is an interdisciplinary science concerned with a collection of technologies relevant to a number of parts of industry: the production of food and medicine, and the treatment of organic waste, for example. Biotechnological principles (for instance, in the manufacture of alcohol from grain, and cheese from milk) have been applied for centuries, but major advances in bioengineering have been made in recent decades.

Microbiological and biochemical contributions centre around the metabolic processes of living organisms, especially complex molecules (enzymes) produced by their cells. While some higher plant and animal cells and tissues are used, the principal agents of biotechnological activity are micro-organisms. They are the main source of productivity in, and of catalytic enzymes for, biotechnological reactions that involve both natural and man-made organic molecules. As there are many kinds of micro-organisms (although relatively few have, to date, been used by biotechnologists), they possess a large gene pool. They are also capable of procreating quickly, and are among the most efficient producers of protein. The latter ability means that single-celled organisms may be valuable in the future production of this essential material (SCP). The majority of micro-organic enzymes are confined within their bodies, but the most widely used in biotechnology are those discharged from them to a substrate where reactions occur. Most substrates are liquid, although increasing numbers of solid culture-media are being employed. The majority of biotechnological processes, as Smith (1988) observes, function at low temperatures, consume relatively little energy (notably from fossil fuels) and take place on inexpensive substrates. Raw materials for constructive reactions are termed feedstocks. Synthetic organic chemicals (oil and natural gas, for example) have been extensively used, as has plant and animal biomass, particularly surplus and waste organic matter.

Fermentation is a mainly anaerobic process which involves the decomposition of organic matter by organisms, and is accompanied by energy release. Fermentation technology developed following the discovery of *Penicillium* in 1929. Some 4000 antibiotics are now known, but only about fifty of these (compounds derived from bacteria and fungi, and capable of de-activating other bacteria) are in widespread medical use. Successful fermentation depends upon the presence of an appropriate enzyme catalyst whose effectiveness will not be diminished during the chemical reaction. When plant biomass is used, sugars and starches are first digested or hydrolysed, then fermented to produce acids, alcohols, fats, oils and gases.

In terms of volume, the treatment of household waste water and sewage is the major biotechnological industry in the United Kingdom. Bioreactors for this purpose contain an assortment of micro-organisms, whose overall metabolic capacity is capable of degrading the majority of organic compounds in the waste. The main objective of such a system is the production of effluent that can be safely discharged into the environment (Smith 1988) (see BIO-DEGRADATION; POLLUTION).

The input from genetics mainly involves genetic manipulation by man. This is aimed at raising the output of both essential substances and end products. Radiation and chemicals are employed to induce mutations which lead to the eradication of unwanted traits in the organisms. It is also possible to fuse two non-reproductive cells from different animal species to produce one cell having the components of both nuclei. This technique has produced a number of unique and important anti-bodies. Protoplast (cell material without a wall) of plants and micro-organisms can be fused, and walled cells subsequently generated which possess antibiotic properties. Some micro-organic enzymes assist in fragmenting DNA molecules and others in rejoining them in different combinations. Such recombined DNA can be introduced to a host cell by a carrier. Populations of these cells, consisting of identical individuals, can then be generated. The healing animal hormone insulin and the virus-inhibiting protein interferon have been synthesized via micro-organic activity in this manner.   RLJ

Reading and References
Bull, A.T., Holt, G. and Lilly, M.D. 1982: *Biotechnology: international trends and perspectives*. Paris: OECD.

Bu'Lock, J.D. 1987: Introduction to basic biotechnology. In J. Bu'Lock and B. Kristiansen eds, *Basic biotechnology*. London and New York: Academic Press. Pp. 3–10.
Ginzburg, L.R. 1991: *Assessing ecological risks of biotechnology*. Boston and London: Butterworth–Heinemann.
†Smith, J.E. 1988: *Biotechnology*. 2nd edn. London: Edward Arnold.

biotic isolation The isolation of organisms by hereditary mechanisms which cause the restriction or elimination of interbreeding through processes wholly internal to the organism. The isolation results not from geographical or ecological isolation but from the incompatibility of reproductive structures and genetic isolation. Thus, although populations of the song sparrow (*Melospiza melodia*) and Lincoln's sparrow (*M. lincoleni*) live side by side over considerable areas of the USA, they do not integrate and remain distinct from each other. Most species are isolated by several kinds of mechanisms which are in turn controlled by the action of many different genes. (See also SYMPATRY.)   PAS

Reading
Jones, S.B. and Luchsinger, A.E. 1979: *Plant systematics*. New York: McGraw-Hill.
Stebbins, G.L. 1977: *Processes of organic evolution*. 3rd edn. Englewood Cliffs, NJ: Prentice-Hall.

biotic potential The total reproductive potential of an individual organism or a population; an important concept in the study of plant and animal population dynamics. Essentially it is a measure of the reproductive rate of a given species taking into account the inherited sex ratio, the number of young per female, and the number of generations per unit of time. The biotic potential of many species (e.g. small mammals or bacteria) is enormous and such species would soon swamp the earth if there were no environmental checks on reproduction and on the survival of offspring.   PAS

Reading
Silvertown, J. 1982: *Introduction to plant population ecology*. London: Longman.
Solomon, M.E. 1976: *Population dynamics*. 2nd edn. London: Edward Arnold.

biotope The HABITAT of a BIOCOENOSIS, or a micro-habitat within a biocoenosis. In the first sense the word is synonymous with ECOTOPE, the effective physical environment of a biocoenosis or biotic community. In the second sense it refers to a small, relatively uniform habitat within the more complex community, e.g. although a forest community occupies its own habitat, each layer or stratum within the forest may be regarded as a separate biotope. Likewise, fungi on a fallen log, epiphytic lichens and mosses on tree bark, and animals confined to the forest canopy, each occupy distinct biotopes. (See also NICHE.)   JAM

Reading
Allee, W.C., Emerson, A.E., Park, O., Park, T. and Schmidt, K.P. 1951: *Principles of animal ecology.* Philadelphia and London: W.B. Saunders.
Daubenmire, R. 1968: *Plant communities: a textbook of plant synecology.* New York and London: Harper & Row.

bioturbation The mixing or disruption of the soil zone, lake and marine sediments, or unconsolidated surficial deposits through organic activity, e.g. the excavation of burrows or construction of termite mounds.

bise, bize A cold, dry northerly to north-north-easterly wind occurring in the mountains of Central Europe during the winter months.

blanket bog A type of bog, often composed of peat, which drapes upland terrain and infills hollows, in areas of high precipitation and low evapotranspiration.

blind valley A steep-sided, river-cut valley which terminates in a precipitous cliff. Although blind valleys can be found in any terrain near the source of a river, they are usually produced when a river flows onto limestone bedrock and sinks into subterranean passages, enabling downcutting to occur in the active river valley. This has the effect of depressing the river valley relative to the upstanding limestone rock, thus producing a steep cliff. Blind valleys are favoured sites for cave exploration: caves such as the famous Swildon's Hole, Mendip, south-west England is entered from a blind valley.   PAB

blizzard A snow storm, either of falling snow or deflated snow, usually accompanied by low temperature and high winds.

block faulting The process whereby large regions are tectonically disrupted to form complex systems of troughs and ridges or basins and block mountains. The result of tectonic uplift and subsidence of adjacent blocks of the earth's crust following faulting and fracturing on a grid pattern.

block fields, block streams Spreads or lines of boulders, generally angular, formed by *in situ* shattering by frost of a bedrock surface. They may surround features such as tors and nunataks and other landforms subjected to severe periglacial processes.

blocking An extreme state in which the tropospheric circulation takes the form of large-amplitude stationary waves. Such flow is frequently manifest in stationary anomalies in the weather which may have significant economic repercussions, such as the 1976 drought in the UK. Prediction of the initiation and persistence of blocks is difficult. Hydrology, transfer of energy by both solar and terrestrial radiation, interaction of synoptic-scale weather systems and the larger scale mean flow, and resonance of stationary waves are all possible relevant factors.   JSAG

Reading
Rex, D.F. 1950: Blocking action in the middle troposphere and its effect upon regional climate. I and II. *Tellus* 2, pp. 196–211, and 275–301.
— 1951: The effect of blocking action upon European climate. *Tellus*, 3, pp. 100–11.

blood rain Rain which stains the ground red owing to the incorporation of dust particles carried into the upper atmosphere by wind. Such rain may occur in Europe as a result of outbreaks of Saharan air.

blow-hole Vertical shaft leading from a sea cave to the surface. Air and water may be forced through the hole with explosive force as a rising tide or large waves cause large changes in pressure in the underlying cave.

blowouts Erosional hollows, depressions, troughs or swales within a dune complex (Carter *et al.* 1990). They form readily in vegetated dunes for a variety of reasons: shoreline erosion and/or washover, vegetation die-back and soil nutrient deficiency, destruction of vegetation by animals and fire, and human recreational activities. However, blowout topography need not necessarily arise from erosional processes; it may develop as areas of non-deposition between mobile dune ridges or as gaps in incipient foredunes. Two basic blowout morphologies have been identified: saucer blowouts (shallow, ovoid and dish-shaped with a steep marginal rim) and trough blowouts (deep, narrow, steep-sided with more marked downwind depositional lobes, and marked deflation basins). ASG

Reference
Carter, R.W.G., Hesp, P.A. and Nordstrom, K.F. 1990: Erosional landforms in coastal dunes. In K.F. Nordstrom, N.P. Psuty and R.W.G. Carter eds, *Coastal dunes: form and process*. Chichester: Wiley. Pp. 217–50.

bluehole Like a sapphire set in turquoise, a circular, steep-sided hole which occurs in coral reefs. The classic examples come from the Bahamas (Dill 1977), but other examples are known from Belize and the Great Barrier Reef of Australia (Backshall *et al.* 1979). Although volcanicity and meteorite impact have both been proposed as mechanisms of formation, the most favoured view is that they are the product of karstic processes (e.g. collapse dolines) which acted at times of low sea level when the reefs were exposed to subaerial processes. ASG

References
Backshall, D.G., Barnett, J. and Davies, P.J. 1979: Drowned dolines – the blue holes of the Pompey Reefs, Great Barrier Reef. *BMR journal of Australian geology and geophysics* 4, pp. 99–109.
Dill, R.F. 1977: The blue holes – geologically significant sink holes and caves off British Honduras and Andros, Bahama Islands. *Proceedings of the 3rd International Coral Reef Symposium, Miami* 2, pp. 238–42.

Blytt–Sernander model Provides the classic terminology of the HOLOCENE. It was established by two Scandinavians, A.G. Blytt and R. Sernander, who, in the late nineteenth and early twentieth centuries, undertook various palaeobotanical investigations that revealed vegetational changes from which climatic changes were inferred. They introduced the terms Boreal, Atlantic, Sub-Boreal, and Sub-Atlantic for the various environmental fluctuations that took place. Modern workers recognize that

because factors other than climate affect vegetation change (e.g. man's intervention, soil deterioration through time etc.) the model may be simplistic. ASG

bodden A type of irregularly shaped coastal inlet brought about by the transgression of the sea in an area of undulating terrain, as along the Baltic coastline of East Germany. ASG

bog An area of water-logged ground characterized by thick accumulations of dead but not markedly decomposed plants, particularly mosses, which form acid peat.

bog bursts The sudden disruption of a bog so that there is a release of water and peat which may then flow over a considerable distance.

bogaz Narrow, deep ravines and chasms in karst areas, the products of limestone solution along bedding planes, joints and fissures.

Bølling A short-lived LATE GLACIAL INTERSTADIAL dated at about 12,350–12,750 BP.

bolson A low-lying trough or basin surrounded by high ground and having a playa at its lowest point to which all drainage trends.

bora From Latin *boreas*, the north wind. A dry, cold, gusty, north-east wind which affects the northern part of the Adriatic Sea and the Dalmatian coast (see Jurcec 1981). Peak frequency occurs during the

The classic European Holocene sequence

Period	Zone number	Blytt–Sernander zone name	Radiocarbon years BP
Post Glacial	IX	Sub-Atlantic	post 2450
	VIII	Sub-Boreal	2450–4450
	VII	Atlantic	4450–7450
	VI	Late Boreal	7450–8450
	V	Early Boreal	8450–9450
	IV	Pre-Boreal	9450–12,250
Late Glacial	III	Younger Dryas	10,250–11,350
	II	Allerød	11,350–12,150
	Ic	Older Dryas	12,150–12,350
	Ib	Bølling	12,350–12,750
	Ia	Oldest Dryas	

A cushion bog of Juncaceae in the western Cordillera of Peru. These distinctive habitats have considerable ecological importance and result from the accumulation of organic material in hydrologically favourable areas.

months from October to February and a maximum gust speed of 47.5 m s^{-1} has been recorded at Trieste. Two types of bora are recognized: anticyclonic and cyclonic, according to the atmospheric pressure field. The term bora is now applied to similar winds in other parts of the world, e.g. Oroshi in central Japan (Yoshino 1976).

AHP

Reading and References
†Atkinson, B.W. 1981: *Mesoscale atmospheric circulations.* London and New York: Academic Press.
Jurcec, V. 1981: On mesoscale characteristics of bora in Yugoslavia. In G.H. Liljequist ed., *Weather and weather maps.* Stuttgart: Birkauser Verlag.
Yoshino, M.M. 1975: *Climate in a small area.* Tokyo: University of Tokyo Press.
— 1976: *Local wind bora.* Tokyo: University of Tokyo Press.

bore A tidal wave which propagates as a solitary wave with a steep leading edge up certain rivers. Formation is favoured in wedge-shaped shoaling estuaries at times of spring tides. Other local names include *egre* (England, River Trent), *pororoca* (Brazil), *mascaret* (France). DTP

boreal climate In Köppen's classification scheme a climate which is characterized by a snowy winter and warm summer, with a large annual range of temperature, such as obtains between 60° and 40° north.

boreal forest The northern coniferous zone of the Holarctic. The most northerly section, transitional to the TUNDRA, is synonymous with TAIGA. The southerly sections are frequently made up of dense forest with closed canopies permitting little light to reach the floor, which possesses a variable cover of lichens, mosses and herbaceous plants. Typical hardy and undemanding trees include spruce, fir, and hemlock, with pine providing more open formations and having a denser ground cover. The forest is bounded approximately to the north by the 10°C July average isotherm and to the south by areas with more than four months above 10°C.

PAF

Reading
Larsen, J.A. 1980: *The boreal ecosystem.* New York: Academic Press.

boss A small batholith or any dome-shaped intrusion of igneous rock, especially one exposed at the surface through erosion of the less resistant host rocks.

botryoidal Having a form resembling a bunch of grapes, often applied to aggregate minerals.

bottom-sets Beds of stratified sediment that are deposited on the bottom of the lake or sea in advance of a delta.

Bouguer anomaly A measure of the gravitational pull over an area of the earth after the Bouguer correction to a level datum, usually sea level, has been applied.

boulder clay See TILL.

boulder train A stream of boulders derived by glacial transport from a specific and identifiable bedrock source, and carried laterally in a more or less straight line, thereby permitting former directions of ice movement to be inferred.

boulder-controlled slopes First described by Bryan (1922, p. 43) from the Arizona deserts. Bryan recognized slopes formed on rock with a veneer of boulders and assumed that the angle of the bedrock surface had become adjusted to the angle of repose of the 'average-sized joint fragment'. Measurements have subsequently shown that boulder-covered slopes exist over the range of angles up to about 37° and that the existence of a boulder cover may be due to a number of causes. Melton (1965) suggested that steep angles of the debris (34°–37°) may occur at the angle of static friction of that debris, and that angles of around 26° may be related to its angle of sliding friction. Where the boulders are core stones left by removal of fine-grained saprolite of a former regolith the boulders may lie at any angle up to about 37° and they do not exert any control on the bedrock slope which is essentially a relict weathering front from the base of the regolith (Oberlander 1972). Boulders may also lie upon bedrock which has resulted from the development of a slope angle in equilibrium with the mass strength of the rock (Selby 1982). The idea that the boulders control the angle of the bedrock slope on which they lie is, therefore, open to question in many cases. MJS

Reading and References
Bryan, K. 1922: Erosion and sedimentation in the Papago country, Arizona, with a sketch of the geology. *US Geological Survey bulletin* 730-B, pp. 19–90.

Melton, M.A. 1965: Debris-covered hillslopes of the southern Arizona Desert – consideration of their stability and sedimentation contribution. *Journal of geology* 73, pp. 715–29.

Oberlander, T.M. 1972: Morphogenesis of granitic boulder slopes in the Mojave desert, California. *Journal of geology* 80, pp. 1–20.

Selby, M.J. 1982: Rock mass strength and the form of some inselbergs in the central Namib Desert. *Earth surface processes and landforms* 7, pp. 489–97.

boundary conditions Many physical phenomena can be modelled by mathematical deduction leading to generalized equations; in order to obtain simplified specific solutions to these equations their applicability is deliberately constrained by the definition of particular circumstances known as 'boundary' or 'initial' conditions (Wilson and Kirkby 1975, pp. 206–7 and 222–5). In particular the solution of differential equations requires definition of boundary conditions so that expressions can be found for the arbitrary constants resulting from integration of the equations.

An example is the theoretical derivation of the logarithmic velocity profile in turbulent flow. The rate of change of velocity (v) with height above the bed (y) – the 'velocity shear' – itself decreases with height, according to the differential equation:

$$dv/dy \cdot K/y$$

where K is a constant incorporating the bed shear stress. The variation of velocity with height is obtained by integrating this equation, which gives:

$$v = K \ln y + C.$$

Here, C is a constant of integration. It can be evaluated by specifying the boundary condition that $v = 0$ when $y = y_0$, so that:

$$0 = K \ln y_0 + C$$

and therefore:

$$C = -K \ln y_0.$$

This is actually an initial condition, since the height is defined at which the velocity is zero, and negative velocities are assumed not to occur. The velocity profile equation can now be simplified by inserting the expression for C:

$$v = K \ln y - K \ln y_0$$
$$= K \ln (y/y_0).$$

This describes a curve plotting as a straight line on a graph with a logarithmic height axis and an arithmetic velocity axis; the intercept on the height axis where $v = 0$ is y_0, and the gradient of the line is K (Richards 1982, pp. 69–70).

Theoretical models may require multiple boundary conditions, some of which are necessarily dynamic in order to provide realistic solutions. The slope profile shape characteristic of different types of slope process is modelled by a partial differential equation in which the rate of change of local slope sediment transport with distance from the divide (x) equals the rate of change of local slope surface elevation (y) with time (t). To solve this equation and derive the characteristic form of profile as a graph of y against x, the initial conditions are: (a) the divide is fixed at $x = 0$, and sediment transport is zero at the divide; and (b) an initial slope shape ($y = f(x)$) is defined from which the characteristic form evolves.

A boundary condition then defines the slope foot; this base level is fixed at x_1, and although a fixed base level elevation ($y = 0$ at $x = x_1$) may be set, a dynamic boundary condition may be established in which $y(x_1) = f(t)$. Quite different characteristic profiles will emerge from the solution of the initial general equation according to the nature of this boundary condition, which models the behaviour of basal erosion at the slope foot.   KSR

Reading and References
Richards, K.S. 1982: *Rivers: form and process in alluvial channels.* London: Methuen.

Sumner, G.N. 1978: *Mathematics for physical geographers.* London: Edward Arnold.

Wilson, A.G. and Kirkby, M.J. 1975: *Mathematics for geographers and planners.* Oxford: Oxford University Press.

boundary layer When a fluid and a solid are in relative motion the boundary layer is the zone in the fluid closest to the solid surface within which a velocity gradient develops because of the retarding frictional effect of contact with the solid. The fluid is at rest relative to the solid immediately adjacent to the surface, but with distance from the surface the frictional effect diminishes and velocity increases, at a rate dependent on the local flow characteristics. The velocity gradient of the boundary layer occurs in overland flow on hillslopes, river flow in channels, the swash and backwash of beaches, airflow over a desert dune, but also immediately adjacent to a sand grain falling through the water or air, where it is the *relative* motion producing the boundary

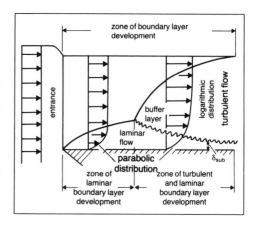

layer, rather than the fluid motion over a static solid surface. The diagram illustrates the development of a boundary layer over a surface parallel to the direction of motion within deep water flow. At the entry point where flow begins over the surface, a laminar boundary layer forms, but at some distance downstream, this is replaced by a turbulent boundary layer if the flow conditions are appropriate; that is, if the REYNOLDS NUMBER exceeds about 2000. The boundary layer is 'fully developed' if the velocity profile extends to the surface of the flow, which is normal in rivers. In a deep fluid layer, however, the motion at some distance from the surface is unaffected by the boundary influence and the velocity is that of a free, or external, fluid stream.

Within a laminar boundary layer viscous forces within the flow are pronounced, and adjacent fluid layers are affected by the molecular interference of the fluid viscosity. The velocity increases with distance from the solid in an approximately parabolic curve (Allen 1970, pp. 36–9). In a turbulent boundary layer, the pattern of velocity increase with distance from the bed is very complex. Close to the bed the fluid is sufficiently retarded for viscous effects to be pronounced and laminar flow occurs; this is the very thin 'laminar sublayer'. If grains on a sedimentary surface are smaller in diameter than the thickness of the laminar sublayer the flow is 'hydrodynamically smooth', and the grains are protected against entrainment. In the HJULSTRÖM CURVE, threshold velocities are seen to be higher for silt and clay sizes than for sand sizes. Above the laminar sublayer is a buffer zone before the true turbulent velocity profile is reached. In the turbulent boundary layer interference between fluid elements occurs at a scale controlled by the depth of eddy penetration, and measured velocity profiles indicate that velocity increases with the logarithm of distance from the surface (Richards 1982, pp. 68–72). Under the BOUNDARY CONDITIONS defined above it is shown that this relationship takes the form:

$$v = K \ln (y/y_0)$$

where v is velocity, y is distance from the surface, y_0 is the height where the velocity is zero, and K is a constant which is equal to v_*/K. Here, K is the von Karman constant (0.4) <u>and</u> v_* is the 'shear velocity', defined as $\sqrt{\tau_0/\rho_w}$. This is a measure of the steepness of the velocity profile which is dependent on bed shear stress and water density. In hydrodynamically rough conditions where grains are large relative to the thickness of the laminar sublayer, y_0 equals one-thirtieth of the D_{65} grain diameter. If the above equation is converted to common logarithms and the expressions for K and y_0 are inserted, the equation for the logarithmic velocity profile in a hydrodynamically rough, turbulent boundary layer, becomes:

$$v/v_* = 5.75 \, log \, (y/D_{65} + 8.5).$$

This equation may be used to fit to velocity data from the lower 10–15 per cent of the flow in order to project the curve to the bed. The local bed shear stress can then be estimated, as well as the velocity close to the bed sediment at heights where measurement is impractical, especially in the field. Note that both the turbulent fluctuations and the rapid increase of velocity above the bed material, causing strong lift forces, which occur under hydrodynamically rough bed conditions in turbulent flow, are important factors in the entrainment of sediment by the flow.   KSR

Reading and References

Allen, J.R.L. 1970: *Physical processes of sedimentation.* London: Allen & Unwin.

Leeder, M.R. 1982: *Sedimentology: process and product.* London: Allen & Unwin. Pp. 47–66.

†Open University 1978: Oceanography Unit 11 *Introduction to sediments.* Milton Keynes: Open University. Pp. 10–16.

Richards, K.S. 1982: *Rivers: form and process in alluvial channels.* London and New York: Methuen.

bourne A stream or stream channel on chalk terrain that flows after heavy rain.

Bowen's reaction series A series of minerals which crystallize from molten rock

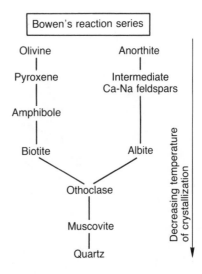

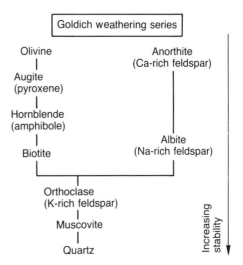

of a specific chemical composition, wherein any mineral formed early in the chain will later react with the melt, forming a new mineral further down the series; the minerals formed under decreasing temperatures of crystallization are more stable in the weathering environment.   ASG

brackish Pertaining to water which contains salt in solution, usually sodium chloride, but which is less saline than seawater.

braided river A river whose flow passes through a number of interlaced branches

A braided river channel in the Rakaia valley, New Zealand. Such multi-thread channels tend to develop in areas with coarse debris, relatively steep slopes and variable discharge.

that divide and rejoin. The term has been applied both to short reaches where a river splits around an island and to very extensive river networks on valley bottoms or alluvial plains, the whole of which may be criss-crossed by rapidly shifting channels with freshly deposited sediment between them. Braiding may be more apparent at some flow levels than at others. For example, single channels of low sinuosity at high flows may assume a braided pattern as channels thread their way between sets of emergent BEDFORMS at low flow. By contrast, single channels may become multiple ones at high flows as inactive channels are reoccupied and developed.

The term 'braided' is applied in a general sense to a whole family of multiple channel river patterns some of which have recently been given separate names. The term applies particularly to 'anastomosing' and 'wandering' rivers. The former, as identified by D.G. Smith in Canada, is a type of stable multi-channel system developed under aggrading conditions with levées and back-swamps. They closely resemble deltaic distributary channel patterns, though they are found in some inland valleys. It is confusing that 'anastomosing' was earlier also used as an alternative for braiding in the general sense. 'Wandering' rivers, first identified in this sense by C.R. Neill, also in

Canada, may consist of alternate stable single channel reaches and unstable multi-channel 'sedimentation zones'. The term has also been used as an alternative for patterns that are transitional between meandering and braided. Examples of these several variants or relations to braided channel patterns are discussed in chapters by D.G. Smith, M. Church, R.I. Ferguson and A. Werritty in Collinson and Lewin (1983).

Braided river patterns – in the general sense of multi-channel systems – appear to be created in various ways. Mid-channel bar development may lead to division of the channel and enlargement of the bar by accretion, possibly with the development of a vegetated island. Alternatively, migratory bars, exposed only at low flows, may simply be exposed bedforms continually shifting at high flows by erosion and accretion. Scour at channel junctions may be important for the local entrainment of sediment which is then redeposited as flows diverge again down-channel. Overbank flood flows also scour out new chute channels.

Particularly in view of the various channel-dividing or multiplying processes involved, not to mention the different kinds of pattern and pattern change, it is not surprising that various conditions and environments have been identified as conducive to the development of braided rivers. These include high energy environments (with steep-gradient channels and high or variable discharges) and high rates of sediment transport. It has been suggested that the braided pattern is one form of channel adjustment to prevailing hydraulic conditions, though it is not yet possible to predict exactly when and why braiding will occur. In general braiding may be found in contemporary ice-marginal gravel rivers (where fluctuating high discharges, high sediment supply rates and steep gradients may be combined) and in some semi-arid sandy rivers (where at least the first two may be common). Localized braiding, in the form of semi-permanent islands, may be widely found. Many of the world's largest rivers have braided lower courses. These include the Amazon, the Brahmaputra and the Hwang He. In North America, braided rivers are found in the arid southwest, on the Great Plains, and in glacier marginal environments in the Rockies and Coast Ranges of Western Canada. In Britain, braided rivers are rare, with a few good examples in Scotland but almost none elsewhere (see CHANNELS, TYPES OF RIVER/STREAM). JL

Reading and References
Collinson, J.D. and Lewin, J. eds. 1983: *Modern and ancient fluvial systems*. Special publication no. 6, International Association of Sedimentologists. Oxford: Blackwell Scientific.
†Richards, K. 1982: *Rivers*. London and New York: Methuen.

brash A mass of fractured rock that has been weathered *in situ*, also applied to a mixture of shattered rock or ice.

braunerde Brown forest soil or brown earth. Dark brown coloured soil with decreasing organic content from the surface downwards. It is usually developed on calcareous parent material and has a high agricultural potential.

breccia A rock that has been greatly fractured into angular fragments, generally less than 2 mm in diameter, by tectonic activity, volcanism or transport over short distances.

broad-crested weir A type of flow gauging WEIR in which the crest is considerably broader than that of a SHARP-CRESTED WEIR. The crest is commonly constructed of concrete, and rectangular, round-nosed and triangular profiles are employed. These weirs are generally used on large rivers, and may be classified according to the shape of the crest or notch into rectangular, compound and other variants. The stage/discharge relationship is frequently established by field rating rather than by using theoretical formulae, but it will take the same general form as that associated with a sharp-crested weir. DEW

Reading
Ackers, P., White, W.R., Perkins, J.A. and Harrison, A.J.M. 1978: *Weirs and flumes for flow measurement*. Chichester: Wiley.
Gregory, K.J. and Walling, D.E. 1976: *Drainage basin form and process*. London: Edward Arnold. Pp. 135–9.

brodel A highly contorted and irregular structure in soils which have been subjected to churning by frost processes.

brousse tigrée Vegetation banding, which may include grassland patterns but which generally consists of bands of more closely spaced trees alternating with bands of sparser vegetation. Its nature and origin have been well described thus by Mabbutt and Fanning (1987, p. 41): 'All are developed in arid or semi-arid areas, in open low woodlands or tall shrublands, with average annual rainfalls of between 100 and 450 mm; they occur on slopes of the order of 0.25%, too gentle for the development of drainage channels, but steep enough to maintain organized patterns of sheetflow; these slopes are mantled with alluvium or colluvium and the patterns are independent of bedrock. The associated soils are earths, and sandier crests or clay flats in the same areas do not have tree bands. The bands of denser vegetation, termed "vegetation arcs" run close enough to the contour to serve as form lines; hence they tend to be convex downslope on interfluves and convex upslope in shallow drainage ways. In drier areas the banding may be restricted to the better-watered depressions, but it is commonly best-developed on low interfluves, with the intervening depressions marked by uniformly dense tree cover. Such tracts of more concentrated sheetflow have been named "water lanes".

The bands commonly occur in fairly regular sequences or ladder-like "tiers" downslope, the tiers being bounded by water lanes. Tree bands may extend up to a kilometre, or more along the contour, but in detail they are commonly slightly irregular, "burgeoning here and becoming attenuated there; dying out and succeeding one another *en echelon*".

The downslope distance between bands ranges from 70–500 m, although it is mainly between 100 and 250 m and the interband intervals are commonly between two and four times as wide as the bands.'

ASG

Reference
Mabbutt, J.A. and Fanning, P.C. 1987: Vegetation banding in western Australia. *Journal of arid environments* 12, pp. 41–59.

Brückner cycle A series of cold, wet seasons followed by a series of hot, dry ones which recur regularly over a period of about 35 years.

brunizem A prairie soil developed under grassland in temperate latitudes. Characteristically a brown surface zone overlies a leached horizon which grades into a brown subsoil on non-calcareous bedrock.

Bruun rule An empirical equation designed to predict absolute horizontal shoreline recession (r) arising from absolute sea level rise (s).

$$r = ls/h$$

where l and h are the length and height of the equilibrium cross-shore profile from beach crest to offshore. Shoreline recession is caused by the upward and landward movement of the cross-shore profile as it moves to readjust to the disequilibrium caused by a rise in sea level. Measurements of l and h are difficult as the seaward edge of the equilibrium profile has to be established. This position (closure depth) should relate to the start of onshore sediment transport by waves, but is usually regarded as variable depending on arbitrary definitions of the maximum wave causing such transport. The closure depth has in recent years come to be defined by some multiple of significant storm-wave height associated with a return period of n-years (Hands 1983).The original study (Bruun 1962) related to measured profile recession of Florida barrier-island shorelines over twenty years. Schwartz (1967) thought that the equation had universality sufficient to indicate the status of a rule, but subsequent work indicates that this is an overstatement. Widespread use of the rule to establish building set-back lines on eroding coasts by coastal managers has been controversial. Doubts have been expressed about the universal validity of such cross-shore profile analysis when beach changes are dominated by long-shore sediment supply. The rule has been championed by workers with experience mainly of the open barrier beaches (i.e. USA). Recession of gravel-dominated beaches on closed or crenellate coasts does not conform to this absolute rule, though r and s are positively correlated when cited as rates of change (Orford *et al.* 1991). It is important to realize that sea level *per se* does not cause recession; it is merely the datum upon which waves and tides, which do the work of profile alteration, operate.

JO

References
Bruun, P. 1962: Sea level rise as a cause of shore erosion. *Journal of the Waterways and Harbour Division of the American Society of Civil Engineers* 88, pp. 117–30.

Hands, B. 1983: The Great Lakes as a test model for profile responses to sea level changes. In P.D. Komar ed., *Handbook of coastal processes and erosion.* Boca Rotan, Florida: CRC Press, pp. 167–89.

Orford, J., Carter, R.W.G. and Forbes, D.L. 1991: Gravel barrier migration and sea level rise: some observations from Story Head, Nova Scotia, Canada. *Journal of coastal research* 7, pp. 477–88.

Schwartz, M.L. 1967: The Bruun theory of sea level rise as a cause of shoreline erosion. *Journal of geology* 73, pp. 528–34.

Bubnoff units Provide a means for quantifying rates of slope retreat or ground loss (perpendicular to the ground surface). A unit equals 1 mm per 1000 years, equivalent to $1 \text{ m}^3 \text{ km}^{-2}$ (Fischer, 1969).

ASG

Reference
Fischer, A.G. 1969: Geological time-distance rates: the Bubnoff unit. *Bulletin of the Geological Society of America* 80, pp. 594–652.

buffer A solution to which large amounts of acid or alkaline solutions may be added without markedly altering the original hydrogen ion concentration (pH).

bush encroachment The advance of indigenous shrub and tree species into an area of land previously cleared of trees for the purposes of agriculture. Also the establishment of shrubs and tree species in an area which was formerly grassland, as a result of reduced frequency of grass fires. It is a particular environmental problem in some savannah areas and may be accelerated by overgrazing of palatable grasses by domestic animals.

bushveld The savannah lands of sub-Saharan Africa, ranging from open grassland, through parkland with scattered trees to dense woodland.

butte A small, flat-topped and often steep-sided hill standing isolated on a flat plain. Often attributed to erosion of an older landsurface, the butte representing a remnant or outlier.

butte temoin A flat-topped, erosional outlier on the scarp side of a cuesta.

Buys Ballot's law An observer in the northern hemisphere, standing with his back to the wind, will have low pressure to his left and high pressure to his right; the converse is true in the southern hemisphere. This law was formulated in 1857 by the Dutch meteorologist Buys Ballot. (See also CORIOLIS FORCE; GEOSTROPHIC WIND.)

BWA

bysmalith A plutonic plug or mass of igneous rock which has been forced up into the overlying rocks causing them to dome up and fracture.

C

caatinga A form of thorny woodland found in areas such as north-east Brazil, and characterized by many xerophytic species.

caballing The mixing of two water masses of identical *in situ* densities but different *in situ* temperatures and salinities, such that the resulting mixture is denser than its components.   ASG

Cainozoic (Cenozoic) A geological era spanning the Palaeocene, the Eocene, the Oligocene, the Miocene, the Pliocene and the Pleistocene. It was a time of climatic decline, possibly associated with the breaking of the super-continent of Pangaea into the individual continents we know today, which moved into high latitudes so that ice caps could develop.   ASG

calcicole A plant which flourishes with a large amount of exchangeable calcium in the soil. Examples include wood sanicle (*Sanicula europaea*) and traveller's joy (*Clematis vitalbs*). Plants which clearly cannot tolerate such conditions are calcifuge; examples include common heather (*Calluna vulgaris*) and most other ericaceous

Subdivisions of the Cainozoic era

	Date of beginning in millions of years
Pleistocene	1.8
Pliocene	5.5
Miocene	22.5
Oligocene	36.0
Eocene	53.5
Palaeocene	65.0

Source: Berggren, W.A. 1969: Cainozoic stratigraphy, planktonic foraminiferal zonation and the radiometric time-scale. *Nature* 224, pp. 1072–5.

plants. The effect of pH on mineral nutrition appears to be the operative factor.   KEB

Reading
Crawley, M.J. ed. 1986: *Plant ecology*. Oxford: Blackwell Scientific.

calcifuge Any plant which grows best on acidic soils, e.g. bracken.

calcrete A terrestrial material composed predominantly, but not exclusively, of calcium carbonate, which occurs in states ranging from powdery and nodular to highly indurated, as the result of displacive and/or replacive introduction of vadose carbonates into greater or lesser quantities of soil, rock or sediment within a weathering profile. The term, which is synonymous with caliche and kunkur, does not embrace cave deposits (SPELEOTHEM), spring deposits (for which TUFA or travertine are accepted terms), marine deposits (such as BEACH ROCK) or lacustrine algal stromatoliths. Calcretes, the profiles of which may exceed 40 m in thickness, are widespread in semi-arid areas such as the Kalahari, Western Australia, Tunisia and the High Plains of the USA, where precipitation is in excess of evapotranspiration, so that there is a tendency for carbonate (derived from bedrock, dust, surface run-off, ground-water etc.) not to be fully leached out of the system. Local translocation and accumulation predominate.

The average chemical composition is $CaCO_3$ 78 per cent; SiO_2 12 per cent; Al_2O_3 2 per cent; Fe_2O_5 2 per cent; and MgO 3 per cent. In addition, calcretes may contain clay minerals like palygorskite, sepiolite, and concentrations of uranium ore (carnotite).   ASG

Reading
Goudie, A.S. 1983: Calcrete. In A.S. Goudie and K. Pye eds. *Chemical sediments and geomorphology.* London: Academic Press. Pp. 93–131.
Wright, V.P. and Tucker, M.E. eds. 1991: *Calcretes.* Oxford: Blackwell Scientific.

caldera A large, roughly circular, volcanic depression. Calderas usually have a number of smaller vents and can also contain a large crater lake. The distinction between volcanic craters and calderas is essentially one of size, one to two kilometres being the lower limit for the diameter of a caldera. Maximum diameters are in excess of 40 km. Calderas probably form in a variety of ways, but most proposed mechanisms attribute a primary role to collapse or subsidence, which may be related to explosive eruptions.   MAS

Reading
Francis, P. 1983: Giant volcanic calderas. *Scientific american* 248, pp. 60–70.
Williams, H. and McBirney, A.R. 1979: *Volcanology.* San Francisco: Freeman, Cooper.

calms Winds with a velocity of less than one knot and which are represented by a force of zero on the BEAUFORT SCALE.

calving The breaking away of a mass of ice from a floating glacier or ice shelf to form an iceberg or brash ice (small fragments). Large tabular icebergs calve from ice shelves while smaller icebergs and brash ice are commonly produced by valley or outlet glaciers. Most calving is induced by stresses set up within the floating ice mass by ocean swell.   DES

cambering The result of warping and sagging of rock strata which overlie beds of clay. The plastic nature of the clays causes the overlying rocks to flow towards adjacent valleys, producing a convex outline to the hill tops. Classic examples of cambering occurred in the Pleistocene when, under periglacial conditions, great rafts of limestone or sandstone subsided over lias and other clays along the escarpments of southern England. Cambering is often associated with the development of VALLEY BULGES.   ASG

canopy Usually taken to be the uppermost stratum of woodland vegetation, the tree-top layer, though the term may also be used for any extensive above-ground leaf-bearing parts of plants. Despite the obvious

The canopy of a montane forest at around 3000 m on the flanks of Mount Kilimanjaro, East Africa. Beneath the canopy the amount of incoming light is greatly reduced.

importance of this zone in the interception of light and precipitation, and in the production of flowers, relatively little work has been reported, presumably due to practical difficulties. The role of the canopy in the woodland light climate, and therefore in tree regeneration and ground flora, is a vital one: some 80 per cent of incoming radiation may be intercepted in this zone and 10 per cent reflected from the upper leaves and twigs.   KEB

Reading
Crawley, M.J. ed. 1986: *Plant ecology.* Oxford: Blackwell Scientific.
Packham, J.R. and Harding, D.J.L. 1982: *Ecology of woodland processes.* London: Edward Arnold.

capillary forces Essentially SURFACE TENSION and adsorptive forces. Water will rise up a narrow (capillary) tube as a result of adsorptive forces between the water and the tube surface and tension forces at the water surface. These forces bind soil moisture to the soil particles so that it is held in an unsaturated soil at less than atmospheric pressure. This is often called a SUCTION or tension and its strength may be determined using a TENSIOMETER.   AMG

Reading
Baver, L.D., Gardner, W.H. and Gardner, W.R. 1991: *Soil physics* 5th edn. New York: Wiley.
Smedema, L.K. and Rycroft, D.W. 1983: *Land drainage.* London: Batsford.

cap-rock A stratum of hard, resistant rock which overlies less competent strata and protects them from erosion.

capture (or river capture) The capture of part of one drainage system by another system during the course of drainage pattern evolution. Interpretation of drainage networks in terms of river capture was an integral feature of the Davisian CYCLE OF EROSION and the distance to base level or sea level, the exposure of easily eroded rocks, or the effects of discharge increase following climatic change could all be reasons why one river was able to erode more rapidly and so capture the headwaters of another. A terminology was developed for several aspects of the streams involved in capture, as shown in the diagram. The beheaded stream becomes a misfit stream as it is now too small for the valley. River capture has certainly featured prominently in the evolution of world river systems and, for example, the easternmost tributary of the Indus was captured by the Ganges in geologically recent times, transferring drainage from a large area of the Himalayas from Pakistan to India. Knowledge of the sequence of river capture is sometimes necessary in the location of placer deposits which are alluvial deposits containing valuable minerals. Placer deposits from ore deposits may occur in an area (such as *x* on diagram B) no longer directly connected to the drainage system with ores outcropping (at *y*) in the headwaters (Schumm 1977).   KJG

Reference
Schumm, S.A. 1977: *The fluvial system*. Chichester: Wiley.

carapace *a.* The upper normal limb of a recumbent fold.
b. A soil crust which is exposed at the surface, especially a surficial calcrete.

carbon cycle The 'life' cycle, carbon being one of the three basic elements (with hydrogen and oxygen) making up most living matter. Over 99 per cent of the earth's carbon is locked up in calcium carbonate rocks and organic deposits such as coal and oil, both being the result of millions of years of carbon fixation by living organisms on land and in the oceans. The biotic cycle is similarly split into terrestrial and oceanic subsystems. Photosynthesis by pigmented plants fixes the carbon dioxide from air and water; almost half is returned by plant respiration, the rest builds up as plant materials. The carbon is then returned to the atmosphere via animal respiration or plant decomposition. Fossil fuel consumption has increased atmospheric CO_2 fairly dramatically in the past few decades and is the basis of current concern over a GREENHOUSE EFFECT.   KEB

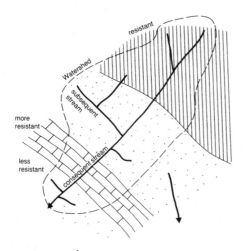

A

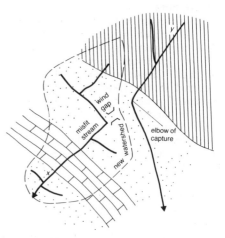

B

Reading
Bach, W., Crane, A.J., Berger, A.L. and Longhetto, A. eds 1983: *Carbon dioxide.* Dordrecht, Holland: D. Reidel.
Bradbury, I.K. 1991: *The biosphere.* London: Belhaven Press.
Goudie, A.S. 1993: *The human impact.* 4rd edn. Oxford: Basil Blackwell.

carbon dating (also called C14 dating, and radiocarbon dating) An isotopic dating technique, developed in 1949, based on the principle that radioactive elements such as C14 are subject to decay through time. The rate of decay being known for a particular element, the time interval may be assessed between the present and the time when the particular parent material was fixed and its decay began. This technique, formerly used mainly for organic carbonate (peat, wood, charcoal etc.) is now being extended to a wider range of materials, especially soil carbonates, bones and mollusca. Some laboratories can now bring material 75,000 years old within dating range.   ASG

Reading
Worsley, P. 1981: Radiocarbon dating: principles, application and sample collection. In A.S. Goudie ed., *Geomorphological techniques.* London: Allen & Unwin. Pp. 277–83.

carbon dioxide problem One of the most important environmental issues facing the human race. The burning of fossil fuels, the destruction of forests, and the reduction of humus levels in soils under agricultural treatment are increasing the levels of carbon dioxide in the atmosphere above their recent natural levels of around 270 ppm. If the levels continue to increase it is possible that world temperatures will increase because of the GREENHOUSE EFFECT, setting in train such consequences as changes in the positions of rainfall belts, increases in the growing seasons in high latitude areas, and melting of ice caps, etc. Particular concern has been voiced about the possibility of severe droughts in areas such as the Great Plains of the USA. On the other hand, elevated CO_2 levels tend to promote plant growth and reduce plant water needs. The forecasting of future CO_2 levels is problematical in view of such factors as uncertainties surrounding future energy consumption, and a dearth of knowledge about how much of the gas will be absorbed by oceans and vegetation.
                                ASG

Reading
Hansen, J., Johnson, D., Lacis, A., Lebedeff, S., Lee, P., Rind, D. and Russell, G. 1981: Climatic impact of increasing atmospheric carbon dioxide. *Science* 213, pp. 957–66.
Houghton, J.T., Jenkins, G.J. and Ephraums, J.J. eds 1990: *Climate change: the IPCC scientific assessment.* Cambridge: Cambridge University Press.

carbonation The reaction of minerals with dissolved carbon dioxide in water. The process is dominant in the weathering of limestone, since rainwater contains a small proportion of carbon dioxide (0.03 per cent by weight) and thus acts as weak acid dissolving limestone rock. The conventional chemical reaction is shown in the following formula:

$$CO_2 + H_2O + CaCO_3 \rightleftharpoons Ca(HCO_3)_2$$

The $Ca(HCO_3)_2$ molecules have never been detected in solution and, while the product of carbonation is well known, the chemical process is not fully explained by this conventional equation (Picknett *et al.* 1976).   PAB

Reference
Picknett, R.G., Bray, L.G. and Stenner, R.D. 1976: The chemistry of cave water. In T.D. Ford and C.H.D. Cullingford eds, *The science of speleology.* London: Academic Press. Pp. 213–66.

carnivore An animal-eating mammal of the order Carnivora, which depends solely on other carnivores or HERBIVORES for its food, and which is located in the higher TROPHIC LEVELS of ecological systems. Carnivores may be predators (e.g. the lion or wolf among the large land animals, many species of beetles, molluscs, centipedes and mites among the smaller); scavengers, such as jackals and seagulls; or animal parasites, including a wide range of bacteria, protozoa, nematodes and winged insects. Excepting the parasites, most carnivores are not restricted to a single species for their food supply; their ranges accordingly tend to be larger than those of the animals on which they depend.   DW

carr *a.* Former bog and marsh land that has been reclaimed by drainage.
b. An isolated islet or rock, especially off the north-east coast of the British Isles.

carrying capacity Represents the population size which the resources of an environment can just maintain without a tendency to decrease or increase. Begon *et al.* (1986, p. 209) explain it thus: 'As population density increases, the per capita birth rate eventually falls and the

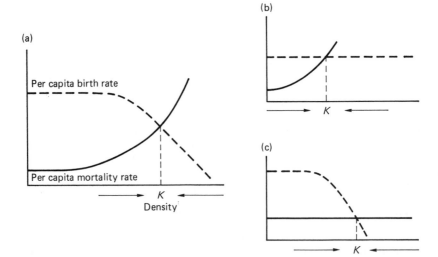

Carrying capacity: Density-dependent birth and mortality rates lead to the regulation of population size. When both are density-dependent (a), or when either of them is (b and c), their two curves cross. The density at which they do so is called the carrying capacity (K). Below this the population increases, above it the population decreases: K is a stable equilibrium. But this figure is a mere caricature of real populations.
Source: *Begon et al. 1986, p. 210. Figure 6.5.*

per capita death rate eventually rises. There must, therefore, be a density at which these curves cross. At densities below this point, the birth rate exceeds the death rate and the population increases in size. At densities above the cross-over point, the death rate exceeds the birth rate and the population declines. At the cross-over density itself, the two rates are equal and there is no net change in population size. This density therefore represents a *stable equilibrium*, in that all other densities will tend to approach it. In other words, intraspecific competition, by acting on birth rates and death rates, can *regulate* populations at a stable density at which the birth rate equals the death rate. This density is known as the *carrying capacity* of the population and is usually denoted by *K*.'   ASG

Reference
Begon, M., Harper, J.L. and Townsend, C.R. 1986: *Ecology: individuals, population and communities*. Oxford: Blackwell Scientific.

carse A flat area of alluvium adjacent to an estuary.

cascading systems See SYSTEMS.

case hardening The feature or process of formation of a hard, resilient crust on the surfaces of boulders and outcrops of soft, porous rock through the filling of voids with natural cement. The cement may consist of a range of different materials, including iron and manganese oxides, silica and calcium carbonate. Beneath the hard surface the rock may be weakened, so that if the crust is breached, cavernous weathering may occur.   ASG

cataclasis The process of rock deformation accomplished by the fracture and rotation of mineral grains, as in the production of a crush breccia.

cataclinal Pertaining to a stream or river which trends in the same direction as the dip of the rocks over which it flows.

catastrophe/catastrophism In general use, the word catastrophe can be applied to any major, normally short-lived and sudden, misfortune leading to widespread change. Catastrophes can be found in both the physical and human environments, physical examples being storm surges, floods and hurricanes. In terms of catastrophe theory, however, catastrophes are more precisely defined events which affect systems and cause their organizations (Thom 1975). This varied use of the word catastrophe springs in part from the changing development of ideas, especially of those relating to earth history.

Catastrophism is a mode of thought that ascribes important change in the physical environment to the action of catastrophic events. It is often placed in direct opposition to the doctrine of UNIFORMITARIANISM which basically ascribes change in the physical environment to small-scale, commonly acting processes (see Gould 1984 for a more comprehensive description of the varied meanings of uniformitarianism).

The origins of catastrophism have often been traced to Baron Georges Cuvier (1769–1832). Cuvier was primarily a palaeontologist and he brought catastrophic ideas to the attention of his fellow geologists. The use of catastrophic episodes to explain earth history was necessitated by current religious and scientific views which held that the history of both rocks and the living world should be subsumed within biblical history. Two important data levels were revealed from the Bible, i.e. the Deluge and the Creation. Cuvier came to the conclusion that in order to incorporate the events revealed in the stratigraphic record, given a relatively short ordained time span of 75,000 years for the total history of the earth as accepted by most scientists at that time, sudden catastrophic changes needed to be invoked. Extinctions and structural discordances could only be explained in these terms. Cuvier's ideas were therefore at variance with those of James Hutton (1726–1797) who is regarded as one of the fathers of uniformitarian views.

Charles Lyell (1797–1875) somewhat unjustly saw catastrophism as being unscientific in its approach (Benson 1984). He was the first main exponent and propagator of uniformitarian views in the geological world and as a result of his persuasive arguments and those of others, uniformitarian views have dominated the study of geology and other earth sciences for over 100 years. Darwin's acceptance of uniformitarian views had great affect on his ideas on the progress of evolution.

Catastrophism and uniformitarianism can therefore be seen in their extreme forms to be two ends of a spectrum of approaches to explaining the physical environment. The development of ideas on the environment is, in part, affected by human history and cultural events which affect the way in which people think and view the world. When war and revolution are common, for example, catastrophic explanations of a variety of phenomena are likely to come into vogue.

In more recent years catastrophism has once again become accepted as a valuable explanatory tool. In palaeontology and biology, for example, the idea of 'punctuated equilibria' (i.e. discrete, sudden changes in species as opposed to gradual evolution) has gained support (Gould and Eldridge 1977) and there have been many explanations of past mass extinctions invoking catastrophic events.

Many geomorphological features can be most satisfactorily explained by recourse to catastrophic ideas (Dury 1980). A classic example of this is provided by the work of Bretz on the channeled scablands of eastern Washington. Bretz suggested in 1923 that these scablands could best be explained by the action of a single gigantic flood over a period of only a few days. Bretz's views did not achieve much recognition when they were published but they have since been shown to be broadly correct. Analogous features have recently been discovered on the surface of Mars and similar catastrophic explanations have been put forward to account for these.

Neocatastrophic views in the earth sciences have been strengthened by the development of catastrophe theory (Thom 1975). This complex, mathematical theory accounts for sudden changes in systems and may be used for modelling. Its potential has been recognized by several geomorphologists (e.g. Graf 1988), but it has received much criticism from mathematicians and its complexity has baffled many earth scientists.

Huggett (1990) provides a useful review of the history of catastrophism and its importance to biology, geology and geomorphology. It is clear that most environmental systems are affected by both catastrophic and gradual changes and the ideas of NON-LINEAR SYSTEMS recently introduced to the earth sciences attempt to include all such changes. Indeed, Huggett goes as far as to suggest that 'over the next few years catastrophic and gradual change will be unified by the theory of non-linear dynamics' (Huggett 1990, p. 200).   HAV

Reading and References
Benson, R.H. 1984: Perfection, continuity and common sense in historical geology. In W.A. Berggren and J.A. Van Couvering eds, *Catastrophes and earth history*. Princeton, NJ: Princeton University Press. Pp. 35–76.

†Berggren, W.A. and Van Couvering, J.A. eds 1984: *Catastrophes and earth history*. Princeton, NJ: Princeton University Press.

Bretz, J.H. 1923: The channeled scablands of the Columbia plateau. *Journal of geology* 31, pp. 617–49.

Dury, G.H. 1980: Neocatastrophism? A further look. *Progress in physical geography* 4, pp. 391–413.

Gould, S.J. 1984: Toward the vindication of punctuational change. In W.A. Berggren and J.A. Van Couvering eds, *Catastrophes and earth history*. Princeton, NJ: Princeton University Press. Pp. 9–34.

— and Eldridge, N. 1977: Punctuated equilibria: the tempo and mode of evolution reconsidered. *Paleobiology* 3, pp. 115–51.

Graf, W.L. 1988: Applications of catastrophe theory in fluvial geomorphology. In Anderson, M.G. ed., *Modelling geomorphological systems*. Chichester: Wiley.

Huggett, R. 1990: *Catastrophism. Systems of earth history*. London: Edward Arnold.

Thom, R. 1975: *Structural stability and morphogenesis: an outline of a general theory of models*. Reading, Mass.: Benjamin.

catchment control The adjustment and arrangement of land use in a catchment so that as far as possible an appropriate quality and quantity of water suitable for distribution throughout the year can be ensured at minimum cost to the community.　　ASG

Reading
Newson, M. 1991: Catchment control and planning: emerging patterns of definition, policy and legislation in UK water management. *Land use policy* 8, pp. 9–15.

catena The catena concept was formulated in British East Africa by the soil scientist G. Milne (1935), and was originally defined as 'a unit of mapping convenience . . . a grouping of soils which while they may fall wide apart in a natural system of classification on account of fundamental and morphological differences, are yet linked in their occurrence by conditions of topography and are repeated in the same relationship to each other wherever the same conditions are met with'. In other words, if an area has well-defined slope units, and if parent materials are similar, then the soil types reflect the differences in leaching, drainage, mass movements etc. on different land surfaces within the area (Fenwick and Knapp 1982).　　ASG

References
Fenwick, I. and Knapp, B.J. 1982: *Soils: process and response*. London: Duckworth.

Milne, G. 1935: Composite units for the mapping of complex soil associations. *Transactions of the Third International Congress of Soil Science* 1, pp. 345–7.

cation exchange The exchange of one species of base attached to the surface of clay colloids for another.

cation-ratio dating A form of surface-exposing dating. It can be classified as a calibrated, biochemical age determination method in which chemical changes in rock varnish are calibrated by established numerical ages for the subaerial exposure of the underlying rock. The cation ratio of $(K^+ + Ca^{2+})/Ti^{4+}$ decreases with time.

ASG

Reading
Dorn, R.I. 1989: Cation-ratio dating of rock varnish: a geographic assessment. *Progress in physical geography* 13, pp. 559–96.

causality The relationship between events in which a second event or configuration (B) can be seen as the product of a prior event (A): in other words A is cause and B is effect. A simple causal relationship is one where B is only and always the result of A: an obvious example is the reaction of litmus paper to the application of an acid solution.

In physical geography – and in historical science in general – causality can rarely be established in a simple experimental fashion, but has to be inferred by repeated observations of A and B. Several problems arise. First, the joint occurrences of A and B may be fortuitous and there may be no physical connection between them. Secondly, both A and B may be responses to some other, truly causal event or variable, C, and the apparently direct causal link between them misleading. Thirdly, A may be a necessary but not a sufficient cause of B, i.e. some further agency or group of agencies is involved.

It is particularly difficult to infer causality with certainty when observations are spatially contiguous or coincident, although similar problems arise with temporal sequences.

The establishment of inferential causal relationships with a high level of reliability depends upon very careful programmes of observation and an exhaustive search for alternative explanations. Two interesting and informative examples from the early literature in geomorphology are Darwin's (1842) establishment of a causal link between the distribution of coral reefs and the nature of volcanic activity; and Ramsay's (1862) identification of a causal relationship between the operation of land ice and the existence of lake basins in the upper portion of river courses.　　BAK

References
Darwin, C.R. 1842: *The structure and distribution of coral reefs*. London: Smith, Elder.

Ramsay, A.C. 1862: On the glacial origin of certain lakes in Switzerland, the Black Forest, Great Britain, Sweden, North America and elsewhere. *Quarterly journal of the Geological Society* 18, pp. 185–204.

causse A term synonymous with karst, derived from the name of the limestone landscape of the Central Massif of France.

cave A natural hole or fissure in a rock, large enough for a man to enter. Although caves can be found in any type of rock, they are most common in limestone regions and are formed by solutional processes of joint enlargement. Caves can be either horizontal or vertical in general form; the latter are usually termed pot-holes. Those produced by solutional processes are normally initiated (i.e. by joint enlargement) in the saturated or phreatic zone. Lowering of the water table allows normal stream or vadose conditions to cut canyons in the more circular phreatic cave tubes. Thus, compound cave cross-sections can result: in this specific case a keyhole-shaped passage is produced. (Indeed, the 20 km cave Agen Allwedd in South Wales is named from the Welsh: Keyhole Cave.) Solutional processes alone do not account for all limestone cave systems; often, when the water table lowers, the overburden of rock, now no longer supported by a water-filled cavity, collapses, producing extensive boulder falls in cave passages.

The general pattern of a cave system depends not only on the processes which have led to its formation but also on the regional jointing, folding and faulting. Caves develop along lines of weakness and the structural geology of the area will dictate the plan and depth of a cave almost as much as fluctuations in the water table.

Solutional caves can also form in rock salt, although such cavities usually form as isolated chambers rather than integrated cave passage networks. Ice too can provide solutional cave systems; some systems can be very long lasting (Bull 1983).

Although caves can also be produced by tectonic activity (which is regularly referred to in textbooks as a viable mechanism of cave formation), in practice they are few and far between. They form as cavities on the limbs and crests of tightly folded rocks but normally only very small recesses are formed, never long cave systems.

The largest of the cave systems formed in non-karstic rocks (PSEUDOKARST) are found in lava. Well-documented, long cave systems exist in Hawaiian lava flows (Wood 1976), sometimes exceeding 10 km in passage length. They are not of course the results of solutional processes but rather products of heat loss at the edges of lava flows, with corresponding continual flowing of molten lava in the core of the flow. Repeated eruptions utilize the same passages to transport their lava along these gently dipping tubes, perpetuating the lava cave system.   PAB

References
Bull, P.A. 1983: Chemical sedimentation in caves. In A.S. Goudie and K. Pye eds, *Chemical sediments in geomorphology*. London: Academic Press. Pp. 301–19.

Wood, C. 1976: Caves in rocks of volcanic origin. In T.D. Ford and C.H.D. Cullingford eds, *The science of speleology*. London: Academic Press. Pp. 127–50.

cavern See CAVE.

cavitation Occurs in high velocity water (above 8–16 m s^{-1}) in irregular channels, when local acceleration causes pressure to decrease to the vapour pressure of water and airless bubbles form. Subsequent local deceleration and increased pressure result in bubble collapse. This process is a manifestation of the conservation of energy; increased kinetic energy during flow acceleration is balanced by decreased pressure energy (see GRADUALLY VARIED FLOW). The bubble implosion generates shock waves that erode adjacent solid surfaces like hammer blows. Cavitation erosion occurs in waterfalls, rapids, and especially in subglacial channels where velocities of 50 m s^{-1} have been observed (Barnes 1956). Typical erosional products are pot-holes and crescent-shaped depressions called *sichelwannen*.   KSR

Reference
Barnes, H.L. 1956: Cavitation as a geological agent. *American journal of science* 254, pp. 493–505.

Cenozoic See CAINOZOIC.

centripetal acceleration Defines the acceleration of a particle moving at constant speed round an arc of a circle. If the particle were unaccelerated it would move in a straight line tangent to the circle, so the acceleration must be towards the centre and some force must exist to provide it. In atmospheric flow of large scale the centripetal acceleration is usually much less than the Coriolis acceleration, to which it is closely related, and accounts to some extent for the slack pressure gradients found in anti-cyclones compared to those found in cyclones.   JSAG

cerrado A form of savannah vegetation, comprising grasses, small trees and tangled undergrowth, found in Brazil.

channel capacity The size of the river channel cross-section to bankfull level, usually expressed as the cross-sectional area in square metres. Various possible definitions of the bankfull capacity have been used (Williams 1978), referring to: the heights of the valley flat, the active floodplain, the benches within the channel, the highest channel bars, the lower limit of perennial vegetation, the upper limit of sand-sized particles in the boundary sediment or the elevation at which the WIDTH-DEPTH RATIO becomes a minimum, where there is a first maximum of the bench index (Riley 1972), or the relation of cross-sectional area to top width changes. Williams concludes that the bankfull level to the active floodplain level is the most useful to the fluvial geomorphologist, whereas the banks of the valley flat are the most important to engineers. Many river

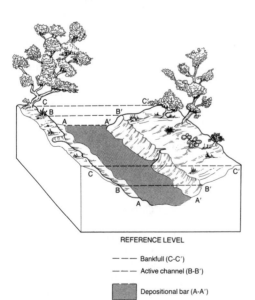

REFERENCE LEVEL

--- Bankfull (C-C')

--- Active channel (B-B')

▨ Depositional bar (A-A')

Channel capacity: *commonly used reference levels.*
Source: *E.R. Hedman and W.R. Osterkamp 1981: Streamflow characteristics related to channel geometry of streams in Western United States.* U.S. Geological Survey water supply paper *2193.*

channels have a compound cross-section in which it is difficult to determine exactly which of the several levels signifies the capacity level which is equivalent to bankfull stage defined elsewhere, and the sharp limits of lichen growth have been shown to allow the consistent identification of channel capacities from one area to another (Gregory 1976). Because some channels are incised into their floodplains and some are compound in cross-section it has been suggested that an active channel can be recognized within the bankfull capacity channel (Osterkamp and Hedman 1981). The upper limit defining the active channel is a break in the relatively steep bank to the more gently sloping surface beyond the channel edge and this break in slope normally coincides with the lower limit of permanent vegetation. The section of channel within the active channel is actively, if not totally, sculptured by the normal process of water and sediment discharge.

A definition of channel capacity is necessary before measurements of cross-sectional area can be related to values of drainage area or discharge. The size of channel at a particular location is related to a range of channel-forming discharges which have often been approximated by a single BANKFULL DISCHARGE or DOMINANT DISCHARGE value. Most recent research has shown how a range of flows acting upon the locally available bed and bank sediment will determine the size of channel capacity. In New South Wales it has been shown that a range of flows (recurrence interval 1.01 to 1.4 years on the annual series) affects the bedforms in the channel and a less frequently occurring range of flows (recurrence interval 1.6 to 4 years) is responsible for the channel capacity (Pickup and Warner 1976). Andrews (1980) has defined effective discharge as the increment of discharge that transports the largest fraction of the annual sediment load over a period of years, and from fifteen gauging stations showed such effective discharges to be equalled or exceeded between 1.5 and 11 days per year and to have recurrence intervals on the ANNUAL SERIES ranging from 1.18 to 3.26 years.

Channel capacity can represent a quasi-equilibrium value at a particular location and relationships between discharge and channel capacity have been used to estimate discharge at ungauged sites. Capacity can change if there is a change in the magnitude and frequency of peak discharges which may be the result of a secular change of climate or which may arise as a consequence of human activity affecting the channel by CHANNELIZATION, by RESERVOIR STORAGE or by gravel extraction, or affecting the drainage area through LAND USE CHANGE including urbanization with the downstream effects of URBAN HYDROLOGY. Channel capacities have been shown to decrease to as little as 15 per cent of their original sizes as a result of reduced flows downstream from reservoirs and to increase to as much as six times their former size as a result of increased peak discharges downstream from urban areas.   KJG

Reading and References
Andrews, E.D. 1980: Effective and bankfull discharges of streams in the Yampa river basin, Colorado and Wyoming. *Journal of hydrology* 46, pp. 311–30.
Gregory, K.J. 1976: Lichen and the determination of river channel capacity. *Earth surface processes* 1, pp. 273–85.
†Gregory, K.J. ed. 1977: *River channel changes.* Chichester: Wiley.
Osterkamp, W.R. and Hedman, E.R. 1981: Perennial-streamflow characteristics related to channel geometry and sediment in Missouri River Basin. *US Geological Survey professional paper* 1242.
Pickup, G. and Warner, R.F. 1976: Effects of hydrologic regime on magnitude and frequency of dominant discharge. *Journal of hydrology* 29, pp. 51–75.
†Richards, K.S. 1982: *Rivers: form and process in alluvial channels.* London and New York: Methuen.
Riley, S. 1972: A comparison of morphometric measures of bankfull. *Journal of hydrology* 17, pp. 23–31.
Williams, G.P. 1978: Bankfull discharge of rivers. *Water resources research* 14, pp. 1141–54.

channel resistance Water flowing in a river channel encounters various sources of resistance which oppose downstream motion and result in energy loss. The potential energy of the water is converted to kinetic energy, and thence to work in overcoming frictional resistance and generating heat, as well as in transporting sediment. Channel resistance, or ROUGHNESS, includes grain resistance controlled by bed material size, internal distortion resistance which encompasses the form resistance of bedforms and flow separation at bends, and spill resistance caused by local acceleration at obstacles. Irregularity of channel form and bank vegetation add to flow resistance. The combined effect of these resistances is summarized by the composite roughness coefficients in the MANNING and CHÉZY EQUATIONS.   KSR

channel storage The volume of water that can be stored along a river channel because of the variations in channel morphology. As a flood HYDROGRAPH travels along a river channel the shape of the hydrograph will change as a result of the storage of water in the channel. Prediction of the character of the hydrograph along the channel is called FLOOD ROUTING.   KJG

channelization (or river channelization) The modification of river channels for the purposes of flood control, land drainage, navigation, and the reduction or prevention of erosion. River channels may be modified by engineering works including realignment or by maintenance measures by clearing the channel. Channelization can influence the downstream morphological and ecological characteristics of river channels through channel erosion giving larger channels, deposition of the sediment released and change in the river ecology (see diagrams). Because of these consequences downstream from channelization schemes and because of the effects that channelization measures can have on the landscape by aesthetic degradation, alternative methods of stream restoration or stream renovation have been suggested.   KJG

Reading
Brookes, A. 1988: *Channelized rivers – perspectives for environmental management.* Chichester: Wiley.
—Gregory, K.J. and Dawson, F.H. 1983: An assessment of river channelization in England and Wales. *Science of the total environment* 27, pp. 97–111.
Keller, E.A. 1976: Channelization: environmental, geomorphic and engineering aspects. In D.R. Coates ed., *Geomorphology and engineering.* Binghamton NY: State University of New York Press, Pp. 115–40.

channels, types of river/stream River channel typology can be based upon morphological features of the network, upon relationship to bed and bank characteristics, upon sedimentary or hydrological processes, upon morphology of the channel or its pattern, or upon stability of the channel. Classification is necessary for consistent description of river channels in different parts of the world, for the analysis and interpretation of river channel genesis, for the investigation of river channel changes and to provide a basis for understanding likely future changes of river channels, especially those taking place after human activity has affected the river channel or the drainage basin. Although channels have been classified by using

NATURAL CHANNEL

Suitable water temperatures:
   adequate shading; good cover for fish
   life; minimal variation in temperatures;
   abundant leaf material input

MAN-MADE CHANNEL

Increased water temperatures:
   no shading; no cover for fish life;
   rapid daily and seasonal fluctuation
   in temperatures; reduced leaf material
   input

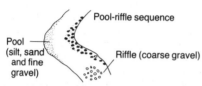

Pool-riffle sequence

Pool
(silt, sand
and fine
gravel)

Riffle (coarse gravel)

Sorted gravels provide diversified habitats
   for many stream organisms

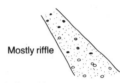

Mostly riffle

Unsorted gravels:
   reduction in habitats; few organisms

POOL ENVIRONMENT

High flow

Diversity of water velocities:
   high in pools, lower in riffles. Resting areas
   abundant beneath undercut banks or behind
   large rocks, etc.

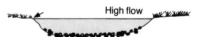

High flow

May have stream velocity higher than
some aquatic life can withstand. Few
or no resting places.

Low flow

Sufficient water depth to support fish and
other aquatic life during dry season

Low flow

Insufficient depth of flow during dry
seasons to supprt diversity of fish
and aquatic life. Few if any pools
(all riffle)

Channelization: *comparison of the natural channel morphology and hydrology with that of a channelized stream, suggesting some possible ecological consequences.*
Source: *A.S. Goudie 1981:* The human impact. *Oxford: Basil Blackwell. Figure 5.6.*

Classification of alluvial channels

Mode of sediment transport and type of channel	Channel sediment (M) (percentage silt clay)	Bed load (percentage of total load)	Channel stability		
			Stable (graded stream)	Depositing (excess load)	Eroding (deficiency of load)
Suspended load	>30	<3	Stable suspended-load channel. Width/ depth ratio <10; sinuosity usually >20; gradient relatively gentle.	Depositing suspended load channel. Major deposition on banks causes narrowing of channel; initial streambed deposition minor.	Eroding suspended-load channel. Streambed erosion predominant; initial channel widening minor.
Mixed load	5–20	3–11	Stable mixed-load channel. Width/depth ratio >10, <40; sinuosity usually <2.0; >1.3; gradient, moderate.	Depositing mixed-load channel. Initial major deposition on banks followed by streambed deposition.	Eroding mixed-load channel. Initial streambed erosion followed by channel widening.
Bed load	<5	>11	Stable bed-load channel. Width/ depth ratio >40; sinuosity usually <1.3; gradient, relatively steep.	Depositing bed-load channel. Streambed deposition and island formation.	Eroding bed-load channel. Little streambed erosion; channel widening predominant.

Source: Schumm 1977.

stream ordering methods, by comparing unstable channels which shift readily with stable channels which do not, and by contrasting alluvial channels with those cut into bedrock, the main bases used for channel classifications have been based either upon the relationship between channel morphology and water sediment transport, or upon river channel planform. These are not the only approaches, however, and a particularly practical way of classifying channels was attempted by Graf (1981) who distinguished degrees of stability of river channels according to the extent to which reaches had shifted as indicated by fifteen topographical map or air photograph surveys of differing dates. Reaches of hazardous instability were obviously inappropriate for bridging and should be avoided during river development. Also related to management is the way in which Palmer (1976) envisaged four zones of a river channel, each having particular management problems. Zone IV the Boulder zone included non-graded

mountain rivers; zone III the Floodway zone had coarse sediment load greater than the transporting capacity of the river so that braiding and islands were common: the Pastoral zone (II) had fine bed-load material and was characterized by deeper, meandering, single thread channels; and the Estuarine zone (I) had periodic gradient reversal. River channel morphology reflects the way in which the water and sediment transported along the channel interact with the available slope and the sediments in, and the vegetation on, the bed and banks. It is the spatial variation of this interaction that accounts for the existence of a range of river channel types. Using the type of sediment load that is moved through the channel, Schumm (1977) developed a tentative classification of alluvial channels as elaborated in the table.

This classification introduces channel pattern but many classifications of river channels have been devised based primarily upon river channel planform. The threshold classification into straight, meandering and

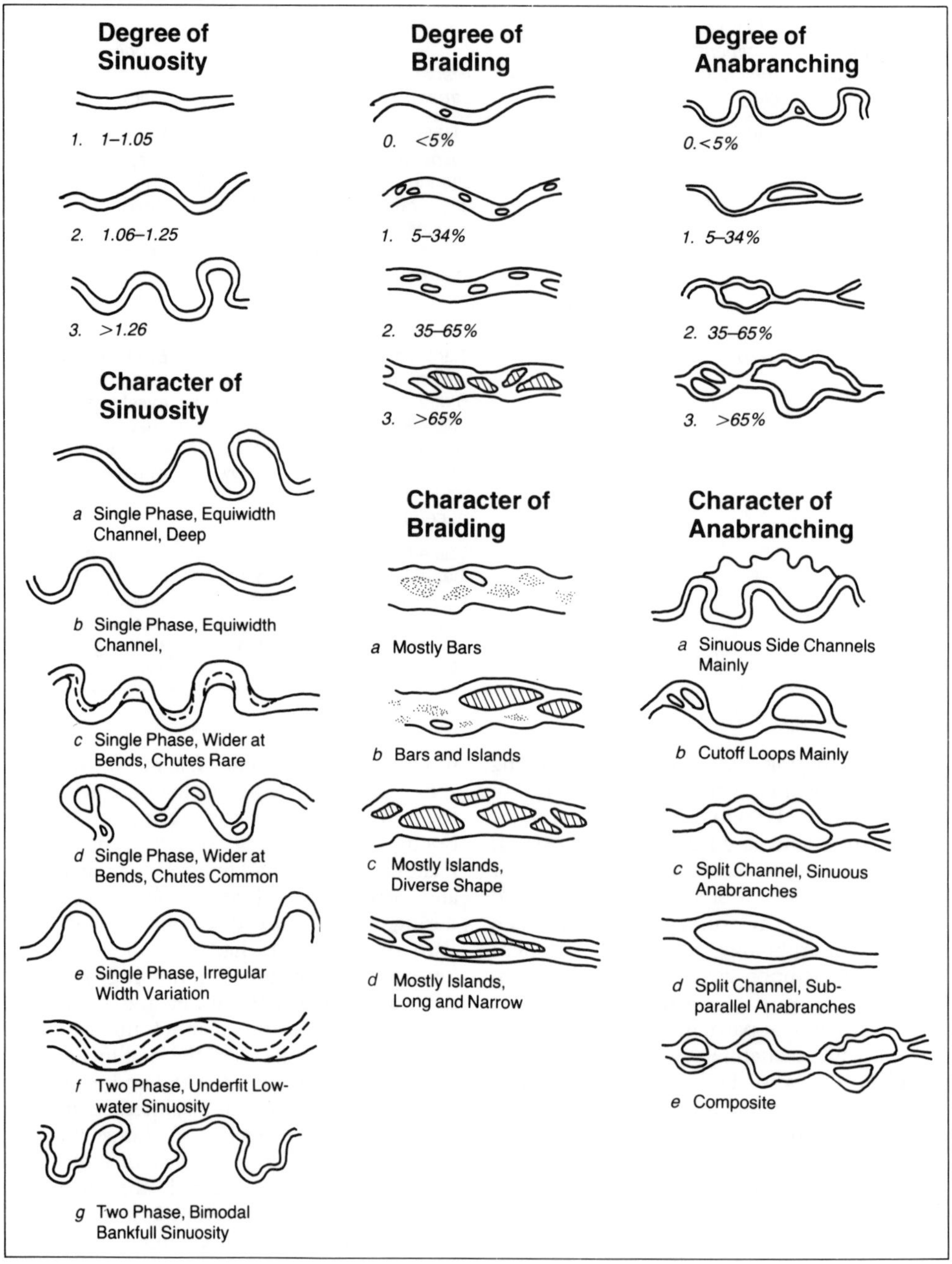

***Channels:* 1.** *Types of channel patterns.*
Source: *Shen* et al. *1981.*

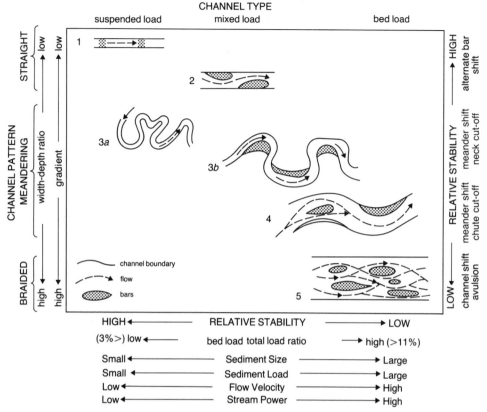

Channels 2. *Channel classification showing relative stability and types of hazards encountered with each pattern.*
Source: *Shen* et al. *1981.*

braided channels proposed by Leopold and Wolman (1957) has been used extensively but Chitale (1970) suggested that a classification into single-thread and multi-thread channels together with an intermediate category was more appropriate because channels are seldom straight for distances greater than about eight channel widths. It is difficult to develop a classification which applies equally effectively to all areas of the world and in the former USSR a classification (Popov 1964) is based upon channel processes represented by the type and arrangement of sand or gravel bars in the channel and the extent to which the channel is constrained by the valley walls. This classification includes:

mid-channel bar (braided);
freely meandering;
embedded (incised non-meandering);
limited (confined) meandering;
side bar (alternate riffle) process;

incomplete meandering (with chute cut-offs).

Also using river activity together with sinuosity Ferguson (1981) classified British river channels as:

active meandering;
actively changing: erosional and depositional activity is not concentrated at bends and point bars, braiding tendencies are common and channels are low sinuosity; not perceptibly migrating: do not contain many exposed point or braid bars at ordinary low flow and include fairly regular sinuous channels as well as irregular and straight ones.

Although some channel patterns have only been described qualitatively, e.g. ANABRANCHING, distributary and reticulate patterns, it is desirable to devise quantitative indices to describe patterns and SINUOSITY has been used most extensively. Brice (1975) developed a descriptive classification of alluvial rivers and embraced the range of channel patterns available by using the properties of degree of sinuosity, braiding and anabranching and the character of meandering, braided and anabranched streams (figure 1). Sinuosity is the ratio of channel length to valley length and a single and double phase pattern is recognized. Two-phase sinuosity refers to a sinuous low-water channel within a wider, less sinuous bankfull channel (sinuosity *f*) or to a bimodal distribution of loop sizes (sinuosity *g*) where there are two wavelength spectra. The degree of braiding is the percentage of the channel that is divided by islands or bars, whereas anabranching is the condition when a river is divided by islands which have a width greater than three times channel width at average discharge. The degree of anabranching is the percentage of river reach length that is occupied by large bars or islands.

The classification of river channels has been advanced by combining classification of channel morphology and process with channel planform in a way which is directly relevant to management. Schumm (1977) developed a channel classification in this way according to the relative stability of the river channels as a basis for recognizing the types of hazards encountered with each pattern and this involved the distinction of five patterns (figure 2). Each pattern is characterized by reference to whether the type of sediment load conveyed through the channel is dominantly suspended load, bed load or a mixture of the two; to width–depth ratio of the channel; to gradient of the reach; and to sediment size, sediment load, flow velocity and stream power. It is on the basis of the five types that relative stability is suggested according to the ways in which channel shifting could take place.   KJG

References
Brice, J.C. 1975: Airphoto interpretation of the form and behaviour of alluvial rivers. Unpublished final report to the US Army Research Office. Quoted in H.W. Shen *et al.* 1981.
Chitale, S.K. 1970: River channel patterns. *Journal of the Hydraulics Division, Proceedings of the American Society of Civil Engineers* 96, pp. 201–21.
Ferguson, R.I. 1981: Channel forms and channel changes. In J. Lewin ed., *British rivers*. London: Allen & Unwin.
Graf, W.L. 1981: Channel instability in a braided, sandbed river. *Water resources research* 17, pp. 1087–94.
Leopold, L.B. and Wolman, M.G. 1957: River channel patterns – braided, meandering and straight. *US Geological Survey professional paper* 282-B.

Palmer, L. 1976: River management criteria for Oregon and Washington. In D.R. Coates ed., *Geomorphology and engineering*. Stroudsburg: Dowden Hutchinson & Ross.
Popov, I.V. 1964: Hydromorphological principles of the theory of channel processes and their use in hydrotechnical planning. *Soviet hydrology*, pp. 188–95.
Schumm, S.A. 1977: *The fluvial system*. Chichester: Wiley.
Shen, H.W. *et al.* 1981: *Federal Highway Administration report* FHWARD–80/160. Washington DC.

chapada A wooded ridge or elevated plateau in the savannah areas of South America, especially Brazil.

chaparral A vegetation type encountered in areas experiencing Mediterreanean climates, characterized by evergreen shrubs with small leathery leaves. (See also MATTORAL.)

chattermarks Crescent-shaped gouges found on the surfaces of rocks and rock particles (even sand grains) either as individual features or as trails. They can be produced either by the grinding of rock-armoured basal ice riding over a rock outcrop to produce crescentic gouge trails (Chamberlain 1888 on rock; Gravenor 1979 on sand grains) or by impaction of sub-rounded grains on other grains in wind or water environments. These latter forms are termed Hertzian cracks in engineering science. Chattermarks on sand grains may also be produced by chemical etching, particularly in a beach environment (Bull *et al.* 1980).   PAB

References
Bull, P.A., Culver, S.J. and Gardner, R. 1980: Chattermark trails as palaeoenvironmental indicators. *Geology* 8, pp. 318–22.
Chamberlain, T.C. 1888: The rock scorings of the great ice invasions. *US Geological Survey seventh annual report*, pp. 147–248.
Gravenor, C.P. 1979: The nature of the Late Paleozoic glaciation in Gondwana. *Canadian journal of earth sciences* 16, pp. 1137–53.

cheiorographic coast The characteristic coastline of areas which have experienced complex, tectonic uplift and subsidence, being made up of alternating deep bays and promontories.

chelation The chemical removal of metallic ions in a rock or mineral by biological weathering. The term derives from the Greek *chela* meaning claw and reflects the process by which the metallic ion is sequestered, held between a pincher-like arrangement of two atoms (a ligand). These ligands most frequently attach themselves to the metal ion through nitrogen, sulphur or oxygen atoms. Ligands are produced by organic molecules of plant, animal and microbial origin and are important, and much neglected, processes of rock disintegration. PAB

Reading
Ehrlich, H.L. 1981: *Geomicrobiology.* New York: Marcel Dekker.

cheluviation Results when water containing organic extracts combines with soil cations to form a chelate. This solution then moves downwards in the soil profile by a process of eluviation, transferring aluminium and iron sesquioxides into lower horizons. ASG

chemosphere A term sometimes applied to the region of the atmosphere, mainly between 40 and 80 km in altitude, in which photochemical processes are important.

chenier ridge A beach ridge which is surrounded by low-lying swamp deposits, and which tends to be made of sand or shell debris. Classic examples occur downdrift from the Mississippi delta, where individual cheniers are up to 3 m high, 1000 m wide, and 50 km long. They are generally slightly curved, with smooth seaward margins but ragged landward margins due to washovers. Conditions conducive to their formation include low wave energy, low tidal range, effective longshore currents and a variable supply of predominantly fine-grained sediment. ASG

Reading
Augustinus, P.G.E.F. ed. 1989: Cheniers and chenier plains. *Marine geology* 90, pp. 219–351.
Hoyt, J.H. 1969: Chenier versus barrier; genetic and stratigraphic distinction. *Bulletin of the American Association of Petroleum Geologists* 53, p. 299–306.

chernozem A black soil rich in humus and containing abundant calcium carbonate in its lower horizons. A soil type characteristic of temperate grasslands, notably the Russian Steppes.

chert A cryptocrystalline variety of silica, e.g. flint, or more specifically a limestone rock in which the calcium carbonate has been replaced by silica.

Chézy equation A FLOW EQUATION developed by Antoine Chézy in 1769 and experimentally tested using data from the River Seine. Its derivation assumes UNI-

FORM STEADY FLOW in which no acceleration or deceleration occurs along a reach, and the resistance to flow must therefore balance the component of the gravity force acting in the direction of the flow (Sellin 1969). The equation is:

$$v = C\sqrt{Rs}$$

where v is mean velocity, R is the HYDRAULIC RADIUS (often taken to be the mean depth in wide, shallow channels) and s is energy or bed slope. C, the Chézy coefficient, is essentially a measure of 'smoothness' or the inverse of channel resistance, and is therefore inversely related to the coefficient in the similar MANNING EQUATION. KSR

Reference
Sellin, R.H.J. 1969: *Flow in channels.* London: Macmillan.

chine A small ravine or canyon which reaches down to the coast, especially in southern England. Chines are well developed in sandstones near Bournemouth.

chinook A warm, dry wind which blows down the eastern slopes of the Rocky Mountains of North America. It is warmed adiabatically during its descent and produces marked increases in temperatures, especially in the spring months. It has some similarity to the föhn winds of Europe.

chlorofluorocarbons (CFCs) A class of compounds, entirely man-made in as much as they are not known to occur naturally, containing atoms of carbon, fluorine and chlorine. Different CFCs contain these atoms in varying proportions. They have been developed to provide aerosol propellants, to use in refrigerators, and for making various types of foam. They are non-toxic, non-flammable and rather inert. They have very long lifetimes in the atmosphere because of their stability, and so their concentrations have been building up inexorably. They have emerged into environmental notoriety because of their role in ozone depletion in the stratosphere, and partly because, molecule for molecule, they are highly effective 'greenhouse' gases (see GREENHOUSE EFFECT). The compositions, lifetimes and concentrations of the five main CFCs are listed in the table. ASG

Reading
Bridgman, H. 1990: *Global air pollution.* London: Belhaven Press.

Characteristics of CFCs

Chemical formula	Abbrev.	Approx. tropospheric lifetime (yr)	1980 Global average amount (ppbv)
CCl_3F	CFC-11	65	0.18
CCl_2F_2	CFC-12	110	0.28
$CClF_3$	CFC-13	400	0.007
CCl_2FCClF_2	CFC-113	90	0.025
$CClF_2CClF_2$	CFC-114	180	0.015
CF_3CF_2Cl	CFC-115	380	0.005

Radiative forcing relative to CO_2 per unit molecule change in the atmosphere

Gas	Relative radiative forcing
CO_2	1
CH_4 (methane)	21
N_2O (nitrous oxide)	206
CFC-11	12400
CFC-12	15800

chott A seasonal lake, often very saline, flooded only during the winter months. Applied especially to the tectonically formed lake basins of North Africa.

chronosequence A temporal relationship between soils and deposits. Vreeken (1975) has recognized four kinds. Postincisive chronosequences are those where there is a sequence of deposits of different age, and the soils related to each deposit have formed from the end of the time of deposition to the present. A pre-incisive chronosequence is a sequence where soil began to form on a particular deposit, but subsequent burial of the soil took place at different times. A time-transgressive chronosequence *without* historical overlap is a vertical stacking of sediments and buried soils, so that the latter record times of non-deposition. A time-transgressive chronosequence *with* historical overlap is the most complicated of the four types, and incorporates aspects of the other three.

ASG

Reading
Vreeken, W.J. 1975: Principal kinds of chronosequences and their significance in soil history. *Journal of soil science* 26, pp. 378–94.

chute A narrow channel with a swift current, applied both to rivers and to the straits between the mainland and islands.

circadian rhythm The approximately 24-hour rhythm of activity exhibited by most living organisms: man, higher animals, insects and plants. The cycle is to some degree independent of day and night cycles and seems to be an important organizing principle in animal and plant physiology. Organisms isolated from external stimuli will continue to display circadian rhythms of temperature, respiration, hormone levels etc. for some time, but may get 'out of phase' and need a diurnal cycle to reset their 'internal clocks'. KEB

Reading
Guthrie, D.M. 1980: *Neuroethology: an introduction*. Oxford: Blackwell Scientific.
Luce, G.G. 1972: *Body time*. London: Temple Smith.

circulation index A numerical measure of properties or processes of the large-scale atmospheric circulation. Indices have been devised to measure the strength of the east–west component of the circulation in middle latitudes – the zonal index, and the north–south component – the meridional index. The indices are usually in terms of differences in the mean pressures of two specified latitudes. The mean pressures are calculated along each of the latitudes. Lamb (1966) suggests that indices can usefully express circulation vigour if measured at points where the main air streams are most regularly developed. Indices were first used in statistical investigation connected with long range forecasting, e.g. Walker's North-Atlantic Oscillation (see Forsdyke 1951) but are now used more widely in studies of climatic change and the general circulation. AHP

References
Forsdyke, A.G. 1951: Zonal and other indices. *Meteorological magazine* 80, pp. 156–60.
Lamb, H.H. 1966: *The changing climate*. London and New York: Methuen.

cirque (also corrie, cwm) A hollow, open downstream but bounded upstream by an arcuate, cliffed headwall, with a gently sloping floor or rock basin. The cirque floor is eroded by glacier sliding while the backwall is attacked by BASAL SAPPING and subaerial rock weathering. Cirques are common in formerly glaciated uplands and have long caught the imagination of physical geographers. They were originally thought to have been formed during the waxing and waning of ice sheet glaciations, but few were occupied by active glaciers during ice sheet withdrawal. Instead it seems likely that they represent many stages during the past few million years when marginal glaciation affected mid-latitude uplands. Most mid-latitude cirques show a preferred orientation towards the north-east in the northern hemisphere and towards the south-east in the southern hemisphere, reflecting mainly the effect of shade in protecting the glacier from the sun but also the effect of wind-drifted snow accumulated by predominantly westerly winds. Preferred orientation is less important in polar and tropical mountains. Cirque altitude is an indication of former snow lines and it is common for basin altitudes to increase with distance from a coast. Cirques have attracted much morphometric analysis and, although the main controls on their morphology remain unclear, it seems that they tend to become more enclosed and deeper with time. DES

Reading
Derbyshire, E. and Evans, I.S. 1976: The climatic factor in cirque variation. In E. Derbyshire ed., *Geomorphology and climate*. New York: Wiley.
Gordon, J.E. 1977: Morphometry of cirques in the Kintail-Affric-Cannich area of north-west Scotland. *Geografiska annaler* 59A. 3–4, pp. 177–94.

cirrocumulus See CLOUDS.

cirrostratus See CLOUDS.

cirrus See CLOUDS.

cladistics (or phylogenetic systematics) The elucidation of the evolutionary history of groups of organisms. Hennig (1966) was responsible for its initial development. Subsequently, it has been refined into a method with more general properties than those stated by Hennig and having a much wider application than he intended (Humphries and Parenti 1986).

Biological systematics seeks to describe and to classify the variation between organisms. Such classifications are often hierarchical and reveal patterns of association (TAXONOMY). Resemblances among the intrinsic properties of organisms (encompassing everything from their chemistry to conduct) can be classified to reveal phylogenetic patterns. As Eldredge and Cracraft (1980) point out, the pattern of similarity of features in organisms has resulted from either evolution by descent or special creation (CREATIONISM). If evolution by ancestry and descent is accepted, together with the fact that novel

characteristics appear in organisms at different times during its course, a hierarchical phylogenetic pattern of clustered intrinsic resemblances (akin to similarities in taxa) can be formulated.

According to Hennig (1966), it is possible to distinguish two principal categories of resemblance in a monophyletic taxon (one having two or more species and including an ancestor and descendants). First, true evolutionary similarities. These may be of two kinds, either those acquired from a distant common ancestor and retained by one or more of its descendants, or those possessed solely by a particular group of organisms which acquired them from a recent common ancestor. The second category is that of false or misleading resemblance and derives from adaptations made during parallel and convergent EVOLUTION. Cladists contend that the intermittent retention of resemblances acquired from an ancient source renders such traits only useful as clustering agents where they can claim novel status. Their methodology also indicates that because ancestors do not appear to have special assemblages of novelties, their definition is difficult.

Accordingly, phylogenetic systematists concentrate upon recently acquired evolutionary resemblances in an attempt to ascertain patterns of novelties and establish proximate relationships between taxa. The intrinsic characteristics of organisms are examined in detail. An attempt is made to

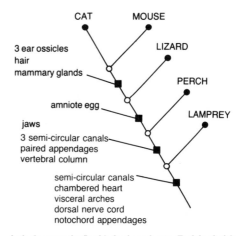

A cladogram for five kinds of vertebrates. Each level of the hierarchy (denoted by branch points) is defined by one or more similarities interpreted as evolutionary novelties.
Source: N. Eldredge and J. Cracraft 1980: *Phylogenetic patterns and the evolutionary process. New York: Columbia University Press. Figure 2.1.*

establish whether these characteristics are primitive or derived. Widespread traits within groups are thought to have originated there, while characters occurring in only a few instances are considered as derived. Significant patterns of phylogenetic resemblance have also emerged after the analysis of homologous characteristics at various stages in the life cycle of organisms. The taxon being investigated is also compared in detail with its nearest relative (sister), as part of the search for character distributions. The aim is to find a universal set (one or more) of similarities in order that a group in which they are all present may be defined. Resemblances which are common to only some of the group can also be used to categorize subsets. Inferred inherited similarities that define subsets at some level within a universal set are known as homologies. Thus homology is synonymous with synapomorphy and refers to novel, shared resemblance endowed by a recent common ancestor (Eldredge and Cracraft 1980).

Data from such analyses are presented as branching diagrams or cladograms, upon which are illustrated clusters of common resemblances regarded as evolutionary innovations. It is often possible to map more than one cladogram from one data set. When compared, these exhibit different patterns that require further investigation. An example provided by Eldredge and Cracraft (1980) will illustrate the principles and problems of cladistic analysis (see p. 91). The cat, mouse, lizard, perch and lamprey are vertebrates with a universal set of resemblances which include semicircular canals in their heads and a chambered heart. Subgroups can be made of the perch and lamprey as they lack an amniote egg, and the lizard, perch and lamprey which do not possess hair or mammary glands. The apparent conflict in the relationships between these organisms can be clarified with reference to the stages in their life cycles. Each has a substantial amount of cartilage in at least one stage of their ontogeny. Cartilage is superseded by bone in the advanced stages of development of the cat, mouse, lizard and perch. Thus bone is diagnostic of a subgroup (each member of which is also part of a bigger group including the lamprey), whose members possess a vertebral column, paired appendages, jaws and semicircular canals. A further subgrouping emerges with reference to an amniote egg, which is possessed by the cat, mouse and lizard. Consideration of earlier developmental stages of the organisms indicates that all five then had a general vertebrate egg, an amniotic membrane subsequently emerging in the cat, mouse and lizard. The cat and mouse can be combined in isolation from the lizard as they are endowed with hair, mammary glands and three ear ossicles, these again being adaptations from more widespread vertebrate traits. Such relationships can be expressed in the cladogram shown here.

The next stage in the cladistic analysis of these data is to look outside the group of five organisms for others with equivalent shared resemblances. In this context, two other chordates, the lancelet amphioxus and the tunicates are relevant. They, however, do not have the characteristics of the cat, mouse, lizard and perch grouping, nor of the large group which includes the lamprey. Thus the amphioxus, tunicates and lamprey comprise outgroups and the cat, mouse, lizard and perch a subgroup. The latter can be further subgrouped in that cats, mice and lizards possess amniote eggs but perch do not, their eggs being analogous to those of the outgroups. Finally, cats and mice are the exclusive possessors of hair, mammary glands and three ear ossicles, qualifying as a subgroup by virtue of these evolutionary novelties.

A progression in this type of analysis, whereby more complex and accurate hypotheses concerning ancestor–descendent relationships can be formulated, comes via additional information and takes the form of a phylogenetic tree, which can also be represented diagrammatically. At the end of this sequence is the phylogenetic scenario which consists of a tree with an overlay of adaptational narrative (Eldredge 1979).

Cladistics has caused considerable controversy and acrimony among systematists. Some cladistic viewpoints conflict, often sharply, with certain of those of the other major biological taxonomists – the evolutionary systematists and the numerical taxonomists. Eldredge and Cracraft (1980) state that cladistic strategy involves investigation of the structure of nature, and the formulation of hypotheses concerning this which contain the basis for their own evaluation. This approach differs from that of the Darwinian school of evolutionary taxonomy which, they contend, invokes acknowledged processes in the explanation of form, such that its conclusions are not amenable to disproval. The cladists do not claim to prove anything, but to carry out pattern analysis without fixed ideas concerning causal processes, and within a framework that accepts that evolutionary patterns and processes are related. They also accept that morphological and taxonomic diversity are related but not synonymous, stating that evolutionary systematists believe that they are equivalent. The principles of cladistics also have a bearing on speciation. Cladists recognize species as distinctive entities to the extent that the emergence of a new one interferes with the pattern of ancestry and descent. Darwinists explain long-term intrinsic modifications in species populations by means of natural selection and adaptation. Cladism leads to the notion that such gradual speciation is unlikely. Cladists believe that species remain unmodified for a period, and are quite suddenly replaced by others. Speciation occurs in areas of various dimensions. When these areas are defined, they can be related to the phylogenetic histories of the relevent species as depicted by cladograms (VICARIANCE BIOGEOGRAPHY).

Cladists disagree with numerical taxonomists whom, they contend (mainly on the basis of computerized cluster analysis), feel that evolutionary history defies reason. They are at variance with both evolutionary systematists and numerical taxonomists over a fundamental tenet of their method. This relates to cladistic assemblages which comprise novelties rather than retentions, the latter being the usual characteristics employed by alternative approaches. In consequence, the cladists' interpretation of the evolutionary status of certain taxa differs from that of their fellow systematists.　RLJ

Reading and References
Eldredge, N. 1979: Cladism and common sense. In J. Cracraft and N. Eldredge eds, *Phylogenetic analysis and paleontology.* New York and Guildford: Columbia University Press. Pp. 165–97.
†Eldredge, N. and Cracraft, J. 1980: *Phylogenetic patterns and the evolutionary process.* New York and Guildford: Columbia University Press.
†Hennig, W. 1966: *Phylogenetic systematics.* Urbana: University of Illinois Press.
Humphries, C.J. and Parenti, L.R. 1986: *Cladistic biogeography.* Oxford: Clarendon Press.

clast A coarse sediment particle, usually larger than 4 mm in diameter; clast sizes include pebbles, cobbles and boulders.

clastic Composed of or containing fragments of rock or other debris that have originated elsewhere.

clathrate A solid where a compound or element (the guest) is held within a cage of a crystalline lattice of another compound or element (the host). There is no chemical bonding between guest and host but the structure is maintained by the closeness of molecular fit. At low temperatures and pressures, where the host is water ice, the guest may be methane (crystalline hydrate). Such gas hydrates are a possible source of natural gas in permafrost areas.　WBW

Reading
Makogon, Yu. F. 1988: Gas-hydrate accumulations and permafrost development. *Permafrost, Fifth International Conference Proceedings.* Tapir, Trondheim, Norway. Vol. 1, pp. 95–101.

clay A sedimentary rock composed of aluminium silicates and other minerals which are derived from the chemical breakdown of other rocks. Any material made up of particles less than about 2.0–4.0 μm (0.002–0.004 mm) in diameter.

clay dunes Dunes formed of aggregates of clay-sized material rather than of the more normal quartz sand-sized grains. They generally consist of about 30 per cent of clay-sized material, and may also contain some quartz grains, silt, and evaporite crystals. They form as a result of the deflation of lagoonal and dried lake sediments, which then accumulate as LUNETTES on the lee sides of the depressions. They are widespread in southern Australia, the coastal plain and the High Plains of Texas, the CHOTTS of North Africa, coastal plain of Argentina, and in the Kalahari, especially in areas where the mean annual precipitation is between 200 and 500 mm (Bowler 1973).　ASG

Reference
Bowler, J.M. 1973: Clay dunes: their occurrence, formulation and environmental significance. *Earth science reviews* 9, pp. 315–38.

clay–humus complex Consists of a mixture of clay particles and decaying organic material that attracts and holds the cations of soluble salts within the soil profile.

claypan A stratum of compact but not cemented clayey material found within the soil zone.

clay-with-flints An admixed deposit of clay and gravel, predominantly flint, occurring locally in depressions in the chalk uplands of southern England. Probably the insoluble components of the chalk and/or reworked Tertiary deposits.

clear water erosion The erosion caused by rivers whose sediment load has been removed by the construction of a dam and reservoir. With a reduced sediment load, incision occurs rather than aggradation. This can create serious problems for bridges and other man-made structures downstream.

cleavage Of minerals, the plane along which a crystal can be split owing to its internal molecular arrangement. Also those planes of weakness in fissile rocks which are not related to jointing or bedding.

CLIMAP In 1971 a consortium of scientists from many institutions was formed to study the history of global climate over the past million years, particularly the elements of that history recorded in deep-sea sediments. This study is known as the CLIMAP (Climate, Long-range Investigation, Mapping and Prediction) project, and is part of the United States National Science Foundation's International Decade of Ocean Exploration Programme. Its administrative centre is at Lamont-Doherty Geological Observatory, Columbia University, New York. One of CLIMAP's goals is to reconstruct the earth's surface at particular times in the past, and a good example is contained in CLIMAP Project Members (1976). These reconstructions can serve as boundary conditions for atmospheric general circulation models, as in Gates (1976).   JGL

References
CLIMAP Project Members 1976: The surface of the ice-age earth. *Science* 191, pp. 1131–7.

Gates, W.C. 1976: Modelling the ice-age climate. *Science* 191, pp. 1138–44.

climate The long-term atmospheric characteristics of a specified area. Contrasts with weather. These characteristics are usually represented by numerical data on meteorological elements, such as temperature, pressure, wind, rainfall and humidity. These data are frequently used to calculate daily, monthly, seasonal and annual averages, together with measures of dispersion and frequency.

Climatic statistics have been published for a huge number of stations, and there are some useful publications which have summarized tables of major climatic phenomena. Among the most useful are Landsberg (1969), Wernstedt (1972), Müller (1982) and Pearce and Smith (1984).   ASG

References
Landsberg, H.E. ed. 1969 onwards: *World survey of climatology.* 15 vols. Amsterdam: Elsevier.

Müller, M.J. 1982: *Selected climatic data for a global set of standard stations for vegetation science.* The Hague: Junk.

Pearce, E.A. and Smith, C.G. 1984: *The world weather guide.* London: Hutchinson.

Wernstedt, F.L. 1972: *World climatic data.* Lemont, Penn.: Climatic Data Press.

climate modelling See GENERAL CIRCULATION MODELLING.

climate modification See WEATHER MODIFICATION.

climatic change Evidence of past climates from a large variety of both instrumental and proxy records indicates that they have varied on many time-scales. The dominant characteristic of essentially all climatic time series, covering time-scales from tens of thousands of years down to a few weeks, is the continuous red distribution of the variance spectrum without the occurrence of prominent peaks. An exception to this is that palaeoclimatic spectra are terminated on the low-frequency side by a peak at a period of 10^5 years, reflecting the growth and decay of the major temperate latitude ice sheets. Also apparent in palaeoclimatic spectra are weak peaks, barely statistically significant above the continuum, near periods of 2×10^4 and 4×10^4 years, which have been attributed to variations in the earth's orbit in accordance with MILANKOVITCH'S HYPOTHESIS.

The long-term palaeoclimatic history of the earth is not well known, but it appears that in the very remote past the earth evolved into a temperature regime differing little in its very broad features from that of the present day. It is believed that the pattern of climate with warm tropical regions, cool poles and periodic ice ages has altered scarcely at all, at least during the past 10^9 years. During the past 500-million-year period the earth was warmer than it is today and during more than 90 per cent of

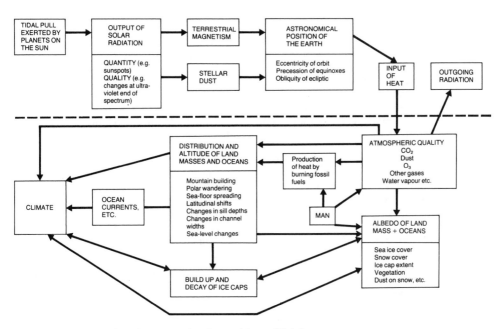

Climatic change: a schematic representation of some of the possible influences.

the time the poles were ice free. Numerous geological records make it clear that about 55 million years BP the global climate began a long cooling trend leading to the formation of the Antarctic ice sheet, and eventually the middle latitude ice sheets. For at least the past million years the earth's climate has been characterized by an alternation of glacial and interglacial episodes, marked in the northern hemisphere by the waxing and waning of continental ice sheets and in both hemispheres by periods of rising and falling temperatures.

Recent climatic changes include a marked cold period from about AD 1430 to 1850 known as the Little Ice Age and a marked warming during the first half of the present century. Future temperatures are expected to increase because of the increasing amount of CO_2 in the atmosphere caused by burning fossil fuels.   JGL

Reading
Frakes, L.A. 1979: *Climates throughout geologic time.* Amsterdam: Elsevier.

Goudie, A.S. 1992: *Environmental change.* 3rd edn. Oxford: Oxford University Press.

Lamb, H.H. 1977: *Climate: present, past and future. 2: Climatic history and the future.* London and New York: Methuen.

— 1982: *Climate, history and the modern world.* London and New York: Methuen.

Pittock, A.B., Frakes, L.A., Jenssen, D., Peterson, J.A. and Zillman, J.W. 1978: *Climatic change and variability: a southern perspective.* Cambridge: Cambridge University Press.

climatic classification Climates may be classified quantitatively into broad types depending on statistics, such as the annual levels of temperature and precipitation, and also on the seasonal variations of these parameters. The climate of a locality is mainly governed by factors such as latitude, position relative to continents and oceans, position relative to large-scale atmospheric circulation patterns, altitude, and local geographical features. A very broad classification may be made into CONTINENTAL CLIMATE, which is found mainly in the interior and eastern parts of continents and is characterized by low precipitation and large seasonal ranges of temperature, and maritime climate, typical of oceanic islands and the western parts of continents and characterized by high precipitation and relatively uniform seasonal temperature. Further divisions may be made into tropical, temperate and polar climates.

No single climatic classification can serve more than a limited number of purposes satisfactorily and many different schemes have therefore been developed. Many climatic classifications are concerned with the relationships between climate and vegetation or soils. The work of W. Köppen is the prime example of this type of classification, and his work has been used extensively in geographical teaching. The key features of Köppen's final classification are temperature and aridity criteria, based on the study of vegetation groups.

Energy and moisture budgets can be used as a basis for climatic classification. The most important of these is Thornthwaite's 1948 classification which is based on the concept of potential EVAPOTRANSPIRATION (calculated from temperature) and the water budget (see WATER BALANCE). Budyko (1974) has developed a similar, but more fundamental approach, using net RADIATION rather than temperature. He related the net radiation available for evaporation from a wet surface to the level required to evaporate the mean annual precipitation, the resulting ratio being used to define major vegetation zones. JGL

References
Budyko, M.I. 1974: *Climate and life*. New York: Academic Press.
Thornthwaite, C.W. 1948: An approach towards a rational classification of climate. *Geographical review* 38, pp. 55–94.

climatic geomorphology This subject developed during the period of European colonial expansion and exploration at the end of the nineteenth century, when unusual and spectacular landforms were encountered in newly discovered environments like deserts and the humid tropics. In the USA, W.M. Davis recognized 'accidents', whereby non-temperate climatic regions were seen as deviants from his normal cycle and introduced, for example, his arid cycle (Davis 1905). Some regard Davis as one of the founders of climatic geomorphology (see Derbyshire 1973), though the French (e.g. Tricart and Cailleux 1972) have criticized him for his neglect of the climatic factor in landform development. Much important work on dividing the world map into climatic zones (morphoclimatic regions) with distinctive landform assemblages was attempted both in France (e.g. Tricart and Cailleux 1972) and in Germany (e.g. Büdel 1982). This geomorphology described itself as geographical (Holzner and Weaver 1965).

In recent years certain limitations have become apparent:

1 Much climatic geomorphology has been based on inadequate knowledge of rates of processes and on inadequate measurement of process and form.
2 Many of the climatic parameters used for morphoclimatic regionalization have been seen to be meaningless or crude.
3 The impact of climatic changes in Quaternary times has disguised the climate–landform relationship.
4 Climate is one step removed from process.
5 Many supposedly diagnostic landforms are either relict features or have a form which gives an ambiguous guide to origin.
6 Climate is but one factor in many that affect landform development.
7 Climatic geomorphology tends to concentrate on the macroscale rather than investigating the details of process.
8 Macroscale regionalization has little inherent merit. ASG

Reading and References
†Birot, P. 1968: *The cycle of erosion in different climates*. London: Batsford.
Büdel, J. 1982: *Climatic geomorphology*. Princeton, NJ: Princeton University Press.
Davis, W.M. 1905: The geographical cycle in an arid climate. *Journal of geology* 13, pp. 381–407.
Derbyshire, E. ed. 1973: *Climatic geomorphology*. London: Macmillan.
†— 1976: *Climate and geomorphology*. London: Wiley.
Holzner, L. and Weaver, G.D. 1965: Geographical evaluation of climatic and climato-genetic geomorphology. *Annals of the Association of American Geographers* 55, pp. 592–602.
†Stoddart, D.R. 1969: Climatic geomorphology. In R.J. Chorley ed., *Water, earth and man*. London: Methuen. Pp. 473–85.
Tricart, J. and Cailleux, A. 1972: *Introduction to climatic geomorphology*. London: Longman.

climatic optimum See ALTITHERMAL.

climato-genetic geomorphology An attempt to explain landforms in terms of fossil as well as contemporary climatic influences. Büdel (1982) recognized that landscapes were composed of various relief generations and saw the task of climato-genetic geomorphology as being to recognize, order and distinguish these relief generations, so as to analyse today's highly complex relief. Among the most important relief generations are those of the 'tropicoid palaeo-earth' (produced by conditions of a seasonal tropical type between the Cretaceous and the Mid-Pliocene) of the Late Pliocene, and the three main epochs of the

Quaternary: the Earliest Pleistocene, the Ice Age Pleistocene and the Holocene. ASG

Reference
Büdel, J. 1982: *Climatic geomorphology*. Princeton, NJ: Princeton University Press.

climatology Traditionally concerned with the collection and study of data that express the prevailing state of the atmosphere. The word study here implies more than simple averaging: various methods are used to represent climate, e.g. both average and extreme values, frequencies of values within stated ranges, frequencies of weather types with associated values of elements. Modern climatology is concerned with explaining, often in terms of mathematical physics, the causes of both present and past climates. JGL

Reading
Barry, R.G. and Chorley, R.J. 1992: *Atmosphere, weather and climate*. 6th edn. London: Routledge.
Lockwood, J.G. 1979: *Causes of climate*. London: Edward Arnold.

climax vegetation Mature vegetation in a steady state equilibrium with prevailing environmental conditions. Climax vegetation is stable in the sense that the only major change is the replacement of its component plant populations. It may be seen as the self-perpetuating terminal stages of SUCCESSION, which develop eventually on sites where environmental conditions remain unchanged for a sufficiently long period.

Climax vegetation is considered by many to be a mosaic of distinct plant communities. Others, notably R.H. Whittaker, consider that it is made up of a continuum of plant populations which intergrade in response to complex environmental gradients. Although Whittaker's 'climax pattern hypothesis' has come to replace both MONOCLIMAX and POLYCLIMAX theories, the recognition of climax vegetation and the validity of climax concepts remain controversial. JAM

References
Whittaker, R.H. 1953: A consideration of climax theory: the climax as a population and pattern. *Ecological monographs* 23, pp. 41–78.
— 1974: Climax concepts and recognition. In R. Knapp ed., *Vegetation dynamics. Handbook of vegetation science*. Vol 8. The Hague: Junk. Pp. 137–54.

climbing dunes Sand accumulations which climb gentle slopes.

climogram Usually refers to two types of climatic diagram, a climograph and a hythergraph, which were introduced by T. Griffith Taylor in 1915. The diagrams depict graphically the annual range of climatic elements at a particular location, emphasis being placed on the way in which these elements affect the comfort of man. In a climograph mean monthly WET-BULB TEMPERATURE is plotted as the ordinate and mean monthly relative humidity as the abscissa. The points are then joined to form a 12-sided polygon. In a hythergraph mean monthly temperature is plotted against mean monthly precipitation. Data for several stations, representing different climatic zones, are usually plotted on a single climogram, for comparative purposes. Many other types of climogram have been devised (Monkhouse and Wilkinson 1971). DGT

Reference
Monkhouse, F.J. and Wilkinson, H.R. 1971: *Maps and diagrams*. London and New York: Methuen. Pp. 246–50.

cline A gradation in the range of physiological differences within a species which result from the adaptions to different environmental conditions.

clinographic curve A curve which is representative of the slope or slopes of the surface of the earth.

clinometer An instrument for measuring angles in the vertical plane, particularly those of dipping rocks and hillslopes.

clinosequence A group of related soils that differ in character as a result of the effect of slope angle and position.

clint The ridge or block of limestone between the runnels (grikes) on a rock outcrop. Clints form on remnant features caused by the solution of the limestone by water under soil or drift cover, etching weaknesses or joint patterns in the rock. Clint is an English term (helk is also used) but the features are called KARREN in German. The most famous clints and grikes can be found at Malham Tarn, Malham, Yorkshire. PAB

clisere The series of climax plant communities which replace one another in a particular area when CLIMAX VEGETATION is subjected to a major change in climate, such as a change from glacial to interglacial conditions. A postclisere is the sequence of plant communities in response to a favourable climatic change; a preclisere is produced where an unfavourable climatic change occurs. In all cases, the POTENTIAL CLIMAX may be found in adjacent regions. (See also ALLOGENIC SUCCESSION.)   JAM

Reading
Davis, M.B. 1981: Quaternary history and the stability of forest communities. In D.C. West, H.H. Shugart and D.B. Botkin eds, *Forest succession: concepts and applications*. New York and Heidelberg: Springer.

clitter Massive granite boulders, especially those found on Dartmoor. Clitter probably represents a type of blockstream or blockfield which may have resulted from the fashioning of tors.

cloud dynamics The role played by air motions in the development and form of clouds. Extensive sheets or layers of stratiform cloud are an expression of the gentle, uniform and widespread ascent of deep layers of moist air. In contrast, localized convective or cumuliform cloud bears witness to the presence of more vigorous but much smaller-scale ascent of moist bubbles of air.

Widespread ascent is associated with the extensive low-level convergence and frontal upgliding that characterizes middle latitude depressions, and involves vertical velocities of a few cm s^{-1} which can reach 10 cm s^{-1} in the vicinity of the fronts. Cirrostratus, altostratus and stratus are products of this kind of ascent, although they can be modified if they occur in thin layers and are not shielded by any higher level cloud. A stratus layer cools at its top by radiation to space and is warmed at its base by absorbing infrared radiation emitted from below. This process destabilizes the layer, leading to overturning and a dappled appearance as the stratus is gradually transformed into stratocumulus. Cirrocumulus and altocumulus can also form in this way.

Localized ascent is associated generally with strong surface heating in an unstable atmosphere such as occurs occasionally over summertime continents or in polar air outbreaks across oceans. The rising convective bubbles transport heat upwards in a central core of air which becomes visible as a cumulus cloud once condensation occurs. When these clouds occur on settled days in the form of fair weather

cumulus they tend to be only 1–2 km deep and display updraughts of up to 5 m s^{-1}. In unsettled weather convection can extend to the tropopause and these cumulonimbus clouds exhibit updraughts often stronger than 5 m s^{-1} and exceptionally up to 30 m s^{-1}. These convective clouds are surrounded by clear sky in which the air is sinking gently.

Cumulonimbus clouds go through a unique dynamical life cycle which is most developed at maturity with an organized up- and downdraught, the latter of which flows out at the surface as a gust front. The dissipating stage is dominated by downdraughts associated with extensive rainshafts and evaporative cooling of the raindrops.   RR

cloud forest Occupies those zones in mountainous terrain where clouds occur sufficiently regularly to provide moisture to support the growth of forest, which is often of broad-leaved evergreen type.

cloud microphysics Deals with the physical processes associated with the formation of cloud droplets and their growth into precipitation particles. Liquid water droplets and ice crystals in clouds all possess a nucleus; the former have hygroscopic nuclei (e.g. dust, smoke, sulphur dioxide and sodium chloride) and the latter freezing nuclei which form most commonly between $-15°C$ and $-25°C$ (for example very fine soil particles). The hygroscopic nuclei vary in diameter between 4×10^{-7} m and 2×10^{-5} m and exhibit concentrations of about 10^8 m^{-3} in oceanic air and 10^9 m^{-3} in continental air. Microphysics is concerned with explaining how cloud droplets which have diameters from less than 2×10^{-6} m up to 10^{-4} m grow to precipitation particle diameters of up to 5×10^{-4} m (drizzle) and above (rain).

It is observed that at a given temperature below $0°C$, the relative humidity above an ice surface is greater than over liquid water and that saturation vapour pressure over water is greater than over ice, notably between $-5°C$ and $-25°C$. The result is that in clouds where ice crystals and supercooled water droplets coexist, the cystals grow preferentially by SUBLIMATION at the expense of the droplets. This Bergeron–Findeisen mechanism is important in precipitation production especially in the extra-tropics where a good deal of clouds have substantial layers colder than 0°C. This means that most rain and drizzle in these areas starts off as ice crystals or snowflakes high in the parent cloud.

In natural clouds the droplet size varies and those that have grown on giant nuclei are large and exhibit the greatest fall speeds. These droplets grow by colliding with slower moving smaller droplets and this process of coalescence is efficient enough to produce drizzle size or sometimes raindrop size particles. It is most favoured in deep clouds with prolonged updraughts and is fairly common in clouds of tropical maritime origin. Cloud electrification increases the efficiency of coalescence.   RR

Reading
Mason, B.J. 1962: *Clouds, rain and rainmaking*. Cambridge: Cambridge University Press.

cloud streets Rows of cumulus or stratocumulus clouds lying approximately along the direction of the mean wind in the layer they occupy. A single cloud street may form downwind of a persistent source of thermals, but extensive areas of parallel streets are common, being associated with a curvature of the vertical profile of the horizontal wind. The axes of adjacent rows are separated by distances which are approximately three times the depth of the layer of convection. They are especially frequent over the oceans during outbreaks of cold arctic or polar air. The air motion within the cloud streets is one of longitudinal roll vortices, with air undergoing spiralling motions.   KJW

Reading
Agee, E.M. and Asai, T. eds 1982: *Cloud dynamics*. (Advances in earth and planetary science.) Tokyo: Terra Scientific; Dordrecht: D. Reidel.
Atkinson, B.W. 1981: *Mesoscale atmospheric circulations*. London and New York: Academic Press.
Stull, R.B. 1988: *An introduction to boundary layer meteorology*. Amsterdam: Kluwer.

clouds Both the most distinctive feature of the earth viewed from space (see SATELLITE METEOROLOGY) and the most transient element of the climate system. At any time about half of the globe is covered by clouds composed of water droplets or ice crystals and occurring at altitudes throughout the TROPOSPHERE. The dynamic nature of cloud processes (see CLOUD DYNAMICS) is most readily viewed on a bright summer day when vertical development is vigorous. Clouds are formed when air is cooled below its saturation point (see CLOUD MICROPHYSICS). This cooling is usually the result of ascent accompanied by ADIABATIC expansion. Different forms of VERTICAL MOTION give rise to different cloud types: vigorous local CONVECTION causes convective clouds; forced ascent of stable air

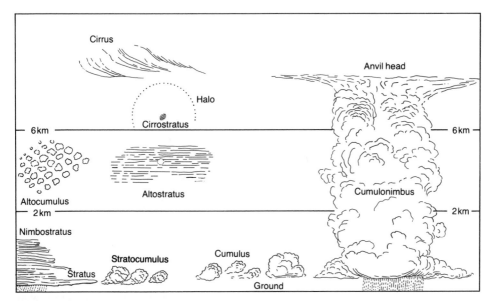

Cloud genera showing typical heights in middle lattiudes.
Source: J.G. Harvey 1976: *Atmosphere and ocean. London: Artemis. Figure 3.5.*

Among the best known of cloud forms are towering cumulus. These examples occurred in Texas, USA.

(usually over an adjacent AIR MASS) produces layer clouds and forced lifting over a topographic feature gives rise to orographic clouds, which may be convective or stable in character. Clouds can also form as a result of other processes, such as the cooling of the lowest layers of the atmosphere in contact with a colder surface in, for instance, the formation of radiative FOG as a result of radiative cooling of the surface at night and advective fog resulting from movement of warm, moist air across a cold surface. The International Classification scheme (WMO, 1956) groups clouds into four basic categories and uses the latin names given to them by an English chemist Luke Howard in 1803; *cumulus* – a heap or pile; *stratus* – a layer; *nimbus* – rain; and *cirrus* – a filament of hair. Additionally the height of the cloud can be indirectly identified by the use of the terms: *strato* – low level; *alto* – middle level; and *cirro* – high. Thus, middle level, cumuliform cloud, often known as a 'mackerel sky', is called altocumulus. Clearly many different combinations of character and height descriptors are possible. The passage of a warm sector depression system often includes rain from nimbostratus clouds while conditions in summer frequently lead to cumuliform development, perhaps culminating in a fully fledged cumulonimbus cloud and a THUNDERSTORM.   AH-S

Reference
WMO 1956: *International cloud atlas.* Geneva: World Meteorological Organization.

cluse Originally a French term, now in general use to describe specifically a steep-sided valley which cuts through a mountain ridge in the Jura mountains.

coastal classification Attempts to classify coasts have been many and varied. In line with Davisian evolutionary concerns in geomorphology, Johnson (1919) based his classification on the history of relative sea-level movement, and so had submergent coasts, emergent coasts, neutral coasts and compound coasts. Shepard (1963) developed a classification based on the relative importance of coastal and non-coastal processes; his primary coasts were those that were largely unmodified by coastal processes, while his secondary coasts were modified by coastal processes, whether erosional, depositional or organic. Another process-orientated classification was made by Davies (1964) and based on the relative strengths of wave activity, producing categories such as high latitude, storm wave environments, low latitude, swell wave environments, and coasts of low energy type (e.g. enclosed seas or those protected by ice). Finally, in line with the developing interest in plate tectonics (see diagram), attempts have been made to formulate a coastal classification on the basis of the position of a particular coastline relative to particular types of plate boundary (Inman and Nordstrom 1971).

ASG

Reading and References
Davies, J.L. 1964: A morphogenetic approach to world shorelines. *Zeitschrift für Geomorphologie* 8, pp. 127–42.
Inman, D. and Nordstrom, C. 1971: On the tectonic and morphologic classification of coasts. *Journal of geology* 79, pp. 1–21.
Johnson, D.W. 1919: *Shore processes and shoreline development.* New York: Wiley.
Shepard, F.P. 1963: *Submarine geology.* New York: Harper & Row.

coastal dunes Aeolian dunes that are found above the high-water marks of sandy beaches, particularly on windward, sediment-rich coastlines. They may contain substantial amounts of calcareous material (AEOLIANITE) but are more often quartz rich. 'The role of vegetation in fixing, trapping and accumulating sediment, and the role of waves in shaping the beach, replenishing sediment sources and reworking the foredunes distinguish coastal dunes from the majority of active dunes in arid areas' (Carter *et al.* 1990, p. 3). Coastal

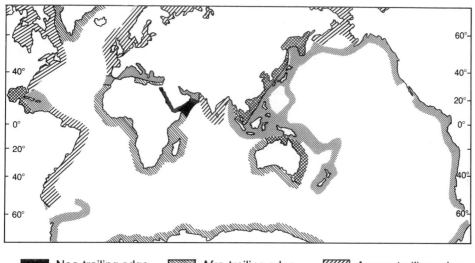

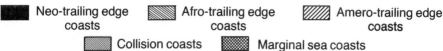

Neo-trailing edge coasts   Afro-trailing edge coasts   Amero-trailing edge coasts

Collision coasts   Marginal sea coasts

A geophysical coastal classification in terms of plate tectonics with some amendments and additions. The outer coast of Baja California does not fit into any of the listed categories.
(i)   Collision coasts: *formed where two plates converge.*
    (a) Continental collision coasts: *where a continental margin is located along the zone of convergence.*
    (b) Island arc collision coasts: *where no continental margin is located along the zone of convergence.*
(ii) Trailing edge coasts: *where a plate-embedded coast faces a spreading zone.*
    (a) Neo-trailing edge coasts: *where a new zone of spreading is separating a land mass.*
    (b) Afro-trailing edge coasts: *where the opposite continental coast is a collision coast.*
(iii) Marginal sea coasts: *where a plate-embedded coast faces an island arc.*
Source: *J.L. Davies 1972:* Geographical variation in coastal development. *London: Oliver & Boyd. Figure 2.*

dunes, which occur in all climatic environments including the humid tropics, range in size and longevity from tiny ephemeral forms to massive persistent dunefields. Unstable forms often migrate as parabolic, transverse or barchan forms. They are an important protection against coastal erosion and valuable wildlife habitats, but are susceptible to human pressures.   ASG

Reference
Carter, R.W.G., Nordstrom, K.F. and Psuty, N.P. 1990: The study of coastal dunes. In Nordstrom, K.F., Psuty, N.P. and Carter, R.W.G. eds, *Coastal dunes: form and process.* Chichester: Wiley. Pp. 1–14.

cockpit karst The scenery produced by the solution of limestone resulting in a hummocky terrain of conical residual hills surrounded by DOLINES or SINKHOLES. They are very characteristic of Jamaican limestone scenery and are often associated with humid tropical karst landscapes. The hills are also called KEGELKARST or MOGOTES.   PAB

Reading
Sweeting, M.M. 1972: *Karst landforms.* London: Macmillan.

coefficient of permeability See PERMEABILITY.

coevolution The evolutionary interaction between two species so that over a long period of time they become co-adapted to each other through a series of selective influences. The predator–prey relationship in animals provides some of the clearest examples. A predator such as a cheetah has evolved to outrun prey such as gazelles, whose defence is to run faster themselves, and to evolve more sensitive detection mechanisms, forcing the cheetah to stalk more quietly and unobtrusively. Coevolution between insects and plants has been the subject of much recent work. Besides the numerical relationship between the ecological 'conspicuousness' of plants such as the oak tree in England, with about 300 insect species associated with it, contrasting with the ash, with less than fifty associated insects, there are some very precise and beautiful examples which are less obvious. Chemical defences may be used by plants against particular insects – e.g. oak produces leaves with more tannin as a defence against defoliating caterpillars – or else mimicry may be employed. An example of this is the extraordinary way in which the leaves of *Passiflora* species in central America mimic the leaves of other rainforest species in an attempt to deceive *Heliconiid* butterflies who otherwise may lay their eggs on the leaves. Some *Passiflora* species even produce small swellings on the leaves to mimic butterfly eggs and hence escape predation! Coevolution must also have occurred between larger herbivores and grazed vegetation: the production of spines, of tough unpalatable leaves and of basal shoots as in grasses, are all responses to grazing pressure.

Coevolution can also lead to SYMBIOSIS or mutualism, as in lichens, and other relationships which may be looked upon as controlled or 'beneficial' parasitism such as that between mycorrhizal fungi and their host plants. The whole field of pollination ecology also throws up many examples of coevolution between insects and plants.

KEB

Reading
Howe, H.F. and Westley, L.C. 1986: Ecology of pollination and seed dispersal. In Crawley, M.J. ed, *Plant ecology*. Oxford: Blackwell Scientific.
Krebs, C.J. 1985: *Ecology: the experimental analysis of distribution and abundance.* 3rd edn. New York: Harper & Row.
Putman, R.J. and Wratten, S.D. 1984: *Principles of ecology*. Beckenham, Kent: Croom Helm.

cohesion The attraction of particles to each other (usually clay minerals in soils) which is not governed directly by a FRICTION law (i.e. it is independent of STRESS) but does provide a measure of the strength of a material. Thus sands do not exhibit cohesion but what is termed (intergranular) friction, while clays, or soils which contain clays, show cohesion. The strength is supplied by the structure of the clay minerals and the way in which chemical bonding is produced in these structures. It can be measured, as in soil mechanics, by the MOHR–COULOMB EQUATION.   WBW

col A pass or saddle between two mountain peaks or a narrow belt of low pressure separating two areas of high pressure.

cold front A frontal zone in the atmosphere where, from its direction of movement, rising warm air is being replaced by cold air. It has an average slope of about 1 in 60 over the lowest several hundred metres rather like the forward bulge of a density bore. Above this level both warm and cold fronts have similar slopes. A cold front is often found to the rear of an EXTRATROPICAL CYCLONE marking a sudden change from warm, humid and cloudy weather to clear, brighter weather. Satellite images usually show cold fronts clearly, as a thick band of cloud spiralling away from the depression centre. Where uplift within the warm air is rapid, rainfall may be heavy along the cold frontal zone. As it passes, temperatures fall suddenly together with a rapid veering of the wind.   PS

cold pole Can be defined as the location of lowest mean monthly temperature, or of lowest mean annual temperature, or of the coldest air in the TROPOSPHERE. The last case is usually indicated by the area of lowest THICKNESS on the chart of 1000–500 mb thickness, and in January is found in the northern hemisphere over north-east Siberia, and in July over the north pole. In the southern hemisphere the lowest 1000–500 mb thickness is found over the Antarctic continent. The lowest surface air temperatures in the northern hemisphere are found in north-east Siberia at Verhojansk and Oymjakon, where they have reached $-68°C$. The lowest surface air temperatures in the southern hemisphere are recorded in Antarctica, where they have reached $-88°C$.   JGL

colk A pot-hole in the bed of a river.

colloid Any non-crystalline, partially solid substance. Often applied to individual, lens-shaped particles of clay.

colluvium Material that is transported across and deposited on slopes as a result of wash and mass movement processes. It is frequently derived from the erosion of weathered bedrock (ELUVIUM) and its deposition on low-angle slopes, and can be differentiated from ALLUVIUM, which is deposited primarily by fluvial agency.

Colluvium may be many metres thick, often contains fossil soil layers which represent halts in deposition, shows some crude bedding downslope, and is generally made of a large range of grain sizes. Cut and fill structures may represent phases when stream incision has been more important than colluvial deposition, and in southern Africa (Price Williams *et al.* 1982) many colluvial spreads have suffered from intense DONGA (gully) formation. ASG

Reference
Price Williams, D., Watson, A. and Goudie, A.S. 1982: Quaternary colluvial stratigraphy, archaeological sequences and palaeoenvironment in Swaziland, Southern Africa. *Geographical journal* 148, pp. 50–67.

colluvium-filled bedrock depression
Elongated gully-like features, cutting into and extending down the substrate surface beneath the regolith. Some are associated with topographic depressions in the overlying ground surface, but many have no topographic expression on the ground surface and may occur on planar slide slopes, spurs and ridges. A wide range of processes may produce the bedrock depression (e.g. landsliding, fluvial dissection, solifluction, percoline erosion) into which colluvium accumulates. ASG

Reading
Crozier, M.J., Vaughan, E.E. and Tippett, J.M. 1990: Relative instability of colluvium-filled bedrock depressions. *Earth surface processes and landforms* 15, pp. 329–39.

colonization The occupation by an organism of new areas or habitats thereby extending either its geographical or its ecological range. For colonization to succeed three main phases are usually necessary; effective dispersal or migration, germination or breeding and establishment, and, finally, survival in the new site through time. Most colonizing species are subject to R-SELECTION. Species which are able to increase their numbers exponentially on arrival in a new area have strong advantage (e.g. annual and biennial plants), while animals exhibiting behavioural adaptability and a marked ability to learn new behaviour tend to be natural colonizing species. Man has proved a potent aid to many colonizers, spreading them to new areas where they have subsequently flourished (e.g. *Opuntia stricta* in Australia) or opening up new habitats ripe for colonization. (See also ALIENS; DISPERSAL; R- AND K-SELECTION.) PW

Reading
Elton, C.S. 1958: *The ecology of invasions by animals and plants*. London: Methuen.
Krebs, C.J. 1978: *Ecology: the experimental analysis of distribution and abundance*. New York: Harper & Row.
MacArthur, R.H. 1972: *Geographical ecology*. New York: Harper & Row.

combe, coombe A small, often narrow valley. Frequently applied to the dry or seasonal stream valleys of chalk country.

comfort zone The range of meteorological conditions within which the majority of the population, when not engaged in strenuous activity, will feel comfortable. It is often expressed in terms of effective temperature (ET), which combines dry-bulb and WET-BULB TEMPERATURES and air movement into a single biometeorological index. The comfort zone varies with climatic zone and season of the year. Personal factors such as age, clothing, occupation and degree of acclimatization are also important. In the tropics, the comfort zone is often taken to range from 19 to 24.5°C ET. In the UK, comparable values are 15.5–19°C ET in summer and 14–17°C ET in winter (Air Ministry 1959). The effects of direct exposure to solar radiation are not normally included in a consideration of the comfort zone. DGT

Reading and Reference
Air Ministry 1959: *Handbook of preventive medicine*. London: HMSO. P. 164.
†Terjung, W.H. 1966: Physiologic climates of the conterminous United States: a bioclimatic classification based on man. *Annals of the Association of American Geographers* 56, pp. 141–79.

commensalism The weakest type of association of species living together in SYMBIOSIS. One or more 'guest' species benefits from living in association with a 'host' species, but the latter neither benefits nor is harmed by the presence of the guests. Mites which live in human hair represent a case in point, as also does the pitcher plant (*Nepenthes*), which can host up to nineteen species. On a world scale commensalism is most common in oceans, especially in the Indian and West Pacific oceans, where at the generic level it attains 59 per cent among caridean shrimp populations (Vermeij 1978). DW

Reference
Vermeij, G.T. 1978: *Biogeography and adaption: patterns of marine life*. Cambridge, Mass. and London: Harvard University Press.

comminution The reduction of a rock or other substance to a fine powder, often as a result of abrasion.

community Any assemblage of populations of living organisms in a prescribed area or habitat. The term 'community' is an ecological unit used in a more general, broad, collective sense to include groups of various sizes and degrees of integration. The community comprises a typical species composition that has resulted from the interaction of populations over time.

Botanists and zoologists have defined the term community in widely differing ways. Three main ideas are involved in community definitions, which claim to have one or more of the following attributes:

    co-occurrence of species,
    recurrence of groups of the same species,
    homeostasis or self-regulation.

Two opposing schools have developed in ecology over the question of the nature of the community. The *Organismic school* suggests that communities are integrated units with discrete boundaries. The *Individualistic school* suggests that communities are not integrated units but collections of populations that require the same environmental conditions.   AP

Reading
Krebs, C.J. 1985: *Ecology: the experimental analysis of distribution and abundance*. New York: Harper and Row.

compaction In engineering terms, the expulsion of air from the voids of a soil. This is usually achieved by artificial means, e.g. with various types of roller in road and dam construction. It differs from CONSOLIDATION which is usually a natural process, although the latter may involve some air expulsion. The aim of compaction is to achieve a high bulk density or lower VOID RATIO, so improving the overall strength of the soil.   WBW

compensation flows Designated river discharges that must continue to flow in a river below a direct supply reservoir or abstraction point to allow for other riparian activities and interests downstream. Such flows are legally established in different ways in different countries and at different times, and may be necessary to maintain water quality, to satisfy the needs of wildlife or recreation, or to provide water for other abstractors or users.   JL

competence The competence of flowing water is the maximum particle size transportable by the flow. A related concept is the threshold flow, which is the minimum flow intensity capable of initiating the movement of a given grain size. When the largest grains on a stream bed are just mobile the flow is the threshold (critical) flow for those grains, and their size represents the competence of the flow. It is possible to predict theoretically the threshold flow conditions for non-cohesive grains coarser than 0.5–0.7 mm, by balancing moments caused by fluid drag and particle immersed weight (Carson 1971, p. 26; Richards 1982, pp. 79–84). The threshold shear stress increases linearly with the grain diameter, while threshold velocity increases with the square root of grain diameter or the one-sixth power of its weight (the so-called 'sixth-power law'). These relationships can be used to define the competence of a given shear stress or velocity (see HJULSTRÖM CURVE). However, the competence to *maintain* particle motion is usually greater than that to *initiate* motion, so large grains may be carried *through* a reach in which they cannot be entrained.

In flowing water grain motion is normally caused by fluid stresses. In air, however, an impact threshold also occurs. The kinetic energy of a sand grain travelling by SALTATION in air allows it to move grains up to six times its size. The concept of fluid competence is therefore less applicable to aeolian transport.   KSR

References
Carson, M.A. 1971: *The mechanics of erosion*. London: Pion.
Richards, K.S. 1982: *Rivers: form and process in alluvial channels*. London and New York: Methuen.

competition The inevitable interactions between members of the same and different species attempting to secure limited resources from the environment, and leading to increases and decreases in the populations of more and less successful organisms. Competition takes many forms: between plants for light, water, space and nutrients; between animals for mates, food, shelter and for social position within a hierarchy. While many of the principles are the same for both plants and animals, there are several obvious practical differences, such as the ability of plants to spread and reproduce asexually. Competition is also a very complex phenomenon, with many of the precise mechanisms by which it occurs being poorly understood. The literature tends, therefore, to be over-reliant upon a relatively small number of classic case studies. More recently, there has been increasing recognition of the role of other factors in the distribution of species and the structure of plant and animal communities, particularly herbivory, predation and habitat disturbance.

The classic experiments concluded that species grown together in the same environment did not do as well as when grown separately, for one species tended to triumph and the other became extinct. This led to the 'competitive exclusion principle' and quantitative expressions such as the Lotka–Volterra equations (Begon *et al.* 1990, pp. 247–51), based on the logistic curve of POPULATION DY-NAMICS. All discussion of competition also involves consideration of the concept of the NICHE, the position of an organism within the community and its inter-relationships with the organisms around it, as extensively reviewed by Putman and Wratten (1984).

Reviews of the influence of competition on community structure tend to stress the importance of other factors, while not doubting that competition has a major role (e.g. Begon *et al.* 1990), though some see the predictions of a simple theory of competition for essential resources as the main explanation of the observed patterns (Tilman, 1986), at least in plant communities.   KEB

Reading and References
Begon, M., Harper, J.L. and Townsend, C.R. 1990: *Ecology*. 2nd edn. Oxford: Blackwell Scientific.

Putman, R.J. and Wratten, S.D. 1984: *Principles of ecology*. Beckenham, Kent: Croom Helm.

Tilman, D. 1986: Resources, competition and the dynamics of plant communities. In Crawley, M.J. ed., *Plant ecology*. Oxford: Blackwell Scientific.

complex response The term introduced by Schumm (1973) to describe the variety of linked changes which may occur within a drainage basin in response to the single passage of a geomorphical THRESHOLD. These changes may well be both spatially and temporally separated and it may, in consequence, be difficult *ex post facto* to establish that all are ultimately due to a single event.   BAK

Reference
Schumm, S.A. 1973: Geomorphic thresholds and the complex response of drainage systems. In Marie Morisawa ed., *Fluvial geomorphology*. Publications in geomorphology. Binghampton: State University of New York. Pp. 299–310.

compressing flow Compressing and extending flows describe longitudinal variations in velocity in a moving medium. The concept is widely used in glacial geomorphology to describe longitudinal variations in flow along the length of a glacier (Nye 1952). Compressing flow refers to a reduction in the length of a unit of glacier ice in a downstream direction, and extending flow refers to an increase in length in the same direction. Such changes must be accompanied by a corresponding variation in glacier width or height. As valley walls often prevent lateral expansion the variations in a glacier are commonly accompanied by a change in depth, with compressing flow associated with thickening and extending flow with thinning. Any such thickening must be associated with an upward component of ice flow while thinning is associated with a downward component of ice flow. When such vertical flow is superimposed on normal down-glacier flow it can have important geomorphological effects by redistributing rock debris. Compressing flow can bring basal debris to the surface and its later deposition by the melting of underlying ice can lead to complex hummocky moraine landforms. Extending flow can carry surface and englacial material to the glacier bottom, thus replenishing the supply of erosive tools. Such vertical movements are usually taken up by ice deformation or CREEP, but where stresses are sufficiently high thrusting and faulting may occur. The association of compressing and extending flows with changing ice thickness leads to predictable spatial associations. Extending flow is characteristic of the accumulation zone where the glacier mass increases downstream, while compressing flow is common in the ABLATION zone where the glacier thins downstream, and especially near the snout where thinning is most marked. Extending flow is also associated with a bed convexity which steepens downstream, as in the case of an ICE FALL, while compressing flow occurs as the glacier crosses a bed concavity. Variations in the rate of BASAL SLIDING along the length of a glacier can also induce compressing and extending flows. For example, a change from warm- to cold-based thermal regime or a thinning of the basal water layer in a downstream direction favours compressing flow, while the opposite conditions favour extending flow. DES

Reference
Nye, J.F. 1952: The mechanics of glacier flow. *Journal of glaciology* 2.12, pp. 82–93.

compressional Pertaining to descending atmospheric air masses, which as a result of pressure increases are warm and dry. Also pertaining to geological faults and fractures which are the product of lateral increases of pressure and to earthquake waves, specifically P-waves.

conchoidal fracture A fracture in a rock or mineral which is shaped like a shell, i.e. concave down. The characteristic fracture of siliceous rocks and minerals.

concordant Within the same plane. Applied particularly to the mountain summits of a region, the concordance of summits attesting to the existence of a plateau prior to incision and dissection of the landsurface.

concretion A solid lump or mass of a substance, or aggregates of these, incorporated within a less competent host material.

condensation The process of the formation of liquid water droplets from atmospheric water vapour mainly onto 'large' hygroscopic nuclei with diameters between 2×10^{-7} m and 10^{-6} m. The cooling necessary to produce condensation in the atmosphere is effected either by ADIABATIC expansion during ascent, by contact with a colder surface or by mixing with colder air. The LATENT HEAT released during this process can be a significant source of heat in weather systems. RR

conductance, specific A measure of the ability of an aqueous solution to conduct an electrical current. It is reported in units of microsiemens per centimetre at $t°C$ (μS cm^{-1} $t°C$). Pure water has a very low specific conductance, but the conductance will increase with an increasing concentration of charged ions in solution.

Individual ions exhibit different conductance values and the level of conductance recorded for a given total ion concentration will therefore vary according to its ionic composition. For most dilute natural waters an increase in temperature of $1°C$ will increase the conductance by approximately 2 per cent and values of specific conductance are normally corrected to a reference temperature of 20 or $25°C$.

Because specific conductance (SC) measurements are simple and rapid to make, and reflect the total ion concentration, they have been widely used as a means of estimating the total dissolved solids (TDS) concentration of water samples. The relationship between SC and TDS for waters of different chemical composition is generally determined empirically and commonly takes the form:

$$TDS = KSC$$

where K varies between 0.55 and 0.75. This relationship may be complicated by changing ionic composition and by the presence of non-ionized material in solution (e.g. SiO_2 and dissolved organic matter) which will contribute to TDS but not SC. DEW

Reading
Foster, I.D.L., Grieve, I.C. and Christmas, A.D. 1982: The use of specific conductance in studies of natural waters and solutions. *Hydrological sciences bulletin* 26, pp. 257–69.

conductivity, hydraulic The term given to the parameter K in the equation defining DARCY'S LAW. It is concerned with the physical properties of both the fluid and the material through which it flows, reflecting the ease with which a liquid flows and the ease with which a porous medium permits it to pass through it. Hydraulic conductivity has the dimensions of a velocity and is usually expressed in m s^{-1}. Since hydraulic conductivity may vary according to direction, K_x, K_y and K_z can be used to represent the hydraulic conductivity values in the x, y and z directions. Hydraulic conductivity should be distinguished from the term PERMEABILITY (or INTRINSIC PERMEABILITY) k which refers only to the characteristics of the porous medium and not to the fluid which passes through it.

The saturated hydraulic conductivity of soils and other sediments may be measured in the laboratory with a PERMEAMETER. If samples are essentially undisturbed, the results are point values representative of the field conditions. However, the hydraulic conductivity of an aquifer (see GROUND-WATER) is best determined by field methods. Piezometer tests can be used to determine *in situ K* values in a porous material around a piezometer tip. Pumping tests at a well provide measurements representative of a much larger aquifer volume (Freeze and Cherry 1979).   PWW

Reference
Freeze, R.A. and Cherry, J.A. 1979: *Groundwater.* Englewood Cliffs. NJ: Prentice-Hall.

cone of depression The shape of the depression in the WATER TABLE surface around a well that is being actively pumped for GROUNDWATER. Pumping results in a lowering (or DRAW DOWN) of the water table that is greatest in the well itself, but reduces radially with distance from the well (see diagram). In a rock with horizontal PERMEABILITY that is uniform in every direction, the cone of depression will be symmetrical about the well. But where groundwater flow occurs more readily in one direction than in another (perhaps because of the influence of dominant joints), the cone of depression will be asymmetrical and elongated in the direction of greatest permeability. Cones of depression can develop in wells pumping both unconfined and confined aquifers.

PWW

Reading
Freeze, R.A. and Cherry, J.A. 1979: *Groundwater.* Englewood Cliffs. NJ: Prentice-Hall.
Linsley, R.K., Kohler, M.A. and Paulhus, J.L.H. 1988: *Hydrology for engineers* 3rd edn. New York: McGraw-Hill.

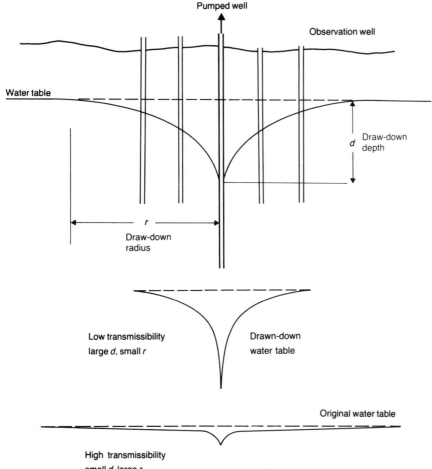

Cone of depression.

confined groundwater See GROUND-WATER.

congelifluction A general term applied to the movement of rock and earth, usually down hillslopes, as a result of freezing and thawing of ice, especially in permafrost regions.

congelifraction The weathering of rock by freeze–thaw process.

congeliturbation A general term for the heaving of the ground surface as a result of the freezing and thawing of ice.

conglomerate A rock which is composed of or contains rounded or water-worn pebbles and cobbles more than about 2 mm in diameter.

connate water Water trapped in the interstices of sedimentary rocks at the time of their deposition. It is usually highly mineralized and is not involved in active GROUNDWATER circulation, although connate waters may be expelled from their original location by compaction pressure and migrate, accumulating in more permeable formations. (See also JUVENILE WATER; METEORIC WATER.)   PWW

conscious weather modification See WEATHER MODIFICATION.

consequent stream A stream which flows in the direction of the original slope of the landsurface.

conservation Until recent years the word conservation was used very largely in connection with PROTECTED ECOSYSTEMS and LANDSCAPES, i.e. in the context of preservation of the status quo. It was used for the preservation of wild animals and plants in nature reserves, the passing of legislation to prevent the direct killing of particular species, and the seclusion of valued landscapes (be they natural or man-made) from the normal processes of economic development. In the 1960s it frequently included also the protection of wild land, primarily for the purpose of outdoor recreation. In economics it has generally been used in the context of temporal distribution, i.e. the costs and benefits of not using a resource now but leaving it for future usage. In the first sense

the word was largely sectional in applying to particular resources such as wild plants and animals, analogous to crop plants or minerals; in the second it was more of an attitude implying that it was better to be promised jam tomorrow than to eat bread today. The idea that conservation is an attitude to all resources (and, in a material sense, non-resources as well) is currently the strongest and underpins, for example, the WORLD CONSERVATION STRATEGY. It is translated operationally into a way of management of the biosphere so that it may yield the greatest sustainable benefit to present generations while maintaining its potential to meet the needs of future generations. It becomes therefore a positive (developing) set of policies rather than a negative (preserving) set, embracing the protection, maintenance and enhancement of the environment, emphasizing the sustainability of the various processes developed (i.e. SUSTAINED YIELD) and the avoidance of irreversible actions, especially where living resources are concerned. Conservation is, in general, anthropocentric (though some writers claim that it would be better if it were biocentric or even focused on global systems). It may be summed up in the saying that 'we have not inherited the earth from our parents, we have borrowed it from our children.'   IGS

Reading
Ehrenfeld, D.W. 1976: The conservation of non-resources. *American scientist* 64, pp. 648–56.

IUCN/UNEP/WWF 1980: *World conservation strategy.* Gland: IUCN.

Jones, G.E. 1987: *Conservation of ecosystems and species.* London: Croom Helm.

Soulé, M.E. and Wilcox, B.A., eds 1980: *Conservation biology.* Sunderland, Mass.: Sinaner.

consociation A natural vegetation community which is dominated by one species.

consolidation In engineering terms, the expulsion of water from the void spaces (see VOID RATIO) of a soil. This is achieved by natural burial processes as sediment accumulates or when a structure is erected on a soil. A normally consolidated soil (usually clay) is one which has never been subjected to an overburden pressure (i.e. load on top) greater than that which it currently has. An over-consolidated soil is one which has had a greater overburden pressure. This last effect is usually achieved by denudation of the overlying material: lodgement TILL is over-consolidated because of the pressures

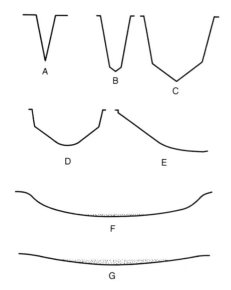

Constant slope: Wood's diagram of the slope cycle. A = free face only; B and C = constant slope forms; D and E = waning slope develops; F = waxing slope forms, waning slope rises up side of constant slope, alluvial filling represented by dots; G = constant slope has been consumed, alluvial fill deepens, slopes gradually flatten and approach a peneplain.
Source: *Wood 1942*.

previously applied by the glacier above. Over-consolidated clays have higher strengths than an otherwise equivalent normally consolidated clay, measured by an over-consolidation ratio. WBW

constant slope A term used by Wood (1942) to define the straight part of a hillside surface, lying below the free face, and having an inclination determined by the angle of repose of the TALUS material forming it. The constant slope extends upwards until it buries, or replaces the free face and so eliminates the supply of fresh talus debris. The WANING SLOPE of lower inclination is formed of weathered talus and eventually extends upwards and replaces the constant slope. The idea of a 'constant' slope is a theoretical construct with constancy of slope length and slope angle being of limited duration in a geological time span. MJS

Reading and Reference
Wood, A. 1942: The development of hillside slopes. *Proceedings of the Geologists' Association* 53, pp. 128–40.
†Young, A. 1972: *Slopes*. Edinburgh: Oliver & Boyd.

continental climate A type of climate characteristic of the interior of large land masses of middle latitudes. The major effects of continentality are to produce extreme seasonal variations of temperature and to depress the mean annual temperature below the latitudinal average. Since the oceans are the main sources of atmospheric moisture the continental interiors also tend to be dry, allowing the subtropical deserts to extend into temperate latitudes. The most extreme form of continental climate is found in eastern Asia, where there is an annual range of temperature of about 60°C. JGL

continental drift The movement of continents relative to each other across the earth's surface. Although it was a subject of speculation by numerous early workers, the first comprehensive case for continental drift was presented by Alfred Wegener in 1912. Wegener cited various lines of evidence in support of his notion of a super-continent (Pangaea) which gradually separated into a northern (Laurasia) and a southern (Gondwanaland) landmass before finally splitting into the continents of the present day. The evidence included the matching configuration of opposing continental coastlines, the similarity of geological structures on separate continental masses, the anomalous location of ancient deposits, indicating specific climatic conditions, and the distribution of fossil species through time. The theory was rejected at first by most geologists and geophysicists, largely because of a lack of a viable mechanism, but further support for continental drift was provided during the 1950s and 1960s through evidence from PALAEOMAGNETISM. In the late 1960s the idea was incorporated into the PLATE TECTONICS model. MAS

Reading
Du Toit, A.L. 1937: *Our wandering continents*. Edinburgh: Oliver & Boyd; New York: Hafner.
Hallam, A. 1973: *A revolution in the earth sciences: from continental drift to plate tectonics*. Oxford: Clarendon Press.
McElhinny, M.W. ed. 1977: *Past distribution of the continents*. Amsterdam: Elsevier.
Wegener, A. 1966: *The origin of continents and oceans*. Translated by John Biram from the fourth revised German edition. New York: Dover.
Wilson, J.T. ed. 1976: *Continents adrift and continents aground*. San Francisco: W.H. Freeman.

continental islands Occur in close proximity to a continent, to which they are also geologically related, and which are detached from the mainland by a relatively narrow, shallow expanse of sea. They were formerly united to the mainland. Oceanic ISLANDS, on the other hand, are and have been, geographically isolated and rise from the floors of the deep ocean basins.

continental shelf A portion of the continental crust below sea level, consisting of a very gently sloping, rather featureless, surface forming an extension of the adjacent coastal plain and separated from the deep ocean by a much more steeply inclined continental slope. The gradient of most continental shelves is between 1 and 3 m km^{-1}. Some shelves reach a depth of over 500 m but 200 m is often conveniently used as a depth limit. Their mean width is around 70 km but there is a marked variation between different coasts. It is estimated that continental shelves cover approximately 5 per cent of the earth's surface.   MAS

Reading
Kennett, J.P. 1982: *Marine geology.* Englewood Cliffs, NJ and London: Prentice-Hall.
Shephard, F.P. 1973: *Submarine geology.* 3rd edn. New York: Harper & Row.

continental slope Lies to the seaward of the CONTINENTAL SHELF and slopes down to the deep sea floor of the abyssal zone.

continuity equation A statement that certain quantities such as mass, energy or momentum are conserved in a system, so that for any part of the system the net increase in storage is equal to the excess of inflow over outflow of the quantity conserved.

The most familiar example of a continuity equation is perhaps the STORAGE equation in hydrology which describes the conservation of mass for water. Continuity equations are equally applicable to the conservation of total mass of earth materials, or of mass for particular chemical elements or ions. In these contexts the continuity equation is one fundamental basis for models of hillslope or soil evolution (e.g. Kirkby 1971). The equation is usually used in these models as a partial differential equation, which formally connects rates of change at a site over time to rates of change over space at a given time. One simple form of the continuity equation is:

$$\partial s/dt + \partial Q/dx = a$$

where s is the amount stored (e.g. as elevation of rock and soil materials), Q is the rate of flow (e.g. as total transport of earth materials downslope), a is the net rate of accumulation (e.g. as wind-deposited dust), and x and t are respectively distance downslope and time elapsed.

An equation of this type cannot normally be solved without specifying the relevant processes which control the rates of flow, accumulation, etc. Continuity equations are equally relevant to the conservation of energy in micro-meteorological studies and in many other physical and chemical systems. The concept of continuity dates back to Leonardo da Vinci and remains a fundamental principle underlying our understanding of the physical world.   MJK

Reading and Reference
†Davidson, D.A. 1978: *Science for physical geographers.* London: Edward Arnold.
Kirkby, M.J. 1971: Hillslope process-response models based on the continuity equation. *Transactions of the Institute of British Geographers, special publication 3*, pp. 15–30.

contour A line on a map which joins areas of equal height or equal depth.

contraction hypothesis Suggestion that as the earth's interior cooled, so it would contract. It was felt, especially in the nineteenth century, that the resulting compression of the earth's crust would create fold mountains.

contrail An abbreviation for condensation trail which is a thin line of water droplets or ice crystals that condense after emission in the exhaust gases of aircraft engines. At a given height each engine type has a critical air temperature above which a contrail will not form; in the case of the British Isles they are seldom observed below about 6000 m in winter and 8500 m in summer. Once formed they generally spread in width and if persistent indicate slow EVAPORATION in a relatively humid upper TROPOSPHERE.   RR

contributing area The area of a catchment which is, or appears to be, providing water for storm run-off.

The term was first used by Betson (1964) and may be calculated as the stream discharge divided by the rainfall (or net rainfall) intensity, usually expressed as a proportion or percentage of total catchment area. This ratio is meaningless when it is not raining, and is usually calculated over a total storm period, in the form total storm run-off (normally calculated as QUICKFLOW) divided by total storm rainfall. For many well-vegetated small catchments, the contributing area is less than 5 per cent of the catchment, but is commonly many times greater where vegetation is sparse (within the area of storm rainfall).   MJK

Reference
Betson, R.P. 1964: What is watershed runoff? *Journal of geophysical research* 69, pp. 1541–52.

control structures Structures in the form of weirs or flumes which are installed in a river cross-section to enable river discharge to be measured. In each case a formula relates discharge to the depth of water in or above the control structure. (See also DISCHARGE.)   KJG

convection In general, mass movement within a fluid resulting in transport and mixing of properties of that fluid. In the atmosphere a class of fluid motion in which warmer air goes up while colder air goes down. Unfortunately fluid dynamicists sometimes use the word in place of ADVECTION. In the case of fair-weather convection, air at low levels is slowly warmed by sunshine absorbed at the ground until it can lift away from the surface to be replaced suddenly by comparatively cool air from above. If the air is sufficiently moist cumulus cloud may form above the updraught as an indicator of the process. *Cumulonimbus* convection is more complex (see CLOUDS). Air rising from the surface is usually slightly cooler than its immediate environs but it is very moist, so that when it has risen above cloud base and made use of the latent energy of the water vapour by condensing to cloud droplets it becomes very buoyant. This air then accelerates vigorously, dragging the reluctant air near the surface after it. Slantwise convection occurs when the warm and cold air are initially side by side. This mechanism accounts for sea breezes, valley winds and weather systems (see MESOMETEOROLOGY).   JSAG

Reading
Ludlam, F.H. 1980: *Clouds and storms*. University Park, Pa.: Pennsylvania State University Press.

convergence See DIVERGENCE.

coordination number Refers to the packing of objects around a given body. The packing of sand particles can be represented by a coordination number which refers to the number of contacts a given particle makes with other particles. The higher the coordination number the greater the interparticle friction and interlocking and the higher the SHEAR STRENGTH of the material. The value also affects (but is not a substitute for) POROSITY and PERMEABILITY of a material.

In chemical structures the coordination is also important in determining the properties of materials; this is especially true with clay minerals. There are two basic building blocks in clays: first a silicon atom with four (larger) oxygen atoms around it in the form of a tetrahedron to give four-fold coordination; secondly, an atom of magnesium or aluminium surrounded by six hydroxyl radicals in an octahedron to give six-fold coordination.   WBW

Reading
Whalley, W.B. 1976: *Properties of materials and geomorphological explanation*. Oxford: Oxford University Press.

coprolite Fossilized animal dung and excreta. Phosphatic nodules.

coquina A carbonate rock which consists largely or wholly of mechanically sorted, weakly to moderately cemented fossil debris (especially shell debris) in which the interstices are not necessarily filled with a matrix of other material.   ASG

coral algal reef A marine structure, containing colonies of scleractinian corals. Normally composed mainly of calcium carbonate, REEFS are also complex and productive ecosystems. Skeletal carbonates provide much of the reef framework and organisms also assist in reef erosion and sediment production. The distribution of coral algal reefs is controlled by environmental factors, notably water temperature, clarity and salinity. Most reef-forming corals prefer sea temperatures between 17 and 33°C, salinities of between 30 and 38 parts per thousand, and clear water. Light is also important, and coral growth is usually restricted to the upper 25 or 30 metres. Because of these factors, coral algal reefs are found mainly between latitudes 30° N and S on mud-free coastlines, particularly in western parts of the Pacific, Indian and Atlantic Oceans.

Coral reefs, like this example from the Maldive Islands in the Indian Ocean, are among the world's most diverse and productive ecosystems. Reefs are restricted to low latitudes where sea-water temperatures are high and may occur as atolls, fringing reefs and barrier reefs.

There are three main types of coral algal reef, i.e. fringing reefs, BARRIER REEFS and ATOLLS. Bank reefs and ridge reefs (found in the Red Sea) have also been described. Reefs formed on continental coastlines are often complex, multiple features which are difficult to categorize. Several theories have been proposed to explain the genesis of barrier reefs and atolls, invoking subsidence, sea-level change and sea-floor spreading.

The coral algal reef ecosystem has very different environmental conditions from the surrounding sea, encouraging a flourishing of life. Zooxanthellate acleractinian corals form much of the primary reef framwork. These corals live in symbiotic association with unicellular algae, the *Zooxanthellae*, and are colonial organisms with the ability to reproduce sexually or asexually. Calcareous, encrusting organisms (such as coralline algae, corals, bryozoans, gastropods

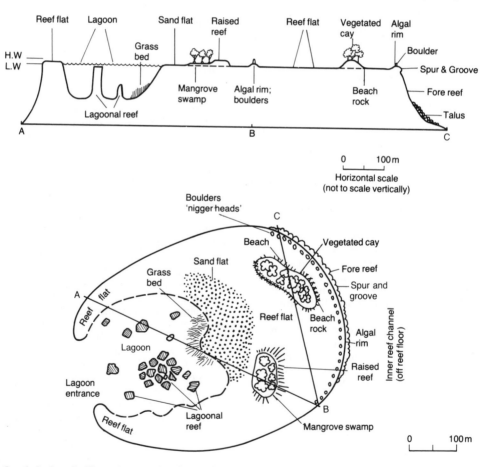

Coral algal reefs: *Plan and cross-section of an ideal coral reef, showing the major environments.*

and serpulid worms) attach themselves to cavities within the primary reef framework, forming a secondary structure.

Reef ecosystems are vulnerable to catastrophic events, such as hurricanes and bleaching episodes, which may cause mass mortality of corals. Human stresses, such as pollution and increase of sediment load, have damaged reefs in many areas. Recently, there has been much speculation over the future impacts of climatic warming on coral reefs (see Stoddart 1990). The geomorphology of coral algal reefs is controlled by the interplay of growth and erosion, producing a reef front (often with spur and groove topography) on the ocean side, a reef flat, and a back reef zone. HAV

Reading and Reference
Guilcher, A. 1988: *Coral reef geomorphology*. Chichester: Wiley.

Hopley, D. 1982: *The geomorphology of the Great Barrier Reef*. New York: Wiley Interscience.

Stoddart, D.R. 1990: Coral reefs and islands and predicted sea-level rise. *Progress in physical geography* 14, pp. 521–36.

coral bleaching A new and unexplained phenomenon which threatens to destroy coral reefs. Corals are bleached when the colourful symbiotic algae they house are lost. The algae can re-enter their hosts if conditions are favourable, and bleached reefs have recovered. When the algae are absent for any length of time, the coral dies. The extent of bleaching varies with depth: the shallower the water, the worse the bleaching. Scientists have hypothesized that the cause of the bleaching is stress, brought on by unusually warm water, changes in salinity, excessive exposure to ultraviolet radiation or extreme climatic changes. Most corals thrive when the water is between 25 and 29°C, and it is believed that algae die when water temperatures exceed the upper limit. ASG

Reading
Brown, B.E. ed. 1990: Coral bleaching. *Coral reefs* 8, pp. 153–232.

— and Ogden, J.C. 1993: Coral bleaching. *Scientific American*, 268, pp. 44–50.

core The intensely hot (2700°K) inner part of the earth. It begins at around 2900 km from the surface at the Gutenberg Discontinuity. The outer portions of the core may be liquid and the inner solid.

corestone A cobble or boulder of relatively unweathered rock which is or has been incorporated within the weathered rock which surrounds it.

Coriolis force Also known as the geostrophic force. An apparent force on moving particles in a frame of reference which itself

A snow cornice in the Gurkha Himal on the north-east ridge of Himalchuli. Snow plays a major geomorphological role in high altitude situations through nivation processes. When cornices collapse avalanches may occur.

is moving, usually rotating. Such a force is required if Newton's laws of motion (see EQUATIONS OF MOTION) are to be applied in the rotating framework. The Coriolis force is of major importance to the movement of both oceanic waters and air. In meteorology the Coriolis force per unit mass of air arises from the earth's rotation and is equal to $-2\Omega \times V$, where Ω and V are vectors representing respectively the angular velocity of the earth and the velocity of the air relative to the earth. In practical terms the force 'deflects' air particles to the right in the northern hemisphere and to the left in the southern hemisphere. It affects only the direction, not the speed of the wind. (See also GEOSTROPHIC WIND and WIND.)

BWA

Coriolis parameter Twice the component of the earth's angular velocity (Ω) about the local vertical, $2\Omega \sin \phi$, where ϕ is latitude.

cornice An overhanging accumulation of wind-blown snow and ice found on a ridge or a clifftop, usually on the lee side.

corniche An organic protrusion growing out from steep rock surfaces at about sea level, and providing a narrow pavement or sidewalk-like path at the foot of sea cliffs. Comparable rock ledges caused by erosional processes and coated with organic material are termed trottoirs. Corniches are often formed of calcareous algae. They are largely intertidal, being best developed in the inlets of exposed coasts and generally protrude about 0.2–2.0 m. Vermetids and serpulids may contribute to their development.

ASG

corrasion Mechanical erosion of rocks by material being transported by water, wind and ice over and around them.

corrie See CIRQUE.

corrosion The process of solution by chemical agencies of a rock in water as distinct from the mechanical wearing away of rock by water or its bed load (CORRASION).

cost–benefit ratio The ratio of the expected costs of an anticipated project to the expected benefits expressed in monetary terms. Projects are usually considered warranted if the cost–benefit ratio exceeds unity. All costs, including values assigned to qualitative concepts, that can be reasonably anticipated are compared to all reasonably expected benefits calculated in a similar manner (Sewell 1975, p. 10).

Costs in a cost–benefit analysis usually include material, labour, and associated direct costs of construction. For many long-term capital projects a major consideration is the amount of interest cost incurred on construction funds. Land acquisition costs also tend to be an important component, especially in urban areas or for water projects requiring extensive reservoirs. The costs associated with environmental degradation are usually included, but they are frequently difficult to quantify and are controversial items. Social costs including dislocation of homes and businesses and lost opportunities are also difficult to assess but none the less important to the decision-maker.

Benefits to be compared to costs usually account for increased productivity and increased land values near to the project. More difficult to define but of equal importance are enhanced environments that are more useful for human purposes than the original natural environment and increased opportunities in a social sense.

Although the mathematics of cost–benefit analysis are clearly defined (Thuesen *et al.* 1971), the approach suffers from two major weaknesses. First, the process gives little attention to those aspects which are difficult to convert to monetary values, yet to the decision-maker these may be the most important considerations. Secondly, the process tends to emphasize the short-term considerations because they are the most easily assessed, while the long-term considerations may be the most important.

WLG

Reading and References
Sewell, G.H. 1975: *Environmental quality management.* Englewood Cliffs, NJ: Prentice-Hall.
†Thuesen, H.G., Fabrycky, W.J. and Thuesen, G.J. 1971: *Engineering economy.* 4th edn. Englewood Cliffs, NJ: Prentice-Hall.

coulée A flow of volcanic lava which has cooled and solidified.

couloir A deep gorge or ravine on the side of a mountain, especially in the Alps.

Coulomb equation See MOHR–COU-
LOMB EQUATION.

cover, plant The proportion, or percentage, of ground occupied by the aerial parts
of a plant species or group of species. With
the overlapping of species in most plant
communities, the combined percentage will
nearly always exceed 100, except in very
open vegetation. Special scales have been
devised for estimating both the degree and
character of plant cover in quadrats of a
chosen size, such as the Domin scale and
the Braun–Blanquet scales. Some methods
involve sampling, like the number of points
touching a given species, whereas other
systems are simply estimates by eye.   PAS

Reading
Kershaw, K.A. 1973: *Quantitative and dynamic plant ecology.*
2nd edn. London: Edward Arnold.
Willis, A.J. 1973: *Introduction to plant ecology.* London: Allen
& Unwin. Especially ch. 11.

coversand A generally thin cover of
sandy material of aeolian origin that may
have been modified by subsequent reworking by fluvial, periglacial and other processes. DUNE forms may be relatively
undistinct or absent altogether. Such
deposits cover large areas of the lowlands
of Europe on the margins of the great
Pleistocene ice caps, and have also been
recognized in certain parts of Britain (Catt
1977).   ASG

Reading and Reference
Catt, J.A. 1977: Loess and coversands. In F.W. Shotton
ed., *British Quaternary studies.* Oxford: Clarendon Press. Pp.
221–9.
Koster, E.A. 1988: Ancient and modern cold-climate
aeolian sand deposition: a review. *Journal of Quaternary
science 3, pp. 69–83.*

crab-holes Small abrupt depressions in
the ground surface which vary in diameter
from a few centimetres to more than a
metre, and in depth from *c.* 5 to 60 cm.
They occur in sediments which are prone to
vertical cracking and horizontal piping.
ASG

Reading
Upton, G. 1983: Genesis of crabhole microrelief at Fowlers
Gap, western New South Wales. *Catena 10, pp. 383–92.*

crag-and-tail A streamlined landform
comprising a rock obstruction with glacial
deposits in its lee. One of the best known
examples is in the centre of Edinburgh
where the crag, capped by Edinburgh
Castle, is an old volcanic plug and the

High Street, which slopes steadily eastwards, for over a kilometre, marks out the
tail. The landform is one of a family of
streamlined subglacial forms such as the
DRUMLIN. The material deposited in the lee
may include both till and water deposits.
DES

crater A depression at the crest or on the
flanks of a volcanic cone where a pipe or
vent carrying gases and lava reaches the
surface. Also the impact scar left by a
meteorite when it impacts on the surface
of a planet or moon.

craton A continental area that has experienced little internal deformation since the
Precambrian (about 570 million years ago).
These areas can be divided into a very stable
core, known as a SHIELD, and a marginal
platform zone where gently tilted or flat
sedimentary rocks bury the Precambrian
basement. Crustal movement in cratons is
largely vertical and results in the formation
of broad domes and basins. The term
craton is also used in an alternative sense
to refer to the very ancient core of shield.
MAS

Reading
Spencer, E.W. 1977: *Introduction to the structure of the earth.*
2nd edn. New York: McGraw-Hill.
Windley, B.F. 1984: *The evolving continents* 2nd edn.
London and New York: Wiley.

creationism The theory that attributes
the origin of all species of organisms (and
indeed all matter) to special creation as
opposed to evolution. Creationists maintain
that plants and animals were brought into
existence, in their present form, by the
direct intervention of divine power. Some
authorities, such as John Ray (1627–1705)
and William Paley (1743–1805), emphasized the adaptation of organisms to their
environment as evidence of the 'wisdom of
God'. Creationists usually suggest that all
organisms were created at the same time, or
over a very short period (as in the Genesis
accounts), although Louis Agussiz (1807–
1873) entertained the possibility of 'multiple creations'.

'Creation science' or 'scientific creationism' attempts to demonstrate the literal
truth of accounts of the creation, such as
that in the book of Genesis, using the
techniques of modern science. Bone beds,
such as that of the Rhaetic in Britain, are
attributed to a catastrophic destruction of

organisms caused by the deluge. Creation scientists have drawn attention to imperfections in the fossil record, and the rarity within it of links between the major groups of organisms, arguing that these invalidate the theory of evolution. They have stated that there are problems in explaining the evolution of complex structures, such as the vertebrate eye, by natural selection, for, creation scientists argue, an 'incompletely evolved' eye would be of no value to the organism. They have also taken certain recent scientific doubts about natural selection as a *principal* mechanism of evolution as calling into question the whole of evolutionary doctrine.

In certain jurisdictions, creation scientists have sought (and on occasions won) the right to place their ideas alongside those of evolutionary theory before school students. Unquestionably, creationists have caused textbook writers to be much more careful in their wording, but most evolutionists would argue that many of their ideas are based upon the misunderstanding, or in some cases, the deliberate distortion, of scientific evidence.   PHA

Reading
Baker, S. 1976: *Is evolution true?* Welwyn. Herts. and Grand Rapids, Mich: Evangelical Press.
Ruse, M. 1982: Creation science: the ultimate fraud. In J. Cherfas ed., *Darwin up to date*. London: New Science Publications. Ch. 2, pp. 7–11.

creep The gradual downslope movement of soil and rock debris on hillsides or glacier ice under the influence of gravity.

crescentic bar See BAR, COASTAL.

crevasse A chasm or deep fissure in the surface of an ice sheet or glacier. Also a breach in a levée along the bank of a river.

critical erosion velocity For fluid flows over an erodible bed, that velocity which is just able to entrain the bed material. In practice this is not so easy to define because bed material may be mixed in size, cohesive or well packed, so that it is not possible simply to relate bed material incipient motion to a precise velocity value. There is also the problem of designating which velocity: is it mean velocity in the vertical, as has been used for streamflow, or that very near the bed? For material moving at the bed, different velocities may be required for sliding and for overturning.

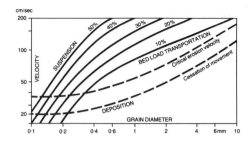

Relation between flow velocity, grain size and state of sediment as proposed in curves devised by Hjulström 1935 and modified by A. Sundborg 1956: The river Klaralven, a study of fluvial processes. Geografiska Annaler 38, pp. 127–316.

Despite these complications, empirical relationships between sediment size and entrainment velocity have been established, notably by Hjulström in the mean velocity relations and in more recent studies involving near-bed velocity studies. Incipient motion is also specified in terms of other hydraulic measures such as stream power and bed or dimensionless shear stress.   JL

Reading
Richards, K. 1982: *Rivers*. London and New York: Methuen.

critical load A concept in pollution studies which involves the idea that there is a certain pollution load level above which harmful effects on biological systems, such as decline and disappearance of fish populations, will occur.   ASG

Reading
Brodin, Y-W. ed. 1992: The critical-load concept: an instrument to combat acidification and nutrient enrichment. *Ambio* 21, pp. 332–87.

cross-bedding The arrangement of laminae and beds in sedimentary strata at different angles from the principal planes of stratification. The pattern of cross-bedding provides evidence of the environment and modes of deposition.

cross-lamination Thin layers of sediment, often only a few millimetres thick, that dip obliquely to the main bedding plane. The cross-laminae represent individual sedimentation units resulting from small-scale fluctuations in velocity and in rates of supply and deposition of silts and sands forming the lee, stoss and trough laminae of migrating ripples.   JM

cross-profile, valley/river channel A profile may be surveyed at right-angles to the river flow direction across a river channel or across the valley in which the river channel occurs. Information from contours or topographic maps is often sufficiently detailed to draw valley cross-profiles. (See also CHANNEL CAPACITY.)

KJG

crumb structure Pertains to those soils which have their fine particles accumulated in the form of aggregates or crumbs, so that they have a more open, coarser and workable texture.

cryergic Periglacial in its broadest sense. Pertaining to the periglacial features and processes which occur in those areas not immediately adjacent to glaciated areas.

cryoplanation The flattening and lowering of a landscape by processes related to the action of frost. (See also ALTIPLANATION.)

ASG

Reading
Priesnitz, K. 1988: Cryoplanation. In M.J. Clarke ed., *Advances in periglacial geomorphology*. Chichester: Wiley. Pp. 49–67.

cryostatic pressures Freezing-induced pressures thought to develop in the ACTIVE LAYER in pockets of unfrozen material which are trapped between the downward migrating freezing plane and the perennially frozen ground beneath. Although recorded in experimental studies, the existence of substantial cryostatic pressures in the field has yet to be convincingly demonstrated. Generally, the presence of voids in the soil, the occurrence of frost cracks in winter, and the weakness of the confining soil layers lying above, prevent pressures of any magnitude from forming. Nevertheless, cryostatic pressures are often invoked to explain various forms of patterned ground and mass displacements (cryoturbations) in the active layer. HMF

Reading
French, H.M. 1976: *The periglacial environment*. London and New York: Longman, Esp. pp. 40–4.
Mackay, J.R. and Mackay, D.K. 1976: Cryostatic pressures in non-sorted circles (mud hummocks), Inuvik. *Canadian journal of earth sciences* 13, 889–97.
Washburn, A.L. 1979: *Geocryology: a survey of periglacial processes and environments*. New York: Wiley. Esp. p. 167.

cryoturbation The process whereby soils, rock and sediments are churned up by frost processes to produce convolutions or involutions. The process is especially active in the zone above permafrost which is subject to seasonal freezing and thawing – the active layer. (See also FROST HEAVE.)

ASG

Reading
Vandenberghe, J. 1988: Cryoturbations. In M.J. Clarke ed., *Advances in periglacial geomorphology*. Chichester: Wiley. Pp. 179–98.

cryovegetation Consists of plant communities comprising such types as algae, lichens and mosses, which have adapted to life in environments where there is permanent snow and ice.

cryptovolcano A small, roughly circular area of greatly disturbed strata and sediments which though suggestive of volcanism does not contain any true volcanic materials.

cuesta A ridge which possesses both scarp and dip slopes.

cuirass An indurated soil crust which mantles the landsurface protecting the underlying unconsolidated sediments from erosion.

cumec A measure of discharge, being an abbreviation for cubic metre per second.

cumulative soil profiles receive influxes of parent material while soil formation is still going on; i.e. soil formation and deposition are concomitant at the same site. Their features are thus partly sedimentological and partly pedogenic. Among topographic sites favourably situated for their formation are areas receiving increments of loess, river floodplains, and colluvial and fan deposits at the base of hillslopes. ASG

cumulonimbus See CLOUDS.

cumulus See CLOUDS.

cupola A dome-shaped mass of igneous rock which projects from the surface of a batholith.

current bedding Layering produced in accumulating sediment by fluid flow which is oblique to the general stratification. Laminae (less than 1 cm) or thicker strata in tabular or wedge-shaped units may comprise near-parallel sets bounded by plane, curved or irregular surfaces, with boundaries representing accretionary limits or erosional truncations. Such features are produced through the development of fluid BEDFORMS varying in size from CURRENT RIPPLES to BARS, and their form may give indications of the direction of current flow, flow regime and sediment supply. Also called cross-bedding (a term now preferred by most sedimentologists and subdivided into cross-lamination and cross-stratification) or false bedding, this phenomenon may help in the identification of sedimentary environments. For example, large-scale current bedding may relate to delta growth, the progradation of aeolian dune slip faces or to river channel point bar sedimentation. Smaller-scale features derive from ripple development, and different kinds of ripples (e.g. straight-crested or linguoid) may produce contrasted current bedding patterns.   JL

Reading
Allen, J.R.L. 1970: *Physical processes of sedimentation*. London: Allen & Unwin.

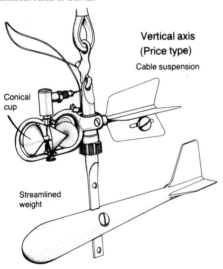

Vertical axis
(Price type)

Cable suspension

Conical cup

Streamlined weight

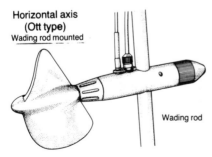

Horizontal axis
(Ott type)
Wading rod mounted

Wading rod

— 1982: *Sedimentary structures*. Vol. 1. Amsterdam: Elsevier.
Collinson, J.D. and Thompson, D.B. 1982: *Sedimentary structures*. London: Allen & Unwin.

current meter An instrument for measuring the velocity of flowing water in freshwater and marine environments. Many principles have been employed and available types include rotating, electromagnetic, optical and pendulum current meters. The rotating current meter is the most widely used for river measurements and consists of a propeller (horizontal axis type) or a rotor formed by a series of cups (vertical axis type) which rotates at a speed proportional to the flow velocity. The revolutions are counted over a fixed period of time, and velocity is computed from calibration data. The meter body may be mounted on a wading rod or suspended on a cable.   DEW

Reading
Buchanan, T.J. and Somers, W.P. 1969: Discharge measurements at gaging stations. *US Geological Survey techniques of water resources investigations*. Book 3, Ch. A8.

current ripples Small-scale wave-like undulations developed by fluid flow over a sandy or coarse silty bed. Their spacing or wavelength is usually less than 50 cm and the height difference between trough and crest is seldom more than 3 cm. Larger sand features may be called dunes or sandwaves and larger dynamically related features in coarser sediment may be termed BARS.

Current ripples may be described in terms of their plan and profile characteristics. They may be straight, sinuous or indented (linguoid, cuspate, lunate) in plan and may have peaked, rounded and asymmetrical crests. Ripple development occurs through flow separation from the bed; dimensions may be related to applied fluid shear stress.   JL

Reading
Allen, J.R.L. 1968: *Current ripples*. Amsterdam: North-Holland.

currents, nearshore In the nearshore environment water particles beneath wind-driven waves develop elliptical orbits when the water depth is less than half the wave length, and an oscillatory bed current develops normal to the wave front. This increases in velocity as water depth decreases, and gains in geomorphological effectiveness when it is competent to move sediment. The wave approach is often oblique to the shore at angles of 10–20°,

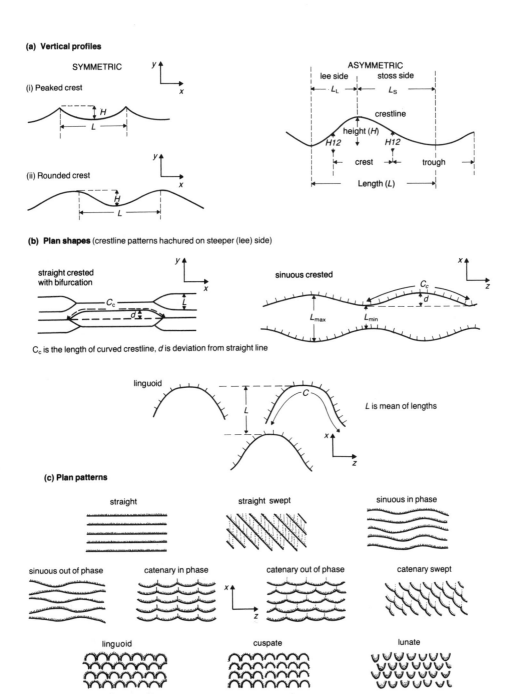

(a) Vertical profiles

SYMMETRIC

(i) Peaked crest

(ii) Rounded crest

ASYMMETRIC

lee side   stoss side

L_L L_S

crestline

height (*H*)

$H12$ $H12$

crest   trough

Length (*L*)

(b) Plan shapes (crestline patterns hachured on steeper (lee) side)

straight crested with bifurcation

sinuous crested

C_c d

L_{max} L_{min}

C_c is the length of curved crestline, *d* is deviation from straight line

linguoid

L

C

L is mean of lengths

(c) Plan patterns

straight   straight swept   sinuous in phase

sinuous out of phase   catenary in phase   catenary out of phase   catenary swept

linguoid   cuspate   lunate

Current ripples: *definition diagrams for many of the descriptive terms used in the description of ripples. Most of the terms can also be applied to larger ripple-like bedforms. The reference axes are x parallel to the current, y vertical and z horizontal and perpendicular to the current.*
Source: *J.D. Collinson and D.B. Thompson 1982:* Sedimentary structures. *London: Allen & Unwin. Figure 6.1.*

so littoral drift of sediment occurs in the swash and backwash when the wave breaks. The wave-normal current may be resolved into shore-normal and longshore component current (Pethick 1984, pp. 34 ff). The longshore current is accentuated by the need to satisfy continuity, since an outward flow is required to balance the shoreward flow of water. This takes the form of a rip current in a clearly defined channel running parallel to the beach, then forming a localized stream passing through the breaker zone at speeds of up to 1 m s^{-1} to disperse eventually about 300 m offshore. Rip currents form part of a series of cell-like circulation systems related to EDGE WAVES. KSR

Reference
Pethick, J.S. 1984: *An introduction to coastal geomorphology.* London: Edward Arnold. Pp. 34–5.

currents, ocean Drift currents in the oceans are driven by the major global wind systems, and therefore contribute to the net poleward energy transfer necessitated by latitudinal imbalance in solar radiation receipt. Near the equator dominantly east-west currents are driven by the trade winds, and an easterly directed countercurrent at the equator completes a circulation known as 'gyre'. Within the ocean basins currents follow the continental margins. In the North Atlantic a clockwise circulation includes the

eastern and western 'boundary currents', the Canaries Current and the Gulf Stream. Drift currents may have clearly defined temperature and density boundaries with the surrounding water, and migrating meanders which intensify then break away to form cells or rings. With increasing depth beneath the surface successive deflection of the current occurs relative to that above, to the right in the northern hemisphere and to the left in the southern, because of the Coriolis force which results from the earth's rotation. This increasing deflection is the EKMAN SPIRAL. KSR

currents, river River currents vary both across the section within which the flow is confined, and with depth because of the bed friction that controls the vertical velocity profile. In addition, SECONDARY FLOWS occur with components directed across the channel. The main longitudinal current is generally fastest in the deepest part of the cross-section (the thalweg), but shifts position with changes in river discharge. In a meander bend at low discharge a high sinuosity current (the 'pool current') follows the thalweg closely, but at high discharge a 'bar-head current' of lower sinuosity and steeper slope develops across the point bar. Sediments deposited by river currents have grain sizes which reflect current strength, and structures which

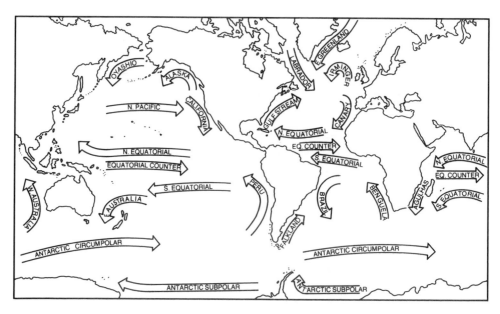

Surface currents of the world during the northern hemisphere winter.

record the current direction. Coarse sediments and their structures tend to display more consistent directional properties, whereas finer sediments are deposited by currents of more varied direction which are forced to conform to the bed topography created by the higher flows.   KSR

currents, tidal Tidal currents are horizontal water movements asociated with the tidal rise and fall of the sea surface. In the open ocean standing wave tides are associated with continuous currents of about 0.3–0.5 m s^{-1} which change direction through 360° in one tidal cycle (clockwise in the northern hemisphere). At the coast, and especially in estuaries, the tide is more like a progressive wave, and is associated with stronger, reversing, currents which flood as the tidal wave crest approaches the land, and ebb as the tidal wave trough approaches. Slackwater periods occur at high and low tides, and maximum tidal current velocities occur at mid-tide. In an estuary, as the tide wave progresses inland its amplitude is decreased by frictional effects and it becomes asymmetric as the crest of the tidal wave entering the deeper water at the mouth and travelling faster catches the slower moving trough ahead of it. Ebb and flood current velocities are influenced by the tidal range and asymmetry, and by the estuary morphology and freshwater discharge (Pethick 1984, pp. 59 ff). The flood is often shorter than the ebb, so higher velocities are needed to discharge the same volume of water through the same cross-section. More sediment may thus be moved landward than seaward, and the head of an estuary becomes a sediment sink.   KSR

Reference
Pethick, J.S. 1984: *An introduction to coastal geomorphology.* London: Edward Arnold. Pp. 58–63.

cusp, beach A ridge made of sands and gravel on a foreshore or beach. Cusps tend to be uniformly spaced and to run at right angles to the water's edge where they peter out.

Reading
Komar, P.D. 1983: Rhythmic shoreline features and their origins. In R.A.M. Gardner, and H. Scoging, eds, *Megageomorphology.* Oxford: Oxford University Press.

cut-off An abandoned reach of river channel, produced particularly where a meander loop has become detached from

A cut-off meander developed in Minnesota, USA. Such cut-offs are a common feature of many river floodplains.

the active river channel because the neck of the loop has been breached. The abandoned reach may be occupied by an OXBOW lake which gradually fills with sediment. Different cut-off processes are possible: with highly sinuous channels adjacent reaches may impinge on each other directly (neck cut-offs), while in other cases the scouring out of longer short-circuiting channels across the inside of meander bends during floods may produce cut-offs at much lower sinuosities (chute cut-offs). The accumulation of sediment in BARS may also lead to the detachment of channel reaches. The term is sometimes applied to the new channel itself as well as to the abandoned one or to the process in general.   JL

cutan A thin coating of clay on soil particles or lining the walls of a void in the soil zone.

cutter Linear slots cut in bedrock by solution along a guiding structual element, they are equivalent to the British term, grike (White 1988).   ASG

Reference
White, W.B. 1988: *Geomophology and hydrology of karst terrain.* New York: Oxford University Press.

cycle, biochemical See BIOGEOCHEMICAL CYCLES.

cycle, climatic A recurrent climatic phenomenon. The term is best reserved for changes of strictly periodic origin such as the annual temperature cycle, but it is often used loosely to describe many changes of climate which occur at approximately fixed intervals.

The cycle of major glaciations with a peak about every 10^5 years is a good example. Various short period cycles have been described by numerous authors, but they are often the subject of some controversy. Examples are the supposed climatic cycles related to solar activity in general and SUNSPOTS in particular. (See MILANKOVITCH HYPOTHESIS.)   JGL

cycle of erosion The sequence of denudational processes and forms which, in theory, exist between the initial uplift of a block of land and its reduction to a gently undulating surface or peneplain close to BASE LEVEL. The concept was codified and popularized by the American geographer, William Morris Davis (1850–1934), who termed the sequence the geographical cycle (1899).

In reality, the process described is not a cycle at all, but a one-way movement of mass from higher to lower elevation. The true cycle involved is that first described by James Hutton (1788) as the sequence of uplift, denudation, sedimentation and lithification, followed by renewed uplift. This *geological* cycle is the most basic of the numerous biogeochemical cycles in which the earth's atoms are repeatedly recombined into new chemical, geological and biological compounds. In all these cycles, there are many different pathways which matter may follow over time periods of enormously different lengths.

The Davisian concept of the cycle of erosion is thus mis-named. It is also over-simple, since it emphasizes a generally relentless progression of material from highlands to the oceans or enclosed basins, and the creation of a characteristic sequence of landforms during the process. The early, intermediate and late stages of the cycle are termed youth, maturity and old age. While it was recognized that changes in base level might interrupt the progression, leading to rejuvenation, the general tenor of Davis's ideas led to the belief that such interruptions were unusual.

Davis worked originally upon the sequence of forms produced by fluvial action in areas of humid climate: this was termed the 'normal' cycle. Later, supposedly distinctive, sequences were added by Davis and others to represent the development of arid, glacial, karst, coastal and periglacial regions, together with a recasting of Darwin's theory of coral reef development (1842) in a more explicitly cyclic form.

The reality and the theoretical utility of the cycle of erosion as a concept have been increasingly challenged. In particular, the frequent and intense changes in global climate which appear to have occurred during the Pleistocene – to say nothing of associated fluctuations in base level – render it improbable that any area will contain the simple sequence of forms and, even if these occur, it is difficult to see how they can be the products of any simple, unidirectional set of processes. Therefore, while it is true that 'what goes up must come down', it does not seem useful to ascribe the descent to the operation of Davisian cycles of erosion.   BAK

Reading and References
†Chorley, R.J. 1965: A re-evaluation of the geomorphic system of W.M. Davis. in R.J. Chorley and P. Haggett eds, *Frontiers in geographical teaching*. London: Methuen.
Darwin, C.R. 1842: *The structure and distribution of coral reefs*. London: Smith, Elder.
†Davis, W.M. 1899: The geographical cycle. *Geographical journal* 14, pp. 481–504.
Hutton, J. 1788: Theory of the earth: or an investigation of the laws observable in the composition, dissolution and restoration of land upon the globe. *Transactions of the Royal Society of Edinburgh* 1, pp. 209–304.

cyclogenesis See CYCLONE.

cyclone (or depression) A region of relatively low atmospheric pressure, typically one to two thousand kilometres across, in which the low level winds spiral counterclockwise in the northern hemisphere and clockwise in the southern. Cyclones are common features of surface weather maps and are frequently associated with windy, cloudy and wet weather.

Cyclogenesis (or the formation of cyclones) occurs in preferred areas and is usually most vigorous in wintertime; such areas in the northern hemisphere are the western North Atlantic, western North Pacific and Mediterranean Sea. The birth and subsequent movement of cyclones is closely linked to the presence of the planetary scale ROSSBY WAVES in the atmosphere. They form frequently in the downstream or eastern limb of these large-scale troughs and move as features embedded in the deep generally poleward flow.

The inward spiralling air ascends within the system and flows out in the upper troposphere. In any atmospheric column, if more mass is exported aloft than is imported at low levels, the surface pressure will fall, while surface pressure will rise and the low will fill if there is a net gain of mass in the column.

Cyclones are highly transient features which are associated with disturbed weather and, if frontal, with strong horizontal gradients of TEMPERATURE and HUMIDITY, and sharp changes in cloud cover and type. Across the extra-tropical ocean basins they carry out very important heat transport in a meridional direction which offsets to some extent the persistent equator to pole imbalance in the RADIATION budget. In frontal cyclones warm air is transported poleward and cooled while cold air moves towards the equator and is warmed.   RR

Reading
Palmén, E. and Newton, C.W. 1969: *Atmospheric circulation systems*. New York and London: Academic Press.

cyclostrophic A term which relates to the balance of forces in atmospheric systems in which the flow is tightly curved, e.g. near the centre of a hurricane. In this case the centrifugal force is substantially larger than the CORIOLIS FORCE and the cyclostropic wind (V) is:

$$V = \frac{P_n^{1/2}}{R_T}$$

where P_n is the horizontal pressure gradient force and R_T the local radius of curvature of the isobars.   RR

cyclothem A uniform sequence of sedimentary strata repeated several times through a stratigraphic succession and indicative of repetitive cycles of sedimentation under similar environmental conditions.

cymatogeny The warping of the earth's crust over horizontal distances of tens to hundreds of kilometres with minimal rock deformation, producing vertical movements of up to thousands of metres. The term, introduced by L.C. King in 1959, describes crustal movements intermediate between EPEIROGENY and OROGENY and applies not only to the formation of broad domal uplifts but also to the linear vertical movements represented by mountain ranges such as the Andes. Uplift is assumed to be induced by vertical movements associated with processes active within the earth's MANTLE and not to arise from the large-scale horizontal movements proposed in the PLATE TECTONICS model.   MAS

Reading and Reference
King, L.C. 1959: Denudational and tectonic relief in southeastern Australia. *Transactions of the Geological Society of South Africa* 62, pp. 113–38.

†—1967: *The morphology of the earth*. 2nd edn. Edinburgh: Oliver & Boyd; New York: Hafner.

D

dalmatian coast A coastline characterized by chains of islands running parallel to the mainland, deep bays and steep shorelines, being the product of subsidence of an area of land with mountain ridges running parallel to the coast.

Daly level Named after R.A. Daly, a student of coral reef development who first put forward the theory that coral reefs were affected by cold periods during the Pleistocene when colder temperatures prevailed and there were lower sea levels. Daly suggested that during these phases the coral would be destroyed and wave-cut platforms, formed by marine planation, would develop on the dead coral. These platforms are, according to Daly, evidenced by present-day lagoon floors. Subsequent sea-level rise would encourage regrowth of corals at the edges of these platforms. If Daly's theory is correct lagoon floors should be flat and occur at similar levels throughout the world. HV

Reading
Stoddart, D.R. 1973: Coral reefs: The last two million years. *Geography* 58, pp. 313–23.

dambo A linear depression without a well-marked stream channel which is characteristic of old, relatively gently sloping landsurfaces, especially in the tropics. Frequently forming reticulate networks, they are often marked by the presence of termitaria, small pan-like depressions, and different vegetation assemblages from the surrounding woodland and savannah. They have been recognized especially in Africa, notably in Sierra Leone, Ghana, Nigeria, Zambia, Zimbabwe and Mozambique (Mäckel 1974). ASG

Reading and Reference
Boast, R. 1990: Dambos: a review. *Progress in physical geography* 14, pp. 153–77.
Mäckel, R. 1974: Dambos: a study in morphodynamic activity on the plateau regions of Zambia. *Catena* 1, pp. 327–65.

dams The damming of rivers by artificial structures to create reservoirs has been one of the most dramatic and widespread deliberate impacts that humans have had on the natural environment. Such structures change river hydrology, sediment loads, riparian vegetation, patterns of aggradation and erosion, the migration of organisms, seismic activity, etc. ASG

Reading
Petts, G.E. 1984: *Impounded rivers – perspectives for ecological management.* Chichester: Wiley.

Darcy's law Defines the relationship between the discharge of a fluid through a saturated porous medium and the gradient of HYDRAULIC HEAD (figure 1). It is the most important law of GROUNDWATER hydrology.

Henri Darcy was a French engineer. In 1856 he published the results of an experiment undertaken to determine the nature of water flow through sand (see figure 2). He found the outflow discharge from the sand to be directly proportional to the loss in hydraulic head (h) after a given length (l) of flow through the sand. This relationship may be written as:

$$v = -K\frac{dh}{dl}$$

where v is the specific discharge, dh/dl is the hydraulic gradient, and K is a constant of proportionality known as the hydraulic CONDUCTIVITY. Thus if dh/dl is held constant, $v \propto K$.

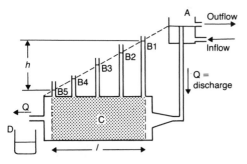

Darcy's law: 1. *Apparatus showing the relationships expressed in Darcy's law.*
A, *Constant head device*; B, *manometers*; C, *porous medium being measured:* D, *outflow, with discharge Q.*
Note the change in head h *with distance* l.
Source: *D.I. Smith, T.C. Atkinson and D.P. Drew. In T.D. Ford and G.H.D. Cullingford eds 1976: The science of speleology. London and New York: Academic Press. Figure 6.2.*

The specific discharge is sometimes termed the macroscopic velocity, the Darcy velocity or Darcy flux. It is the volume rate of flow through any cross-sectional area perpendicular to the flow direction (Freeze and Cherry 1979). It has the dimension of a velocity, but it should be clearly distinguished from the microscopic velocities of water passing through individual intergranular spaces during flow through the porous medium.

Darcy's law is valid for groundwater flow in any direction, including when it is being forced upwards against gravity in circuitous groundwater flow paths. It may also be used to describe the flow of moisture in soil. However, there are limits to its validity. Freeze and Cherry (1979) point out that if it were universally valid, a plot of specific discharge v against hydraulic gradient dh/dl

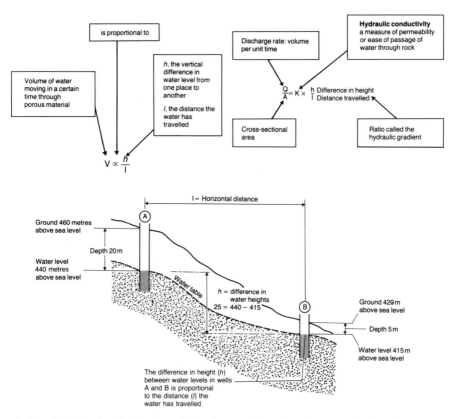

Darcy's law: 2. *The law formulated by Darcy is given in terms of the volume of water moving through any opening in a given amount of time, essentially a velocity term, and the geometry of the general flow, or the ratio of the vertical to the horizontal distance. Darcy reasoned that the permeability of a rock is what more or less slowed down the flow for a given drop of height (h) in a certain distance (l) and so made this into an equation by multiplying the right-hand side by a proportionality factor, called K. Darcy identified K as a measure of the permeability of the rock, or in other words, how easily it transmits water. From this equation, we can either determine the velocity of flow or, knowing the velocity, the hydraulic conductivity.*
Source: *F. Press and R. Siever 1978: Earth. San Francisco: Freeman. Box 6-1.*

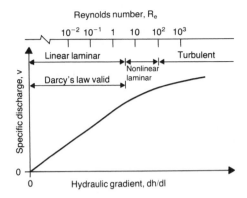

Darcy's law: 3. *Range of validity.*
Source: *Freeze and Cherry 1979.*

Reading and Reference
†Castany, G. 1982: *Principes et méthodes de l'hydrogeologie.* Paris: Dunod.
Freeze, R.A. and Cherry, J.A. 1979: *Groundwater.* Englewood Cliffs, NJ: Prentice-Hall.
Todd, D.K. 1980: *Groundwater hydrology.* 2nd edn. New York and Chichester: Wiley.

would reveal a straight line relationship for all gradients between zero and infinity. This is not the case. There appear to be both upper and lower limits to its validity.

Darcy's law is only applicable under conditions of laminar flow. But at the upper limits to laminar flow where the Reynolds number is in the range of 1 to 10, the law breaks down (see figure 3). Hence in the non-linear laminar flow regime, Darcy's law is not valid. Evidence is less conclusive at the lower limit, but some work suggests that there may be a threshold hydraulic gradient below which flow does not occur.

The above equation describing Darcy's law is one-dimensional in form, but it may also be developed to describe three-dimensional flow (Freeze and Cherry 1979). In three-dimensions, specific discharge v is a vector with components v_x, v_y and v_z. Hydraulic conductivity K may not be the same in each direction. Thus a three-dimensional generalization of Darcy's law may be written as:

$$v_x = -K_x \frac{dh}{dx}$$

$$v_y = -K_y \frac{dh}{dy}$$

$$v_z = -K_z \frac{dh}{dz}.$$

PWW

Darwinism The biological theory of evolution by natural selection as propounded by Charles Darwin (1809–1882) and set out in *Origin of Species* (1859) and *The Descent of Man* (1871). In demonstrating the mutability of species Darwin succeeded in setting the whole idea of scientific enquiry free from theological constraints. Thus the biblical notion of successive deluges modifying the landscape (CREATIONISM) was superseded by a theory of 'evolution' whereby random variations in fauna and flora would be selectively preserved and inherited by subsequent generations. Darwinism is sometimes taken to refer to any kind of evolution (the term did not appear in *Origin of Species* until the fifth edition), indeed to any kind of evolutionary theory which relies on the natural selection principles but rejects the doctrine of inheritance of acquired characteristics as suggested by Lamarck. Specifically, however, the idea of adaptation to the environment and selective change at the species level revolutionized the scientific community of the nineteenth century so much that the development of geography as a science became possible. Darwin's theory allowed the disciplines geomorphology, pedology, ecology and natural history more generally to calculate more *time* in which the sequential progression of development occurred; dispelled were the time-limited ideas of Bishop Ussher and the Anglican Church that the world was a divine creation formed 4004 BC. The pre-Darwinian ideas of landform studies in the Linnaeus taxonomic form produced only confusion in the early nineteenth-century scientific community. This was due in the main to the failure of taxonomic organization to provide any unifying principle which would allow scientists to order the myriad of landform types which had been recognized.

W.M. Davis, utilizing the Darwinian principle of *evolution* through time, provided a cycle of landscape evolution from youth to maturity and finally to old age. This was a direct analogue of the Darwinian idea of a plant or animal undergoing sequential change through time. Such change embodies a second important quality of Darwinism, the idea of organization in change. The importance of this in nineteenth-century scientific thinking was the rejection of the preconception of Platonic thinking of immutability of form. Dominant in the rationale of geologists at this time was the idea of changeless ideas or forms. Indeed it was generally held that change and variation were no more than illusions and that genuine reality was of fixed types permanently distinguished from one another. Such ideas delayed the recognition of evolution in nature and specifically prevented the science of ecology from developing. Indeed the idea of the inter-relationship between fauna and flora and their environment is a basic tenet of ecological understanding. Darwinism became the underlying principle of the subject.

Darwinism also provides geography with the idea of struggle and selection. Cause–effect relationships preoccupied pre-Darwinian thought at the expense of *process*. Subsequent Darwinian disciples have stressed the importance of the environmental influence. In a broad sense these 'Darwinian impacts' which include Social Darwinism, Darwinism and its influence upon politics, theology, philosophy, psychology, anthropology, literature and even music all affect our general viewpoint within the specific field of physical geography. All these facets of the 'Darwinian revolution', as it is now called, relied initially (as did geology and geomorphology) on the idea of evolution. Darwin's important message, which lay relatively unheeded until the 1930s, was the idea of randomness or chance variation in nature. Indeed natural selection of species was effected (according to Darwin) by this mechanism of chance.

Many objections to Darwinism have been raised; the most serious concerned the method of inheritance. In 1867 Jenkin pointed out that favourable variations in species would soon disperse when inter-bred with the 'normal' non-variant types within that species. This damaging comment could, however, have been answered with reference to the work of Mendel who recognized the structure of inheritance, but Darwin however, reverted to a Larmarckian stance, a theory which he termed Pangenesis.

Darwinian ideas of evolution were in general widely accepted as a doctrine, but the natural selection idea was rejected. By the end of the nineteenth century only two important scientists, August Weismann and Alfred Russel Wallace, believed in the random or chance variation idea. Weismann argued that inheritance of acquired characteristics was wrong, indeed impossible and Wallace, the often overlooked co-originator of Dawinism, maintained an unswerving loyalty to the doctrine. With Mendel's work and subsequent development of genetics, Darwin's theory remains the only plausible explanation of life. In the geographical sciences, however unintended, Darwinism provides evolution through time, organization and relationships between plants and animals and the environment, struggle and selection and, to a limited extent, chance and randomness. Generally, Darwinism provides the science of geography with a framework free of the static view of nature and landforms, free of the Linnaean taxonomic viewpoint and free of the Greek view of changeless ideas and forms. Alfred Russel Wallace wrote of Darwin in *Natural selection and tropical nature* (1895):

Nature and Nature's laws lay hid in night. God said, 'Let Darwin be' and all was light. PAB

References

Darwin, C.R. 1859: *On the origin of species by means of natural selection; or, the preservation of favoured races in the struggle for life.* London: John Murray.

— 1871: *The descent of man and selection in relation to sex.* London: John Murray.

Green, J.C. 1980: The Kuhnian paradigm and the Darwinian revolution in natural history. In G. Gutting, ed., *Paradigms and revolutions.* Notre Dame, Indiana: University of Notre Dame Press.

Oldroyd, D.R. 1983: *Darwinian impacts.* Milton Keynes: Open University.

Stoddart, D.R. 1966: Darwin's impact on geography. *Annals of the Association of American Geographers* 56, pp. 683–98.

Wallace, A.R. 1895: *Natural selection and tropical nature.* London: Macmillan.

daya A small, silt-filled solutional depression found on limestone surfaces in some arid areas of the Middle East and North Africa.

Reading
Mitchell, C.W. and Willimott, S.G. 1974: Dayas of the Moroccan Sahara and other arid regions. *Geographical journal* 140, pp. 441–53.

dead ice topography See STAGNANT ICE TOPOGRAPHY.

débâcle The breaking up of ice in rivers in spring.

deciduous forest An area in which the dominant life form is trees, whose leaves are shed at a particular time, season or growth stage. Deciduousness is a protective mechanism against excessive transpiration and represents an alternative strategy to being evergreen. Leaf shedding occurs most commonly in either cold or dry conditions, when the availability of soil water to roots is reduced. If transpiration were to continue unchecked during these periods, deciduous trees with a full leaf canopy would suffer serious water deficiencies. During the resting season buds are enclosed in tough, protective scales, but the timing of bud formation varies for each different tree species. PAF

Reading
Cousens, J. 1974: *An introduction to woodland ecology.* Edinburgh: Oliver & Boyd.
Reichle, D.E. ed. 1973: *Analysis of temperate forest ecosystems.* New York: Springer-Verlag.

décollement A feature resulting from the detachment of strata from underlying beds during folding, with the result that the upper strata slip forward.

decommissioning The process of dismantling and disposing of old nuclear reactors and other integral components of nuclear plant following final reactor shutdown. As the first generation of nuclear reactors (e.g. from nuclear submarines and Britain's Magnox power stations) come to the end of their operational lives, the costs and problems of decommissioning take on increasing significance. The actual process of dismantling reactors will be delayed by the presence of high levels of radioactivity, so that one solution is to entomb reactors under a mound of sediment until radioactivity levels have decayed to a point where dismantling can more safely take place. Large amounts of nuclear waste will need to be disposed of. The events involved in decommissioning are as follows (from Mounfield 1991, p. 368):

(a) defuelling and shipment off the site of all spent fuels, a major task that could take several years for a gas-cooled reactor;
(b) decontamination (chemical cleaning) of accessible pipework and, if necessary, of parts of the site;
(c) packaging and removal of operational solid radwastes;
(d) dismantling and removal of all buildings and structures outside the biological shield, including contaminated boilers, circulatory systems and fuel ponds;
(e) 'mothballing' or 'entombment' of the reactor for a period of time (determined partly by the decay rate of cobalt-60);
(f) dismantling and removal of the reactor itself and the biological shield;
(g) site decontamination to enable free entry and reuse. ASG

Reference
Mounfield, P. 1991: *World nuclear power.* London: Routledge.

decomposer An organism which helps to break down dead or decaying organic material, and so aid the recycling of essential nutrients to plant producers. Bacteria and small fungi are the major decomposer groups. Digestive enzymes are released from the bacterial cells or fungal filaments, which turn the organic material into a soluble form capable of being ingested, and at the same time CO_2 and H_2O are released back into the environment. In natural ecological systems, many thousands of decomposer organisms ensure the efficient operation of the detritus FOOD CHAIN. DW

deep weathering The rotting and decomposition of rocks at the earth's surface to depths of several tens of metres (or more) under the influence of percolating meteoric water and other surface processes. The products of deep weathering are unlikely to accumulate if rates of erosion are high, and they tend to be a feature of old erosion surfaces in the tropics. They may be associated with DURICRUSTS.

deforestation The removal of trees from a locality. This removal may be either temporary or permanent, leading to partial or complete eradication of the tree cover. It can be a gradual or rapid process, and may occur by means of natural or human agencies, or a combination of both.

Spurr and Barnes (1980) enumerate the major causes of deforestation. Natural tree removal is of relatively little significance on a global scale. Its mechanisms often lead to partial and temporary clearance that is followed by secondary succession, as a result of which forest develops again. The major natural cause of tree removal is fire resulting from lightning strike. Such burns are an essential part of certain forest ecosystems (e.g. in some types of pine forest). Gales may cause trees to be broken or uprooted in what is termed windthrow. Disease can also lead to the elimination of forest trees. Native animals (e.g. elephants in savannah woodland) can also damage trees by removing foliage and bark, and by trampling and uprooting them. Temporary severe weather (such as cold or drought) can lead to tree death (DIE-BACK), while secular modifications of climate may contribute to deforestation. In the latter instance, several millennia of reduced temperatures and increased precipitation may, for example, accelerate the accumulation of soil organic matter and inhibit forest growth by preventing tree regeneration.

The principal cause of deforestation is human activity. Forests are often permanently cleared on a large scale for a variety of agricultural and urban–industrial purposes using cutting and burning techniques. Some 33 per cent of the biosphere ($c.4 \times 10^9$ ha) is at present forested, compared with approximately 42 per cent about a century ago. Although this represents a pronounced recent decline, it is merely part of a process which has been operational in certain parts of the world for millennia.

As Lamb (1979) notes, while some trees continue to be removed, major deforestation is no longer a feature of most developed countries. There are some exceptions (e.g. parts of the boreal coniferous forests), but, in general, these areas have already been effectively deforested, and their emphasis is now upon woodland preservation. In north-west Europe, for instance, postglacial pollen records indicate the presence of an extensive mixed deciduous forest from $c.7–5 \times 10^3$ years BP, after which time it was progressively cleared by human activities. The forest soils, also found fossil (PALAEOSOLS), were fertile, and hence enticing to prehistoric and later agriculturalists.

Some developed countries have a considerably shorter history of major deforestation. These (such as the USA, Australia and New Zealand) were settled by Europeans, who, together with their introduced animals, brought about substantial tree losses in hundreds rather than thousands of years.

This is illustrated by New Zealand which was over two-thirds forested at the time of European colonization, and now has less than a 20 per cent tree cover.

The most significant deforestation at present is in the less-developed countries. Tropical tree cover has been removed for millennia by native inhabitants as an aid to hunting, for fuel, itinerant agriculture and settlement. As population increases, the cleared area is getting larger, and secondary succession to forest rarer, with lasting changes being brought about in the vegetation. Tropical rain forest is also being subjected to the constant, organized, commercial removal of its hardwood timber for export to developed countries. If, as is increasingly the case, such felling is entire (clear), regeneration of similar forest vegetation is unlikely.

Forests are complex ecological structures. They represent the optimum sites for photosynthesis within the biosphere, and thus contain a substantial proportion of the earth's biomass. Moreover, they possess considerable biotic diversity, which makes them important gene pools (see GENE-COLOGY). Their photosynthetic activity means that they assimilate a considerable amount of atmospheric CO_2, the concentration of which is increasing as a result of fossil fuel burning. Increased CO_2 concentration may be a contributor to a global warming of climate (GREENHOUSE EFFECT). Forest removal could therefore ampify this trend, because less CO_2 will be taken up by the trees. Additionally, the burning of wood releases CO_2 to the atmosphere. Tree burning also depletes atmospheric oxygen, and destroys an important source of oxygen (see PHOTO-SYNTHESIS).

As well as destroying trees, deforestation eliminates dependent animal habitats in the forest ecosystem. Deforestation involving fire may also kill animal as well as plant life. As Tivy and O'Hare (1981) observe, tree removal causes changes in the light, temperature, wind and moisture regimes of an area. A greater quantity of light is able to reach the ground, where temperatures will also be increased. A consequence of this is an accelerated rate of organic matter decomposition. Both temperature ranges and windspeeds are greater after forest clearance. There is also a higher intensity of rainfall, and a lower evapotranspiration rate. Hence, more run-off occurs, as does enhanced leaching and soil erosion. Studies at the Hubbard Brook Experimental Catchment in the USA (Likens *et al.* 1977) have revealed details of the biogeochemistry (BIOGEOCHEMICAL CYCLES) resulting from deforestation. For example, compared with a forested area, run-off from one that was deforested increased, and had a higher concentration of dissolved matter. The increased run-off was principally the result of the diminution in evapotranspiration, while the greater quantity of dissolved matter was derived mainly from that not taken up by plants as nutrients, and from substances released by the higher rate of organic matter breakdown on the former forest floor.   RLJ

Reading and References
Lamb, R. 1979: *World without trees*. London: Wildwood House.

Likens, G.E., Bormann, F.H., Pierce, R.S., Eaton, J.S. and Johnson, N.M. 1977: *Biogeochemistry of a forested ecosystem*. New York, Berlin and Heidelberg: Springer-Verlag.

Richards, J.F. and Tucker, R.P. eds 1988: *World deforestation in the twentieth century*. Durham, North Carolina: Duke University Press.

Spurr, S.H. and Barnes, B.V. 1980: *Forest ecology*. 3rd edn. New York and Chichester: Wiley.

Tivy, J. and O'Hare, G. 1981: *Human impact on the ecosystem*. Edinburgh and New York: Oliver & Boyd.

deflation The process whereby the wind removes fine material from the surface of a beach or a desert. It is a process which contributes to the development of LOESS, DUST storms and some STONE PAVEMENTS.   ASG

deformation A general geological term for the disruption of rock strata, folding, faulting and other tectonic processes.

De Geer moraines Landforms of glacial deposition consisting of 'swarms of small ridges orientated perpendicular to the ice flow direction' (Larsen *et al.* 1991, p. 263). They are thought to form either marginally to an ice body by glacier pushing at the grounding line or subglacially by material being squeezed up into basal crevasses.

ASG

Reference
Larsen, E., Longva, O. and Follestad, B.A. 1991: Formation of De Geer moraines and implications for deglaciation dynamics. *Journal of Quaternary science* 6, pp. 263–77.

deglaciation The process by which glaciers thin and withdraw from an area. The usual cause is climatic amelioration

which reduces snow accumulation or increases ABLATION, but sea-level rise relative to the land can also increase calving and thus the rate of ablation. The literature is full of arguments about whether the dominant process of deglaciation is thinning (downwasting) or snout retreat (backwasting). Since both must occur together it is probably helpful to regard the relative dominance of one or other as varying according to glacier type, location and the nature of the climatic or sea-level change. (See also STAGNANT ICE TOPOGRAPHY.)

DES

degradation The lowering, and often flattening, of a landsurface by erosion.

degree day For a given day the difference between the mean temperature and a given threshold, normally 15.5°C for a heating degree day:

$$\text{degree days} = 15.5 - \frac{\left(\begin{array}{c}\text{maximum}\\\text{temperature} +\\\text{minimum}\\\text{temperature}\end{array}\right)}{2}$$

If the mean temperature is greater than 15.5°C, negative degree days can be used to estimate air conditioning needs. For space heating, two other formulae have been devised for the days when the maximum temperature is above 15.5°C.

If the daily maximum temperature is above 15.5°C by a lesser amount than the daily minimum temperature is below 15.5°C, then:

degree days $= \frac{1}{2}$ (15.5 − minimum temperature) $- \frac{1}{4}$ (maximum temperature − 15.5).

If the daily maximum temperature is above 15.5°C by a greater amount than the daily minimum temperature is below 15.5°C, then:

degree days $= \frac{1}{4}$ (15.5 − minimum temperature).

Generally speaking, degree days are used for heating purposes and ACCUMULATED TEMPERATURE is used for agricultural purposes.   JET

delayed flow The part of the streamflow which lies below an arbitrary cut-off line drawn on the hydrograph, representing the more slowly responding parts of the catchment.

The division between QUICKFLOW and delayed flow is usually made by a line which rises from the start of the hydrograph rise at a gradient of $0.55\ 1\ s^{-1}/km^2$. h until the line meets the falling limb of the hydrograph. This procedure was suggested by Hewlett (1961) as an arbitrary but objective basis of separating the hydrograph peaks associated with each storm. It was earlier proposed as a replacement for older methods of hydrograph separation (e.g. by Linsley *et al.* 1949) which were considered to have only a spurious physical basis (see BASE FLOW).

MJK

References

Hewlett, J.D. 1961: Soil moisture as a source of base flow from steep mountain watersheds. *US Department of Agriculture: Southeastern Forest Experimental Station paper* 132.

Linsley, R.K., Kohler, M.A. and Paulhus, J.L.H. 1949: *Applied hydrology.* New York: McGraw-Hill.

dell A small, well-wooded stream or river valley.

deltas Accumulations of river-derived sediment deposited at the coast when a stream enters a receiving body of water, which may be an ocean, gulf, lagoon, estuary or lake. Deltas result from the interaction of fluvial and marine (or lacustrine) forces. Sediment accumulations at the mouths of rivers are subject to reworking by waves and tidal currents. Development of deltas involves the progradation of river mouths and delta shorelines producing a subaerial deltaic plain surmounting delta-front deposits which have accumulated to seaward.

Deltas exhibit appreciable variability of morphology and depositional patterns, reflecting global, regional and local variations in such controlling factors as river discharge, marine energy, tidal regime and geological structure and dynamics (Coleman and Wright 1975; Wright 1978, 1982). Although processes such as ocean currents may be important in a particular case, it is generally agreed that the interaction of river, wave and tide regimes is the major factor influencing delta morphology and sediment types (Galloway 1975). The diagram shows a tripartite classification using selected major deltas to illustrate variations in deltaic characteristics

associated with the relative contribution of each process.

Deltas consist of subaqueous and subaerial regions. The former constitutes the foundation of a delta and can be subdivided into a prodelta sedimentary unit and a delta-front sedimentary unit. The prodelta is typically fine grained whereas the delta front often possesses a bar morphology consisting of coarser-grained material (Wright 1982). These bars may result from a number of processes operating at or adjacent to river mouths (Wright 1978). The subaerial portion can be divided into two subregions: a lower delta plain which extends to the inland limit of tidal influence, and an upper delta plain dominated by fluvial processes. Distributary channels, natural levees, overbank crevasse splays, tidal channels and flats, interdistributary depressions (lakes, swamps, marshes, lagoons and bays), beaches, cheniers and dunes are landforms which may be found on many deltas of the world. Over time the active portion of a delta dominated by riverine processes may be abandoned because of the development of a new distributary system. In the abandoned sector marine, estuarine and paludal processes will modify the primary delta landscape.

BGT

References

Coleman, J.M. and Wright, L.D. 1975: Modern river deltas: variability of processes and sand bodies. In M.J. Broussard ed., *Deltas: models for exploration*, Houston, Texas: Houston Geological Society.

Galloway, W.E. 1975: Process framework for describing the morphologic and stratigraphic evolution of deltaic depositional systems. In M.L. Broussard ed., *Deltas: models for exploration.* Houston, Texas: Houston Geological Society.

Wright, L.D. 1978: River deltas. In R.A. Davis Jr ed., *Coastal sedimentary environments.* New York: Springer;Verlag.

— 1982: Deltas. In M.L. Schwartz ed., *The encyclopedia of beaches and coastal environments.* Stroudsburg: Hutchison and Ross.

demoiselle A pillar of earth or other unconsolidated material that is protected from erosion by a capping boulder.

dendroecology The study of the width of the annual growth rings of trees in order to interpret specific ecological events that resulted in changes in a tree's ability to photosynthesize and fix carbon.

Today studies involve a variety of coniferous and deciduous tree species, along with some shrubs, in a variety of areas, particularly in arctic and alpine

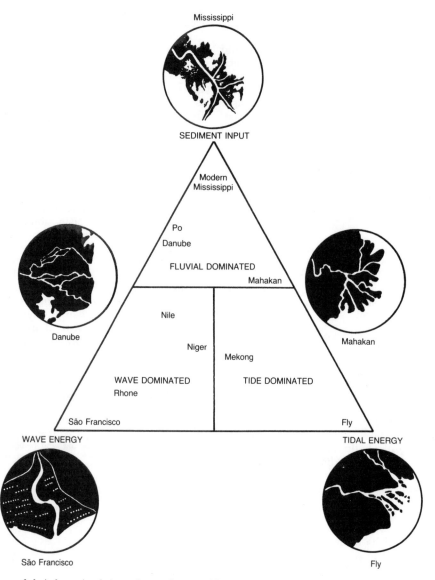

Delta *morphological types in relation to three environmental factors.*
Source: *Galloway 1975.*

areas where temperature limits tree growth, and in specific localities where events such as floods or glacial movements also restrain growth. The isotopic composition of the wood, rather than the width of the ring, has recently yielded information about past temperatures. Data provided by dendro-ecology are used by all sciences involved in palaeoecological reconstruction, e.g. geography, geology, archaeology, hydrology, forestry, biology, limnology, and ecology.

RHS

Reading
Douglas, P.E. 1919: Climatic cycles and tree growth. Vol. 1. *Carnegie Institute of Washington Publication* 289.
Fritts, H.C. 1976: *Tree rings and climate.* New York: Academic Press.

density The frequency of a phenomenon or mass of a substance per unit or volume. Stream density, for example, refers to the number of stream channels per unit area of a landscape.

density current A descending body of air or water with a high suspended sediment load. Turbidity currents on lake-bed and sea-floor slopes, clouds of falling volcanic ash and dust storms are all examples of density currents.

density dependence The action of environmental factors controlling the growth of populations of organisms, which vary with the density of the population. In contrast, factors that affect population size independently of the number of individuals present are described as density independent. Climatic and physical factors, such as floods or earthquakes, are usually classed as density independent factors. Resources (food, shelter) are often considered to be density dependent, as are biotic factors such as competition and parasites.

The relative importance of density dependent and independent factors in regulating population numbers has been a subject of considerable debate. Andrewartha and Birch (1954) argued that natural populations are controlled primarily by density independent factors and emphasized the ultimate role of severe climatic conditions. An alternative view stressed the importance of density dependent factors involving competition (Lack 1954). These differences in viewpoint may also be influenced by contrasts in the ecosystems analysed, which range from small organisms, such as insects, in arid areas in the case of Andrewartha and Birch, to larger organisms, such as birds, in temperate landscapes in the case of Lack.

The regulation of populations may often involve the interaction of density independent and dependent factors. Moreover, density dependent factors may sometimes vary in their effectiveness from one population density to another. In some cases predation may control population size at low prey densities but become ineffective at high densities.   ARH

Reading and References
†Andrewartha, H.G. and Birch, L.C. 1954: *The distribution and abundance of animals*. Chicago: University of Chicago Press.

†Lack, D. 1954: *The natural regulation of animal numbers*. Oxford: Oxford University Press.

†Odum, E.P. 1971: *Fundamentals of ecology*. 3rd edn. Philadelphia and London: W.B. Saunders. Ch. 7, pp. 162–233.

denudation Literally, the laying bare of underlying rocks or strata by the removal of overlying material. It is usually defined as a broader term than 'erosion', to include weathering and all processes which can wear down the surface of the earth. While some denudational processes may continue to operate below sea level, most discussions are concerned with subaerial denudation. Studies of other planets – notably Mars – make it clear that denudation is also a factor to be considered in the explanation of non-terrestrial topography (Gornitz 1979: Francis 1981).

The principal agent of subaerial denudation on the earth is water, as Playfair clearly recognized as long ago as 1802. Other important agents are related to stresses generated by pressure (atmospheric, gravitational and crustal); and the actions of organisms, including man. The impact of bodies such as meteorites, although of limited importance on the modern earth, assumes great significance on Mercury and the moon.

The rate at which denudation proceeds is fundamentally dependent upon two things: first, the intensity with which the different

Denudation caused by vegetation clearance and grazing in dry tropical forest on the eastern slopes of the Andes. Gullying is a characteristic feature of such badland scenery.

agents operate, singly or collectively; secondly, the ability of the ground surface and the underlying materials to withstand the stresses generated. It is this intimate connection between the nature of the applied force and the actual resistance of earth materials which makes it difficult to produce realistic generalizations about denudation rates, either over space or through time. However, four factors are clearly important in determining the denudational environment.

1 Surface geometry
As denudation ultimately depends upon the movement of material, the three-dimensional form of the surface is apt to constrain the operation of all processes which are in essence dependent on gravitational forces. Flat surfaces with little potential energy of elevation (i.e. close to local or regional BASE LEVEL) will lose mass very slowly, since only land ice, groundwater circulation, wind and organisms are effectively capable of removing material. In contrast, strongly dissected surfaces with steep slopes, which stand high above local base level, will expose large volumes of material to attack and should, other things being equal, produce rapid denudation.

2 Properties of materials
The chemical constituents of earth materials and the way in which they have been combined by igneous, sedimentary or metamorphic processes obviously guide the effectiveness of different denudational agents. The proportion of elements and compounds which are soluble in the weak acid solution of rainwater, and the way in which these components are disposed in relation to other constituents of the strata, will clearly govern the efficiency of solutional attack. The number of voids or pores, their sizes and geometries, will influence the internal frictional strength of the material, as well as governing the way in which water can move within the rock and the pressures which can be generated. The sheer weight of the strata, per unit volume, will produce stresses; and so forth. In reality, landsurfaces are made up of different materials, piled rather higgledy-piggledy, at various attitudes and in varying thicknesses. The actual operation of denudational agencies is, therefore, closely linked to the sequence and attitude of layers of material of different kinds (Selby 1993). The 'resistance' of a horizontal layer of impermeable, massive rock may be irrelevant, in denudational terms, if it is perched at the head of a slope composed of finely stratified rocks of low mechanical strength which dip steeply towards the slope surface and allow substantial flow of groundwater out onto the surface. By and large, thick, massive layers of dense, fine-grained rocks – such as quartzites or basalts – have the greatest all-round resistance to subaerial denudation; while thin layers of dipping or contorted, weakly cemented sedimentary or low-grade metamorphic rocks tend to prove most susceptible to a variety of attacks.

3 Tectonic setting
Both the geometry of the ground surface and the nature and attitude of the materials of which it is formed are, to a degree, dependent upon both the geological history and the present tectonic setting of any area. Large expanses of the stable crustal units known as cratons – such as the Canadian Shield, or western Australia – tend to be substantially flat surfaces, near base level and composed of old, resistant rocks: modern denudation rates are correspondingly low. In contrast, active mountain belts, where crustal movement and uplift are continuing, not only exhibit steep, high relief, but also tend to contain folded beds of sedimentary and metamorphic rocks: the Southern Alps of New Zealand and the Himalayas both provide vivid examples of the intense denudation which can result in such circumstances. However, the role of tectonism is more subtle and complex than this (Ollier 1981). Tectonic forces not only determine the pressure and temperature regimes under which strata are lithified, but also create changes in the environment of strata after lithification. These variations, in turn, affect the stability both of the minerals (Sparks 1971) and of the rock masses. Equally important, in much of the world, are the stresses generated by crustal movements. The role of earthquakes in denudation is easy to underestimate if one does not live in a region where they commonly occur; the magnitude and frequency of their effects in areas such as Papua New Guinea are formidable (Simonett 1967). Other things being equal, the intensity of modern tectonic activity tends to be directly related to the overall speed of denudation.

4 Climate

The distribution and variation in precipitation and insolation receipts and their interactions, together with the pattern of global and local wind systems, produce the moasic of climates which are of immediate and vital importance in controlling denudation rates. The climatic elements not only supply energy to the denudational agents directly, but also through their influence on plant and animal life and human activities help to control the detailed patterns of applied force and resistance at the ground surface. The climatic control over vegetation cover is of particular significance (Langbein and Schumm 1958; Schumm 1965): for any given temperature regime, there will be a precipitation level at which there is insufficient water available to maintain a complete blanket of vegetation, yet enough to cause substantial erosion. At lower precipitation totals, sediment yields decline, because of inadequate water supply; while at higher amounts, more complete vegetation cover inhibits denudation. A further major climatic control over denudation is linked to the distribution of ice sheets and glaciers: this depends not only on the overall global heat budget, but also on delicate local balances between precipitation and insolation receipts. At both levels the distribution of relief is also a very important factor. On the whole denudation is most rapid where the climate is semi-arid and warm; and slowest where it is either wet and cool, or dry and very cold. The removal or modification of vegetation cover, however, can make such a generalization highly misleading. In the absence of a complete cover of vegetation it is the warm, wet climatic zones which produce staggeringly high denudation rates, especially where slopes are steep.

The interaction of the four controlling factors outlined above is exceedingly complex, even under natural conditions. In the modern world, of course, denudation rates are being increasingly influenced on the local and regional scale by the pattern of human activities.

Many attempts have been made to produce global maps of denudation rates, two of the most comprehensive being those by Fournier (1960) and Strakhov (1967); these show interesting points of disagreement, especially in the rates they estimate. Both agree, however, that some of the highest relative rates of erosion are found in south-east Asia (hot, wet, tectonically active and very intensively altered by man) and some of the lowest in true arid areas – both hot and cold – as well as the flat, cold tundra areas of northern North America and Eurasia. The fact that these authors' estimates of actual rates for large regions often vary by at least an order of magnitude indicates the need for caution. A better sense of the way in which denudation actually proceeds can be gained by an examination of a classic local study, such as that by Rapp (1960).

At whatever scale, it is very important to realize that denudation is fundamentally *episodic* in its operation, rather than continuous. As the interaction of force and resistance is also non-uniform in space, denudation also tends to be concentrated in relatively discrete zones. To quote an 'average' denudation rate of x mm 1000 year^{-1} tends to give the false impression that very thin sheets of material are being pared off the whole landsurface at an infinitely slow, but steady rate. Only on the very longest of geological time-scales would the calculation of average rates seem to be appropriate (and even then, massive over-generalization must be borne in mind). On a historical scale, the emphasis is better focused on the identification of zones of high and low denudation and, increasingly, on the recognition of those areas where human intervention may cause abrupt acceleration of natural rates.   BAK

Reading and References

Fournier, F. 1960: *Climat et érosion: la relation entre l'érosion du sol par l'eau et les précipitations atmosphériques*. Paris: Presses Universitaires de France.

Francis, P. 1981: *The planets*. London: Penguin Books.

Gornitz, V. ed. 1979: *Geology of the planet Mars*. Stroudsberg, Pa.: Dowden, Hutchison & Ross.

†Langbein, W.B. and Schumm, S.A. 1958: Yield of sediment in relation to mean annual precipitation. *Transactions of the American Geophysical Union* 39, pp. 1076–84.

†Ollier, C.D. 1981: *Tectonics and landforms*. London: Longman.

Playfair, J. 1802: *Illustrations of the Huttonian theory of the earth*. London: Cadell & Davies.

Rapp, A. 1960: Recent development of mountain slopes in Kärkevagge and surroundings, northern Scandinavia. *Geografiska annaler* 42, pp. 73–200.

Schumm, S.A. 1965: Quaternary palaeohydrology. In H.E. Wright and D.G. Frey eds, *The Quaternary of the United States*. Princeton, NJ: Princeton University Press.

†Selby, M.J. 1993: *Hillslope materials and processes*. 2nd edn. Oxford: Oxford University Press.

Simonett, D.S. 1967: Landslide distribution and earthquakes in the Bewani and Torricelli Mountains, New Guinea. In J.N. Jennings and J.A. Mabbutt eds, *Landform studies from Australia and New Guinea*. Canberra: ANU Press.

Sparks, B.W. 1971: *Rocks and relief*. London: Longman.

Strakhov, N.M. 1967: *Principles of lithogenesis*, Vol. 1. Trans. from the Russian edn of 1964. Edinburgh: Oliver & Boyd.

denudation chronology An attempt by geomorphologists to reconstruct the erosional history of the earth's surface. The original definition of geomorphology that emerged in the US Geological Survey in the 1880s saw the new science of geomorphology as being in all essentials equivalent to what is now termed denudation chronology.

In the mid-nineteenth century planed-off surfaces had been identified by British geomorphologists in areas of complex structure and lithology, such as mid-Wales, while, with the exploration of the walls of the Colorado Canyon by Major Powell and his colleagues, great unconformities were recognized, leading to the concepts of BASE LEVEL and peneplain. In Europe, Suess postulated that planation surfaces might be susceptible to correlation on a worldwide basis as a result of worldwide (eustatic) changes of sea level in the geological past. Denudation chronology therefore arose as a prime focus of geomorphology, the aim of which was to use the study of erosional remnants to reconstruct the history of the earth where the stratigraphic record was interrupted or unclear. Techniques were developed to help in the identification of such erosional remnants, including superimposed contours, altimetric frequency curves, etc. (Richards 1981) and particular energy was expended on trying to fathom out whether surfaces were the product of marine or subaerial denudation. Crucial in such an analysis was the degree of adjustment of streams to structure; streams on subaerial peneplains were thought to be better adjusted than those developed on marine planation surfaces. Much of the evidence for denudation chronology was morphological, with all the implications that this has for its reliability (Rich 1938). The tectonic warping of small remnants rendered height correlation difficult. Supposedly accordant summits might have been greatly lowered by erosion, areas of flat ground might be susceptible to a whole range of different interpretations of their origin, and adjustments of streams to structure might be affected by a variety of tectonic factors, including antecedence (Jones 1980). The more successful attempts at denudation chronology were able to supplement the morphological evidence with information gained from deposits resting on the planation surfaces. In southern England, for example, Wooldridge and Linton (1955) were able to use the Lenham Beds to establish the presence of the supposed marine Calabrian Transgression of Plio-Pleistocene times. Other notable studies include those of Baulig (1935) in France, D.W. Johnson (1931) in the USA, and E.H. Brown (1960) in Wales. In the 1960s, as geomorphology became less concerned with evolution and more concerned with process studies, morphometry, and systems, considerable dissatisfaction was expressed about denudation chronology as a basis for the discipline (Chorley 1965) but with developments in plate tectonics, in our knowledge of the importance of Pleistocene events, in the amount of information that can be gained from a study of submarine deposits in basins like the North Sea, and with improvements in dating techniques, it remains a viable branch of study.   ASG

References

Baulig, H. 1935: *The changing sea level*. London: Philip.

Brown, E.H. 1960: *The relief and drainage of Wales: a study in geomorphological development*. Cardiff: University of Wales Press.

Chorley, R.J. 1965: The application of quantitative methods to geomorphology. In R.J. Chorley and P. Haggett eds, *Frontiers in geographical teaching*. London: Methuen. Pp. 148–63.

Johnson, D.W. 1931: *Stream sculpture on the Atlantic Slope: a study in the evolution of Appalachian rivers*. New York: Columbia University Press.

Jones, D.K.C. ed. 1980: *The shaping of southern England*. London: Academic Press.

Rich, J.L. 1938: Recognition and significance of multiple erosion surfaces. *Bulletin of the Geological Society of America* 49, pp. 1695–722.

Richards, K.S. 1981: Geomorphometry and geochronology. In A.S. Goudie ed., *Geomorphological techniques*. London: Allen & Unwin. Pp. 38–41.

Wooldridge, S.W. and Linton, D. 1955: *Structure, surface and drainage in south-east England*. 2nd edn. London: Philip.

denudation rates Provide a measure of the rate of lowering of the landsurface by erosion processes per unit time and are expressed in millimetres per 1000 years (mm 1000 year^{-1}) or in the direct equivalent of cubic metres per square kilometre per year (m^3 km^{-2} year^{-1}). These rates are commonly calculated using information on sediment (physical denudation) and solute yields (chemical denudation) from drainage basins, coupled with an estimate of soil or rock density. As such they are an index of the rate of denudation in the upstream catchment area. Maximum reported denudation rates are probably those for the island of Taiwan which exceed $10,000$ m^3 km^{-2} year^{-1} (Li 1976).

There are, however, several important problems in the derivation and interpretation of denudation rates based on measurements of river loads. For example, in the case of suspended sediment yields, it must be recognized that only a proportion of the eroded sediment will be transported to the basin outlet and that the associated denudation rate may be an underestimate. With the dissolved load, however, a large proportion may reflect non-denudational sources and should not be included in the calculation (Janda 1971). Furthermore, it may be unrealistic to convert values of river load to a uniform rate of surface lowering over the entire catchment and to assume that current river loads are representative of past conditions (Meade 1969).   DEW

References

Janda, R.J. 1971: An evaluation of procedures used in computing chemical denudation rates. *Bulletin of the Geological Society of America* 82, pp. 67–80.

Li, Y.H. 1976: Denudation of Taiwan Island since the Pliocene epoch. *Geology* 4, pp. 105–7.

Meade, R.H. 1969: Errors in using modern stream-load data to estimate natural rates of denudation. *Bulletin of the Geological Society of America* 80, pp. 1265–74.

depletion curve (or recession curve or base-flow recession curve) Represents gradual drainage of water from storage in a drainage basin and it is often possible to construct a master depletion curve for a particular site on a river system. This depletion curve will usually provide a very reliable means of predicting the decline of base-flow discharge at a site during dry conditions, although some drainage basins may exhibit seasonal variations in the form of the curve as a result of differences in loss of stored water through evapotranspiration (Federer 1973).

The depletion curve may be derived by producing a composite curve from the recession limbs of storm HYDROGRAPHS at a gauging site (see diagram) and one of a number of functions may be fitted to the depletion curve to describe it and to allow quantitative comparison of curves from different sites within a drainage basin, different drainage basins or to check the consistency of curves produced from different hydrograph recession limbs at the same site. One widely applied depletion function is:

$$Q_t = Q_o e^{-\alpha t}$$

where Q_o is flow at any time in the period of depletion or base flow, Q_t is flow after time t from flow Q_o, α is the recession coefficient, and e is the base of natural logarithms.

This function will plot as a straight line on semi-logarithmic graph paper and a modification of the method of depletion curve construction shown in the diagram is to plot the recession limbs on semi-logarithmic graph paper so that they define a straight line. However, in practice a perfect straight line is rarely found and this is probably partly a result of the fact that a number of stores are contributing to the depletion curve, all of which produce a recession flow at different rates.

The sources of discharge contributing to the depletion curve are likely to be very different in different drainage basins. Traditionally, it has been assumed that depletion flow is derived from effluent seepage from an aquifer and so the recession coefficient (α) has often been called the aquifer coefficient, implying that it is a simple parameter of the drainage

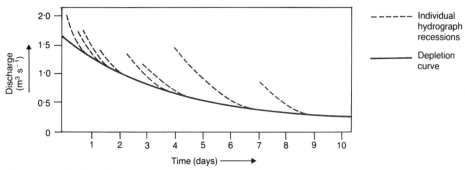

Construction of a depletion curve.

characteristics of a single aquifer. However, in practice such a simple situation is highly unlikely and the number of stores contributing to base flow will vary with the structure and size of the catchment. Many drainage basins are not underlain by efficient aquifers and yet they exhibit depletion flow which is largely generated from soil moisture storage. Higher discharges on a depletion curve are almost certainly produced by drainage from soil moisture and from the unsaturated zone of aquifers as well as from the saturated zone. Drainage basins underlain by more than one aquifer may experience effluent seepage from different aquifers at different points on the drainage network and may even lose flow by influent seepage at some locations. As a result, the depletion curve should not be expected to have a simple form because every store contributing to flow should have its own depletion curve which should produce a complex composite curve (see RECESSION LIMB OF HYDROGRAPH).   AMG

Reading and References
Bako, M.D. and Owoade, A. 1988: Field application of a numerical method for the derivation of baseflow recession constant. *Hydrological processes* 2, pp. 331–6.
Federer, C.A. 1973: Forest transpiration greatly speeds streamflow recession. *Water resources research* 9, pp. 1599–605.
Hall, F.R. 1968: Base-flow recessions – a review. *Water resources research* 4, pp. 973–83.
Linsley, R.K., Kohler, M.A. and Paulhus, J.L.H. 1988: *Hydrology for engineers*. 3rd edn. New York: McGraw-Hill.

depression See CYCLONE.

depression storage Consists of water trapped in small surface depressions or hollows during a rainfall event. Depression storage must be filled before overland flow can occur and this component of the total volume of rainfall will eventually be evaporated or will infiltrate the soil. (See also SURFACE DETENTION; SURFACE STORAGE.)   AMG

Reading
Sneddon, J. and Chapman, T.G. 1989: Measurement and analysis of depression storage on a hillslope. *Hydrological processes* 3, pp. 1–13.

depth–area curve Relates rainfall amount to the area covered by the rainfall. Average rainfall is inversely related to the area covered by that average rainfall amount and a depth–area curve may be defined for individual storm events or for all storm events of a particular duration in an area (see below). The depth–area curve is constructed using observations from a dense network of rain-gauges and by plotting the area enclosed by each isohyet against the mean precipitation within the isohyet.   AMG

Reading
Dunne, T. and Leopold, L.B. 1978: *Water in environmental planning*. San Francisco: W.H. Freeman.
Nicks, A.D. and Igo, F.A. 1980: A depth–area–duration model of storm rainfall in the southern Great Plains. *Water resources research* 16, pp. 939–45.
Shaw, E.M. 1988: *Hydrology in practice*. 2nd edn. New York: Van Nostrand Reinhold.

depth–duration curve Relates the magnitude of rainfall to its duration. This type of curve is usually constructed to relate the magnitude of extreme rainfall events to their duration at a single site or over an area.   AMG

Reading
Dunne, T. and Leopold, L.B. 1978: *Water in environmental planning*. San Francisco: W.H. Freeman.
Niemczynowicz, J. 1982: Areal intensity–duration–frequency curves for short term rainfall events in Lund. *Nordic hydrology* 13, pp. 193–204.
Shaw, E.M. 1988: *Hydrology in practice*. 2nd edn. New York: Van Nostrand Reinhold.

desalinization (or desalination) The production of freshwater from saline brines, especially seawater, by distillation or any other process.

desert An area in which the vegetation cover is sparse or absent and where the ground surface is thus exposed to the atmosphere and the associated physical forces. Generally deserts are areas where precipitation is small in amount and of infrequent and irregular occurrence. There is a deficit of moisture because potential losses from evapotranspiration exceed the moisture received from precipitation. One classification of deserts based on moisture deficit was that of Thornthwaite (1948) who employed a moisture index where:

$$\frac{s - 0.6d}{e} \times 100$$

where s is the sum of monthly surpluses of precipitation above estimated potential evaporation, d is the sum of monthly deficits, and e is the estimated annual evaporation based on mean monthly values of temperature, with adjustment for season of rainfall and including a factor for soil-moisture storage. Under this scheme areas

A crusted and cracked desert surface in the Atacama showing raindrop impact pits. The flowers are of Nolana plants. Clay crusting can make desert surfaces relatively impermeable and cause substantial run-off generation.

with a moisture index of between −20 and −40 were termed semi-arid, and those with below −40 arid.

The nature of deserts is not solely determined by climatic conditions for, like other major climatic zones, great differences are caused by changes in rock type, tectonic conditions, etc. ASG

Reading and Reference
Cooke, R.U. Warren, A. and Goudie, A.S. 1993: *Desert geomorphology*. London: UCL Press.

Goudie, A.S. and Wilkinson, J.C. 1977: *The warm desert environment*. Cambridge: Cambridge University Press.

Thornthwaite, C.W. 1948: An approach toward a rational classification of climate. *Geographical review* 38, pp. 55–94.

desert pavement See STONE PAVEMENT.

desert varnish A thin patina (just a few microns thick) of shiny dark stain that develops on rocks in arid areas and some other environments. It is composed primarily of iron and manganese oxides which may either be derived from *in situ* weathering of the underlying rock or a result of inputs (probably aeolian) of materials from outside the rock. Although geomorphologically insignificant, it may have some potential for dating purposes. ASG

Reading
Whalley, W.B. 1983: Desert varnish. In A.S. Goudie and K. Pye eds, *Chemical sediments and geomorphology*. London: Academic Press.

desertification (desertization) Can be defined, following Rapp (1974), as 'the spread of desert-like conditions in arid or semi-arid areas, due to man's influence or climatic change'. In this definition lies one of the problems of desertification – its cause. The question has been asked whether this process is caused by temporary drought periods of high magnitude, some long-term climatic change towards aridity (see DESICCATION), man-induced climatic change, or the result of human

Desert varnish developed on a rock slope in the dry Hunza valley in the Karakorams of Pakistan. Note the contrast in tone between the varnish and the rock underlying it as revealed by the petroglyphs.

action through man's degradation of biological environments in arid zones. Interminable conferences have been held on this theme and it is now generally accepted that it is largely a combination of man's activities with occasional runs of dry years that leads to presently observed desertification. Depletion of vegetation cover, by firewood collection, overgrazing and cultivation, has set in train such insidious processes as water erosion and deflation, and these in turn limit the ability of the environment to support ever increasing human population levels. If a drought occurs the environment is even less able to support population. The problem of desertification was brought sharply into focus by the severe problems encountered in the Sahel Zone of West Africa in the late 1960s and early 1970s (United Nations 1977).   ASG

Reading and References
Goudie, A.S. ed. 1990: *Techniques for desert reclamation.* Chichester: Wiley.

Grainger, A. 1990: *The threatening desert: controlling desertification.* London: Earthscan.

Rapp, A. 1974: A review of desertization in Africa – water, vegetation and man. *Secretariat for International Ecology, Stockholm* report 1.

United Nations 1977: *Desertification: its causes and consequences.* Oxford: Pergamon.

desiccation A concept that has been current at least since the mid-nineteenth century. It is the belief that portions of the earth's surface are drying up, leading to the spread of deserts, the depletion of groundwater reserves, the dwindling of rivers, and the decline of settlements. By some it has been regarded as the result of progressive climatic deterioration in postglacial times, while others have seen it as a result of human mismanagement of the environment.   ASG

Reading
Goudie, A.S. 1972: The concept of post-glacial progressive desiccation. *Oxford School of Geography research papers series* 4.

design discharge The discharge which a structure or development is designed to resist or to cope with. A dam across a river must be designed to retain a flood of a particular size and a flood prevention scheme will also be designed to convey a flood of a particular magnitude. Although very large recent events may provide the experience against which schemes may be designed, a design discharge is usually

selected by reference to a specific RECURRENCE INTERVAL and this has to be such that it will provide a reasonable expectation of protection and yet not be too costly. Therefore, for land drainage works or for flood prevention schemes, estimated recurrence intervals in the range 50–75 years, or occasionally 100 years, may be used.   KJG

desquammation Onion-weathering. The disintegration of rocks, especially those in desert areas, by peeling of the surface layers.

Devensian See LATE GLACIAL.

dew, dewpoint Dew is formed as a condensate from moist air which is cooled by contact with a surface which loses heat by RADIATION. It occurs once the chilling has lowered the air's TEMPERATURE to its dewpoint which is the temperature at which an air sample becomes saturated by cooling at constant PRESSURE and absolute HUMIDITY. The numerical value of dewpoint is obtained by using humidity tables in conjunction with measurements of dry-bulb and wet-bulb temperature.   RR

Monteith, J.L. 1957. Dew. *Quarterly Journal of the Royal Meteorological Society* 83, 322–41.

diabatic See ADIABATIC.

diachronous Pertaining to a sedimentary unit of a single facies which belongs to two or more units of geological time.

diaclinal Describes those rivers whose courses cross the strike of geological structure at right angles.

diagenesis Post-depositional changes which have altered a sediment, particularly cementation and compaction.

diamictite, diamicton Terms proposed by the American geologist R.F. Flint for non-sorted terrigenous deposits and rocks containing a wide range of particle sizes, regardless of genesis. Examples of diamictites include till and mudflow deposits.

diapir An anticlinal fold which has resulted from the upward movement of mobile rocks, such as halite, lying beneath more competent strata. Sometimes a surface dome produced by such movements.

diastrophism Tectonic processes which produce dramatic changes in the shape of the earth's surface, such as orogenies, faulting and folding.

diatreme The general term for vents and pipes which have been forced through sedimentary strata by the forces of underlying volcanism. Kimberlite pipes are examples of diatremes.

die-back Mortality beginning at the extremities of plants. One or more phenomena lead to stress which manifests itself in the decline and death of leaves, shoots and roots. Plants susceptible to die-back are often woody and possess restricted environmental tolerances. In deciduous trees, a characteristic first sign of it is the premature discoloration and fall of leaves on outermost banches, with no subsequent leaf regrowth in that area (Norton 1985).

There appears to be a number of possible causes of die-back. Extreme and lasting drought may initiate it, as seems to have occurred in the shallow-rooting sugar maple (*Acer saccharum*) in the eastern United States during pronounced dry spells of the first, third and sixth decades of this century (Westing 1966). Recent die-back of sugar maple groves in Ontario, Quebec and Vermont could be linked to ACID PRECIPITATION (Norton 1985). Drought may also heighten the risk of disease which leads to die-back. For instance, fungi which do not harm white ash (*Fraxinus americana*) growing under normal conditions have probably been responsible for canker formation during dry spells in New York State over the past half-century (Silverborg and Ross 1968). Water abundance may be related to die-back. For example, salt-marsh substrate in southern England possessed reducing conditions that caused the build-up of sulphide, the toxic effect of which may have led to die-back in cord-grass (*Spartina townsendii*) (Goodman and Williams 1961).

Progressive die-back lowers the resistance of plants to other environmental factors (insect attack, for example) unfavourable to their survival, and thus can combine with such factors to bring about death (Norton 1985).   RLJ

Reading and References
Goodman, P.J. and Williams, W.J. 1961: Investigations into 'die-back' in *Spartina townsendii* agg. III. Physiological correlates of 'die-back'. *Journal of ecology* 49, pp. 391–8.

Norton, P. 1985: Decline and fall. *Harrowsmith* 9, pp. 24–43.
Silverborg, S.B. and Ross, E.W. 1968: Ash die-back disease development in New York State. *Plant disease reporter* 52, pp. 105–7.
Westing, A.H. 1966: Sugar maple decline: an evaluation. *Economic botany* 20, pp. 196–212.

diffluence The process of a glacier overflowing its valley into an adjacent one. Also the separation in airflow which occurs when an airstream decelerates.

diffusion equation A flow equation for *transient* flow through a saturated homogeneous, porous rock in which HYDRAULIC CONDUCTIVITY is the same in each direction. *Steady* flow through a porous medium requires the rate of fluid flow into a given volume of the rock to be equal to the rate of flow out of it. This is expressed mathematically by the equation of continuity:

$$-\frac{dv_x}{dx} - \frac{dv_y}{dy} - \frac{dv_z}{dz} = 0$$

where v_x, v_y, v_z are specific discharges in directions x, y, and z. If the porous rock is homogeneous and hydraulic conductivity is the same in each direction, then another equation incorporating DARCY'S LAW can be written to express the steady state saturated flow through it:

$$\frac{d^2h}{dx^2} + \frac{d^2h}{dy^2} + \frac{d^2h}{dz^2} = 0$$

where h is the HYDRAULIC HEAD and x, y, and z are a coordinate system defining position. This partial differential equation is known as *Laplace's equation*. It is incorporated into another equation termed the *diffusion equation*, in order to describe *transient* saturated flow through a porous medium with similar homogeneous properties, as follows:

$$\frac{d^2h}{dx^2} + \frac{d^2h}{dy^2} + \frac{d^2h}{dz^2} =$$

$$\frac{pg(\alpha + n\beta)}{K} \frac{dh}{dt}$$

where p is density, g is gravitational acceleration, α is the vertical compressibility of the aquifer, β is the compressibility of water, n is porosity, and K is hydraulic conductivity.

Since $p\,g\,(\alpha + n\beta) =$ specific storage S_s, the right-hand side of the equation may be simplified to $S_s/K\,dh/dt$.

The specific storage is the volume of water that a unit volume of aquifer releases from storage per unit decline in h. The solution $h\,(x, y, z, t)$ describes the value of the hydraulic head at any point in a flow field at any time (Freeze and Cherry 1979).   PWW

Reference
Freeze, R.A. and Cherry, J.A. 1979: *Groundwater*. Englewood Cliffs, NJ: Prentice-Hall.

digital image processing A REMOTE SENSING technique involving the handling and modification of images that are held as discrete units, e.g. the sampling, correction and enhancement of a Landsat/MSS image can be achieved by this method.

A discrete image comprises a number of individual picture elements known as pixels, each one of which has an intensity value and an address in two-dimensional image space. The intensity value of a pixel, which is recorded by a digital number (DN), is dependent upon the level of ELECTROMAGNETIC RADIATION received by the sensor from the earth's surface and the number of intensity levels that have been used to describe the intensity range of the image. There are three stages in the processing of a discrete image. First, the images which are stored on computer-compatible tapes are read into a computer; next, the computer manipulates these data; then results of these manipulations are displayed. These three stages can be performed using a range of computers; workstations and the larger personal computers are currently the most popular among physical geographers.

There are many techniques for the processing of digital images, and physical geographers tend to concentrate on six of them: image restoration and correction, image enhancement, data compression, colour display, image classification and the development of geographic information systems.

1 Image restoration and correction form the first stage in any image-processing sequence and include, first, the restoration of the image by the removal of effects whose magnitudes are known, like the non-linear response of a detector or the curvature of the earth and, secondly, the correction of the image by the suppression of effects whose magnitudes can only be estimated, such as atmospheric scatter or sensor wobble.

2 Image enhancement involves the 'improvement' of an image in the context of a particular application. The most popular image enhancements are the selective increase in image contrast (stretching), ratioing wavebands against each other to display differences between wavebands, and digital filtering to smooth or sharpen edges within an image.

3 Data compression involves the reduction of many images into one image for ease of interpretation.

4 Colour display involves the combination of images with colour, again for ease of interpretation.

5 Image classification can be achieved by several techniques, notably the density slicing of one image or the supervised classification of several images. Density slicing involves the grouping of image regions with similar DN, either automatically or interactively. Supervised classification involves the careful choice of wavebands, the location of small but representative training areas, the determination of the relationship between object type and DN in the chosen wavebands, the extrapolation of these relationships to the whole image data set and the display and accuracy assessment of the resultant images.

6 Geographic information systems involve the combination and use of any spatial data that can be referenced by geographic coordinates. The three processing steps for this operation are, first, data encoding where spatial data are broken into polygons or grids; secondly, data management, where these data are spatially filed and, thirdly, data manipulation where these data are retrieved, transformed, analysed, measured, composited or modelled.   PJC

Reading
Curran, P.J. 1985: *Principles of remote sensing.* Harlow: Longman Scientific and Technical.

Mather, P.M. 1987: *Computer processing of remotely-sensed images: an introduction.* Chichester: Wiley.

Moik, J.G. 1980: *Digital processing of remotely sensed images.* Washington, DC: National Aeronautics and Space Administration.

Richards, J. 1986: *Remote sensing digital image analysis: an introduction.* Berlin: Springer-Verlag.

dikaka Accumulation of dune and sand covered by scrub or grass vegetation, extended to include plant-root cavities in dune sediments (calcified root tubules).

dilation (or dilatation) Describes the action of PRESSURE RELEASE in a rock mass by the removal of overlying material by erosional processes. Severe glaciation may cause dilation joints to open in glaciated terrain, while in granite areas the opened joints on many INSELBERGS may be a result of pressure release following the removal of the overlying sedimentary or metamorphosed rocks. The joints often approximately parallel the ground surface configuration.   ASG

dilution effect A term used to describe the behaviour of those solute concentrations in a stream which decrease during a storm run-off event. This decrease is ascribed to the dilution of solute-rich base flow by additional inputs of storm run-off which, in view of its shorter residence time within the drainage basin, possesses lower solute concentrations. Some solute concentrations in a stream may, however, increase during periods of storm run-off.   DEW

dilution gauging A method of measuring river discharge by introducing a tracer into the river channel and timing the passage of the tracer over a known length of channel. (See also DISCHARGE.)

Reading
Water Research Association 1970: *River flow measurement by dilution gauging.* Water Research Association technical paper TP74. Medmenham: Water Research Association.

White, K.E. 1978: Dilution methods. In R.W. Herschy ed., *Hydrometry: principles and practices.* Chichester: Wiley.

diluvialism The belief in the role of Noah's flood, as reported in the book of Genesis, in shaping the landscape. Before the true origin of glacial drift was recognized such materials were ascribed to a great deluge, when 'waves of translation' covered the face of the earth. The heterogeneous and unsorted character of the drift seemed ample proof that it had been laid down in the turbulent waters of a universal flood, and Dean Buckland of Oxford termed such material 'diluvium' to distinguish it from the 'alluvium' formed by rivers. By the 1830s recognition of the often complex stratigraphy of the drift, and the Ice Age, greatly weakened the diluvial viewpoint. As the catastrophic interpretation of landscape and geological history gave way to UNIFORMITARIANISM, diluvialism became obsolete.   ASG

Reading
Davies, G.L. 1969: *The earth in decay.* London: Macdonald.

dimensionless unit hydrograph See UNIT HYDROGRAPH.

dimethylsulphide (DMS) The most abundant volatile sulphur compound in seawater is produced by planktonic algae and bacterial decay. It oxidizes in the atmosphere to form a sulphate aerosol that is a major source of cloud-condensation nuclei. Because of this, DMS may have an important climatic impact through its impact on cloud albedo and the earth's radiation budget (Charlson *et al.* 1987). Increasing cloud production and albedo caused by increased planktonic productivity resulting from global warming could act as a negative feedback in the climate system. This is a hypothesis that merits careful consideration.   ASG

Reference
Charlson, R.J., Lovelock, J.E., Andreae, M.O. and Warren, S.G. 1987: Oceanic phytoplankton, atmospheric sulphur, cloud albedo and climate. *Nature* 326, pp. 655–61.

dip The angle between the inclination of sedimentary strata and horizontal.

dipslope The more gentle slope of a cuesta; the slope of the landsurface that approximates the dip of the underlying sedimentary rocks.

dirt cone A conical hill or dome of ice that is completely mantled by till or rock fragments. It owes its existence and form to the mantle of debris which retards the rate of surface lowering resulting from the melting of the ice.

disasters Geophysical events of an extreme nature which exact a heavy toll upon life and property. These are usually called 'natural disasters' or 'natural hazards' and refer to extreme events whose energy levels are such that human societies have no control over them. Predictability of these events is poor in the short and medium terms. Examples are those resulting from crustal processes, such as earthquakes, volcanic eruptions and subsequent tidal waves (tsunami); those associated with weather events, such as floods and landslides, wind damage from tornadoes and hurricanes; the results of unexpected heavy snowfalls and avalanches; those which are climatic in origin, such as periods of drought extending over several years; and fire in ecosystems which are inhabited by permanent residents, such as Californian chaparral and Australian bush. Research into the perceptions that many people have shows that they consider the risks to be lower than objectively calculated probabilities indicate and fail to adapt their structures and lives to them, e.g. earthquake-damaged towns are rebuilt on the same sites in the same materials. It is a moot point, therefore, whether 'natural hazard' is the right term since a large component of the disaster is caused by human behaviour, i.e. a failure to recognize the hazard and act accordingly. This is a facet of behaviour apparently not confined to natural hazards. IGS

Reading
Alexander, D., 1993: *Natural disasters*. London: UCL Press.
Burton, I., Kates, R.W. and White, G.F. 1978: *The environment as hazard*. New York: Oxford University Press.

discharge The volume of flow of water or fluid per unit time. It is usually expressed in cumecs which are cubic metres per second ($m^3 s^{-1}$) but for small discharges it may be more conveniently expressed as litres per second ($1 s^{-1}$). In Imperial Units the cusec (cubic feet per second) was originally used and is still employed in the USA. It is necessary to measure discharge and to obtain continuous records of discharge variation for the investigation of the HYDROLOGICAL CYCLE.

Discharge may be measured in a number of different ways and the method adopted at a particular gauging station will depend upon the size of the river, the stability of the channel, the variability of the flow and of the sediment transported, and the length and accuracy of the record required. The major methods of discharge measurement are:

1 *Volumetric gauging* involves collection of the total volume of flow over a specified period of time. It is the most accurate method but can only be used where it is easy to collect the discharge in a large container and to time the increases in water level. Can therefore be used to measure flow from small plots or experimental areas.

2 *Control structures* are structures installed in the cross-section of the stream channel which includes weirs and flumes. Both types of control structure have a formula which relates depth of water to discharge. A weir is a structure placed across the channel and may be sharp crested, in which case the plate inserted in the channel cross-section has a sharp edge on the V notch and the angle at the centre of the V may be 90° or 120° or other angles. Alternatively, the weir may be broad crested and this is preferred for large basins and has various forms which include flat-V and Crump types and often need a rating curve established to relate the depth of water above the weir and the velocity of water. Where the gauging station needs to measure a range of flows it may be necessary to construct a compound weir where a V notch may occur in the centre of a rectangular cross-section, for example. The flume is an artificial channel constructed by raising the channel bed into a hump or by contracting the sides of the channel, or by combining both. The cross-section of the flume is adapted to suit the range and magnitude of discharge and it may be rectangular, triangular or trapezoidal. Flumes have the advantage that silt and debris are easily carried through whereas it could collect upstream of a weir. Several types of flume exist and in a standing wave flume or critical depth flume there is a direct relationship between depth of water upstream of the throat of the flume and discharge. In other cases (Parshall and Venturi flumes) head is measured within and upstream of the throat and the difference between the two values is directly related to discharge.

3 *Velocity-area technique* is the most frequently used method for discharge measurement and depends upon the fact that discharge is:

$$Q = Va$$

where V is velocity and a is water cross-sectional area. The velocity is usually measured by current meter and this is used in each of several verticals across the channel and the spacing between the verticals should not exceed 5 per cent of the channel width. At many gauging stations the velocity and hence discharge is measured at a range of flows, and a rating curve is constructed relating discharge to depth of water or stage. This rating curve can then be the basis for converting continuous records of river stage into discharge values.

4 *Dilution gauging* is a method of discharge measurement depending upon calculation of the degree of dilution by the flowing water of an added tracer solution which may be sodium chloride (NaCl) or sodium dichromate (Na-Cr_2O_3). The tracer may be injected either at a constant rate or by gulp injection. In the latter case an amount of the tracer solution is introduced instantaneously into the stream and the passage of the 'Ionic Wave' or slug past a downstream site at a known distance downstream is measured usually using a conductivity meter.

5 *The slope–area method* of estimating discharge is effected by using a flow equation whereby velocity (V) can be estimated by surveying water surface slope (S), hydraulic radius (R) and estimating roughness (n) using an equation such as the Manning equation in which:

$$V = \frac{R^{\frac{2}{3}}S^{\frac{1}{2}}}{n}$$

6 *Electromagnetic gauging* is particularly useful where there is no stable-discharge relationship or where weed growth impedes flow. An electric current passed through a large coil buried beneath the river bed induces an electromotive force in the water and the force recorded by probes at each side of the river is directly proportional to the average velocity through the cross-section.

7 *Ultrasonic gauging* can be used where water flow is not hampered by vegetation or sediment. The time taken for acoustic pulses beamed from transmitters on one side of the river to travel to sensors on the other side is recorded and gives mean velocity at a specified depth.

Most of the above methods of discharge measurement will provide a value for a single moment. To obtain continuous records of discharge it is usual to employ a stage recorder which will give continuous records of water depth which can subsequently be converted to discharge values.

The discharge record obtained from a gauging station has to be expressed in a form which can be analysed in relation to controlling parameters. This can be done by establishing the general character of the discharge record and by calculating daily, monthly or annual flows for a specific period, usually a year. A further way is to calculate the total run-off volume for a specified period and this is usually expressed as depth of run-off from the entire catchment area and calculated by dividing the total volume of water which passes the gauging station by the surface area of the drainage basin. The run-off (R) can then be compared directly with precipitation (P) over the basin during the same time period and the ratio gives the run-off percentage ($R/P \times 100\%$). If the average flow is plotted for the year a diagram can be drawn to show the river regime. Such regime diagrams reflect the broad influence of climate and some river regimes will show a major concentration during a short period, for example in an area affected by snowmelt, whereas other areas will have a fairly uniform distribution of discharge throughout the year.

A discharge record may also be analysed for a short period and individual HYDROGRAPH events may be identified either by extracting specific parameters from a single hydrograph or by generalizing the hydrographs as UNIT HYDROGRAPHS.

Variation in river discharge reflects the pattern of climate over the basin, particularly the incidence and intensity of precipitation, the DRAINAGE BASIN CHARACTERISTICS and also the effects of change in the drainage basin including LAND USE changes. KJG

Reading
Gregory, K.J. and Walling, D.E. 1976: *Drainage basin form and process.* London: Edward Arnold.

Herschy, R.W. ed. 1978: *Hydrometry, principles and practices.* Chichester: Wiley.

disclimax A stable plant community resulting from the disturbance of CLIMAX VEGETATION. The word is normally used for communities disturbed to such an extent by the activities of man or domesticated animals that the former climax has been largely replaced by new species. The species may even be introduced, as in the case of the prickly pear cactus, which has formed a disclimax over wide areas in Australia. It is similar to the term plagioclimax, which has been used for many plant communities of the English landscape produced by grazing, cutting or burning over many centuries. (See also SERE; SUBCLIMAX; SUCCESSION.)   JAM

Reading
Oosting, H.J. 1956: *The study of plant communities.* San Francisco: W.H. Freeman.
Tansley, A.G. 1949: *The British Islands and their vegetation.* Vol. I. Cambridge: Cambridge University Press.
Vogl, R.J. 1980: The ecological factors that produce perturbation-dependent ecosystems. In J. Cairns Jr ed., *The recovery process in damaged ecosystems.* Ann Arbor: Ann Arbor Science.

disconformity An unconformity in a geological sequence which is not represented by a difference in the inclination of the strata above and below.

discordance An unconformity. A difference in the inclination of two strata which are contiguous.

dishpan experiments Laboratory experiments which use rotating, fluid-filled dishpans to simulate large-scale atmospheric motions. They usually take the form of a shallow annulus which rotates about its vertical axis at a rate which may be varied. The rim of the pan represents the 'equator' while the centre represents the 'pole' and the radial or 'latitudinal' temperature gradient may be changed by heating the rim and/or cooling the centre, for example.

The basic aim is to study how the fluid in the dishpan transports heat from source (rim) to sink (centre) and how the flow regimes change as the rotation rate and radial temperature difference are varied.

For a fixed thermal pattern a low rotation rate is characterized by a simple overturning convective cell (HADLEY CELL) with rising motion at the rim, inward motion at the top of the fluid, sinking at the centre and outward motion at the bottom of the fluid. This cellular circulation is a simple model of the tropical Hadley cell.

If the spin rate is increased the fluid motion breaks into unstable wave-like perturbations which play a crucial role in the inward heat transport. This turbulent motion in which transient EDDIES effect the transfer is called the Rossby regime (see ROSSBY WAVES) and is reminiscent of the disturbed regions of the extra-tropical latitudes.   RR

disjunct distribution A geographical distribution pattern in which two or more populations of an organism are exceptionally widely separated for the organism concerned, thus creating a major discontinuity, and may involve now isolated relict populations, exceptional long-range migration or the separation of populations through continental movement. The southern hemisphere beeches (*Nothofagus*), for example, are thought to have been formerly linked on one continental landmass, Gondwanaland, but today they occur widely disjunct in South America, New Zealand, Australia and New Guinea.   PAS

Reading
Stott, P.A. 1981: *Historical plant geography: an introduction.* London: Allen & Unwin.

dispersal The mechanism of migration, by which plants and animals are disseminated over the surface of the earth. In plants, dispersal involves the transport of any spore, seed, fruit or vegetative portion which is capable of producing a new plant in a new locality. These propagules or diaspores may be carried by air or water, on or inside animals, by the movement of soil or rock, or they may be exploded from the parent plant. In animals, dispersal depends on the mechanisms for movement, which may range from facilities for flying to swimming and running. For both plants and animals, man is a potent agent of dispersal. (See also ALIENS; MIGRATION.)   PAS

Reading
Seddon, B. 1971: *Introduction to biogeography.* London: Duckworth. Especially ch. 8.

dissection The destruction of a relatively flat landscape through incision and erosion by streams.

dissipative beach Beaches can be classified into two basic types: dissipative and reflective. A third type, termed intermediate, represents those beach states which contain elements of the two basic types (Wright *et al.* 1979, 1982).

Under dissipative conditions, incident waves break and lose much of their energy before reaching the beach face. Broken waves or dissipative bores form the resulting surfzone with bore height decreasing in amplitude towards the shore. Depending on incident wave height, the surfzone may be as wide as 500 m under fully dissipative conditions.

Dissipative beaches are characterized by a wide low gradient beach face extending from the foot of the foredune into the surfzone. Multiple bars or breaker zones may be present across the surfzone. Water circulation in the surfzone is dominated by strong onshore flow in the upper water column (alongshore if incident wave approach is oblique to the shoreline), and offshore towards the bed (Wright *et al.* 1982). BGT

References
Wright, L.D., Chappell, J., Thom, B.G., Bradshaw, M.P. and Cowell, P. 1979: Morphodynamics of reflective and dissipative beach and inshore systems: Southeastern Australia. *Marine geology* 32, pp. 105–40.
Wright, L.D., Guza, R.T. and Short, A.D., 1982: Dynamics of a higher energy dissipative surfzone. *Marine geology* 45, pp. 41–62.

dissolved load Material in solution transported by a river and including both inorganic and organic substances. There is no clear boundary between a true solution and the presence of material as fine colloidal particles and all material contained in a water sample which has been passed through a 0.45 μm filter is conventionally regarded as dissolved. Sources of the dissolved load include rock weathering, atmospheric fallout of aerosols of both oceanic and terrestrial origin, atmospheric gases, and decomposition and mineralization of organic material. In general the total concentration (mg l^{-1}) of material in solution in a river will decline as discharge increases due to a DILUTION EFFECT, but the load transported (kg s^{-1}) will increase.

On a global basis, the dissolved loads of perennial rivers and streams range from <1.0 t km^{-2} year^{-1} to a maximum of about 500 t km^{-2} year^{-1}, although even higher levels may exist in streams draining saline deposits. Meybeck (1979) has estimated the mean dissolved load transport by rivers from the landsurface of the globe to the oceans at 37.2 t km^{-2} year^{-1}, and Walling and Webb (1983) have pointed out the importance of mean annual run-off and lithology in controlling the global pattern of dissolved load transport. DEW

References
Meybeck, M. 1979: Concentration des eaux fluviales en éléments majeurs et apports en solution aux océans. *Revue de géologie dynamique et de géographie physique* 21, pp. 215–46.
Walling, D.E. and Webb, B.W. 1983: The dissolved loads of rivers: a global overview. In *Dissolved loads of rivers and surface water quantity/quality relationships.* IAHS publication no. 141. Pp. 3–20.

dissolved oxygen Will be present in most natural waters, but interest in this water quality parameter has focused largely on rivers, because of its importance for fish and other aquatic life and the potential for significant spatial and temporal variation. The dissolved oxygen content of streamflow primarily reflects interaction with the overlying air, since oxygen from the atmosphere is dissolved in the water. Assuming equilibrium conditions, the dissolved oxygen concentration is essentially a function of the water temperature, which influences its solubility, and of the atmospheric pressure, which reflects the partial pressure of the gas. The solubility of oxygen at 0°C and 760 mm atmospheric pressure is 14.6 mg l^{-1} and this will decrease with increasing temperature to 7.6 mg l^{-1} at 30°C. Deviations from the equilibrium may occur in both a positive and a negative direction. Supersaturation can result from the production of oxygen within the water body through photosynthesis by macrophytes and algae during daylight hours. Oxygen will be consumed by the respiration of aquatic organisms and by the biochemical oxidation of organic material and pollutants, and reduction in dissolved oxygen concentration will occur if this consumption exceeds the rate of atmospheric reaeration. The potential oxygen consumption associated with an organic pollutant is expressed by its biochemical oxygen demand or BOD value. DEW

dissolved solids The total concentration of dissolved material in water is frequently expressed as a total dissolved solids (TDS) concentation. This parameter can be determined by evaporating to dryness a known volume of water. The evaporation

Average composition of world river water

	Concentration (mg l^{-1})								
	Ca^{2+}	Mg^{2+}	Na^+	K^+	Cl^-	SO_4^{2-}	HCO_3^-	SiO_2	Total
Livingstone 1963	15.0	4.1	6.3	2.3	7.8	11.2	58.4	13.1	118.2
Meybeck 1979	13.4	3.35	5.15	1.3	5.75	8.25	52.0	10.4	99.6

is normally carried out at a temperature of 103–105°C, but some standard procedures specify a temperature of 180°C. The value obtained should be viewed as only approximate, as there is a possibility of loss of dissolved material by volatilization or of incomplete dehydration of the residue. The residue will contain both inorganic and organic material. An alternative procedure involves summation of the results obtained from analysis of individual constituents, although these need to be expressed in terms of an anhydrous residue (e.g. bicarbonate will exist as carbonate) in order to ensure comparability. There may be poor correspondence between results from the two methods, through the problems outlined above and the general lack of data for dissolved organic material.

Because of their generalized nature, measurements of TDS concentration are of limited value in water quality assessment and pollution studies but they are frequently employed in investigations of DISSOLVED LOAD transport by rivers and in associated assessments of chemical DENUDATION RATES.

On a global scale, discharge-weighted TDS concentrations encountered in rivers range between minima of approximately 5–8 mg l^{-1} recorded in several tributaries of the Amazon, and maxima of 5000 mg l^{-1} or more associated with streams draining areas of saline deposits. Values are, however, typically in the range 30–300 mg l^{-1} and Meybeck (1979) cites a mean inorganic TDS concentration of 99.6 mg l^{-1} for world river water. Dissolved organic matter generally constitutes only a small proportion of the total dissolved solids, and concentrations commonly fall in the range 2–40 mg l^{-1}, with a mean of approximately 10 mg l^{-1}. The inorganic component is dominated by a limited number of major elements and Ca^{2+}, Mg^{2+}, Na^+, K^+, Cl^-, HCO_3^-, SO_4^{2-} and SiO_2 generally account for 99 per cent of the material, with a number of lesser constituents comprising the remainder. Several workers have

attempted to define a world average river composition in terms of major inorganic constituents and those provided by Livingstone (1963) and Meybeck (1979) are listed in the table. In the latter case, an attempt has been made to deduct anthropogenic contributions.　　　　　　　　　　DEW

References
Livingstone, D.A. 1963: Chemical composition of rivers and lakes: data of geochemistry, Chapter G. *US Geological Survey professional paper* 440G.
Meybeck, M. 1979: Concentrations des eaux fluviales en éléments majeurs et apports en solution aux océans. *Revue de gééologie dynamique et de géographic physique* 21, pp. 215–46.

distributary A stream channel which divides from the main channel of a river. One of the channels a river subdivides into when it becomes braided or when it reaches its delta.

distribution graph A graph first employed by Bernard in 1935 to represent the unit hydrograph (see UNIT RESPONSE GRAPH) in percentage form. In a distribution graph ordinates are expressed as the percentage of total run-off. According to unit hydrograph theory the storm discharge response is directly proportional to the effective rainfall, and therefore the percentage of run-off in each time period will remain the same regardless of effective rainfall amount if the rainfall duration is constant.　　　　　　　　　　AMG

Reading and Reference
Bernard, M. 1935: An approach to determine stream flow. *Transactions of the American Society of Civil Engineers* 100, pp. 347–95.
†Wilson, E.M. 1969: *Engineering hydrology*. London: Macmillan.

diurnal tides Occur in a limited number of parts of the world and consist of only one high and one low tide in each 24 hour period. They only occur where the coastline has the correct configuration.

divagation The lateral movement of a river's channel into an area of previously deposited alluvium.

divergence The phenomenon of air flowing outwards from an air mass being replaced by air descending from above, also the splitting of oceanic currents, often as a result of offshore winds, allowing upwelling of cold water from the depths.

Written as div ∨ it is given by:

$$\text{div} \vee = \frac{\partial u}{\partial x} + \frac{\partial v}{\partial y} + \frac{\partial w}{\partial z}$$

where ∨ is the three-dimensional velocity vector, having x, y and z components u, v and w; units are $(\text{time})^{-1}$. Often in meteorology the term is used to mean horizontal divergence, being the instantaneous fractional rate of change of an infinitesimal horizontal area within the fluid, given by:

$$\text{div} \vee_H = \frac{\partial u}{\partial x} + \frac{\partial v}{\partial y}.$$

Typical values of large-scale horizontal divergence in the atmosphere are about 10^{-5}s^{-1}. Negative divergence is called convergence.

Local variations of density in the atmosphere are relatively small, so that horizontal divergence and convergence usually result in VERTICAL MOTION. KJW

Reading
Atkinson, B.W. ed., 1981 *Dynamical meteorology: an introductory selection*. London and New York: Methuen.

divergent erosion A term introduced by the German geomorphologist H. Bremer (1971) to describe the difference between erosion in the subtropics, where chemical weathering is most effective on substantial surfaces and weakest on steeper slopes, and the mid-latitudes, where erosion is weakest on horizontal surfaces and most effective on slopes. ASG

Reference
Bremer, H. 1971: Flüsse, Flachen-und Stufenbildung in der feuchten Tropen. *Würzburger Geographische Arbeiten* 35.

diversity Usually the number and relative abundance of species in a biological community (see ECOSYSTEM). However, the resource and habitat diversity of ecological systems can also be measured (Magurran 1988). A small proportion of the species in a community usually accounts for a considerable number of its members (see DOMINANT ORGANISM), while a large

proportion will be represented by relatively few individuals. Such a situation leads to the presence of a common and rare species component in a community, with the latter category mainly responsible for its diversity (Odum 1971). There are two types of species diversity. Local (alpha) diversity relates to species variety within a particular community, while regional diversity (see BETA DIVERSITY) is that which exists between communities in juxtaposition (Whittaker 1975; Ricklefs 1980).

Species diversity may be either totally enumerated or, more commonly, sampled. Its most basic measure is given by the ratio between the number of species present (S) and the number of individuals (N). The number of species per unit area can also be ascertained. Estimation of diversity from the number of individuals provides no insight into the commonness or rarity of species. Communities with the same S/N ratio may possess major differences in species relative abundance. Among measures employed to accommodate both number and relative abundance, an index of diversity measured by the Shannon–Weiner function is favoured. This takes the form:

$$H = -\sum_{i=1}^{S} (p_i)(\log_2 p_i)$$

where H = index of species diversity; S = number of species; p_i = proportion of the total sample belonging to the ith species (Krebs 1985).

While such measures provide indices of diversity, the calculated values do not relate to the frequency distribution of the sample. As Hutchinson (1978) states, four main distributions have been recognized in relation to numbers of species and individuals. The most commonly encountered seems to be log-normal, although declining geometrical, log-series and random distributions have been observed.

Important species diversity variations include those on the equatorial-polar latitudinal gradient (high to low respectively) and on islands of increasing size (low to high respectively). Possible causes of diversity differences are numerous, complex, interactive and imperfectly understood. They have been discussed by Pielou (1975) and Krebs (1985).

Greater speciation, hence a more diverse biota, seems to be aided by the combination of physical environmental equability and stability, together with habitat variety, which is characteristic of low rather than high latitudes. Competition for resources also appears to be greater in tropical than other ecosystems, so that there tends to be more niche differentiation in low latitudes. The latter, in turn, helps promote species diversity. On islands, species abundance appears to approximately double when insular area is ten times as great. Here, species diversity is also a function of parameters such as habitat variety and competition. The number of species present is determined by an equilibrium between immigration to, and extinction within, the area (see ISLAND BIOGEO-GRAPHY).

Until recently, it was widely assumed that species diversity fostered stability in ecosystems. However, mathematical modelling of ecosystem components has prompted May (1981) to postulate the converse. He argues that high diversity levels lead to increased susceptibility to disruptive effects, so that a complex ecosystem will become unstable if its balance is disturbed.

Over the short term, communities appear to diversify with time. At present, it is usual for mature communities to contain more species than youthful ones. Longer-term patterns of diversity, deduced from fossil taxa, are more debatable. The uncertainty is largely due to the uneven distribution of geological evidence in both space and time. Gould (1981) discusses the two major hypotheses relating to long-term diversity trends. The first states that diversity has increased throughout geological time. The second contends that the available ecological niches were filled earlier in biospheric history. Since that time, there has been a steady-state in respect of diversity, with variations in the equilibrium number of species occurring in response to plate-tectonically controlled environmental fluctuations.

Species diversity measures can provide a pointer to the welfare of ecosystems, and are of potential use in conservation strategy and environmental monitoring (Magurran 1988). Species diversity is often regarded as synonymous with biodiversity. However, the latter also includes the assessment of genetic diversity, which fosters evolution, and of the diversity of ecosystems, within which there are links between biotic and abiotic components (World Conservation Monitoring Centre 1992).   RLJ

Reading and References

Gould, S.J. 1981: Palaeontology plus ecology as palaeobiology. In R.M. May ed., *Theoretical ecology: principles and applications*. 2nd edn. Oxford and Boston: Blackwell Scientific. Pp. 295–317.

Hutchinson, G.E. 1978: *An introduction to population ecology*. New Haven and London: Yale University Press.

†Krebs, C.J. 1985: *Ecology: the experimental analysis of distribution and abundance*. 3rd edn. New York and London: Harper & Row.

†Magurran, A.E. 1988: *Ecological diversity and its measurement*. London and Sydney: Croom Helm.

May, R.M. 1981: Patterns in multi-species communities. In R.M. May ed., *Theoretical ecology: principles and applications*. 2nd edn. Oxford and Boston: Blackwell Scientific. Pp. 197–227.

Odum. E.P. 1971: *Fundamentals of ecology*. 3rd edn. Philadelphia: Saunders.

†Pielou, E.C. 1975: *Ecological diversity*. New York and London: Wiley.

Ricklefs, R.E. 1980: *Ecology*. 2nd edn. London: Nelson.

Whittaker, R.H. 1975: *Communities and ecosystems*. 2nd edn. New York and London: Macmillan.

World Conservation Monitoring Centre 1992: *Global biodiversity: status of the Earth's living resources*. London and New York: Chapman and Hall.

diversivore An animal which may vary its food intake to include both a plant-eating and an animal-eating pattern. The best examples are human beings for, although on a world scale humans are probably more herbivore than anything else, consuming large amounts of cereals in particular, they also ingest meats and fish (notably for protein), and choose to eat some decomposers too (e.g. fungi). There are several other well-known examples: in Welsh rivers the net-spinning caddis fly (*Hydropsyche*) feeds on detritus, green algae, mayflies and midges, and the stonefly (*Perla*) lives off a similar diet (Jones 1949). Such organisms do not fit easily into the 'normal' structure of FOOD CHAINS and TROPHIC LEVELS. In the real world, food chains with a high degree of diversivory are rare, except for insect–host parasite systems (Yodzis 1981).   DW

References

Jones, J.R.E. 1949: A further ecological study of a calcareous stream in the Black Mountain district of South Wales. *Journal of animal ecology* 18, pp. 142–59.

Yodzis, P. 1981: The stability of real ecosystems. *Nature* 289, pp. 674–6.

doab A term used in the Indian subcontinent to describe the low alluvial plain between two converging rivers.

doldrums A zone of light, variable winds, low atmospheric pressure, high humidity and temperature, and frequent cloudy and unsettled weather located near or slightly north of the equator. The doldrums are bounded to the north by the north-east trade winds of the northern hemisphere and to the south by the south-east trade winds of the southern hemisphere (see INTER-TROPI-CAL CONVERGENCE ZONE). They shift north and south with the seasons, being farthest north in June to October. In a

broader sense, the term doldrums refers to the lethargic, monotonous, warm, humid weather of summer and to the listless, often despondent, human response to it.   WDS

doline A roughly conical depression formed essentially by the solution and/or collapse of underlying limestone strata. Dolines are often, but not necessarily, sites of sinking water (hence the name SINKHOLE

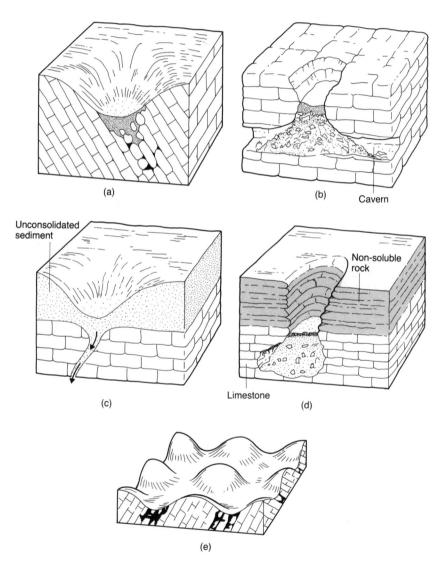

Doline: *five different types of limestone solution hollow: (a) solution doline produced by widening of joints;(b) collapse doline produced by a cave roof falling in;(c) subsidence doline produced by removal of unconsolidated sediment down a fissure or joint (d) subjacent karst collapse doline;(e) cockpits (intersecting star-shaped dolines), typical of tropical areas.*
Source: *A.S. Goudie 1984:* The nature of the environment. *Oxford and New York: Basil Blackwell.*

or swallow hole) and can be mantled by subsequent glacial drift deposits. When limestone dissolves and collapses beneath gritstone (as on the north crop of the South Wales limestone) dolines can form on rock other than limestone – they are then called interstratal karst dolines (Thomas 1974). They can be surface expressions of collapsed cave passages and are common sites for cave digging activities. The age of formation of many of the British dolines has been questioned; they may be considerably older than was once thought (Bull 1980).

PAB

References
Bull, P.A. 1980: The antiquity of caves and dolines in the British Isles. *Zeitschrift für Geomorphologie Supplementband* 36, pp. 217–32.
Thomas, T.M. 1974: South Wales interstratal karst. *British Cave Research Association* 3, pp. 131–52.

dolocrete A form of calcrete in which magnesium carbonate is a major component, and which probably forms as a groundwater precipitate near the water table of a brackish water body. ASG

Reading
El-Sayed, M.I., Fairchild, I.J. and Spiro, B. 1991: Kuwaiti dolocrete: petrology, geochemistry and groundwater origin. *Sedimentary geology* 73, pp. 59–75.

dome dune A low, circular or oval mound formed where dune height is inhibited by unobstructed strong winds. It generally lacks a slip face.

domestication The manipulation of the life and reproduction of plants and animals by man to produce genetic changes that result in new races. The domestication of biota represents a change in human economies from food collecting (e.g. hunting, fishing, gathering) to food production, sometimes called the Neolithic revolution. The transition from hunting and gathering to food production must always have been a gradual process, although once inaugurated it may often have spread quickly. The recognition of the process of domestication in archaeological contexts is often difficult and may be apparent only when it is complete, so that the beginnings of such processes are often lost beyond recovery. The reasons why domestication started in such early foci as south-west and south-east Asia, and Meso-America are still under discussion: the innovation and stimulus derived from a city, climatic change and 'oasis' formation, ecosystem models, environmental potential models, and demographic pressure have all been invoked. Once established, food-producing societies raise the carrying capacity of the land (the CARRYING CAPACITY for late Pleistocene hunter-gatherers was 0.1 person $km \gg fb^2$, for early dry farms 1–2 persons $km \gg fb^2$, early irrigated lands 6–12 persons $km \gg fb^2$), encourage the development of sedentary societies, cause changes in the structure of society, allow craft specialization, the development of surpluses and very probably the growth of civilization as we usually define it. Whether food production allows rapid population growth or is called forth by it, there is clearly a close link between the two. IGS

Reading
Bender, B. 1975: *Farming in prehistory: from hunter-gatherer to food-producer*. London: Arthur Barker.
Harris, D.R. Alternative pathways towards agriculture. In C.A. Reed ed., *Origins of agriculture*. The Hague: Mouton. Pp. 179–243.

dominant discharge The discharge to which the average form of river channels is related. The dominant discharge that determines the size of a river channel cross-section or the size of river channel pattern at a particular location will also depend on the character and quantity of sediment transported and also on the composition of the bed and bank materials. The dominant discharge will not be a single value but a range of flows. (See also CHANNEL CAPACITY.) KJG

dominant organism An organism of principal importance in either the whole or part of a community. Dominance may be physiognomic (of a particular form, such as a tree – see LIFE FORM), taxonomic (of an evolutionary category, such as a species), or ecological (of quantity and function, such as the amount of standing crop (see BIOMASS), the competitive ability of a producer or consumer). An organism can exhibit more than one type of dominance. This is especially so in the case of physiognomy and taxonomy (Shimwell 1971).

Both plants and animals may dominate. However, plants comprise most of the earth's biomass and are essential for the survival of animals (see ECOSYSTEM). Thus, strictly speaking, plants always dominate communities (Odum 1975). Nevertheless, as Daubenmire (1968) points out, there are instances (in agricultural ecosystems, for example) in which animals assume an important role. Similar regard may be given to the current ascendancy of man in the majority of world ecosystems.   RLJ

Reading and References
Daubenmire, R.F. 1968: *Plant communities: a textbook of plant synecology.* New York and London: Harper & Row.
Odum, E.P. 1975: *Ecology: the link between the natural and social sciences.* 2nd edn. London and New York: Holt Reinhart & Winston.
Shimwell, D.W. 1971: *Description and classification of vegetation.* London: Sidgwick & Jackson.
†Willis, A.J. 1973: *Introduction to plant ecology.* London: Allen & Unwin.

dominant wind The wind which plays the most significant role in a particular local situation, in contrast to the PREVAILING WIND.

donga Derived from the Nguni word *Udonga*, meaning a wall, the term used in southern Africa to describe a gully or badland area caused by severe erosion. Dongas are especially prevalent in colluvium and in weathered bedrock in areas where the mean annual rainfall lies between 600 and 800 mm. Where the materials in which they are developed have high exchangeable sodium contents, they may have highly fluted 'organ pipe' sides (Stocking 1978). At Rorke's Drift they provided a snare for unwary troops in the battle between the British and Zulus. ASG

Reference
Stocking, M.A. 1978: Interpretation of stone lines. *Southern African geographical journal* 60, pp. 121–34.

dormant volcano A volcano which, though not currently or perhaps even recently active, is not extinct since it is likely to erupt in the future.

double mass analysis A plot of cumulative values of one variable, or values from one site against cumulative values of another variable or from another site. It has often been used to plot the values of precipitation recorded at one station against the records of another station or against the average of several other stations. Unless the double mass curve which is plotted has a slope of 1.0 there is a difference between the sites and this may be interpreted in terms of instrument siting or of temporal variations which affect one variable but not the other.   KJG

Reading
Dunne, T. and Leopold, L.B. 1978: *Water in environmental planning.* San Francisco: W.H. Freeman.

downwelling The process of accumulation and sinking of warm surface waters along a coastline. A change of airflow of the atmosphere can result in the sinking or downwelling of warm surface water. The resulting reduced nutrient supply near the surface affects the ocean productivity and meteorological conditions of the coastal regions in the downwelling area.   ASG

draa A large-scale accumulation of aeolian sand dunes geographically distinct from other sand fields but smaller than a sand sea.

drain A man-made channel constructed to allow the free flow of water from an area.

drainage May refer either to the natural drainage of the landsurface or to the system of LAND DRAINAGE introduced by human activity. Natural drainage of the landsurface is organized in drainage basins which are those areas in which water is concentrated and flows into the DRAINAGE NETWORK. The drainage basin is usually defined by reference to information on surface elevation, for example, from contours on topographic maps although the position of the WATERSHED on the ground surface, which is the line separating flow to one basin from that to the next, may not correspond to the PHREATIC DIVIDE beneath the surface. The pattern of natural drainage has been studied in relation to the DRAINAGE DENSITY and DRAINAGE BASIN CHARACTERISTICS which can be quantified and used in rainfall–run-off modelling and in the interpretation of river discharge; to the nature of the drainage network including the pattern of the drainage and also the stream ORDER; and to the evolution of the drainage pattern. In the course of drainage evolution the details of the several patterns such as trellis or rectangular (see diagram for DRAINAGE NETWORK) may be related to geological structure such as the alternation of hard and soft rocks or to the presence of joints or faults. Where drainage patterns are discordant with the structure and cross folds or faults, for example, it has been suggested that either the drainage has been superimposed from a cover rock that originally occurred above the rocks at present exposed in the landscape, or the drainage was antecedent and the drainage pattern was maintained as the structures were developed by endogenetic uplift giving the folded and/or faulted structures.   KJG

drainage basin characteristics The geological, topographical, soil and vegetational/land use characteristics of the drainage basin unit. Topographic characteristics, include measures of basin size, relief, shape, and internal character including DRAINAGE DENSITY of the stream network. Many models constructed to estimate water or sediment yield from drainage basins include parameters to represent climatic indices and also drainage basin characteristics. It is therefore necessary to be able to express drainage basin characteristics (or catchment characteristics) in quantitative terms.   KJG

Reading
Gregory, K.J. and Walling, D.E. 1976: *Drainage basin form and process*. London: Edward Arnold.

drainage density Is calculated by dividing the total length of stream channels in a basin (ΣL) by the drainage basin area (A_d) as:

$$D_d = \frac{\Sigma L}{A_d}$$

Drainage density is a very significant measure of drainage basin character because the values of D_d reflect the climate over the basin and the influence of other DRAINAGE BASIN CHARACTERISTICS including rock type, soil, vegetation and land use, and topographic characteristics. Drainage density also has an important

influence upon streamflow because water flow in channels is faster than water flow over or through slopes. The higher the drainage density the faster the hydrograph rise and the greater the peak discharge. Although drainage density is a very significant parameter it must be used with careful attention given to the extent of the drainage network which can be determined from topographic maps of different scales, from remote sensing sources, or from field survey; and also to the composition of the DRAINAGE NETWORK because networks are composed of channels which have flows for different periods of the year.   KJG

Reading
Gregory, K.J. 1967: Drainage networks and climate. In E. Derbyshire ed., *Geomorphology and climate*. Chichester: Wiley. Pp. 289–318.

drainage network The system of river and stream channels in a specific basin or area (see diagram). Whether a particular headwater stream is included in the drainage network depends upon CHANNEL TYPE, and networks have been analysed qualitatively and quantitatively. Qualitative classifications of drainage networks depend upon the way in which drainage patterns reflect either underlying geological structure (rectangular, parallel, dendritic, trellised), prevailing regional slope (radial, centripetal), geomorphological history (deranged), or some combination of these (e.g. annular). Quantitative analysis has focused upon drainage NETWORK structure founded upon stream ORDER and upon DRAINAGE DENSITY. When viewed according to channel process the drainage network can be regarded as composed of PERENNIAL, INTERMITTENT and EPHEMERAL streams.   KJG

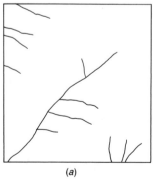 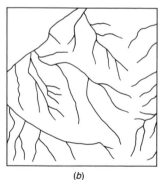 

(a)   (b)   (c)

Drainage density *measurements can distinguish between networks which are coarse (a), with a low to medium density (b) and to fine (c) which have a high drainage density value.*

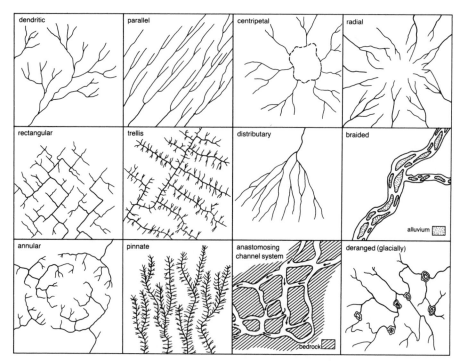

Drainage network: *Types of drainage pattern.*
Source: *H.F. Garner 1974:* The origin of landscapes. *New York: Oxford University Press. Pp. 60–1.*

draw down The extent to which the WATER TABLE is reduced in elevation as a result of pumping water from a well. The amount of draw down diminishes logarithmically with distance from the site of pumping, and this determines the shape of the CONE OF DEPRESSION in the water table. Under ARTESIAN conditions, draw down may also occur in the potentiometric surface should heavy pumping occur at a bored well. In a coastal AQUIFER, the reduction in HYDRAULIC HEAD caused by draw down will encourage salt water intrusion beneath the well (see GHYBEN–HERZBERG principle). PWW

Reading
Freeze, R.A. and Cherry, J.A. 1979: *Groundwater.* Englewood Cliffs, NJ: Prentice-Hall.
Linsley, R.K., Kohler, M.A. and Paulhus, J.L.H. 1988: *Hydrology for engineers*, 3rd edn. New York: McGraw-Hill.

dreikanter A pebble found on the surface in desert regions; it has three distinct facets on its upper surface as a result of wind abrasion.

drift potential A measure, in vector units, of the sand-moving power of the wind. It is derived from reduction of surface wind data throughout a weighting equation and usually represents one year. Resultant drift potential involves treating vector unit totals from various directions on a sand rose as vector quantities, which are then resolved trigonometrically to a resultant, the magnitude of which is the resultant drift potential, and the direction of which is the resultant drift direction. ASG

Reading
McKee, E.D. 1979: A study of global sand seas. *US Geological Survey professional paper* 1052.

dripstone Any accumulation of water-soluble salts on the roofs, walls and floors of caves. A general term for stalactites and stalagmites.

drought The condition of dryness because of lack of PRECIPITATION. The deficit is normally sufficiently persistent for EVAPORATION to lead to a substantial decrease in the moisture content of soils and in other hydrological parameters such as groundwater flow and streamflow.

The perception and therefore the definition of a drought varies with climate. In the

British Isles, for example, an absolute drought is a period of at least fifteen consecutive days during which no day reports more than 0.2 mm rain. A partial drought is a spell of at least twenty-nine days during which there may be some days that experience slight rain but for which the mean daily rainfall does not exceed 0.2 mm. Absolute droughts occur about once a year on average in the lowlands of south-east England. In contrast, more prolonged drought is in a sense a regular annual feature of some tropical climates where one rainy season is separated by a long dry season – found characteristically across areas at the limit of the poleward excursion of the INTERTROPICAL CONVERGENCE ZONE.

Droughts in middle latitudes are associated with the unusual persistence of anticyclonic conditions and specially with the presence of a BLOCKING anti-cyclone. Under such a regime the rain-bearing frontal systems are steered around the flanks of the stationary high pressure area so that anomalous dryness in one place is linked to anomalous wetness in others. For example the period from May 1975 to August 1976 was one of extreme drought stretching from Scandinavia to western France, with southern England recording only 50 per cent of the long-term average precipitation for such a 16 month period. In August 1976 the persistent high pressure meant that much of the same area recorded less than 50 per cent of the average rainfall, while both Iceland and the northern Mediterranean reported over 150 per cent.

The Sahelian drought of 1969 to the mid-1970s is believed to have been the result of a variety of factors including the anomalous south-eastward expansion of the Azores Anti-cyclone, lower sea surface temperature across parts of the eastern Atlantic and increased desertification due to overgrazing. This latter artificial change leads to an increase in surface albedo which on balance will be associated with enhanced subsidence. RR

Reading
Doornkamp, J.C. and Gregory, K.J. eds. 1980: *Atlas of drought in Britain 1975–6*. London: Institute of British Geographers.

drumlin A depositional landform, normally consisting of glacially derived debris, which has been streamlined by the passage of overlying ice. The typical form is blunt upstream and tapered downstream. Length to breadth ratios are commonly between 2:1 and 4:1, while heights vary from 5 to 50 m. The thrust of recent work is to envisage localized lodgement beneath moving ice associated with strength variations within the TILL layer or the enhanced frictional effects of an obstacle such as a bedrock bump or resistant patch of sand and gravel. The streamlining reflects erosional and depositional processes as the ice passes over the lodged material. DES

Reading
Chorley, R.J. 1959: The shape of drumlins. *Journal of glaciology* 3.25, pp. 339–44.

Menzies, J. 1979: The mechanics of drumlin formation with particular reference to the change in porewater content of the till. *Journal of glaciology* 22.87, pp. 373–84.

Menzies, J. and Rose, J. 1987: *Drumlin symposium*. Rotterdam: Balkema.

dry deposition The process by which pollutant gases or particles are transferred directly from the atmosphere on to liquid and solid surfaces. Particularly important for the deposition of SO_2, but deposition of NO_2 is slow and dry deposition is not a major removal process for atmospheric NO_x. Deposition is primarily through turbulent transfer and the velocity of deposition depends upon relative humidity, pH and, especially in urban areas, aerodynamic resistance, whereby surfaces exposed to the wind experience greatest deposition. Deposition is independent of rainfall, and maps of Britain produced by the Review Group on Acid Rain (1987) show dry deposition of sulphur to be more widespread and loadings to be considerably higher than those for wet-deposited non-marine sulphur. Particulate deposition tends to concentrate near to source areas and particulates deposited between rains can be concentrated in the early stages of surface run-off creating an acid surge. Dry deposition can be increased by land use changes, for example, afforestation, which increase aerodynamic resistance and/or surface area. BJS

Reference
Review Group on Acid Rain 1987: *Acid deposition in the United Kingdom*. 2nd report. Warren Spring Laboratory. London: HMSO.

dry valley A valley which is seldom, if ever at the present time, occupied by a stream channel. These valleys are widespread on a variety of rock types, including sandstones, chalk and limestone in southern

England, but they are also known from many other parts of the world, including the coral reefs of Barbados. An enormous range of hypotheses has been put forward to explain why they are generally dry.

Hypotheses of dry valley formation
Uniformitarian
1 Superimposition from a cover of impermeable rocks or sediments
2 Joint enlargement by solution through time
3 Cutting down of major through-flowing streams
4 Reduction in catchment area and groundwater lowering through scarp retreat
5 Cavern collapse
6 River capture
7 Rare events of extreme magnitude

Marine
1 Non-adjustment of streams to a falling Pleistocene sea level and associated fall of groundwater levels
2 Tidal scour in association with former estuarine conditions

Palaeoclimatic
1 Overflow from proglacial lakes
2 Glacial scour
3 Erosion by glacial meltwater
4 Reduced evaporation caused by lower temperatures
5 Spring snowmelt under periglacial conditions
6 Run-off from impermeable permafrost.

The uniformitarian hypotheses require no major changes of climate or base level, merely the operation of normal processes through time; the marine hypotheses are related to base-level changes; and the palaeoclimatic hypotheses are associated primarily with the major climatic changes of the Pleistocene. Dry valleys show a considerable range of shapes and sizes, from mere indentations in escarpments, to great winding chasms like Cheddar Gorge in the Mendips. ASG

Reading
Goudie, A.S. 1993: *The nature of the environment*, 3rd edn. Sect. 4.15. Oxford and New York: Basil Blackwell.

dry weather flow A term used to describe low flow in a river as an alternative term to BASE FLOW. It is also used to refer to the total daily rate of flow of domestic and trade waste sewage in a sewer in dry weather. The daily total of sewage dis-

charge is used to represent dry weather flow because both domestic and trade waste sewage vary greatly in quantity with 24 hour periods. The dry weather flow is the background flow in the sewer and may be 'measured after a period of seven consecutive days during which rainfall has not exceeded 0.25 mm' (Bartlett 1970).

Reading and Reference
Bartlett, R.E. 1970: *Public health engineering-design in metric sewerage*. Oxford: Elsevier.
†Linsley, R.K., Kohler, M.A. and Paulhus, J.H.L. 1982: *Hydrology for engineers*. 3rd edn. New York: McGraw-Hill.

du Boys equation In 1879 Paul du Boys developed one of the earliest bed-load transport equations from consideration of the TRACTIVE FORCE of flowing water. The bed-load transport weight per unit width of channel per unit time (g_b) is expressed as a function of the mean bed shear stress (τ_0) in excess of the threshold shear stress (τ_{oc}) required to initiate particle motion:

$$g_b = A(\tau_0 - \tau_{oc})\tau_0$$

The constant A and the threshold mean bed shear stress (τ_{oc}) are both regarded as functions of sediment particle size alone. Derivation of the equation is simply based on the assumption that bed sediment moves in discrete layers whose velocity decreases linearly from the bed surface to zero at a depth dependent on the fluid shear stress at the bed (Embleton and Thornes 1979, pp. 238–9). KSR

Reading and Reference
†du Boys, P.F.D. 1879: *Études du régime du Rhône et l'action exercée par les eaux sur un lit à fond de graviers indefinement affouilable*. *Annales des Ponts et Chaussées* Ser. 5.18, pp. 141–95.
Embleton, C. and Thornes, J.B. eds 1979: *Process in geomorphology*. London: Edward Arnold.

dune An accumulation of sand (sometimes also clay, gypsum, carbonate grains, etc.) deposited by the wind and aerodynamically shaped by aeolian processes. They are landforms, often mobile, which are characteristic of some deserts and sea coasts. They occur in many forms (see figure), and have many names (see AKLE; ANTIDUNE; BARCHAN; BEDFORMS; CLAY DUNE; DOME DUNE; DRAA, FULJE; IMPEDED DUNE; LUNETTE; PARABOLIC DUNE; REVERSING DUNE; SEIF; STAR DUNE; TRANSVERSE DUNE). The classic work on dunes is that of Bagnold (1941), while a comprehensive global survey of

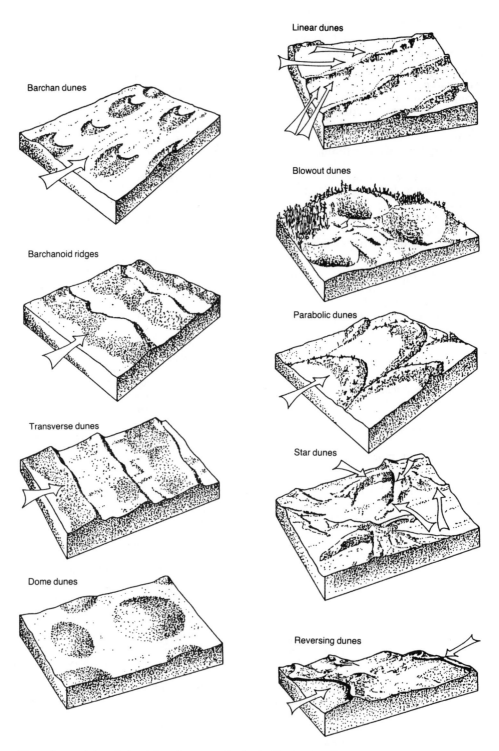

Barchan dunes

Linear dunes

Barchanoid ridges

Blowout dunes

Transverse dunes

Parabolic dunes

Dome dunes

Star dunes

Reversing dunes

Dune: *the morphology of selected desert dune types after the classification of McKee 1979.*

Sand dunes and succulent plants near Chala in the Atacama Desert, Peru. Dunes that accumulate around vegetation are often called nebkhas.

Classification of dune types controlled largely by vegetation, topographical features or localized sediment sources

Type	Form and position	Mode of development
Blowout	Circular rim around depression	Localized deflation
Parabolic dune	'U' or 'V' shape in plan view with arms opening upwind to enclose a blowout	Deposition of sand locally deflated upwind; arms are usually fixed by vegetation
Lunette	Crescent-shaped opening upwind	Accumulation downwind of localized sediment source such as desiccated lake basin or pan
Shrub-coppice dune (nebkha)	Roughly elliptical to irregular in plan, streamlined downwind	Accumulation around and downwind of vegetation clump
Lee dune	Elongated downwind from topographical obstruction	Accumulation on protected lee side of obstacle
Foredune	Roughly arcuate with arms extending downwind either side of obstruction	Accumulation in zone of disrupted airflow immediately windward of obstacle
Climbing dune	Irregular accumulation rising up windward side of large topographical obstruction	Accumulation in zone of disrupted airflow on windward side of obstacle
Falling dune	Irregular accumulation descending leeward side of large topographical obstruction	Accumulation in zone of disrupted airflow on upwind side of obstacle
Echo dune	Elongated ridge roughly parallel to, and separated from, windward side of topographical obstruction	Accumulation in zone of rotating airflow upwind from large obstacle

Source: Summerfield, M.A. 1991: *Global geomorphology*. London and New York: Longman Scientific and Technical and Wiley. Table 10.3.

forms and materials is that of McKee (1979). A survey of processes, forms, sediments, and ancient aeolian sand bodies is provided by Brookfield and Ahlbrandt (1983).   ASG

Reading and References
Bagnold, R.A. 1941: *The physics of blown sand and desert dunes.* London: Chapman and Hall.
Brookfield, M.E. and Ahlbrandt, T.S. eds 1983: *Eolian processes and sediments.* Amsterdam: Elsevier.
McKee, E.D. 1979: A study of global sand seas. *United States Geological Survey professional paper* 1052.
Pye, K. and Tsoar, H. 1990: *Aeolian sand and sand dunes.* London: Unwin Hyman.

durability A term used with reference to building materials, which gives an indication of the service life of the material (e.g. a building stone). It is not an absolute quality nor can it be easily quantified, as it must be related to tests appropriate to the type of deterioration which the structure is likely to undergo (e.g. various weathering phenomena). Such tests can be standardized. WBW

Reading
Frohnsdorff, G. and Masters, L.W. 1980: The meaning of durability and durability prediction. *American Society for Testing and Materials special technical publication* 691, pp. 17–30.

duricrust A hard crust formed at or near the landsurface during the processes of weathering of rocks and soil formation, usually in tropical or arid regions (see illustration). The main types include alcrete, calcrete, ferricrete (laterite) and silcrete.

Reading
Goudie, A.S. 1973: *Duricrusts of tropical and subtropical landscapes.* Oxford: Clarendon Press.

duripan A synonym for silcrete. An indulated and cemented silica-rich accumulation within the soil zone or zone of weathering.

dust Consists of solid particles of varying character and size that are carried in suspension in the atmosphere. The prime sources of such particles are volcanic eruptions, the deflation of silt and clay sized materials from desert surfaces and from unvegetated glacial deposits, and the pollution produced by industrial and domestic combustion and by fires. When present in appreciable quantities it causes haze, and if the visibility is reduced to less than 1000 m dust storm conditions are said to exist. In some desert areas dust storms occur for some tens of days in each year and have probably been more frequent during the cold, dry, windy interpluvials of the Pleistocene (Goudie 1983). One product of the deposition of dust is the formation of LOESS. Dust, especially through its backscattering effect on solar radiation, may contribute to climatic change, and it is also a significant environmental hazard (Péwé

Duricrust: An iron-rich duricrust (laterite) cap-rock above basalt at Panchgani, Maharashtra, India. Note the barren nature of the plateau surface and the cambering of laterite blocks down the hillside.

1981). It is also one result of drought, desiccation and desertification (Worster 1979).   ASG

Reading and References
Goudie, A.S. 1983: Dust storms in space and time. *Progress in physical geography* 7, pp. 502–30.

Péwé, T.L. ed. 1981: Desert dust: origin, characteristics and effect on man. *Geological Society of America special paper* 186.

Pye, K. 1987: *Aeolian dust and dust deposits.* London: Academic Press.

Worster, D. 1979: *Dust bowl: the southern plains in the 1930s.* New York: Oxford University Press.

dust veil index (DVI) A quantitative method developed by Lamb (1970) for comparing the magnitude of volcanic eruptions. The formulae use observations either of the depletion of the solar beam, temperature lowering in the middle latitudes, or the quantity of solid matter dispersed as dust. The reference dust veil index is 1000, assigned to the Krakatoa 1883 eruption, and the index is calculated using all three methods, where the information is available, for statistical comparison purposes.   ASG

Reading
Lamb, H.H. 1970: Volcanic dust in the atmosphere with a chronology and assessment of its meteorological significance. *Philosophical Transactions of the Royal Society of London, Series A* 266, pp. 425–533.

dyke A sheet-like intrusion of igneous rock, usually orientated vertically, which cuts across the structural planes of the host rocks. The wall or trough formed by differential weathering of such an intrusion when exposed at the landsurface.

dynamic equilibrium A term so widely used in physical geography that it has lost almost all strictness of definition and seems to mean a situation which is fluctuating about some apparent average state, where that average state itself is also changing through time. In Huggett's view (1980) the problems with the definition arise because an idea from thermodynamics (which allows for micro-scale chaotic movement of molecules while the whole system is in equilibrium) has been transposed to apply to a macro-scale. Following Huggett, one can agree that dynamic equilibrium in physical geography today 'is taken either as synonymous with "steady state" or with some false equilibrium in which the system appears to be in equilibrium (*qua* steady state) but in reality is changing very, very slowly with time'.

In this broad sense the term has been neatly substituted for a number of earlier, equally ill-defined concepts, most notably GRADE and climax. Unfortunately, such substitution cannot be said to have resulted in clarification. On the macro-scale, the problem is a statistical one, which applies to any attempt to equate a belief that a long-term, directed change is occurring with observations which indicate departures from some theoretically persistent condition. In such a case, how can the fluctuations be separated from the underlying trend? Conversely, how should it be decided if some of the fluctuations are too great to be 'permissible' elements of a dynamic equilibrium, but instead indicate a departure from that equilibrium state and the onset of a relaxation period leading to some new equilibrium? The choice seems to be entirely a product of the spatial and temporal scale under discussion.

In practice it is virtually impossible to decide whether any part of the physical environment *is* in dynamic equilibrium, since it is equally difficult to decide that any part is *not* in that state. Unless the term is defined with clarity its use adds very little except confusion to any discussion of change over time.   BAK

Reference
Huggett, R. 1980: *Systems analysis in geography.* Oxford: Clarendon Press.

dynamic source area The changing part of a catchment which is physically contributing overland flow at any time. The SOURCE AREA of a catchment is envisaged as changing in response to subsurface flow conditions before and during storms. This concept is important to an understanding of hillslope hydrological response.   MJK

Reading
Kirkby, M.J. ed. 1978: *Hillslope hydrology.* Chichester: Wiley.

dynamical meteorology Study of the forces acting on, and the subsequent motion of air in weather systems; compare SYNOPTIC METEOROLOGY; SATELLITE METEOROLOGY. The control of most of the important physical processes in the atmosphere, such as condensation, depends critically on air motion (see WIND). Hence dynamical meteorology is seen by many to represent the core of meteorology.   JSAG

Reading
Aktinson, B.W. ed. 1981: *Dynamical meteorology: an introductory selection*. London and New York: Methuen.

dynamics As distinct from statics, the study of the forces on, and accelerations of, moving bodies. The application of this study to the atmosphere gives rise to DYNAMICAL METEOROLOGY. The basic principles can, of course, be applied to all aspects of the natural environment which involve moving bodies, e.g. oceanography, hydrology, slope mechanics.   JSAG

Reading
Atkinson, B.W. ed. 1981: *Dynamical meteorology: an introductory selection*. London and New York: Methuen. Esp. pp. 1–20.

E

---

earth hummocks A form of patterned ground characterized by rounded hummocks which produce an irregular net pattern over the landsurface.

earth pillar A pinnacle of soil or other unconsolidated material that is protected from erosion by the presence of a stone or boulder at the top.

earthflow See MASS MOVEMENT TYPES.

earthquake A series of shocks and tremors resulting from the sudden release of pressure along active faults and in areas of volcanic activity. The shaking and trembling of the earth's surface associated with subterranean crustal movements. (See tables 1 and 2.)

easterly wave A shallow disturbance in the trade winds, which moves westwards causing convergence and associated storms.

Table 1 Modified Mercalli scale of earthquake intensity

I	Not felt except by a very few under especially favourable circumstances.		Noticed by persons driving motor cars.
II	Felt only by a few persons at rest, especially on upper floors of buildings. Delicately suspended objects may swing.	VIII	Damage slight in specially designed structures; considerable in ordinary substantial buildings, with partial collapse; great in poorly built structures. Panel walls thrown out of frame structures. Fall of chimneys, factory stacks, columns, monuments, walls. Heavy furniture overturned. Sand and mud ejected in small amounts. Changes in well-water levels. Disturbs persons driving motor cars.
III	Felt quite noticeably indoors, especially on upper floors, but many people do not recognize it as an earthquake. Standing motor cars may rock slightly. Vibration like passing truck.		
IV	During the day felt indoors by many, outdoors by few. At night some awakened. Dishes, windows, doors disturbed; walls make creaking sound. Sensation like heavy truck striking building. Standing motor cars rocked noticeably.	IX	Damage considerable in specially designed structures, well-designed frame structures thrown out of plumb; great in substantial buildings, with partial collapse. Buildings shifted off foundations. Ground cracked conspicuously. Underground pipes broken.
V	Felt by nearly everyone; many awakened. Some dishes, windows, etc. broken; a few instances of cracked plaster; unstable objects overturned. Disturbances of trees, poles, and other tall objects sometimes noticed. Pendulum clocks may stop.	X	Some well-built wooden structures destroyed; most masonry and frame structures destroyed with foundations; ground badly cracked. Rails bent. Landslides considerable from river banks and steep slopes. Shifted sand and mud. Water splashed over banks.
VI	Felt by all, many frightened and run outdoors. Some heavy furniture moved; a few instances of fallen plaster or damaged chimneys. Damage slight.	XI	Few, if any, masonry structures remain standing. Bridges destroyed. Broad fissures in ground. Underground pipelines completely out of service. Earth slumps and land slips in soft ground. Rails bent greatly.
VII	Everybody runs outdoors. Damage negligible in buildings of good design and construction; slight to moderate in well-built ordinary structures; considerable in poorly or badly designed structures; some chimneys broken.	XII	Damage total. Waves seen on ground surfaces. Lines of sight and level distorted. Objects thrown upward into the air.

Source: US Geological Survey.

Table 2 *Energy equivalents of earthquakes compared with the Richter scale*

Earthquake magnitude	TNT equivalent
1.0	0.17 kg
1.5	0.9 kg
2.0	5.9 kg
2.5	28 kg
3.0	179 kg
3.5	450 kg
4.0	5.5 t
4.5	29 t
5.0	181 t
5.3	455 t
5.5	910 t
6.0	5.7×10^3 t
6.3	14.4×10^3 t
6.5	28.7×10^3 t
7.0	181×10^3 t
7.1	228×10^3 t
7.5	910×10^3 t
7.7	1811×10^3 t
8.0	5706×10^3 t
8.2	$11,421 \times 10^3$ t
8.5	$28,711 \times 10^3$ t
9.0	$181,999 \times 10^3$ t

echo dunes Dunes that exist as single sand ridges formed parallel to a vertical cliff.

ecliptic The plane within which the earth orbits about the sun. The path the sun appears to take as it moves across the sky.

ecogeographical rules Describe common patterns of change in the measurable characteristics (size, colour and so forth) in animal populations produced by regular gradients of climatic factors, notably temperature and moisture. Examples include ALLEN'S RULE and BERGMANN'S RULE.   ASG

ecological energetics The study of energy fixation, transformation and movement within ecological systems. The three most important types of energy found are solar energy, chemically stored energy, and heat energy. Short-wave energy from the sun, in the visible light spectrum of from 0.36 to 0.76 μm, is fixed within the cells of green plants, blue-green and other algae, and phytoplankton, by transformation, through the process of PHOTOSYNTHESIS, into chemical energy, following which it is passed down either the grazing or the decomposer FOOD CHAIN. The long-wave heat energy associated with plant and animal

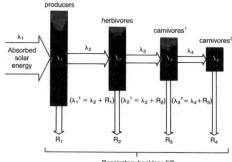

RESPIRATION may leave ecological systems at any time. Heat is not directly available for reuse as an energy source by organisms. Both plant and animal communities at high altitudes (above 2000 m) may also receive small but significant amounts of ultraviolet radiation, with wavelengths of less than 0.36 μm, which is potentially damaging to cell tissue, and against which chemical and physical defences are required.

To avoid confusion, and since all forms of ecologically useful energy can be converted into heat, the calorie or kilocalorie is used as the standard unit of measurement in ecological energetics. One calorie is equivalent to the amount of heat required to raise the temperature of 1 g of water by 1°C; and 1 kilocalorie (kcal) = 10^3 calories. The joule (J), a measure of mechanical or work energy, is now increasingly utilized: the amount of work needed to raise 1 g weight against gravity to a height of 1 cm = 981 erg; 10^7 erg =1 J; 4.2 J = 1 cal. The conversion factor between mechanical and heat energy was not discovered until the mid-nineteenth century.

Studies in ecological energetics are inevitably complex, not least because of the large numbers of different organisms found within most biological systems, the uncertainty as to the precise roles of some of these in respect of energy transfer, and the often complicated technology required to unravel the intricacies of energy flow, and their changing patterns in time and space. Even so, some understanding of the essential energy controls of individual land-based organisms had been gained from A.L. Lavoisier's studies in France on respiration, as early as 1777. And from the 1920s, investigations into the energy flow of biologically simple lake systems were under way, particularly in North America.

But the major theoretical step forward was provided by R.L. Lindemann in 1942. It was he who first formalized the concept of TROPHIC LEVELS, and the energy relationships between them. He suggested that the standing crop (BIOMASS) of each trophic level might be measured, and the result then translated into energy equivalents. He hypothesized that the mean energy flow between different levels could be represented by simple equations, which took into account the ultimate controls of the first and second laws of thermodynamics. If the energy content of any trophic level is given as Λ along with a subscript to indicate the food-chain role of the organisms within it (Λ_1 = producers; Λ_2 = herbivores; Λ_3 = carnivores, etc.), and the passage of energy between any two levels is designated by λ, then that which is transferred from Λ_n to Λ_{n+1} will be λ_{n+1}. Further, the heat loss in respiration may be termed R_1, R_2, etc, according to the particular trophic level under review. The heat loss R, coupled with the quantity of energy in transfer between any two trophic levels may jointly be described as λ_n, and is the amount which, at the lower trophic level, does not go into biomass production. From this, the rate of energy exchange in any trophic level may be expressed as:

$$\frac{\Delta \Lambda_n}{\Delta_t} = \lambda_n + \lambda_n{}',$$

or in other words, the rate at which energy is taken up by that level (λ_n) to form new biomass, minus the rate at which energy is lost from it ($\lambda_n{}'$) (see diagram).

In many respects Lindemann's theory marked a turning point in the study of ecological energetics, for it provided a framework for all later research. It encouraged Slobodkin (1959) to undertake a series of laboratory experiments on *Daphnia* populations, from which he deduced that the mean transfer of energy from one trophic level to another would be *c.*10 per cent (it is now known that it can vary widely from this: see TROPHIC LEVELS). Others (e.g. Odum 1971) have accumulated a vast amount of data on the patterns of energy transfer in both natural and agricultural systems. Still others (e.g. Kozlovsky 1968) have suggested alternative methods of examining the efficiency of energy transfer

between trophic levels, as for example that related to consumption efficiency:

Consumption efficiency =

$$\frac{\text{intake of food at trophic level } n}{\text{net productivity at trophic level } n-1} \times 100$$

This may increase slightly from the lowermost trophic levels, but seems in general to fall between 20 and 25 per cent; and the majority of the remainder goes into the decomposer food chain.

Overall, ecological energetics has brought home the importance of small and microscopic organisms in the patterns of energy transfer everywhere, and the need to maintain the presence of these in balanced systems. A good deal is now known about the fixation of solar energy into plant cells at the primary producer level, but much still remains to be learnt about energy transfer and fixation further down the food chain (see BIOLOGICAL PRODUCTIVITY; NET PRIMARY PRODUCTIVITY)   DW

Reading and References
Kozlovsky, D.G. 1968: A critical evaluation of the trophic level concept. I. Ecological efficiencies. *Ecology* 49, pp. 48–60.
Lindemann, R.L. 1942: The trophic-dynamic aspect of ecology. *Ecology* 23, pp. 399–418.
†Odum, E.P. 1971: *Fundamentals of ecology.* 3rd edn. Philadelphia: W.B. Saunders.
†Phillipson, J. 1966: *Ecological energetics.* London: Edward Arnold.
Slobodkin, L.B. 1959: Energetics in *Daphnia pulex* populations. *Ecology* 40, pp. 232–43.

ecological explosions Ecological events marked by an enormous increase in the numbers of some kind or kinds of organism. The term was carefully defined by Elton (1958) and was employed to indicate the bursting out from control of populations that were previously held in restraint by other forces. Classic examples of such explosions are found in the epidemics of infectious viruses and bacteria, such as influenza and bubonic plague, and in the rapid spread of the North American grey squirrel (*Sciurus carolinensis*) after its introduction to Britain in the nineteenth century.

Many organisms subject to such population outbursts are serious agricultural pests, such as the desert locust (*Schistocerca gregaria*), swarms of which are described in the Old Testament. In most locust species it is possible to identify two types of distributional area, namely a highly localized *outbreak area*, where the species persists permanently and where swarms form, and an *invasion area*. The causes of the devastating plagues appear to be related to weather conditions, particularly to moisture, operating in and through the process of phase transformation, in which the locusts exhibit polymorphism, changing from solitary forms (*solitaria*) to gregarious or swarming forms (*gregaria*). The causes of many ecological explosions remain far from clear and it is interesting to observe that many species which are rare in their normal habitat experience such bursts of population when spread by man to new areas and environments. PAS

Reading and Reference
Drake, J.A., Mooney, H.A., di Castri, F., Groves, R.H., Kruger, F.J., Rejmánek, M. and Williamson, M. eds 1989: *Biological invasions: a global perspective*. Chichester: Wiley.

Elton, C.S. 1958: *The ecology of invasions by animals and plants*. London: Methuen.

†Krebs, C.J. 1978: *Ecology: the experimental analysis of distribution and abundance*. 2nd edn. New York: Harper & Row. Esp. ch. 16.

ecological transition A concept (Bennett 1976) which concerns the reduction in the farmer's dependence on his land that often accompanies his incorporation into the cash economy. The economic opportunities of industry or urban life gradually provide viable social alternatives to rural life and individual farmers can afford to be less concerned about the possibility of long-term decline in the productivity of their land. Over-cultivation may result from this reduced ecological sensitivity in rural populations. ASG

Reference
Bennett, J.W. 1976: *The ecological transition: cultural anthropology and human adaptation*. Oxford: Pergamon.

ecology A branch of science which studies the relations of plants and animals with each other and with their non-living environment. In conception it is epistemologically holistic, but in fact reductionist methodology is most common because of the complexity of the systems (ECOSYSTEMS) involved. No consensus has evolved about the position of humanity in relation to ecological sciences. Is mankind a component of the ecology of a region, albeit often dominant (in which case perhaps the term human ecology is permissible) or is our species an outsider which must have an impact upon ecological relations which are the product of natural selection? The type of thinking of which ecology is a modern representative is of considerable antiquity, and Glacken traces the lineage from Greek teleology through concepts such as the great chain of being. In the modern period the word ecology is generally attributed to Ernst Haeckel; in the twentieth century its study has been widespread in educational and research establishments and its influence upon geography quite strong. In outline there seems to have been a phase of largely ecological inventory concerned with mapping and classifying plant and animal communities, and with study of the ecological relationships of individual species of plants and animals (autecology). In the search for the pattern of the whole community (synecology) workers have looked closely at the functional relations between the components of an ecological system. Thus SUCCESSION, POPULATION DYNAMICS and BIOLOGICAL PRODUCTIVITY have all become important ecological studies, bringing together the outcome of the flows of energy and matter through an ecosystem. The relevance of the study of ecology in MAN–ENVIRONMENT RELATIONS has been profound at an intellectual level, though its influence at policy level is much more limited than that of economics, for example. But we have learned a great deal about the interconnections of phenomena and the unpredictability (because of the complexity of the systems involved) of certain types of change, in particular those resulting from human impact upon biotic communities and their habitats. This has given rise to many attempts at ecological modelling, with stochastic processes being described by a series of equations, normally with the aid of a computer. One outcome is the realization that the ECOSPHERE is not, as nineteenth-century science would have it, like a billiard table but more like a mobile, in which a small touch on one dangling element causes the whole to change its balance. Another powerful ecological idea has been that of CARRYING CAPACITY, in which it is observed that any species' population is kept within limits by its interaction with other biota, energy and nutrient supplies. This idea has fostered the conviction that the earth has a carrying capacity for humanity in terms of the numbers that can be supplied with food and other resources; the concept is sometimes applied on smaller scales as well. (An additional element is that *Homo sapiens* may be subject to a psychological carrying capacity too, because of certain territorial features of our behaviour.) The carrying capacity idea is behind the environmentalist school of thought in man–environment relations. In turn this has led to the use of the term ecology for a set of attitudes concerned with social behaviour ('the Ecology Centre') or political advocacy ('the Ecology Party'). In the form of the 'Greens', this view had some influence in national politics in France and Germany in the 1960s and 1970s.   IGS

Reading
Glacken, C.J. 1967: *Traces on the Rhodian shore*. Berkeley and Los Angeles: University of California Press.

Gorz, A. 1980: *Ecology as politics*. Boston: South End Press; London: Pluto Press.

Krebs, C.J. 1978: *Ecology: the experimental analysis of distribution and abundance*. 2nd edn. New York: Harper & Row.

Odum, E.P. 1971: *Fundamentals of ecology*. 3rd edn. Philadelphia: W.B. Saunders.

ecosphere A synonym for biosphere, but also used as a definition of the zone between Venus and Mars where life as we know it may be possible.

ecosystem The North American ecologist E.P. Odum defines an ecosystem as 'any unit that includes all of the organisms in a given area interacting with the physical environment so that a flow of energy leads to . . . exchange of materials between living and non-living parts of the system'. This amplifies the earlier definition given by A.G. Tansley, who coined the term in 1935, and confirms the concept of the ecosystem in an aggregative hierarchy (see Tansley 1946). Individuals aggregate into populations, populations come together in communities, and a community plus its physical environment comprise an ecosystem. In many ways the concept is independent of scale, for the definition is valid for a drop of water with a few micro-organisms in it or for the whole of planet earth, but in usual practice the term is used for units below the scale of the major world BIOMES. Although an ecosystem may be characterized in synoptic terms, i.e. by an inventory of its components, both biotic and physical, the essential features of the term are (*a*) that it implies a functional and dynamic relation between the components, going beyond a frozen mosaic of species distribution, and (*b*) that it is holistic, implying that the whole possesses emergent qualities which are not predictable from our knowledge of the constituent parts. The study of functional relationships in ecosystems has usually concentrated on phenomena which can be accurately measured and which are common to both biotic and abiotic parts of the system: energy, water, and mineral nutrients are frequent examples. Energy flow through the various TROPHIC LEVELS of a system and its dissipation into heat can be used to see how the system has in its evolution partitioned the energy, and how efficiently it is passed from level to level. Studies of nutrients have often revealed mechanisms for keeping them within the ecosystem: under natural conditions relatively little of the nutrient capital of the system is lost in run-off or animal migration. In arid areas, the use of water by the system may be similarly conserved by a variety of adjustments within the ecosystem, as well as the physiological responses of individual plants and animals. The temporal dimensions of the system are also amenable to study; for example, population numbers through time are often collected. For most species, each ecosystem has a CARRYING CAPACITY, an optimum level for a particular population, which may be a simple number or subject to fluctuations of various kinds. Again, the changes in species composition and physiognomy of an ecosystem through time may be studied, as in the SUCCESSION from bare ground left by a glacier through various types of vegetation to a stable, self-reproducing forest. When succession has apparently terminated at an ecosystem type which sustains itself and gives way, under natural conditions, to no other then this is said to be a mature or climax ecosystem. The applied side of the concept is evident in the idea of BIOLOGICAL PRODUCTIVITY, which is the rate of organic matter production per unit area per unit time, a rate which can be used to compare natural ecosystems with those affected by human activity or indeed totally man-made. The concept of STABILITY is important in the human–biophysical interface because it relates to the resilience of an ecosystem to human-induced perturbation. If we perform a particular act of environmental manipulation, will an ecosystem recover its former state (given the cessation of the impact) or will it perhaps break down irreversibly? The concept itself may also apply to ecosystems with a large man-directed component, such as agriculture (the term agro-ecosystem is sometimes used), pastoralism, fisheries, forestry and even cities themselves, as in the work on Hong Kong by K. Newcombe and colleagues (Boyden *et al.* 1981), in which the urban area is seen as a functional ecosystem with inputs and outputs of energy and matter. IGS

Reading and References
†Boyden, S., Millar, S., Newcombe, K. and O'Neill, B. 1981: *The ecology of a city and its people: the case of Hong Kong.* Canberra: ANU Press.
Odum, E.P. 1969: The strategy of ecosystem development. *Science* 164, pp. 262–70.
Putman, R.J. and Wratten, S.D. 1984: *Principles of ecology.* London: Croom Helm.
Tansley, A.G. 1946: *Introduction to plant ecology.* London: Allen & Unwin.

ecotone The transition on the ground between two plant communities. It may be a broad zone and reflect a gradual blending of two communities or it may be approximated by a sharp boundary line. It may coincide with changes in physical environmental conditions or be dependent on plant interactions, especially COMPETITION, which can produce sharp community boundaries even where environmental gradients are gentle. It is also used to denote a mosaic or interdigitating zone between two more homogeneous vegetation units. They have special significance for mobile animals through edge effects (such as the availability of more than one set of HABITATS within a short distance). JAM

Reading
Allee, W.C., Emerson, A.E., Park, O., Park, T. and Schmidt, K.P. 1951: *Principles of animal ecology.* Philadelphia and London: W.B. Saunders.
Daubenmire, R. 1968: *Plant communities: a textbook of plant synecology.* New York and London: Harper & Row.

ecotope The physical environment of a biotic community (BIOCOENOSIS). It includes those aspects of the physical environment that are influences on or are influenced by a biocoenosis. Together with its biocoenosis, the ecotope forms an integral part of a BIOGEOCOENOSIS. There are two major component parts of the ecotope: the effective atmospheric environment (climatope) and the soil (edaphotope). (See also BIOTOPE; HABITAT.) JAM

ecotoxicology 'The science which includes all studies carried out with the intention of providing information to further our understanding of the effects that chemicals and radiations (that become bioavailable as a direct or indirect result of man's activities) exert on organisms in their natural habitats' (Depledge 1990, p. 251). ASG

Reading and Reference
Depledge, M.H. 1990: New approaches in ecotoxicology: can individual physiological variability be used as a tool to investigate pollution effects? *Ambio* 19, pp. 251–2.
Levine, S.A., Harwell, M.A., Kelly, J.R. and Kimball, K.D. 1988: *Ecotoxicology: problems and approaches.* Berlin: Springer-Verlag.

ecotype Coined by Turesson (1922) to describe populations of organisms within a single species that exhibit genetically produced differences in morphology or physiology which have adapted to a particular habitat, but can interbreed with other ecotypes (ecospecies) of the same species without loss of fertility. Well-known examples come from plants that, because of their low mobility, exhibit evolutionary isolation of subpopulations on small geographic scales. In many cases, these differences amongst habitats result from developmental responses, the phenotypes of populations are fixed but vary among individuals from place to place, which may limit the exchange of genes. Such region and habitat differences in adaptations broaden the ecological tolerance ranges of many species by dividing them into smaller subpopulations, each differently adapted to consistent local environmental conditions. AP

Reading
Krebs, C. 1985: *Ecology: the experimental analysis of distribution and abundance.* 3rd edn. New York.: Harper and Row.
Ricklefs, R.E. 1990: *Ecology.* 3rd edn. San Francisco: W.H. Freeman.
Turesson, G. 1922: The genotypical response of the plant species to the habitat. *Hereditas* 6, pp. 147–236.

edaphic A term referring to environmental conditions that are determined by soil characteristics. Edaphic factors include the physical, chemical and biological properties of soils such as pH, particle size distribution and organic matter content. Plants and animals may be influenced by soil characteristics, although in many instances these edaphic factors interact with other aspects of the HABITAT.

Areas underlain by serpentine rock provide a well-defined example of the influence of edaphic factors on plant distribution. Serpentine soils are low in major nutrients but contain very high levels of chromium, nickel and magnesium. These sites are occupied by distinctive plant communities adapted to these unusual soil conditions. ARH

Reading
Watts, D. 1971: *Principles of biogeography.* New York and London: McGraw-Hill. Ch. 4, pp. 175–84.

edaphology The science that deals with the influence of soils on living things, particularly plants, including man's use of land for plant growth.

eddies Like beauty, eddies depend on the eye of the beholder. In any model or data set we tend to see a slowly varying signal (the 'drift' or the 'climate') plus resolved fluctuations (eddies) and unresolved fluctuations (TURBULENCE and perhaps noise). In a useful record or model these signals and fluctuations define distinct physical processes. When visualizing the nature of the eddy flow it is desirable to subtract the mean flow by plotting the flow relative to the eddy. (Imagine trying to describe the mechanism of a motor-car engine relative to coordinates fixed in a pedestrian.) When this is done there are usually some closed STREAMLINES which are taken by some authors to define an eddy. There is also a tendency to notice eddies only in flow plotted relative to the ground. This makes the definition more pictorial and less mechanistic. Many eddies are noticed because they are the finite-amplitude form of mathematically well-defined solutions to a linearized equation, e.g. Kelvin–Helmholtz. A popular class of eddies is found to the lee of, or propagating downwind of, bluff bodies like steep mountains or islands, such as Jan Mayen or Madeira, where accompanying cloud features can be seen on satellite pictures.   JSAG

Reading
Atkinson, B.W. 1981a: *Dynamical meteorology; an introductory selection.* London and New York: Methuen.
— 1981b: *Meso-scale atmospheric circulations.* London and New York: Academic Press.

eddy diffusivity (eddy viscosity) A conceptual device designed to represent the mixing effect of eddies. By analogy with the kinetic theory of gases the flux of a quantity is supposed to be proportional to the mean gradient of the quantity, the distance over which the air moves and the speed with which it does so. Molecular mean free path of 10^{-7} m multiplied by molecular speed of 300 m s^{-1} gives the kinematic molecular diffusivity of 3×10^{-5} m^2 s^{-1}. Eddy radius of 10 m multiplied by a fluid velocity of 1 m s^{-1} gives kinematic eddy diffusivity of 10 m^2 s^{-1}.   JSAG

Reading
McIntosh, D.H. and Thom, A.S. 1969: *Essentials of meteorology.* London: Wykeham Publications.

edge wave Edge waves are oscillations moving along the shore as a result of instability in wave set-up alongshore. They are surface WAVES with crests perpendicular to the shore, occurring where energy is trapped against the shore. Zero-order edge waves have an amplitude maximum at the shoreline. Higher-order edge waves have one or more zero crossings, with decreasing maxima and minima between them in the offshore direction. They can be standing or progressive waves. The former often occur in bays when reflection takes place. Edge waves are common on beaches steeper than 1 in 10. Sediment is deposited at the nodes of the edge waves, often resulting in rhythmic topography and an even pattern of rip currents. Longer period edge waves form crescentic bars, while shorter ones can form cusps. Rip current spacing is equal to edge wave length. (See p. 170.)   CAMK

Reading
Bowen, A.J. and Inman, D.L. 1971: Edge waves and crescentic bars. *Journal of geophysical research* 76, p. 8662–71.
Huntley, D.A. and Bowen, A.J. 1975: Field observations on edge waves and their effect on beach material. *Geological Society of London journal* 131, pp. 69–81.
Komar, P.D. 1976: *Beach processes and sedimentation.* Englewood Cliffs, NJ: Prentice-Hall.

edge-line The representation on a map of a major topographical break-in-slope by a thick line.

effective rainfall (ER) The proportion of the total PRECIPITATION that is available for a specific purpose. This definition can be refined only when the scale and object of a particular study are known. An agroclimatologist (see AGROMETEOROLOGY), for instance, calculating the availability of water for plant growth, may simply assume that it equals the amount reaching the soil surface and that intercepted (see INTERCEPTION) by vegetation, i.e. referring to the diagram (on p. 171):

$$ER = S + ITC$$

or he or she may take into account RUN-OFF and drainage losses, i.e.

$$ER = S + ITC - DRA - RO.$$

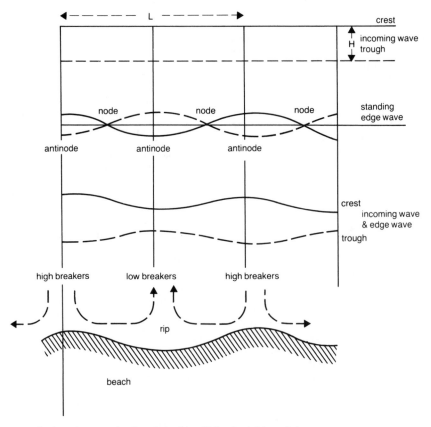

Edge wave *plus incoming curve showing relationship with breaker height and rip current.*

A hydrologist examining discharge may consider it necessary to include run-off and GROUNDWATER contributions, i.e.:

$$ER = RO + GWflow$$

whereas a hydrogeologist may be interested only in the amounts of water remaining in aquifers, i.e.:

$$ER = GWstor$$

CTA

effective stress A concept related to total stress in the MOHR–COULOMB EQUATION, which takes the effects of pore water pressure into account when the strength of a soil is calculated. The pressure exerted by water in soil pore spaces acts upon the grains, tending to force them apart when saturated (positive pore water pressure) or pull them together when the pore spaces have only a small amount of water in them (negative pore

water pressure). The normal stress (σ) is modified by the subtraction of the pore water pressure (u) so that $\sigma' = (\sigma - u)$, where σ' is the effective normal stress. The Mohr–Coulomb equation or failure criterion then becomes:

$$s = c' + \sigma' \tan \phi$$

where the primes denote that effective stresses have been considered. Positive values of u thus tend to decrease the soil strength but negative values (suction) increase it.

WBW

Reading
Mitchell, J.K. 1976: *Fundamentals of soil behaviour.* New York: Wiley.

Smith, G.N. 1974: *Elements of soil mechanics for civil and mining engineers.* London: Crosby Lockwood Staples.

Statham, I. 1977: *Earth surface sediment transport.* Oxford: Clarendon Press.

Whalley, W.B. 1976: *Properties of materials and geomorphological explanation.* Oxford: Oxford University Press.

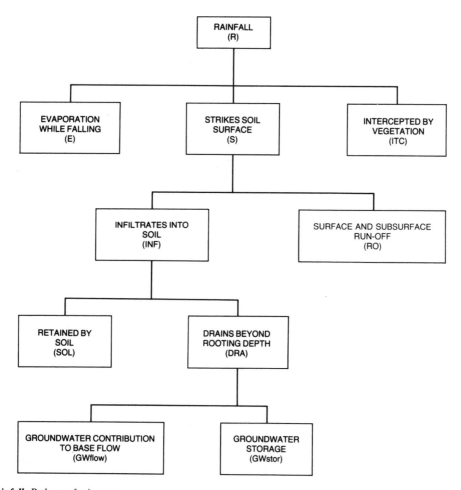

Rainfall: Pathway of rainwater.
Source: *N.G. Dastane 1974: Effective rainfall in irrigated agriculture. Irrigation and drainage paper 25. Rome: FAO.*

effluent The polluted water or waste discharged from industrial plants.

effluent stream A stream which flows from a lake or the small distributary of a river.

egre A tidal bore.

Eh See REDOX POTENTIAL.

Ekman spiral An idealized mathematical description of wind distribution in the atmospheric BOUNDARY LAYER. It is an equangular spiral which forms the locus of the end points of the wind vectors as a function of height, all having a common origin. The GEOSTROPHIC WIND is its limit point. The assumed conditions under which this description is valid are that EDDY DIFFUSIVITY and density are constant within the layer, and the geostrophic wind is constant and unidirectional.

If the x-direction is taken along the geostrophic wind direction, the equations for the components of the wind vectors, u and v, in the x- and y-directions, respectively, at any level z are:

$$u = v_g(1 - e^{-z/\beta} \cos^{z/\beta}); \; v = v_g e^{-z/\beta} \sin^{z/\beta}$$

where v_g is the geostrophic wind speed and $\beta = (2K_M/f)^{\frac{1}{2}}$ where K_M is the eddy viscosity and f is the CORIOLIS PARA-

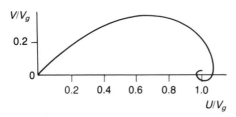

Hodograph of Ekman spiral solution.

METER. The form of the spiral is shown in the diagram.

Through most of the layer the wind blows across the isobars towards lower pressure at an angle which is maximum at the surface at 45°.

The theory of the spiral was developed by Ekman in 1902 for the variation of wind-driven ocean current with depth below the surface.　　KJW

Reading
Holton, J.R. 1972: *An introduction to dynamic meteorology.* New York and London: Academic Press.

el Niño effect The appearance of un-usually warm water off the coasts of Peru, Ecuador and Chile. This is normally a region of upwelling of cold, nutrient-rich water brought up from the lower depths of the ocean to replace surface water driven westward by the trade winds. From late December to March the upwelling usually weakens and is replaced to some extent by warm water moving in from the west and north. This water is low in nutrients, adversely affecting the fish and bird population of the region. El Niño occurs irregularly about fourteen times a century when the influx of warm water is especially intense and widespread. It is associated with heavy rains along the normally dry Peruvian coast and with major shifts in the GENERAL CIRCULATION OF THE ATMOSPHERE.　WDS

Reading
Enfield, D.B. 1989: El Niño, past and present. *Review of geophysics* 27, pp. 159–87.
Pickard, G.L. 1979: *Descriptive physical oceanography.* New York: Pergamon.

elastic rebound theory The contention that the faulting of rocks results from the sudden release, through movement, of the elastic energy that has accumulated owing to pressure and tension in the earth's crust. Sudden movement dissipates the energy.

electrical resistance block This method for determining soil moisture content is based on the principle that the electrical resistance of a block of porous material buried in the soil and in moisture equilibrium with the soil is inversely proportional to the soil moisture content. Electrical resistance blocks can be made of gypsum or nylon and contain two electrodes between which the resistance is measured. Although electrical resistance blocks are used to estimate changes in soil moisture content, the contained moisture is in equilibrium with the matric or capillary potential of the surrounding soil rather than its moisture content. (See also CAPILLARY FORCES; TENSIOMETER.)　AMG

Reading
Curtis, L.F. and Trudgill, A. 1974: The measurement of soil moisture. *British Geomorphological Research Group technical bulletin* 13. Norwich: Geo Abstracts.

electromagnetic radiation Energy that is propagated through space or through a material in the form of an interaction between electric and magnetic wave fields. This is the link between the earth's surface and the majority of sensors used in REMOTE SENSING. The three measurements used to describe these waves are: wavelength (λ) in micrometres (μm), which is the distance between successive wave peaks; frequency (γ) in Hertz (Hz), which is the number of wave peaks passing a fixed point in space per unit time; and velocity (c) in m s^{-1}, which within a given medium is constant at the speed of light. As wavelength has a direct and inverse relationship to frequency, an electromagnetic wave can be characterized either by its wavelength or by its frequency.

Electromagnetic radiation occurs as a continuum of wavelengths and frequencies from short wavelength, high-frequency cosmic waves to long wavelength, low-frequency radio waves (see diagram).

The wavelengths that are of greatest interest in remote sensing are visible and near infrared radiation in the waveband 0.4–1 μm, infrared radiation in the waveband 3–14 μm and microwave radiation in the waveband 5–500 mm.　PJC

Reading
Curran, P.J. 1985: *Principles of remote sensing.* Harlow: Longman Scientific and Technical.
Drury, S. 1990: *A guide to remote sensing: interpreting images of the Earth.* Oxford: Oxford University Press.
Monteith, J.C. and Unsworth, M.H. 1990: *Principles of environmental physics.* 2nd edn. London: Edward Arnold.

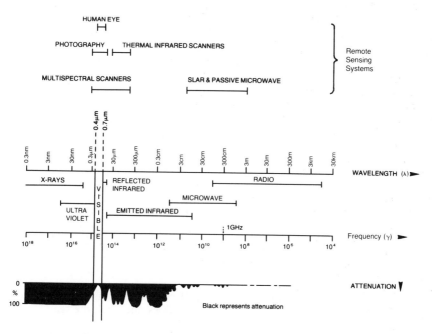

electron spin resonance (ESR) A dating technique applicable to a wide variety of Quaternary materials. It has been used to date materials such as speleothems, molluscs, corals and tooth enamel. However, it is still considered by many workers as an experimental rather than routine dating technique.

The principle of ESR dating in Quaternary studies is based on the relative amounts of induced energy states in electrons produced by the rupturing of electron bonding during radioactive decay. Such radiation results from the natural radio-isotopes of uranium, thorium and potassium contained by the sample and by the enclosing sediment. The sample acts as a natural dosimeter recording the cumulative radiation dose received at the sample site since deposition. Electrons become trapped at charge deficit sites associated with defects and impurities within the crystal lattice. These trapped electrons form paramagnetic centres or radicals whose density may be measured directly, repeatedly and non-destructively by ESR.

ESR simply provides a means of measuring the cumulative effects of this radiation on the sample, calibrating the effects of this radiation on the sample, and of calibrating the sensitivity of the sample of radiation. It is particularly useful as it uses small samples (<2g) of materials which cannot be dated using other techniques and whose age is greater than the applicable range of methods. The dating range of this technique in theory should be applicable to the whole of the Quaternary time-scale.   AP

Reading
Ikeya, M. 1985: Dating methods of Pleistocene deposits and their problems: IX electron spin resonance. In Rutter, N.W. ed., *Dating methods of Pleistocene deposits and their problems*. Geoscience Canada Reprint Series 2, Geological Association of Canada. Pp. 73–87.
Smart, P.L. 1991: Electron spin resonance (ESR) dating. In Smart, P.L. and Frances, P.D. eds, *Quaternary dating methods: a user's guide*. Technical Guide 4. Quaternary Research Association, Cambridge. Pp. 128–60.

eluviation The movement of soil materials through the soil zone resulting in depletion and accumulation in different horizons. It occurs *in situ* but components of the soil may be moved laterally or vertically.

eluvium The material which is produced through the rotting and weathering of rock in one place. *In situ* weathered bedrock.

endangered species One of a group of terms used to describe the status of wildlife. The following definitions are those employed by the Species Survival Commission of the World Conservation Union (IUCN) and are accepted for use by international bodies such as the Convention on Trade in Endangered Species of Flora and Fauna.

extinct: species not definitely located in the wild during the past fifty years.

endangered: species in danger of extinction, whose survival is unlikely if the causal factors continue operating.

vulnerable: species believed likely to move into the 'endangered' category in the near future if the causal factors continue operating.

rare: species with small world populations that are not at present 'endangered' or 'vulnerable'.

indeterminate: species known to be 'endangered', 'vulnerable' or 'rare', but not enough information is available to determine which category is appropriate.

insufficiency known: species that are suspected of belonging to one of the above categories but are not definitely known to be due to lack of information.

threatened: a general term used to denote species which are in any of the above categories.   ASG

endemism The confinement of plant and animal distributions to one particular continent, country or natural region. Thus the white spruce (*Picea glauca*) is endemic to North America, the coast redwood (*Sequoia sempervirens*) to coastal California and Southern Oregon, and the genus *Dasynotus* to a few square kilometres of Idaho. Organisms confined to one small island, one mountain range or just a few restricted localities are termed 'local' or 'narrow' endemics, whereas those with more substantial distributions are called 'broad' endemics.

The number of endemics in the northern hemisphere is lower than that in the southern hemisphere, although their count is reduced in both hemispheres where the lands were occupied by the cap of the continental glacier during the Pleistocene. Endemism is most marked on islands, such as Darwin's famous Galápagos Islands, particularly in the warmer regions of the world. No less than 90 per cent of the native plants of the Hawaiian Islands are endemic to the island group. Endemism is also marked on isolated mountain tops, especially in the tropics. There is usually a close relationship between the number and type of endemics and the geological age of the habitats they occupy. Naturally, the study of endemism is one of the main ways of characterizing the different faunal and floristic regions of the world. (See also ISLAND BIOGEOGRAPHY; REFUGIA; VICARIANCE BIOGEOGRAPHY.)   PAS

Reading
Daubenmire, R.F. 1978: *Plant geography, with special reference to North America.* New York and London: Academic Press.
Richardson, I.B.K. 1978: Endemic taxa and the taxonomist. In H.E. Street ed., *Essays in plant taxonomy.* London and New York: Academic Press. Pp. 245–62.
Stott, P. 1981: *Historical plant geography.* London: Allen & Unwin.
Williamson, M. 1981: *Island populations.* Oxford: Oxford University Press.

endogenetic Pertaining to the forces of tectonic uplift and disruption originating within the earth and to the landforms produced by such processes (see EXOGENETIC).

endoreic Pertaining to a drainage system which does not reach the sea.

endrumpf A peneplain. A landsurface that has been reduced to a flat plain or gently undulating landscape by erosive processes.

energy flow May be loosely defined as the energy transformations which occur within the planet earth system. Energy is the capacity to do work and, for present purposes, includes mechanical, chemical, radiant and heat energy. Mechanical energy may be further subdivided into kinetic and potential energy. Kinetic energy or free 'useful' energy is possessed by a body by virtue of motion and is measured by the amount of work required to bring the body to rest. Potential energy is stored and becomes useful only when converted into the free form and can do work. This includes movement, friction, and the expenditure of heat. On earth, energy sources are solar radiation, rotational energy of the solar system and radiogenic heat involving geothermal heat flow. It is no exaggeration to claim that physical geography is the science of energy flow from these sources through the atmospheric, oceanic, aeolian, fluvial (hydrological cycle), glacial, biological, human, tectonic and geothermal systems as described by the laws of thermodynamics and as defined by complex circulation patterns and thermal gradients.

The main circulations are atmospheric circulation, heat and moisture balance, photosynthesis and ecological energetics, the hydrological cycle and tectonics. The important gradients are latitudinal, altitudinal, seasonal, daily, heat gradients and across system contrasts (e.g. land–sea).

DB

Reading
Bloom, A.L. 1969: *The surface of the earth.* Englewood Cliffs, NJ: Prentice-Hall.
Caine, N. 1976: A uniform measure of sub aerial erosion. *Bulletin of the Geological Society of America* 87, pp. 137–40.
Chapman, D.S. and Mach, A.N. 1975: Global heat flow: a new look. *Earth and planetary science letters* 28, pp. 23–32.

energy grade line Water flowing down a channel loses energy because of the work done in overcoming FRICTION, both internally and with the channel perimeter, and in transporting sediment. The rate of energy loss per unit length of channel is measured by the energy gradient, which is the slope of the energy grade line. This line may be plotted above the water surface at a distance equal to the velocity head ($v^2/2g$), and therefore measures the variation of the total energy (potential and kinetic) of the flow (see GRADUALLY VARIED FLOW). In UNIFORM STEADY FLOW the energy grade line, water surface and channel bed are parallel. In gradually varied flow, the three slopes differ but the energy grade line always slopes downwards in the direction of flow.

KSR

energy in man–environment relations As far as science is concerned, it is energy rather than love that makes the world go round, for apart from the spin imparted to the globe at some stage of its birth, all processes both of a purely biophysical nature and of a cultural kind are suffused with the gathering, use and transformation of energy, and may sometimes indeed be classified by the quantities of energy which flow through them. The types of energy which animate the biological processes of the ECOSPHERE are gravitational, solar, geothermal or magnetic in origin. Solar radiation may be transformed into chemical energy by the process of photosynthesis, which is confined to green plants. Human use of energy may be of two types: (*a*) metabolic, i.e. those uses of energy which sustain individuals as animal organisms in the form of a nutritional intake sufficient for daily life and for successful reproduction; and (*b*) cultural, i.e. energy uses which are not apparently immediately necessary for survival but which are an integral part of a culture, whether it be making beads in a simple society or motor vehicles in a complex one. Thus access to energy can be seen to classify contemporary societies: the less developed have a low per capita consumption of energy for cultural purposes (indeed, for some the metabolic energy is below the optimum as well), whereas the developed use great quantities of energy for heating, lighting, transport, manufacture, services and a myriad of other needs.

Human societies which subsisted on HUNTING AND GATHERING had access to solar energy, mostly in a recently fixed chemical form, as the plants and animals they ate and transformed into somatic energy, which might be intensified in its application by tools, e.g. the blowpipe which takes the energy of one set of muscles and directs it down a single channel (the bow, sling and slingstick work on similar principles). In addition the release of energy from organic sources through FIRE gave such groups a potent weapon to change whole ECOSYSTEMS and indeed to manipulate them for higher energy yield. The ability to produce and control fire may also have given early men the ability to colonize non-tropical regions of the earth. The development of domesticated plants and animals intensifies the use of solar energy because the crop which yields the energy is all in one place, whether it be a field of cereals or a herd of animals. Such societies are likely to produce an energy surplus which can be devoted to non-subsistence purposes, e.g. the maintenance of a large population of city-dwellers. Agriculturalists are also likely to have access to wind energy (derived from solar radiation) as in windmills and sailing ships, and gravitational energy, as in watermills and even a few tidal mills. A huge change came in the nineteenth century with widespread access to stored photosynthetic energy in the form of fossil fuel, which enabled societies to develop steam power and then electric power. The increased energy availability gave access in turn to many more materials (especially iron and other metals) leading to the development of industrial societies which now use the full gamut of energy sources other than, perhaps, the magnetic fields of the earth.

To these has been added since 1945 the energy of the atomic nucleus, which can be unleashed by either fission or fusion. This energy can be used to generate electrical power or in weapons. In the latter form sufficient destructive energy, including that of radioactivity, could be released to destroy not only man-made structures but most ecosystems and their living components as well. 'Energy is eternal delight', said William Blake, but he put the words into the mouth of the Devil.   IGS

Reading
Cottrell, F. 1955: *Energy and society*. Westport, Conn.: Greenwood Press.
Holdgate, M.W., Kassas, M. and White, G.F. eds 1982: *The world environment 1972–1982*. Dublin: Tycooly International for UNEP. Ch. 12.
Odum, H.T. and Odum, E.C. 1981: *Energy basis for man and nature*. 2nd edn. New York: McGraw-Hill.

englacial Describes conditions within the body of a glacier and is therefore to be distinguished from the *subglacial* environment which is beneath a glacier, the *supraglacial* environment on the glacier surface and the *proglacial* environment in front of the glacier margin.

entrainment A concept most widely used in environments including fluids. It is most conveniently explained using an example from the atmosphere. Entrainment means the mixing of environmental air into a pre-existing organized air current so that the environmental air becomes part of the current. Such a process occurs at the edges of cumulus clouds where clear air is entrained into the cloud. The concept has ready applicability to all physical geographical cases involving fluids.   BWA

entropy A concept in thermodynamics which describes the quantity of heat supplied at a given temperature. The entropy of a unit mass of substance remains constant in the process of adiabatic expansion, i.e. when no additional heat is supplied to the mass. The term has been transferred by analogy to other fields (notably information theory) and introduced into geomorphology by Chorley (1962) and Leopold and Langbein (1963) as a measure of energy unavailable to perform work (high entropy equals greater unavailability of energy). This analogue use has been widely criticized as inappropriate.   BAK

References
Chorley, R.J. 1962: Geomorphology and general systems theory. *United States Geological Survey professional paper 500-B*.
Leopold, L.B. and Langbein, W.B. 1963: The concept of entropy in landscape evolution. *United States Geological Survey professional paper 500-A*.

environmental assessment An attempt objectively to evaluate the quality of the environment in terms of biophysical attributes and/or aesthetic value. It is used to give the natural environment credibility and comparability with socioeconomic data in planning decisions. The concept of environmental assessment has been important since the 1960s with the gradual recognition that problems such as loss of habitat and genetic variety, pollution, population growth and an increasing reliance on non-renewable energy and mineral resources cannot be solved by economic growth or technology and that finite limits exist within the environment.   SMP

environmental economics The aim to conserve, maintain, use and reuse natural resources so that the quality of life is retained without excessive waste. It recognizes that the major cause of environmental problems in market systems is failure of the incentives generated by these markets to lead towards efficient use of resources.

At the beginning of the twentieth century Alfred Marshall suggested that the basis of orthodox economics – the market – presented an oversimplified idea of reality and introduced the concept of external costs and benefits. These describe instances where one fiscally independent economic unit directly affects another, without intervention of the market. On the cost side the instance of a locomotive igniting an adjacent field is commonly quoted; a good example of a beneficial externality is that of bees pollinating an orchard owner's apples.

The 1960s and 1970s witnessed the rapid development of these ideas owing to an increasing awareness of environmental problems, particularly pollution, and the notion that both free enterprise and Marxist economics encourage the squandering of natural resources. A profound asymmetry had developed in the effectiveness and efficiency of the market system which works well in stimulating the exploitation, processing and distribution of basic resources, but almost completely fails in the efficient disposal of residuals to common property assets.

The natural environment has clearly been largely ignored in the conventional economic account and commonly the earth is regarded as a bottomless rubbish dump. Boulding (1966, 1970) describes this as the 'cowboy economy', in which success is measured in terms of the amount of material turned over by the factors of production. He compares this to the 'spaceship economy' in which maintenance of existing capital stocks within limits is the criterion for success.

The practice of economics is certainly still the single most important feature governing the relationship between man and the environment. However, environmental economics acknowledges its own limitations for regulating current human resource requirements. Price cannot always be equalled with value; for instance, unspoilt countryside may have intrinsic value but cannot have an accurate price fixed upon it. The interest of the individual in the market is often not the same as the general interest (Hardin 1968), indeed some resources (e.g. clean air) are not within the market so are not subject to the choices normally available. Furthermore, market economies are not well suited to respond to problems which suddenly force themselves upon resource managers or the public (e.g. disease causing a harvest to collapse) nor to cope with the long timelags within which complex technology needs to develop substitutes.

The ultimate goal of environmental economics is to reach the steady state with little or no economic growth in the industrialized nations (Boulding 1966). In so doing it attempts to investigate that which now governs price and supply of materials and, in particular, the role of energy, the economic relations of rich and poor countries and the construction of new measures of human welfare in terms of the whole resource process, rather than just a portion of it, which has for so long been the sole concern of economists.   SMP

Reading and References
Boulding, K.E. 1966: Ecology and economics. In F. Fraser Darling and J.P. Milton eds, *Future environments of North America*. New York: Natural History Press.
— 1970: The economics of the coming spaceship earth. In H. Jarrett ed., *Environmental quality in a growing economy*. Baltimore: Johns Hopkins University Press.
†Cottrell, A. 1978: *Environmental economics*. London: Edward Arnold; New York: Halsted Press.
Hardin, G. 1968: The tragedy of the commons. *Science* 162, pp. 1243–8.
†Knese, A.V. 1977: *Economics and the environment*. London and New York: Penguin.
Marshall, A. 1930: *Principles of economics*. 8th edn. London: Macmillan.

environmental engineering geomorphology The application of geomorphological techniques and analyses to the solution of planning, environmental management, engineering or similar problems. It includes resource surveys, terrain analysis, evaluation of surface form and material properties, hazard surveys, process monitoring, laboratory analysis and experiment, hardware models of engineering sites and computer modelling. It is the fastest growing area of geomorphology, owing to an increasing awareness by decision makers of the complexity of environmental conditions, the significance of natural hazards for new projects in developing areas, and new standards for the application of geoscience knowledge to engineering. New methods of funding applicable research, government contracts, and public awareness have provided impetus. Work depends on the provision and use of quantitative data bases, including hydrology, soil mechanics, structural geology and subsurface detail. A major benefit is the growth of interdisciplinary studies and the development of a comprehensive geoscience–geotechnics programme of investigation.   DB

Reading
Brunsden, D., Doornkamp, J.C. and Jones, D.K.C. 1978: Applied geomorphology: a British view. In D. Embleton *et al.* eds, *Geomorphology: present problems, future prospects*. Oxford: Oxford University Press.
Cooke, R.U. 1976: Urban geomorphology. *Geographical journal* 142, pp. 59–65.
Cooke, R.U. and Doornkamp, J.C. 1990: *Geomorphology in environmental management*, 2nd edn. Oxford: Oxford University Press.
Douglas, I. 1976: Urban hydrology. *Geographical journal* 142, pp. 65–72.
Hails, J. 1977: *Applied geomorphology*. Amsterdam and New York: Elsevier.
Jones, D.K.C. 1980: British applied geomorphology: an appraisal. *Zeitschrift für Geomorphologie* Suppl. 36, pp. 48–73.
Thornes, J.B. 1979: Research and application in British geomorphology. *Geoforum* 10, pp. 253–9.

environmental impact A net change, either positive or negative, in man's health and well-being and in the stability of the ecosystem on which man's survival depends. The change may result from an accidental or a planned action and can affect the change in balance either directly or indirectly.

Direct impacts are generally premeditated and planned and are commonly felt soon after environmental modification. The effects are often long term but normally reversible and include alterations such as land use changes, various constructional and excavational activities, the direct ecological impact of agricultural practices and the direct effects of weather modification programmes.

In contrast, indirect effects are normally unplanned and are often socially, if not economically, undesirable. Effects are often delayed until well after the original impact and depend upon the sensitivity of the system to change, the existence of threshold conditions, and interaction between different side-effects of the initial impact. Many impacts are long term, cumulative and irreversible, difficult to identify and almost impossible to predict, and include the introduction of DDT and other toxic elements into the environment and the subsequent accumulation of those into food chains over a wide area, triggering long-term and possibly long-range climatic modifications by particulate and gaseous pollution and indirect local climatic effects associated with changed land surface configuration or material composition.

Many impacts are caused by pressures related to the rapid increase in population

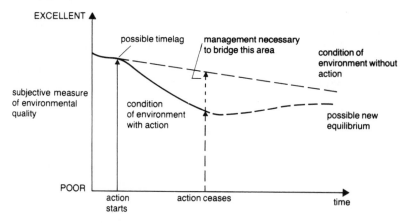

1. *Basic resource process.*

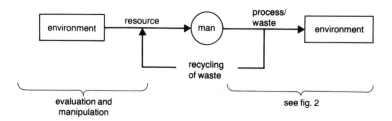

1 management may be preventative

2 management may be too late

2. *Breakdown of effect of man's actions.*

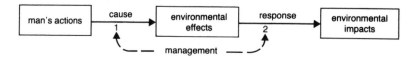

3. *Changing environmental quality resulting from man's activity.*

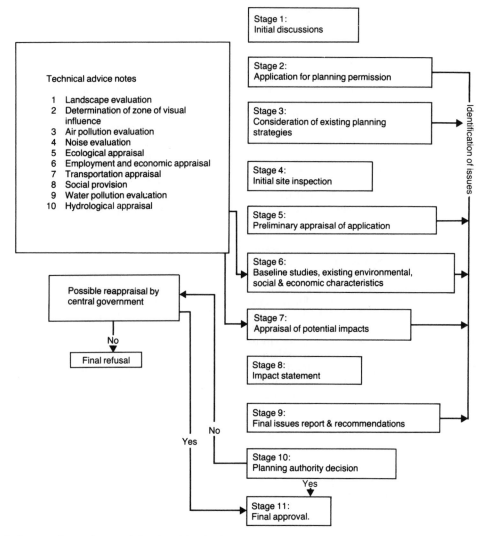

Stage 1:
Initial discussions

Stage 2:
Application for planning permission

Stage 3:
Consideration of existing planning strategies

Stage 4:
Initial site inspection

Stage 5:
Preliminary appraisal of application

Stage 6:
Baseline studies, existing environmental, social & economic characteristics

Stage 7:
Appraisal of potential impacts

Stage 8:
Impact statement

Stage 9:
Final issues report & recommendations

Stage 10:
Planning authority decision

Stage 11:
Final approval.

Identification of issues

Technical advice notes

1 Landscape evaluation
2 Determination of zone of visual influence
3 Air pollution evaluation
4 Noise evaluation
5 Ecological appraisal
6 Employment and economic appraisal
7 Transportation appraisal
8 Social provision
9 Water pollution evaluation
10 Hydrological appraisal

Possible reappraisal by central government

No

Final refusal

No

Yes

Yes

Yes

4. *An approach to project appraisal under existing development control procedures.*
Source: *Clark 1976.*

growth, especially on physical resources – land, food, water, forests, metals – and on the biological environment, whose ability to remove and recycle human waste and provide an important set of functions such as pest control or fish production is being severely strained. In addition there are pressures on society's ability to dispense services – education, medicine and law administration – and in personal values such as privacy, freedom from restrictive regulations and the opportunity to chase a lifestyle.

It is not only population growth that causes these pressures. The consumption of materials and energy per person have grown simultaneously, linked with the type of technology that is being employed to facilitate consumption, and the economic, political and social forces influencing decision-making are contributory factors. The ability of both individuals and governments to react has not kept pace.

The environmental impact may be the final stage of a basic resource process, whereby man takes a resource from the environment, uses it in some fashion and then returns it to the environment (figure 1). Man's actions, whether they be legislative proposals, policies, programmes or operational procedures may well set into motion or accelerate environmental effects – which are, for example, dispersal of pollutants, soil erosion or displacement of persons – unless some form of preventative management is initiated (figure 2). If management at this stage is ineffectual or avoided, however, an environmental impact will probably occur, and any subsequent management may be too late, too expensive or merely palliative. Unfortunately most knowledge is still concerned with what is put into the system rather than with how the environment responds to it, and it is this response that defines the nature and magnitude of the environmental impact.

If, at a later stage, man completes his action the response may be a lessening of the environmental impact (figure 3) owing to natural homeostasis in the system. A good example is a colliery spoil heap which gradually revegetates sometime after mining has ceased. It is rare, however, for the new equilibrium even to approach the original environmental quality. There are two main approaches to the assessment of environmental impacts. In the first the problem of resources receives particular attention and some of the more subjective ecological and aesthetic aspects of the environment, although recognized as important, are not included. One attempt (Ehrlich *et al.* 1977) tried to relate the problem in terms of population and consumption:

environmental impact = population × consumption of goods per person × environmental impact per quantity of goods consumed or
environmental impact = population × affluence × technology.

However most forms of consumption give rise to many forms of environmental impact; changes in technology might reduce some impacts but increase others and different impacts associated with alternative technologies (e.g. oil spills and coal mining) are difficult to compare, with the result that the above measure is almost impossible to quantify.

The second approach considers the probable consequences of human intervention in the natural environment with the goal of minimizing environmental damage, while developing resources. Known as Environmental Impact Assessment (EIA) it was developed by ecologists and economists in the USA and became a legal requirement under S102(2)(C) of the Natural Environmental Policy Act 1969 for all major federal actions affecting the human environment.

In Europe, EIA is used early in existing planning procedure and does not replace it. It identifies adverse effects, suggests alternatives to proposals and considers short-term local environmental use in relation to long-term productivity. It imposes limitations, therefore, on the magnitude and impact of resource processes.

The flowchart (figure 4) outlines a general impact evaluation methodology. The critical part is stage five – preliminary appraisal – as the value of any assessment of impact will depend heavily on the thoroughness with which the appraisal work is undertaken. Both the specific characteristics and the potential interactions between the existing local situation and that in the development proposal are evaluated and are normally identified using an impact matrix, which acts as a framework and checklist.

Most commonly used is the Leopold matrix which lists all possible environmental, social and economic parameters and attempts to evaluate the importance and magnitude of each impact identified. Others include the component interaction matrix, normally used for biological, physical and climatic dependencies, the disruption matrix, which measures environmental disturbance for each alternative proposal and the Sorenson matrix, a three-stage computer analysis which may also suggest corrective action or control mechanisms.

Baseline studies (stage six) are used to provide information for the matrix and both are considered to appraise the potential impacts, and produce an ENVIRONMENTAL IMPACT STATEMENT.   SMP

Reading and References
Clark, B. 1976: Evaluating environmental impacts. In T. O'Riordan and R. Hey eds, *Environmental impact assessment*. Farnborough: Saxon House.

Ehrlich, P.R., Ehrlich, A.H. and Holdren, J.P. 1977: *Ecoscience*. San Francisco: W.H. Freeman.

†O'Riordan, T. 1981: *Environmentalism*. 2nd edn. London: Pion.

†Park, C.C. 1980: *Ecology and environmental management*. Studies in physical geography 3. Folkestone, Kent and Dawson, Colorado: Westview Press.

†Simmons, I.G. 1981: *The ecology of natural resources*. 2nd edn. London: Edward Arnold.

environmental impact statement (EIS) The summary of all the information gathered on each potential ENVIRONMENTAL

IMPACT that might be realized for a given development proposal. It is an integral part of the procedure of environmental impact analysis.

The EIS discusses as succinctly as possible: (1) a brief description of a proposed action; (2) the likely impact of the proposed action on the environment; (3) any predicted adverse or beneficial effects as a result of the proposal; (4) whether the impacts are likely to be long term or short term; (5) whether the impacts will be reversible or irreversible; (6) the range of direct and indirect impacts associated with the proposed action; and (7) whether the impacts are likely to be of local and/or national strategic significance. In addition, the prospects for the area under consideration if the development does not take place are outlined so that decision-makers can compare the potential effects of approving the application with the implications of the 'no change' alternative.

Most EISs do not need non-environmental questions (such as economic impact) to be included, although many agencies voluntarily include such information. The aim of the EIS is to identify and develop methods and procedures and use empirical information to ensure that unquantified environmental amenities and values are given appropriate consideration in decision-making along with economic and technical considerations. An EIS may, therefore, include rankings or hierarchies to differentiate the degree of importance or magnitude of impacts, although this procedure can lead to a false sense of objectivity.   SMP

Reading
O'Riordan, T. and Hey, R. eds 1976: *Environmental impact assessment*. Farnborough: Saxon House.
Wathern, P. ed. 1990: *Environmental impact assessment: theory and practice*. London: Routledge.

environmental lapse rate See LAPSE RATE.

environmental management Provides resources from the bioenvironmental systems of the planet but simultaneously tries to retain sanative, life-supporting ecosystems. It is therefore an attempt to harmonize and balance the various enterprises which man has imposed on natural environments for his own benefit. To achieve this, long-term strategies are evolved, based on reducing stress on ecosystems from contamination or overuse. In addition, environmental management pursues short-term strategies that are sufficiently flexible to preserve the long-term options: in other words, no resource process that brings about irreversible environmental changes should be allowed to develop.

This temporal scale is of major importance in environmental management; for example, engineering solutions may be necessary in the short term to check localized coastal erosion, but in the long term conservation of the entire coast with acceptance of slow erosion, accretion or movement may be required. Extended time perspectives may also have to accommodate extreme events.

Different approaches are similarly evident with the spatial element. Using the coastal example again: mangement may consider a single beach profile or a complete sedimentary cell involving supply of material from a river mouth, movement, storage in beaches and abstraction to dunes or marine deeps. In a predominantly agricultural area the maintenance of a wild population of predatory birds for scientific interest or pest control is not possible with an isolated nature reserve, but requires a whole network of protected areas.

Environmental management is, therefore, dealing with the rationalization of the resource process – the flows of material (and energy) from natural states through a period of contact with man to their ultimate disposal. It has much in common with environmental planning and those two are sometimes used interchangeably. Strictly, planning approaches the natural resource problem with a cultural or demand bias, whereas in environmental management the emphasis is on the resources themselves. It is a dynamic discipline which does not urge the preservation of resources at all costs but attempts to identify or specify major groups of resources, consider the way each changes, evaluate and resolve conflicting demands upon them and finally conserve the resources. It believes that the very process by which renewable resources are produced can be manipulated, but that the production of non-renewable resources is virtually unmanageable because of the timescale under which they develop.

Traditionally, environmental management has considered that, generally speaking, environmental problems need more adjustment to socioeconomic systems and has been more concerned with the maintenance of the ecological and geomorphological balance – for instance the study of and control of movements of pollutants and pesticides in food chains, overtrampling of ecologically interesting swards – leading to the preparation of conservation strategies. However, it is becoming more socially aware as renewable resource systems become so thoroughly altered by man that they can seldom be left to produce or even function in a stable condition without interdisciplinary management involving biological and physical, economic and political, and scientific and aesthetic approaches.

In terms of values environmental management is rather ambivalent towards economic growth, recognizing that there is an absolute limit to materials and the surface area of the planet, but seeing little reason to prevent resource use unless ecological stability is threatened. In any natural ecosystem the overall limiting factor must be the amount of incident solar energy, but within this context other boundaries may operate. Man may alleviate a critical unit, e.g. by using chemical fertilizers, or may introduce a new lower limit, e.g. untreated sewage in coastal waters may decrease the light reaching littoral and sublittoral vegetation, limit productivity and so reduce the recreational potential of the system.

Different societies have differing attitudes to environmental management determined by their own order of priorities. In the USA and Europe the dominant purpose of environmental management in the past has been to obtain useful materials, an emphasis which is decreasing in favour of more concern about the life-supporting role of the ecosystem and the aesthetic value of the environment given greater impacts upon it. In contrast, the struggle to obtain food has always dominated environment management in countries such as Egypt, where the gathering of useful materials has taken a secondary role and is largely regarded only as a basic development aimed at export markets, and the care of wildlife and aesthetic preservation of the environment is of peripheral interest with little value.

It is unrealistic to pretend that totally successful environmental management is currently much more than a concept, except perhaps in the relatively simple situation of Antarctica, where resource processes are readily identifiable and man's intrusion is limited. Some process response reactions, e.g. the avoidance of flood hazard, might be regarded as environmental management at a local scale, but there are critical problems within the concept. One of the most important is the dualism between ecology and economics which have completely different resource and time perspectives. It is also quite difficult in many cases to distinguish between changes which have been brought about exclusively by man and those which are at least partly natural. A further difficulty is that problems of environmental management are not the same worldwide on account of differing attitudes of wealth and its distribution; sociopolitical systems, population growth rates, and the implementation of western 'developed' culture leading to rapid urbanization and industrialization, which takes no account of natural environmental processes. Indeed, in many instances environmental problems are caused by unbalanced or over-rapid development rather than a complete disregard for environmental management.

SMP

Reading
Blacksell, M. and Gilg, A.W. 1981: *The countryside: planning and change.* Resource Management series 2. London and Boston: Allen & Unwin.

Douglas, I. 1983: *The urban environment.* London: Edward Arnold.

Edington, J.M. and Edington, M.A. 1977: *Ecology and environmental planning.* London: Chapman & Hall; New York: Wiley.

Goldsmith, F.B. and Warren, A., 1993 *Conservation in progress.* London and New York: Wiley

O'Riordan, T. 1981: *Environmentalism.* 2nd edn. London: Pion.

Park, C.C. 1980: *Ecology and environmental management.* Studies in physical geography 3. Folkestone, Kent and Dawson, Colorado: Westview Press.

Simmons, I.G. 1981: *The ecology of natural resources.* 2nd edn. London: Edward Arnold.

epeiric sea A shallow body of marine water on the continental shelf which is connected with an ocean.

epeirogeny The warping of large areas of the earth's crust without significant deformation. It can be contrasted with OROGENY, which is associated with linear zones of uplift. Epeirogenic uplift can affect regions thousands of kilometres across and is the major form of uplift in most CRATONS. The causes of the predominantly vertical movements associated with epeirogeny are uncertain but may be related to expansion resulting from localized heating within the crust or at the base of the LITHOSPHERE, possibly in conjunction with phase changes in the MANTLE.   MAS

Reading
Crough, S.T. 1979: Hot spot epeirogeny. *Tectonophysics* 61, pp. 321–33.
Ollier, C.D. 1981: *Tectonics and landforms*. London: Longman.

ephemeral, plant A species with short life cycles, such as many so-called annual plants which may complete a number of life cycles (from seed to seed) within a year. Such plants are fairly extreme *r*-strategists (see *R*- AND *K*- SELECTION) with a high investment in reproductive capacity and a high rate of dispersal (e.g. the dandelion), few defences against predators and seeds capable of long dormancy periods – several years in many cases.   KEB

Reading
Begon, M., Harper, J.L. and Townsend, C.R. 1990: *Ecology: individuals, populations and communities*. 2nd edn. Oxford: Blackwell Scientific.
Solbrig, O.T. ed. 1980: *Demography and evolution in plant populations*. Botanical Monographs, vol. 15. Oxford: Blackwell Scientific.

ephemeral stream A stream which is often one of the outer links of a DRAINAGE NETWORK and which contains flowing water only during and immediately after a rainstorm which may be fairly intense. As the water flows along the ephemeral stream channel it may infiltrate into the channel bed as a transmission loss by influent seepage and therefore the peak discharge may decrease downstream along the ephemeral channel by as much as 5 per cent per km of channel. In arid and semi-arid areas of the world ephemeral streams are very extensive and represent the major CHANNEL TYPE.   KJG

Reading
Renard, K.G. and Laursen, E.M. 1975: Dynamic behaviour model of ephemeral streams. *Journal of the Hydraulic Division of the American Society of Civil Engineers* 101, pp. 511–28.
Thornes, J.B. 1977: Channel changes in ephemeral streams: observations, problems and models. In K.J. Gregory ed., *River channel changes*. Chichester: Wiley. Pp. 317–35.

epicentre The point on the earth's surface which lies directly above the focus of an earthquake.

epilimnion The surface layer of water of a lake or sea. The water which lies between the surface and the thermocline.

epipedon A diagnostic surface horizon which includes the upper part of the soil that is darkened by organic matter, or the upper eluvial horizons, or both.

epiphyte A plant which grows upon the surface of another plant but does not obtain food from the host plant.

epoch A unit of geologic time equivalent to a series, a division of a period.

equation of state A relationship between properties of a material, and/or forces acting on it, in a state of equilibrium. The most familiar examples of equations of state in physical geography refer to the balance of forces acting on a body or within a material in equilibrium. Stability analysis for landslides is, for instance, based on such a balance. Equations of state may also refer to other properties, and the term has a special significance in thermodynamics as the unique relationship between temperature, pressure and volume for a body of fluid.   MJK

equations of motion Expressions governing the motion of a body or a material under the action of a force or forces. The equations most commonly take the form:

$$\text{force} = \text{mass} \times \text{acceleration}$$

in either its linear form or as moments (torques) about a centre. Since force is a vector, the equation is a vector equation, and may be resolved to give up to three component equations in directions which are mutually at right angles. Equations of motion are one example of expressions which control the rate of a process, usually subject to the constraints of the CONTINUITY EQUATION. The term was originally used in the context of solid bodies, but may also be used to describe the motion of a fluid such as water or air, either travelling with a physical body of fluid or describing motion at a fixed point.   MJK

equatorial rain forest A lowland evergreen TROPICAL FOREST lying approximately 5° north and south of the equator in near-continuous rainfall climates, over

2000 mm year^{-1}, and not limited by low temperatures. The forests are multilayered, over 30 m tall, shallow rooted and often buttressed, containing a profusion of climbing plants and epiphytes and the greatest diversity and abundance of plants and animals of any terrestrial biome. Their BIOLOGICAL PRODUCTIVITY also heads the league for terrestrial biomes. Although the main global formations are comparable in structure, life forms and animal adaptations, the biological evolution and species composition is profoundly different in each area.

PAF

Reading
Golley, F.B., Lieth, H. and Werger, M.J.A. eds 1982: *Tropical rain forest ecosystems*. Amsterdam: Elsevier.
Longman, K.A. and Jenik, N. 1987: *Tropical forest and its environment* 2nd edn. London: Longman.

equatorial trough A narrow, fluctuating belt of unsteady, light, variable winds, low atmospheric pressure, and frequent small-scale disturbances. It is located near the equator between the trade wind belts of the two hemispheres. However its position, breadth, and intensity are constantly changing. From time to time it disappears completely, especially over the continents. Along its meandering position occur most of the frequent, heavy showers for which the tropics are so well known. The equatorial trough includes the prevailing calms of the DOLDRUMS and is frequently referred to as the INTERTROPICAL CONVERGENCE ZONE (ITCZ).

WDS

Reading
Rumney, G.R. 1968: *Climatology and the world's climates*. New York: Macmillan.

equifinality Arises when a particular morphology (e.g. a landform) can be generated by a number of alternative processes, process assemblages, or process histories. Under such circumstances the morphology alone cannot be used as a basis for reconstructing the process of origin of a feature. For example, a central assumption in CLIMATIC GEOMORPHOLOGY is that landforms differ significantly between climatic zones because of variation in the climatic factors that control weathering, run-off, erosion and deposition. However, specific landforms may originate in different ways, and are therefore not restricted to single climate zones. U-shaped valleys are characteristic of glaciated highlands, but also occur in high-relief sub-

tropical areas where basal sapping maintains steep valley sides after intensive chemical weathering at the water table (Wentworth 1928). Tors are also features produced by quite distinct sets of processes in different areas; both deep chemical weathering with subsequent stripping of the weathered mantle and frost-shattering with mass movement generate similar morphological features. Thus the supposed characteristic forms of particular process assemblages and climatic regimes may in fact have diverse origins which display equifinality, and a simple correlation between form, process and climate cannot be demonstrated.   KSR

Reference
Wentworth, C.K. 1928: Principles of stream erosion in Hawaii. *Journal of geology* 36, pp. 385–410.

equilibrium A concept commonly applied to environmental open systems, that is, systems in which the quantities of stored energy or matter are adjusted so that input, throughput and output of energy or matter are balanced. For example, the earth receives short-wave solar radiation at the top of the atmosphere. Of the total receipt (263 kcal cm^{-2} year^{-1}), 31 per cent are reflected and 69 per cent (181 kcal cm^{-2} year^{-1}) are absorbed. Input and output are balanced by the earth maintaining an equilibrium mean temperature such that it emits 181 kcal cm^{-2} year^{-1} of long-wave radiation. In a river system equilibrium is often defined as a balance of erosion and deposition. This is achieved by morphological adjustments which maintain sediment transport continuity. If a short river reach experiences more bed-load input from upstream than output downstream, the excess is deposited and the slope is steepened and the cross-section shallowed. The bed-load input can then be transported through the reach more effectively, and the output is increased to balance the input. Any local particle detachment (bed erosion) is balanced by deposition. This equilibrium is maintained by negative feedback; the deposition of excess load changes reach morphology so that the transport capacity increases and further deposition is prevented. The inputs to an open system vary through time, for example on a seasonal basis, but so long as average annual input is constant the system state is constant, and the equilibrium is one in which the relationship between form and process is stationary. This is a STEADY STATE (Chorley and

Kennedy 1971, pp. 201–3). If the annual average input is changing through time sufficiently slowly for the system to adjust the condition is a DYNAMIC EQUILIBRIUM. Technically, however, there is always a lag between the change in the process input variable and the internal morphological adjustment of the system, so the term *quasi-equilibrium* is sometimes used in this case.   KSR

Reference
Chorley, R.J. and Kennedy, B.A. 1971: *Physical geography: a systems approach*. London: Prentice-Hall.

equilibrium line A notional line describing some sort of balance between process and form. The notion can be applied widely, e.g. to the profile of a slope or plan of a beach, and is closely bound up with concepts of EQUILIBRIUM in natural systems. (See DYNAMIC EQUILIBRIUM; EQUILIBRIUM SHORELINE.)

equilibrium line of glaciers A notional altitudinal line on a glacier where ABLATION balances accumulation. In most situations this means that the summer is just warm enough to melt the snow and ice that has accumulated during the previous winter. Above the equilibrium line on a glacier is the accumulation zone where accumulation exceeds ablation each year, while below the equilibrium line is the ablation zone where ablation exceeds accumulation each year. The amount of snow and ice melted at the equilibrium line each year is a measure of the activity of a glacier with high values implying high velocities (Andrews 1972). Glaciers are most active in mid-latitude, temperate areas and become less active towards continental interiors and the poles. The equilibrium line altitude varies in a similar way. Where there is high winter accumulation the summer temperature must be high in order to melt the snow and ice. Thus equilibrium line altitudes tend to be low near maritime coasts and to rise towards continental interiors in response to precipitation gradients. This pattern is brought out by the distribution of both present-day mountain glaciers and abandoned glacial CIRQUES.

In temperate environments where the glacier ice is at the pressure melting point the equilibrium line may coincide with the FIRN line which marks the line separating bare ice from snow at the end of the ablation season. But on cold glaciers snowmelt may percolate down and freeze onto the glacier as SUPERIMPOSED ICE. Under these circumstances the positions of the firn line and equilibrium line on a glacier may differ.
                                    DES

Reading and Reference
Andrews, J.T. 1972: Glacier power, mass balances, velocities and erosion potential. *Zeitschrift für Geomorphologie* NF 13, pp. 1–17.
†Paterson, W.S.B. 1981: *The physics of glaciers*: Oxford and New York: Pergamon.

equilibrium shoreline A hypothetical state that actual shorelines may or may not approximate. It is a dynamic state in which the geometry of the beach reflects a balance between materials, processes and energy levels (climate). The ideal EQUILIBRIUM beach has curvature and sand prism characteristics which are adjusted so closely that the energy available transports the detritus supplied, over a period to be measured in years rather than months, days or seconds.   ASG

Reading
Tanner, W.F. 1958: The equilibrium beach. *Transactions of the American Geophysical Union* 39, pp. 889–91.

equipotentials Lines on a GROUNDWATER map joining points of equal fluid potential (or HYDRAULIC POTENTIAL). Fluid potential at any point is the product of HYDRAULIC HEAD and acceleration due to gravity. Consequently, since gravitational acceleration is practically constant, a WATER TABLE contour map with equal contour intervals is a potentiometric map in the horizontal plane. The distance between the contours (equipotentials) depicts the gradient of the potential. Hence hydraulic gradient varies inversely with contour spacing (Ward and Robinson 1990). In accordance with DARCY'S LAW, water always flows in a down gradient direction perpendicular to the equipotentials, the path followed by a particle of water being known as a streamline. A mesh formed by a series of equipotentials and corresponding streamlines is known as a flow net.

Equipotentials may also be constructed in the vertical plane. If fluid potential increases with depth, groundwater flow will be towards the surface, but if it decreases vertically flow will be downward (Hubbert 1940).   PWW

Reading and References
†Freeze, R.A. and Cherry, J.A. 1979: *Groundwater*. Englewood Cliffs, NJ: Prentice-Hall.

Hubbert, M.K. 1940: The theory of groundwater motion. *Journal of geology* 48, pp. 785–944.

Ward, R.C. and Robinson, M. 1990: *Principles of hydrology*. 3rd edn. Maidenhead, Berks: McGraw-Hill.

era The largest unit of geological time, being a span of one or more periods.

erg A sand desert. A desert area characterized by sand sheets and dunes. A sand sea.

ergodic hypothesis As used in geomorphology, suggests that under certain circumstances sampling in space can be equivalent to sampling through time. Geomorphologists have sometimes sought an understanding of landform evolution by placing such forms as regional valley-side slope profiles and drainage networks in assumed time sequences. The concept of the cycle of erosion was based to a large extent on ergodic assumptions, as was Darwin's model of coral reef development. Chorley *et al.* (1984, p. 33) point to certain dangers in ergodic reasoning: landforms may be assembled into assumed time sequences simply to fit preconceived theories of denudation; there is always a risk of circular argument; and form variations may result from factors other than their position in time. ASG

Reference
Chorley, R.J., Schumm, S.A. and Sugden, D.E. 1984: *Geomorphology*. London and New York: Methuen.

erodibility The degree to which a rock or sediment is susceptible to erosion by wind, water or ice. (See also EROSIVITY.)

erosion The group of processes whereby debris or rock material is loosened or dissolved and removed from any part of the earth's surface. It includes weathering, solution, corrasion and transportation.

erosion surface A term commonly used in Britain to describe a flattish plain resulting from erosion. Since, strictly speaking, erosion surfaces may be far from flat, it is probably more helpful to use the term *planation* surface instead.

Planation surfaces assume a central role in a geomorphological approach concerned with the evolution of landscape since they are generally regarded as the end product of either a cycle of erosion – the peneplain in the Davisian sense (see CYCLE OF EROSION) or of a particular blend of surface processes, for example the pediplain in a semi-arid environment (see PEDIMENT), the etchplain

An erosion surface truncating intensely deformed and resistant strata of quartzite on the Gibb River Road, Kimberley, western Australia. That such surfaces exist is indisputable.

in a humid tropical environment, or the wave-cut platform in the coastal environment.

In the first half of this century, when the study of landscape evolution was the prime goal of geomorphology, much attention was devoted to the identification of present and relict planation surfaces as the key to understanding DENUDATION CHRONOLOGY (King 1950; Wooldridge and Linton 1955). In many parts of Britain relict planation surfaces (summit planes) are common and comprise master features of the total landscape (see, e.g., Brown 1960; Sissons 1967). A major difficulty was the problem of dating, and the evolutionary approach in geomorphology became unfashionable in the 1960s and 1970s. Now, with new forms of radiometric dating and stratigraphic evidence available from offshore sediments, there are signs of a new lease of life for studies of landscape evolution and real prospects of understanding the significance of relict planation surfaces (e.g. Lidmar-Bergstrom 1982). (See also BASE LEVEL.) DES

Reading and References
Brown, E.H. 1960: *The relief and drainage of Wales*. Cardiff: University of Wales Press.

Brown, E.H. and Clayton, K., eds. *The geomorphology of the British Isles*. (A series of regional volumes; varying dates.) London: Methuen.

King, L.C. 1950: The study of the world's plain lands. *Quarterly journal of the Geological Society* 106, pp. 101–31.

Lidmar-Bergstrom, K. 1982: Pre-Quaternary geomorphological evolution in southern Fennoscandia. *Sveriges Geol Unders. Series C*, 785.

Ollier, C.D. 1979: Evolutionary geomorphology of Australia and Papua-New Guinea. *Transactions of the Institute of British Geographers* ns 4.4, pp. 516–39.

Sissons, J.B. 1967: *Evolution of Scotland's scenery*. Edinburgh: Oliver & Boyd.

Wooldridge, S.W. and Linton, D.L. 1955: *Structure, surface and drainage in south-east England*. 2nd edn. London: George Philip.

erosivity A measure of the potential ability of a soil to be eroded by a given geomorphological agency. For given soil and vegetation conditions the effects of a storm, for example, can be compared with another storm quantitatively and a scale of erosivity values can be produced. The *erodibility* of a soil is the vulnerability of a soil to erosion; for given rainfalls one soil can be compared with another and a scale of values produced. Erodibility can be considered as the effects of the basic (edaphic) characteristics of the soil, plus the way in which a soil is treated or managed. Thus EROSION can be considered as a function of both erosivity and erodibility and they are related in the universal soil loss equation, details of which are given in Morgan (1986). WBW

Reading
Morgan, R.P.C. 1986: *Soil erosion and conservation*. London: Longman.

Selby, M.J. 1993: *Hillslope materials and processes*. 2nd edn. Oxford: Oxford University Press.

erratic (glacial) A rock or boulder that has been carried to its present location by the action of a glacier. In the English-speaking world in the first half of the nineteenth century erratics were commonly attributed to ice rafting in the universal flood, although in Switzerland, Germany and Norway their true origin had already been appreciated. The significance of erratics in demonstrating former widespread glaciers in Britain was shown by Agassiz (1840). DES

Reference
Agassiz, J.L.R. 1840: Glaciers, and the evidence of their having once existed in Scotland, Ireland and England. *Proceedings of the Geological Society* 3, pp. 321–2.

eruption A discharge or eruption of volcanic material, either gaseous, liquid or solid, at the earth's surface.

escarpment The steeper slope of a cuesta. Often used as a synonym for a cuesta.

esker A sinuous ridge of coarse gravel representing the deposits of a MELTWATER stream normally flowing subglacially. Eskers may be hundreds of kilometres in length and 100 m in height. In many situations they are beaded, which means that mounds occur along their length, particularly at points where they change direction. It is common to find that eskers form complex patterns of tributaries and distributaries and that sometimes ridges are discontinuous or linked by rock-cut meltwater channels. Most eskers are the channel deposits of subglacial meltwater rivers and their orientation is usually parallel to that of overall ice flow (Shreve 1972). The deposits are related to both closed channel flow and open channel flow within a conduit (Bannerjee and McDonald 1975; Saunderson 1977). DES

References
Bannerjee, I. and McDonald, B.C. 1975: Nature of esker sedimentation. In A.V. Jopling and B.C. McDonald eds, *Glaciofluvial and glaciolacustrine sedimentation*. Society of

Economic Paleontologists and Mineralogists special publication 23. Tulsa, Oklahoma. Pp. 132–54.

Saunderson, H.C. 1977: The sliding bed facies in esker sands and gravels: a criterion for full-pipe (tunnel) flow? *Sedimentology* 24, pp. 623–38.

Shreve, R.L. 1972: Movement of water in glaciers. *Journal of glaciology* 11, 62. pp. 205–14.

estuary The sections of a river which flow into the sea and which are influenced by tidal currents. Estuaries form transition zones between freshwater rivers and salt-water oceans, with fluctuations in water level, salinity, temperature and velocity. They are constantly modified by erosion and deposition, resulting in tidal flats and salt marshes, deltas, spits and lagoons. A funnelling of tidal currents may produce powerful periodic waves or estuarine bores. Shallow sedimentary estuaries are rich in nutrients and very high in BIOLOGICAL PRODUCTIVITY, providing nurseries for fish and other animals. Deep estuaries such as fiords are colder, less productive and less biologically diverse.

Estuaries can be classified into a number of types on the basis of chemical characteristics (Dyer 1986): salt wedge estuaries in tideless seas, partially mixed estuaries where there are appreciable tidal movements, and well-mixed estuaries where the strength of the tidal currents is strong relative to river flow. They can also be classified on the basis of their tidal range. This determines the tidal current and residual current velocities and therefore the amount and source of sediments. *Microtidal estuaries* occur where tidal range is less than 2 m and so are dominated by freshwater inflow upstream of the mouth and by wind-driven waves seaward of the mouth. They often contain a fluvial delta and spits and bars at the seaward margin. In *mesotidal estuaries* (tidal range *c.* 2–4 m) tidal currents are of greater importance but because of the still somewhat modest tidal range tidal flow does not extend very far upstream. Thus most mesotidal estuaries are relatively stubby. In the case of *macrotidal estuaries*, tidal ranges in excess of 4 m produce a situation where tidal influences extend far inland. Such estuaries have long, linear sand bars parallel to the tidal flow, but their most distinguishing characteristic is

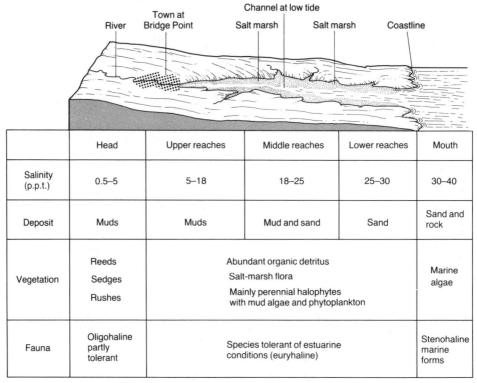

	Head	Upper reaches	Middle reaches	Lower reaches	Mouth
Salinity (p.p.t.)	0.5–5	5–18	18–25	25–30	30–40
Deposit	Muds	Muds	Mud and sand	Sand	Sand and rock
Vegetation	Reeds Sedges Rushes		Abundant organic detritus Salt-marsh flora Mainly perennial halophytes with mud algae and phytoplankton		Marine algae
Fauna	Oligohaline partly tolerant		Species tolerant of estuarine conditions (euryhaline)		Stenohaline marine forms

The estuarine environment. Zones of transition separate the different environments shown.
Source: *P. Furley and W. Newey 1983:* Geography of the biosphere. *London: Butterworth. Figure 13.12.*

their trumpet-shaped flare. The Severn Estuary in Britain, the Delaware Estuary in the USA, and the Plate Estuary in Latin America are prime examples of this type.   PAF/ASG

Reading and Reference
Dyer, K.R. 1986: *Coastal and estuarine sediment dynamics.* Chichester: Wiley.
Ketchum, B.H. ed. 1983: *Estuaries and enclosed seas.* Amsterdam: Elsevier.
McLusky, D.S. 1981: *The estuarine environment.* Glasgow: Blackie.

etchplain A wide erosional surface, often inselberg-studded, that is a feature of shield areas in the tropics. Their formation has been accounted for by Büdel (1982, p. 36) as follows: 'Along the basal surface of weathering, chemical decomposition works ceaselessly downwards, while in the rainy season finely worked material is correspondingly removed from above by highly effective sheet wash. This mechanism of double planation surfaces is alone responsible for creating these etchplains over long geologic periods.'   ASG

Reference
Büdel, J. 1982: *Climatic geomorphology.* Princeton: Princeton University Press.

etesian wind The prevailing wind over the Aegean sea in summer (from the Greek *etesios*, annual). It blows steadily from the north with moderate force, bringing dry cold continental air and clear sky from mid-June to the beginning of October.   JSAG

Reading
Meteorological Office. 1962: *Weather in the Mediterranean.* Vol. I. London: HMSO.

eugeogenous rock Rock which produces a large amount of debris when it is decomposed by weathering.

eulittoral zone The portion of the coastal zone which extends seawards from high-water mark down to the limit of attached plants (generally at a depth of 40–60 m).

euphotic zone The surface layer of a body of water in which photosynthesis can take place because of the availability of light.

eustasy A term which embraces sea-level changes of a worldwide nature. Local changes of sea level complicate the eustatic pattern and are caused by local factors such as ISOSTASY, OROGENY and EPEIROGENY. During the first decades of the twentieth

century, following on in part from the work of Suess, a number of workers including De Lamothe, Deperet, Baulig, and Daly, proposed that most sea-level oscillations and strandlines of the Quaternary were glacio-eustatic (see Guilcher 1969). They believed, correctly, that sea level oscillated in response to the quantity of water stored in ice caps during glaciations and deglaciations.

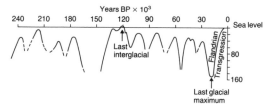

Eustasy: The nature of world sea-level change over the past quarter of a million years.
Source: A.S. Goudie 1984: The nature of the environment. Oxford and New York: Basil Blackwell.

They suggested that there was a suite of characteristic levels in Morocco and elsewhere around the Mediterranean which could be related to different glacial events:

1 Sicilian (80–100 m);
2 Milazzian (55–60 m) – between the Günz and the Mindel;
3 Tyrrhenian (30–35 m) – between the Mindel and the Riss;
4 Monastirian (15–20 and 0–7 m) – between the Riss and the Würm; and
5 Flandrian – the present post-Würm transgression.

The transgressions of the interglacial were succeeded by regressions during glacials and the height of the various stages declined during the course of the Pleistocene. Total melting of the two main current ice caps – Greenland and Antarctica – would raise sea level a further 66 m if they both melted. Deep-sea core evidence, however, does not suggest that in previous interglacials of the Pleistocene these two ice caps did disappear, and without a general melting of them, sea level would only have been a few metres higher than now in the interglacials. This fact does not tie in too happily with the simple glacio-eustatic theory of progressive sea-level decline during the Pleistocene. Some factors other than glacio-eustasy must be responsible for the proposed high sea-levels of early Pleistocene times. Because other, sometimes local, factors have played a role, few people now seriously believe that through height alone one can correlate shorelines over wide areas on the basis of a common interglacial age.

Nevertheless, low Quaternary sea levels brought about by the ponding up of water in the ice caps were quantitatively extremely important. Donn *et al.* (1962) on the basis of theoretical considerations from known ice volumes, reckon that in the Riss, possibly the most extensive of the glaciations, sea levels might have been lowered by 137–159 m below current sea level. In the last glacial (Würm–Wisconsin–Weichsel) they give a figure for lowering of rather less: 105–123 m. On the basis of isotopic dates for coral and associated material in the Great Barrier Reef (Australia), California, and the south-east sea areas, Veeh and Veevers (1970) favour the conclusion that towards the end of the last glaciation, sea level dropped universally to at least −175 m, some 45 m deeper than hitherto suspected.

Although orogeny is normally regarded as being an essentially local factor of sea-level change, and eustasy as being worldwide, there is one class of process, *orogenic eustasy*, whereby a local change can have worldwide effects. It therefore acts as some sort of a link between these two main types of change, and is a worldwide change of sea level produced by changes in the volumes of the ocean basins resulting from orogeny (mountain building) (Grasty 1967).

In recent years the importance of a third type of eustatic change has been identified, notably by Mörner (1980). This is termed *geoidal eustasy*. The shape of the earth is not regular, and at present the GEOID (caused by the earth's irregular distribution of mass) has a difference between lows and highs of as much as 180 m. The ocean surface reflects this irregularity in the geoid surface, which varies according to forces of attraction (gravity) and rotation (centrifugal), and will respond by deformation to a change in these controlling forces. The possible nature of such changes is still imperfectly understood, but they include fundamental geophysical changes within the earth, changes in tilt in response to the asymmetry of the ice caps, changes in the rate of rotation of the earth, and redistribution of the earth's mass caused by ice-cap waxing and waning.

Although glacio-eustasy is the most important of the eustatic factors that have affected world sea levels during the course of the Quaternary, it is worth looking at some of the other minor eustatic factors which have played a role, especially over the long term. The infilling of the ocean basins by sediment, for example, would tend to lead to a sea-level rise. Higgins (1965) estimated that this could lead to a rise of 4 mm 100 years^{-1}. This is equivalent to a rise of 40 m in a million years. Two very minor factors are the addition of juvenile water from the earth's interior and the variation of water level according to temperature. The latter could raise the level of the sea by about 60 cm for each 1°C rise in temperature of the seawater. The former could probably add about 1 m of water in a million years. The evaporation and desiccation of pluvial lakes, some of which had large dimensions, would be unimportant in affecting world sea levels, adding a maximum of about 10 cm to the level of the sea, if they were all to be evaporated to dryness at the same time (Bloom 1971).

Another cause of eustatic changes of sea level, especially in the Holocene, is the process called 'isostatic decantation'. Isostatic uplift in the neighbourhood of the Baltic basin and of Hudson Bay has led to a reduction of the volume of these seas; and the water from them has thus been decanted into the oceans to affect worldwide sea levels. A comparison of the area and volume of the late-glacial precursor of Hudson Bay with Hudson Bay itself suggests that the volume of water decanted could only be sufficient to cause a rise in world sea level of about 0.63 m. The contribution of the Baltic Sea would be even less. This factor can thus be largely ignored.   ASG

Reading and References

Bloom, A.I., 1971: Glacial eustatic and isostatic controls of sea level since the Last Glaciation. In K.K. Turekian, ed., *The Late Cenozoic glacial ages.* New Haven: Yale University Press. Pp. 355–79.

Donn, W.L., Farrand, W.R. and Ewing, M. 1962: Pleistocene ice volumes and sea level changes. *Journal of geology* 70, pp. 206–14.

†Goudie, A.S. 1992: *Environmental change.* 3rd edn. Oxford: Clarendon Press. Ch. 6.

Grasty, R.L. 1967: Orogeny, a cause of world wide regression of the seas. *Nature* 216, p. 779.

Guilcher, A. 1969: Pleistocene and Holocene sea level changes. *Earth science reviews* 5, pp. 69–98.

Higgins, C.G. 1965: Causes of relative sea-level changes. *American scientist* 53, pp. 464–76.

Mörner, N.A. 1980: *Earth rheology, isostasy and eustasy.* New York: Wiley.

Veeh, H. and Veevers, J.J. 1970: Sea level at −175 m off the Great Barrier Reef, 13,600 to 17,000 years ago. *Nature* 226, pp. 526–7.

eutrophic Pertaining to lakes and other freshwater bodies which abound in plant nutrients and which are therefore highly productive. Lakes tend to become more eutrophic as they become older, and eutrophication can also result from the addition of nutrients as a result of pollution. This can cause phenomena such as algal blooms. (See also NUTRIENT STATUS.)

eutrophication The addition of mineral nutrients to an ECOSYSTEM, generally raising the NET PRIMARY PRODUCTIVITY. It is usually used of man-induced additions of elements such as nitrogen and phosphorus to salt and freshwater, which are naturally low in those elements, but it also occurs in terrestrial systems and may be a natural phenomenon. In current usage, it very often relates to the loads of N_2 and P in fresh and offshore waters heated by such effluents as sewage, fertilizer run-off and detergents. The effects are often algal blooms, deoxygenation of water through consequent bacterial activity and, in the sea, rapid growth of small organisms called dinoflagellates which are implicated in 'red tides'. Eutrophication is usually a local or regional problem at most, though enclosed seas like the Mediterranean may be more vulnerable than open oceans. Its main drawback is the loss of expensively gained nutrients which then have to be replaced, since they cannot be economically retrieved from the water in which they have become diluted.   IGS

Reading
Whitton, B.A. ed. 1975: *River ecology.* Berkeley and Los Angeles: University of California Press.

evaporation The diffusion of water vapour into the atmosphere from freely exposed water surfaces. This includes water losses from lakes, rivers, even clouds and saturated soil and plant surfaces but it does not incorporate transpiration losses from plants. It is imperative therefore to distinguish between the process of evaporation which concerns only free-standing water bodies and that of EVAPOTRANSPIRATION. While the process of evaporation is generally understood, its accurate measurement has long proved difficult. Before considering the various instruments and formulae available for this task it is necessary to introduce the variables that govern rate of water loss.

The rate of evaporation is partly controlled by solar radiation, which supplies the energy required to transform liquid water into water vapour, i.e. the latent heat of vaporization or 2.44×10^6 J kg^{-1} at a temperature of 25°C. The proportion of net radiation received by the earth (Q^*) that is available for this process (QLE) depends not only upon the transmission, absorption and reflection of the earth's atmosphere and surface, but also the amounts employed for heating the atmosphere (QH), and for heating the ground (QS), i.e. (Oke, 1987):

$$Q^* = QH + QLE + QS$$

where Q^* is the net radiation balance (W m^{-2}), QH is the sensible heat flux, QLE is the latent heat flux, and QS is the ground heat flux.

The humidity of the air above an evaporating surface will eventually increase until, when the air becomes saturated, evaporation will cease unless these layers are dispersed. The atmospheric 'demand' for moisture is therefore controlled not only by the radiation balance but also by humidity and windspeed. The rate of evaporation is influenced in addition by the characteristics of the water body itself, that is depth, extent and water quality (Ward and Robinson 1990). Understanding evaporation rates is further complicated by the need to distinguish between potential and actual evaporation.

Actual evaporation is the observed rate of water loss, whereas the potential rate is the evaporation that would occur from a free-standing freshwater surface due to atmospheric demand if there were no other limiting factors. These two rates could be equal where the water supply is plentiful, e.g. a large freshwater lake, but potential rates may not be reached when the evaporating surface does not have the same characteristics, e.g. over soils. In addition potential rates may be exceeded over small water bodies when relative humidities are low. Potential evaporation is therefore a theoretical concept referring only to the atmospheric demand for moisture over a large freshwater evaporating surface.

Several instruments have been designed to measure the volume of water loss directly from saturated surfaces and thus provide an estimate of potential rates. Atmometers are commonly employed because they are cheap and simple to use. The device consists of an inverted measuring cylinder with a porous plate at its base. As most of the energy used in evaporating water is derived from the surrounding air rather than direct solar radiation they are over-sensitive to humidity and windspeed (Jackson 1989). The instrument in most widespread use is the evaporating pan, in effect a small reservoir of water approximately 1.2 m across and 25 cm high (US Weather Bureau class 'A'). It is simple and effective but the loss of water is influenced by its exposure, design, material and colour. In addition, estimation of the depth of water is liable to significant observer error.

Potential evaporation rates can be obtained by surrounding these instruments by extensive saturated surfaces or by using empirical coefficients to convert

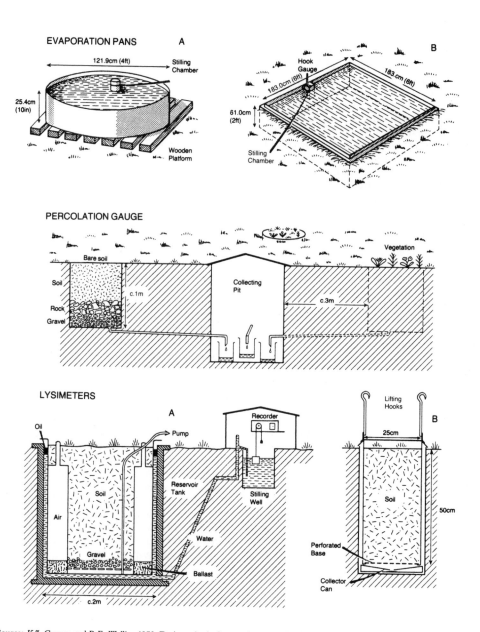

Source: *K.J. Gregory and D.E. Walling 1973:* Drainage basin form and process. *London: Edward Arnold.*

observations. Doorenbos and Pruitt (1975) for example, provide coefficients for evaporating pans but information is also required on the windspeed, relative humidity and exposure.

The measurements of actual evaporation from soils can be estimated by lysimeters, consisting of isolated columns of soil placed in large containers thus enabling the observation of changes in soil moisture and drainage. Evaporation rates can then be deduced by a tabulation of the water balance:

$$E = P - D - RO \pm \Delta S$$

where E is evaporation, P is precipitation, D is drainage, RO is run-off, and ΔS are changes in soil moisture. Unfortunately these devices are difficult to construct and maintain, and the container walls can distort lateral movements of soil moisture. Despite their potential high level of accuracy, lysimeters therefore tend to be used only in research studies.

The water balance approach can be used for water bodies of any size but an alternative method is to measure directly the diffusion of water vapour through the atmosphere, rather than estimating the various inputs and outputs of the hydrological cycle. Rosenberg (1974) lists two main approaches, the eddy correlation approach and profile techniques. Eddy correlation is based on the instantaneous measurement of humidity and vertical windspeed at a selected height when at time (t):

$$QLE(t) = \rho L q(t) W(t)$$

where $QLE(t)$ is the evporation flux (strictly the vertical transport of latent heat), $q(t)$ is the water vapour content, $W(t)$ is the instantaneous windspeed, L is the latent heat of vaporization, and ρ is the density of dry air.

The usefulness of this equation depends upon the instantaneous measurement of humidity and heat fluxes (Moore *et al.* 1976) resulting in expensive and highly sophisticated instrumentation. A device, the HYDRA, has been developed by the Institute of Hydrology, Wallingford and used in a study of evaporation at Lake Toba, Indonesia (Sene *et al.* 1991).

The profile method relies on either the measurement of vertical humidity gradients (Sellers 1965) or simultaneous changes in temperature and humidity. This approach is based on the equation:

$$QLE = \rho L K_w \left(\frac{\Delta q}{\Delta z}\right)$$

where K_w is the eddy diffusivity of water vapour in the air, Δq is the atmospheric humidity difference between two heights Δz apart, and Δz is the height increment.

Unfortunately the diffusivity of water vapour is a variable, especially when the air is buoyant above the boundary layers, and the measurements of humidity need to be precise. An alternative approach called Bowen's ratio is therefore often used, based on the assumption that the thermal diffusivity and the diffusivity of water vapour are equal, then:

$$QH = \rho c_p K_h \frac{\Delta T}{\Delta z}$$

where QH is the sensible heat flux, c_p is the specific heat of air at constant pressure, K_h is the thermal diffusivity, and $\Delta T / \Delta z$ is the temperature gradient in the vertical. Then Bowen's ratio is:

$$B = \frac{QH}{QLE} = \frac{c_p \Delta T}{L \Delta q}$$

and if

$$Q^* = QLE + QH + QS,$$

and

$$QLE = (Q^* - QS)/(1 + B).$$

Unfortunately the profile approach, like the eddy correlation technique, requires exact and instantaneous measurements, in addition to steady state conditions. Hence it is mainly used for research applications, where accurate results over short time periods are required (e.g. Stannard and Rosenberry 1991). Routine measurements of evaporation tend to rely upon the water balance approach but this method is limited by data reliability. Atmometers and pans are widely used because of low cost and simplicity but they both suffer from errors. CTA

References

Doorenbos, J. and Pruitt, K.C. 1975: *Crop water requirements*. FAO Irrigation and Drainage Paper 24, Rome.

Jackson, I. 1989: *Climate, water and agriculture in the Tropics*. 2nd edn. New York: Longman.

Oke, T. 1987: *Boundary layer climates*. 2nd edn. London: Routledge.

Moore, C.J., McNeil, D.D. and Shuttleworth, W.O. 1976: *A review of existing eddy correlation sensors*. Institute of Hydrology Report 32, Wallingford.

Rosenberg, N.J. 1974: *Microclimate: the biological environment*. New York: Wiley.

Sellers, W.D. 1965: *Physical climatology*. London: University of Chicago Press.

Sene, K.J., Gash, J.H.C. and McNeil, D.D. 1991: Evaporation from a tropical lake: comparison of theory with direct measurements. *Journal of hydrology* 127, pp. 193–217.

Stannard, D.I. and Rosenberry, D.O. 1991: A comparison of short term measurements of lake evaporation using eddy correlation and energy budget methods. *Journal of hydrology* 122, pp. 15–22.

Ward, R.C. and Robinson, M. 1990: *Principles of hydrology*. 3rd edn. London: McGraw-Hill.

evaporite A water-soluble mineral (or rock composed of such minerals) that has been deposited by precipitation from saline water as a result of evaporation, especially in coastal sabkhas or in salt lakes of desert areas. Among the most common minerals are sodium chloride (halite) and calcium sulphate (gypsum and anhydrite).   ASG

Reading
Warren, J.K. 1989: *Evaporite sedimentology*. Englewood Cliffs, NJ: Prentice-Hall.

evapotranspiration The diffusion of water vapour into the atmosphere from a vegetated surface. Evapotranspiration is influenced by climate, soil hydrology and plant characteristics and this term has been criticized for combining these elements when they can respond in different ways to energy and moisture variations. Nevertheless, it is convenient to consider a homogeneous vegetated surface for many hydrological applications and this concept is widely used.

Plant stomata remain open for the processes of photosynthesis and respiration, which are essential for growth. This exposes the moist interiors of the stomata and encourages the diffusion of water vapour, that is, transpiration. In addition, water can be lost through small openings in the corky tissue covering stems and twigs. This water loss establishes a water deficit within the plant that may be transmitted to the roots. Whether or not this occurs depends upon a number of factors, e.g. plant physiology, moisture content, size of deficit, etc. (Begg and Turner 1976). Providing water is not limiting, transpiration rates will be largely controlled by atmospheric conditions and the character-

istics of the vegetated surface, i.e. extent, structure and stomatal behaviour. Hence the concept of potential evapotranspiration relates to a specific plant cover, in addition to climate. Several definitions have been proposed and Ward and Robinson (1990) quote an approach by Penman that has gained widespread acceptance: 'Evaporation from an extended surface of a short green crop, actively growing, completely shading the ground, of uniform height and not short of water.'

Potential evapotranspiration (PE_t) is frequently calculated for a grass cover, but these values can be converted to any cropped surface by the crop coefficients provided by Doorenbos and Pruitt (1975) and Wright (1982) where:

$$\frac{PE(crop)}{PE_t} = \text{crop coefficient.}$$

The crop coefficient is assumed to be a constant, depending upon ground cover, i.e. state of development of the plant and prevailing meteorological conditions.

Actual evapotranspiration (AE_t) rates may exceed potential rates because of an increase in the extent of the evaporating surface, e.g. in forested areas. The reverse frequently occurs, due to water deficits. Immediately after rain has fallen the evaporation component of evapotranspiration will be high, but it will decrease markedly as soil and plant surfaces dry out. Thereafter, transpiration will be the main contributor relying upon soil moisture supplies, but water may only be absorbed by the roots when the soil has a higher moisture potential. Hence transpiration rates will be reduced as soil moisture contents decrease, but the exact relationship between evapotranspiration and soil moisture is complex. Hagan and Stewart (1972) present a comprehensive tabulation of the soil moisture contents at which crops start to undergo stresses that will ultimately reduce yields. While useful for irrigation scheduling, this study does not indicate the rates at which transpiration will decrease beyond these threshold values. Jackson (1989) shows that AE_t responses to a limiting moisture supply are complex, being influenced by PE_t rates and plant physiological adaptations including the rooting system and stomatal opening.

Actual evapotranspiration is then a combination of atmospheric conditions, soil moisture content and plant characteristics. Rates of actual water loss can be measured directly by monitoring plant water status, soil moisture changes or by using a lysimeter (see EVAPORATION). The practical difficulties of routinely monitoring such moisture changes over an extensive area has resulted in studies inferring AE_t rates from estimates of PE_t.

Various empirical formulae have been devised to measure potential evapotranspiration, based on the assumption that it is controlled by the earth's energy balance which is reflected in mean monthly temperatures, e.g. Blaney-Criddle and Thornthwaite. These formulae rely upon empirical coefficients that limit their applicability to regional monthly water balance computations. Penman (1948) devised an alternative formula by applying understanding of the diffusion of water vapour from a moist surface to incorporate both energy (H) and aerodynamic (EA) functions in the following form:

$$PE_t = \left(\frac{\Delta}{\gamma}H + EA\right) / \left(\frac{\Delta}{\gamma} + 1\right)$$

where PE_t is the potential evapotranspiration (mm of water day^{-1}), Δ is the slope of the saturation vapour pressure curve at mean air temperature (mb °C^{-1}), and γ is the psychometric constant (0.66 mb °C^{-1}).

Observations of temperature, windspeed, vapour pressure and either duration of daylight or the net radiation balance are required on a daily basis. The equation does contain a large number of constants but fortunately values have been calculated for most parts of the world and hence it has been widely employed in numerous studies. Unfortunately, two of the assumptions upon which the formula is based, that the moisture and heat fluxes from the ground are constant to a height of 1 m and that the heat stored in the ground is negligible, may become violated for time periods of less than one day. Shuttleworth (1988) suggests that such semi-empirical equations are only reliable over periods of ten days or more.

The Penman formula has become the standard approach for the measurement of PE_t for over forty years because of its general reliability and modest data requirements. However, in 1990 (Monteith 1991) the FAO decided to phase out the use of PE_t and crop coefficients to calculate crop water requirements through dissatisfaction with this empirical approach. In future the 1965 Penman–Monteith revision (Monteith 1981) will be used. Monteith adopts a more rational process approach by incorporating the aerodynamic resistances created by the vegetated surface with the surface resistances to the diffusion of water vapour exerted by the transpiring plant. Thus AE_t rates can be obtained with knowledge of these resistances and this approach has been used in the MORECS (Thompson et al. 1981) system of AE_t calculation in the UK.

An alternative approach, the CRAE, summarized by Morton (1983) has emphasized the importance of areal estimates and rejects the sole use of point meteorological measurements because of the complementary relationship between PE_t and AE_t. That is, when water is scarce, AE_t will be low, relative humidities will be low, temperatures will be high and consequently PE_t will be high. Calculations of areal AE_t are made using knowledge of water availability and PE_t rates (Kovacs 1987), but criticisms have been levelled concerning processes in the convective boundary layer of the atmosphere. Lemeur and Zhang (1990) have compared estimates of AE_t by the CRAE and Penman–Monteith methods and concluded that the latter was the most appropriate but there remains much debate over the most suitable method for AE_t determination.   CTA

References

Begg, J.E. and Turner, N.C. 1976: Crop water deficits. *Advances in agronomy* 28, pp. 161–217.

Doorenbos, J. and Pruitt, W.O. 1975: *Crop water requirements.* FAO Irrigation and Drainage Paper 24, Rome.

Hagan, R.M. and Stewart, J.I. 1972: Water deficits, irrigation design and programming. *Proceedings of the American Society of Civil Engineers Irrigation and Drainage Division* 98, pp. 215–35.

Jackson, I. 1989: *Climate, water and agriculture in the Tropics.* 2nd edn. New York: Longman.

Kovacs, G. 1987: Estimation of average areal evapotranspiration. *Journal of hydrology* 95, pp. 227–40.

Lemeur, R. and Zhang, L. 1990: Evaluation of three evapotranspiration models in terms of their applicability for an arid region. *Journal of hydrology* 114, pp. 395–411.

Monteith, J.L. 1981: Evaporation and surface temperature. *Quaterly Journal of the Royal Meteorological Society* 107, pp. 1–27.

— 1991: Weather and water in the Sudano-Sahelian zone. *Proceedings of Niamey Workshop on Soil Water Balance in the Sudano-Sahelian Zone.* IAHS Publ. 199, pp. 11–29.

Morton, F.J. 1983: Operational estimates of areal evapotranspiration and their significance to the science and practice of hydrology. *Journal of hydrology* 66, pp. 1–76.

Penman, H.L. 1948: Natural evaporation from open water, bare soil and grass. *Proceedings of the Royal Society, Series A* 193, pp. 120–45.

Shuttleworth, J.W. 1988: Macrohydrology; the new challenge for process hydrology. *Journal of hydrology* 100, pp. 31–56.

Thompson, N. Barrie, I.A. and Ayles, M. 1981: The Meteorological Office rainfall and evaporation calculation system: MORECS. *Hydrological Memoir 45.* Meteorological Office, Bracknell.

Ward, R.C. and Robinson, M. 1990: *Principles of hydrology.* 3rd edn. London: McGraw-Hill.

Wright, J.L. 1982: New evaporation crop coefficients. *Proceedings of the American Society of Civil Engineers Irrigation and Drainage Division* 108, pp. 57–74.

evergreen Plants which have long-lasting leaves. If the length of the photosynthetic or growing season is short, or if nutrients are in very short supply, evergreen forms are favoured since they do not have to use resources in building up their photosynthetic capacity each year. The relative dominance of evergreen trees and shrubs in high latitude and high altitude areas, and of evergreen dwarf shrubs in peatlands, are examples of this response to climate and nutrient availability.   KEB

Reading
Crawley, M.J. ed. 1986: *Plant ecology.* Oxford: Blackwell Scientific.

Larsen, J.A. 1982: *Ecology of the northern lowland bogs and conifer forests.* New York and London: Academic Press.

evolution The relatively gradual change in the characteristics of successive generations of a species or race of an organism, ultimately giving rise to species or races different from the common ancestor. Evolution reflects changes in the genetic composition of a population over time.

The word evolution was used in the eighteenth century in connection with embryological development. It was Charles Lyell, in 1832, who applied it to the notion of organic transmutation, in his discussion of the ideas of Jean-Baptiste de Lamarck (1744–1829). Charles Darwin (1809–1882) did not use the word extensively: he preferred the phrases 'descent with modification' and 'transmutability of species'.

Lamarck was the first to develop a coherent theory of evolutionary change. In 1800 he put forward the notion that the most primitive forms of life had been generated spontaneously, and that from these simple types all others had been successively produced. Lamarck later explained the mutability of species in terms of two factors: (*a*) a somewhat obscure 'power of life' which was assumed to be responsible

The Berlin specimen of Archaeopteryx. *This fossil has claims to be considered intermediate between reptiles and birds. It possessed a long tail, teeth, and a somewhat reptilian bone structure, yet it also had feathers. The exact status of* Archaeopteryx *fossils is questioned by some authorities.*

for the increasing complexity of the various classes, and (*b*) the influence of the environment.

It was Lamarck's belief that animals responded to changes in their surroundings by developing new habitats: these were associated with changes in the organisms' structure, which could be inherited – 'the inheritance of acquired characteristics'.

The theory of evolution by natural selection was developed independently by Charles Darwin and by Alfred Russel Wallace (1823–1913). Darwin probably had some vague awareness of the possible transmutability of species during his sojourn on the *Beagle* (1831–6), but he did not put pen to paper on the subject until early 1837, some months after his return to England from his voyage around the world. His ideas were well formulated by the time he wrote his *Essay of 1844* (see De Beer 1958) but he did not publish them. Wallace, who had spent some ten years travelling and collecting natural history specimens in South America and south-east Asia, sent a paper summarizing his ideas to Darwin in 1858, and a number of Darwin's associates arranged for a joint presentation to the Linnean Society of London on 1 July 1858 (see De Beer 1958). Darwin's *On the Origin of Species* followed in 1859. In this book Darwin argued that the transmutation of species had occurred as the result of 'natural selection or the preservation of favoured races in the struggle for life'.

The essential mechanism of natural selection may be summarized as follows:

1 In most species of organisms far more young are produced than can possibly survive to adulthood to reproduce themselves.
2 Organisms vary: individuals in a population differing from one another.
3 There is therefore a 'struggle for existence', or competition between organisms, in which those most adapted to their environment survive – 'the survival of the fittest'.
4 Because variations are heritable, organisms possessing the best adaptive characteristics will ultimately predominate in the population: as Wallace put it there is a 'tendency for varieties to depart indefinitely from the original type'. Darwin and Wallace thus explained adaptive change as the result of natural selection operating over long periods on the quite small variations present in populations of animals and plants. (They did not exclude the possibility of the existence of other mechanisms.)

Post-Darwin palaeontological work has, to a limited extent, supported the evolutionary hypothesis through the discovery of 'missing links' such as *Archaeopteryx* and certain alleged sequences of fossils in some groups (e.g. the ammonites). Nevertheless, important gaps remain. (S.J. Gould and N. Eldredge (1977) have stressed 'punctuated equilibrium', the doctrine that major species change is concentrated into very short bursts, alternating with long periods of relative stability.)

Despite increasing acceptance of Darwin's ideas throughout the last decades of the ninteenth century the mechanism by which characters were transferred from generation to generation remained obscure. Even when Gregor Mendel's theory of particulate inheritance (first put forward in the 1860s) began to be appreciated early

in the twentieth century some scientists saw it as an alternative to Darwin's ideas. The *Modern Synthesis* was achieved in the 1930s and 1940s through the cooperation of specialists from a variety of disciplines (Huxley 1942). Field workers provided evidence from studies of speciation (particularly in small, more or less closed environments such as remote islands), geographical variation and adaptation, emphasizing that a species should be considered as a population, rather than an ideal type.

Experimentalists demonstrated the importance of MUTATIONS, showing that they could be inherited in accordance with Mendel's rules, and that selection could work on continuous variations, ultimately altering the character of the population.

Other mechanisms for evolution, besides natural selection, have also been suggested. Darwin himself mentioned *sexual selection* – the preference by some female animals for males with elaborate ornamentation or colouring. Peacocks with their decorative tails, he argued, had 'the power to charm females'. In time these exaggerated characteristics, and the elaborate courtship rituals sometimes associated with them, would be accentuated in the population through the mechanism of inheritance, and the failure of less well-endowed males to find mates.

More recently (Dover 1982), *molecular drive*, a mechanism for evolution working independently of either selection or GENETIC DRIFT has been suggested. Genetic changes that result from mutations, it is argued, spread through internal mechanisms or 'turnover' of DNA inside the cell nucleus. A number of related changes can occur 'in unison', and as the genes of different individuals are intermixed in a sexually breeding population, the changes could spread rapidly through that population.

The term *convergent evolution* is frequently used to describe the process by which members of unrelated or distantly related groups become superficially similar to each other, usually through a common adaptation to a specialized environment or way of life, through the action of similar selection mechanisms. Examples include the evolution of an elongate, streamlined body and the elimination or reduction of limbs by fish, ichthyosaurs, Cetacea and birds such as penguins; the development of similar membranous wing structures by both bats and pterodactyls; and strong similarities between certain marsupials and placental mammals having a similar mode of life: wolf and thylacine; mole and marsupial mole. COEVOLUTION is the contemporaneous and associated evolution of two organisms (or groups of organisms) that are ecologically linked – such as flowering plants and their pollinators (insects, birds, or bats).

Evolutionary theory profoundly influenced thinking on a number of aspects of physical geography, such as the CYCLE OF EROSION and SUCCESSION.   PHA

Reading and References
†Cherfas, J. ed. 1982: *Darwin up to date.* London: New Science Publications.
Darwin, C.R. 1859: *On the origin of species.* London: John Murray. Many subsequent editions.
De Beer, G. ed. 1958: *Evolution by natural selection: papers by Darwin and Alfred Russel Wallace.* Cambridge: Cambridge University Press.
Desmond, A. and Moore, S. 1992: *Darwin.* London: Michael Joseph.
Dover, G. 1982: Molecular drive: a cohesive model of species evolution. *Nature* 299, pp. 111–17.
Gould, S.J. and Eldridge, N. 1977: Punctuated equilibria: the tempo and mode of evolution reconsidered. *Palaeobiology* 3, pp. 115–51.
Huxley, J. 1942: *Evolution: the modern synthesis.* London: Allen & Unwin.
Kohn, D. ed. 1985: *The Darwinian Heritage.* Princeton, NJ and Guildford: Princeton University Press.
†Mayr, E. 1976: *Evolution and the diversity of life: selected essays.* Cambridge, Mass. and London: Harvard University Press.
— and Provine, W.B. 1980: *The evolutionary synthesis: perspectives on the unification of biology.* Cambridge, Mass. and London: Harvard University Press.
†Stebbins, G.L. 1971: *Processes of organic evolution.* Englewood Cliffs, NJ: Prentice-Hall.

evorsion The erosion of rock or sediment in a river bed through the action of eddies and vortices.

exaration The process by which glaciers pluck or quarry bedrock. Abrasion is not involved.

exfoliation Onion-weathering or desquamation. The weathering of a rock by peeling off of the surface layers.

exhaustion effects Encountered frequently in detailed studies of the variation of suspended sediment concentrations in rivers and streams during storm run-off events, and attributable to a progressive reduction or exhaustion in the availability of sediment for mobilization and transport. They may occur during an individual event,

where they will be reflected in the occurrence of maximum suspended sediment concentrations and loads before the hydrograph peak, or during a closely spaced sequence of hydrographs, where they will give rise to a progressive decrease in the sediment concentrations associated with similar levels of water discharge.   DEW

exhumation The exposure of a subsurface feature through the removal by erosion of the overlying materials.

exogenetic Pertaining to processes occurring at or near the surface of the earth and to the landforms produced by such processes. By contrast ENDOGENETIC processes originate within the earth.

exotic A term normally used to describe a plant or animal which is kept, usually in a semi-natural or artificial manner, in a region outside its natural provenance. A wide range of garden plants and aviary birds are exotics, having been introduced by horticulturalists and bird fanciers from widely differing regions of the world. The term tends to be used in a rather restricted sense to describe especially spectacular species collected from markedly different climatic regimes. Exotic should be contrasted with the term ALIENS, which is the commoner designation for introductions which are more generally naturalized in their new locations.   PAS

Reading
Whittle, T. 1970: *The plant hunters.* London: Heinemann.

expansive soils Soils which shrink and swell according to their moisture content. Under the SEVENTH APPROXIMATION soil classification scheme they are vertisols. Soils and rocks containing sodium montmorillonite clays are especially susceptible, though alkali soils and pyrites also exhibit volume change properties. Areas with marked seasonality of climate (e.g. semi-arid areas) are especially prone to the effects of expansive soils because of the great changes in moisture status that take place during the year. Changes in vegetation cover may influence moisture status and in this way man can alter the character of these materials. The presence of expansive soils can cause severe damage to structures, and while the problem exists in Britain, especially during drought years like 1976, it is in areas with a more severe climate that the problem becomes acute. Holtz (1983) has estimated that in the USA the damage caused by expansive soils at 1982 prices is $6000 million per annum, and that the losses exceed those caused by earthquakes, tornadoes, hurricanes and floods combined. Expansive soils are therefore one of the most serious natural hazards.　　　ASG

Reading and Reference
†Driscoll, H. 1983: The influence of vegetation on the swelling and shrinkage of clay soils in Britain. *Géotechnique* 33, pp. 93–126.
Holtz, W.G. 1983: The influence of vegetation on the swelling and sinking of clays in the United States of America. *Géotechnique* 33, pp. 159–63.

experimental catchment A small drainage basin used for detailed investigations of hydrological and geomorphological processes. During the UNESCO International Hydrological Decade (1965–74) the term was frequently used in a more limited context to refer to small (<4 km²) catchment studies, where the vegetation cover or land use was deliberately modified in order to study the hydrological impact of such changes. These studies were seen as experiments. Various strategies involving single, paired and multiple watershed experiments have been employed in order to decipher the nature and magnitude of changes in catchment behaviour.　　　DEW

Reading
Rodda, J.C. 1976: Basin studies. In J.C. Rodda ed., *Facets of hydrology*. Chichester: Wiley.
Toebes, C. and Ouryvaev, V. eds. 1970: *Representative and experimental basins: an international guide for research and practice*. Paris: UNESCO.

extending flow See COMPRESSING FLOW.

extinction The disappearance of an individual organism, a group of organisms or a local population from existence. Although many recorded extinctions are directly attributable to the actions of man, the fossil record clearly reveals that nearly all lineages have become extinct without leaving descendants and that evolutionary conservatism will frequently lead to extinction in the face of significant environmental changes, such as climatic change, the development of new competitors, predators or diseases, or the loss of a major food supply. Many classic mass extinctions, such as that of the dinosaurs, remain enigmas unexplained by modern science, although there is never a dearth of speculation,

scientific or otherwise, on the causes of their demise. The process of extinction continues, however, and over the past 200 years or so it is estimated that some fifty-three birds and seventy-seven mammals have become extinct.

One particularly relevant principle in the discussion of extinction is that of 'competitive exclusion' or the 'ecological replacement principle'. When one group disappears, another appears to take its place in a corresponding or identical habitat. This theory therefore asserts that when an ecological resource is used simultaneously by more than one kind of organism and when this resource is insufficient to furnish all their needs, then the better or best adapted organism will eventually eliminate the competitors. It has even been argued that new forms must always replace old ones and that for every species that evolves, another becomes extinct. It is, however, clear that competition is not always present in the process of extinction, which may simply reflect evolutionary convervatism in relation to the speed and direction of environmental change. Indeed, competition may resolve itself by producing diverging adaptations, leading to overall diversification and not to extinction. Other factors involved in extinctions include overspecialization, reduced mutation, loss of behavioural versatility and changes in community patterns.

Whatever the causes, the palaeontological record clearly indicates that extinction has paralleled evolution throughout the entire history of the plant and animal kingdoms, a fact which has yet to be faced by the modern CONSERVATION movement, which so often aims to maintain all forms, whatever the cost and against all odds, in the battle between birth and death. (See also ISLAND BIOGEOGRAPHY.)　　　PAS

Reading
Cox, C.B. and Moore, P.D. 1993: *Biogeography: an ecological and evolutionary approach*. 5th edn. Oxford: Blackwell Scientific.
Krebs, C.J. 1978: *Ecology: the experimental analysis of distribution and abundance*. New York: Harper & Row.
MacArthur, R.H. 1972: *Geographical ecology*. New York: Harper & Row.
Stanley, S.M. ed. 1987: *Extinction*. New York: Scientific American Library.

extra-tropical cyclone An area of low pressure which develops in the westerly wind belts of middle latitudes and is associated with characteristic weather patterns. Also known as a depression or low. At the surface the extra-tropical cyclone consists of an area of low pressure surrounded by winds blowing in an anticlockwise direction in the northern hemisphere and in a clockwise direction in the southern hemisphere. Central pressure values vary greatly from about 930 hPa to 1015 hPa. The most intense cyclones usually occur in winter and, on average, those of the southern hemisphere are deeper than those in the northern hemisphere. The diameter of the system may range from about 500 km to over 3000 km. The associated weather is unsettled with cloud, strong winds, periods of rain and sudden changes of

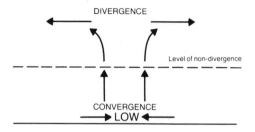

Figure 1.

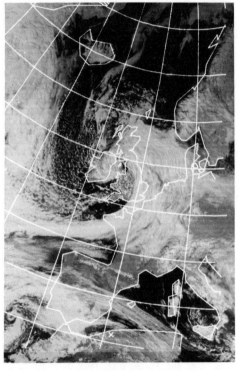

An extra-tropical cyclone is centred over eastern Ireland with dense cloud forming the characteristic spiral away from the centre. Ahead of the system the cloud boundary is rather diffuse with evidence of some cirrus cloud over southern Norway. The sudden change in cloud type at the cold front is visible down the centre of England where convective shower clouds in the cold air behind the front contrast with the layer cloud ahead of the front. Over the Atlantic Ocean to the west of Ireland, patterns are clearly visible indicating variable convection. Another cyclone is seen to the west of Spain together with jet stream clouds over North Africa. The mountains of Norway, the Alps and Pyrenees stand out because of their snowcover.
Photograph taken at 14.55 GMT on 17 March 1980.

temperature as the frontal zones pass. Satellite photographs show the beauty of the cloud patterns around the cyclones. They usually consist of a large 'comma-shaped' mass of cloud near the centre of the low with bands of frontal cloud near the warm and cold fronts streaming away towards the equator on a curved path (see illustration). Many differences of detail are found between cyclones but almost all follow this basic pattern. Over time the cyclones tend to move eastwards or north-eastwards, but some will remain almost stationary or follow unusual tracks depending upon air circulation in the middle atmosphere.

Cyclones are dynamic features of our atmosphere undergoing a sequence of changes from initiation to decay. The start of cyclone development involves a weak surface trough or front and a wave in the upper-level westerlies. Intensification of the surface low takes place as a result of divergence in the upper westerly wave which leads to the formation of a cyclonic circulation along the surface front (figure 1). Establishment of the cyclonic circulation allows the flow of warm air to occur to the east of the cyclone centre and cold air transfer to the west of the centre. This temperature modification changes the distribution of uplift from one in which there is a general area of ascent and low-level convergence to one in which ascent takes place mainly east and poleward of the surface low and descent west and equatorward of the low. The circulation associated with the cyclone causes the cold front to move eastward and equatorward and the warm front to move eastward and poleward. This stage of development is shown in figure 2(a) below. Further advection of warm and cold air together with changes in the upper airflow produce a gradual (or sometimes rapid) change in the appearance

(a)   (b)

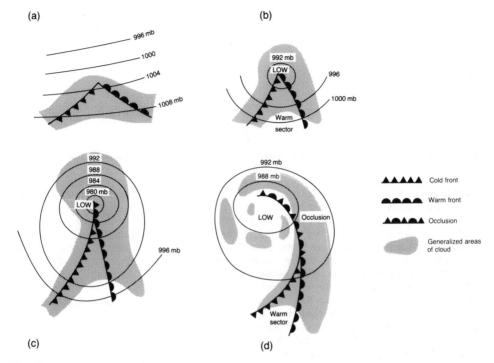

(c)   (d)

Figure 2.

of the surface low. By stage (d) the storm has reached its maximum intensity and started to occlude. Cold air is present at all levels over the cyclone centre. As the system decays, advection over the centre almost ceases. Surface friction and internal dissipation ensure that this eddy in the westerly circulation loses its identity.

In the northern hemisphere troughs in the westerlies have favoured positions of location because of the mountain ranges and the distribution of land and ocean. As a result cyclones have preferred areas of formation: off Newfoundland and north-east of Japan for example. In the southern hemisphere, the westerly flow is more symmetrical about the pole and longitudinal variations in cyclone formation are less pronounced. (See also CYCLONES.)   PS

Reading
Carlson, T.N. 1988: *Mid-latitude weather systems*. London: Harper-Collins.
Palmén, E. and Newton, C.W. 1969: *Atmospheric circulation systems*. New York: Academic Press.

extrusion flow of glaciers A view that the lower layers of a glacier are squeezed out by the weight of the overlying ice and move faster than it. The concept has been abandoned as physically impossible.

extrusion, volcanic A feature produced by rocks which have been deposited at the earth's surface after eruption, in a molten or solid state, from volcanic vents and fissures.

exudation basin A depression occurring at the head of glaciers emanating from the Greenland ice cap.

eye The centre of a cyclone where air descends from the upper atmosphere filling the low pressure zone. A calm area around which cylonic winds blow.

eyot An islet in a river or lake.

F

fabric The three-dimensional arrangement of the particles of a sediment. It is a bulk property and can be specified at a variety of scales and in a number of ways. It is possible to refer to the 'clay fabric', the organization and arrangement of clays over a volume of a few millimetres. Most usually, clay fabrics are related to the engineering properties of the soil; especially to the response to STRAIN; this is similar to the analysis of strain and fabrics of metamorphic rocks. The clay fabric may have a rather different (microfabric) from the overall fabric, which takes much larger particles into account and covers a much greater volume. The volume taken is critical to measurement and interpretation of fabrics. The fabrics of stones (clasts) are often used as indicators of palaeocurrent direction, streams or ice sheet movement (TILL FABRIC ANALYSIS). They can also be used to help to identify a specific geomorphological mechanism, e.g. imbrication of pebbles in a stream, as well as to help explain the mass behaviour of a sediment, e.g. preferred permeability direction. Fabrics are sometimes determined in two dimensions but ideally three dimensions should be used. Where stones are employed, a preferred orientation is determined from analysis of a number (more than fifty) of individual measurements which are statistically analysed to give a measure of the fabric for that position. A clast long axis (a) needs to be determined and its orientation (or azimuth), with respect to, say, true north and its plunge (with respect to an angle below the horizontal) determined. These two values can be plotted on an equal area projection (or net) and contoured for all the clasts taken. Alternatively, the maximum projection (a/b) plane can be determined and the orthogonal to this (the c axis) measured with respect to azimuth and plunge. This again can be plotted. Values of azimuth and plunge (of a line) or dip (of a plane) can be used to calculate mean orientations directly by various statistical measures and tested against several types of spherical distribution.   WBW

Reading
Goudie, A.S. ed. 1990: *Geomorphological techniques* 2nd edn. London: Allen & Unwin.
Whalley, W.B. 1976: *Properties of materials and geomorphological explanation.* Oxford: Oxford University Press.

facet A flat surface on a rock or pebble produced by abrasion.

facies The characteristics of a rock or sediment which are indicative of the environment under which it was deposited. A distinct stratigraphic body which can be distinguished from others on the basis of appearance and composition. Lateral variations in the nature of a stratigraphic unit.

falling dune Sand accumulation (sloping at the angle of repose of dry sand) to the lee of a cliff or mountain side as sand is blown off the top.

false-bedding The stratification of sediments in several units inclined to the general stratification. The product of fluvial, littoral and aeolian sedimentation.

fan The accumulation of sands and gravels deposited by debris flows, streams and rivers, usually in arid and periglacial regions when they issue from confined channels and dissipate their energy and sediment load.   ASG

Reading
Nielsen, T.H. and Moore, T.E. 1984: *Bibliography of alluvial fan deposits.* Norwich: Geo Books.

fanglomerate Indurated alluvial fan gravel.

fatigue failure Fracture as a result of repeatedly applied cyclic stresses, at levels far below the instantaneously determined strength of a material: it is widely recognized as a major contributor to failure in metals. In aircraft, regular vibrations have led to disastrous fracturing. Experimental studies of cyclic stresses in rocks have shown that fatigue fracturing occurs at 80–60 per cent of the ultimate strength and after a number of cycles which range from 10^3 to 10^6. The importance of fatigue failure in rocks exposed to repeated wetting and drying, thermal expansion and contraction, and salt crystallization and dissolution, is not known. It may, however, be a significant cause of several forms of rock fracturing which have not yet been explained satisfactorily.   MJS

Reading
Selby, M.J. 1993: *Hillslope materials and processes.* 2nd edn. Oxford: Oxford University Press. Ch. 8.

fault A crack or fissure in rock, the product of fracturing as a result of tectonic movement. The line along which displacement of formerly adjacent rocks has taken place as a result of earth movements. Normal faults develop under a pattern of stress which is predominantly tensional. A down-faulted block between a pair of more or less normal faults is known as a graben, and the up-faulted block is called a horst. By contrast, reverse faults are normally associated with zones of compression. Where the angle of dip is low the term thrust fault is used. When the mean compressive stress is vertical, strike-slip faults are formed (also called wrench faults and transcurrent faults). Where both horizontal and vertical movements are significant, the term oblique-slip fault is applied.   ASG

faunal realms The largest divisions into which the faunas of the world are customarily grouped; also called faunal kingdoms or empires. The earliest enduring classification of world faunas was drawn up by Sclater in 1858 and dealt with the distribution of birds, a group already well known by that date. This was then extended by Günther to include reptiles and finally consolidated into a widely accepted and applicable system by Alfred Russel Wallace, particularly in his early classic of zoogeography, *The Geographical Distribution of Animals* (1876: see WALLACE'S REALMS). In this, he subdivided each of Sclater's original six regions into four subregions. The Sclater–Wallace division is still in use today, despite

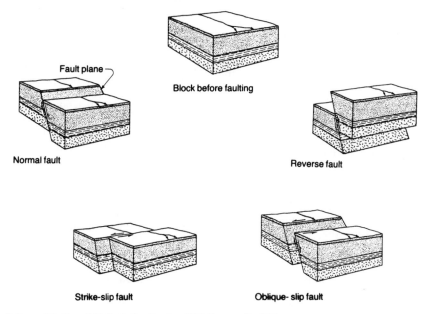

Source: *F. Press and R. Siever 1978: Earth. San Francisco: W.H. Freeman. Fig. 20.33.*

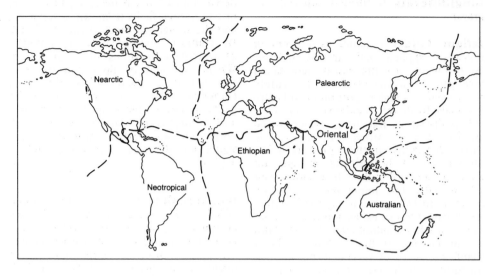

Source: *A.S. Goudie 1984:* The nature of the environment. *Oxford and New York: Basil Blackwell.*

attracting continuous criticism, and it is constantly under revision. Smith (1983), for example, has recently attempted a logical and statistical derivation of world mammal faunal regions.

The main faunal realms normally recognized are:

1 The *Nearctic*, comprising virtually the whole of North America (e.g. the common raccoon, the American bison).
2 The *Neotropical*, including Central America and South America (e.g. sloths, opossums, vicuña, guanaco).
3 The *Palearctic*, embracing Europe and Asia north of the tropics (e.g. the European bison).
4 The *Ethiopian*, comprising Africa, mainly south of the Sahara (e.g. the aardvark, lemurs).
5 The *Oriental*, covering tropical Asia (e.g. the orang-utan, the tarsier, tree-shrews).
6 The *Australian*, a complex realm including parts of South-East Asia, Oceania, Australia and New Zealand (e.g. the Australian marsupials, such as the kangaroo, the Tasmanian wolf, etc.).

Because they have many closely related species or animals in common, the *Nearctic* and the *Palearctic* are usually linked together in one realm known as the *Holarctic* (e.g. lynx, wolf, fox, brown bear, weasel). This larger unit is, in turn, frequently joined with the *Ethiopian* and the *Oriental* realms to create one major single division known as *Arctogaea*, a concept first suggested by Huxley in his system. In Huxley's classification, the *Neotropical* realm was called *Neogaea* and the *Australian* realm *Notogaea*, terms still in use today.

The boundary zones between some of these realms are far from clear-cut and controversy has particularly surrounded that between the Australian and Oriental realms in the region known as Wallacea (see WALLACE'S LINE). Essentially, the realms reflect the present-day distribution of animals and are basically defined longitudinally by the great oceanic barriers and latitudinally by the subtropical warm temperate dry belt. It is also apparent that the different faunas are the product of continental drift, brought about through the processes of plate tectonics and sea-floor spreading, and that the distinctive marsupial fauna of the Australian realm, for example, survived because this landmass was separated from the rest of the world before placental mammals were able to invade it and oust the more primitive forms. Left in isolation, the marsupials were able to radiate by ADAPTIVE RADIATION into a large number of different types. (See also FLORISTIC REALMS.)   PAS

Reading and Reference
†Illies, J. 1974: *Introduction to zoogeography.* London: Macmillan.

†Schmidt, K.P. 1954: Faunal realms, regions, and provinces. *Quarterly review of biology* 29, pp. 322–31.

Smith, C.H. 1983: A system of world mammal faunal regions. 1. Logical and statistical derivation of the regions. *Journal of biogeography* 10, pp. 455–66.

†Udvardy, M.D.F. 1969: *Dynamic zoogeography*. New York: Van Nostrand Reinhold.

feather edge The thin edge of a wedge-shaped sedimentary rock that tapers out and eventually disappears as it abuts an area where no deposition has taken place.

feedback See SYSTEMS.

feedback loops or homeostasis See SYSTEMS.

felsenmeer See BLOCK FIELDS, BLOCK STREAMS.

fen A low-lying area partially inundated by water which is characterized by accumulations of non-acid peat.

feral relief Describes the landscape occurring in areas where the sides of the main valleys are dissected by insequent streams. It is a product of rapid run-off with intense dissection.

ferrallitization A combination of intense weathering and efficient removal of the more soluble weathering products under warm, wet conditions. There are three basic aspects to the process: intensive and continuous weathering of parent material involving the release of iron and aluminium oxides and silica, as well as of bases; translation of soluble bases and silica; and formation of 1:1 kaolin-type clays.

Ferrel cell A thermally indirect, weak, MERIDIONAL CIRCULATION in middle latitudes of the atmosphere, in which air rises in the colder regions around 60° latitude and descends in the warmer regions around 30° latitude (subtropic high pressure belts). This cell, in which the surface winds are predominantly westerly, forms part of a three-cell (in each hemisphere) mean meridional circulation pattern, the other two cells being the Hadley and polar cells.   KJW

Reading
Holton, J.R. 1992: *Introduction to dynamical meteorology*. New York and London: Academic Press.

Ferrel's law States that all moving masses of air move to the right (as they proceed) in the northern hemisphere and to the left in the southern hemisphere. This is because air, in moving from higher to lower pressure, is subject to the effect of the earth's rotation (see CORIOLIS FORCE).

KJW

Reading
Palmén, E. and Newton, C.W. 1969: *Atmospheric circulation systems*. New York and London: Academic Press.

ferricrete An iron-pan or a near-surface zone of iron oxide cementation. An accumulation of iron oxides and hydroxides within the soil zone as a result of weathering or soil-forming processes such as laterization. It is a type of duricrust and tends to be associated with deep weathering profiles.

fetch The extent of the area of sea or ocean over which waves have developed through the effect of broadly unidirectional winds.

fiard A river valley which has been drowned by the sea owing to a rise in sea level or subsidence of the land.

field capacity The condition reached when a soil holds the maximum possible amount of water in its voids and pores after excess moisture has drained away. A measure of the volume of water a soil can hold under these conditions.

field drainage The process of artificially accelerating the movement of water through soil. This is undertaken to lower levels of saturation, to reduce the extent of surface water ponding or run-off and to minimize the periods during which saturation or surface water exists on agricultural land. This process is alternatively called under-drainage, in contrast to the surface and arterial drainage which involve ditching and the modification of existing water courses to accelerate channelized run-off.

Field drainage may be implemented by excavating a network of trenches and inserting plastic piping or tubing, earthenware pipes or permeable rock fill, and then backfilling the trenches again. In heavy cohesive soils an extra network of mole drainage may be added. This is undertaken by drawing a former (the 'mole') through the soil at the required depth, the hole remaining open for a number of years afterwards. Deep ploughing ('subsoiling') may also accelerate field drainage.   JL

Reading
Green, F.H.W. 1978: Field drainage in Europe. *Geographical journal* 114, pp. 171–4.
Robinson, M. 1990: Impact of improved drainage on river flows. *Institute of Hydrology, Wallingford, Report*, 113.

Finger Lakes A group of semi-parallel lakes in New York State. The name originated in the Indian legend of the Great Spirit placing his hand on the earth causing the finger-like indentations. An alternative explanation is that they were cut by glacial erosion.   DES

Reading
Clayton, K.M. 1965: Glacial erosion in the Finger Lakes region (New York State, USA). *Zeitschrift für Geomorphologie* 9, pp. 50–62.
Coates, D.R. 1966: Discussion of K.M. Clayton: Glacial erosion in the Finger Lakes region (New York State, USA). *Zeitschrift für Geomorphologie* 10, pp. 469–74. (Also in the same issue pp. 475–7, Clayton, K.M. Reply to Professor Coates.)

fiord (fjord) A glacial trough whose floor is occupied by the sea. Fiords are common in uplifted mid-latitude coasts, for example in Norway, East Greenland, eastern and western Canada and Chile. Historically, the origin of fiords has generated a great deal of interest, for example an extreme view championed by Gregory (1913) that they are essentially tectonic forms. Today it seems easiest to think of them as essentially glacial with such features as steep sides, overdeepened rock basins and shallow thresholds at the coast as characteristic of erosion by ice streams or valley glaciers exploiting either pre-existing river valleys or underlying weaknesses in the bedrock (Holtedahl 1967).   DES

Reading and References
Gregory, J.W. 1913: *The nature and origin of fjords.* London: John Murray.
Holtedahl, H. 1967: Notes on the formation of fjords and fjord valleys. *Geografiska annaler* 49A, pp. 188–203.
†Loken, O.H. and Hodgson, D.A. 1971: On the submarine geomorphology along the east coast of Baffin Island. *Canadian journal of earth sciences* 8.2, pp. 185–95.
Syvitski, J.P.M., Burrell, D.C. and Skei, J.M. 1987: *Fjords: processes and products.* New York: Springer-Verlag.

fire One of the greatest forces employed by man for the modification of the environment; it may have been used deliberately as long as 1.4 million years ago (Gowlett *et al.* 1981). It has been used for a great variety of reasons: to clear forest for agriculture; to improve grazing land for domestic animals

or to attract game; to deprive game of cover and to drive them from cover; to kill or drive away predatory beasts and pests; to repel enemies; for cooking; to expedite travel; to protect settlements or encampments from great fires by controlled burning; to satisfy the sheer love of fires as spectacles; to make pottery, smelt ores, harden spears, etc; to provide warmth; and to prepare charcoal. Although man is responsible for many fires, they can arise naturally because of lightning, spontaneous combustion, or volcanic activity. Fire has many ecological consequences, affecting, for example, the diversity of pine forests, the distribution of prairie, and the origin of savannah. In either its natural or its man-made forms it has been an influence in nearly every BIOME, with the possible exception of the moist tropical forest. Among other components of ecosystems it affects the flow of water. (See also FIRES, HYDROLOGICAL EFFECTS.)   ASG

Reading and References
†Daubenmire, R. 1968: Ecology of fire in grassland. *Advances in ecological research* 5, pp. 209–66.
Gowlett, J.A.J., Hairns, J.W.K., Walton, D.A. and Wood, B.A. 1981: Early archaeological sites, hominid remains and traces of fire from Chesowanja, Kenya. *Nature* 294, pp. 125–9.
†Kozlowski, T.T. and Ahlgren, C.E. eds 1974: *Fire and ecosystems.* New York: Macmillan.
Levine, J.S. ed. 1991: *Global biomass burning. Atmospheric, climatic and biospheric implications.* Cambridge, Mass.: MIT Press.
†Stewart, O.C. 1956: Fire as the first great force employed by man. In W.L. Thomas ed., *Man's role in changing the face of the earth.* Chicago: University of Chicago Press. Pp. 115–33.

fires, hydrological effects Fire affects the hydrological cycle by destroying vegetation and by radically altering soil conditions. These changes cause adjustments in the processes of infiltration, run-off, erosion, and sedimentation. Fire affects the infiltration and run-off process in at least five ways.

1 The characteristics of litter on the ground surface are changed as organic material is burned away. The result is that rainsplash erosion and the mobilization of sediment are likely to be more significant after a fire than previously.

2 The removal of vegetation by fire increases the extent of exposed soil in a given area, so that run-off is more effective in entraining sediment.

3 The heat from the fire alters the structure of the soil layer, frequently resulting in the loss of cohesion which

contributes to increased sediment yield from the burn area.

4 Fire usually leads to a decrease in porosity of the soil layer as grains are fused together and as the amount of fine-grained materials increases, also contributing to increased run-off.

5 Solute concentrations in run-off may be dramatically changed by fire as chemical changes occur in fire debris and soils.

The overall effect of these changes is to decrease infiltration and to increase run-off 0–95 per cent, depending on initial conditions (Branson *et al.* 1981, p. 65). The degree of impact is related to severity of the burn, to textural class of the surface materials, and to the slope involved (Settergren 1967). Sediment yield increases 5–13 times after a forest fire in mountainous areas (Anderson 1975), amounting to an increase of as much as 500 t ha^{-1}, though usually the range of increase is in the order of 2–10 t ha^{-1} in less steep terrain (White and Wells 1979).

WLG

References
Anderson, H.W. 1975: Sedimentation and turbidity hazards in wildlands. In *Watershed management*. Symposium Proceedings of the Irrigation and Drainage Division of the American Society of Civil Engineers.

Branson, F.A., Gifford, G.F., Renard, K.G. and Hadley, R.F. 1981: *Rangeland hydrology*. 2nd edn. Dubuque, Iowa: Kendall Hunt.

Settergren, C.D. 1967: Reanalysis of past research on effects of fire on wildland hydrology. *Missouri Agricultural Experiment Station research bulletin 954.*

White, W.D. and Wells, S.G. 1979: Forest fire devegetation and drainage basin adjustments in mountainous terrain. In D.D. Rhodes and G.P. Williams eds *Adjustments of the fluvial system*. Dubuque, Iowa: Kendall Hunt.

firn The term generally applied to snow which has survived a summer melt season and which has not yet become glacier ice.

firn line See EQUILIBRIUM LINE.

fission track dating Based on the principle that traces of an isotope, ^{238}U, occur in minerals and glasses of volcanic rocks, and that this isotope decays by spontaneous fission over time, causing intense trails of damage called tracks. These narrow tracks, between 5μ and 20μ in length, vary in their number according to the age of the sample. By measuring the numbers of tracks an estimate can be gained of the age of the volcanic minerals.   ASG

Reading
Miller, G.H. 1981: Miscellaneous dating methods. In A.S. Goudie ed., *Geomorphological techniques*. London: Allen & Unwin. Pp. 292–7.

fissure eruption The eruption of volcanic gases, lavas and rocks through a large crack or chasm in the earth's surface rather than through a pipe or vent.

A fissure developed in a lava platform in northern Iceland. Note the volcanic cone in the background. Many of the world's great spreads of lava were extruded from such fissures.

Flandrian transgression Occurred as the ice sheets of the last glacial melted in the Holocene and Late Pleistocene. World sea levels rose, possibly by as much as 170 m. The continental shelves were flooded, and river and glacial valleys were transformed into rias and fiords respectively. This transgression was more or less complete by 6000 years ago, and may locally have reached a few metres above the present level. It caused the separation of Ireland from mainland Britain, and of Britain from the continent of Europe. In areas of very gentle slope, like the Arabian Gulf, the rapid lateral spread of the transgression (by as much as 100 m year^{-1} at its peak) may have contributed to the biblical story of the flood. ASG

Reading
Kidson, C. 1982: Sea-level changes in the Holocene. *Quaternary science reviews* 1, pp. 121–51.

flash A local name from north-western England, especially Cheshire, for water-filled depressions produced by ground subsidence associated with salt dissolution. Many of them have resulted from salt mining and brine pumping. ASG

Reading
Bell, F.G. 1975: Salt and subsidence in Cheshire, England. *Engineering geology* 9, pp. 237–47.

flashiness The rapidity with which the stage discharge increases at a stream cross-section. The hydrograph (a graphical plot of water discharge versus time) of a flashy stream shows a rapid increase in discharge over a short period, with a quickly developed high peak in relationship to normal flow (Ward 1978, p. 10). On the other hand, a sluggish stream develops peaks more slowly, and the peaks are relatively lower than those developed on flashy streams.

The term flashy depends on personal judgement and is rarely quantified. It is frequently applied to streams in arid or semi-arid regions where the normal flow is zero. Flash floods cause waves or bores to descend the stream course, so that hydrographs show nearly instantaneous rises from zero to high values of discharge in extreme cases of flashiness (Schick 1970).

Flashiness results from the actions of several possible factors that intensify flood peaks, including accelerated run-off rates caused by land use changes, basin shapes that become progressively more narrow in the downstream direction, especially intense rainfall events, and rapid snowmelt. WLG

References
Schick, A.P. 1970: Desert floods. In *Symposium on the results of research on representative and experimental basins*. International Association of Scientific Hydrology publication. No. 96. Pp. 479–95.

Ward, R. 1978: *Floods: a geographical perspective*. London: Macmillan; New York: Wiley.

flatiron Steep triangular cliff facets resulting from the presence of a capping of rock resistant to erosion which protects the underlying more readily eroded rocks.

ASG

Reading
Schmidt, K-H. 1989: Talus and pediment flatirons – erosional and depositional features on dryland cuesta scarps. *Catena* Suppl. 14, pp. 107–18.

float recorder A device for recording the water level in a river channel, lake, well or similar situation. A flexible cable or steel tape connecting a float and counterweight passes over a pulley which transmits vertical movements of the float to the recording mechanism. Several principles are employed. For example, in a vertical recorder the pulley moves the pen vertically across a

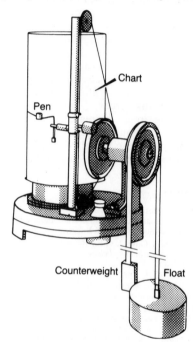

1. Vertical float recorder.

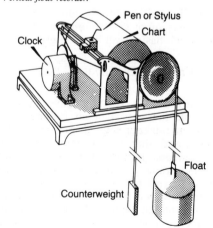

2. Horizontal float recorder.

moving chart, whereas with a horizontal recorder the pulley rotates the chart against the pen, which is driven across the chart by a clock. Digital recording devices also exist (see DISCHARGE). DEW

Floating bogs Refer to any lake-fill bog or kettle-water wetland with a quaking *Sphagnum* mat. Such bogs, which are common in parts of North America, are generally referred to as 'schwingmoor' or 'kesselmoor' in Europe. ASG

Reading
Warner, B.G. 1993: Palaeoecology of floating bogs and landscape change in the Great Lakes drainage basins of North America. In F.M. Chambers ed., *Climate change and human impact on the landscape*. London: Chapman & Hall. Pp. 237–45.

flocculation The process of aggregating into small lumps, applied especially to clays.

flood A high water level along a river channel or on a coast that leads to inundation of land which is not normally submerged. River floods which involve inundation of the FLOODPLAIN may be caused by:
Precipitation when storm precipitation is very intense; when precipitation is very prolonged and follows a period of wetter than average conditions; when the snowpack melts and snowmelt floods are an annual feature of many river regimes; when rain falls on snow and accelerates snowmelt; or when ice and snowmelt are combined and melting river ice produces breakup floods.
Collapse of dams which may be natural where a landslide temporarily blocks a river or a log jam dams the river channel; or which may occur when a man-made dam bursts and the impounded water flows down the river valley as a flood wave or when a landslide into a reservoir leads to a wave overtopping the dam wall.
Drainage of ice-dammed lakes which can lead to the release of great volumes of water (in Iceland JOKULHLAUPS are flood waves which roar down-valley as a result of the failure of a lake dammed by, or contained within, a glacier).
In addition to being effected by the above causes coastal floods may also be produced by:
High tides – especially in combination with river floods.
Storm surges – abnormally high sea levels at about the time of spring tides.

Tsunami – large waves produced by submarine earthquakes, volcanic eruptions, landsliding or slumping.
The FLOOD FREQUENCY in a particular area may vary over time, especially as a result of human activity, and in some areas the damage produced by floods has increased despite expenditure on flood protection simply because there is a tendency to assume that flood protection measures have completely eliminated the flood hazard. Responses to flood hazard include:
Flood protection by structural measures along the river or coast to reduce the effects of flooding; these include walls and embankments, river diversion schemes, flood barriers.
Flood reduction by taking action in the drainage basin by afforestation, controlled vegetation or agricultural changes; or by the construction of small or large dams.
Flood adjustment by adjusting to the hazard by accepting it; by land use zoning; by taking emergency measures when floods occur; or by floodproofing so that flooding will damage structures and buildings as little as possible. KJG

Reading
Ward, R.C. 1978: *Floods: a geographical perspective*. London: Macmillan.

flood frequency An analysis using data from the period of hydrologic records to establish a relationship between discharge and return period or probability of occurrence as a basis for estimating discharges of specific recurrence intervals. Several methods are available for calculating the recurrence intervals or return periods (T) but one most frequently employed ranks the N years of record from the highest (rank $m = 1$) to the lowest (rank $m = N$) and uses:

$$T = \frac{N+1}{m}.$$

To obtain the N years of record use of the highest discharge from each year of record is the basis for the ANNUAL SERIES, whereas using the N highest independent discharges provides the ANNUAL EXCEEDANCE SERIES. The recurrence intervals may be plotted against the discharges for a specific gauging station and the curve fitted by eye or by fitting an appropriate theoretical probability distribution. A range of distributions are available and the Gumbel extreme value type 1 based on GUMBEL EXTREME VALUE THEORY is one frequently used. The flood frequency plot of discharge against recurrence interval can be the basis for estimating the recurrence interval of discharge including the BANKFULL DISCHARGE and for giving the DESIGN DISCHARGE for a particular return period when a structure is being planned. When a regional flood frequency analysis is required using the data from a series of gauging stations within the region it is necessary to undertake a homogeneity test which identifies a homogeneous region according to the ten-year flood estimated from the probability curve for each gauging station.   KJG

Reading
NERC 1975: *Flood studies report.* 5 vols. London: Natural Environment Research Council.
Newson, M.D. 1975: *Flooding and flood hazard in the United Kingdom.* Oxford and New York: Oxford University Press.
Ward, R.C. 1978: *Floods: a geographical perspective.* London: Macmillan.

flood geomorphology 'The study of the role of floods in shaping the landscape, including the analysis of flood causes, flood processes, resistance factors to flood-induced landscape change, and changes in flood-related processes and forms through time. Geomorphologists are also interested in how the landscape affects flood processes. Indeed, over varying time-scales, in varying climatic and physiographic settings, floods and riverine landscapes constitute complex mutually interactive systems. The study of those systems, with the goal of understanding thresholds for response, feedback elements, and other intricacies, constitutes the core of flood geomorphic research' (Baker *et al.* 1988, p. ix).   ASG

Reference
Baker, V.R., Kochel, R.C. and Patton, P.C. eds 1988: *Flood geomorphology.* New York: Wiley Interscience.

flood routing The technique of determining the timing and shape of the flood wave at successive points along a river channel. Techniques for flood peak estimation applicable to small areas are not appropriate for larger basins because as a flood wave is routed along the channel it will change in shape as a result of the storage offered by the channel and the characteristics of the channel morphology. Flood routing may be accomplished either by a theoretical method or by an empirical method of which the MUSKINGUM METHOD is well known.   KJG

Reading
Lawler, E.A. 1964: Flood routing. In V.T. Chow ed., *Handbook of applied hydrology.* New York: McGraw-Hill. Sect. 25–II.

floodplain The low-relief area of valley floor adjacent to a river. This is inundated by water during floods and its surface is normally formed from sediment deposited by the river itself. In detail, floodplains possess some relief in the form of LEVÉES, abandoned channels and BACKSWAMP or floodplain areas. Several metres of relief on floodplains hundreds of metres in width may greatly affect the passage of flood-waters.

Considerable contrasts exist between the morphological and sedimentary characteristics of different floodplains. Many are underlain by considerable thickness of Pleistocene and earlier sediments, including glacial till and outwash, and the present plain may be only a retouched version of what past processes have created. Others, by contrast, may only have a thin veneer of alluvial sediment on a bedrock platform which is being scoured by the present river.

Processes of sedimentation are varied. Some floodplains are created largely by the LATERAL ACCRETION of sediments as meandering streams migrate across the alluvial valley floor; there may be a layer of OVERBANK DEPOSITS produced during floods which extends across the whole floodplain, but this is volumetrically of limited significance. Other floodplains have rivers which remain relatively fixed in position where sedimentation may be dominated by vertical accretion of overbank sediment. In BRAIDED RIVERS, channel migration and switching may rapidly rework near-surface sediments, with a large part of the floodplain being effectively the bed of the stream at any one time.

Floodplain soils are commonly rich agriculturally, but these areas are also liable to the dangers of water-borne pollution and inundation. Floodplains are at one and the same time profitable, hazardous and rapidly changing geomorphological environments.   JL

Reading
Bloom, A.L. 1978: *Geomorphology.* Englewood Cliffs, NJ: Prentice-Hall.
Lewin, J. 1978: Floodplain geomorphology. *Progress in physical geography* 2, pp. 408–37.

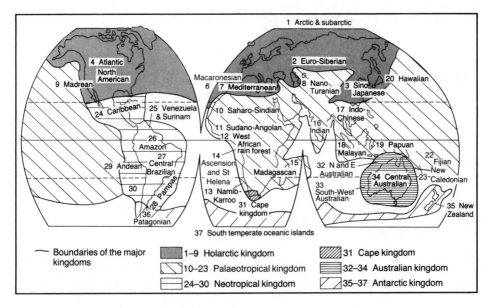

Source: *A.S. Goudie 1993: The nature of the environment. 3rd edn. Oxford and New York: Basil Blackwell.*

Nanson, G.C. and Croke, J.C. 1992: A genetic classification of floodplains. *Geomorphology* 4, pp. 459–86.
Ward, R.C. 1978: *Floods – a geographical perspective.* London: Macmillan.

floristic realms The largest divisions into which the floras of the world are customarily grouped; also called floristic kingdoms or empires. In turn, these may be subdivided into floristic regions, provinces and districts. The earliest substantial system was that of Engler (1879–1882). Floristic realms are mainly characterized by the plant families or substantial sections of families which are endemic to that particular portion of the globe. The main realms normally recognized are: the *Holarctic* floral realm (e.g. most of the Pinaceae), which embraces North America, Greenland, Europe and the northern part of Asia; the *Palaeotropical* realm (e.g. the pitcher plant family, the Nepenthaceae); the *Neotropical* realm (e.g. the Bromeliaceae); and the *Austral* realm (e.g. the Proteaceae), which includes southern South America, southern Africa, Australia and New Zealand. The boundary between the holarctic and the tropical realm is, with a few exceptions, marked by the southward limit of the northern hemisphere coniferous family (Pinaceae) and the northward limit of the palms (Palmae or Arecaceae).

The evolution of these different realms reflects above all the geological history of the earth and the movement of continents brought about by the processes of plate tectonics and sea-floor spreading although, of course, the floristic realms are not confined to specific continents. (See also FAUNAL REALMS; WALLACE'S REALMS.)

PAS

Reading
Good, R. 1974: *The geography of the flowering plants.* 4th edn. London: Longman.
Moore, D.M. ed. 1982: *Green planet: the story of plant life on earth.* Cambridge: Cambridge University Press. Esp. ch. 6.

flow duration curve Shows the percentage of time that given flows are equalled or exceeded at a particular site. They provide a useful summary of flow reliability at a site and, if the flow axis is standardized for variations in catchment area, the graph forms a very good basis for comparing the flow conditions in different drainage basins or at different sites within the same drainage basin.

AMG

Reading
Dunne, T. and Leopold, L.B. 1978: *Water in environmental planning.* San Francisco: W.H. Freeman.
Gregory, K.J. and Walling, D.E. 1976: *Drainage basin form and process.* London: Edward Arnold.

flow equations Define the inter-relationships between velocity, depth, slope and boundary roughness or energy losses in open channel flow. A variety of flow conditions can be defined including UNIFORM STEADY FLOW and GRADUALLY VARIED FLOW. In the former the water surface is parallel to the bed, and width, depth and velocity are constant along a reach. Under such constrained circumstances the velocity formulae developed by Chézy and Manning (see CHÉZY and MANNING EQUATIONS) relate mean velocity to depth, slope and a roughness coefficient. These equations may be used to estimate flood velocities if flood depth and slope are measured and a suitable value of the roughness coefficient is identified. The equations which define point velocity variation with height above the channel bed – that is, the velocity profile shape within the BOUNDARY LAYER – also relate to uniform flow conditions. In gradually varied flow, which is more common in natural river channels, depth and velocity vary from section to section and bed and water surface slopes are not parallel. Under these conditions the flow properties must be modelled by an energy-balance equation based on the principle of conservation of energy between closely spaced sections. This is the BERNOULLI EQUATION, which defines the total energy of the flow as the sum of potential energy, pressure energy, and kinetic energy. KSR

flow regimes Four flow regimes are defined, in terms of the depth and velocity of open channel flow, by the combination of criteria based on the Froude and Reynolds numbers. These describe the relationships between inertial, viscous and gravitational forces acting on the flow. The Froude number distinguishes subcritical ($F_r < 1$) from supercritical (($F_r > 1$) flow, and the Reynolds number identifies laminar ($R_e < 500$) and turbulent ($R_e > 2000$) flow with a transitional state between these limits. The flow regimes are laminar subcritical, laminar supercritical, turbulent subcritical, and turbulent supercritical (see diagram). In river channels, turbulent subcritical conditions are normal, while overland flow on hillslopes is often laminar except when disturbed by rainsplash.

In a SANDBED CHANNEL, the progressive change of flow intensity during the passage of a flood along a reach is associated with a systematic sequence of changing bedforms. In the subcritical regime, as Froude and

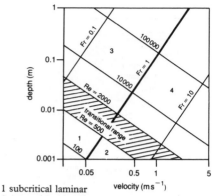

1 subcritical laminar
2 supercritical laminar
3 subcritical turbulent
4 supercritical turbulent

Reynolds numbers increase, small sand ripples up to 4 cm high and 60 cm long are replaced first by dunes with superimposed ripples and then by dunes up to 10 m high and 250 m long, depending on river size. At Froude numbers approaching unity as the flow passes into the upper regime supercritical state, dunes wash out to a plane bed with intense sediment transport, and then antidunes and standing waves form. Thus the flow regimes are closely related to bedform changes, which are in turn associated with changes of sediment transport rate and ROUGHNESS (Simons et al. 1961). KSR

Reading and Reference
Richards, K.S. 1982: *Rivers: form and process in alluvial channels*. London and New York: Methuen.
Simons, D.B., Richardson, E.V. and Albertson, M.L. 1961: Flume studies using medium sand (0.45 mm). *United States Geological Survey water supply paper* 1498A.

flow till See TILL.

fluid mechanics A branch of applied mathematics dealing with the motion of fluids and the conditions governing such motion. Mechanics generally includes kinematics and dynamics, which respectively cover the geometry of motion and the forces involved in motion. Thus the continuity equation which defines channel discharge as the product of cross-section area and velocity is a kinematic statement, while the theoretical velocity profiles of laminar and turbulent boundary layers are the product of dynamical analysis of forces acting upon and within moving fluids. HYDRAULICS deals specifically with the mechanics of *liquid flow*, particularly water. KSR

flume A structure built within a river channel in order to measure the flow. The structure, normally built of concrete, introduces a constriction or contraction into the channel. This produces critical flow conditions for which a unique and stable relationship will exist between the flow depth upstream of the constriction and discharge. The standing wave, critical depth or Parshall flume is the most common design, but there is a wide variety of specialized flume designs which have been produced for specific channel and flow conditions. A flume can provide very accurate measurements of discharge and may be preferred to a WEIR in many situations because sediment is able to pass through the structure. Calibration may involve formulae or field rating. (See also DISCHARGE.) DEW

Reading
Ackers, P., White, W.R., Perkins, J.A. and Harrison, A.J.M. 1978: *Weirs and flumes for flow measurement.* Chichester: Wiley.

fluvioglacial See GLACIOFLUVIAL.

fluviokarst A type of landform developed in limestone areas by a combination of river action and of true KARST processes. Included in this category are gorges and DRY VALLEYS.

flux The rate of flow of some quantity. It is perhaps most easily perceived in the context of fluids (air, water) but it also applies to many other elements of the natural environment, such as mass and energy. Many apparently unchanging forms in the natural world ensue from different fluxes into and out of the system. As such, fluxes are a primary element in natural environmental processes. BWA

flux divergence The variation of fluxes through space. This relatively simple notion has profound implications in the natural environment. As W.M. Davis (1909) noted, landforms are a product of structure, process and stage. It takes but little thought to realize that probably everything in the universe is a function of structure, process and stage – hence the notion can be applied, at least, to all facets of physical geography. In this context we are most usually concerned with the structure and form of a constituent of the natural environment and, more particularly, the processes whereby the structure and form change through time.

In most natural systems fluxes are the essence of process: mass through geomorphological features; water through rivers, oceans, plants; chemicals through soils; air, energy and pollutants through the atmosphere. If the flux of a quantity through a system is constant everywhere, the size and form of the system is frequently unchanging. For example, if the flow of water into and out of a bath is equal, the depth of the water in the bath remains constant; the water itself, of course, is constantly changing. Casual inspection of the water body could easily give the impression that it is unchanging, simply because its depth is constant. In the same way, casual inspection of a scree could easily suggest that it is an unchanging feature whereas in reality significant mass fluxes occur. Clearly, the way to change the depth of the water in the bath is to ensure that input and output differ, that is, to ensure that there is flux divergence in the flow of water through the system. Hence we see that flux divergence is the way in which the state of a feature changes through time and is thus the essence of natural environmental processes.

For many years physical geographers monitored form, with little appreciation of process (and hence fluxes and flux divergences). In the past two decades more emphasis has been put on process and to some extent on fluxes (e.g. river discharge). But there remains much scope for better and more extensive measurements of fluxes and their divergences in the natural environment so as to increase our understanding of the mechanisms of environmental phenomena. BWA

Reference
Davis, W.M. 1909: *Geographical essays.* Boston: Ginn.

flyggberg An asymmetrical hill 1–3 km across and 100–300 m high shaped by the action of overrunning ice. Swedish derivation: *flygg*, a steep rock face; *berg*, mountain.

flysch Deposits of marine sandstones, shales, marls and clays produced during the initial uplift of the Alps by sedimentation and later deformation of materials eroded from the uplifted rocks. The word now refers to any thick succession of alternations of the rocks mentioned above, interpreted as having been deposited by turbidity currents or mass-flow in a deep water environment within a geosynclinal belt.

fog If the air near the ground is cooled enough and its moisture content is high CONDENSATION may occur and a fog will develop. By definition, a fog exists if the visibility in cloudy air is 1 km or less. There are several ways by which moist air may be cooled enough to form fog – loss of RADIATION at night (radiation fog), warm air passing over a cold surface (advection fog), and air flow up an incline (upslope fog). Fog may also form when warm rain falls through cold, but saturated, air (frontal fog) and when cold air passes over warm water (steam fog).   WDS

Reading
Neiburger, M., Edinger, J.G. and Bonner, W.D. 1982: *Understanding our atmospheric environment.* San Francisco: W.H. Freeman.

föhn (chinook in North America) The name given to a strong and gusty down-slope (or 'fall') wind which occurs on the lee side of a mountain range when stable air is forced to flow over the mountains by the large-scale pressure gradient. The air at the foot of the mountains is characteristically dry and warm as a result of the ADIABATIC compression during its descent. In some cases the air may ascend over the range with cloud and precipitation occurring on the windward side, but in other cases a stable layer near the summit level blocks the cross-mountain flow and air descends from the summit level. Föhn onset can give rise to dramatic increases in temperature and such winds are important in accelerating snow-melt in spring along the northern flanks of the Alps and Caucasus and the eastern flanks of the Rocky Mountains.   RGB

Reading
Atkinson, B.W. 1981: *Meso-scale atmospheric circulations.* London and New York: Academic Press.

fold A bend in fomerly planar strata of rock resulting from movement of the crustal rocks. If the strata are flexured in just one plane they are termed a mono-cline. The arching up of strata produces an anticline, while the depression of strata produces a syncline. In a recumbent fold the strata are overturned and both limbs of the folds are nearly horizontal. The

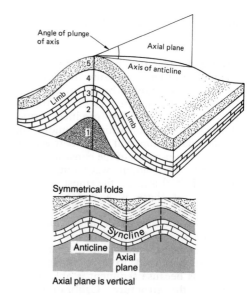

Symmetrical folds

Axial plane is vertical

Fold: 1. Diagrammatic illustration of parts of a fold.

Overturned folds

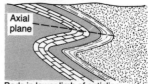

Upper limb of syncline and lower limb of anticline, tilted beyond vertical, dip in same direction

Asymmetrical folds

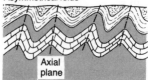

Beds in one limb dip more steeply than those in the other

Recumbent folds

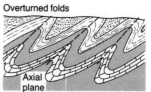

Beds in lower limb of anticline and upper limb of syncline are upside down; axial plane is nearly horizontal

2. Diagrammatic illustration of symmetrical, asymmetrical, overturned and recumbent folds.
Source: *F. Press and R. Siever 1978: Earth. San Francisco: W.H. Freeman. Figures 20.23–24.*

horizontal compression that creates recumbent folds may lead to the shearing of the upper part of the fold along a thrust fault. The strata moved forward over a thrust fault form a nappe. ASG

foliation Fine layering of rocks, usually metamorphic and igneous rock, as a result of parallel orientation of minerals. Slates and schists are typical foliated rocks.

food chain, food web Food chains represent the transfer of food (and, therefore, energy) from one type of organism to another, in sequence and in a linear relationship: for instance, in a marine environment: algae → small fish → squid → man. Normally, however, this would represent an extreme oversimplification of the feeding relationships present in any community, which are much better portrayed as a food web (see diagram).

There are two major types of food chain or food web, termed *grazing* and *detrital*. At the base of the grazing food chain or web are producer organisms which are autotrophs, i.e. they are able to fix incident light energy and so manufacture food from subsequent chemical reactions (see PHOTOSYNTHESIS). These may be green plants, blue-green or other algae, or phytoplankton. All other organisms within it are dependent heterotrophs, i.e. they eat or rearrange existing organic matter. In the overall sequence of food consumption patterns, they may be classed as primary consumers (HERBIVORES), secondary consumers (CARNIVORES, which eat herbivores), tertiary consumers (top carnivores, which eat other carnivores), and so on.

Although essentially similar, detrital food chains begin at producer level with the breakdown of dead or decaying organic material by DECOMPOSER organisms. Consumers then take up the nutrients released by the decomposers, and some of the decomposers themselves; and they in turn are eaten by a range of carnivores. Relationships between the grazing and detrital food chains and webs are complex; in general one may say that much more energy passes through the latter than the former, often in a ratio of *c*.10 to 1 in organically rich communities. DW

Reading
Pimm, S.L. 1982: *Food webs*. London: Chapman & Hall.

foredune A sand dune which has formed on the seaward side of a coastal dune belt (see diagram on p. 216). A dune which accumulates on the upwind side of an obstruction.

foreset beds Layers of sediment which have been laid down on the inclined surface of an advancing deltaic deposit or sand dune.

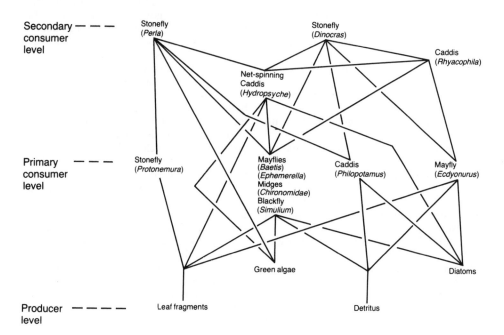

A foredune developed parallel to the shoreline (on the left) at Braunton Burrows, Devon, England. Plants such as marram grass contribute to dune evolution and stabilization. The sand is derived from deflation from a wide beach and intertidal zone.

forest See BOREAL FOREST; DECIDUOUS FOREST; MONSOON FOREST; TROPICAL FOREST.

forest decline A concept that developed particularly in the late 1970s and early 1980s and is related to the German terms *Waldsterben* (forest deaths) and *Waldschaden* (forest damages). The common symptoms include:

1 *Growth-decreasing symptoms*
   Discoloration and loss of needles and leaves.
   Loss of feeder-root biomass (especially in conifers).
   Decreased annual increment (width of growth rings).
   Premature ageing of older needles in conifers.
   Increased susceptibility to secondary root and foliar pathogens.
   Death of herbaceous vegetation beneath affected trees.
   Prodigious production of lichens on affected trees.
   Death of affected trees.
2 *Abnormal growth symptoms*
   Active shedding of needles and leaves while still green, with no indication of disease.
   Shedding of whole green shoots, especially in spruce.
   Altered branching habit.
   Altered morphology of leaves.
3 *Water-stress symptoms*
   Altered water balance.

Increased incidence of wet wood disease.

In many cases it is probably more correct to talk of tree decline in that there are relatively few cases where entire forest eocystems are declining (Innes 1992). The potential causes of the phenomenon, which remain controversial, are poor management practices, ageing of stands, climatic change, severe climatic events, nutrient deficiency, viral, fungal and pest attack, and atmospheric pollution.   ASG

Reference
Innes, J.L. 1992: Forest decline. *Progress in physical geography* 16, pp. 1–64.

forest hydrology The branch of the science of physical geography (Linsley *et al.* 1988, p. 1) that is concerned with the 'study of how forests affect the hydrologic cycle, with particular reference to the regulation of streamflow, water supplies, and erosion control' (Storey *et al.* 1964). The relationships between water and forests have been the subject of scientific and political interest for several centuries. In 1215 King Louis VI of France set out a decree for the management of forests and related waters, and in Switzerland a law of 1342 protected forests from overcutting as an avalanche protection measure (Storey *et al.* 1964). The impacts of deforestation on accelerated soil erosion have been a subject of study and debate in the Mediterranean area for five hundred years. In the New World the US government issued its first report on the hydrological impacts of deforestation in 1849, and in 1891 began setting aside vast areas of forest for the purpose of watershed protection.

From the human perspective forests are considered to have beneficial effects on the hydrological cycle (Kittredge 1948). The forest foliage breaks the fall of raindrops, lessening the erosion caused by drop splash on the surface below. Large quantities of moisture are intercepted by the foliage, temporarily stored on leaves and branches, and then allowed slowly to drop to the surface. This process ensures that precipitation is more likely to be absorbed into the surface instead of being fed to the surface too rapidly for effective infiltration.

The overhead canopy of vegetation in the forest also reduces water loss from the surface by means of evaporation, although this saving is at some cost because the trees transpire large amounts of moisture through their leaves. The forest root system acts as an important soil binder and retards erosion. The litter on the forest floor, organic material that has fallen from the actively growing vegetation, absorbs large amounts of moisture that otherwise would quickly run off into channels. Finally, the forest protects snow cover with shade, preventing rapid melt and flash flooding in downstream areas. Melting takes place over a relatively longer period of time, releasing water later in the melt season when it is more useful for irrigation. WLG

References
Kittredge, J. 1948: *Forest influences*. New York: McGraw-Hill.
Linsley, R.K., Kohler, M.A. and Paulhus, J.L. 1988: *Hydrology for engineers*. 3rd edn. New York: McGraw-Hill.
Storey, H.C., Hobba, R.L. and Rosa, J.M. 1964: Hydrology of forest lands and rangelands. In V.T. Chow ed., *Handbook of hydrology*. New York: McGraw-Hill.

form line A contour line on a map, the precise position of which has not been accurately surveyed but interpolated.

form ratio The ratio of the depth of a stream or river to its width.

fractal dimension (or fractal) A concept put forward by Mandelbrot (1982) which refers to the space filling of curves such that the dimension is between one and two; the trail of Brownian motion is a fractal. The length of a coastline can be measured in different ways which will give values greater than the linear distance between the end points. In extreme cases, the length is very large but it can be characterized by fractals. Complex lines can be analysed by using the ideas of fractal dimension. For a closed loop (such as the outline of a particle) with perimeter P, measured by stepping off a constant length S, a plot of log P versus log S tends to give a constant slope b. The fractal dimension (D) is equal to the slope term plus one $(D = b + 1)$ and log $P \propto (D - 1)$ log S. WBW

Reading and Reference
Mandelbrot, B.B. 1982: *The fractal geometry of nature*. San Francisco: W.H. Freeman.
†Orford, J.D. and Whalley, W.B. 1983: The use of the fractal dimension to quantify the morphology of irregular particles. *Sedimentology* 30, pp. 655–68.

fracture The term given to the splitting of a material into two or more parts; the material is said to have failed. It usually refers to brittle fracture where the stressed body ruptures rapidly with the release of energy to form new surfaces by crack propagation after little or no plastic deformation. Brittle fracture may take place through crystals, along cleavage planes or between grains to give 'inter-granular' fracture. Brittle fracture occurs normal to the maximum applied tensile stress component. This type of failure is usual with hard rocks. Ductile failure is fracture which occurs after extensive plastic deformation and with slow crack formation; this is typical of clays. WBW

Reading
Whalley, W.B. 1976: *Properties of materials and geomorphological explanation*. Oxford: Oxford University Press.
— Douglas, G.R. and McGreevy, J.P. 1982: Crack propagation and associated weathering in igneous rocks. *Zeitschrift für Geomorphologie* 26, pp. 33–54.

fractus Cloud having a broken or shattered appearance, perhaps a convective cloud growing in a place where the wind shear tears apart the incipient cloud, as in fractocumulus. It is a temporary phase of development in which buoyancy and shear are not yet reconciled. JSAG

Reading
Ludlam, F.H. and Scorer, R.S. 1966: *Cloud study*. London: John Murray.

fragipan An acidic, cemented horizon between the base of the soil zone and the underlying bedrock or parent material. Fragipans are normally compact and brittle, and may often be bonded by clays. Other agents may also be involved in the bonding, including silica (Bridges and Bull 1983), iron, aluminium and organic matter. Many are the result of periglacial processes, and they are widespread in Europe and North America. ASG

Reading and Reference
Bridges, E.M. and Bull, P.A. 1983: The role of silica in the formation of compact and indurated horizons in the soils of South Wales. *Proceedings of the International Symposium on Soil Micromorphology 1983*, pp. 605–13.
†Grossmann, R.B. and Carlisle, F.J. 1969: Fragipan soils of the eastern United States. *Advances in agronomy* 21, pp. 237–79.

frazil ice Fine spicules of ice in suspension in water, commonly associated with the freezing of seawater.

free face The wall of a rock outcrop that is too steep for debris to rest upon it. It is the portion of a cliff that lies above a scree or talus, and from which, through rockfall and other processes, scree formation may take place.

freeze–thaw cycle A cycle in which temperature fluctuates both above and below 0°C. The amplitude of the temperature change and the period of time over which the fluctuation occurs are important considerations since freezing does not occur instantaneously nor does it always occur at 0°C. A typical diurnal freeze–thaw cycle is one in which the temperature ranges from +0.5°C to −0.5°C within a 24 hour period.

Because of their supposed significance with respect to frost shattering of rock, the frequency and efficacy of freeze–thaw cycles have been the subject of both field and laboratory investigations. Field studies indicate that most freeze–thaw cycles per year ($\approx$40–60) occur in subarctic alpine regions which experience diurnal temperature rhythms. High latitudes experience few cycles on account of the seasonal temperature regimes. In all areas most cycles occur in the upper 0–5 cm of the ground and only the annual cycle occurs at depths in excess of 20 cm. Laboratory studies, in which rock samples are subject to repeated freeze–thaw cycles of varying amplitude and intensity, suggest that the number of freeze–thaw cycles is more important than their intensity as regards rock shattering. If correct, the low frequency of freeze–thaw cycles recorded in present day periglacial environments suggests that frost shattering may be overemphasized as a physical weathering process. Hydration shattering and cryogenic (i.e. frost) weathering in general may be equally if not more important.   HMF

Reading
Washburn, A.L. 1979: *Geocryology: a survey of periglacial processes and environments*. New York: Wiley.

freezing front The boundary between frozen or partially frozen ground and non-frozen ground. During freezing in permafrost regions freezing fronts move downwards from the ground surface and upwards from the permafrost table. In seasonally frozen ground only a downward moving freezing front exists. The freezing front is sometimes equated with the cryofront, the position of the 0°C isotherm in the subsurface, forming the boundary between cryotic (i.e. temperature less than 0°C) and noncryotic (i.e. temperature more than 0°C) ground. The permafrost base, the perma-

frost table, and the top and base of the cryotic portion of the active layer all constitute cryofronts, or freezing fronts. (See also PERMAFROST.)   HMF

Reading
van Everdingen, R.O. 1976: Geocryological terminology. *Canadian journal of earth sciences* 13, pp. 862–7.

freezing index A measure of the combined duration and magnitude of the below-freezing temperatures which occur in a freezing season. It is expressed in DEGREE DAYS.

friction A force resisting relative motion between two solids or between a solid and a fluid. It is a fundamental property in studies of sediment transfer or transport, since frictional resistance to motion must be overcome before masses or particles of sediment can be moved, and sedimentary landforms modified as a result.

The force required to move a solid block on a plane surface is its weight (W) multiplied by the static coefficient of friction (μ) between the block and the underlying surface. If the surface is sloping at an angle $\beta°$, the normal force on the surface is $W \cos \beta$, and the force required to move the block is $\mu W \cos \beta$. The block will slide under its own weight when $\tan \beta = \mu$ (Statham 1977, pp. 12–14). The interlocking friction in a sediment between individual grains is measured by the angle of internal friction (ϕ), which is, however, difficult to define and measure since it is dependent on density, water content, and test conditions (Statham 1977, pp. 41–9). Dry sediment on a slope will slide if the slope angle equals or exceeds the angle of internal friction, which is therefore a limiting or threshold slope angle in the landscape.

Friction also occurs between a solid bed surface and a fluid passing over it. This slows the immediately adjacent fluid to a standstill, but shear in the fluid above allows the development of a velocity gradient within the BOUNDARY LAYER, in which the frictional resistance is successively less effective with distance from the bed. In UNIFORM STEADY FLOW the friction exerted by the bed on the flow is equal and opposite to the drag of the fluid on the bed, and in loose, mobile sediment, this drag (the TRACTIVE FORCE) may overcome the resistance to motion of non-cohesive sediment grains, which is dependent on their weight and the frictional contact between them. KSR

Reference
Statham, I. 1977: *Earth surface sediment transport*. Oxford: Clarendon Press.

fringing reef See REEF.

front A sharp transition zone separating air of different temperatures and origins. The term was introduced by the Bergen School of Meteorology in 1918 as part of their work on EXTRA-TROPICAL CYCLONE structure. The front has a three-dimensional form. It extends into the atmosphere as a gently sloping surface of about 1 in 100 so that the cold, denser air appears as a wedge shape beneath the warmer air. The front lies in a trough of lower pressure accompanied by changes in wind velocity, pressure and temperature as the front passes a site. The intensity of change varies greatly from one front to another.

Horizontal convergence and associated vertical motion are a necessary feature of a well-defined front. According to early ideas the rising of the warm air at a front took place along the frontal surface itself, but recent work, based on Doppler radar, research aircraft, satellite images and rain-gauges, has shown a much more complex structure of air movements near fronts. Basic to the uplift of air at the warm front is a well-defined flow of low-level, moist, warm air within the warm sector which moves parallel to the cold front then ascends above the main warm frontal surface, eventually running parallel to the surface warm front but at a higher level (see figure (b)). Interaction between this flow and that at even higher levels (600 hPa) may trigger off potential instability which produces linear areas of heavier precipitation. Similar rain bands of subsynoptic scale, known as mesoscale precipitation areas (MPA), have been found associated with cold and occluded fronts. Five types of frontal rain bands have been identified: (1) warm frontal; (2) warm sector; (3) cold

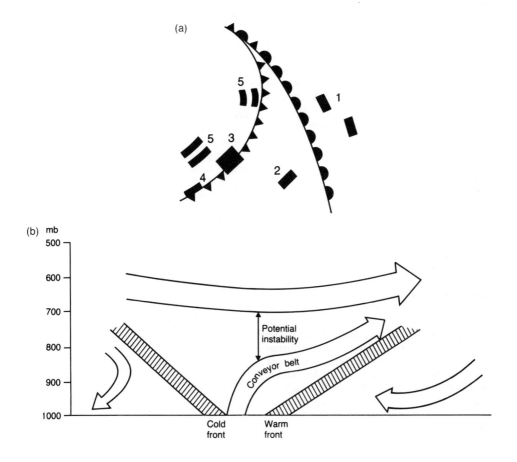

frontal – wide; (4) cold frontal – narrow; and (5) post frontal (see figure (a)). Aircraft observations have confirmed that clouds developing in frontal MPAs do have different liquid water contents and ice particle concentrations from other clouds associated with the fronts. Radar provides a continuous picture of the movement and patterns of the MPAs, but forecasting their initiation is difficult because of the lack of detailed information about atmospheric structure near fronts, and because the horizontal resolution of the present numerical weather prediction models is insufficiently fine to represent the features precisely. PS

Reading
Atkinson, B.W. 1981: *Meso-scale atmospheric circulations*. London: Academic Press.
Browning, K.A. 1982: *Nowcasting*. London: Academic Press.
Carlson, T.N. 1991: *Mid-latitude weather systems*. London: Harper-Collins.

frontogenesis The process of intensification of the thermal gradient at a frontal zone. It takes place mainly by horizontal confluence and convergence when the isotherms are suitably aligned. Since uplift must follow surface convergence, frontogenesis is helped if the upper atmospheric air movements favour the continuation of rising air. The reverse process is known as frontolysis. In this situation the thermal gradient becomes progressively less distinct and the cloud band eventually disperses. PAS

frost (action) The mechanical weathering process caused by alternative or repeated cycles of freezing and thawing of water in pores, cracks, and other openings, usually at the ground surface. The expansion of water upon freezing (approximately 9 per cent by volume) forces material, commonly rock, apart. Termed frost wedging, its efficacy largely depends upon the frequency of FREEZE–THAW CYCLES, the availability of moisture, and the lithological/strength characteristics of the material. Other terms for frost wedging include frost shattering, gelifraction and frost riving. The term 'frost action' is sometimes used to include a wider range of frost-related processes, such as frost heaving, frost creep, thermal contraction cracking and frost weathering. (See also FROST WEATHERING.) HMF

frost creep The ratchet-like downslope movement of the particles as a result of the frost heaving of the ground and subsequent settling upon thawing, the heaving being predominantly normal to the slope and the settling more nearly vertical. Although frost creep is commonly associated with GELIFLUCTION (and is usually included within it in rate measurements because of the difficulty of distinguishing between their contributions to total movement) it is a separate process. Movement associated with frost creep decreases from the surface downwards and depends upon frequency of FREEZE–THAW CYCLES, angle of slope, moisture available for heave, and frost susceptibility of soil. Studies in East Greenland indicate that frost creep exceeds gelifluction by not more, and probably less than 3:1 in most years, and frost creep usually resulted in 30–50 per cent of total annual movement on slopes. In the Colorado Rockies measurements indicate that solifluction is a more effective process than frost creep in the saturated axial areas of lobes, but less effective than frost creep at their edges. HMF

Reading
Benedict, J.B. 1970: Downslope soil movement in a Colorado alpine region: rates, processes and climatic significance. *Arctic and alpine research* 2, pp. 165–226.
Washburn, A.L. 1979: *Geocryology: a survey of periglacial processes and environments*. New York: Wiley.

frost heave The predominantly upward movement of mineral soil during freezing casued by the migration of water to the freezing plane and its subsequent expansion upon freezing. Frost heaving is usually associated with the active layer above permafrost or with seasonally frozen ground. As such, ICE SEGREGATION is an essential component of frost heave. Field studies indicate that heave occurs not only during the autumn freeze-back period but also during winter when ground temperatures are below $0°C$. Frost heaving processes include the upheaving of bedrock blocks, upfreezing of objects, tilting of stones, formation of NEEDLE ICE, and the sorting and migration of soil particles. Frost heaving presents important geotechnical problems in the construction of roads, buildings, pipelines, and airfields in cold environments. HMF

Reading
Mackay, J.R. 1983: Downward water movement into frozen ground, western Arctic coast. *Canadian journal of earth sciences* 20, pp. 120–34.

Slusarchuk, W.A., Clark, J.I. and Nixon, J.F. 1978: Field tests of a chilled pipeline buried in unfrozen ground. *Proceedings of the Third International Conference on Permafrost*, 10–13 July 1978, Edmonton, Alberta. National Research Council of Canada, Ottawa, Vol. 1.

Washburn, A.L. 1979: *Geocryology: a survey of periglacial processes and environments*. New York: Wiley. Esp. pp. 79–96.

frost smoke (also arctic smoke) A fog produced by the contact of cold air with relatively warm water. It is commonly associated with leads of open water which open up in a sea-ice cover but may also occur at the ice edge or over water which is beginning to freeze.

frost weathering A general term used to describe the complex of weathering processes, both physical and chemical, which operate, either independently or in combination, in cold non-glacial environments. The most important physical weathering process is frost wedging which characteristically produces angular fragments of varying sizes. The predominant size to which rocks can be ultimately reduced by frost wedging is generally thought to be silt. Porous and well-bedded sedimentary rocks, such as shales, sandstones and limestones are especially susceptible to frost weathering. Features attributed to frost weathering include extensive areas of angular bedrock fragments (blockfields and blockslopes) and irregular bedrock outcrops termed tors. It has been estimated that frost weathering (i.e. rockfalls induced by frost wedging) over a 50 year period caused steep rock faces in Longyeardalen, Spitsbergen, to retreat at a rate of 0.3 mm year^{-1}.

Many aspects of cold climate weathering are not fully understood. For example, it has been suggested that hydration shattering may be responsible for the large blockfields and blockslopes characteristic of low temperature alpine and polar environments, but this has yet to be proved. Equally, experimental studies in the former USSR indicate that under cold conditions the ultimate size reduction of quartz (0.05–0.01 mm) is smaller than for feldspar (0.1–0.5 mm), a reversal of what is normally assumed for temperate environments. Finally, the current emphasis upon frost wedging in periglacial environments should not obscure the fact that chemical weathering can be significant. The dominance of physical weathering tends to mask

A frost-shattered rock in the Aletschwald, Switzerland. Extreme rates of frost weathering require a combination of frequent temperature cycles around freezing point and the presence of available moisture.

chemical effects. In places, physical and chemical effects combine, as in salt wedging. Like frost wedging, this process breaks up rock into silt-size particles and is particularly effective in cold, arid regions, such as the ice-free areas of Antarctica. Solutional effects in limestone terrain may also be present and karst terrain exists in permafrost regions, further illustrating the inadequacy of a simplistic view of frost weathering. HMF

Reading
van Everdingen, R.O. 1981: *Morphology, hydrology and hydrochemistry of karst in permafrost terrain near Great Bear Lake, NWT.* National Hydrology Research Institute, paper 11, Calgary, Alberta.
French, H.M. 1976: *The periglacial environment.* London and New York: Longman.
Jahn, A. 1976: Contemporaneous geomorphological processes in Longyeardalen, Vestspitsbergen (Svalbard). *Biuletyn peryglacjalny* 26, pp. 253–68.
Washburn, A.L. 1979: *Geocryology: a survey of periglacial processes and environments.* New York: Wiley.
White, S.E. 1976: Is frost action really only hydration shattering? A review. *Arctic and alpine research* 8, pp. 1–6.

frost wedge See ICE WEDGE.

Froude number (F_r) The dimensionless ratio of inertial to gravity forces in flowing water:

$$F_r = v/(\sqrt{gd}),$$

where v is velocity, g is gravitational acceleration and d is depth. The term ($\sqrt{gd}$) is the velocity of a small gravity wave (a surface ripple) and, if $F_r < 1$, the flow is subcritical or tranquil and ripples formed by a pebble dropped into the water travel upstream   because their velocity exceeds that of the stream. If $F_r > 1$ the flow is supercritical, and when $F_r = 1$ the flow is critical. Sudden spatial changes from supercritical to subcritical are HYDRAULIC JUMPS. In sandbed channels, temporal changes of FLOW REGIME during floods cause a consistent sequence of bedform changes, with sand dunes washing out to form antidunes at a local Froude number of approximately unity. However, *mean* Froude numbers in natural channels rarely exceed 0.4–0.5, because the associated rapid energy losses cause bank erosion, channel enlargement, and a reduction of flow velocity and Froude number – an example of negative feedback (see SYSTEMS). KSR

fulgurite A tube in sand or rock produced by the fusing effects of a lightning strike (Withering 1790). Sand fulgurites are especially common in areas of dry, loose, quartz sand typical of deserts. ASG

Reference
Withering, W. 1790: An account of some extraordinary effects of lightning. *Philosophical transactions of the Royal Society of London*, Series D. 80, pp. 293–5.

fulje A depression between barchans or barchanoid sand ridges, especially where dunes are pressing closely on one another. In Australia the term may be used to describe a blowout or small parabolic dune. ASG

fumarole A small, volcanic vent through which hot gases are emitted.

G

gabbro A basic igneous rock composed of calcic plagioclase and clinopyroxene with or without orthopyroxene and olivine. Usually coarse-grained and dark grey to black in colour.

Gaia A concept developed by J.E. Lovelock (1979). He defines Gaia as (p. 11): 'a complex entity involving the Earth's biosphere, atmosphere, oceans and soils, the totality constituting a feedback or cybernetic system which seeks an optimal physical and chemical environment for life on this planet.' He maintains (p. 152) 'that the physical and chemical condition of the surface of the Earth, of the atmosphere and of the oceans has been and is actively made fit and comfortable by the presence of life itself. This is in contrast to the conventional wisdom which held that life adapted to the planetary condition as it and they evolved their separate ways.' ASG

Reference
Lovelock, J.E. 1979: *Gaia – a new look at life on Earth.* Oxford: Oxford University Press.

gallery forest Forest which lines the banks of a river in an area where away from the river's favourable hydrological circumstances such forest does not occur.

GARP (Global Atmospheric Research Programme) A project born out of the World Weather Watch in the late 1960s and early 1970s with the aim of studying the *global* atmospheric circulation by both observational and theoretical means. The observational programme has involved the use of satellites, balloons, radars and ocean-buoys as well as conventional surface and upper air observing systems. The First Garp Global Experiment (FGGE), conducted in the late 1970s was mankind's largest ever scientific experiment. These observational programmes have been complemented by programmes of numerical modelling of the GENERAL CIRCULATION of the atmosphere. BWA

garrigue Xerophytic, evergreen scrubland occurring on thin soils in areas with a dry 'Mediterranean type' of climate. Much of it may result from anthropogenic landscape degradation. It consists of low thorny shrubs and stunted evergreen oaks.

gas laws The thermodynamic laws applying to perfect gases. In particular they relate the pressure, density and temperature of gases in different ways.

The most important, dealing with 'perfect gases', are:

Boyle's law: 'The pressure of a given mass of gas, at constant temperature, is inversely proportional to its volume.'

This may be summarized as $pV = $ constant.

Charles's law: 'The volume of a given mass of gas, at constant pressure, increases by 1/273 of its values at 0°C for every degree Celsius rise in temperature.'

These two laws can be related through the general equation of state for a perfect gas:

$$pV + RT$$

where T is the absolute temperature, p is the pressure and V the volume and R is the gas constant.

Dalton's law of partial pressures: 'A mixture of gases has the same pressure as the sum of the partial pressures of its components.'
WBW

gauging stations Sites at which river flow is determined. The gauging sites may only be equipped to provide point measurements in time or they may provide continuous measurements. The accuracy of the flow estimates will vary according to the gauging technique employed. (See also DISCHARGE, HYDROMETRY.)　　　AMG

geest Ancient alluvial sediments which still mantle the landsurfaces on which they were originally deposited.

gelifluction A type of solifluction occurring in periglacial environments underlain by permafrost. Suitable conditions for gelifluction occur in areas where downward percolation of water through the soil is limited by the permafrost table and where the melt of segregated ice lenses provides excess water which reduces internal friction and cohesion in the soil. Particularly favoured sites include areas beneath or below late-lying snowbanks. Rates of movement, which generally vary between 0.5 and 10.0 cm per year, usually decrease with depth. Frost creep is usually measured as a component of gelifluction. As with solifluction, features related to gelifluction include sheets, stripes and lobes.　　　HMF

Reading
Washburn, A.L. 1979: *Geocryology: a survey of periglacial processes and environment.* New York: Wiley.

gendarmes Pinnacles of rocks projecting vertically from a ridge.

genecology The study of the genetics of populations in relation to habitat; the study of species (and other taxa) through a combination of the methods and concepts of both ecology and genetics. Some species display a range of ECOTYPES – genetic varieties existing in different environments; the investigation of these ecotypes constitutes a part of genecology.　　　PHA

general circulation modelling Also known as climate modelling. Simulation of the large-scale features of the atmospheric circulation by either solving the system of equations that govern atmospheric motion or by reproducing the circulation using laboratory models. The latter are discussed under DISHPAN EXPERIMENTS.

With the advent of highspeed computers it became possible to solve numerically the complete system of basic equations that govern atmospheric motions. Because the equations are highly nonlinear, they must be solved numerically. This is usually done by either rewriting the equations in finite-difference form or by using a combination of finite difference and spectral representation of the variables involved. In either case a time step of 10 to 30 minutes must be used in order for the numerical solution to approximate the exact analytic solution. For a global model with grid points at every 5° of latitude and longitude and at five vertical levels this entails solving the equations at about 65,000 points every time step. Obviously it is only practical to carry out extended integrations of the equations on super computers. Fortunately these are now big enough and fast enough to complete the enormous number of calculations involved in a reasonable, although still considerable, time. A typical run of one model year requires about three days of computer time. This figure varies greatly, of course, depending on the complexity of both the model and the computer.

The basic equations used in most cases are the two horizontal EQUATIONS OF MOTION, the THERMODYNAMIC EQUATION, the water vapour continuity equation, the equation of mass continuity, the HYDROSTATIC EQUATION, and the EQUATION OF STATE. These seven equations involve seven time-dependent variables – pressure (or height), temperature, specific humidity, air density, and the three components of the wind velocity. The first four equations are prognostic, involving time derivatives, and the last three are diagnostic. In some cases one or more of the prognostic equations are made diagnostic by ignoring the time derivatives involved. If this is done in all equations or in all but the energy equation and if the dynamics are either eliminated or greatly simplified, a class of climate models, usually referred to as energy balance models, results. The surface temperature is usually the only independent variable. These models are often further simplified by zonally averaging the equations. They require much less computer time than the complete general circulation models (GCMs) and can therefore be integrated on computers for long periods of time, using a time step, if appropriate, of one day to one month.

In order to solve the basic equations considered above certain boundary conditions are required. The vertical velocity at the top of the atmosphere and normal velocity at the earth's surface are usually set equal to zero. The surface temperature is either specified, especially over the oceans, or determined from the surface energy balance equation. The solar flux at the top of the atmosphere is also specified.

Even in the most complex GCM there are small-scale processes which cannot be explicitly resolved but which are very important in maintaining the overall heat, momentum, and water balances of the system. To achieve satisfactory results these must, therefore, be represented in terms of the distributions of the large-scale properties of the model. Included among such processes are: (1) the transfer of momentum by the effects of viscosity from one part of the atmosphere to another and from the atmosphere to the underlying surface; and (2) diabatic heating due to radiation, the release of latent heat, and heat conduction at the surface.

The development of GCMs is still going on. Very few studies have been made with an interactive dynamic ocean and realistic topography. Most frequently ocean surface temperatures are specified. Even with this simplification, the modelled circulation frequently departs radically from that observed. This seems to be especially true in the southern hemisphere, in which the models often greatly underestimate the strength and breadth of the strong middle latitude surface westerlies. Theoretically, the models should do best in this most zonally symmetrical hemisphere. That they do not suggests that there is something fundamentally wrong either in the physics or in the way in which the models are formulated. WDS

Reading
Berger, A. 1981: *Climatic variations and variability: facts and theories*. Dordrecht: D. Reidel. Esp. pp. 435–59.

Lorenz, E.N. 1967: *The nature and theory of the general circulation of the atmosphere*. Geneva: World Meteorological Organization.

Smagorinsky, J. 1970: Numerical simulation of the global atmosphere. In *The global circulation of the atmosphere*. London: Royal Meteorological Society.

general circulation of the atmosphere Either the totality of atmospheric fluid motions or that part of the circulation associated with the synoptic – and planetary – scale horizontal wind field. The latter is driven either directly or indirectly by horizontal heating gradients in a stably stratified atmosphere and accounts for about 98 per cent of the atmospheric kinetic (wind) energy. The remaining kinetic energy is associated with small-scale motions driven by convective instability.

The energy associated with the planetary – and synoptic-scale – atmospheric disturbances is constantly being syphoned off by small-scale fluid motions which interact among themselves to transfer energy to smaller and smaller scales and ultimately down to random molecular motions. The large-scale circulation is maintained against this frictional dissipation by drawing on the reservoir of POTENTIAL ENERGY inherent in the spatial distribution of atmospheric mass resulting from the gradients of diabatic heating induced primarily by the sun. The conversion from potential to kinetic energy is achieved primarily through vertical motions in the atmosphere.

In the tropics most of the kinetic energy is associated with quasi-steady thermally driven circulations, which include the seasonally varying MONSOONS, the result of land–sea heating contrasts, and the large-scale HADLEY CELLS, especially prominent over the tropical Atlantic and Pacific Oceans. Within these cells the rising currents near the equator, forming the INTERTROPICAL CONVERGENCE ZONE, gain both potential energy and heat energy, the latter through condensation. This energy is carried poleward by the circulation in the upper TROPOSPHERE and lower STRATOSPHERE to latitudes of 30° or 40° where the now descending air is heated by ADIABATIC compression and the potential energy is converted into the kinetic energy of the low latitude trade winds and, to some extent, the middle latitude westerlies.

In middle and high latitudes much of the kinetic energy is associated with moving disturbances called baroclinic waves, which develop within zones of strong horizontal temperature gradient along the POLAR FRONT. The wave energy is gained primarily through the vertical displacement of warm air masses by cold air masses, thus depleting the potential energy of the system. That the middle latitude westerlies can move from west to east faster than the rotating earth is due to the fact that these winds are maintained against frictional retardation primarily by the poleward gradient of temperature and potential energy. WDS

Reading
Lorenz, E.N. 1967: *The nature and theory of the general circulation of the atmosphere.* Geneva: World Meteorological Organization.

Palmén, E. and Newton, C.W. 1969: *Atmospheric circulation systems.* New York: Academic Press.

Smagorinsky, J. 1972: The general circulation of the atmosphere. In D.P. McIntyre ed., *Meteorological challenges: a history.* Ottawa: Information Canada. Pp. 3–42.

Wallace, J.M. and Hobbs, P.V. 1977: *Atmospheric science.* New York: Academic Press.

general system theory Was largely developed by the biologist, Von Bertalanffy, whose basic statement appeared in 1950 and was substantially extended in 1962 (where the term becomes general systems theory). The fundamental proposition of the theory is that systems (defined as structured sets of objects and their attributes) may be identified in all studies of phenomena; and that more may be learnt by the comparison of the ways in which similar (isomorphic) systems function, than by the standard academic concentration on the distinctiveness of their component parts. Thus what is of interest in a comparison of – say – towns, drainage basins and bathtubs, is that they may be considered to share fundamental attributes which relate to their physical boundedness and to the transfers of energy and mass across those boundaries. These transfers serve to integrate the components of each system and this interrelatedness of parts is a diagnostic and important feature.

Since its introduction by Von Bertalanffy, general system theory (or GST) has become a field of study in its own right. The history of the ideas *per se* in physical geography is, however, less than clear and one is forced to admit that GST (as opposed to general ideas of SYSTEMS) has been of little direct moment.

The first explicit use of Von Bertalanffy's ideas in geomorphology was made by Strahler (1950, p. 676), who stated 'A graded drainage system is perhaps best described as an open system in a steady state . . . which differs from a closed system in equilibrium in that the open system has import and export of components.' This and subsequent studies by Strahler and others were then taken up by Chorley (1960) in a major paper which apparently focused on the ideas of GST, rather than simply on concepts of physical systems.

Chorley's discussion is perhaps remarkable in that it nowhere defines GST, although it introduces a series of the key concepts: notably, open and closed systems, entropy, steady state, self-regulation, equilibrium, hierarchial differentiation and organization. In later works (see Chorley and Kennedy 1971; Bennett and Chorley 1978), references to both GST and Von Bertalanffy are exiguous. Similarly Huggett (1980) makes no reference to either GST or its author. Only Chapman (1978, p. 404) is prepared to confess that he is making explicit use of GST.

While it may be argued that the *framework* provided by GST has been widely adopted in physical geography, it seems abundantly clear that the use made has been so derivative that the actual aims and aspirations of GST itself are largely unknown – and therefore irrelevant – to discussions of systems in physical geography. BAK

References
Bennett, R.J. and Chorley, R.J. 1978: *Environmental systems: philosophy, analysis and control.* London: Methuen.

Chapman, G.P. 1978: *Human and environmental systems: a geographer's appraisal.* London: Academic Press.

Chorley, R.J. 1960: Geomorphology and general systems theory. *United States Geological Survey professional paper* 500–B.

— and Kennedy, B.A. 1971: *Physical geography: a systems approach.* London: Prentice-Hall International.

Huggett, R. 1980: *Systems analysis in geography.* Oxford: Clarendon Press.

Strahler, A.N. 1950: Equilibrium theory of erosional slopes approached by frequency distribution analysis. *American journal of science* 248, pp. 673–96; 800–14.

Von Bertalanffy, L. 1950: The theory of open systems in physics and biology. *Science* 111, pp. 23–9.

— 1962: General systems theory, a critical review. *General systems* VII.

genetic drift The effect of sampling error in causing random changes in the relative frequency of genes in a gene pool. Random change in gene frequency occurs in all populations and it is maintained that genetic drift provides a mechanism for evolution. The smaller the population size the greater the possibility of gene loss or fixation: a gene normally present in one in 10,000 individuals may not be present in a population of 100. Particularly important is the 'founder effect': when a small sample of an organism's population is isolated, for example on an island or mountain peak, it may have a different gene frequency from the parent group. This is one reason why organisms on remote islands are frequently ENDEMIC species or subspecies. An animal that has been reduced to a very small population by climatic catastrophe, disease or overhunting is likely to have an impoverished gene pool for some time after recovery.   PHA

Reading
Bonnel, M.L., and Selander, R.K. 1974: Elephant seals: genetic variation and near-extinction. *Science* 184, pp. 908–9.

geo A deep, narrow cleft or ravine along a rocky sea coast which is flooded by the sea.

geocryology The study of frozen, freezing and thawing terrain (but not glaciers) is known as permafrost science or more generally termed geocryology. This widespread term is usually associated with earth materials having a temperature below 0°C. Geocryology seeks to promote an understanding of the dynamics of such environments, especially the study of the origins and history of permafrost.   AP

Reading
Washburn, A.L. 1979: *Geocryology: a survey of periglacial processes and environments.* London: Edward Arnold.

geode A roughly spherical or globular inclusion within a mass of rock. Geodes are hollow and frequently exhibit mineral crystals growing into the central void.

geodesy The determination of the size and shape of the earth by mathematical means and surveys.

geo-ecology See LANDSCAPE ECOLOGY.

geographic information system (GIS) 'Is designed for the collection, storage and analysis of objects and phenomena where geographic location is an important characteristic or critical to the analysis' (Aronoff 1989, p. 1). Many other terms have been used as synonyms for GIS: geobase information system, geographic data system, land information system, cadastral information system, environmental information system and urban information system. Using computers, a GIS (sometimes referred to as 'electronic tracing paper') stores and integrates large amounts of spatially referenced data, and contains the following major components: a data input subsystem, a data storage and retrieval subsystem, a data manipulation and analysis subsystem and a data reporting subsystem. In the 1980s GIS was a major

growth area and received considerable attention from planners, bureaucrats and others (Maguire *et al.* 1991).   ASG

Reading and References
Aronoff, S. 1989: *Geographic information systems: a management perspective.* Ottawa: WDL publications.
Maguire, D.J., Goodchild, M.F. and Rhind, D.W. 1991: *Geographical information systems.* Harlow: Longman Scientific and Technical.
Taylor, D.R.F. ed. 1991: *Geographic information systems.* Oxford: Pergamon Press.

geoid The equipotential surface that would be assumed by the sea surface in the absence of tides, water-density variations, currents and atmospheric effects. It varies above and below the geometrical ellipsoid of revolution by as much as 100 m due to the uneven distribution of mass within the earth. The mean sea-level surface varies about the geoid by typically decimetres, but in some cases by more than a metre.   DTP

geological time-scale The divisions of geological time as listed in the table (p. 228)

Reading
Harland, W.B. 1990: *A geologic time scale.* Cambridge: Cambridge University Press.

geomorphology A term that arose in the Geological Survey in the USA in the 1880s, possibly coined by J.W. Powell and W.J. McGee. In 1891 McGee wrote: 'The phenomena of degradation form the subject of geomorphology, the novel branch of geology.' He plainly regarded geomorphology as being that part of geology which enabled the practitioner to reconstruct earth history by looking at the evidence for past erosion, writing: 'A new period in the development of geologic science has dawned within a decade. In at least two American centres and one abroad it has come to be recognised that the later history of world growth may be read from the configuration of the hills as well as from the sediments and fossils of ancient oceans . . . The field of science is thereby broadened by the addition of a coordinate province – by the birth of a new geology which is destined to rank with the old. This is geomorphic geology, or geomorphology.'

Many scientists had studied the development of erosional landforms (see Chorley *et al.* 1964) before the term was thus defined and since that time its meaning has become broader. Many geomorphologists believe that the purpose of geomorphology goes

Geological time-scale

Era	Sub-era/period/ subperiod/Epoch				Age (Ma BP)
CAINOZOIC	Quaternary		Holocene		0.01
CAINOZOIC	Quaternary		Pleistocene		1.64
CAINOZOIC	Tertiary	Neogene	Pliocene		5.2
CAINOZOIC	Tertiary	Neogene	Miocene	Late	14.2
CAINOZOIC	Tertiary	Neogene	Miocene	Early	23.3
CAINOZOIC	Tertiary	Palaeogene	Oligocene		35.4
CAINOZOIC	Tertiary	Palaeogene	Eocene		56.5
CAINOZOIC	Tertiary	Palaeogene	Palaeocene		65.0
MESOZOIC	Cretaceous		Late		97.0
MESOZOIC	Cretaceous		Early		145.6
MESOZOIC	Jurassic		Late		157.1
MESOZOIC	Jurassic		Middle		178.0
MESOZOIC	Jurassic		Early		208.0
MESOZOIC	Triassic				245.0
PALAEOZOIC	Permian				290.0
PALAEOZOIC	Carboniferous				362.5
PALAEOZOIC	Devonian				408.5
PALAEOZOIC	Silurian				439.0
PALAEOZOIC	Ordovician				510.0
PALAEOZOIC	Cambrian				570.0
PRECAMBRIAN					

Source: Harland 1990.

beyond reconstructing earth history (see DENUDATION CHRONOLOGY), and that the core of the subject is the comprehension of the form of the ground surface and the processes which mould it. In recent years there has been a tendency for geomorphologists to become more deeply involved with understanding the processes of erosion, weathering, transport and deposition, with measuring the rates at which such processes operate, and with quantitative analysis of the forms of the ground surface (morphometry) and of the materials of which they are composed. Geomorphology now has many component branches (e.g. ANTHROPOGEOMORPHOLOGY; APPLIED GEOMORPHOLOGY).   ASG

Reading and References
†Bloom, A.L. 1978: *Geomorphology: a systematic analysis of late Cenozoic landforms*. Englewood Cliffs, NJ: Prentice-Hall.

†Butzer, K.W. 1976: *Geomorphology from the earth*. New York: Harper & Row.

Chorley, R.J., Dunn, A.J. and Beckinsale, R.P. 1964: *The history of the study of landforms*. Vol. 1. London: Methuen.

— , Schumm, S.A. and Sugden, D.E. 1984: *Geomorphology*. London and New York: Methuen.

McGee, W.J. 1891: The Pleistocene history of northeastern Iowa. *Eleventh annual report of the US Geological Survey.* Pp. 189–577.

†Rice, R.J. 1988: *Fundamentals of geomorphology.* 2nd edn. London: Longman.

†Sparks, B.W. 1986: *Geomorphology.* 3rd edn. London: Longman.

Tinkler, K.J. 1985: *A short history of geomorphology.* London and Sydney: Croom Helm.

†Twidale, C.R. 1976: *Analysis of landforms.* Sydney and New York: Wiley.

geophyte A herbaceous plant which has parts beneath the ground surface which survive when the parts above ground die back.

geostrophic wind The geostrophic wind is a wind with a velocity determined by an exact balance of the CORIOLIS FORCE and the horizontal pressure gradient force. This balance results in a configuration of velocity and pressure as described by BUYS BAL-LOT'S LAW, i.e. in the northern hemisphere, when one has one's back to the wind, low pressure lies on the left and high pressure on the right. The converse is true in the southern hemisphere. The geostrophic wind thus blows *along* the isobars and its magnitude is a direct function of the horizontal pressure gradient force, and an inverse function of height and latitude. It is not defined at the equator. At other latitudes the geostrophic wind is a reasonable approximation to the real wind.   BWA

Reading
Atkinson, B.W. 1972: The atmosphere. In D.Q. Bowen, ed., *A concise physical geography.* Amersham: Hulton Educational. Pp. 1–76, esp. pp. 33–48.

Hess, S.L. 1959: *Introduction to theoretical meteorology.* New York: Henry Holt.

geosyncline A very large depression, perhaps several hundred kilometres across and up to 10 km deep, the terrestrial or marine floor of which is built up by sedimentation.

Gerlach trough A sediment trap designed to catch a sample of OVERLAND FLOW and the sediment it carries down a hillside. Troughs with a variety of shapes and sizes have been used to collect slope-wash sediment. They have an upslope lip which is either flush with the surface or inserted beneath the uppermost organic soil layers: disturbance from installing this lip is a major source of error. Sediment settles out and/or is filtered out of water which eventually overflows from the trough,

sometimes through a total water or flow rate meter. The trough is protected by a lid. A second major source of error is the accurate delimitation of the area which yields sediment to the trough. This type of installation is an alternative to direct measurements of slope lowering by erosion pins, etc.   MJK

Reading
Goudie, A. ed. 1990: *Geomorphological techniques.* 2nd edn. London: Unwin Hyman.

geyser A spring or fountain of geother-mally heated water that erupts intermittently with explosive force as a result of increases in pressure beneath the surface.

Ghyben–Herzberg principle This refers to the relationship between freshwater and saltwater in a coastal aquifer. Ghyben and Herzberg were two European scientists who independently investigated this relationship around the turn of the century. They found that since freshwater is less dense than seawater, it rises above underlying intruding saltwater. In unconfined aquifers beneath small islands, a lens of freshwater floats on seawater which surrounds and underlies it, whereas at the edges of larger landmasses there is a sloping interface with freshwater extending to the coast near the surface and seawater penetrating inland at depth (see diagram). The Ghyben–Herzberg principle can be expressed by the equation:

$$Z_s = \frac{\rho f}{\rho s - \rho f} Z_w$$

where Z_s is the depth below sea level to the interface between fresh and saltwaters; ρf and ρs are the density of fresh and saltwaters respectively; and Z_w is the elevation of the water table above sea level. Hence if the density of the freshwater is 1 and that of seawater is 1.025, then under hydrostatic equilibrium the depth Z_s to the interface is 40 times the height of the water table above sea level. Consequently, if pumping from a well in a coastal aquifer results in a draw down of the water table by 1 m, then saltwater will intrude upwards beneath the well by a distance of 40 m (see diagram).

The Ghyben–Herzberg principle simplifies the relationship usually found in nature, because groundwater conditions are usually dynamic rather than static. As a result, the

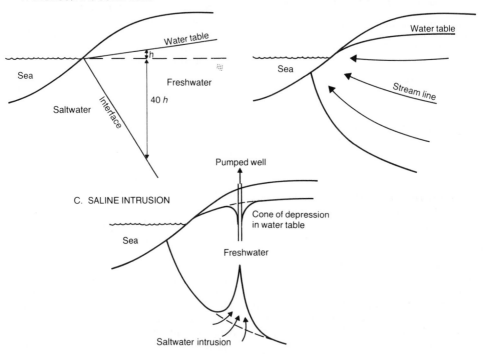

A HYDROSTATIC CONDITIONS

B STEADY SEAWARDS FLOW

C. SALINE INTRUSION

equation usually underestimates the depth to the interface with saltwater, which is commonly seaward of the calculated position. PWW

Reading
Freeze, R.A. and Cherry, J.A. 1979: *Groundwater*. Englewood Cliffs, NJ: Prentice-Hall.
Hubbert, M.K. 1940: The theory of groundwater motion. *Journal of geology* 48, pp. 785–944.
Todd, D.K. 1980: *Groundwater hydrology*. 2nd edn. New York and Chichester: Wiley.
Ward, R.C. and Robinson, M. 1990: *Principles of hydrology*. 3rd edn. Maidenhead: McGraw-Hill.

gibber A desert plain which is mantled with a layer of pebbles or boulders. It is a type of stone pavement.

gilgai The microrelief sometimes resulting from changes in the volume of swelling clays during prolonged expansion and contraction, due to changes in moisture content, especially in less humid areas. It usually consists of a series of microbasins and microknolls in nearly level areas, or of micro-valleys and micro-ridges parallel to the direction of the slope. ASG

Reading
Verger, F. 1964: Mottureaux et gilgais. *Annales de géographie* 73, pp. 413–30.

gipfelflur A plane within which uniform summit levels occur in a mountainous region, especially where the uniformity is neither structural nor the residual portion of a peneplain.

glacial *a.* (adjective) Describes a landscape occupied by glaciers. In this usage the term is similar to *glacierized*, an alternative which has not found general favour. The term *glaciated* describes a landscape which has been covered by glaciers, but normally in the past.

b. (noun) Those occasions during Ice Ages when ice sheets were expanded and average global climates were colder and drier than during the intervening INTERGLACIALS such as exists at present. During many of the seventeen or so PLEISTOCENE glacials ice sheets covered Canada and the northern USA, northern Europe, Britain north of the environs of London, and northwestern Siberia. In addition, the existing ice sheets of Greenland and Antarctica expanded offshore while mountain glaciers throughout the world extended into lower altitudes. Sea ice extended further towards the equator as global ocean temperatures fell (figures 1 and 2). Atmospheric and oceanic circulation was modified. It seems

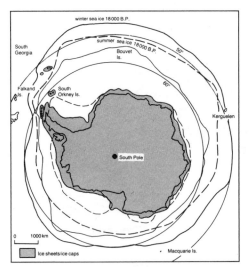

1. The Antarctic during glacial maxima. The thinner lines represent the modern equivalents.
Source: *Sugden 1982. Figure 7.2.*

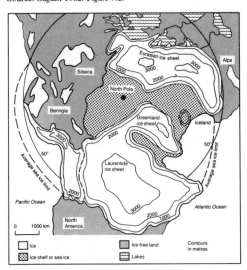

2. The Arctic during glacial maxima.
Source: *Sugden 1982. Figure 7.3.*

likely that the globe as a whole was drier with subtropical deserts extending their equator margins and the equatorial rain forest being restricted to discrete islands by the spread of savannah conditions. Mid-latitude areas in the northern hemisphere saw increased wind action with extensive loess deposits in Europe. China and North America.   DES

Reading
Bowen, D.Q. 1978: *Quaternary geology*. Oxford: Pergamon.

CLIMAP Project Members, 1976: The surface of the Ice-Age earth. *Science* 191, 4232, pp. 1131–7.

Goudie, A.S. 1983: The arid earth. In R. Gardner and H. Scoging eds, *Megageomorphology*. Oxford: Oxford University Press. Pp. 152–71.

Sugden, D.E. 1982: *Arctic and Antarctic*. Oxford: Basil Blackwell; Totowa, NJ: Barnes & Noble.

glacial protectionism The belief that the erosive power of rain and rivers far exceeds that of glacier ice, and that the presence of glaciers in a region protects the landscape from much more effective fluvial attack. Glaciers were thought to rest in depressions like custard in a pie dish, rather than to erode the basins. Proponents of this theory included the British geologists J.W. Judd, T.G. Bonney, E.J. Garwood, and S.W. Wooldridge.   ASG

Reading
Davies, G.L. 1969: *The earth in decay*. London: Macdonald.

glaciation level (also called glacial limit) The altitude above which mountain glaciers occur. Since glacier location is also influenced by topography, in particular by the need for sufficiently gentle slopes on which to form, the commonly used method of fixing the glaciation level is to take the altitude midway between the highest topographically suitable mountain without a glacier and the lowest topographically suitable mountain carrying a glacier (Østrem 1966). Delimited in this way, the glaciation level varies predictably over the globe. It rises from near sea level in high latitudes towards the equator in response to temperature changes, but superimposed on this trend are variations reflecting depression in humid areas, such as the mid-latitude and equatorial regions. The glaciation level also rises from maritime coastal locations towards continental interiors.

DES

Reference
Ostrem, G. 1966: The height of the glacial limit in southern British Columbia and Alberta. *Geografiska annaler* 48A.3, pp. 126–38.

glacier A mass of snow and ice which, if it accumulates to sufficient thickness, deforms under its own weight and flows. If there is insufficient snow to maintain flow, as occurs in some dry polar areas, the glacier may be essentially stagnant and termed a *glacier réservoir* (Lliboutry 1965). If the snowfall is sufficient the snow is transformed to ice and flows from the accumulation zone to the ABLATION ZONE as a *glacier évacateur*.

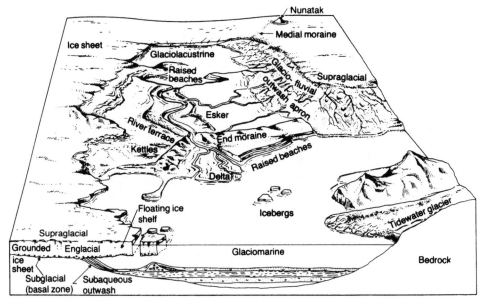

Glacial environments and associated landforms typical of glaciated areas, including glaciofluvial outwash fan, ice-contact end moraine, glacial lake- and sea-bottom deposits, raised beaches and marine delta.
Source: *H.G. Reading ed. 1978:* Sedimentary environments and facies. *Oxford: Blackwell Scientific. Figure 3.12.*

There are three main types of glacier:

1 *Ice sheet or ice cap* where the ice builds up as a dome over the underlying topography. Such domes are often drained radially by outlet glaciers.
2 *Ice shelf* where the ice forms a floating sheet in a topographic embayment and flows towards the open sea.
3 *Mountain glaciers* which are constrained by the underlying topography of the mountains and form a wide variety of types, e.g. cirque, valley, piedmont glaciers. DES

Reading and Reference
Lliboutry, L. 1965: *Traité de glaciologie.* 2 vols. Paris: Masson.
†Sugden, D.E. and John, B.S. 1975: *Glaciers and landscape.* London: Edward Arnold.

glacier milk A popular name given to glacial meltwater with sufficient suspended sediment load to give it a milky-green colour.

glacier table A stone resting on a pillar of ice which protrudes above a glacier surface. The ice has been protected from melting by the presence of the overlying stone.

glacieret A small glacier, such as may develop from a SNOW PATCH.

glacierization The process whereby a landscape is progressively covered by glacier ice.

glacioeustasy See EUSTASY.

glaciofluvial The activity of rivers which are fed by glacial MELTWATER. The main characteristics of such streams are the highly variable discharge and the high sediment loads. Discharge varies markedly on a wide variety of time-scales. Variations over a matter of seconds or minutes relate to the sudden release or closure of basal water pockets as a result of glacier sliding. Diurnal fluctuations respond to high rates of melting by day and produce high flows in the evenings. Fluctuations over a matter of days reflect prevailing weather patterns, whereas a strong seasonal summer flow reflects the effect of a glacier in storing winter precipitation only to release it in the ABLATION season. One particularly sudden seasonal peak in discharge may accompany the rapid emptying of a marginal or subglacial lake (see JÖKULHLAUP). The muddy colour of meltwater streams reflects their high suspended sediment loads and measurements as high as 3800 mg l^{-1} have been measured. In addition the bed load is high and may amount to 90 per cent of the

A classification of glaciofluvial deposits

Dominant sediment	Environment	General form	Relationship to ice	Genetic term
Ice-contact deposits Sand and gravel	Fluvial	Ridge	Marginal, subglacial, englacial, supraglacial	Esker
		Mound		Kame Kame complex
		Spread with depressions	Marginal	Kettled sandur
Proglacial deposits Sand and gravel	Fluvial	Spread	Proglacial	Sandur
Silt and clay	Lacustrine		Proglacial/ marginal	Lake plain
Sand and gravel		Terraces, ridges		Beach
Clay, sand and gravel		Terrace		Kame delta
Silt and clay	Marine	Spread		Raised mud flat
Sand and gravel		Terraces, ridges		Raised beach
Clay, sand and gravel		Terrace		Raised delta

Source: Price, R.J. 1973: *Glacial and fluvioglacial landforms*. Edinburgh: Oliver & Boyd. Table 3, p. 138.

suspended sediment load. Not surprisingly glaciofluvial landforms may reflect prodigious feats of erosion and sedimentation. Formerly glaciated areas, particularly in mid-latitudes, contain abundant erosional evidence in the form of deeply incised meltwater channels and giant pot-holes, subglacial channel courses such as ESKERS and KAMES, and extensive areas of proglacial OUTWASH and lake deposits (glaciolacustrine).   DES

Reading
Elliston, G.R. 1973: Water movement through the Gornergletscher. *Symposium on the Hydrology of Glaciers, Cambridge 9–13 Sept. 1969*. International Association of Scientific Hydrology 95, pp. 79–84.
Østrem, G., Bridge, C.W. and Rannie, W.F. 1967: Glaciohydrology, discharge and sediment transport in the Decade Glacier area, Baffin Island, NWT. *Geografiska annaler* 49A, pp. 268–82.

glaciotectonism Those structures and landforms (e.g. displaced megablocks) produced by deformation and dislocation of pre-existing soft bedrock and drift as a direct consequence of glacier ice movement.   ASG

Reading
Aber, J.S. 1985: The character of glaciotectonism. *Geologie en mijnbouw* 64, pp. 389–95.

glacis A gentle pediment slope, especially in arid and semi-arid regions.

glei, gley A clayey soil rich in organic material that usually develops in areas where the soil is waterlogged for long periods. Various component processes are involved: the reduction of ferric compounds, the translocation of iron as ferrous compounds or complexes, and precipitation of iron as mottles and minor indurations.

Glen's law The relationship between the deformation of ice over time and SHEAR STRESS discovered by J.W. Glen (1955). It has the form:

$$\dot{\epsilon} = A\tau^n$$

where $\dot{\epsilon}$ is the strain rate (deformation rate), τ is the shear stress, A is a constant depending on ice temperature, crystal size and orientation, impurity content and perhaps other factors, and n is a constant whose mean value is normally taken as equal to 3. The relationship models the secondary creep of ice which involves several separate processes, such as the movement of dislocations within crystals, crystal growth, the migration of crystal boundaries and recrystallization.

Glen's law is of fundamental importance in understanding glacier motion. It shows how sensitive glacier ice is to increasing shear stress and, for example, when the shear stress is doubled, the rate of deformation increases 8 times. This inherent sensitivity helps explain the characteristic shallow surface profile of glaciers. It also explains why most internal deformation occurs at the bottom of glaciers and it shows how glaciers move by internal deformation in the absence of BASAL SLIDING.   DES

Reading and Reference
Glen, J.W. 1955: The creep of polycrystalline ice. *Proceedings of the Royal Society of London*, Series A. 228, pp. 519–38.
†Paterson, W.S.B. 1981: *The physics of glaciers*. Oxford and New York: Pergamon.

glint line The escarpment of Palaeozoic rocks which borders the Scandinavian and Laurentide shields and is associated with a line of lakes. Infrequently used today.

global environmental change There are two components to this (Turner *et al.* 1990): systemic global change and cumulative global change. In the systemic meaning, 'global' refers to the spatial scale of operation and comprises such issues as global changes in climate brought about by atmospheric pollution. In the cumulative meaning, 'global' refers to the areal or substantive accumulation of localized change, and a change is seen to be 'global' if it occurs on a worldwide scale, or represents a significant fraction of the total environmental phenomenon or global resource. Both types of change are closely intertwined. For example, the burning of vegetation can lead to systemic change through such mechanisms as carbon dioxide release and albedo change, and to cumulative change through its impacts on soil and biotic diversity.   ASG

Reference
Turner, B.L., Kasperson, R.E., Meyer, W.B., Dow, K.M., Golding, D., Kesperson, J.X., Mitchell, R.C. and Ratick, S.J. 1990: Two types of global environmental change. Definitional and spatial-scale issues in their human dimensions. *Global environmental change* 1, pp. 14–22.

global ocean circulation Surface currents of the oceans, driven by the prevailing winds, have their motion modified by the earth's rotation (see CURRENTS, OCEAN). In the Atlantic, Indian and Pacific Oceans surface circulation patterns are dominated by flow around the subtropical GYRES where the motion is in quasi-geostrophic balance between pressure gradient and Coriolis forces (see GEOSTROPHIC WIND). Here the velocities of western boundary currents, such as the Gulf Stream, can be as great as 3 m s^{-1}. In equatorial regions a more complex pattern, which includes a powerful undercurrent flowing from west to east, results from the weakness of Coriolis effects together with the mean northward displacement of the INTERTROPICAL CONVERGENCE ZONE relative to the geographical equator. The directions of the currents in the northern Indian Ocean reverse seasonally under the influence of the monsoonal winds. At high latitudes in the Atlantic and Pacific Oceans cyclonic (counter-clockwise) subpolar gyres are present; more complex gyratory flows have been deduced for the Arctic. In contrast to the landlocked northern polar regions, a strong eastward-flowing current encircles Antarctica unimpeded by landmasses.

Whereas the surface circulation is relatively accessible to study, below the THERMOCLINE water moves some two or three orders of magnitude more slowly and on a much broader scale, hence less is known with any certainty. This deep or thermohaline circulation originates with the small differences in density between WATER MASSES from different source regions. At high latitudes, because of the low ambient temperatures and the addition of salts expelled in the formation of ice, the density of surface ocean water is increased to the point at which it naturally sinks to deeper levels. The principal source regions of deep ocean water are to be found in the Weddell and Ross Seas in the Antarctic and in the Norwegian Sea in the Arctic. Water from the former regions, known as Antarctic Bottom Water, is the densest of all water masses and flows northwards close to the bottom to cover much of the Indian Ocean and the Pacific and Atlantic Oceans to about 30°N. The presence of the Norwegian and Weddell Sea source regions at the extremities of the Atlantic Ocean, combined with its natural topography, gives rise to a relatively strong meridional flow. The

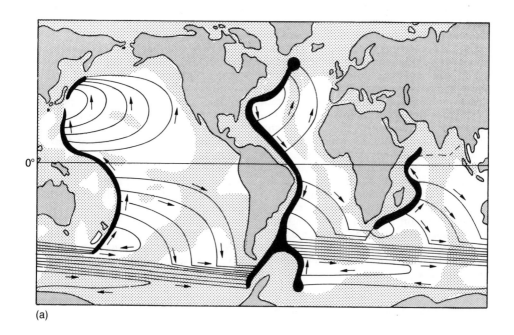

(a)

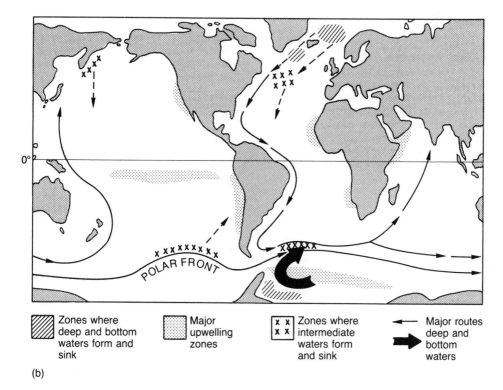

| Zones where deep and bottom waters form and sink | | Major upwelling zones | | x x
x x | Zones where intermediate waters form and sink | ← | Major routes deep and bottom waters |

(b)

Global ocean circulation: (a) Model of deep-water circulation of the world ocean with source regions in the North Atlantic and in the Weddell Sea. (b) Circulation within the deep ocean layers.
Source: *Modified after H. Stommmel in Tolmazin 1985, p. 143.*

southward flow of Atlantic Deep Water from the Norwegian Sea is concentrated in a western boundary current moving at a few cm s^{-1}. Approaching the Southern Ocean it mixes with both Antarctic Bottom Water below and Antarctic Intermediate Water above before being carried eastwards in the strong circumpolar current. The resultant water mass flows northwards as a western boundary current into the Indian and Pacific Oceans, both of which lack significant source regions of deep water. Since there are no obvious zones of upwelling to match the large-scale downwelling of dense water, it is concluded that the return flow to the surface is principally by very slow upward diffusion through the thermocline. The circulation may not be completed until, after perhaps a thousand years, water is returned by wind-driven surface currents to the polar seas.   JEA

Reading
Open University Oceanography Course Team 1989: *Ocean circulation*. Oxford: Pergamon; Milton Keynes: Open University Press.

Pond, S. and Pickard, G.L. 1978: *Introductory dynamic oceanography*. Oxford: Pergamon.

Tolmazin, D. 1985: *Elements of dynamic oceanography*. Boston and London: Allen & Unwin.

global warming The possibility that the earth is or will warm up because of increasing concentrations of certain gases in the atmosphere (see GREENHOUSE EFFECT). Over the past century the earth's mean surface temperature has increased by perhaps 0.3–0.6°C, and it is possible that it may increase by several more degrees during the next hundred years. The degree of warming may, however, be greatest in higher latitudes. The potential impacts of such warming are now the subject of considerable research and include cryospheric melting, sea-level rise, shifts in vegetation and precipitation belts, changes in hurricane characteristics, modifications to land use and human affairs, and changes in disease distribution.   ASG

Reading
Houghton, J.T., Callander, B.A. and Varney, S.K. 1992: *Climate change 1992: the supplementary report to the IPCC Scientific Assessment*. Cambridge: Cambridge University Press.

— , Jenkins, G.J. and Ephraums, J.J. 1990: *Climate change: the IPCC Scientific Assessment*. Cambridge: Cambridge University Press.

Gloger's rule The pigmentation of warm-blooded animal species tends to decrease away from the equator as mean annual temperatures decrease.

GLOSS (Global Sea-Level Observing System) A worldwide network of sea-level gauges, defined and developed under the auspices of the Intergovernmental Oceanographic Commission. Its purpose is to monitor long-term variations in the global level of the sea surface by reporting the observations to the Permanent Service for Mean Sea Level. The levels of the gauges are fixed by GPS: a satellite-based global positioning system, capable of accurately locating points in a three-dimensional geometric framework.   DTP

gnammas Holes produced in rock surfaces, especially igneous rocks and sandstones, by weathering processes. They are a type of rock basin.

gneiss A coarse-grained igneous rock, often a granite, that has been metamorphosed, producing a banded or foliated structure.

goletz terrace (also known as cryoplanation terrace or altiplanation terrace) A hillside or summit bench which is cut in bedrock and transects lithology and structure. It is confined to cold climates.

Reading
Washburn, A.L. 1979: *Geocryology*. London: Edward Arnold.

Gondwanaland A large continent lying predominantly in the southern hemisphere which, it is hypothesized, was rifted apart in the late Palaeozoic. The component blocks of continental rocks now form parts of Africa, Australia, Antarctica, South America and India.

gorge A deep and narrow section of a river valley, usually with near vertical rock walls. More generally a narrow valley between hills or mountains.

gouffres Large pipes or vertical shafts that occur in limestone areas.

graben A valley or trough produced by faulting and subsidence or uplift of adjacent blocks (horsts).

grade, concept of One of the most confusing concepts in geomorphology,

A gnamma developed in granite on Paarl Mountain, Cape Province, South Africa. Two-year-old geomorphologist is shown for scale.

partly because of its inextricable relationship with gradient. Introduced by G.K. Gilbert in 1876, it relates the gradient of a channel to the balance between corrasion (erosion), resistance and transportation. The idea was adopted, adapted and debated by W.M. Davis, J.E. Kesseli and J.H. Mackin in particular and re-surfaces in the influential paper by S.A. Schumm and R.W. Lichty (1965) as the 'graded' time span. It is now generally assumed to be roughly equivalent to DYNAMIC EQUILIBRIUM and, in practical terms, should be viewed – by extension from work on regime theory in alluvial canals – as a state in which channel form is relatively constant despite variations in flow (usually 2–10 years). The application to hillslope form and materials is distinctly problematic.   BAK

References
Gilbert, G.K. 1876: The Colorado Plateau province as a field for geological study. *American journal of science* 3rd series 12, pp. 85–103.
Schumm, S.A. and Lichty, R.W. 1965: Time, space and causality in geomorphology. *American journal of science* 263, pp. 110–19.

graded bedding Comprises sedimentary units that exhibit a vertical gradation in mean grain size. Normal grading is where a fining-upward sequence is present, and may result from deposition in a waning current, from a decrease in sediment supply, or from the progressive sorting or settling out of different size fractions. Inverse grading exhibits an upward-coarsening sequence, and may result from deposition in rising flow conditions or from an increase in sediment supply.   JM

graded slopes Those possessing a continuous regolith cover without rock outcrops. The concept of grade was used by G.K. Gilbert (1876) to indicate a condition of balance between erosion and deposition, brought about by adjustments between the capacity of a stream to do work and the quantity of work that the stream has to do. This definition was formally introduced by W.M. Davis (1899, reprinted in 1954) and applied to hillslopes in the words: 'a graded waste sheet . . . is one in which the ability of the transporting forces to do work is equal to the work they have to do. This is the condition that obtains on those evenly slanting, waste-covered mountain sides which have been reduced to a slope that engineers call *the angle of repose*' (1954, p. 267). Rocky outcrops are not graded because waste can be removed from them faster than it is supplied by weathering. On slopes from which outcrops have been eliminated the 'agencies of removal are just able to cope with the waste that is there weathered plus that which comes from farther uphill' (1954, p. 268). Graded waste slopes decline in angle as the waste becomes finer in texture as a result of weathering 'so that some of its particles may be moved even on faint slopes' (1954, p. 269).

Because of the difficulty of determining the volumetric relationships between weathering and removal, and texture and removal processes, Young (1972, p. 100) has suggested that the term graded slope be used to indicate those lacking outcrops: this definition would make it equivalent to a soil or regolith-covered slope.   MJS

Reading and References
†Davis, W.M. 1899: The geographical cycle. *Geographical journal* 14, pp. 481–504.
†— 1954: The geographical cycle. In D.W. Johnson ed., *Geographical essays*. New York: Dover; London: Constable.
†Gilbert, G.K. 1876: The Colorado Plateau province as a field for geological study. *American journal of science* 3rd series 12, pp. 16–24, 85–103.
†Young, A. 1972: *Slopes*. Edinburgh: Oliver & Boyd.

graded time A time-span intermediate between the longer interval of 'cyclic time' and the shorter period of 'steady time' (Schumm and Lichty 1965). It is defined as 'a short span of cyclic time during which a graded condition or dynamic equilibrium exists' (1965, p. 114), with respect to the landforms and, by reference to Mackin's (1948) discussion of GRADE, it is implied that this 'short span' will be a 'period of years'. In Schumm and Lichty's view the chief practical considerations governing studies on the graded time-scale are that time and initial relief become irrelevant to the enquiry, while the morphology of drainage networks and hillslopes and the hydrologic outputs of drainage basins are dependent variables, contingent upon the independent controls of geology, climate, vegetation, disposition of relief above base level, and the manner in which run-off and sediment are generated within the landscape.

The concept is discussed again by Schumm (1977, pp. 10–13) and, rather confusingly, it is there redefined (figures 1–5) as equivalent to steady-state equilibrium time, with a time span of 100–1000 years. It seems clear that Schumm intends the term to be used to imply an intermediate time-scale in which the focus of investigations is on fluctuations in hydrologic outputs, channel morphology and hillslope form viewed as responses to spatial or temporal patterns of variation in the 'independent' variables listed in the 1965 paper.   BAK

Reading and References
†Mackin, J.H. 1948: Concept of the graded river. *Bulletin of the Geological Society of America* 59, pp. 463–512.
†Schumm, S.A. 1977: *The fluvial system*. New York and London: Wiley.
†— and Lichty, R.W. 1965: Time, space and causality in geomorphology. *American journal of science* 263, pp. 110–19.

gradient wind Results from a balance of horizontal pressure gradient force, CORIOLIS FORCE and the CENTRIPETAL ACCELERATION (or centrifugal force) that exists when air moves in a curved path, such as occurs in a cyclone and an anti-cyclone. BUYS BALLOT'S LAW applies to this wind just as it does to the geostrophic wind, but because more forces are involved, the velocity of the gradient wind is different to that of the GEOSTROPHIC WIND. The one exception to this occurs when airflow is straight, giving a zero centripetal acceleration and hence geostrophic balance and wind.   BWA

Reading
Atkinson, B.W. 1972: The atmosphere. In D.Q. Bowen ed., *A concise physical geography*. Amersham: Hulton Educational. Pp. 1–76, esp. pp. 33–48.
Hess, S.L. 1959: *Introduction to theoretical meteorology*. New York: Henry Holt.

gradually varied flow In most natural river channels the flow is gradually varied because the cross-section and bed slope change downstream and the water surface is not parallel to the bed. Under these conditions the FLOW EQUATIONS for UNIFORM STEADY FLOW are not strictly applicable except locally, and a more detailed analysis of the flow energy is required. So long as the streamlines in a short reach are approximately parallel, and pressure within the flow is therefore HYDROSTATIC, the total energy of a unit mass of water at the bed (e.g. at the upstream in section 1 of the diagram) is the sum of its potential energy, pressure energy, and kinetic energy:

$$E = \rho_w g z_1 + \rho_w g d_1 + \rho_w v_1^2/2$$

or

$$E = \rho_w g(z_1 + d_1 + v_1^2/2g)$$

where the term in brackets is the 'total head' H_1 and:

$$H_1 = z_1 + d_1 + v_1^2/2g$$

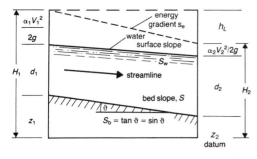

is the BERNOULLI EQUATION. This can be used to define the conservation of energy between two adjacent sections. An energy balance equation between sections 1 and 2 of the diagram states that:

$$z_1 + d_1 + v_1^2/2g = z_2 + d_2 + v_1^2/2g + h_L$$

where h_L is the head or energy loss between the sections.

It is clear from the diagram that in gradually varied flow the water surface, bed slope and energy grade line are not parallel. The energy grade line always slopes downwards in the direction of flow and measures the rate of dissipation of energy (the energy loss) caused by flow resistance and sediment transport. If the water surface slope is gentler than the energy slope the kinetic energy term decreases downstream as the flow decelerates (as in the diagram), whereas a steeper water surface slope would reflect accelerating flow and conversion of energy from potential to kinetic forms (Richards 1982, pp. 72–6). In a POOL AND RIFFLE stream the variations of velocity and depth from section to section reflect this continual conversion of energy from potential to kinetic forms, and vice versa, in response to the changing bed slope. KSR

Reference
Richards, K.S. 1982, *Rivers: form and process in alluvial channels*. London and New York: Methuen.

granite A coarsely crystalline igneous rock composed predominantly of quartz and alkali feldspars. Additional constituents are commonly mica and hornblende.

grasslands Regions in which the natural or the plagioclimax vegetation is dominated by grasses or grass-like plants and non-grass-like herbs. They include temperate grasslands of the steppes, prairies, pampas and veld, tropical grasslands or SAVANNAHS, and smaller zones on mountains, in high latitudes and as patches within other plant formations resulting from fire, soil or drainage controls. Before man's modification of the natural plant and animal cover, grasslands probably occupied around 40–45 per cent of the land surface, a figure increased by the maintenance of grazing land and decreased by conversion to other forms of land use to around 25 per cent today. PAF

Reading
Coupland, R.T. ed. 1979: *Grassland ecosystems of the world*. Cambridge and New York: Cambridge University Press.

gravimetric method A means of soil moisture determination involving taking, weighing, oven drying and reweighing a soil sample and expressing the moisture content (or sample loss in weight) as a percentage of the oven dry weight of the sample. AMG

Reading
Reynolds, S.G. 1970: The gravimetric method of soil moisture determination. Parts I, II and III. *Journal of hydrology* 11, pp. 258–300.

gravity The force imparted by the earth to a mass which is at rest relative to the earth. All masses are attracted to each other according to Newton's Law of Universal Gravitation, but the earth is also rotating and therefore a centrifugal force is also exerted on the mass in question. Hence the force observed, and commonly called gravity, is the combination of the true gravitational force and the centrifugal force. The standard acceleration of gravity at sea level at 45° latitude is 9.80665 m s^{-2}. BWA

gravity faulting An important process that operates in mountainous areas; high available relief enables major rock movements to occur under the influence of gravity, creating hilltop valleys and depressions, and sometimes double summits (*doppelgrate*). ASG

Reading
Beck, A.C. 1968: Gravity faulting as a mechanism of topographic adjustment. *New Zealand journal of geology and geophysics* 11, pp. 191–9.
Paschinger, V. 1928: Untersuchungen über Doppelgrate. *Zeitschrift für Geomorphologie* 3, pp. 204–36.

gravity wave A wave disturbance in which buoyancy acts as a restoring force on fluid parcels displaced from an equilibrium state. The restoring force acts only in the vertical, frequently producing simple harmonic motion around the equilibrium level. The wave ensues because this simple harmonic motion occurs in a horizontal flow. The resultant of the two components of velocity (vertical and horizontal) at any instant gives the velocity of the parcel of air under consideration. A sequence of such velocities throughout one cycle of simple harmonic motion describes the gravity wave form. Examples of gravity waves are lee waves in the atmosphere and water waves. Gravity waves are also known as buoyancy waves for obvious reasons.　　　BWA

grazing, hydrological effects of
From the human perspective grazing affects the hydrological cycle in undesirable ways because it increases the rate of run-off, reduces infiltration rates, and causes erosion and sedimentation problems (Smeins 1975). Despite the importance of the subject, especially in developing countries with agricultural economies, there is a 'general lack of studies which adequately define the hydrologic impact of grazing' (Branson *et al.* 1981).

Despite the lack of widespread research, some impacts of grazing on infiltration have been identified. Grazing animals remove vegetation and sometimes consume so much of the plants that regeneration is lengthy or impossible. Animals also trample the soil, altering its texture and structure. These changes usually result in a reduction in infiltration rates, especially during initial rainfalls. Once saturated conditions are obtained grazing may have no impact on infiltration rates because the rates are so low in natural settings (Lusby *et al.* 1971). In areas of prairie vegetation infiltration rates may be reduced by 85–93 per cent in comparison with areas that are not grazed. Although three to four years may be required to restore infiltration rates to their natural levels in semi-arid regions with favourable conditions, drastically reduced rates may be re-established after only one grazing season.

Decreased infiltration rates on grazed terrain lead to increased run-off, the amount of change being directly related to the intensity of grazing (Dunford 1949). Research in instrumented watersheds in semi-arid conditions has shown that in a two-year period 43 per cent more run-off may be expected on grazed watersheds than on similar ungrazed areas (Lusby *et al.* 1971).

The increased run-off from grazed areas leads to significant impacts on erosion and sediment yield in downstream areas. In the instrumented basins mentioned above soil erosion increased by 45 per cent under grazing (Lusby *et al.* 1971). An extensive analysis in the western USA found that grazing practices were more important in controlling excessive erosion than surface gradient, slope aspect, soil conditions, rodent activity, plant density or plant types.　　　WLG

References
Branson, F.A., Gifford, G.F., Renard, K.G. and Hadley, R.F. 1981: *Rangeland hydrology*. 2nd edn. Dubuque, Iowa: Kendall Hunt.
Dunford, E.G. 1949: Relation of grazing to runoff and erosion on bunchgrass ranges. *US Department of Agriculture, Forest Service research note* RM 7.
Lusby, G.C., Reid, V.H. and Knipe, O.D. 1971: Effects of grazing on the hydrology and biology of the Badger Wash Basin in Western Colorado, 1953–1966. *US Geological Survey, water-supply paper* 1532-D.
Smeins, F.E. 1975: Effect of livestock grazing on runoff and erosion. In *Watershed management*, Proceedings of a symposium held at Logan, Utah, 11–13 August, 1975.

great interglacial A phase of Pleistocene history identified in the classic four-glacial model developed by A. Penck and E. Brückner (see PENCK AND BRÜCKNER MODEL) who believed that the interglacial between the Mindel and the Riss glacials lasted a longer time than any others. Studies of the Pleistocene record in ocean cores tend not to support this view.　　　ASG

greenhouse effect A change to the temperature of the atmosphere brought about by the presence of gases. The Intergovernmental Panel on Climatic Change explains it as follows (Houghton *et al.* 1990, p. xiii): 'Short-wave solar radiation can pass through the clear atmosphere relatively unimpeded. But long-wave terrestrial radiation emitted by the warm surface of the Earth is partially absorbed and then re-emitted by a number of trace gases in the cooler atmosphere above. Since, on average, the outgoing long-wave radiation balances the incoming solar radiation, both the atmosphere and the surface will be warmer than they would be without the greenhouse gases. The main natural greenhouse gases are not the major constituents, nitrogen and oxygen, but water vapour (the biggest contributor), carbon dioxide, methane, nitrous oxide, and ozone in the troposphere (the lowest 10–15 km of the atmosphere) and stratosphere.' The greenhouse effect is 'natural' in that the world is already warmer at its surface by some 33°C than it would be if the natural greenhouse gases were not present. Moreover, measurements of the polar ice cores, going back as far as 160,000 years ago, show that the earth's temperature closely paralleled the amount of carbon dioxide and methane in the atmosphere. However, interest has grown in this phenomenon in recent years because of the fact that concentrations of some of the greenhouse gases are increasing because of human activities. The most important of these is carbon dioxide derived from the burning of fossil fuels and biomass, but also of significance are methane, nitrous oxide and chlorofluoro-carbons (CFCs). Global warming is believed to be a likely consequence of this tendency.   ASG

Reading and Reference
Houghton, J.T., Jenkins, G.J. and Ephraums, J.J. eds 1990: *Climate change: the IPCC scientific statement.* Cambridge: Cambridge University Press.
Jäger, J. and Ferguson, H.L. 1991: *Climate change: science, impacts and policy.* Cambridge: Cambridge University Press.

grey wether See SARSEN.

greywacke A sediment composed of coarse fragments of quartz and feldspar, usually poorly sorted.

grèzes litées Bedded screes of angular rock fragments associated with cold climates and frost shattering. French examples have thicknesses up to 40 m, with layers dipping at as much as 40°, whereas Polish examples are generally only a few metres thick (Dylik 1960). The inclination of the layers parallels that of the slopes, and, in contrast to ordinary gravitational debris slides, the deposits show a striking predominance of fines in their distal parts. Snow patches may play a role in their formation (Guillen 1964) and downwash is an important process. The rhythmic nature of the sediments suggests that under cold conditions the following process occurs: first, freezing of rocks on a cliff face causes disintegration, and the coarse debris thus released slides downward over frozen subsoil; secondly, the following phase of thaw causes a mantle of half-fluid material rich in fines to spread over the stony layer.   ASG

References
Dylik, J. 1960: Rhythmically stratified slope waste deposits. *Biulteyn Peryglacjalny* 8, pp. 31–41.
Guillen, Y. 1964: Les grèzes litées comme depôts cyclothémiques. *Zeitschrift für Geomorphologie Supplementband* 5, pp. 53–8.

grike (gryke) The cleft or runnel in bare limestone pavements which separates the CLINTS. Grikes are formed when limestone is dissolved by water, probably under a soil cover (Trudgill 1972) normally along joint pattern weaknesses. Grikes are called *kluftkarren* in German.   PAB

Reference
Trudgill, S.T. 1972: The influence of drifts and soils on limestone weathering in NW Clare. *Proceedings of the University of Bristol Speleological Society* 13, pp. 113–18.

ground frost Ground, but not necessarily air, below freezing. On cool dry nights there is not enough moisture in the air to stop terrestrial (heat) RADIATION escaping to space. When, as is usual, the earth is a good radiator but a poor conductor of heat, the ground may cool sufficiently to give CONDENSATION (dew). If the air is dry enough the moisture may sublime into crystalline form to make FROST. Such conditions are hazardous for plants whose cells are disrupted by ice crystals. Whether crystals form depends on the amount of 'antifreeze' (sugar and starch) in their tissues.   JSAG

Reading
Monteith, J.L. 1973: *Principles of environmental physics.* London: Edward Arnold.

ground ice A body of more or less clear ice within frozen ground. It takes many forms (see diagram on p. 242), some of the more common being PORE ICE, SEGREGATED ICE, ice veins and ICE WEDGES, PINGO ice, and massive icy beds. Buried glacier ice, buried ICING ice, and buried snowbank ice are sometimes regarded as forms of ground ice even though they are of surface origin. In places, ground ice may constitute more than 50 per cent by volume of the upper 2–3 m of permafrost. Generally speaking, ground ice amounts decrease with increasing depth. Aggradational landforms associated with the formation of ground ice include open- and closed-system pingos, ice wedge polygons, palsas and peat plateaux, and seasonal frost mounds. The degradation of ice-rich permafrost causes THERMOKARST and results in thaw slumping, thaw depressions and lakes, and hummocky unstable topography.   HMF

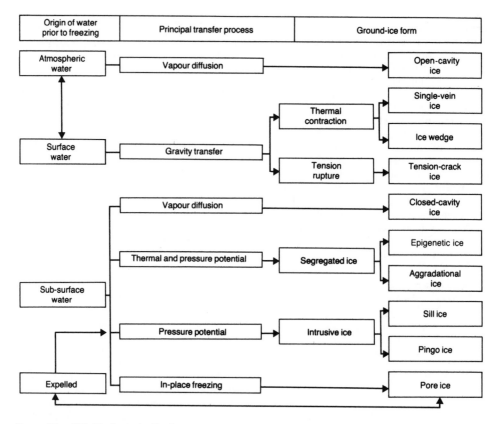

Origin of water prior to freezing	Principal transfer process	Ground-ice form

Ground Ice: J.R. Mackay's classification.
Source: *C. Embleton and J. Thornes 1979:* Process in geomorphology. *London: Edward Arnold. Table 6.1.*

Reading
Mackay, J.R. 1972: The world of underground ice. *Annals of the Association of American Geographers* 62, pp. 1–22.

ground moraine See TILL.

groundwater In its broadest sense includes all subsurface water whether in its liquid, solid or gaseous state, provided it is not chemically combined with the minerals present. In practice it is all subsurface water that participates in the HYDROLOGICAL CYCLE. Only an extremely small proportion of groundwater, such as CONNATE WATER, is prevented from active circulation, although the turnover time of some frozen groundwaters may be very long indeed.

In most textbooks, groundwater is usually taken to be just water that occurs in the permanently saturated (or PHREATIC) zone beneath the WATER TABLE. This is a matter of convenience, and a reflection of the economic importance of water in that zone and of the research that has been done on it. Water in the aerated, unsaturated (or vadose) zone above the water table but beneath the soil is also irrefutably groundwater, but it has attracted relatively little research interest. By convention, soil water is distinguished from groundwater, but logically it is groundwater at the top of the vadose zone, at the interface between surface water and groundwater systems. The complex inter-relationships between the components of these systems is often an important part of the hydrological cycle.

The aerated zone is often dismissed simply as a zone of transmission of water, but recent research, especially in karst terrains, has shown that this zone is often an important store in its own right that can sustain the base flow of subterranean streams above the water table. The uppermost weathered part of the vadose zone directly beneath the soil, but extending vertically for 10 m or so, is the location of greatest vadose storage in karst rocks. The

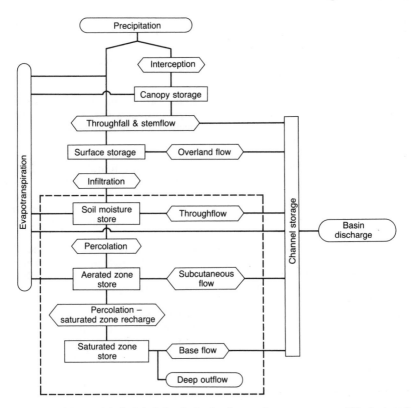

Groundwater: *terrestrial part of the hydrological cycle showing the groundwater components within the dashed lines.*

terms *subcutaneous* or *epikarstic* (French) are used to refer to this zone and its perched aquifer.

The saturated zone in karst terrains is also usually subdivided. Three parts are distinguished, the intermittently flooded (or epiphreatic) zone sometimes being as much as 30 m deep. This draws attention to the fact that the water table is not a static surface marking the top of the saturated zone, but often a highly mobile one that may change not only in vertical position but also in configuration. Nevertheless, the expansion and contraction of the saturated zone in karst rocks is commonly much greater than in other lithologies, due to more efficient recharge and discharge mechanisms via large solution passages.

Groundwater has sometimes been encountered by drilling and mining up to as much as 3000–4000 m below the surface. Its occurrence, however, usually diminishes considerably after a few hundred metres. The low limit of groundwater penetration is determined by hydrostatic pressure and by

Groundwater zones in karst aquifers

1   *Unsaturated (vadose) zone*
1a Soil
1b Subcutaneous (epikarstic) zone
1c Free-draining transmission zone
2   *Saturated (phreatic) zone*
2a Intermittently flooded (epiphreatic) zone
2b Shallow phreatic zone
2c Deep phreatic (bathyphreatic) zone

the availability of voids in the rock, although even if water is present at depth it does not necessarily participate actively in a groundwater circulation. Eventually the pressure exerted by surrounding rocks – lithostatic pressure – closes any developing fissures or voids and renders the rock impervious.

Groundwater is a resource of considerable importance. Subsurface water accounts for about 96.5 per cent of all freshwater on earth, water in the saturated (phreatic) zone alone amounting to 95 per cent. The turnover time of water in these stores is comparatively long compared to surface waters. For example, Nace (1969, 1971) estimates the turnover time (volume in storage/discharge rate) of soil water to be two weeks to a year, that of phreatic water to be two weeks to 10,000 years, depending on location, and that of rivers to average roughly two weeks.

In considering groundwater movement, a distinction should be made between *flow through time* and *pulse through time*. The former is the time taken for water from a given recharge event to pass right through the groundwater system, as measured by a tracer dye for example, whereas the latter is the time taken for a recharge-induced pressure wave to pass through the system. The transmission of a pressure wave is 10^2–10^3 times faster than that of the water itself, groundwater flow usually being very slow and in the laminar regime.   PWW

Reading and References
†Davis, S.N. and De Wiest, R.J.M. 1966: *Hydrogeology*. New York: Wiley.
†Freeze, R.A., and Cherry, J.A. 1979: *Groundwater*. Englewood Cliffs, NJ: Prentice-Hall.
†Hubbert, M.K. 1940: The theory of groundwater motion. *Journal of geology* 48, pp. 785–944.
Nace, R.L., 1969: World water inventory and control. In R.J. Chorley ed., *Water, earth and man*. London and New York: Methuen. Pp. 31–42.
— ed. 1971: Scientific framework of world water balance. *UNESCO technical papers in hydrology* 7.
†Scheller, H. 1962: *Les eaux souterraines*. Paris: Mason.
†Todd, D.K. 1980: *Groundwater hydrology*. 2nd edn. New York and Chichester: Wiley.
†Zötl, J.G. 1974: *Karsthydrogeologie*. Vienna: Springer-Verlag.

growan Decomposed granite or related rock as found on Dartmoor, south-west England, and its environs. The growan may be produced by chemical weathering or metamorphic processes, and it has been suggested that the stripping of growan from corestones of sounder rock may play a role in the formation of tors.

groyne A man-made barrier running across a beach and into the sea which has been constructed to reduce the erosion of the beach by longshore currents. Although groynes may serve to reduce erosion at the place where they are constructed, by reducing the movement of material along the coastline by longshore drift they may cause beach starvation and erosion elsewhere.

grumusol Under the classification of the SEVENTH APPROXIMATION this may be regarded as a vertisol and is in effect a modern American term for a black cotton soil.

grus An accumulation of poorly sorted angular quartz grains and clayey material derived locally from weathered granite.

guano Thick accumulations of bird excrement, usually found on islands where the birds nest. The material is used as a fertilizer as it is rich in phosphates. In some caves (and belfries) there may be a substantial accumulation of bat guano.

gull A fissure or crack, sometimes sediment filled, which opens up on escarpments as a result of the tensions produced by CAMBERING.

gully erosion The pronounced erosion, by ephemeral streams, of soils and other poorly consolidated sediments, producing networks of steep-sided channels.

Gumbel extreme value theory A theory appropriate for the analysis of extreme values; it has been applied particularly to flood frequency analysis where the Gumbel extreme value 1 distribution may be used as the theoretical distribution to fit to the distribution of flood frequency values for a particular gauging station.   KJG

Reading
Gumbel, E.V. 1958: *Statistics of extremes*. New York: Columbia University Press.

gumbo An area of clayey soil which turns to sticky mud when wet. Any damp, sticky clay.

gustiness factor An index of the variations in windspeed. It is calculated from the ratio of the total range of windspeed between gusts and lulls to the mean windspeed in the given period. Gustiness factors are highest in urban areas where the surface roughness is high, and lowest in coastal sites or exposed upland locations where surface friction is small. A gustiness factor may also be defined in terms of wind direction whereby the angular width in radians is the measure of lateral gustiness. For small values it is nearly equivalent to the speed ratio.   PS

guyot A flat-topped mountain on the sea floor, especially in the Pacific, which does not reach the sea surface. A sea mount or drowned island, which is a truncated volcano, formed as a result of subsidence associated with sea-floor spreading. ASG

Reading
Watts, A.B. 1984: The origin and evolution of seamounts. *Journal of geophysical research* 89 (B13), pp. 1106–286.

gypcrete Gypsum crusts, as found in deserts, and comprising loose, powdery or cemented crystalline accumulations dominated by calcium sulphate dihydrate at or near the ground surface. ASG

Reading
Watson, A. 1983: Gypsum crusts. In Goudie, A.S. and Pye, K. eds, *Chemical sediments and geomorphology*. London: Academic Press. Pp. 133–61.

gypsum An evaporite mineral, calcium sulphate dihydrate ($CaSO_4.2H_2O$). A common mineral forming crusts in desert pans and soils.

gyre Great circulatory systems of water in the world's oceans, involving the major surface currents.

gyttja A nutrient-rich peat or organic mud which contains much plankton.

H

habitat The overall environment, but more often specifically the physical environment, in which organisms live. It may be examined in a range of scales which extend from the macroscale (continental, subcontinental), to the mesoscale (regional, local) and microscale, the latter being of particular significance to the numerous small animal species of submicroscopic and microscopic size. All organisms must be morphologically and genetically adapted to the habitats in which they reside for any length of time. Such adaptation usually begins with the acquisition of tolerance to the conditions therein.

For plants on land, the major influential factors of habitat which can limit growth, may be grouped under four headings: climatic, topographic, edaphic and biotic. Of these *climatic* factors especially those related to cold tolerance and the provision of adequate amounts of heat and moisture in the growing season, are normally regarded as being most important, certainly on a continental, subcontinental or regional scale, but they are frequently of much less significance at a local and a microscale level. Climatic influences are those of light (energy), heat, moisture availability and wind. Conditions of light which affect plant growth and survival are its intensity, its wavelength quality, and the photoperiod (period of daylength), all of which can interact either individually or together, at one and the same time. The intensity of light is equivalent to solar energy income, and this can affect the rates of PHOTOSYNTHESIS, the rate of formation of auxins (growth-forming substances) and the vertical structure of vegetation communities which develops in response to their need to utilize radiant energy as efficiently as possible. Vegetation structures are most complex in the energy-rich wet tropics (see TROPICAL FOREST). Some green plants do very well on small amounts of light (sciophytes); but others (heliophytes, e.g. pioneers, weeds, palms) require strong light levels throughout their life cycle. Still others, among them many tree seedlings, demand a sciophytic, followed by a heliophytic phase, to achieve their best growth results. Wavelength quality variations are best exemplified on an altitudinal basis: above 2000 m, augmented levels of ultraviolet energy are received as compared to those of visible light, and many more UV-tolerant species consequently are present at higher altitudes. In contrast, differences in photoperiod are most clearly displayed latitudinally: on a year-round basis, they are minimal in the tropics, and at their maximum at high latitudes. In themselves, they can prevent the successful transfer of, say, mid-latitudinal species into the tropics, and vice-versa. For other features of light-energy control of plants and vegetation, see BIOLOGICAL PRODUCTIVITY; NET PRIMARY PRODUCTIVITY and TROPHIC LEVELS.

Those temperature conditions which influence plant growth and physiology are in most cases a response to latitude, altitude and distance from the sea. Mean annual temperatures are higher at low latitudes, and at low as compared to high altitudes; and both mean diurnal and annual temperature ranges are augmented as one moves away from oceans. All plants adapt to external temperature conditions by adjusting their own internal temperatures to them as much as possible; and they will die if the differences between them become too great. Heat is absorbed by plants either directly by radiation, or by conduction from the layer of heat or water directly over the leaf or stem; and it is lost by conduction or convection from plant surfaces, by the evaporation and transpiration of water as latent heat, by respiration, and by re-radiation of long wavelengths (Oke 1987). Most plants also have a range of further specific adaptations designed to combat possible physiological or morphological damage to them which might result from extremes of heat or cold. Thus a thick, cuticular tissue on leaf surfaces insulates all cell mechanisms from extremes of heat in hot deserts; and dwarf forms may be the only ones capable of surviving in areas of persistent cold, where they are protected by a snow cover for much of the year. In mid-latitudes, the phenomenon of cold-hardiness, which develops each year as growth ceases, and peaks at the height of winter, reduces the danger of frost damage there (Melzack and Watts 1982). In general, most of the higher plants stop growth activity once external temperatures fall to 5°C, so as to enable cold-hardiness to form. There are very few trees in places where mean annual external temperatures are lower than 10°C; and little activity is exhibited by plants generally when the immediate temperature around them exceeds 45°C.

Habitat moisture circumstances (the balance between incoming precipitation and outgoing evapotranspiration, plus the effect of the soil moisture reserve) also vary widely. On a world scale, the atmosphere holds a reserve at any one time of only about ten days' supply of precipitation so that, bearing in mind the large quantities of water required by most plants to survive (see NET PRIMARY PRODUCTIVITY), very efficient methods of gaining and recycling it are necessary in all but the wettest terrestrial biological systems (e.g. bog and marsh land). Within the overall evapotranspirational limits, which are set by energy availability (Penman 1963), the flow of water into plants generally rises with an increase in TRANSPIRATION, which in turn is usually augmented by the maximum opening of stomata under strong light conditions, high temperatures, and above-minimum windspeeds. Most plants can withstand some moisture stress, though if this becomes too great, or continues for too long, they may wilt and die. In arid areas, in which such stresses are severe and pro-longed, xerophytes (i.e. drought-tolerant plants) have an inbuilt range of defences against them. Some may become dormant for long periods, e.g. aloes and sage brush (*Artemesia tridentata*); and others adopt an annual, rapid-growth cycle, springing up and maturing quickly after rain. Still others raise the efficiency with which they can extract water from a dry soil, by physiological means, while at the same time minimizing rates of transpiration through selection either of a small leaf form, leaves with thick cuticular surfaces and few stomata, or extremely narrow leaves. Some families, including Old World euphorbias and New World cacti, become succulents, storing water for relatively long periods in their cells. In contrast, hydrophytes prefer habitats in which their roots are perma-nently placed in water, or in wet soil; and they cannot exist elsewhere. But most plants are mesophytes, tolerant neither of pro-longed excesses, nor deficiencies of water availablity, but of a nice balance of both.

The wind factor in terrestrial habitats influences plant growth in four major respects. First, although of relatively small import in most land biological systems, in which existing vegetation acts as a protec-tive agent itself against the possible adverse consequences to it of strong winds, physical damage may well result even so from the passage of hurricanes or severe storms. Secondly, a combination of strong winds and small particles (dust, sea salt) often produces a blasting effect which is capable of destroying plant cells at heights of 1 m and more above the ground surface; and a bevelling of vegetation occurs, notably along windward coasts. Thirdly, similar effects develop at high altitudes and high latitudes through a wind-chill factor (wind and severe cold temperatures). Fourthly, more generally, but predominantly in mid-latitudes, a sudden rise in windspeed in spring may increase rates of plant transpira-tion to beyond levels at which it can respond, bearing in mind that it is still then in its sluggish, cold-hardiness phase; the condition is termed 'physiological drought', and through it cell damage is initiated through lack of water.

On a more local scale climatic influences of habitat may be overshadowed by factors of topography, and edaphic and biotic controls. Three conditions of *topography* are important. First, there is a direct altitudinal effect, arising from the normal decline in temperature with increase in height of $c.0.6°C$ $100 m^{-1}$ which is prevalent in most parts of the world: and this causes a distinct and well-known zonation of plants and vegetation upslope on mountains, the essential features of which will vary according to latitude and precise location. Secondly, once slope angles have reached c.15°, patterns of vegetation will be further modified, often to include more xerophytes, since run-off will have become that much greater; and, at 35°, they are often so unstable as not to allow the development of vegetation at all. Thirdly, differences in slope aspect in relation to the angle of incident solar radiation can change temperature–water availability patterns (and, in consequence, plant growth patterns too) on alternate sides of the same valley; and the impact of this is greatest between latitudes 35° and 45° north and south of the equator (Holland and Steyn 1975). Among *edaphic controls* are those of soil, soil chemistry (see NET PRIMARY PRODUCTIVITY) and soil water: and any material deficiencies or excesses of soil nutrients or soil water as compared to the mean are likely to restrain plant growth. *Biotic controls* include especially the influence of grazing (see HERBIVORES), of fire, and of man.

These external habitat factors may influence land-based animals too, both directly (temperature, water availability, etc.) and also indirectly, in that they determine to a large extent the nature of local and regional FOOD CHAINS. In consequence of this, the NICHE role of animals is a further important habitat constraint, as also is their precise relationship with other animals (see COMMENSALISM; COMPETITION; PARASITE; SYMBIOSIS). For water-based organisms the main habitat conditions to which they respond are differences in water chemistry, temperature, light penetration and the general state of the food-chain web within the water body.   DW

Reading and References
Holland, P.G. and Steyn, D.G. 1975: Vegetational responses to latitudinal variations in slope angle and aspect. *Journal of biogeography* 2, pp. 179–84.

Melzack, R.N. and Watts, D. 1982: Cold hardiness in the yew (*Taxus baccata*, L.) in Britain. *Journal of biogeography* 9, pp. 231–41.

Oke, T.R. 1987: *Boundary-layer climates*. London: Routledge.

Penman, H.L. 1963: *Vegetation and hydrology*. Commonwealth Agricultural Bureau, Farnham Royal.

Prentice, T.C. 1992: A global biome model based on plant physiology and dominance, soil properties and climate. *Journal of biogeography* 19, pp. 117–34.

†Watts, D. 1971: *Principles of biogeography*. Maidenhead and New York: McGraw-Hill. Ch. 4.

Woodward, F.T. 1987: *Climate and plant distribution*. Cambridge: Cambridge University Press.

haboob 'Derived from the Arabic word *habb*, to blow, refers to any duststorm raised by the action of the wind. More specifically, in a meteorological context, the term refers to a duststorm generated by the evaporative outflow of a parent cumulonimbus, an outflow that can exceed 80 kn in extreme cases. Very turbulent conditions are experienced along the boundary of the cool, dense outflow as it undercuts hot stagnant air leading to vigorous dust raising activity' (Membery 1985, p. 217).   ASG

Reference
Membery, D.A. 1985: A gravity-wave haboob? *Weather* 40, pp. 214–21.

hadal zone Pertaining to the greatest depths (more than 6000 m) of the oceans.   ASG

Hadley cell The name often given to the large-scale thermally driven circulations existing in tropical latitudes and most prominent over the Atlantic and Pacific Oceans. There is one Hadley cell in each hemisphere, heated air rising near the equator in the INTERTROPICAL CONVERGENCE ZONE (ITCZ), flowing poleward aloft, descending at a latitude of 30–40°, especially in the eastern half of the very intense subtropical high pressure areas at these latitudes, and then flowing either poleward or equatorward near the earth's surface. Because of the earth's rotation the equatorward moving currents are deflected towards the west and become the north-east and south-east trade winds of the northern and southern hemispheres, respectively. The poleward moving currents are deflected towards the east and become the middle latitude westerlies. The upper poleward moving branch of the Hadley cell rapidly gains westerly momentum, which is concentrated in the subtropical JET STREAM at a height of 12–15 km above the tropical highs.

The intensity and position of the Hadley cells vary seasonally, the one in the winter hemisphere being stronger and latitudinally more extensive than the one in the summer hemisphere. The boundary between the two cells shifts from an average latitude of about 5°S in February to 10°N in August.

The Hadley cells are a very important source of energy for driving the circulation at higher latitudes. Although they transfer LATENT HEAT and SENSIBLE HEAT energy towards the equator, this is offset by an intense poleward transport of POTENTIAL ENERGY aloft. This transport is enhanced by heat added to the air by the condensation of the water vapour drawn into the ITCZ at low levels.   WDS

Reading
Palmén, E. and Newton, C.W. 1969: *Atmospheric circulation systems*. New York: Academic Press.

haff A coastal lagoon separated from the open seas by a sand pit formed by longshore drifting of sediments.

hagg A channel which separates hummocks in a peat bog.

hail Solid precipitation which falls in the form of ice particles from cumulonimbus clouds. The high concentration of liquid water in these clouds provides an environment which is favourable for the quick growth of ice particles by both coalescence and collision with supercooled water droplets. Hail is commonly spherical and the larger stones are composed of concentric shells of clear and opaque ice with a diameter as large as 10 cm in severe convective storms. Hailstones grow larger by being transported in the strong up- and downdraughts which characterize cumulonimbus clouds.   RR

haldenhang A degrading rock slope which underlies an accumulation of talus or scree.

half-life The time required for 50 per cent of the atoms of a radioactive isotope or substance to decay, with the assumption that it decays in a regular exponential manner.

haloclasty The disintegration of rock as a result of the action of salts, which may result from salt crystallization, salt hydration, or the thermal expansion of salts. It is especially important in arid areas, but may also play a role in building decay in cities. (See also SALT WEATHERING.)   ASG

halons Members of the halogenated fluorocarbon (HF) group of ethane- or methane-based compounds in which H^+ ions are partially or completely replaced by chlorine, fluorine and/or bromine. This group also includes chlorofluorocarbons. Halons are HFs which contain bromine, for example, halon 1211 (CF_2BrCl) and halon 1301 (CF_3Br). They are long-lived and have been implicated in stratospheric ozone depletion, where their damage potential has been estimated at 3–10 times that of equivalent CFC molecules.   BJS

halophyte A plant which flourishes in soils containing sodium chloride.

hamada, hammada A desert region which does not have any surficial materials other than boulders and exposed bedrock.

hamra A red, sandy soil which also contains clay.

hanging valley A tributary valley whose floor is discordant with the floor of the main valley. Hanging valleys are a hallmark of glacial erosion in mountains. The discordance of the valley floors was one of the arguments used in the early nineteenth century to suggest that rivers did not cut valleys and the objection was only removed when the role of glacier activity was appreciated. The valley cross-sections were adjusted to the glaciers they held and it is likely that the glacier surfaces met concordantly (Penck 1905).   DES

Reference
Penck, A. 1905: Glacial features in the surface of the Alps. *Journal of geology* 13, pp. 1–17.

hardness The resistance of a material to scratching. Moh's scale, which is not quantitative, depends upon the ability of one mineral in a series (usually 1 to 10, but also 1 to 8 or 1 to 12) to scratch others below it but not those above, e.g. quartz (7) will scratch feldspar (6) but not topaz (8). More quantitative measures, e.g. those of Brinell, Rockwell, Vickers, depend on the load applied to a standard indenter applied for a given time to produce a given pattern on the material under test. The Schmidt hammer test used in concrete research gives an approximate value of hardness via the 'rebound value' (in reality a measure of the coefficient of restitution of the material). This can be used to determine the compressive strength of the material. WBW

hardpan A compacted or cemented subsurface horizon within the soil zone, sometimes termed a duripan.

harmattan A dry, north-easterly wind of West Africa which blows in the winter months. Blowing out of the Sahara, it frequently carries much sand and dust.

harmonic analysis The representation of tidal variations as the sum of several harmonics, each of different period, amplitude and phase. The periods fall into three *tidal species*, long-period, diurnal and semidiurnal. Each tidal species contains *groups* of harmonics which can be separated by analysis of a month of observations. In turn, each group contains *constituents* which can be separated by analysis of a year of observations. In shallow water, harmonics are also generated in the third-diurnal, fourth-diurnal and higher species.   DTP

Hawaiian eruption A type of volcanic eruption characterized by the emission of large quantities of highly fluid, basic lava issuing from vents and fissures. Explosive eruptions of material are rare.

hazard See NATURAL HAZARD.

haze Small particles of dust or salt suspended in the air, invisible to the naked eye. They are AEROSOLS usually of less than one micron (10^{-6} m) and cause preferential scattering of light adding to the colours of sunrise and sunset. Hence they can obscure visibility, especially if as condensation nuclei they attract moisture and grow in size. Haze is distinguished therefore from mist which does not scatter light preferentially due to the larger droplet sizes.   JET

head See SOLIFLUCTION.

headcut The upslope limit of a gully system, characterized by a steep wall which is cut back, migrating upslope, as further erosion occurs.

headward erosion The processes involved in the upslope migration of the head of a gully or source of a stream.

headwater A stream which forms the source and upper limit of a river, especially a large one.

heat budget Heat is a form of energy and it defines in a general way the aggregate internal energy of motion of the atoms and molecules of a body. It may be taken as being equivalent to the specific heat of a body multiplied by its absolute temperature in degrees Kelvin and by its mass, where the specific heat of a substance is the heat required to raise the temperature of a unit mass by one degree. Temperature is the condition which determines the flow of heat from one substance to another, the direction being from high to low temperatures. So long as only one object is considered its temperature changes represent proportional changes in heat content. The definition of heat content suggests that when a variety of masses and types of material are considered the equivalence of heat and temperature disappears. Often a small hot object will contain considerably less heat than a large cool one, and even if both have the same mass and temperature their heat contents can differ because of differing specific heats.

The transfer of heat to or from a substance is effected by one or more of the processes of conduction, CONVECTION or RADIATION. The common effect of such a transfer is to alter either the temperature or the state of the substance or both. A heated body may acquire a higher temperature (sensible heat) or change to a higher state (liquid to gas, or solid to liquid) and therefore acquire latent or hidden heat. Conduction is the process of heat transfer through matter by molecular impact from regions of high temperature to regions of low temperature without the transfer of the matter itself. It is the process by which heat passes through solids but its effects in fluids (liquids and gases) are usually negligible in comparison with those of convection. In contrast, convection is a mode of heat transfer in a fluid, involving the movement of substantial volumes of the substance concerned. Conduction is the main method of heat transfer in solid rocks and the soil, while the convection process frequently operates in the atmosphere and oceans.

A dry surface with no atmosphere will assume a very simple heat balance, such that:

$$R_N = R_T(1 - \propto) - \epsilon\sigma T^4 = H$$

where R_N is the net radiation, R_T is the global (solar plus sky) radiation, $\propto$ is the albedo, ϵ is the infrared emissivity, $\epsilon\sigma T^4$ is the long-wave radiation loss from a surface at temperature T K, and H is the heat flux into/out of the soil.

Energy fluxes towards the soil surface are taken as being positive. If there is no heat source other than the global radiation, then over long time periods the net radiation R_N will be zero, since the radiative energy gained must equal the radiative energy lost. The flow of heat H into or out of the soil can produce a small imbalance in the short term, but in general H is very small compared with the magnitude of the other energy fluxes. It follows therefore that in general the temperature T will change in close accord with the daily march of incoming radiation, and that it will vary greatly between day and night.

The temperature of the surface for a given global radiation flux depends on its ALBE-DO, on its infrared emissivity and on its thermal inertia. The higher the albedo, the more radiation is reflected and the lower the surface temperature. While the emissivity of many natural substances approaches unity, it is usually not exactly equal to unity but a few per cent less. It is the surface emissivity which controls the infrared radiation loss by the surface to space. Surfaces with a low emissivity will lose heat by radiation more slowly than surfaces with higher emissivities at the same temperature. Similarly, to lose the same amount of heat by radiation, a surface with a low emissivity will have to be at a higher temperature than a surface with a high emissivity.

The flux of sensible heat into the surface is controlled by the thermal properties of the surface and the subsurface. A natural parameter which expresses the thermal properties of a soil is the thermal inertia $(\rho c \lambda)^{1/2}$, where ρ, c and λ are the density, specific heat and thermal conductivity respectively. If the thermal inertia is large, the subsurface absorbs a large amount of heat during the day and then conducts a large amount of heat to the radiating surface during the night. Temperature variations of the surface will in consequence be moderate. In contrast, soils of poor thermal inertia conduct little heat to the subsurface, and attain high temperatures in the day-time and low temperatures at night. Obviously there are no dry surfaces of this type on earth but the lunar surface is of this nature.

The surface heat balance equation with the atmosphere present is the same as before, except that the atmosphere is capable of radiating energy R_L and advecting heat S. Thus:

$$R_N = R_T(1 - \alpha) - \epsilon \sigma T^4 + R_L = H + S.$$

The atmosphere will also modify the incoming solar beam by scattering and absorbing it, so not all the radiant energy will come from the direction of the sun. Since the atmosphere also absorbs and re-radiates infrared radiation there is an infrared flux towards the surface as well as away from it. Two general classes of dry landsurfaces can be found on earth. The first is fairly common and is found in the tropical deserts, where it is dominated by the global radiation flux. The second sometimes occurs over dry surfaces at high latitudes in winter and is dominated by the atmospheric heat flux S.

In the tropical deserts there is, because of the clear skies, a large solar radiation input, which leads to high surface temperatures. Since the winds are normally light, horizontal heat transfer by the atmosphere is small, and as a consequence the main heat loss from the surface is by infrared radiation. The infrared radiation losses can be large, leading to relatively small net radiation values despite the large solar radiation input. The surface temperature follows the variations of incoming solar radiation closely and is controlled by it. Great variations of temperature occur under these conditions and numerous stations in the Sahara have recorded maximum temperatures above 45°C and minimum temperatures below 0°C. Even higher maximum temperatures would be recorded but for the fact that the albedo of desert surfaces tends to be high and therefore a fair amount of incoming radiation is reflected.

During winter in high latitudes the incoming solar radiation (S) is small, and under these conditions the heat transported by the atmosphere is considerably greater than the incoming dominated by the sensible heat flux(es). Under these conditions, with moderate winds, the surface temperature follows the air temperature which in turn depends on the prevailing synoptic conditions. While wet surfaces of this nature are common, extensive dry surfaces are rare.

Wet surfaces are the most common type of surface found in nature and the surface energy balance may be described by the equation in the previous section together with the addition of the energy lost during EVAPORATION (LE_a):

$$R_N = R_T(1-\alpha) - \epsilon\sigma T^4 + R_L$$

$$= H + LE_a + S$$

where L is the latent heat of vaporization, and E_a is the actual evaporation. A wet surface in this context is one from which the evaporation is controlled solely by the prevailing radiational and meteorological conditions, that is to say the evaporation is independent of the rate of water supply. Such a surface could be one which is physically wet, such as an ocean or a landscape after rain, or it could be one in which evaporation is occurring freely from water moving through plants (transpiration).

The study of the heat budget of a natural surface falls into two stages. First, there is the study of the radiation balance, which leads to an estimation of the available net radiation. Secondly, the net radiation has to be divided among the sensible heat flows to the soil and the atmosphere, and the latent heat flow to the atmosphere, to produce the full heat balance. The ratio of the sensible heat flow to the atmosphere to the latent heat flow is known as the Bowen ratio, β, and can be written as:

$$\beta = \frac{\text{sensible heat loss to atmosphere } (S)}{\text{latent heat loss to atmosphere } (LE_t)}$$

In the absence of atmosphere advection, β can vary between $+\alpha$ for a dry surface with no evaporation to zero for an evaporating wet surface with no sensible heat loss. If there is atmospheric heat advection, β may become negative, indicating a flow of heat from the atmosphere to the surface.　　　　　　　　　　　JGL

Reading
Budyko, M.I. 1974: *Climate and life.* New York: Academic Press.
Houghton, J.T. 1977: *The physics of atmospheres.* Cambridge: Cambridge University Press.
Lockwood, J.G. 1979: *Causes of climate.* London: Edward Arnold.
Miller, D.H. 1981: *Energy at the surface of the earth.* International geophysical series 27. New York and London: Academic Press.
Sellers, W.D. 1965: *Physical climatology.* Chicago: University of Chicago Press.

heat island See URBAN METEOROLOGY.

heathland An area of evergreen sclerophyllous shrubland where heath families are present though not necessarily dominant (e.g. Diapersiaceae, Empetraceae, Epacridaceae, Ericaceae, Grubbiaceae, Prionotaceae. Vacciniaceae, etc.). Heathlands develop on areas where soil is low in nutrient status. The stand is generally less than 2 m high. The heathland flora evolved early in the Mesozoic in the southern part of Gondwanaland and is now widespread in many parts of the world, including the *fynbos* of South Africa.　　　　　　ASG

Reading
Sprecht, R.L. 1979: *Heathlands and related shrublands.* Amsterdam: Elsevier Scientific.

helical flow Instability in fluid motion often produces cross-currents superimposed on the dominant flow direction which result in a continuous spiral motion known as helical or helicoidal flow. The SECONDARY FLOWS form a roller vortex with its longitudinal axis parallel to the main flow. In desert airflow these parallel vortices cause lateral transport of sand to zones between the circulation cells, and longitudinal dunes are created (Cooke, Warren and Goudie 1993, p. 382). In rivers helical flow is associated with meander formation and maintenance (Leliavsky 1955, pp. 122–9), and arises because water is superelevated on the outside of a bend. This results in a transverse water slope that generates a current at the bed directed to the inside of the bend. Erosion of the bank occurs at the outside of the bend where the current plunges, and deposition occurs on the inner, point-bar bank where it rises. Helical flow created in one bend may be transmitted downstream to encourage further bend formation.　　　　　KSR

References
Cooke, R.U., Warren, A. and Goudie A. 1993: *Desert geomorphology.* UCL Press.
Leliavsky, S. 1955: *An introduction to fluvial hydraulics.* London: Constable.

helictite A small calcium carbonate SPELEOTHEM which grows in curved or spiralling forms in limestone caves. It grows by the slow accretion of calcium carbonate crystals from the top of the structure, fed by an internal canal. When the crystal-growth forces exceed that of the normal hydraulic force of the flowing water, the crystals can grow away from the vertical. Helping this erratic crystal growth, there may often be clay particle blockages in the feeder canal or even draughts within the cave passage. Helictites are quite common in limestone caves of all environments and can often produce intricate tangled structures. Famous ones can be found in the Kango Caves of South Africa.   PAB

helm wind A strong, cold north-easterly wind that occasionally blows down the western slope of the Cross Fell range into the Vale of Eden in north-west England. It occurs when the horizontal component of airflow is virtually perpendicular to the hills, which restricts directions to the north-east, and when a stable layer of air lies about 600 m above the summit of the range. These conditions accord with the requirements for mountain and LEE WAVES in the atmosphere. The resultant airflow resembles that of water flowing over a weir. The classic study of the helm wind is by Manley (1945).   BWA

Reading and Reference
†Atkinson, B.W. 1981: *Meso-scale atmospheric circulations.* London and New York: Academic Press. Esp. pp. 25–108.
Manley, G. 1945: The helm wind of Cross Fell 1937–1938. *Quarterly journal of the Royal Meteorological Society.* 71, pp. 197–215.

hemera A zone or any other period of geological time as determined from an assemblage of fossils.

herbivore A plant-eating organism; an animal which feeds directly on photosynthetic species. Other than the plant itself, it is the organism which lies at the base of the food chain. Most herbivores are very small, e.g. the aphid species which live off cultivated plants; the largest are the grazing and browsing mammals of Africa (such as the giraffe or elephant). The herbivore group also includes man's domesticated animals. On land the smaller the size of the herbivore, the more likely it is to be linked with one plant species alone. Overall, in terrestrial environments, the ecological effect of herbivores under normal conditions is greatest within grasslands, where up to 35 per cent of production may be consumed; within forests, the equivalent figure is customarily 4–7 per cent (Hughes 1971; Whittaker 1975). Overstocking may result in the unbalanced and excessive removal of vegetative resources in all land habitats. In aquatic systems herbivores are relatively unimportant, most lighter plants being eaten by decomposers, moulds and bacteria.   DW

References
Hughes, M.K. 1971: Tree biocontent, net production and litter falls in a deciduous woodland. *Oikos* 22, pp. 62–73.
Whittaker, R.H. 1975: *Communities and ecosystems.* 2nd edn. New York: Macmillan.

heterosphere The outer portion of the earth's atmosphere beyond the hemosphere.

heterotrophs Organisms which depend upon organic foods in order to obtain energy. Most bacteria, fungi and animals fall into this category.

hiatus A gap in the stratigraphic record or in geological time which is not represented by any sediments.

high energy window The suggestion, first introduced by Neumann (1972), that in the mid-Holocene on tropical coasts there was a period when wave energy was higher than at present. This occurred during the phase when the present sea level was being first approached by the Flandrian (Holocene) transgression and prior to the protective development of coral reefs. The 'window' may have operated on a more local scale on individual reefs with waves breaking not on margins of an extensive reef flat as at the present time, but more extensively over a shallowly submerged reef top prior to the development of the reef flat (Hopley 1984).   ASG

References
Hopley, D. 1984: The Holocene 'high energy window' on the central Great Barrier Reef. In B.G. Thom, ed., *Coastal geomorphology in Australia.* Sydney: Academic Press. Pp. 135–50.
Neumann, A.C. 1972: Quaternary sea level history of Bermuda and the Bahamas. *American Quaternary Association Second National Conference Abstracts*, pp. 41–4.

hillslope flow processes The mechanisms and routes of flow followed by precipitation down hillsides to a stream channel.

Some precipitation is intercepted before reaching the ground, and if not evaporated will reach it directly or by stemflow, with some delay and with some spatial redistribution, particularly concentrating round tree-trunks, for example. If rainfall intensity exceeds the INFILTRATION capacity, or if the soil is saturated, surface depressions will fill and overflow to give OVERLAND FLOW which normally reaches the stream channels but may infiltrate into the soil downslope, especially if a rainstorm is very brief or very local, or if the soil thickens appreciably.

Infiltration occurs most readily into soil crumbs or peds, but also overflows down soil cracks and other structural voids, thereby bypassing the surface soil layers and usually infiltrating into peds farther down but in some cases connecting directly to the groundwater table.

Within the soil flow is mainly vertical in the unsaturated zone, flowing under the predominant influence of hydraulic potential gradients. Soil porosity normally decreases with depth, though there are many exceptions from this general rule. If downward flow is impeded a saturated zone may develop, either in contact with the groundwater body as a whole or perched above it. In such a layer hydraulic potential gradients are negligible and saturated subsurface flow or THROUGHFLOW predominates. As throughflow accumulates downslope the level of saturation in the soil rises towards the surface, or the SATURATION DEFICIT falls, and flow may re-emerge on the surface as return flow. The area of zero saturation deficit defines the SOURCE AREA for a rapid streamflow response to subsequent rain by overland flow production as described above.   MJK

Reading
Kirkby, M.J. ed. 1976: *Hillslope hydrology.* Chichester: Wiley.

Hjulström curve In 1935 the Swedish geomorphologist Filip Hjulström presented an empirical curve defining the threshold flow velocities required to initiate motion of grains of different sizes on a stream bed. By displaying this curve in conjunction with curves of the settling velocity or depositional velocity, the conditions required for transport, traction and deposition of sediment can be defined (see diagram). The threshold flow velocity is at a minimum for well-graded sand particles in the 0.2–0.5 mm size range, and higher velocities are necessary to entrain both finer and coarser sediments. Finer sediment includes cohesive silt and clay show whose entrainment is inhibited by particle aggregation and by submergence in the laminar sublayer (see BOUNDARY LAYER). Coarser sediments (gravel, pebbles and cobbles) require higher threshold velocities simply because they are heavier. However, there is no unique mean velocity at which grains of given size first move, because the bed velocity is a more critical control of entrainment by flowing water. For a particular threshold

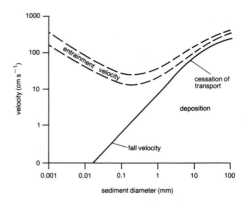

bed velocity the mean velocity is greater in deeper water. Also, the particle size distribution affects the threshold velocity, since coarse grains in a poorly sorted sediment may shelter finer grains and prevent their entrainment. A similar shape of curve defines the fluid threshold in air (Bagnold 1941), but the different fluid densities result in different entrainment velocities in the two media.   KSR

References
Bagnold, R.A. 1941: *The physics of blown sand and desert dunes.* London: Chapman & Hall.
Hjulström, F. 1935: Studies of the morphological activities of rivers as illustrated by the River Fyris. *Bulletin of the Geological Institute, University of Uppsala* 25, pp. 221–527.

hoar frost The deposition of ice crystals usually by the direct sublimation of water molecules, but also by the freezing of dew, on a solid surface exposed to the atmosphere. The surface, if cooled at night by loss of radiation, in turn cools the water vapour immediately above it to its frost point and sublimation begins. The crystals are usually white.   JET

hodograph A scheme for representing windspeeds and directions at different heights at a given time by plotting them on a polar coordinate diagram. Successive winds are plotted as vectors so that they all 'blow' from the respective directions towards the central point, which represents the station at which the observations were made. The hodograph is then constructed by drawing a series of vectors which join the successive end points of the plotted winds.

For example, the three winds plotted in the diagram for 700, 500 and 400 mb (broken lines) are shown with the hodograph construction (solid lines) which represents the shear vectors for the 700–

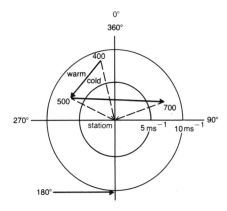

0°
360°
400
warm
cold
500
700
270° — 90°
station 5 ms⁻¹ 10 ms⁻¹
180°

Hodograph: The blank diagram comprises compass directions and a series of concentric circles, the radii of which represent windspeed. The centre of all the circles represents the observing station. Three winds are shown (dashed lines) for the levels 400, 500 and 700 mb. The wind at 700 mb, for example, blows from a direction of about 70° with a speed of about 8 ms⁻¹. The solid arrows show the size and direction of the vector difference of the winds at the top and bottom of any layer. This vector difference is known as the thermal wind.

500 mb and 500–400 mb layers. Assuming the observed winds to be geostrophic, the shear vectors can be considered to be THERMAL WINDS and therefore to provide information about the nature of temperature ADVECTION.

Between 700 and 500 mb the wind backs with height and within this layer the air is blowing from a relatively cold region (cold air lies to the left of thermal wind). So winds that back with height signify cold advection and those that veer (500–400 mb here) mean warm advection. The thermal winds can be measured directly from the scale on the polar diagram.   RR

Reading
Atkinson, B.W. 1968: *The weather business.* London: Aldus Books.

hogback A long ridge of rock dipping steeply on both sides that is the exposure of a stratum or rock which has been tilted until the originally horizontal beds are almost vertical.

Holocene The second series of the Quaternary period, following the Pleistocene. Often called the postglacial, it has extended from *c*. 11,000 years ago until the present and has been characterized by interglacial conditions. It has been marked by various climatic fluctuations (see ALTITHERMAL; BLYTT–SERNANDER MODEL;

HYPSITHERMAL; LITTLE CLIMATIC OPTIMUM; NEOGLACIAL etc.) and by rapid sea-level rise (see FLANDRIAN TRANSGRESSION). The term was introduced during the 1860s, and means 'entirely modern'.   ASG

Reading
Roberts, N. 1989: *The Holocene: an environmental history.* Oxford: Blackwell.

holokarst A term coined by Cvijic to describe the coastal Dinaric karst belt of the former Yugoslavia where the full suite of karst landforms is found and limestone solution processes dominate the landscape. Cvijic contrasted this landscape with the less well-developed karst landscape further inland which he termed merokarst. The term holokarst has been generally used to describe any limestone landscape with a fully developed range of karst features.
                                      HAV

Reading
Cvijic, J. 1893: Das Karstphänomen. *Geographische Abhandlung* 5, pp. 217–329.
Cvijic, J. 1925: Types morphologiques des terrains calcaires. Le holokarst. *Comptes Rendus de l'Académie des Sciences* 180, pp. 592–4.

homoclines Locations which experience the same types of climatic regimes.

homoiothermy (homeothermy) The maintenance by an organism of a relatively constant body temperature, independent of, and usually somewhat above, that of the organism's immediate surroundings. Mammals and birds are homoiothermic or 'warmblooded'.   PHA

hoodoo An unusually shaped pillar or outcrop of rock produced by erosion (see diagram on p. 256).

horizon, soil A stratum or level within the soil zone which is chemically or physically distinct from the levels above and below. A level within the soil zone which is affected by specific soil-forming processes.

horn, glacial A pyramidal peak with three or more distinct faces steepened by glacial undercutting. The classic situation occurs when cirque glaciers encroach on a mountain from all sides.

horse latitudes The latitude belts over the oceans at approximately 30–35° north and south where winds are predominantly

Rock pinnacles (hoodoos) in the Wasatch Formation at Bryce Canyon, USA, resulting from the severe erosion of very gently dipping sedimentary rocks.

calm or very light and the weather often hot and dry. These latitudes represent the normal axes of the subtropical high pressure belts. The origin of the name is uncertain but it is suggested that the crews of sailing ships carrying horses to the West Indies had occasionally when becalmed either to jettison their live cargo as fodder ran out or eat it. The term 'subtropical high pressure areas' is now frequently used in preference to 'horse latitude'.   AHP

Reading
Boucher, K. 1975: *Global climate*. London: English University Press.

horst An upstanding block of the earth's crust that is bounded by faults and has been uplifted by tectonic processes. The down-faulted areas which bound horsts are called graben.

Horton overland flow model Forecasts overland flow per unit area as rainfall intensity minus the INFILTRATION capacity, where this difference is positive. In Horton's original version of the model (1933),

infiltration capacity was calculated as an exponential decay of the form:

$$f = f_c + (f_0 - f_c) \exp(-ct)$$

where f is the infiltration capacity at time t, and f_c, f_0 and c are constants, which depend on the soil and its moisture distribution at the start of infiltration. At the start of a rainstorm all rainfall is considered to infiltrate until the infiltration capacity falls to the current rainfall intensity. Subsequently the infiltration curve determines how much water enters the soil, and any excess is diverted as OVERLAND FLOW. Total overland flow discharge is obtained by routing this flow downslope, combined with the overland flow produced at each site. This overland flow is then considered to supply the sharply rising and falling peak of stream hydrograph response to rainstorms. The model was first developed for agricultural soils in an area of very intense rainfalls, and it works very well under these conditions. At the simplest level the model appears to imply a uniform rate of overland flow production so that flow discharge increases linearly downslope. Allowance for wetter soils downslope, however, forecasts increasing production downslope, so that Horton's model is able to perform well as a forecasting tool, even in humid forested environments where its physical basis is undermined by the rarity of observable overland flow, and where PARTIAL AREA MODELS are now preferred by many researchers.

Criticisms of the Horton model have been made at several levels. Attempts to improve the model may be made by improving the infiltration equation used, or by using an equation which takes into account the reduced rate at which infiltration capacity falls if rainfall intensity is at less than the maximum capacity. More fundamental criticisms note the existence of storm hydrograph peaks in the absence of observable overland flow, particularly in forested catchments, and have to take greater account of THROUGHFLOW within the soil as a direct source of stormflow and to forecast saturated SOURCE AREAS in which overland flow production is concentrated.   MJK

Reading and Reference
Horton, R.E. 1933: The role of infiltration in the hydrological cycle. *Transactions of the American Geophysical Union* 14, pp. 446–60.

†Kirkby, M.J. ed. 1978: *Hillslope hydrology. Chichester:* Wiley.

Horton's laws

Two laws of drainage composition were suggested by R.E. Horton (1945): the law of stream *numbers* whereby an inverse geometric series related number of streams of a particular ORDER *and order*, and the law of stream *lengths* which was based on a geometric series between mean lengths of streams of each order *and order*. These two laws were subsequently complemented by three others giving an inverse geometric series relation between mean slope of streams of a particular order *and order*, a geometric series relating mean basin area of streams of a particular order *to order*, and the law of contributing areas which was the logarithmic relation of drainage areas of each order and the total stream lengths which they contained and supported. Although modifications to these five laws were made with the advent of different systems of stream ordering and some useful comparisons were made between areas using the five relationships, it was realized by 1970 that the 'laws' were at least in part a consequence of the definition of stream ordering and therefore largely statistical relationships. Recent research has demonstrated that other parameters defined by MORPHOMETRY are more useful to relate to indices of drainage basin process and that the structure of drainage networks can be illuminated by topology of the NETWORK.   KJG

Reading and Reference
†Gardiner, V. 1976: *Drainage basin morphometry.* British Geomorphological Research Group technical bulletin 14. Norwich: Geo Abstracts.
Horton, R.E. 1945: Erosional development of streams and their drainage basins; hydrophysical approach to quantitative morphology. *Bulletin of the Geological Society of America* 56, pp. 275–370.

hot spot

A small area of the earth's crust where an unusually high heat flow is associated with volcanic activity. Of approximately 125 hot spots thought to have been active over the past 10 million years most are located well away from plate boundaries (see PLATE TECTONICS). The major theory of hot spot formation involves the effects of a PLUME of hot MANTLE rising to the surface. Some researchers consider that hot spots are sufficiently stationary with respect to the mantle to provide a reference frame for determining plate motions and CONTINENTAL DRIFT.   MAS

Reading
Burke, K.C. and Wilson, J.T. 1976: Hot spots and the earth's surface. In J.T. Wilson ed., *Continents adrift and continents aground.* San Francisco: W.H. Freeman, Pp. 58–69.
Duncan, R.A. 1981: Hot spots in the southern oceans – an absolute frame of reference for motion of the Gondwana continents. *Tectonophysics* 74, pp. 29–42.
Morgan, W.J. 1972: *Plate motions and deep mantle convection.* Geological Society of America memoir 132, pp. 7–22.
Vink, G.E., Morgan, W.J. and Vogt, P.R. 1985: The earth's hot spots. *Scientific American* 252, pp. 32–9.

hot spring

An emission of hot, usually geothermally heated, water at the landsurface.

hum

A residual hill in limestone country formed through surface lowering of the surrounding landsurface.

humate

A collective term for the dark-brown to black gel-like humic substances formed as a result of the decomposition of organic matter in soils and sediments. It may be translocated down the profile by vadose water. Subsurface accumulations of this material occur most commonly in podzols. If the humate dries out it may harden sufficiently to create a humicrete.
                ASG

Reading
Pye, K. 1982: Characteristics and significance of some humate-cemented sands (humicretes) at Cape Flattery, Queensland, Australia. *Geological magazine* 119, pp. 229–42.

humidity

A term which relates to the water vapour content of the atmosphere and is expressed in a variety of ways.

Thus the relative humidity of an air sample is expressed as a percentage which is found by relating the observed VAPOUR PRESSURE at a given temperature to the saturation value of vapour pressure at that temperature. It is evaluated by using tables or a special slide rule in conjunction with dry-bulb and wet-bulb temperature readings. Relative humidity commonly displays a daily cycle which has a phase opposite to that of temperature so that the highest values are observed near dawn and the lowest during the afternoon.

Absolute humidity indicates the actual amount of water vapour present in a sample of air. For example, the humidity mixing ratio of an air parcel is defined to be the ratio of the mass of water vapour to the mass of dry air with which the water vapour is associated. At the surface it may range between near zero in frigid polar areas to

25 g kg^{-1} in very humid tropical air. Absolute humidities can be deduced from wet-bulb temperature readings. RR

humus Partially decomposed organic matter which accumulates on and within the soil zone.

hunting and gathering An economic mode of subsistence characterized by food collecting rather than food production. It is generally applied to societies who have not adopted agriculture but live by hunting mammals, collecting fruits and seeds, and fishing in both fresh and marine waters. Before the coming of agriculture in late Pleistocene societies, all human societies were sustained in this way but now only a few groups in isolated places live like this and they have mostly been affected by industrial societies elsewhere, e.g. by the provision of iron tools. In the present context, interest in these societies is perhaps two-fold. The human animal has been for most of its evolutionary existence (i.e. at least 90 per cent) a hunter-gatherer, so we may wonder whether there are aspects of our physiology and psychology which are adapted to that role rather than that of, for example, urban-industrial living. Naturally, others think that we are sufficiently plastic, in a cultural sense, for this history to be unimportant. There has been for years the idea that hunter-gatherers were passive members of their ecosystems and did not manage the resources of them. This has been shown to be untrue, with examples of the use of fire to manage plant resources in Australia, animal populations in pre-contact North America and mesolithic England, and to alter watercourses in Australia. Even so, it appears that many hunter-gatherer groups did not consciously manage their environments at all. Some, however, were careful to husband their resources; others apparently did not do so, with variable consequences depending upn the resilience of the local environment.
IGS

Reading
Lee, R.B. and DeVore, I. eds. *Man the hunter*. Chicago: Aldine Press.

hurricane See TROPICAL CYCLONE.

hydration The process whereby an anhydrous mineral, one not containing water molecules within its crystal structure, takes up water to form a crystallographically distinct mineral. The partial decomposition of rocks by water.

hydraulic autogeometry Over long stretches of river the channel dimensions or the width, depth and velocity of flow within the channel at a constant discharge frequency change systematically in trends described by the power-function relations of the downstream hydraulic geometry. More locally within a reach, however, these characteristics of channel and flow vary in relation to the POOL AND RIFFLE or MEANDERING patterns. At this scale the properties of successive cross-sections spaced about one channel width apart are spatially correlated with those of the closest sections upstream. There is thus autocorrelation between nearby locations, and this pattern of variation can be referred to as hydraulic autogeometry, and can be described using appropriate STOCHASTIC MODELS. KSR

hydraulic conductivity See CONDUCTIVITY, HYDRAULIC.

hydraulic diffusivity (D) The ratio of the hydraulic CONDUCTIVITY K to the specific storage S_s or transmissivity T to storativity S, as follows:

$$D = K/S_s = T/S$$

The square root of hydraulic diffusivity is proportional to the velocity of HYDRAULIC HEAD transmission in the aquifer.

The specific storage is the volume of water that a unit volume of aquifer releases from storage per unit decline in hydraulic head, whereas the dimensionless parameter storativity is the volume of water it releases per unit surface area per unit decline in hydraulic head above the surface. Transmissivity (or transmissibility) is the product of the hydraulic conductivity and aquifer thickness.

The hydraulic diffusivity is therefore an aquifer parameter that combines the transmission characteristics and the storage properties. PWW

Reading
Freeze, R.A. and Cherry, J.A. 1979: *Groundwater*. Englewood Cliffs, NJ: Prentice-Hall.

Ward, R.C. and Robinson, M. 1990: *Principles of hydrology*. 3rd edn. Maidenhead: McGraw-Hill.

hydraulic force The water flowing in a channel reach exerts a hydraulic force on the bed and may move grains of unconsolidated bed material. This is usually known as the TRACTIVE FORCE, and is the effective component of the weight of water acting parallel to the bed in the direction of flow. It is usually expressed in relation to the bed area over which it acts, and is then called the unit tractive force, or the bed shear stress (force per unit area). The roughness of the channel bed imposes a frictional drag on the flow (see BOUNDARY LAYER), and in a reach where no flow acceleration occurs, this is equal and opposite to the hydraulic force of the flow on the bed. KSR

hydraulic geometry The hydraulic geometry of alluvial channels was defined by Leopold and Maddock (1953) as: (1) the at-a-station adjustment of flow character- istics such as discharge (w), depth (d) and velocity (v) within a section as discharge (Q) varies; and (2) the general downstream adjustment of these flow properties as discharge increases at a constant flow frequency. Both at-a-station and down- stream, these adjustments are commonly described by the power-functions:

$$w = aQ^b; d = cQ^f; v = kQ^m$$

and since $wdv = Q$, it follows that $ack = 1$ and $b + f + m = 1$. The overall flow geo- metry of a river system is represented by relationships for an upstream and a down- stream section at both a high and a low frequency flow (see diagram). However, the downstream trends are only approximate because factors other than discharge, such as bed and bank sediment properties, control the changing shape of the channel within which the flow then adjusts at discharges which are not competent to change the section.

At-a-station flow geometry varies between rectangular sections with steep cohesive banks in humid areas, and parabolic sections in sandy sediments in semi-arid regions. In the former, $b = 0.05$, $f = 0.45$ and $m = 0.50$, and in the latter $b = f = m = 0.33$. Downstream the adjust- ment of width and depth accommodates more of the increase in discharge, and $b = 0.50$, $f = 0.40$ and $m = 0.1$ (Richards 1982, pp. 148–59). Downstream exponents vary according to the trends in ROUGHNESS and long-profile slope, but their general similarity reflects the tendency of all rivers to create forms which allow equal dissipa- tion of energy and minimize total work.

At the local reach scale hydraulic autogeometry describes a further pattern in flow and channel form. KSR

Reading and References
Ferguson, R.I. 1986: Hydraulics and hydraulic geometry. *Progress in physical geography* 10, pp. 1–31.
Leopold, L.B. and Maddock, T. 1953: The hydraulic geometry of stream channels and some physiographic implications. *United States Geological Survey professional paper* 252.

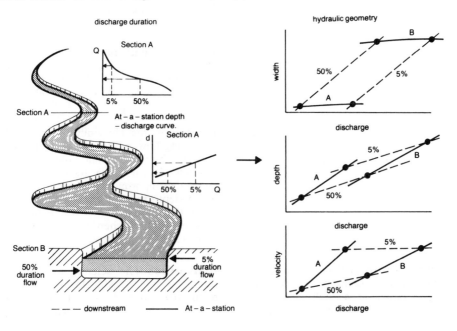

Richards, K.S. 1982: *Rivers: form and process in alluvial channels*. London and New York: Methuen.

hydraulic gradient If piezometric tubes are inserted into flowing water (soil throughflow, groundwater flow, pipe flow or open channel flow) water will rise to the hydraulic grade line. The slope of this line is the hydraulic gradient. So long as conditions remain hydrostatic, which requires that no extreme flow curvature occurs and that slopes are less than 1 in 10, the hydraulic gradient is the water table or water surface slope, or the piezometric surface slope in the case of confined groundwater flow or pipe flow.   KSR

hydraulic head (*h*) is the sum of the elevation head (z) and the pressure head (*hp*). Each of these parameters has the dimension of length and is measured in metres above a convenient datum, usually taken as sea level. The change of hydraulic head with distance (*l*) is known as the hydraulic gradient (*dh/dl*) and is an important term in the equation defining DARCY'S LAW.   PWW

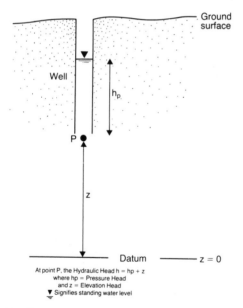

At point P, the Hydraulic Head h = hp + z
where hp = Pressure Head
and z = Elevation Head
▼ Signifies standing water level

Reading
Freeze, R.A. and Cherry, J.A. 1979: *Groundwater*. Englewood Cliffs, NJ: Prentice-Hall.

hydraulic jump A given discharge in an open channel of constant width can be conveyed either at a high velocity and shallow depth or at a low velocity in deep water; a sudden change between these states along a channel is a hydraulic jump. The FROUDE NUMBER F_r passes from $F_r > 1$ (supercritical flow) to $F_r < 1$ (subcritical flow) during this transition. A hydraulic jump may be caused naturally by a sudden reduction of bed slope, or a large submerged obstacle (a boulder or bank slump), and is deliberately generated by hydraulic engineers to dissipate energy over weirs and dams, to ensure that upstream flow is critical in measuring installations such as flumes, and below sluices to ensure that they are not drowned (Sellin 1969, pp. 52–68).   KSR

Reference
Sellin, R.J.H. 1969: *Flow in channels*. London: Macmillan.

hydraulic potential or fluid potential at any point in a porous substance is the product of the HYDRAULIC HEAD and the acceleration due to gravity. It is the mechanical energy of water (in this case) per unit mass. Since gravitational acceleration is almost constant near the earth's surface, hydraulic potential is very closely correlated with hydraulic head (Freeze and Cherry 1979). In accordance with DARCY'S LAW, subsurface water flows from points of high hydraulic potential towards points of low hydraulic potential. This occurs down the hydraulic gradient perpendicular to the EQUIPOTENTIALS.   PWW

Reference
Freeze, R.A. and Cherry, J.A. 1979: *Groundwater*. Englewood Cliffs, NJ: Prentice-Hall.

hydraulic radius The hydraulic radius *R*, or the hydraulic mean depth, is the ratio of wetted perimeter (*P*) to the cross-sectional area of flow in a channel (*A*):

$$R = A/P.$$

It measures the efficiency of a section in conveying flow, as it is the area of flow per unit length of water–solid contact. In wide, shallow channels the mean depth is a convenient approximation for *R*; when the width–depth ratio of a rectangular channel is 50, this approximation results in a 4 per cent error, but when it is 10 the error increases to 17 per cent.   KSR

hydraulics The science in which the principles of FLUID MECHANICS are applied to the behaviour of water flowing in pipes or open channels over rigid or loose boundaries. Basic hydraulic theories relating to water flow deal with the development and refinement of FLOW EQUATIONS which define the flow velocity in terms of depth, slope and boundary roughness, with assessment of the forces, momentum, and energy losses within the flow, and with analysis of velocity variations in two- and three-dimensional BOUNDARY LAYERS. Some of the most difficult problems in hydraulics occur in the area where it impinges most obviously on physical geography – namely, in the analysis of flow over loose boundaries, when the flow transports sediment at the bed, creates bedforms, and therefore affects its own flow resistance in a complex feedback process.

Applied hydraulics is concerned with the economic development of water supply, and therefore deals with the storage, conveyance and control of water. Storage problems include assessment of the ability of the physical supply within an area to cope with estimated and forecast demand, as well as the difficulties of maintenance of supply – for example, because of over-exploitation of groundwater, or reservoir sedimentation and loss of storage capacity. Hydraulic design is concerned with dams, spillways, and conveyance and diversion systems such as canals and irrigation ditches. There are close links between REGIME THEORY, which considers the design of stable sediment-bearing artificial channels, and the geomorphological study of natural river morphology. The design of control and measurement systems is a further important aspect of hydraulic analysis, in which spatially varied flow including free overfalls and HYDRAULIC JUMPS is deliberately manipulated to provide control conditions for the purpose of flow measurement in weirs and flumes.   KSR

Reading
Chow, V.T. 1959: *Open-channel hydraulics.* Tokyo: McGraw-Hill.
Sellin, R.H.J. 1969: *Flow in open channels.* London: Macmillan.

hydrodynamic levelling The transfer of survey datum levels by comparing mean sea level at two sites, and adjusting them to allow for gradients on the sea surface due to currents, water density, winds and atmospheric pressures.   DTP

hydrofracturing The rupture of rock caused when water enters cracks in rock

which may be little wider than the combined diameters of a few molecules of water. Under cold conditions if ice seals off the end of such a crack, liquid water may be forced towards the tip of the crack and so extend it.   ASG

hydrogeological map A cartographic representation of subterranean water resources information with associated surface water and geological data. Hydrogeological maps may be prepared from continental to local scales, depending on their purpose. Continental and regional scale maps commonly depict the major aquifers and identify water bearing lithologies and GROUNDWATER basins. Major SPRINGS and their discharges are also frequently depicted. Water balance information is sometimes contoured. Larger-scale local maps, in addition, usually show WATER TABLE (or piezometric) contours (or EQUIPOTENTIALS in the case of ARTESIAN aquifers) and may provide information on the direction of groundwater flow, aquifer thickness, borehole locations, recharge areas, and saltwater intrusions. Water quality data are also sometimes presented.   PWW

hydrogeomorphology The study of water and its effects on physical environment. It includes the domains of saltwater as coastal hydrogeomorphology and of freshwater as fluvial hydrogeomorphology. The term was introduced to embrace geomorphological studies of form and process relationships which are intermediate between fluvial geomorphology and coastal geomorphology (which tended to concentrate upon landforms and landform chronology) and hydrology and coastal hydraulics (which were concerned with the mechanisms of fluvial and coastal processes).   KJG

Reading
Gregory, K.J. 1979: Hydrogeormorphology: how applied should we become? *Progress in physical geography* 3, pp. 83–100.
Scheidegger, A.E. 1973: Hydrogeomorphology. *Journal of hydrology* 20, pp. 193–215.

hydrographs Show discharge plotted against time for a point in a drainage basin. The discharge is usually monitored at a site on the stream channel network (stream hydrograph) but it may be measured at a section on a hillslope (hillslope hydrograph).

Stream hydrographs represent the pattern of total discharge through time and are made up of flow from different stores within the drainage basin (see HYDROLOGICAL CYCLE). During dry periods discharge gradually recedes and the hydrograph at a particular site may often be represented by a standard DEPLETION CURVE or base-flow recession curve. During rainfall events, the stream discharge responds in the form of a storm hydrograph (see diagram). The shape of the storm hydrograph is affected by both the spatial and temporal pattern of the rainfall, ANTECEDENT MOISTURE conditions, and by the drainage basin characteristics, including catchment geometry, rock and soil types, vegetation and land use pattern and drainage network structure.

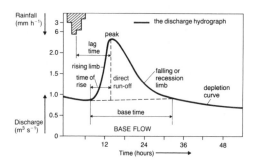

Storm hydrograph.

Hydrograph analysis often begins with a separation of the hydrograph into components. Traditional hydrograph separation subdivides the hydrograph into surface run-off or direct run-off, base flow and sometimes interflow. The subdivision is essentially empirical with straight lines joining either the beginning of the storm hydrograph rise or a point beneath the hydrograph peak to one or more points on the falling limb of the hydrograph. These points on the falling limb may be identified from the curvature of the falling limb, from the point of divergence of the falling limb from the depletion curve or by identifying breaks of slope on semi-logarithmic plots of the hydrograph (Barnes 1939; Gregory and Walling 1976). Hibbert and Cunningham (1967) avoided the generic terms for describing the components of the separated storm hydrograph by suggesting a separation into quick flow and delayed flow by a straight line from the point of hydrograph rise to the falling limb of the hydrograph with a slope of $0.551s^{-1}$ km^{-2} h^{-1}. Hydrograph separation based upon water quality variations (e.g. Pinder and Jones 1969;

Pilgrim *et al.* 1979) show that straight line separations do not accurately reflect the way in which storm hydrographs are composed of water from different source areas.

Whichever separation technique is used, parameters of the storm hydrograph may be analysed to represent hydrograph size and shape and some widely used parameters include the time of rise, base time, run-off volume, peak discharge, percentage run-off and hydrograph peakedness. The unit hydrograph technique (Sherman 1932; see UNIT RESPONSE GRAPH) is often applied to separated storm hydrographs since it provides a simple means of summarizing and predicting the storm hydrograph at a site.   AMG

Reading and References
Barnes, B.S. 1939: The structure of discharge recession curves. *Transactions of the American Geophysical Union* 20, pp. 721–5.
Gregory, K.J. and Walling, D.E. 1976: *Drainage basin form and process*. London: Edward Arnold.
Hibbert, A.R. and Cunningham, G.B. 1967: Streamflow data processing opportunities and applications. In W.E. Sopper and H.W. Lull eds, *International symposium on forest hydrology*. Oxford and New York: Pergamon.
Linsley, R.K., Kohler, M.A. and Paulhus, J.L.H. 1988: *Hydrology for engineers* 3rd edn. New York: McGraw-Hill.
Pilgrim, D.H., Huff, D.D. and Steele, D.M. 1979: Use of specific conductance and contact time relations for separating flow components in storm runoff. *Water Resources Research* 15, pp. 329–39.
Pinder, G.F. and Jones, J.F. 1969: Determination of the groundwater component of peak discharge from the chemistry of total runoff. *Water resources research* 5, pp. 438–45.
Robson, A.J. and Neal, C. 1990: Hydrograph separation using chemical techniques: an application to catchments in Mid-Wales. *Journal of hydrology* 116, pp. 345–63.
Sherman, L.K. 1932: Stream flow from rainfall by the unit-graph method. *Engineering news record* 108, pp. 501–5.

hydroisostasy The reaction of the earth's crust to the application and removal of a mass of water, as when eustatic sea-level changes have affected the depth of water over the continental shelves causing the crust to be depressed at times of high sea level and to be elevated at times of low sea level. The same process can operate in lake basins like Lake Bonneville in response to climatically induced fluctuations in lake level, and may create warped shorelines.

ASG

Reading
Bloom, A.L. 1967: Pleistocene shorelines: a new test of isostasy. *Bulletin of the Geological Society of America* 78, pp. 1477–94.

hydrolaccolith A pingo, a dome of ice just beneath the surface in areas of permafrost. The product of the freezing of artesian spring water as it nears the surface.

hydrological cycle The hydrological cycle describes the continuous movement of all forms of water (vapour, liquid and solid) on, in and above the earth's surface and it is the central concept of HYDROLOGY.

The hydrological cycle includes the condensation and freezing of water vapour in the atmosphere to form liquid or solid precipitation, the movement of water from precipitation through one or more of a range of conceptual stores, including SURFACE STORAGE, soil moisture storage, groundwater storage, stream channels and the oceans until at some stage the water returns to the atmosphere as water vapour through the processes of evaporation and transpiration. Figure 1 schematically presents the drainage basin hydrological cycle or the possible routes that water may follow within a drainage basin to form part of stream discharge at the basin outlet or to be lost to the basin through evapotranspiration. The time taken for water to pass to and from a store, the capacity of the store and the volume of water contained in each store at a particular time will strongly influence the discharge response of the drainage basin to a rainfall event. Figure 2 summarizes the

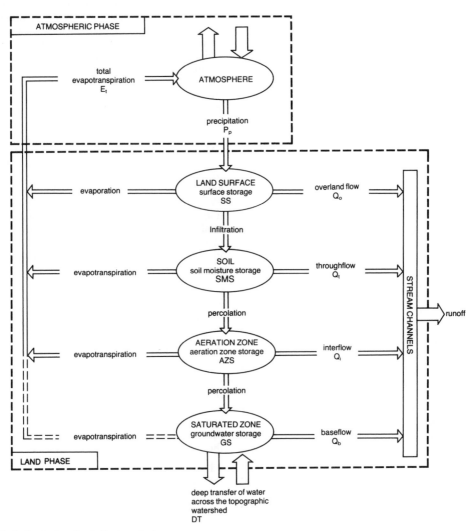

Hydrological cycle: 1. Drainage basin.

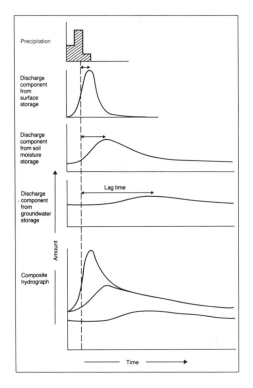

Precipitation

Discharge
component
from
surface
storage

Discharge
component
from soil
moisture
storage

Discharge
component
from
groundwater
storage

Lag time

Amount

Composite
hydrograph

Time

Hydrological cycle: 2. Schematic representation of the way in which discharge from different stores might combine to produce a composite storm hydrograph in a small drainage basin.

way in which the different contributions to run-off might be generated in a drainage basin in response to a storm.

Figure 2 illustrates, first, that more inaccessible stores will respond more slowly to rainfall, if they generate a measurable response. For example, water must infiltrate the soil and percolate to the water table before groundwater storage can be increased and so produce increased base flow to the stream. All these processes may take a long time and so introduce a long lag before the peak of base flow passes the gauging station on the stream at the catchment outlet. Kirkby (1975), for example, suggested orders of magnitude for drainage times in hours to surface, soil and groundwater storage as 0, 5 and 500 h respectively, and drainage from these stores downslope to channel storage as 0.5, 30 and 200 h respectively. Once in the channel network, travel time to the gauging station may also be substantial in large drainage basins. Secondly, larger stores, containing

more water, will produce a more attenuated response to a particular input of water. Thus, in figure 2, outflow from surface storage is less attenuated than that from soil moisture storage which in turn is less attenuated than that from groundwater storage, and this reflects the condition of the stores in the particular drainage basin considered in this example. Finally, if a store is full of water, it can only absorb as much as can drain from the store in a fixed time period and, as a result, excess water may build up in the preceding store in the cycle. An example of this which has important effects on the basin discharge characteristics is the effect of soil saturation on overland flow. ANTECEDENT conditions will control the extent to which areas of the soil in a drainage basin are saturated and so the dynamic area of the soil in a drainage basin will affect the area of the basin over which saturation overland flow will occur during a storm. This example also illustrates the fact that each of the conceptual lumped stores of figure 1 may be represented by a large number of distributed stores in a real drainage basin.

The hydrological cycle provides a conceptual framework for hydrological studies and it may be extended to include the movement of solutes and sediments, heat and biota as they are transported and stored within the water cycle. The cycle presents a logical structure for the measurement, analysis and modelling of hydrological processes and for the identification of the impact of man on the quality, quantity, routing and storage of water.   AMG

Reading and Reference
Kirkby, M.J. 1975: Hydrograph modelling strategies. In R. Peel, M. Chisholm and P. Haggett eds, *Processes in physical and human geography*. London: Heinemann.
†Kirkby, M.J. ed. 1978: *Hillslope hydrology*. Chichester: Wiley.
Shaw, E.M. 1988: *Hydrology in practice*. 2nd edn. New York: Van Nostrand Reinhold.

hydrological maps Maps of the characteristics of the components of the hydrological cycle. McKay (1976) listed the commonly mapped characteristics of the hydrological elements as follows:

1 Networks of instruments.
2 Means or medians.
3 Departures from the mean.
4 Variability.
5 Total amounts for a specific event or
   duration.

6 Extremes, including the mean extreme and variability.
7 Number of days or months with specified condition.
8 The time of beginning, duration and ending of phenomena.
9 Intensity with a specified frequency of occurrence.
10 Ratios for different durations or frequencies.
11 Combinations of several elements or components.
12 As graphs, variability in time, intensity-duration-frequency, annual regimes and mean directional array.

Precipitation, evapotranspiration and runoff are the most frequently mapped hydrological variables because they are complementary, forming the three components of a simpified WATER BALANCE which, therefore, give a good indication of the water resources of any area.   AMG

Reference
McKay, G.A. 1976: Hydrological mapping. In J.C. Rodda ed., *Facets of hydrology*. Chichester: Wiley.

hydrological year See WATER YEAR.

hydrology The science concerned with the study of the different forms of water as they exist in the natural environment. Its central focus is the circulation and distribution of water as it is expressed by the WATER BALANCE and HYDROLOGICAL CYCLE. Hydrology embraces not only the study of water quantity and movement but also the degree to which these are affected by man's activities, including deliberate management of water resources and the inadvertent effects of man on hydrological processes. It is often subdivided into physical and applied hydrology. Physical hydrology includes the detailed measurement and analysis of information on hydrological processes to improve understanding of the functioning of the hydrological system and also the refinement of statistical and mathematical methods of predicting and modelling these physical processes. Applied hydrology is concerned with the application of the understanding of hydrological processes to their modification and management. 'Water resources and pollution on the one hand and flooding and erosion on the other, are the chief concerns of the hydrologist' (Rodda *et al.* 1976).

The study of hydrology is at least as old as the ancient civilizations of Egypt, because the provision of a reliable water supply is essential to the survival of man. However, the development of plausible theories concerning the circulation of water in the hydrological cycle did not appear until the seventeenth century. These were largely based on observations of rainfall and river flow in the Seine basin by Pierre Perault and Edmé Mariotte and on the ideas of Edmund Halley, who simulated evaporation from the Mediterranean and concluded that this could account for all surface drainage. Since then the science of hydrology has developed rapidly and at the international level several events are of particular significance. In 1919 the International Union of Geodesy and Geophysics (IUGG) was established and in 1922 the International Association of Scientific Hydrology (IASH; renamed in 1973 the International Association of Hydrological Sciences, IAHS) was created as a hydrological section of the IUGG. The period from 1965 to 1974 was named the International Hydrological Decade (IHD) which was suggested by IASH members and was organized under the auspices of UNESCO. The IHD was followed by the International Hydrological Programme (IHP) and both the IHD and IHP had broad objectives which have resulted in enormous international research efforts. The objectives of the IHD and IHP have been summarized by Nemec (1976), and see INTERNATIONAL HYDROLOGICAL DECADE and INTERNATIONAL HYDROLOGICAL PROGRAMME.

International cooperation in hydrology is also encouraged by the World Meteorological Organization (WMO), whose Commission for Hydrology was established in 1959 and focuses particularly on operational aspects of hydrology including hydrological measurements and hydrological forecasting. WMO has cooperated with the IHD and IHP as well as introducing its own Operational Hydrology Programme (Walling 1981).

Hydrology is not only a long-established science which has been given major impetus at the international level, it is also an interdisciplinary subject bringing together specialists from an enormous range of disciplines including biology, chemistry, civil engineering, environmental planning, forestry, geology, geomorphology, mathematics and physics. AMG

Reading and References
†Dunne, T. and Leopold, L.B. 1978: *Water in environmental planning.* San Francisco: W.H. Freeman.
†Gregory, K.J. and Walling, D.E. 1976: *Drainage basin form and process.* London: Edward Arnold.
†Kirkby, M.J. 1978: *Hillslope hydrology.* Chichester: Wiley.
†Linsley, R.K., Kohler, M.A. and Paulhus, J.L. 1982: *Hydrology for engineers.* 3rd edn. New York: McGraw-Hill.
Nemec, J. 1976: International aspects of hydrology. In J.C. Rodda ed., *Facets of hydrology.* Chichester: Wiley.
—, Downing, R.A. and Law, F.M. 1976: *Systematic hydrology.* London: Butterworth.
†Rodda, J.C. 1976: *Facets of hydrology.* Chichester: Wiley.
Shaw, E.M. 1988: *Hydrology in practice.* 2nd edn. New York: Van Nostrand Reinhold.
†Walling, D.E. 1981: Physical hydrology. *Progress in physical geography* 5, pp. 123–31.
Ward, R.C. and Robinson, M. 1990: *Principles of hydrology.* 3rd edn. Maidenhead, Berks: McGraw-Hill.

hydrolysis *a.* The disintegration of organic compounds through their reaction with water.
b. The formation of both an acid and a base by a salt when it dissociates with water. This is an important mechanism of chemical weathering of rock.

hydrometeorology The application of meteorology to hydrological problems. Hydrometeorology is concerned particularly with the atmospheric part of the HYDROLOGICAL CYCLE. This involves the input of water to the surface through PRECIPITATION and outputs by EVAPORATION and TRANSPIRATION. Precipitation input to a catchment area depends upon atmospheric factors which can be examined at both short and long time-scales. Rainfall from an individual cloud will be determined by *microphysical processes* within the CLOUD as well as by the synoptic controls on cloud development. The extent to which suitable atmospheric conditions for cloud growth and precipitation formation prevail will be influenced by atmospheric circulation changes not only on a daily scale but also at the seasonal and annual time-scales. One of the main applications of hydrometeorology is for flood prediction. Statistical techniques are used to analyse existing precipitation data in order to calculate the return of flood events, i.e. how often a particular flood level is likely to occur, to estimate the PROBABLE MAXIMUM PRECIPITATION over an area, or even to simulate river-level variations. Some engineering design problems need similar information to provide cost-effective solutions when building reservoirs, bridges, sewers and planning irrigation schemes. PS

Reading
Shaw, E.M. 1988: *Hydrology in practice.* 2nd edn. London: Van Nostrand Reinhold.

hydrometry The measurement of water flow in channels. Accurate flow measurements are difficult to achieve, particularly when the cross-section of the flowing water is large or when the channel is shifting, irregular or subject to heavy weed growth. The techniques of flow or discharge measurement fall into three groups. Since discharge at a cross-section is equal to the product of the mean velocity of flow and the cross-sectional area of the flowing water, methods of gauging the flow are based upon either the direct measurement of discharge or on the product of an indicator of water cross-sectional area (usually through water depth or stage at a fixed site) and a means of estimating velocity of flow. In addition, measurements of the discharge may be just point measurements in time or they may be continuous. Measurements of stage, velocity or discharge are usually undertaken at fixed sites known as gauging stations. The table summarizes the main methods of flow measurements and recent developments of these techniques are discussed in IAHS (1982) (see DISCHARGE). AMG

Reading and Reference
†Gregory, K.J. and Walling, D.E. 1976: *Drainage basin form and process.* London: Edward Arnold.
†Herschy, R.W. 1976: New methods of river gauging. In J.C. Rodda ed., *Facets of hydrology.* Chichester: Wiley.
IAHS 1982: *Advances in hydrometry.* Proceedings of the Exeter symposium. International Association of Hydrological Sciences publication 134.

hydromorphy A process that occurs in soils which produces gleying and mottling as a result of the intermittent or permanent presence of excess water.

hydrophyte A herbaceous plant which has parts beneath water which survive when the parts above water die back.

hydrosphere The earth's water, which exists in both fresh and saline form and may occur in a liquid, solid or gaseous state. Land, sea and air each contribute to the total volume of water, which is conveyed between various locations and transformed from one state to another (HYDROLOGICAL

Hydrometry: Methods for river flow measurement

Variable	Point measurement	Continuous measurement	Comments
Stage	1 Stage board 2 Float and counter-weight 3 Tape and electrical contact 4 Pressure bulb 5 Crest stage gauge (for peak flow only)	1 Float and counter-weight or pressure bulb attached to continuous recorder 2 Bubble gauge and recorder	Float and counterweight or tape and electrical contact both require a level water surface and so are usually used with a stilling well
Velocity	1 Current meter 2 Floats 3 Tracer (e.g. NaCl or $NaCr_2O_3$) velocity 4 Electromagnetic flowmeter 5 Ultrasonic flowmeter	Electromagnetic flowmeter and recorder (average velocity in cross-section) Ultrasonic flowmeter and recorder (average velocity at one or more depths in the cross-section)	
Discharge	1 Current meter (mean or mid section method)	Continuously monitored stage with established stage – discharge relationship	Choice of wading or cableway or moving boat methods. Smooth, straight section required.
	2 Tracer dilution (gulp or continuous injection)		Suitable for turbulent sections or where high velocities or irregular bed make the use of current meter difficult
	3 Electromagnetic gauging	Continuous measurement of velocity and stage	Suitable for sections with aquatic growth and unstable bed
	4 Ultrasonic gauging	Continuous measurement of velocity and stage	Suitable for very wide rivers and estuaries
	5 Structures (weirs and flumes)	Continuous stage measurement coupled with a stage – discharge relationship	Suitable for small rivers
	6 Volumetric		Suitable for measuring very low flows
	7 Slope area method (e.g. Manning equation)		Suitable for estimating flows at an ungauged site

CYCLE). The overall quantity of water in the hydrosphere remains more or less constant. An insignificant increment accrues from volcanic water vapour, while dissociation of water vapour in the upper atmosphere (photodissociation) represents a minor loss.

About 70 per cent of the earth's surface is occupied by water. Some 97.3 per cent of its volume is currently in the oceans (1350×10^{15} m^3), the maximum extent of which is in the southern hemisphere. Of the 2.7 per cent of terrestrial water, most is polar snow and ice (29×10^{15} m^3). Groundwater (the majority below soil level) accounts for 8.4×10^{15} m^3, lakes (its main superficial terrestrial location) and

rivers 0.2×10^{15} m^3, and water in living organisms 0.0006×10^{15} m^3. Water vapour is the most important variable constituent of the atmosphere, but only accounts for 0.013×10^{15} m^3 of the total amount of water in the hydrosphere (Peixoto and Ali Kettani 1973). RLJ

Reading and Reference
†Chorley, R.J. ed. 1969: *Water, earth and man: a synthesis of hydrology, geomorphology and socio-economic geography.* London: Methuen.
†Cloud, P. 1968: Atmospheric and hydrospheric evolution on the primitive earth. *Science* 160, pp. 729–36.
Peixoto, J.P. and Ali Kettani, M. 1973: The control of the water cycle. In F. Press and R. Siever eds., *Planet earth (readings from Scientific American).* San Francisco: W.H. Freeman.

hydrostatic equation This equation is as follows:

$$\frac{\partial p}{\partial z} r = \rho g \qquad (1)$$

where p is pressure, z is height, ρ is air density, and g is the acceleration of gravity. Alternatively it may be written as:

$$-\frac{1}{\rho}\frac{\partial p}{\partial z} = g \qquad (2)$$

Equation 2 reveals that the upward pressure gradient force (left-hand side) is balanced by the downward force of gravity (right-hand side). Such a balance is known as hydrostatic equilibrium. The assumption that the atmosphere is a hydrostatic equilibrium has proved very useful in the analysis of large-scale motion, because at that scale, buoyancy forces may safely be ignored. At smaller scales, e.g. motion within clouds, we cannot validly assume that hydrostatic equilibrium exists.　　BWA

Reading
Hess, S.L. 1959: *Introduction to theoretical meteorology.* New York: Henry Holt.

hydrostatic pressure The fluid pressure p exerted by the underlying column of water in a body of water when at rest. It is given by Pascal's law as:

$$p = \rho g \psi + p_0$$

where ρ is the mass density of the water, g is the acceleration due to gravity, ψ is the pressure head (or vertical depth of water) and p_0 is atmospheric pressure at the surface of the water. In groundwater hydrology practice, the latter is usually set equal to zero and p is calculated in pressure above atmospheric and expressed in $N\,m^{-2}$ or Pa.　　PWW

Reading
Freeze, R.A. and Cherry, J.A. 1979: *Groundwater.* Englewood Cliffs, NJ: Prentice-Hall.

hydrothermal alteration Alteration of rocks due to earth pressures and temperatures, which are milder than those which produce metamorphism. It may give rise to chemical changes which are similar to weathering. Generally, weathering effects decrease from the surface downwards, while hydrothermal alteration increases downwards towards the fluids which give rise to

the activity. Pneumatolysis is similar but relates to gases (including steam) at still higher temperatures and pressures.　　WBW

hyetograph A self-recording instrument for measuring rainfall continuously. A chart showing the distribution of rainfall over a region.

hygrograph A self-recording instrument for measuring atmospheric humidity.

hygrometer A device for measuring the relative humidity of the atmosphere. Those which give a continuous record are called hygrographs.　　ASG

hypabyssal rock Igneous rock which is intrusive but has consolidated in a zone above the base of the earth's crust and hence has distinct structural characteristics.

hypogene Term applied to mineral and ore deposits formed by water rising towards the surface, in contrast to supergene.

hypolimnion The lowest layers of cold water which occur at the bottom of an ocean, sea or lake whereas the EPILIMNION consists of the warmer, less dense upper layers. The intermediate zone is called the thermocline.　　ASG

hypsithermal A term introduced by Deevey and Flint (1957) for a warm HOLOCENE phase covering four of the traditional Blytt–Sernander pollen zones, embracing the Boreal through to the Sub-Boreal (8950–2550 BP). It is broadly equivalent in meaning to ALTITHERMAL or climatic optimum.　　ASG

Reference
Deevey, F.S. and Flint, R.F. 1957: Post-glacial hypsithermal interval. *Science* 125, pp. 182–4.

hypsographic Pertaining to the branch of geography which deals with altitudes.

hypsographic curve A generalized profile of the surface of the earth and the ocean floors. A curve or graph which represents the proportions of the area of the surface at various altitudes above or below a datum.

hypsometric curve See HYPSOGRAPHIC CURVE.

hypsometry The measurement of the elevation of the landsurface or sea floor above or below a given datum, usually mean sea level.   ASG

Reading
Cogley, J.G. 1985: Hypsometry of the continents. *Zeitschrift für Geomorphologie Supplementband* 53, pp. 1–48.

hysteresis A term borrowed from the study of magnetism and used to describe a bivariate plot, which evidences a looped form and therefore a different value of the dependent variable, according to whether the independent variable is increasing or decreasing. Examples include relationships between matric potential and soil moisture content, river discharge and water stage, and suspended sediment and solute concentrations and river discharge. In the latter case different concentrations are associated with a given level of discharge on the rising and falling limbs of a hydrograph, and this hysteresis effect may be ascribed either to a difference in timing or lag between the response of the two parameters or to asymmetry in either relative behaviour. Both clockwise and anticlockwise hysteresis loops have been reported for relationships between suspended sediment and solute concentrations and discharge. From a study of the River Rother in the UK, Wood (1977) suggested that the precise form of the hysteresis loop associated with the relationship between suspended sediment concentration and discharge for specific events could be related to the period between successive flow events and to the duration and intensity of each event.   DEW

Reference
Wood, P.A. 1977: Controls of variation in suspended sediment concentration in the River Rother, West Sussex, England. *Sedimentology* 24, pp. 437–45.

I

ice Normally the solid form of water formed by: (a) the freezing of water; (b) the condensation of atmospheric water vapour directly into ice crystals; (c) the sublimation of solid ice crystals directly from water vapour in the air; or (d) the compaction of snow. Each form of ice has important implications for the physical geography of the earth.

(a) Ice derived from the freezing of water involves SEA ICE and lake ice which contribute a characteristic morphology to water surfaces and shorelines in high and mid-latitudes, also GROUND ICE which forms a significant component of permafrost landscapes (see also ICE WEDGE; PINGO), and the freezing of water within a snow pack to form ice lenses or, in the case of a glacier, SUPERIMPOSED ICE. Repeated freezing and thawing is also an efficient means of WEATHERING and can cause rapid rock breakdown.

(b) Ice crystals in the atmosphere begin as minute ice particles which form round condensation nuclei. These nuclei may be dust particles with favourable molecular structure or even crystals of sea salt. They will grow so long as they exist in an atmosphere with an excess supply of water vapour and will form an ice cloud or, if conditions are suitable near the ground surface, an ice fog. If the ice crystals reach a critical size they begin to fall as snow. RIME can contribute to the growth of the falling ice crystal and also directly to the ground surface whenever supercooled water droplets freeze on impact with a solid object.

(c) The sublimation of ice crystals directly from water vapour in the air can take place within a snowpack in response to strong temperature gradients and causes the growth of fragile ice crystals known as depth hoar. The process also produces hoar frost, the ice equivalent of dew.

(d) The compaction of snow to form GLACIER ice involves a number of metamorphic processes whose overall effect is to increase the crystal size and eliminate air passages. Snow which has survived a summer melt season and begun this process of transformation is known as FIRN. When consolidation has proceeded sufficiently far to isolate the air into separate bubbles then the firn becomes glacier ice. Carbon dioxide ice occurs extra-terrestrially and is of geomorphological importance, e.g. on Mars (Lucchitta 1981).   DES

Reading and Reference
†LaChapelle, E.R. 1973: *Field guide to snow crystals*. 3rd edn. Washington: University of Washington Press.
Lucchitta, B.K. 1981: Mars and earth: comparison of cold-climate features. *Icarus* 45, pp. 264–303.
†Paterson, W.S.B. 1981: *The physics of glaciers*. Oxford and New York: Pergamon.

ice age A period in the earth's history when ice sheets were extensive in mid and high latitudes. Such conditions were often accompanied by the widespread occurrence of SEA ICE, PERMAFROST, mountain glaciers in all latitudes and sea level fluctuations.

Ice ages have affected the earth on many occasions. There are records of major glaciations in the Precambrian (on several occasions), the Eocambrian (650–700 million years ago), the Ordovician (c. 450 million years ago), the Permo-Carboniferous (250–300 million years ago) and during the Cainozoic (the past 15 million years or so). On occasions the evidence of even the ancient ice ages is splendidly clear and ice directions can be inferred from striations and grooves, basal temperatures and basal water conditions from the type of TILL and sea level fluctuations from rhythmic marine sediments. The freshness of the striations cut by the Ordovician ice sheet in what is now the Sahara is quite stunning to the visitor.

Within an ice age there are sharp fluctuations of ice extent. During the Cainozoic, mid-latitude ice sheets built up on many occasions (see GLACIAL) only to disappear during interglacials. The fluctuations within the ice age can be related to cyclic variations in the amount of solar radiation received by the earth. Cycles of varying radiation receipt were postulated by Milankovitch (see MILANKOVITCH HYPOTHESIS) in 1924 and related to variation in the earth's orbital eccentricity, tilt and precession. Evidence from deep sea cores reveals a similar cyclic variability of ice build-up and decay (Hays *et al.* 1976) and current research is trying to understand the link between the two.

It is not clear why ice ages have occurred during part of the earth's history only. Perhaps an ice age needs a continent to be positioned in an approximately polar position so that ice can build up permanently and chill the earth as a whole.   DES

Reading and Reference
Hays, J.D., Imbrie, J. and Shackleton, N.J. 1976: Variations in the earth's orbit; pacemaker of the Ice Ages. *Science* 194. 4270, pp. 1121–32.

†Imbrie, J. and Imbrie, K.P. 1979: *Ice ages.* London: Macmillan.

†John, B. ed. 1979: *The winters of the world.* Newton Abbot: David & Charles.

ice blink A mariners' term for the white glare on the underneath of clouds indicating the presence of pack ice or glacier ice.

ice cap A dome-shaped GLACIER with a generally outward and radial flow of ice. The difference between an ice cap and an ice sheet is normally taken to be one of scale with the former being less than 50,000 km^2 in area and the latter larger. The marginal regions of the ice cap may be drained by OUTLET GLACIERS which flow beyond the ice cap in rock-walled valleys.   DES

ice contact slope A slope, formerly banked up against an ice mass, which has experienced slumping as a result of ice melting. Such slopes are commonly associated with KAME TERRACES and ESKERS.   DES

ice cored moraine A ridge or spread of glacial rock debris which contains a buried ice core. Protected from melting by the overlying debris, the ice core can survive for considerable periods of time.   DES

ice dam A blockage of drainage caused by ice which leads to periodic and/or rapid fluctuations in meltwater drainage. Large ice dams occur subglacially and are particularly catastrophic. The drainage outbursts are usually triggered when the glacier internal drainage network developing during the ablation season taps the subglacial lake. An initially small outflow melts open a very large passage in a matter of hours. The Icelandic word JÖKULHLAUP is often used to describe the increased discharge associated with the breaching of a subglacial ice dam.

In High Arctic regions streamflow occurs in late spring when valleys are choked with snow. Channel development begins with a saturation of the valley snowpack and water movement within or through the snow. The ponding and subsequent release of water behind snow dams formed by drifts is a common occurrence.   HMF

Reading
Sugden, D.E. and John, B.S. 1976: *Glaciers and landscape.* London: Edward Arnold. Esp. pp. 295–7.

Woo, M-K. and Sauriol, J. 1980: Channel development in snow-filled valleys, Resolute, NWT, Canada. *Geografiska annaler* 62A, pp. 37–56.

ice dome A term used to describe the main form of an ice sheet or ice cap and to distinguish it from streaming flow associated with OUTLET GLACIERS.   DES

ice edge A nautical term for the boundary between open water and floating ice.

ice fall A heavily crevassed area of a glacier associated with flow down a steep rock slope. The zone is one of EXTENDING FLOW and is marked by arcuate rotational slumps.   DES

ice field An approximately level area of ice which is distinguished from an ice cap because its surface does not achieve a dome-like shape and because flow is not radial outwards.   DES

ice floe A piece of floating sea or lake ice which is not attached to the land. In the Arctic and Antarctic ice floes are commonly from tens of metres to several kilometres across and 2–3 m thick. Some ice floes form in one winter and melt the following

Immense ice floes photographed from the mast-head of the Imperial Trans-Antarctic Expedition 1914–16.

summer. Where they survive from one year to the next they are known as multi-year ice and tend to be tougher and, in places where they have been rafted on top of one another or crushed together, thicker. (See also SEA ICE.)   DES

Reading
Nansen, F. 1897: *Farthest north*. 2 vols. London: Constable.

ice flow The movement of ice by internal deformation or BASAL SLIDING. Most studies of ice flow concern glaciers and relatively little is understood, for example, about the flow of debris-rich ground ice in permafrost areas.

A glacier flows in response to shear stresses set up in the ice mass by the force of gravity. These vary according to the thickness of the glacier and its surface slope and can be calculated from the equation: $\tau = \rho gh \sin \alpha$ where τ is the shear stress, ρgh is the weight of overlying ice and α is the slope of the ice surface. Internal deformation of the glacier takes place mainly through the action of CREEP which is modelled for glaciers by GLEN'S LAW. This shows that glacier flow is highly sensitive to an increase in shear stress and this is why most internal deformation occurs at the bottom of a glacier. Near the glacier bottom bedrock obstacles set up locally high stresses and enhanced basal creep is the mechanism by which they are

passed. In situations where the base of the glacier is at the pressure melting point and a film of water exists at much of the ice–rock interface, basal sliding contributes to glacier flow and rates may equal or exceed the rate of internal deformation for the glacier as a whole. Where glaciers overlie saturated soft sediments it has been discovered that deformation of the sediment may contribute to the forward movement of the glacier (Boulton 1979).

Most glaciers flow at the rate of 10–100 m per year but where rates of basal sliding are high, as in the case of SURGING GLACIERS, flow rates may exceed several kilometres per year.   DES

Reading and Reference
Boulton, G.S. 1979: Processes of glacier erosion on different substrata. *Journal of glaciology* 23.89, pp. 15–38.
†Paterson, W.S.B. 1981: *The physics of glaciers*. Oxford and New York: Pergamon.

ice fog A suspension of minute ice crystals in the air reducing visibility at the earth's surface. The optimum conditions for ice fog build-up are temperatures below −30°C and a supply of water vapour. Such conditions are common in and around Arctic settlements in winter, where an inversion causes low air temperatures and vehicles and heating plants contribute more water vapour than can be absorbed without condensing. Such fogs are associated with severe pollution.   DES

Reading
Benson, C.S. 1969: The role of air pollution in Arctic planning and development. *Polar record* 14.93, pp. 783–90.

ice front The vertical cliff forming the seaward edge of an ice shelf or floating glacier.

ice jam A blockage caused by the accumulation of pieces of river ice or sea ice in a narrow channel.

ice rind A stage in the growth of sea ice when the accumulating ice crystals coagulate to form a brittle skin.

ice segregation Sometimes called ice lensing, ice segregation is the process by which water freezes, thereby causing heave of the ground surface. Primary (i.e. capillary) and secondary heave can be distinguished. In primary heave, the critical conditions for the growth of segregated ice are:

$$P_i - P_w = \frac{2\sigma}{r_{iw}} < \frac{2\sigma}{r}$$

where P_i is the pressure of ice, P_w is the pressure of water, σ is the surface tension of ice to water, r_{iw} is the radius of the ice–water interface, and r is the radius of the largest continuous pore openings.

Secondary heave is not clearly understood but may occur at temperatures below $0°C$ and at some distance behind the freezing front. Porewater expulsion from an advancing freezing front is another mechanism for ice segregation, especially massive ice bodies, provided that porewater pressures are adequate to replenish groundwater that is transformed into ice.   HMF

Reading
Miller, R.D. 1972: Freezing and heaving of saturated and unsaturated soils. In *Frost action in soils*. National Academy of Sciences – National Academy of Engineering highway research record 393.
Washburn, A.L. 1979: *Geocryology: a survey of periglacial processes and environments*. New York: Wiley. Esp. pp. 68–70.
Williams, P.J. 1977: General properties of freezing soil. In P.J. Williams and M. Fremond eds, *Soil freezing and highway construction*. Ottawa: Paterson Centre, Carleton University.

ice sheet A large dome-shaped glacier (over 50,000 km^2 in area) with a generally outward and radial flow of ice. On a continental scale such ice sheets can exceed a thickness of 4 km, as they do in Antarctica. A simple model can be used to approximate the surface slope of an ice sheet, assuming the ice is perfectly plastic. The profile equation is:

$$h = 3.4\,(L - x)^{\frac{1}{2}}$$

where L is the distance from ice sheet centre to edge and h is the ice thickness at a distance $L - x$ from the edge and both quantities are in metres (Paterson 1981).   DES

Reading and Reference
†Nansen, F. 1890: *The first crossing of Greenland*. London: Longman.
Paterson, W.S.B. 1981: *The physics of glaciers*. Oxford and New York: Pergamon. P. 163.
†Robin, G. de Q., Drewry, D.J. and Meldrum, D.T. 1977: International studies of ice sheet and bedrock. *Philosophical transactions of the Royal Society London* Series B. 279, pp. 185–96.

ice shelf A floating sheet of ice attached to an embayment in the coast. It is nourished by snow falling onto its surface and by land-based glaciers discharging into it. The seaward edge is a sheer cliff rising some 30 m above sea level. The shelf surface is virtually flat although it rises slightly inland with an increase in ice thickness in the same direction. Freed of basal friction associated with land-based glaciers, ice velocities are high and commonly 1–3 km year^{-1}. Periodically calving removes huge tabular ICEBERGS from the front. Thirty per cent of the Antarctic coastline is fringed by ice shelves. The Ross Ice Shelf extends 900 km inland and is 800 km across.   DES

Reading
Robin, G. de Q. 1975: Ice shelves and ice flow. *Nature* 253, pp. 168–72.
Thomas, R.H. 1974: Ice shelves: a review. *Journal of glaciology* 24.90, pp. 273–86.

ice stream A relatively narrow zone of swiftly moving ice within an ice sheet or ice cap, often bordered by spectacular crevasses. The high velocities probably reflect high sliding velocities associated with a basal water film. Ice streams often form the heads of OUTLET GLACIERS.   DES

Reading
Drewry, D.J. 1983: Antarctic ice sheet: aspects of current configuration and flow. In R. Gardner and H. Scoging eds, *Megageomorphology*. Oxford: Oxford University Press.

ice wedge A massive, generally wedge-shaped, ground ice body composed of foliated or layered, vertically orientated ice which extends below the permafrost table. Large ice wedges may be 1–2 m wide near the top and extend downwards for 8–10 m. They form in cracks in polygonal patterns originating in winter by thermal contraction of the ground into which water from melting snow penetrates in the spring. Repeated annual contraction and subsequent cracking of the ice in the wedge, followed by freezing of water in the crack, lead to an increase in width and depth of the wedge. Ice wedges require PERMAFROST for their formation and existence. They often give a distinct polygonal micro-relief to the tundra surface.   HMF

Reading
Lachenbruch, A. 1966: Contraction theory of ice-wedge polygons: a quantitative discussion. In *Proceedings of the Permafrost International Conference 1963*. National Academy of Science – National Research Council of Canada publication 1287. Washington, DC.
Mackay, J.R. 1974: Ice-wedge cracks. Garry Island. NWT. *Canadian journal of earth sciences* 11, pp. 1336–83.

iceberg A mass of floating ice which has broken away from a floating glacier. Calving of ice from the glacier is induced by flexure under the influence of ocean swell. Once free, iceberg drift is mainly in response to ocean currents though relatively minor deviations result from the influence of wind. Ice shelves produce large tabular icebergs whose area may attain several thousand square kilometres. OUTLET GLACIERS produce smaller, less regular icebergs.   DES

Reading
Proceedings of the Conference on the Use of Icebergs: scientific and practical feasibility, Cambridge, UK, 1–3 April 1980. *Annals of glaciology* 1, 1980.

icing A mass of surface ice formed during the winter by successive freezing of sheets of water that may seep from the ground, from a river or from a spring. Icings are widespread in periglacial areas.   DES

igneous rock Rock formed when molten material, magma, solidifies, either within the earth's crust or at the surface.

illuviation The precipitation and accumulation of material within the B horizon of a soil after the material has been leached from the surface or overlying soil horizons.

imbrication The condition exhibited by flattened river pebbles or tabular thrust sheets when they overlap each other.

impeded dunes, as opposed to free dunes, are related in their position and form directly to vegetation, a topographical barrier or a localized source of sediment.
                                                 ASG

impermeable Having a structure or texture which does not allow the movement of water through a rock or soil material under the natural conditions in the groundwater zone.

impervious Impermeable. Having a texture which does not allow the movement of water, oil or gas through a rock or soil material. Under certain conditions a rock may have an impervious texture though a stratum of the rock is permeable owing to joints and fractures.

in and out channel The name given to a small, discontinuous channel produced by meltwater flow from a glacier onto the adjacent hillside.

incised meander See MEANDERING.

inconsequent stream A stream not apparently related to landsurface features or major geological controls, but following minor surface features without being developed into an organized pattern overall. The term was used by G.K. Gilbert (1877). It became largely superseded by the synonymous 'insequent' of W.M. Davis (1894) and is now used hardly at all.   JL

Reading and References
Davis, W.M. 1894: Physical geography as a university study. *Journal of geology* 2, pp. 66–100.

†— 1899: The geographical cycle. *Geographical journal* 14, pp. 481–504.

Gilbert, G.K. 1877: *Report of the geology of the Henry Mountains*. Washington: US Geographical and Geological Survey of the Rocky Mountain Region.

indeterminacy The situation in which the results of an investigation remain open to two or more conflicting interpretations. For example, if we are investigating the form of valley cross-sections in the mid-latitudes, in east–west trending basins where strata dip steeply to the north or south, we are likely to find that the cross-sections are asymmetric. There are numerous potential causes for asymmetry (see Kennedy 1976) and it may be possible to eliminate some – such as glaciation – from serious consideration. If, however, the steeper valley sides are found to be those where the strata dip into the slope, we will find that we are left with two possible explanations for our findings: either that the asymmetry is due to structural control; or (since north- and south-facing hillsides in mid-latitudes inevitably receive different amounts of solar radiation) that it is due to microclimatic differences that are reflected in varying rates or kinds of hillslope processes. It may well be, of course, that both factors are at work. However, in terms of providing a convincing explanation of the pattern of hillside forms, the result of such an investigation would be indeterminate.

Given the multiplicity of potential causal variables of interest to the physical geographer and their tendency to interact in a non-trivial fashion, it is all too common for studies to produce indeterminate answers. It is for this reason that it is desirable to conduct structured or controlled investigations, making use of the principles of experimental design (see Cox 1958) and of the method of multiple working hypotheses (Chamberlin 1890). (Experimental design is, in effect, a formalized way of putting the method of multiple working hypotheses into practice.) What does this mean?

First, we must be as clear and precise as possible in defining the phenomenon whose variations are of interest: it may be the distribution of pingos, or the pH of rain, or the number of species of herbaceous plants, but in each case we must be absolutely clear what constitutes a pingo, rain or a herb. This may sound laboured, but in fact indeterminacy is not uncommonly produced by a failure to appreciate that two or more variants of phenomena are under study, not one: for example, it is now generally accepted that pingos are of two distinct types (open-system, or Greenland; and closed-system, or Mackenzie; see Washburn 1979) with very different relationships to topography and subsurface water bodies. Failure to recognize this distinction at the outset could produce an indeterminate answer to the question of controls on the distribution of ice-cored hills.

Secondly, we must set out as many potential sources of variation as we can imagine, within the bounds of physical possibility. If we are studying acid rain we are likely to consider that the prevalent wind direction is important; but are other factors, too, of potential significance? Intensity and duration of precipitation, synoptic conditions, effect of local topography, season: all or any may influence the patterns we observe. At this point it is essential to examine the results of previous studies very carefully for information about the nature of relationships established by other workers.

Thirdly, we must design the network of observations to be made, whether these will be field measurements over space or time; or investigations based on maps, remotely sensed data, or archival material. Our object here is to represent, within our data, all major potential sources of variation in such a way that their individual and collective influences will remain distinguishable. This sometimes may be achieved by the use of 'control': for example if we wish to examine changes in herbaceous species produced by the elimination of grazing pressures, we should need to examine not only spatial variations which might affect the situation, but also natural, temporal fluctuations and this could be achieved by leaving control plots. In other cases we may have to proceed by very carefully arranged surveys (see Chorley and Kennedy 1971, pp. 146–9 and figure 4.16).

Fourthly, we must analyse our data scrupulously. This means that we must examine the influence of all the major factors identified at the outset, look for interactions between them, and keep a very wide eye open for the potential role of factors we might initially have overlooked. This analysis may be quantitative or qualitative: if the former, the analysis of variance, developed by R.A. Fisher (see Fisher 1928 and 1935) provides a flexible and powerful tool with the advantage that it was initially designed to cope with spatial data.

Finally, we have to interpret our findings. It is at this point that the recognition of indeterminacy is the most common, most painful and most problematic. To be forced to admit that a careful investigation has failed to provide a clear-cut answer is psychologically disappointing. However, if the investigation really *was* a careful one, it is very unlikely indeed that no possibilities have been eliminated. What is to be done about those which remain? They must all be clearly stated, for a start, and should be given equal and fair weight in the concluding discussion. More important, the investigator should then attempt to outline further, critical observations which might resolve the indeterminate solution. If, on the other hand, the indeterminacy seems all-pervasive and incapable of resolution then that, too, should be made explicit: at the least, it may deter other investigators from banging their heads against that particular brick wall.

Indeterminacy is, then, a common fact of life in many investigations of natural phenomena. It is not unique to physical geography, nor even to the earth sciences, but it is unusually common throughout the historical sciences in general (see Simpson 1963). Given the potential problem, the most useful method for proceeding seems to be to recognize this at the outset of any study and to take the steps outlined above so that the nature and extent of any indeterminate solution are clear to future workers.   BAK

References
Chamberlin, T.C. 1890: The method of multiple working hypotheses. *Science* old series 15.92. Reprinted 1965, new series 148, pp. 754–9.

Chorley, R.J. and Kennedy, B.A. 1971: *Physical geography: a systems approach*. London: Prentice-Hall International.

Cox, D.R. 1958: *Planning of experiments*. London: Wiley.

Fisher, R.A. 1928: *Statistical methods for research workers.* Edinburgh: Oliver & Boyd.

— 1935: *The design of experiments.* Edinburgh: Oliver & Boyd.

Kennedy, B.A. 1976: Valley-side slopes and climate. In E. Derbyshire, ed., *Geomorphology and climate.* London: Wiley. Pp. 171–201.

Simpson, G.G. 1963: Historical science. In C.C. Albritton ed., *The fabric of geology.* Stanford: W.H. Freeman. Pp. 24–48.

Washburn, A.L. 1979: *Geocryology.* London: Edward Arnold.

index cycle A roughly cyclic variation in the zonal index, the average length of which is about six weeks, with individual cycles varying from three to eight weeks duration. The cycle is most marked in winter in the northern hemisphere. It was Namias (1950) who first described this circulation cycle varying at its extremes between a zonal type of high index circulation, and a meridional type, low index, circulation. The latter type is often referred to as a BLOCKING situation. The existence of the true cyclic nature of changes in index is now severely questioned. AHP

Reading and Reference
Namias, J. 1950: The index and its role in the general circulation. *Journal of meteorology* 7, pp. 130–9.

†— and Clapp, P.F. 1951: Observational studies of general circulation patterns. In *Compendium of meteorology.* Washington, DC: American Meteorological Society. Pp. 561–7.

induration The process of hardening through cementation, desiccation, pressure or other causes, applied particularly to sedimentary materials.

infiltration The process by which water percolates into the soil surface. Two main zones can be observed in the soil when infiltration is proceeding at its maximum rate, that is from a surface which is saturated with a thin layer of standing water. There is an upper transmission zone with an almost constant moisture content close to saturation. Below this is a sharp wetting front in which the moisture content declines rapidly towards its pre-infiltration value. Within the transmission zone, flow is driven mainly by gravitational forces. Across the wetting front there is a strong hydraulic or tension gradient, tending to push the water into the dryer soil in the way that water is drawn into a fine capillary tube. This hydraulic gradient advances the wetting front down into the soil and so allows additional water to infiltrate at the surface. The rate at which water can infiltrate under these ideal circumstances is called the infiltration capacity and it decreases as the wetting front advances deeper into the soil.

Infiltration capacity may be expressed either in terms of time since the process began, or in terms of current moisture storage. The main advantage of the storage expressions is that they may remain valid during infiltration at less than the capacity rate, as commonly occurs at the start of a rainstorm when infiltration capacity tends to be very high. One example of a widely used empirical infiltration equation is that used in the HORTON OVERLAND FLOW MODEL. Another equation with a better theoretical basis was put forward by Philip (1957/8):

$$f = A + Bt^{\frac{1}{2}}$$

where f is the infiltration capacity at time t and A and B are constants which depend on the soil and its initial moisture distribution.

In this expression the constant term A mainly represents the steady rate of infiltration under gravitational potential (i.e. the weight of the water) and the time-dependent term is due to the hydraulic potential gradient at the advancing wetting front. It may be seen that the infiltration capacity is initially very high, and decreases steadily towards a constant rate, which is usually achieved within one to two hours.

An example of a storage-based infiltration equation is the Green and Ampt (1911) equation:

$$f = A + C/S$$

where f and A are as above, S is a soil water storage value, and C is a constant of the soil and its initial moisture.

In the original formulation of this expression the storage S was the total amount of water infiltrated since the start, but an alternative is to budget S as representing a store of infiltrated water which leaks at steady rate A. This version has the advantage that if converted to a time-dependent form under conditions of surface ponding, it is exactly equivalent to the Philip equation above (with $B = (2C)^{\frac{1}{2}}$). During a rainstorm of constant or varying intensity, this kind of storage-based model allows estimation of the infiltration capacity at any time in terms of the water that has actually entered the soil previously, which may have been at any rate less than (or equal to) the current infiltration capacity.

Infiltration may be likened to the process of pouring water into a bottle: it may fail to get in either because it is being poured in too fast or because the bottle is already full. On a hillside, saturated THROUGHFLOW may increase downslope or in areas of flow concentration until the SATURATION DEFICIT is zero: in other words the bottle may be full or almost full. A second criterion for infiltration may therefore also be expressed in storage terms in that ponding will occur when soil water storage reaches a critical level, S_c.

In areas of high rainfall intensities and low infiltration capacity, infiltration capacity is commonly exceeded and the Horton overland flow model is generally applicable to the estimation of streamflow. This includes areas which are naturally or artificially clear of dense vegetation, that is to say semi-arid areas and (seasonally) cultivated fields, as will be seen below. In areas of low rainfall intensity and/or dense vegetation cover, including forested areas and much of the humid-temperate zone, infiltration capacities are seldom exceeded, and PARTIAL AREA MODELS, which estimate the areas of saturation and the volumes of THROUGHFLOW, are generally more appropriate for estimating streamflow volumes and HILLSLOPE FLOW PROCESSES.

Rates of infiltration are usually compared in terms of the steady long-term rate (A in the equations above). This rate responds to some extent to soil texture, typically ranging from 0–4 mm^{-1} h^{-1} on clays through to 3–12 mm^{-1} h^{-1} on sands where the soils are initially wet and unvegetated. Vegetation cover and protection by coarse particles shields the surface from raindrop impact which is otherwise liable to break down the top layer of soil peds and pack the resulting soil grains down into the next layer as a thin, impermeable crust. On a crusted surface infiltration capacity is commonly very low except in extremely intense storms which may break the crust: where crusting is prevented, capacities are much higher. Within the soil, structure has a greater influence on infiltration capacity than does texture, and vegetation and its associated organic soil have a strong influence on soil structure. Thus vegetation cover increases infiltration capacity in two ways, so that steady rates may be 50–100 mm^{-1} h^{-1} under a good cover, compared to less than 10 mm on bare crusted soil.

Where soil structural voids are marked and soil textural pores fine, as for example for a cracked clay soil, soil water may bypass much of the soil mass. At the surface soil peds allow infiltration at maximum capacity, and excess rainfall overflows down the structural voids. In the largest voids water flows as a film down each wall and infiltrates into additional peds below the surface. The advance of water down each void is limited by this infiltration into the walls. In most cases rainfall enters the soil peds within a few tenths of a metre of the surface, but along a highly convoluted wetting front which follows the geometry of the largest voids. The resulting pattern may then be conceived either as a greater average depth of penetration of infiltrated water; or as shallow penetration to match the current rainfall intensity. In extreme cases water may bypass the soil as a whole and make direct contact with groundwater via bedrock fissures, etc. It should be pointed out that almost all real soils show some structural voids, but that local bypassing is only important where there is a very marked contrast between the textural and structural pore size distributions.

The importance of infiltration in physical geography lies in its role within the catchment and hillslopes hydrological cycle in partly determining the flow routes of precipitation to the streams and so determining the timing of streamflow response. Infiltration also plays an important part in separating groups of hillslope processes. Water which travels as overland flow takes little part in supplying soil moisture for plant growth and, because it comes into little and rather brief physical contact with mineral soil, picks up little solute load except from the litter layer. OVERLAND FLOW therefore tends to dilute stream solute concentrations, which therefore tend to be lower during intense storms when overland flow is greatest. Overland flow also erodes and transports all surface wash/soil erosion. The infiltrated water flows through the soil as throughflow which is able to carry negligible amounts of suspended material, but is in intimate contact with mineral soil grains from which it is effective in leaching solutes and promoting chemical weathering. The infiltrated water is also responsible for providing water for plant growth and for establishing patterns of hydraulic potential which have a powerful influence on slope stability in the context of mass movement.

In other words the process of infiltration is a critical regulator of the landscape system in both the short, hydrological term and in longer erosional time spans.   MJK

Reading and References
†Goudie, A. ed. 1990: *Geomorphological techniques*. 2nd edn. London: Unwin Hyman.
Green, W.H. and Ampt, G.A. 1911: Studies on soil physics I: the flow of air and water through soils. *Journal of agricultural science* 4.1, pp. 1–24.
†Horton, R.E. 1945: Erosional development of streams and their drainage basins: hydrophysical approach to quantitative morphology. *Bulletin of the Geological Society of America* 56, pp. 275–370.
†Knapp, B.J. 1978: Infiltration and storage of soil water. In M.J. Kirkby ed., *Hillslope hydrology*, Chichester: Wiley. Pp. 43–72.
Philip, J.R. 1957/8: The theory of infiltration. *Soil science* 83, pp. 345–57 and 435–48; 84, pp. 163–77 and 257–64; 85, pp. 278–86 and 333–7.

inflorescence The flowering shoot of a plant.

influent Either a tributary stream or river, or a term applied to a stream which supplies water to the groundwater zone.

infrared imagery See THERMAL INFRARED LINESCANNER.

infrared thermometer/thermometry Bodies reflect, emit and sometimes transmit radiation falling upon them. The efficiency of the surface in generating contributions are known, respectively, as reflectivity, emissivity and transmissivity. If the latter is zero (i.e. an opaque material) then reflectivity + emissivity = 1. The wavelengths near, but longer than, the visible spectrum are termed infrared and constitute most of the radiation from a 'hot' body. This radiation can be detected by an instrument which does not require contact with the body; when allowance is made for the emissivity of the surface its temperature can be measured. A THERMAL INFRARED LINESCANNER is a more complex instrument used in remote sensing.   WBW

ingrown meander See MEANDERING.

inheritance See DARWINISM; EVOLUTION.

inlier An outcrop of rock which is completely surrounded by younger formations, frequently the result of erosion of the crest of an anticline.

inselberg (German for island hill) A general class of large residual hill which usually surmounts an eroded plain. Small residual rock masses tend to be called TORS,

A spectacular granite inselberg, the Spitzkoppje, rising from the gravel plains of the Namib Desert in central Namibia.

large domed residuals tend to be called domed inselbergs or bornhardts, while large accumulations of boulders in the form of a hill are called koppies.

Inselbergs of resistance are those that are left as prominent landforms as a result of their superior resistance (brought about by the jointing density of the rock or its mineralogical composition), while inselbergs of position remain as prominent features because they are on divides farthest from lines of active erosion.

There has been considerable debate as to inselberg origin, and three main mechanisms have been proposed: that they are produced by scarp retreat across bedrock; that they are a result of scarp retreat across deeply weathered rocks; or that they result from differential weathering followed by stripping of the regolith.

Inselbergs occur in a wide range of rock types, but the most common lithologies appear to be sandstones and conglomerates (e.g. Ayers Rock in Australia or Meteora in Greece) or gneisses and granites, especially those that have widely spaced joints and a high potassium content.   ASG

Reading
Bremer, H. and Jennings, J. eds 1978: Inselbergs. *Zeitschrift für Geomorphologie Supplementband* 31.

King, L.C. 1975: Bornhardt landforms and what they teach. *Zeitschrift für Geomorphologie*, NF 19, pp. 299–318.

Pye, K., Goudie, A.S. and Thomas, D.S.G. 1984: A test of petrological control in the development of bornhardts and koppies on the Matopos Batholith, Zimbabwe. *Earth Surface processes and landforms* 9, pp. 455–67.

Thomas, M.F. 1974: *Tropical geomorphology*. London: Macmillan.

Twidale, C.R. 1982: The evolution of bornhardts. *American scientist* 70, pp. 268–76.

insequent stream A drainage network that has developed as a result of factors which are not determinable.

insolation a term used in two senses:

1 The intensity at a specified time, or the amount in a specified period, of direct solar radiation incident on unit area of a horizontal surface on or above the earth's surface.

2 The intensity at a specified time, or the amount in a specified period, of total (direct and diffuse) solar radiation incident on unit area of a specified surface of arbitrary slope and aspect.

In general, insolation depends on the solar constant, calendar date, latitude, slope and aspect of surface, and degree of transparency of the atmosphere.   JGL

insolation weathering (or heating and cooling weathering) The disintegration of rock in response to temperature changes setting up stresses. Early travellers heard, or reported they heard, sounds like pistol shots as rocks cooled in the evening, and thus arose a classic process in desert geomorphology. Experimental work in the twentieth century and recognition that weathering appears to be far more effective in the presence of moisture have led to pure insolation weathering being related to a position of relatively lowly importance (Schattner 1961).   ASG

Reference
Schattner, I. 1961: Weathering phenomena in the crystalline of the Sinai, in the light of current notions. *Bulletin of the Research Council of Israel* 10G, pp. 247–65.

instantaneous unit hydrograph (IUH) See UNIT HYDROGRAPH.

intact strength The strength of a rock sample which is free of large-scale structural discontinuities such as joints, fissures or foliation partings. It is usually expressed as a measure of unconfined compressive strength.   MJS

intensity of rainfall The rate at which rain falls in unit time, usually expressed in mm h^{-1}. It is measured using autographic (or recording) rain gauges and is important because of its impact on run-off characteristics. More intense rainfall will produce a more peaked run-off response. Rainfall intensity is inversely related to rainfall duration at a site and the intensity–duration–frequency analysis of rainfall records provides a good indication of the potential run-off characteristics of an area. (See also DEPTH–AREA CURVE; DEPTH–DURATION CURVE; RAIN GAUGE.)   AMG

Reading
Dunne, T. and Leopold, L.B. 1978: *Water in environmental planning*. San Francisco: W.H. Freeman.

interbasin transfers, effects of A method of water supply whereby the natural or regulated flow from one river system is transferred usually by pumping to another river system. This method of water supply is likely to increase in future decades and many schemes have already been evaluated such as the possibility of the southward diversion of flow from Siberian rivers such as the Lena and the Ob and the transfer of water from the Chiang Jiang (Yangtze) to the Huang He (Yellow River) basin in China. Developments have been approached cautiously because of the possible ENVIRONMENTAL IMPACTS of such interbasin transfers and it is known that in addition to changes of the water balance, especially in the evaporation term, there may also be substantial changes in the morphology and ecology of the river channels themselves because of the increase or decrease in streamflows.   KJG

interception The process by which precipitation is trapped on vegetation and other surfaces before reaching the ground. Interception loss is the component of intercepted precipitation which is subsequently evaporated, although this is also frequently described as interception. The character of the intercepting surfaces has a major impact on the amount of precipitation that is intercepted and then lost through evaporation. (See also STEM FLOW and THROUGHFLOW.)   AMG

Reading
Courtney, F.M. 1981: Developments in forest hydrology. *Progress in physical geography* 5, pp. 217–41.
Crockford, R.H. and Richardson, D.P. 1990: Partitioning of rainfall in a eucalyptus forest and pine plantation in southeastern Australia (four papers). *Hydrological processes* 4, pp. 131–88.
Durocher, M.G. 1990: Monitoring spatial variability of forest interception. *Hydrological processes* 4, pp. 215–29.

interflow A component of streamflow which responds to rainfall more slowly than surface run-off and more rapidly than BASE FLOW. In the HORTON OVERLAND FLOW MODEL streamflow was initially separated into overland flow and groundwater flow, and various procedures were used for partitioning the stream hydrograph into these components. Interflow was originally introduced as an intermediate hydrograph component which fell between the other two. It was considered to represent groundwater which re-emerged as overland flow; or else it was considered to be shallow groundwater flow. Some literature now uses the term interflow interchangeably with subsurface soil flow or THROUGHFLOW, but its original physical identification was rather tenuous.   MJK

Reading
Linsley, R.K., Kohler, M.A. and Paulhus, J.L.H. 1949: *Applied hydrology*. New York: McGraw-Hill.

interfluve The area of high ground which separates two adjacent river valleys.

interglacial A phase of warmth between glacials when the great ice sheets retreated and decayed, and tundra conditions were replaced by forest over the now temperate lands of the northern hemisphere. The Holocene is an interglacial, but some of the Quaternary interglacials may have been slightly warmer than today. Iversen (1958) identified various stages in a glacial–interglacial cycle in north-western Europe. Development begins with a *protocratic* phase, characterized by rising temperature, raw, basic or neutral mineral soils, favourable light conditions, and a pioneer vegetation of small plants, with exacting requirements for both nutrients and light. In the following *mesocratic* phase, comprising the climax of the interglacial, there are maximum temperatures, brown forest soils, mull plants, dense climax forest, and a vegetation that while still demanding nutrients is tolerant of shade. The *oligocratic* phase arises as a result of soil development, and involves more acid soils and more open vegetation. The *telocratic* phase marks the end of interglacial forest development. Heaths expand, and bogs develop in response to climatic deterioration. The climatic deterioration culminates in a *cryocratic* phase when cold conditions and soil instability are hostile to tree growth.   ASG

Reference
Iversen, J. 1958: The bearing of glacial and interglacial epochs on the formation and extinction of plant taxa. In O. Hedberg ed., *Systematics of today*. Acta Universitets Upsaliensis 1958, pp. 210–15.

intermittent spring A natural outflow point of underground water, that sometimes dries up completely (see SPRINGS). Normally such a spring is intermittent because the WATER TABLE upstream of the spring has fallen to or below the elevation of the spring, with the result that the hydraulic gradient leading to the spring is zero.   PWW

intermittent stream A stream is classified as intermittent if flow occurs only seasonally when the water table is at the maximum level. The drainage network is composed of ephemeral, intermittent, and perennial streams and the network expands during rainstorms and extends to limits affected by antecedent conditions especially antecedent moisture. Flow may occur along intermittent streams for several months each year but will seldom occur when the water table is lowered during the dry season. KJG

international hydrological decade

(IHD) A ten-year period (1965–1974) which was inaugurated in order to remedy deficiences in the understanding of hydrology by reference to particular themes and to specific areas. The major objectives were stated as:

1 The collection of basic data, particularly needed for the development of other components of the programme.
2 The production of inventories and water balances.
3 Improved knowledge for water resources development.
4 Exchange of information to meet scientific and practical needs of all participating countries.
5 Broad education and training programmes for professional hydrologists and hydrological technicians designed particularly for the needs of developing countries. (See also HYDROLOGY.) KJG

Reading
Nemec, J. 1976: International aspects of hydrology. In J.C. Rodda ed., *Facets of hydrology*. Chichester: Wiley. Pp. 331–62.

international hydrological programme

(IHP) A sequel to the INTERNATIONAL HYDROLOGICAL DECADE was inaugurated in 1975 because of the success of the IHD and the need to continue some programmes of investigation and to develop others. The major aims are listed as:

1 To provide a scientific framework for the general development of hydrological activities.
2 To further the study of the hydrological cycle and to improve the scientific methodology for assessment of the water resources throughout the world thus contributing to their rational use.
3 To evaluate the environmental implications of changes introduced by man's activities in the hydrological cycle.
4 To promote exchange of information in hydrological research and on new developments in hydrology.
5 To promote education and training in hydrology.

6 To assist member states in the organization and development of their national hydrological activities. KJG

Reading
Nemec, J. 1976: International aspects of hydrology. In J.C. Rodda ed., *Facets of hydrology*. Chichester: Wiley. Pp. 331–62.

interpluvial

A relatively dry phase interspersed with the wetter phases (pluvials) of the Pleistocene and Holocene. In many parts of the tropics the period at the end of the Late-Glacial Maximum (between *c.*18,000–13,000 years ago) was dry enough to cause lake levels to fall and dune fields to expand. ASG

Reading
Goudie, A.S. 1983: The arid earth? In R.A.M. Gardner and H. Scoging eds, *Mega-geomorphology*. Oxford: Oxford University Press.

interstadial

There is as yet no universally accepted definition which differentiates an interstadial from an interglacial. However, it may be defined as a relatively short-lived period of lesser glaciation and relatively greater warmth and thermal improvement during the course of a major glacial phase. During such phases conditions were not of sufficient magnitude and/or duration to permit the development of temperate deciduous forest of the full interglacial type. Information about interstadial environment conditions have been obtained and assessed using faunal and floral evidence, the timing of which have been obtained mainly from radiocarbon dating where this technique permits. AP

Reading
Goudie, A.S. 1992: *Environmental change*. 3rd edn. Oxford: Clarendon Press.

Lowe, J.J. and Walker, M.J.C. 1984: *Reconstructing Quaternary environments*. London: Longman.

interstices

Voids such as pores and fissures that occur within a rock. They can be classified according to their origin, shape and size. Primary interstices are those formed when the rock was created, such as intergranular pores in a sandstone, whereas secondary interstices are the result of later tectonic activity or weathering, such as fault planes and voids left by the differential corrosion of minerals. Interstices are most often small and interconnected, although large isolated interstices termed *vugs* also sometimes occur. Both small primary and large secondary interstices can be present simultaneously in a rock. For example, well-jointed and bedded sedimentary rocks, especially if rich in carbonate, may have a geometrical lattice of large interstices produced by solution along the fissures system and a fine intergranular POROSITY within the main body of the rock.   PWW

intertropical convergence zone (ITCZ) A band of nearly continuous low pressure, light and variable winds, high humidity, and intermittent heavy rain showers found near the equator. The name is derived from its location between the tropics of the two hemispheres and from the fact that it represents a narrow zone along which the trade winds of the two hemispheres converge. The ITCZ is often clearly visible on satellite photographs, especially over the oceans. It appears as a narrow, well-defined cloud band. Occasionally two ITCZs will be visible, most frequently in the eastern tropical Pacific Ocean shortly after the equinoxes. At that time the mean position of the ITCZ is near the equator. In the eastern Pacific cool surface water, produced by upwelling, splits the convergence zone into a northern and southern branch.

The ITCZ meanders with both longitude and season. The extreme positions occur in February and August when temperatures are highest in the respective summer hemisphere. In February its location ranges from 18°S over South Africa and Australia to 7°N in the eastern Pacific Ocean and in August from 3°N over India and South-East Asia. The MONSOON rains of the latter region and also of the Sahel region of Africa are triggered by the ITCZ.   WDS

Reading
Gedzelmen, S.D. 1980: *The science and wonders of the atmosphere*. New York: Wiley.
Riehl, H. 1979: *Climate and weather in the tropics*. New York: Academic Press.

intrazonal soil A soil group comprising well-developed soils, the main characteristics of which can be attributed more to local factors such as relief, drainage or parent material than to climate factors.

intrenched meander A meander or bend in a river channel that has become incised into the surrounding landscape as a result of local tectonic uplift (see MEANDERING).

intrinsic permeability (or specific permeability) A measure of the capacity of a rock or soil under natural conditions to transmit fluids. It depends upon the physical properties of the porous medium, such as pore size, shape and distribution. It is measured in m^2 or darcy units (one darcy is approximately equal to 10^{-8} cm^2). Intrinsic permeability k is related to hydraulic CONDUCTIVITY K (which takes into account the physical properties of the liquid as well as the rock) as follows:

$$K = \frac{k\rho g}{\mu}$$

where ρ mass density and μ dynamic viscosity are functions of the fluid alone, and g is acceleration due to gravity (Freeze and Cherry 1979).   PWW

Reference
Freeze, R.A. and Cherry, J.A. 1979: *Groundwater*. Englewood Cliffs, NJ: Prentice-Hall.

introductions, ecological Introductions, usually deliberate, of an organism into new regions lying outside the range of its natural occurrence, by which it is hoped to bring about some specific ecological condition or control in the receiving areas or habitats. Although any introduction of a species, accidental or otherwise, into a new environment is likely to have considerable ecological repercussions, the concept of ecological introductions is mostly used to describe carefully planned introductions made with specific ecological intentions in mind. For example, where aquatic weeds are an increasing problem, water authorities may decide to introduce the herbivorous grass carp (*Ctenopharyngdon idella*) to bring them under control. In many instances, the aim is to create or recreate some particularly effective food chain or ecological relationship.

A famous example of such an introduction is afforded by the deliberate spread of an Australian ladybird, *Novius (Vedalia) cardinalis*, to California. Around 1968, the fluted or cottony cushion scale insect (*Icerya purchasi*), another native of Australia, appeared in California to threaten the famous citrus groves of the area. Within a few years, this threat had been overcome by the numerous descendants of the 139 specimens of the ladybird introduced from Australia, where it was a natural enemy of the fluted scale insect and from where it had been introduced to California to control the alien pest. As Charles Elton (1958) has written, this was a 'miracle of ecological healing' in which 'Australia administered the poison, but it also supplied the antidote.' The success was repeated in many other countries, e.g. New Zealand and Egypt, where the fluted scale became a problem. It is a perfect example of an ecological introduction, in which one alien species is controlled by a natural predator from its own ecosystem, thus using and re-establishing a natural ecological chain. Another pest, this time from the plant kingdom, required the introduction of the cinnabar moth, which cleared vast areas of the prickly pear (*Opuntia* species) in Australia.

In most cases of predator control, the number of introductions need not be great, partly because of the speed of breeding, but also because of the fundamental principle of the ecological pyramid, in which predators are virtually always scarcer than their prey. Many ecological introductions, however, fail and some may even go badly wrong, leading to new and more serious problems including ECOLOGICAL EXPLOSIONS, the cure proving worse than the disease. Finally, there have been attempts to reintroduce formerly native animals and plants into their old habitats in order to try to establish the past ecological order, e.g. the wolf, a top carnivore, in certain forest areas in the west of Germany. (See also ALIENS; BIOLOGICAL CONTROL.)   PAS

Reading and Reference
Elton, C.S. 1958: *The ecology of invasions by animals and plants*. London: Methuen.
†Emden, H.F. van 1974: *Pest control and its ecology*. London: Edward Arnold.
†Samways, M.J. 1981: *Biological control of pests and weeds*. London: Edward Arnold.
†Simmons, I.G. 1979: *Biogeography: natural and cultural*. London: Edward Arnold.

intrusion A mass of igneous rock that has penetrated older rocks through cracks and faults before cooling (see diagram on p. 284). The process of emplacement of such a mass of rock (see diagram on p. 284).

inversion of temperature An increase of temperature with height, the inverse of the normal decrease of temperature with height that occurs in the TROPOSPHERE. Temperature inversion layers are very stable and greatly restrict the vertical dispersion of atmospheric pollutants. They can form in

several different ways. (1) Radiative cooling of the air near the ground at night. These inversions are very common on clear nights, but dissipate rapidly after sunrise. (2) Advective cooling of warm air passing over a cold surface. These persistent inversions may be accompanied by thick fog if the air is moist. (3) A cold air mass undercutting a warm air mass along a FRONT. These frontal inversions act as an invisible barrier between the two AIR MASSES. (4) Radiative heating in the upper atmosphere. The STRATOSPHERE and thermosphere are examples of this type of inversion. Thunderstorm clouds and pollutants rarely penetrate far into the stratosphere because of its great stability. (5) Descent and ADIABATIC heating of air from the upper troposphere. These subsidence inversions are most common near and to the east of ANTICYCLONES. They may be very intense and persist for days, trapping noxious pollution in a thin air layer near the ground.   WDS

Reading
Battan, L.J. 1979: *Fundamentals of meteorology*. Englewood Cliffs, NJ: Prentice-Hall.

inverted barometer effect Adjustment of sea level to changes in barometric pressure; in the case of full adjustment, an increase in barometric pressure of 1 mb corresponds to a fall in sea level of 0.01 m. If there is full adjustment, the observed pressures at the sea bed are unchanged.
     DTP

inverted relief The condition resulting from the erosion of areas of high relief, such as anticlines, to produce low-lying areas, such as valleys, which simultaneously results in the originally low-lying inclines becoming hills. Equally the deposition of resistant lag gravels, lava streams or duricrusts in river valleys may cause them to be left upstanding in a subsequent phase of erosion.

involution The refolding of large nappes, producing complex structures of more recent nappes within old nappes. Also a term synonymous with cryoturbation.

ion concentrations Studies of the DISSOLVED SOLIDS content of precipitation, run-off and water from other phases of the hydrological cycle will frequently consider the concentrations of individual constituents. With the exception of dissolved silica

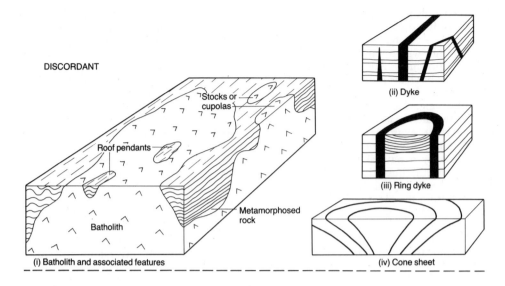

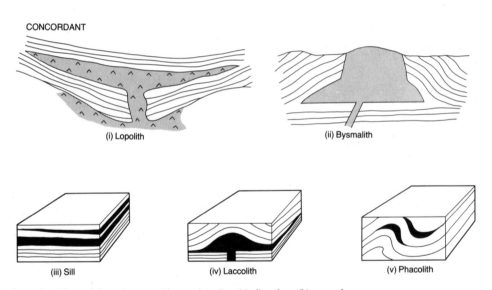

Intrusion: *Some of the major types of igneous intrusion:* (a) *discordant;* (b) *concordant.*
Source: *A.S. Goudie 1993:* The nature of the environment, third edn. *Oxford and Cambridge MA: Blackwell Publishers p. 262.*

and small quantities of dissolved organic matter, the dissolved material is largely dissociated into charged particles or ions and water analyses are therefore generally expressed in terms of concentrations of individual ions (mg l^{-1} or milliequivalents l^{-1}). The major cations (positively charged ions) in natural waters are Ca^{2+}, Mg^{2+}, Na^+, and K^+; and the major anions (negatively charged ions) are HCO_3^- and CO_3^{2-}, SO_4^{2-}, Cl^-, F^-, and NO_3^-. DEW

ionic wave technique See DILUTION DISCHARGE.

ionosphere Region above a height of about 50 km in which the gas density is so low and the temperature so high that positive and negative ions can move with some degree of independence. Electric currents so generated cause daily fluctuations in the earth's magnetic field, affect the propagation of radio waves and respond to solar flares. Other planets seem to have similar layers at similar values of pressure.   JSAG

Reading
Goody, R.M. and Walker, J.C.G. 1972: *Atmospheres*. Englewood Cliffs, NJ: Prentice-Hall.

island arc A chain of islands, mostly volcanic in origin, with a characteristic arcuate planform, rising from the deep ocean and associated with an ocean trench. Island arcs, such as those of the south-western Pacific, are generally located fairly near to continental masses and their curvature is typically convex towards the open ocean. Some island arcs comprise an inner arc of active volcanoes and an outer arc of non-volcanic origin formed from sediments thrust up from the ocean floor. According to the PLATE TECTONICS model, island arcs form as a result of the volcanism induced by subduction of oceanic LITHO-SPHERE.   MAS

Reading
Karig, D.E. 1974: Evolution of arc systems in the western Pacific. *Annual review of earth and planetary sciences* 2, pp. 51–75.
Sigimura, A. and Uyeda, S. 1973: *Island arcs: Japan and its environs*. Amsterdam: Elsevier.

island biogeography In general terms, the study of the distribution and evolution of organisms on islands; more narrowly, the examination of MacArthur and Wilson's equilibrium theory of island biogeography. Island floras and faunas have always fascinated biogeographers and biologists, and all the great nineteenth-century biologists were intrigued by the highly distinctive plants and animals and geographically isolated environments they found on these 'natural laboratories'. In a famous lecture to the British Association given in 1866, J.D. Hooker outlined the main characteristics and origins of island floras, while Darwin wrote of his famous Galápagos finches (the Geospizinae) and in 1880 A.R. Wallace published his classic book entitled *Island life or the phenomena and causes of insular faunas and floras*.

In the main, island biota are more polar in character than their neighbouring mainland counterparts and there are usually fewer species than on a similar-sized continental area. Moreover, the species mix tends to be disharmonic, being different from that on the mainland and often seeming out of harmony ecologically. For example, there are sometimes no top carnivores. The mix is usually an assemblage of taxas noted for their capacity to accomplish long-range dispersal and migration. In many instances, the island progeny of these able dispersers have lost their dispersal ability, a phenomenon well exemplified by the ill-fated dodo (*Raphus cucullatus*) of Mauritius, a large flightless bird related to the pigeon. The small populations on islands are subject to a range of distinctive ecological pressures and under these conditions evolution is accelerated, with new forms developing through ADAPTIVE RADIATION and hybridization.

Islands are therefore noted for their endemic organisms and for the large and abnormal percentage of their floras and faunas which are endemic. For example, some 45 per cent of the birds of the Canaries are endemic and no less than 90 per cent of the flora of the Hawaiian Islands, the most isolated of all floristic regions. Islands are also the homes of relict organisms, which have survived there, but have become extinct elsewhere. This is especially the case on islands which were once part of a continental system, as with Crete, a former remnant of an old mountain system that connected the Balkans with southern Anatolia. Extinction, however, is also known in island biota and it is the balance between immigration and extinction which is at the heart of the now much discussed equilibrium theory of island biogeography.

This theory was first published by R.H. MacArthur and E.O. Wilson in 1963 as the equilibrium theory of island zoogeography, but was widened to include plants in their book of 1967, simply entitled *The theory of island biogeography*. The theory was mainly stimulated by thoughts on the distributions found across oceanic islands in the Pacific.

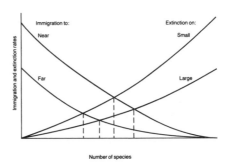

Number of species

The core of the theory is simple. It is argued that the number of species on an island is determined by a balance between immigration (which is a function of the distance of the island from the mainland) and extinction (which is a function of island area). The theory assumes a pool of species P, which can immigrate to the island and which is the number of species on the neighbouring landmass. The theory is illustrated by a widely used diagram which is given on p. 285.

In the diagram, the equilibria occur at the intersections of the extinction and immigration curves, points which are indicated by a series of letters referring to different types of island. The theory can also be presented in the form of an equation:

$$S_{t+1} = S_t + I + V - E,$$

where S_t is the number of species at time t, I is the number of immigrants by time $t + 1$, V the number of new species evolving *in situ* on the island, and E the number of extinctions.

This simple and seemingly logical theory has come under a wide range of criticism. First, it deals only with the number of species and not the number of individuals of the species on the island. In other words, it ignores population numbers. Secondly, it does not really deal with evolution, although in chapter 7 of the book the theory is tentatively extended to include this process. Thirdly, it appears to ignore historical factors which might, for example, mean that many organisms are relicts, subject only to extinction and with no potential for immigration. Fourthly, the theory lumps together all species and treats them as functioning in a similar manner. Fifthly, there are serious problems in defining both immigration and extinction and many argue that it is unacceptable to make immigration solely a function of distance. Finally, of course, it must not be assumed that all islands are in 'equilibrium', for in many this has yet to be reached, even in terms of the theory. Yet, despite all these criticisms, the theory has stimulated a new and invigorated interest in island biogeography.

It should also be noted that the theory has received a much wider application than its use on oceanic islands and that it has been related to the fauna and flora of biological 'islands' on continental areas, such as the remaining relicts of tropical rain forest. It is now being used to help determine the minimum size of viable conservation areas in which the local populations will be able to maintain themselves in some form of equilibrium. In general, it is true to say that the main tenets of island biogeography also apply to a wide range of such biological islands, including isolated mountain tops, ponds and lakes, and tracts of woodland surrounded by agriculture.

One thing is certain, namely, that the arrival of man on many isolated islands has seriously disrupted and altered their ecosystems. Extinction rates have increased markedly with his presence, especially where he has created ecological disharmony by the introduction of alien species (see ALIENS). The Atlantic island of St Helena, for example, has seen the demise of its endemic St Helena ebony (*Trochetia melanoxylon*) which was destroyed by goats, first introduced in 1513, and by the deforestation of the island for fuel. All that is left is a barren landscape with a few relict fragments of the original biota persisting on cliffs and ridges. On the other hand, many species introduced by man have themselves begun to change and form distinctive island races. It is believed that the special forms of the long-tailed field mouse (*Apodemus sylvaticus*) found on the Scottish and Scandinavian islands have developed from ancestors brought to these scattered locations by the Vikings. The house mouse (*Mus musculus*) on the island of Skokholm off Wales is some 30 per cent different in form from the mainland populations in Pembrokeshire. Yet it was probably only introduced to the island around 1900 by rabbit catchers and farmers. Thus, islands continue to be wonderful laboratories for the study of immigration, extinction and evolution and they will remain at the centre of biogeography for a long time to come. (See also ALIENS; ENDEMISM; EXTINCTION; REFUGIA; SPECIES AREA CURVE.)   PAS

Reading
†Brown, J.H. 1971: Mammals on mountain tops: non-equilibrium insular biogeography. *American naturalist* 105, pp. 467–78.

†Carlquist, S. 1974: *Island biology*. New York: Columbia University Press.

†Gilbert, F.S. 1980: The equilibrium theory of island biogeography: fact or fiction? *Journal of biogeography* 7, pp. 209–35.

†Gorman, M. 1979: *Island ecology*. Outline studies in ecology. London: Chapman & Hall.

†Lack, D. 1947: *Darwin's finches*. Cambridge: Cambridge University Press.

†— 1976: *Island biology illustrated by the land birds of Jamaica*. Oxford: Blackwell Scientific.

MacArthur, R.H. and Wilson, E.O. 1967: *The theory of island biogeography.* Princeton, NJ: Princeton University Press.

†Stoddart, D.R. 1977, 1983: Biogeography. *Progress in physical geography* 1, pp. 537–43; 7, pp. 256–64.

†Williamson, M. 1981: *Island populations.* Oxford: Oxford University Press.

islands Landsurfaces totally surrounded by water and smaller in size than the smallest continent (Australia). Oceanic islands are built up from the ocean floor and are part of the basal structure, not attached to continents, as in the example of the Hawaiian group. Continental islands are part of the neighbouring continental geological structure, as exemplified by the British Isles. The dispersal and colonization of plants and animals to islands is related to the distance from the species source. Hence oceanic islands tend to be occupied by a smaller number of species, highly adapted to the available HABITAT or NICHE and frequently ENDEMIC. PAF

Reading
Gorman, M. 1979: *Island ecology.* London: Chapman & Hall.

MacArthur, R.H. and Wilson, E.O. 1967: *The theory of island biogeography.* Princeton, NJ: Princeton University Press.

Menard, H.W. 1986: *Islands.* New York: Scientific American Books.

Nunn, P. 1993: *Oceanic islands.* Oxford: Blackwell.

isochrones Lines joining points on the earth's surface at which the time is the same. Lines joining points which experienced a seismic wave at the same time.

isocline A fold which is so pronounced that the strata forming the limbs of the fold dip in the same direction at the same angle.

isolation, ecological The ecological or habitat separation of one population from another so that interbreeding is normally prevented, even though the organisms involved may have overlapping geographical ranges (see SYMPATRY). Thus, although two closely related organisms live in the same region, interbreeding does not take place because they occupy different habitats. A classic example is afforded by two sympatric African species of *Anopheles* mosquito, the one, *A. melas,* confined to brackish water habitats, the other, *A. gambiae,* to freshwater. PAS

Reading
Ross, H.H. 1974: *Biological systematics.* Reading, Mass.: Addison-Wesley.

isopleths Lines drawn on maps connecting points which are assumed to be of equal value (e.g. contours on a topographical map). Among specific types of isopleth are those shown in the table.

isostasy The condition of hydrostatic equilibrium between sections of the LITHOSPHERE with respect to the underlying

Some isopleth types

Type	Connects up points of equal
Isobar	Barometric pressure
Isobase	Uplift or subsidence during a specified time period
Isobath	Distance beneath the surface of a water body
Isobathytherm	Temperature at a given depth below sea level
Isocheim	Mean winter temperature
Isoflor	Floral character
Isoglacihypse	Altitude of the firn line
Isohaline	Salinity in the oceans
Isohel	Recorded sunshine hours
Isohyet	Rainfall amount
Isomer	Mean monthly rainfall expressed as a percentage of the mean annual rainfall
Isoneph	Cloudiness
Isonif	Snow depth
Isopach	Rock-stratum thickness
Isoryme	Frost intensity
Isotach	Wind or sound velocity
Isothere	Mean summer temperature
Isotherm	Temperature
Isothermobath	Seawater temperature at a given depth

Units of the comparatively rigid outer layer of the earth in effect 'float' in the more mobile and denser material at greater depth. Isostatic adjustment was originally thought to occur by vertical movements of the crust with respect to the underlying MANTLE, but some isostatic models now assume that adjustment occurs through the movement of the lithosphere, which comprises not only the crust but an underlying zone of comparatively rigid mantle.

Two models of isostasy were proposed during the nineteenth century. G.B. Airy noted that the gravitational attraction of the Himalayan mountains was less than could be explained if the range was simply above a radially homogeneous crust and mantle. The gravity anomaly was explained by Airy as a result of crustal blocks of equal density but diffferent thicknesses.

The thickest blocks form the highest topography and are supported by roots of light crust, which have displaced the denser underlying mantle. At a depth equal to, or greater than, the thickness of the crust, the pressure in the mantle is constant and hydrostatic equilibrium is attained. An alternative model proposed by J.H. Pratt attempted to explain isostasy by variations in the density rather than the thickness of crustal blocks. In this model the crust is assumed to be of equal thickness and areas of high elevation are associated with low density crust, which is more buoyant with respect to the underlying mantle than adjacent areas of denser crust. Although these models make unrealistic assumptions about the fluid nature of the mantle and the ability of crustal blocks to move independently, they describe adequately the gross variations in gravitational attraction over the earth's surface. The crustal thickness model is particularly applicable to most continental mountain systems, whereas the density model provides a more adequate explanation of the relief of the mid-ocean ridges.

While the lithosphere has a tendency to attain isostatic equilibrium, several factors may prevent this from occurring. For example, temperature and density variations associated with convection in the mantle can lead to marked gravity anomalies indicative of a lack of isostatic equilibrium. Another factor is the rigidity of the lithosphere, which means that variations in surface loading over a small area may not promote isostatic compensation, while compensations to a really extensive change

in loading may produce vertical movements well beyond the zone of loading itself. This is especially well illustrated by glacial isostasy, the response of the lithosphere to the loading and unloading of the surface by ice. When an ice sheet develops the surface is depressed by an amount proportional to the ratio between the density of the ice and that of the mantle, but the zone of depression may extend beyond the actual area of the ice sheet by up to 300 km. Melting of the ice sheet leads to isostatic rebound, the speed of which provides important evidence concerning the RHEOLOGY of the mantle. Rates of uplift estimated from raised shorelines and other features may exceed 20 mm year^{-1}. Much slower rates of isostatic compensation are associated with erosional unloading of the continents, although these rates are sustained over much longer periods and are consequently an important factor in continental uplift.   MAS

Reading
Andrews, J.T. ed. 1974: *Glacial isostasy* . Stroudsburg, Pa.: Dowden, Hutchinson & Ross.

Lyustikh, E.N. 1960: *Isostasy and isostatic hypotheses.* New York: American Geophysical Union Consultants Bureau.

Mörner, N-A. ed. 1980: *Earth rheology, isostasy and eustasy.* Chichester: Wiley.

Smith, D.E. and Dawson, A.G. 1983: *Shorelines and isostasy.* London: Academic Press.

isotope A form of an element which, while always having the same number of protons in the nucleus, has another form or forms with differing numbers of neutrons. Thus carbon with an atomic number (number of protons = Z) of 6 has isotopes of mass (protons + neutrons) 12, 13 and 14. Of these ^{14}C decays radioactively (see CARBON DATING). Lead (^{210}Pb) is also an important decay dating method for recent sedimentary sequences. Other isotopes are stable. Oxygen has isotopes mass 16 and 18 which do not decay radioactively. Though chemically the same, their physical properties are slightly different. During natural cycles of evaporation and condensation a natural fractionation takes place which is temperature dependent. There are thus changes in the isotopic composition of water. The value of the ratio ^{16}O/^{18}O (δ^{18}O; measured in parts per thousand) is used in ice and marine core determinations of palaeotemperature. Hydrogen/deuterium (δD) variations are also used for a similar purpose often together with δ^{18}O. Other stable isotope ratios have environmental importance. WBW

Reading
Faure, G. 1986: *Principles of isotope geology*. 2nd edn. New York: Wiley.
Goudie, A.S. ed. 1990: *Geomorphological techniques*. 2nd edn. London: Unwin Hyman.

isthmus A narrow strip of land which connects two islands or two large land masses.

ITCZ See INTERTROPICAL CONVERGENCE ZONE.

J

jet stream A band of fast-moving (>30 m s^{-1}) air usually found in middle latitudes in the upper TROPOSPHERE, and associated with strong horizontal gradients of density and temperature below (BARO-CLINITY). Mean zonal cross-sections show a subtropical and polar-front jet. These probably define the equatorial and pole-ward limit of the excursions of a single jet distorted by successive weather systems.

JSAG

Reading
Ludlam, F.H. 1980: *Clouds and storms.* University Park: Pennsylvania State University.

joint probability estimates Estimates of probabilities of extreme sea levels and currents based on the probabilities of independent occurrence of the contributing tidal and surge events. By separating the individual statistics it is possible to calculate more reliable estimates from the available observations.

DTP

jökulhlaup An expressive Icelandic term for catastrophic drainage of a subglacial or ice-dammed lake. The lake may build up seasonally or over several years only to drain in a matter of hours when conditions are suitable for meltwater to open up tunnels in the glacier, mainly by frictional heating. In Iceland some jökulhlaups may be triggered by volcanic activity.

DES

Reading
Björnsson, H. 1992: Jökulhlaups in Iceland: prediction, characteristics and simulation. *Annals of glaciology* 16, pp. 95–106.

Röthlisberger, H. 1972: Water pressure in intra- and subglacial channels. *Journal of glaciology* 11.62, pp. 177–203.

Thorarinsson, S. 1953: Some new aspects of the Grimsvötn problem. *Journal of glaciology* 2.14, pp. 267–74.

juvenile water Water that originates from the interior of the earth and has not previously existed as water in any state. Consequently, it has not previously partici-pated in the hydrological cycle. The term was coined by Meinzer (1923), who contrasted juvenile with METEORIC WATER.

PWW

Reference
Meinzer, O.E. 1923: Outline of ground-water hydrology. *US Geological Survey water-supply paper* 494.

K

K- and r-selection See r- AND K-SELECTION.

K-cycle The name given to a concept of landscape development involving the cyclic erosion of soils on upper hillslopes during unstable climatic phases and soil development during stable phases. The term is much used in Australia.

kame An irregular mound of stratified sediment associated with GLACIOFLUVIAL activity during ice stagnation. It is a Scottish term for a landform much prized for the variety it adds to golf courses. DES

kame terrace A terrace formed between a hillside and a glacier by glaciofluvial activity. The landform is commonly associated with the former presence of stagnant ice down-wasting in valleys.

kamenitza A generally small solutional basin developed on the surfaces of soluble rocks such as limestones. They are a type of LAPIÉ. ASG

kaolin A clay mineral, mainly hydrated aluminium silicate, or any rock or deposit composed predominantly of kaolinite. China clay or other material from which porcelain can be manufactured.

karren (singular karre) A collective name describing small limestone ridges and pool structures which have developed as a result of the solution of rock by running or standing water. There are many types of karren, all differentiated by morphology (Bögli 1960). The commonest are rillenkarren (sharp ridges between rounded channels). In Britain the best examples are on Hutton Roof Crag, Kirkby Lonsdale (Lancs), while spectacular karren scenery can be found at Lluc in Mallorca. The term is German in origin; the French equivalent is *lapiés*. PAB

Reference
Bögli, A. 1960: Kalklosung und Karrenbildung. *Zeitschrift für Geomorphologie Supplementband* 2, pp. 4–21.

One class of karren comprises the narrow vertical solution flutes called rillenkarren. These examples, which are 2 cm across, are from Mallorca.

Classification of solutional microforms developed on limestone

	Form	Typical dimensions	Comments
Forms developed on bare limestone — *Developed through areal wetting*	Rainpit	<30 mm across, <20 mm deep	Produced by rain falling on bare rock. Occurs in fields on gentle rather than steep slopes. Can coalesce to give irregular, carious appearance
	Solution ripples	20–30 mm high; may extend horizontally for >100 mm	Wave-like form transverse to downward water movement under gravity. Rhythmic form implies that periodic flows or chemical reactions are important in their development
	Solution flutes (rillenkarren)	20–40 mm across, 10–20 mm deep	Develop due to channelled flow down steep slopes. Cross-sectional form ranges from semi-circular to V-shaped but is constant along flute
	Solution bevels	0.2–1 m long, 30–50 mm high	Flat, smooth elements usually found below flutes. Flow over them occurs as a thin sheet
	Solution runnels (rinnenkarren)	400–500 mm across, 300–400 mm deep, 10–20 m long	Down runnel increase in water flow leads to increase in cross-sectional area. May have meandering form. Ribs between runnels may be covered with solution flutes
Developed through concentration of run-off	Grikes (kluftkarren)	500 mm across, up to several metres deep	Formed through the solutional widening of joints or, if bedding is nearly vertical, of bedding planes
	Clints (flackkarren)	Up to several metres across	Tabular blocks detached through the concentration of solution along near-surface bedding planes in horizontally bedded limestone
	Solution spikes (spitzkarren)	Up to several metres	Sharply pointed projections between grikes
Forms developed on partly covered limestone	Solution pans	10–500 mm deep, 0.03–3 m wide	Dish-shaped depressions usually floored by a thin layer of soil, vegetation or algal remains. CO_2 contributed to water from organic decay enhances dissolution.
	Undercut solution runnels (hohlkarren)	400–500 mm across, 300–400 mm deep 10–20 m long	Like runnels but become larger with depth. Recession at depth probably associated with accumulation of humus or soil which keeps sides at base constantly wet
	Solution notches (korrosionkehlen)	1 m high and wide, 10 m long	Produced by active solution where soil abuts against projecting rock giving rise to curved incuts
Forms developed on covered limestone	Rounded solution runnels (rundkarren)	400–500 mm across, 300–400 mm deep 10–20 m long	Runnels developed beneath a soil cover which become smoothed by the more active corrosion associated with acid soil waters
	Solution pipes	1 m across, 2–5 m deep	Usually become narrower with depth. Found on soft limestones such as chalk as well as mechanically stronger and less permeable varieties

Note: The commonly encountered German terms are given in parentheses.

Source: Summerfield, M.A. 1991: *Global geomorphology*. London and New York: Longman Scientific and Technical and Wiley. Table 6.6. Based largely on discussion in Jennings, J.N. 1985: *Karst geomorphology*. Oxford: Blackwell. Pp. 73–82.

karst Generally, the term given to limestone areas which contain topographically distinct scenery, including CAVES, SPRINGS, BLIND VALLEYS, KARREN and DOLINES. Specifically, Karst is a region of limestone country between Carniola and the Adriatic coast, which is characterized by typical limestone topography.

Karst regions are typified by the dominant erosional process of solution, the lack of surface water and the development of stream sinks (dolines), cave systems and resurgences or springs. Indeed, the process of stream sinking is known as karstification. All the resultant landforms associated with karst scenery depend upon this phenomenon of stream sinking.

The most abundant rock type which exhibits karst features is limestone and the best karst scenery can be found when that limestone is pure, very thick in areas of upstanding relief, in an environment which provides enough water for solutional processes. Other calcareous rocks such as chalk fit many but not all of the above prerequisites. Chalk is often too soft to give rise to distinctive karst scenery.   PAB

Reading
Ford, D. and Williams, P. 1989: *Karst geomorphology and hydrology*. London: Unwin Hyman.
Ford, T.D. and Cullingford, C.H.D. eds 1976: *The science of speleology*. London: Academic Press.
Jennings, J.N. 1985: *Karst geomorphology*. Oxford: Basil Blackwell.

katabatic flows Downslope winds, often coupled with, or induced by, the large-scale atmospheric circulation. These flows may reach surrounding lowlands as dry warm or cold winds, blowing at speeds in excess of 50 m s^{-1} for several days. Examples of warm katabatic wind are the FÖHN on the north slopes of the Alps in Europe and the Chinook on the east slopes of the Rockies in the USA. These strong winds derive their warmth either from ADIABATIC compression during descent or from heat released by condensation on the windward slopes of the mountains or from both mechanisms together. This heat can increase the temperature of the air by 20°C or more. Warm katabatic winds occur most frequently during the cooler months and when there is a rapid sea-level pressure from the highlands to the lowlands. Many people become depressed or irritable when these winds blow.

Cold katabatic winds occur when a large pool of cold air, forming perhaps over a mountain glacier or on ice caps, becomes so deep that it spills over into the highlands. Heat release by condensation is not involved here, so the air remains cold. The glacier winds of Greenland and Antarctica, the BORA along the Adriatic coast of Yugoslavia, and the MISTRAL along the French Mediterranean coast are good examples of cold katabatic winds.   WDS

Reading
Atkinson, B.W. 1981: *Meso-scale atmospheric circulations*. New York and London: Academic Press.
Gedzelman, S.D. 1980: *The science and wonders of the atmosphere*. New York: Wiley.

kata-front Any front at which the warm air is subsiding relative to the cold air. As a result frontal activity is weak with only a belt of shallow stratiform cloud marking its presence. The change from an ana- to a kata-frontal structure can be seen sometimes on satellite images by their different cloud characteristics. Kata-fronts tend to develop at some distance from the cyclone centre where uplift in the warm air is less marked.   PAS

kavir A playa, sabkha or similar desert depression which is occasionally filled with water. The term is used in Iran and neighbouring areas.

kegelkarst Groups of residual conical-shaped limestone hills produced by limestone solution in adjoining DOLINES or SHAKEHOLES. The remnant limestone blocks are steep-sided and often heavily vegetated. They are also called cone-karst, COCKPIT KARST or MOGOTES.   PAB

Reading
Sweeting, M.M. 1972: *Karst landforms*. London: Macmillan.

kelvin wave A long wave in the oceans whose characteristics are altered by the rotation of the earth. In the northern hemisphere the amplitude of the wave decreases from right to left along the crest, viewed in the direction of wave travel. DTP

kettle, kettle hole An enclosed depression resulting from the melting of buried ice. Kettle holes are characteristic features of STAGNANT ICE TOPOGRAPHY.   DES

khamsin A hot, dry wind which blows from the desert to the south across the north African coast. Also called the sirocco and ghibli.

kinematic wave Consists of zones of high and low density which travel through a medium at a velocity which is generally different from that of the medium as a whole.

One set of solutions of the CONTINUITY EQUATION for material in a medium is generally dominant under certain circumstances. The theory was originally developed by Lighthill and Whitham (1955) and has been applied in a number of contexts within physical geography. The continuity equation in its differential form may be rewritten, using the same notation, as:

$$c\partial s/dx + \partial s/dt = a.$$

In this formulation, c is defined as the kinematic wave velocity, equal to $[dQ/dS]$ evaluated at x, and not necessarily or usually a constant. For the simplest case where c is constant and $a = 0$, the complete solution to the above equation is:

$$S = f(x - ct)$$

for an arbitrary function f. This represents a wave travelling without change of form at velocity c. In more complex cases different parts of the wave travel at different velocities, but the concept of a wave remains.

Kinematic waves have been applied to glacier response to climatic changes (Nye 1960), to the movement of riffles in stream beds (Langbein and Leopold 1968), to the movement of stream knick-points and valley-side terraces, and to the routing of river and overland and flow flood peaks.

MJK

Reading and References
†Freeze, R.A. 1978: Mathematical models of hillslope hydrology. In M.J. Kirkby ed., *Hillslope hydrology*. Chichester: Wiley.

Langbein, W.B. and Leopold, L.B. 1968: River channel bars and dunes – theory of kinematic waves. *US Geological Survey professional paper* 122L, L1–209.

Lighthill, M.J. and Whitham, G.B. 1955: On kinematic waves I: flood movements in long rivers. *Proceedings of the Royal Society* Series A. 229, pp. 281–316.

Nye, J.F. 1960: The response of glaciers and ice sheets to seasonal and climatic changes. *Proceedings of the Royal Society* Series A. 256, pp. 559–84.

kinematics The branch of mechanics dealing with the description of the motion of bodies without reference to the force producing the motion.

BWA

Reading
Petterssen, S. 1956: *Weather analysis and forecasting*. New York: McGraw-Hill. Esp. chs 2 and 3.

kinetic energy Energy due to the translational movement of a body. It is not so definitive as it looks. It depends on the frame of reference, e.g. an object lightly tossed from a rapidly moving railway train has potentially lethal energy for a bystander and vice versa and also upon scale, e.g. 'temperature' of a gas represents the kinetic energy of individual molecules of the gas.

JSAG

kingdoms of animals and plants The simple classic division of all living things (except the non-cellular, problematical viruses) into two categories of plants and animals (*plantae* and *animalia*) is no longer found entirely satisfactory and has been abandoned by most life scientists. The principal problem with the traditional twofold classification is that it groups organisms that are very unlike one another under the same heading. A few organisms lie uneasily in either category.

No classification of kingdoms is entirely satisfactory; the following is in widespread use.

Monera
Not having a well-defined nucleus: bacteria and blue-green algae. Acellular organisms, i.e. those lacking clear division into cells.

Protista
Acellular organisms mostly lacking chlorophyll: flagellates (some of which do possess chlorophyll), amoebae, foraminifera, sporozoans, ciliates. Some of these form colonies, being incipiently multi-celled.

Fungi
All kinds of fungi, including slime moulds. This group has long been included with the plants, but its members have a long, quite distinct evolutionary history. Non-photosynthetic organisms, mostly with definite cell walls.

Plantae
Six phyla (major groups) of photosynthetic organisms with cell walls, ranging from acellular forms (some algae) to much more complex organisms with highly developed organs and systems of organs (ferns, flowering plants).

Animalia
Multi-celled, non-photosynthetic organisms without cell walls. Some classifications recognize over 300 phyla, varying from simple forms with few cells, the mesozoa, through sponges which have partly differentiated tissues, to a great diversity of complex metazoans with well-developed organs and organ systems, e.g. worms, insects, molluscs, vertebrates. (See also FAUNAL REALMS; FLORISTIC REALMS.)

PHA

klippe An outcrop of rock that is separated from the rocks upon which it rests by a fault. It may represent an erosional remnant of a nappe or may have been emplaced by gravity sliding.

knickpoint A break in profile, generally in the long profile of a river. This was especially thought of as the product of REJUVENATION, where a steeper-gradient lower reach is receding headward as a result of local or general lowering of base level, and this steeper profile intersects with a gentler upper one. Some knickpoints may be sharply defined, as in a waterfall, or they may be much less distinguished and only apparent after detailed field survey of stream profiles.

The term is also applied to any profile irregularity, as at tributary confluences or associated with lithological or structural controls, and not just those produced following rejuvenation. Furthermore, rejuvenation may itself be generated in alternative ways – extensively by eustatic sea level change, or by isostatic and tectonic movements, or by changes in river discharge or sediment load. The early geomorphologists W.M. Davis and W. Penck (writing of *Knikpunkte* in German) were among the foremost in developing studies of such phenomena, though they used the word differently and in more restricted senses than are now generally adhered to. Emphasis is now placed on the interaction of many factors in stream system development, so that several possibilities for breaks in long profile would need to be explored.

JL

Reading
Small, R.J. 1970: *The study of landforms*. Cambridge and New York: Cambridge University Press.

knock-and-lochan topography A landscape of ice-moulded rock knobs with intervening lochans which have been eroded along lines of structural weakness. The type site is in the north-western highlands of Scotland (Linton 1963), but it is also characteristic of much of the Canadian and Scandinavian shields (Sugden 1978).

DES

Reading and References
†Gordon, J.E. 1981: Ice-scoured topography and its relationships to bedrock structure and ice movement in parts of northern Scotland and West Greenland. *Geografiska annaler* 63A. 1–2, pp. 55–65.
Linton, D.L. 1963: The forms of glacial erosion. *Transactions of the Institute of British Geographers* 33, pp. 1–28.
Sugden, D.E. 1978: Glacial erosion by the Laurentide ice sheet. *Journal of glaciology* 20.83, pp. 367–91.

koniology (coniology) The scientific study of atmospheric DUST together with its solid pollutants, such as soot, pollen, microbial spores, etc.

kopje A hillock or rock outcrop, applied especially in South Africa.

Köppen's climatic classification A system of climatic differentiation based upon TEMPERATURE and PRECIPITATION linked to vegetation zones. It is one of the most widely used methods of classification but has undergone numerous modifications since it was devised about 1900. Five major categories subdivide the earth's climates: tropical forest, dry, warm temperate rainy, cold forest and polar. Further subdivisions are made on the basis of the rainfall regime, temperature characteristics and any other special features. Each region can then be identified on the basis of a sequence of letters, e.g. Csb indicates a coastal Mediterranean climate with a mild winter and a dry but warm summer.

PS

Reading
Barry, R.G. and Chorley, R.J. 1992: *Atmosphere, weather and climate*. 6th edn. London: Routledge.

krotovina Infilled animal burrows or filaments found in soils and sediments such as loess.

krummholz From the German, meaning crooked wood, refers to the stunted and gnarled woodlands characteristic of forest margins at high altitudes and high latitudes. The dwarfing, distortion and, in extreme conditions, the prostrate habit of trees is a result of the combined effects of wind and cold. Such features are common in the transition zone between the sub-alpine forest and alpine tundra in high latitudes, or in the elfin woods of high tropical elevations, bearing a heavy cover of epiphytes and a ground layer cushioned by mosses, herbaceous plants and grasses. The word is also used to describe dense, tangled thickets in tropical forest.   PAF

kumatology A neglected term developed by the great British amateur geographer, Vaughan Cornish, for the study of wave-like forms encountered in nature.   ASG

kunkar See CALCRETE.

kurtosis (particle size) A measure, as in statistics, of the peakedness of distribution; in the case of sediments it relates to both sorting (standard deviation) and differences from a normal distribution (where a normal distribution would hve a kurtosis value of 1.0). A flat (platykurtic) distribution would be found in a poorly sorted sediment such as a till; a peaked (leptokurtic) distribution would be found in a well-sorted sediment such as a wind-blown sand. The usefulness of this measure has been questioned.   WBW

Reading
Briggs, D. 1977: *Sediments*. London: Butterworth.
Tucker, M.E. 1981: *Sedimentary petrology: an introduction*. Oxford: Blackwell Scientific.

L

laboratory experiments, fluvial and hydrological Investigations, normally involving the testing of hypotheses or properties, of river- and water-related phenomena under controlled environments. Though normally conducted indoors, some involve very large outdoor equipment, as in the 'Rainfall-Erosion Facility' 9 × 15 m container at Colorado State University (Schumm 1977). Some field sites have also been considered 'natural laboratories', particularly where rates of change are rapid and where some amount of experimental control is possible to a degree approaching that which can be established in an outdoor laboratory.

Laboratory experiments can be very varied in scope (see Goudie 1990, table 1.2): they may involve the hardware of slopes or river channels in flumes, experimental simulation of weathering or soil moisture movement under controlled conditions on field-sampled materials, or tests to establish the strength of materials. JL

Reading
Goudie, A.S. ed. 1990: *Geomorphological techniques*, 2nd edn. London: Unwin Hyman.
Schumm, S.A. 1977: *The fluvial system*. New York: Wiley.
†Yoxall, W.H. 1983: *Dynamic models in earth-science instruction*. Cambridge and New York: Cambridge University Press.

laccolith A mass of intrusive rock which though concordant with the host rocks has domed up the overlying strata. The base of the laccolith is either horizontal or convex downward.

Reading
Corry, C.E. 1988: Laccoliths: mechanisms of emplacement and growth. *Geological Society of America special paper* 220.

lacustral See PLUVIAL.

lacustrine Of or pertaining to lakes.

lag gravel An accumulation of coarse rock fragments at the landsurface which has been produced by the removal of finer particles, generally by deflation.

lag time The period elapsing between the occurrence of a causative phenomenon and its resulting effect, as in the time difference between peak storm rainfall and the later peak in stream discharge that results from it. This may be an important consideration in many physical processes. For example, BEDFORMS in rivers may be related to river flows, but such flows vary over time and it also takes a finite time for the bedforms themselves to develop. Thus there may be delayed response between development and a change in flow. JL

Reading
Allen, J.R.L. 1974: Reaction, relaxation and lag in natural sedimentary systems: general principles, examples and lessons. *Earth science reviews* 10, pp. 263–342.

lahar A mass movement feature on the flank of a volcanic cone. The volcanic ash may move either as a mudslide when saturated with water or as a dry landslide as a result of earth tremors.

lake breeze See SEA/LAND BREEZE.

lakes See LIMNOLOGY.

laminar flow A type of flow in which the movement of each fluid element is along a specific path with uniform velocity, with no diffusion between adjacent 'layers' of fluid. Injected dye maintains a straight, coherent thread. The shear stress between adjacent layers increases from zero at the surface to a maximum at the fluid–solid contact (e.g. the river bed), and the flow velocity increases parabolically with height above the bed. Fluid motion is laminar if viscous forces are so strong relative to inertial force that the fluid viscosity significantly influences flow behaviour. The viscosities of air and water are so low that laminar flow is rare in these fluids. Laminar flow is characterized by a REYNOLDS NUMBER of below 500, e.g. in the very shallow water of overland flow on hillslopes, and then only when the water is undisturbed by raindrop impact.   KSR

land breeze See SEA/LAND BREEZE.

land capability A measure of the value of land for agricultural purposes. The capability unit is described as a group of soils that are nearly alike, based on an interpretation of soil data. The soils in each will: (1) produce similar kinds of cultivated crops and pasture plants with similar management practices; (2) require a similar conservation treatment and management under the same kind and condition of vegetative growth; and (3) have comparable potential productivity.

Subclasses are defined according to their limitations for agricultural use and hazards to which they are exposed. In the USA four general limitations are recognized: erosion hazard, wetness, rooting zone limitations and climate, of which all but the last are closely related to geomorphology. In the UK another subdivision – gradient or soil pattern – is also added.

For more than twenty years the Soil Surveys of England and Wales and Scotland have been preparing Land Use Capability maps based on soil profile characteristics, which divide land into seven categories according to suitability for agriculture. Class 1 is described as land with very minor or no physical limitations; Class 7 as land with extremely severe limitations that cannot be rectified. One of the stated aims of these surveys is to provide detailed information for land use planning, but although maps can provide general strategic guidelines, they are insufficiently accurate or detailed to provide specific data to determine whether or not individual developments should be allowed.   SMP

Reading
Bibby, J.C. and Mackney, D. 1969: *Land use capability classification.* London: Soil Survey.

land drainage The removal of excess water from the land for a variety of purposes. Drainage may be necessary to improve the productivity of the land by increasing crop growth rates and by allowing a wider range of cultivation practices. Land drainage is often used in areas affected by flooding both to remove excess water quickly when a flood occurs and to improve infiltration during storms. In improving infiltration of rainfall, land drainage may also have the effect of reducing soil erosion generated by overland flow. Land drainage may also be used for controlling soil salinity.

Land drainage systems consist of three components: the field system, the main system and the outlet. The main system usually consists of a network of ditches which convey the drained water to an outlet where the water discharges into the natural drainage network of the area. The character of the field system depends upon whether the excess water results from a high water table or from permeability variations in the soil leading to impeded drainage at or near the surface.

Impeded drainage at or near the surface of the soil may be alleviated by shallow surface ditches and, where possible, by the cambering of the surface between ditches. Alternatively, mole drainage systems may be applied, particularly where the soil has a high clay content. Subsoiling and deep ploughing may also be used to loosen the soil and promote good drainage.

A high water table may be lowered using pipes or ditches within the field drainage system. The depth, layout and spacing of the pipes and ditches controls the water-table level and two common patterns of field drainage are the herringbone system and the parallel grid system. In the herringbone pattern the field drains are aligned across the slope at a slight angle to the contours to promote drainage to the collector drains which are aligned down the slope. In the parallel grid system, the field drains meet the collector drain at right angles and flow in the field drains is promoted by increasing the depth of each drain in the direction of the collector (see CHANNELIZATION; WETLAND DRAINAGE).   AMG

Reading
Armstrong, A.C. and Garwood, E.A. 1991: Hydrological consequences of artificial drainage of grassland. *Hydrological processes* 5, pp. 157–74.
Green, F.H.W. 1979: Field under-drainage and the hydrological cycle. In G.E. Hollis ed., *Man's impact on the*

hydrological cycle in the United Kingdom. Norwich: Geo Books.

Hill, A.R. 1976: The environmental impact of agricultural land drainage. *Journal of environmental management* 4, pp. 251–74.

Smedema, L.K. and Rycroft, D.W. 1983: *Land drainage.* London: Batsford Academic.

land systems Subdivisions of a region into areas having within them common physical attributes which are different from those of adjacent areas. Any one land system normally has a recurring pattern of topography, soils and vegetation, reflecting the underlying rocks (geology), erosional and depositional processes (geomorphology) and the climate under which these processes operate. A land unit, the detailed component of a land system, is particularly useful in evaluating land for agricultural and engineering purposes and in devising problem-orientated classifications. The resultant land systems maps are easily interpreted, and both rapid and economical to produce.   SMP

Reading
Veateh, J.O. 1983: *Agricultural land classification and land types of Michigan.* Michigan Agricultural and Experimental Station special bulletin 544.

land use, hydrological effects of
Land use affects the hydrological cycle by altering the natural rates of infiltration, run-off, erosion, and sedimentation. Rural land-use practices alter natural vegetation diversity by limiting the number of species and forms, and agricultural activities change

soil conditions and leave the surface barren for parts of the year in many cropping schemes. In urban areas impervious surfaces replace porous natural ones, and drainage networks are altered to include many artificial channels. These rural and urban changes have definable consequences for the hydrological cycle and for the products of its operation, especially the integrative measure of sediment yield.

The connections between land use, run-off, and sediment yield were demonstrated for rangelands by Noble (1965) who measured run-off and soil loss under three conditions of vegetation cover. Under good range conditions with diverse cover he found that only 2 per cent of the precipitation became run-off and that soil loss amounted to only 12 t km^{-2}. When land use of the area reduced the diversity and cover to fair conditions run-off increased to 14 per cent and soil loss to 122 t km^{-2}. Finally, when land use of the area resulted in the elimination of some species and the development of areas without vegetation cover, run-off increased to 73 per cent and soil loss to 1349 t km^{-2}.

Wolman (1967) has shown that when land use of a given humid area passes through a series of changes, sediment production also changes in a predictable fashion. Under conditions essentially unaltered by human activities sediment production of the example area was relatively low. The introduction of primitive farming resulted in a marked increase of

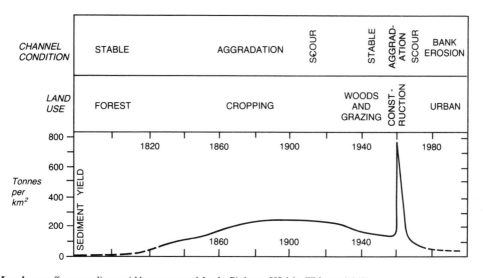

Land use: *effects on sediment yield as constructed for the Piedmont USA by Wolman (1967).*

sediment production. Mechanized farming without conservation practices resulted in further increases that raised the sediment production to a level several times the natural rate. Conservation practices reduced the sediment yield, and the abandonment of the fields which allowed the return of shrubs, dense grasses, and trees also reduced sediment yields to levels approaching their natural values. When the area became urbanized construction activities increased sediment yields to levels 50 or more times the natural levels. This brief peak was followed by a precipitous decline as the impervious surfaces of the city reduced sediment yield to very low levels. Similar studies by Ursic and Dendy (1965) in the southern USA showed that sediment yield for abandoned fields was ten times the yield for nearby pine forests, while the yield for cultivated fields was more than 100 times that of the pine forests.

Conservation practices can reduce excessive soil loss by as much as 90 per cent (Baird 1964). Remedial measures include crop rotation to include species that do not leave large areas of the surface exposed to erosion for long periods. Strip planting can be used to alternate protective crops with those that are more damaging. By leaving some fields fallow for some periods of time the land manager can reduce erosion by increasing the amount of organic material in the soil. On sloping lands terraces may be useful in reducing the gradient of field surfaces, thereby reducing the amount and velocity of run-off and its accompanying erosion.

In urban areas the development of large areas of impervious surfaces and the installation of artificial drainways ensure that run-off increases in amount and collects more quickly than in unaltered environments. Leopold (1968) showed a systematic relationship among these variables in urban areas. The consequence is that downstream from urban areas flood peaks rise more rapidly and are higher than in natural circumstances, resulting in increased flood damage. The flashy discharges occasionally erode newly expanded floodplains that accumulate during the high sediment yield period of construction (Graf 1975).

Control of the urban run-off problem may take either structural or non-structural approaches. The structural alternatives include the construction of retention basins in upstream areas to retard the rate of accumulation in main streams, or the construction of channel improvements in the downstream areas to speed the water out of the area. Unfortunately, this latter strategy merely transfers the problem to other areas downstream. Non-structural alternatives include the setting aside of the undeveloped areas in the upstream sections to act as sinks for run-off, and the avoidance of flood hazard zones downstream through building restrictions and zoning. (See also URBAN HYDROLOGY.)   WLG

References
Baird, R.W. 1984: Sediment yields from Blackland watersheds. *Transactions of the American Society of Agricultural Engineers* 7, pp. 454–6.

Graf, W.L. 1975: The impact of suburbanization on fluvial geomorphology. *Water resources research* 11, pp. 690–3.

Leopold, L.B. 1968: Hydrology for urban land planning: a guidebook on the hydrologic effects of urban land use. *US Geological Survey circular* 554.

Noble, E.L. 1965: Sediment reduction through watershed rehabilitation. *Proceedings of the Federal Interagency Sedimentation Conference*, US Department of Agriculture miscellaneous publication 970, pp. 114–23.

Ursic, S.J. and Dendy, F.E. 1965: Sediment yields from small watersheds under various land uses and forest covers. *Proceedings of the Federal Interagency Sedimentation Conference*, US Department of Agriculture miscellaneous publication 970, pp. 47–52.

Wolman, M.G. 1967: A circle of sedimentation and erosion in urban river channels. *Geografiska Annaler* 49A, pp. 385–95.

land-bridge An isthmus or other connection between two land masses across which animals and plants move to colonize a new environment.

landfill or sanitary landfill A method of disposing of refuse on land with the intention of creating the minimum nuisance or hazard by confining the refuse to the smallest practical area, reducing it (e.g. by compaction) to the smallest practical volume, and covering it with a capping. Landfill sites can include natural depressions and old man-made depressions, specially dug trenches or, in flat areas, artificial mounds. Sites need to be lined to prevent leaching of chemicals and organic wastes into groundwater. Problems also result from biochemical degradation which can lead to subsidence, the formation of disagreeable odours, and the formation of potentially explosive gases (e.g. methane).

ASG

Reading
Costa, J.E. and Baker, V.R. 1981: *Surficial geology: building with the earth*. New York: Wiley.

Landsat See UNMANNED EARTH RE-
SOURCES SATELLITES.

landscape ecology A term introduced
by the German geographer Carl Troll who
also later used the term geo-ecology. It has
two components (Vink 1983): an approach
to the study of the landscape which
interprets it as supporting natural and
cultural ecosystems; and the science which
investigates the relationships between the
biosphere and anthroposphere and either
the earth's surface or the abiotic components.   ASG

Reference
Vink, A.P.A. 1983: *Landscape ecology and land use*. London:
Longman.

landscape evaluation The classification of rural landscapes so that appropriate
planning action may be taken for their
future management. The approach has
developed since the early 1960s as the
problem of defining the aesthetic quality of
the landscape, so that it has real parity with
social and economic factors, has been
evident.

The evaluation operation involves more
than the simple identification and mapping
of land use changes, although many early
subjective attempts used empirical landscape components, which were then scored
and aggregated (Linton 1968) to identify
spatial variations in landscape characteristics. It also incorporates the values that
people attach to landscapes, which causes
complications as these values are likely to be
very subjective.

The development of objective techniques
during the 1970s was a major advance in the
field. These techniques attempt to weight
factors according to their contribution to or
detraction from landscape quality in a given
area (Coventry Subregional Study 1971).
Although involving computer analysis,
many initial decisions concerning factor
scores are still subjective.

There is a range of preference techniques
which identifies people's reactions to landscape by asking them to rank photographs in
order of landscape quality (Fines 1968).
However, finding representative individuals
to undertake the ranking, standardizing of
photographic quality and translating of
preferences into planning decisions cause
many problems.

The social basis for evaluation has
recently received attention. In this approach value is defined in terms of social
indicators rather than landscape components (Penning Rowsell *et al.* 1979). Unlike
most other methods these are not merely
theoretically based, but aim to solve
resource management problems.   SMP

Reading and References
Coventry, Solihull, Warwicks Subregional Study 1971:
Journal of the Town Planning Institute 57, pp. 481–4.
Fines, K.D. 1968: Landscape evaluation: a research project
in East Sussex. *Regional studies* 2, pp. 41–55.
Linton, D.L. 1968: The assessment of scenery as a natural
resource. *Scottish geographical magazine* 84, pp. 219–38.
Penning Rowsell, E.C., Gullet, G.H., Seale, G.H. and
Witham, S.A. 1979: *Public evaluation of landscape quality*.
Planning research group report 13. Middlesex Polytechnic.
†*Transactions of the Institute of British Geographers* 66 (old
series), 1975. Special edition on the topic.

landslide A landslip. The movement
downslope under the influence of gravity
of a mass of rock or earth (see diagram on p.
302). A mass of rock or earth that has
moved in such a way.

lapié See KARREN.

lapse rate The rate of *decrease* of a
quantity with height, usually applied to
temperature but sometimes also to the
mixing ratio (see HUMIDITY) of water
vapour to air. The typical rate of temperature change in the TROPOSPHERE is
6.5 K km^{-1} whereas for the temperature
of dry air displaced adiabatically it is
10 K km^{-1} (dry adiabatic lapse rate). It
follows that the troposphere is usually stable
to dry adiabatic processes.   JSAG

Reading
Ludlam, F.H. 1980: *Clouds and storms*. University Park, Pa.:
Pennsylvania State University Press.

Late Glacial A term used for the span of
time between the maximum of the Last (in
Britain, the Devensian) Glacial (*c.*18,000
BP) and the beginning of the Holocene
interglacial (*c.*11,000–10,000 BP). It was
marked by various minor stadials and
interstadials (e.g. see ALLERØD).   ASG

latent heat That part of the thermal
energy involved in a change of state, like
the $2.4 \times 10^6 \text{ J kg}^{-1}$ of energy released
when water vapour condenses to liquid.
This process makes rising cloudy air cool
less rapidly than does the ambient air with
height and hence the cloudy air becomes
buoyant. EVAPORATION of liquid water at
the ground 'saves up' solar energy in latent

Type of
movement

Falls

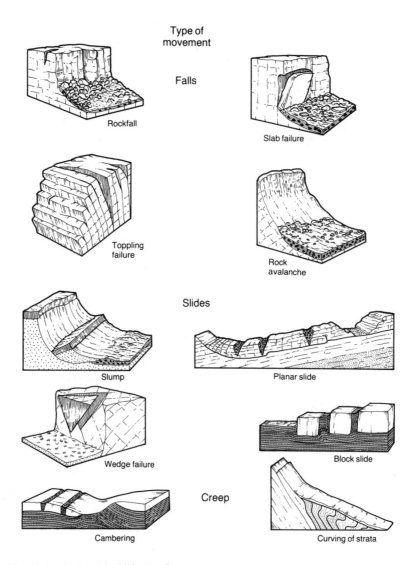

Rockfall

Slab failure

Toppling
failure

Rock
avalanche

Slides

Slump

Planar slide

Wedge failure

Block slide

Creep

Cambering

Curving of strata

Landslides: *A classification of landslides in rock.*
Source: *M.J. Selby 1982:* Hillslope materials and processes. *Oxford: Oxford University Press. Figure 7.18.*

form until it can be released in cloudy convection. Typical English thunderstorms rain French water. JSAG

Reading
Ludlam, F.H. 1980: *Clouds and storms*: University Park, Pa.: Pennsylvania State University Press.

lateral accretion The process by which bed sediments accumulate at the side of a channel as it shifts laterally. The term applies notably to the sediment accumulating in POINT BAR DEPOSITS, but lateral accretion can also occur in BRAIDED RIVERS as channels shift and bars enlarge. Such deposits may contrast with the vertical accretion of sediments deposited from suspension which are usually finer in size and may accrete beyond the confines of channels, and they may be particularly important volumetrically among the near-surface sediments of many FLOODPLAINS. Such sediments may possess a distinctive type of cross-bedding, called epsilon cross-bedding by J.R.L. Allen, in which the dipping or sigmoid-shaped beds represent successive increments of accretion developed at right-angles to the general flow direction.   JL

Reading
Allen, J.R.L. 1970: *Physical processes of sedimentation.* London: Allen & Unwin.
Collinson, J.D. and Thompson, D.B. 1982: *Sedimentary structures.* London: Allen & Unwin.
Reading, H.G. ed. 1986: *Sedimentary environments and facies.* 2nd edn. Oxford: Blackwell Scientific.

lateral flow Applied particularly to subsurface near-horizontal or ground slope-aligned water movement, in contradistinction to vertical water movement. Water flow along permeable soil horizons may be an important mechanism in the transfer of water from soils and hillslopes into streams without such flow taking place over the surface of the ground or through deep percolation to groundwater. (See also INTERFLOW; THROUGHFLOW.)   JL

lateral migration The movement of stream channels across valley floors through bank erosion and accompanying deposition on the opposite bank. This may proceed at varying rates on different rivers, on some amounting to several metres a year and on others none at all.   JL

Reading
Osborn, G. and Du Toit, C. 1991: Lateral planation of rivers as a geomorphic agent. *Geomorphology* 4, pp. 249–60.

lateral moraine See MORAINE.

laterite A surface accumulation of the products of rigorous chemical selection, developing where conditions favour greater mobility of alkalis, alkali earths and silica than of iron and aluminium. Bauxite is a laterite rich in aluminium. Laterite was originally defined by its ability to harden rapidly and irreversibly on exposure to the air, a property which led to its use as building bricks in southern India (Buchanan 1807). The term has been extended to include related materials (MacFarlane 1983) which were hard or contained hard parts, even though this induration may be an original result of iron segregation rather than of exposure.

Laterite profiles vary enormously in scale. The laterite may be a few centimetres to tens of metres thick, and below this the saprolite, leached or unleached of iron, varies from a few centimetres to over 100 m. Thick profiles develop on low angle slopes (Goudie 1973), and laterites can be divided into those that result from relative accumulation of iron and aluminium sesquioxides, and those that result from absolute accumulation.

Relative accumulations owe their concentration to the removal of more mobile components, and absolute accumulations to the physical addition of materials.   ASG

References
Buchanan, F. 1807: *A journey from Madras through the countries of Mysore, Kanara and Malabar.* London: East India Company.
Goudie, A.S. 1973: *Duricrusts of tropical and subtropical landscapes.* Oxford: Clarendon Press.
MacFarlane, M. 1983: Laterites. In A.S. Goudie and K. Pye eds, *Chemical sediments and geomorphology.* London: Academic Press.

latosol A lateritic soil.

Laurasia The northern part of Pangaea, a super-continent thought to have been broken up by continental drift. The southern continent, Gondwanaland, was separated from it by the Tethys Ocean.

lava Molten rock material which is extruded from volcanoes and volcanic fissures. (See p. 304.)

le Chatelier principle Named after the French inorganic chemist, H.L. le Chatelier (1850–1936), it defines a condition of a system in STABLE EQUILIBRIUM in which a change in any one of the governing forces will cause the equilibrium to shift so that the original condition is restored. In other words, the initial change sets up an internal reaction which is equal and opposite to that change and there is no net alteration in the system.

As used by physical geographers this concept is generally conflated with ideas of homeostasis, negative feedback and self-regulation.

In its original form and related to the thermodynamics of (strictly, isolated) systems, it states: 'If the temperature of a system in equilibrium be raised, the equilibrium will shift in favour of the reaction in which heat is absorbed (endothermic reaction); the converse also applies.' This shows how weathering of rocks, more or less in equilibrium in the lithosphere, tends towards exothermic (heat evolving) reactions. Oxidation, hydration and carbonation are volume-increasing

Bedded sheets of lava of phonolitic type on Grand Canary.

reactions in weathering typically of this kind.   BAK/WBW

leaching The downward movement of water through the soil zone which results in the removal of water-soluble minerals from the upper horizons and their accumulation in the lower soil zone or groundwater.

leaching requirement The fraction of irrigation water that must be leached through the root zone to maintain the soluble salt level in the soil at an acceptable level in relation to the salt tolerance of the proposed crop.

lee depression Region of low pressure found downwind of a mountain chain, representing the large-scale part of the drag of the ground on the air. The cyclonic circulation is due to the interplay of the vorticity and divergence of the air to the lee of the obstacle. Occasionally such features appear to develop and move away, as in cyclogenesis in the lee of the European Alps, which results in a depression over northern Italy that forces water up the Adriatic sea. This occasionally results in the flooding of Venice.   JSAG

Reading
Harwood, R.S. 1981: Atmospheric vorticity and divergence. In B.W. Atkinson ed., *Dynamical meteorology: an introductory selection.* London and New York: Methuen. Pp. 35–54.

McIntosh, D.H. and Thom, A.S. 1969: *Essentials of meteorology.* London: Wykeham Publications.

lee eddy A closed circulation (primarily in the vertical plane) often found on the downward side of steep obstacles. On the smaller scale, it defines a good place for scenic picnics; on the larger scale, it identifies places prone to recirculation of pollutants.   JSAG

lee waves Waves in the atmosphere of about 6 km wavelength (see GRAVITY) extending downwind of an obstacle in trains that may be up to 400 km long. Often made visible (remarkably so on satellite pictures) by alternating bands of clear and cloudy air (lenticular CLOUDS), they are conceptually important as identifying some aspects of the irreversibility of atmospheric processes.   JSAG

Reading
Atkinson, B.W. 1981: *Meso-scale atmospheric circulations.* London and New York: Academic Press.

Ludlam, F.H. 1980: *Clouds and storms.* University Park, Pa.: Pennsylvania State University Press.

lessivage Involves the translocation of silicate clays in colloidal suspension in a soil profile without any change in their chemical composition. It contrasts with podzolization in which the clay materials decompose and the hydrous oxides of iron and aluminium are mobilized.

levée A broad, long-crested ridge running alongside a FLOODPLAIN stream or intertidal inlet, composed generally of coarse sand to silt grade suspended sediment deposited by floodwaters as they overtop channel banks. The levée may slope gently away from the river and consist of progressively finer sediment as distance from the channel increases. Rather different features of the same name may also be created on steeper slopes by debris flows: here they may comprise boulders or coarse material. Levées may also be artificially created or raised for flood protection.   JL

Reading
Reading, H.G. ed. 1986: *Sedimentary environments and facies*. 2nd edn. Oxford: Blackwell Scientific.

lichenometry A technique for dating Holocene events that was developed in the 1950s. It is especially useful for dating glacial fluctuations over the past 5000 or so years. It is believed that most glacial deposits are largely free of lichens when they are formed, but that once they become stable, lichens colonize their surfaces. The lichens become progressively larger through time. By measurement of the largest lichen thallus of one or more common species, such as *Rhizocarpon geographicum*, an indication of the date when the deposit became stable can be obtained.   ASG

Reading
Innes, J.L. 1985: Lichenometry. *Progress in physical geography* 9, pp. 187–254.
Worsley, P. 1990: Lichenometry. In A.S. Goudie ed., *Geomorphological techniques*. 2nd edn. London: Unwin Hyman. Pp. 422–8.

life cycle analysis (alternatively environmental life cycle analysis or product life analysis). The evaluation of the environmental burdens associated with a product, process or activity. It involves the quantification of the amounts of energy and materials used and the wastes released to the environment during the entire life of product, process or activity, from extraction and processing of raw materials through to transport, manufacturing, maintenance, recyling and final disposal. The purpose of the analysis is to identify opportunities to implement improvements.   ASG

life form The body shape of an organism at maturity, most commonly used in reference to plants. RAUNKIAER'S classification, for example, is based mainly on the nature of perennating buds and their position and protection. This distinguishes plants over or below 25 cm from the ground, at soil level, below ground or lying in mud or water. Other life-form features of plants include the length of shoots or the nature and density of the root system. Animal classifications also employ life-form attributes which result from morphological adaptations to the environment.   PAF

lightning A luminous discharge associated with a thunderstorm. Several types can be distinguished, principally:

1 cloud discharges (also called sheet lightning) occur between different parts of a thunderstorm, giving a diffuse illumination;
2 ground discharges (also called forked lightning) occur between cloud and ground along a tortuous path with side branches from a main channel;
3 air discharges occur between a part of the cloud and the adjacent air, but otherwise are similar in appearance to a ground discharge (see illustration on p. 306); and
4 ball lightning, reported to have the appearance of a moving, luminous globe-discharge about 20 cm in diameter, sometimes disappearing in a violent explosion.

In addition to being a significant natural hazard, lightning is an important agent in fire ecology and may also have miscellaneous geomorphological effects.   KJW

Reading
Golde, R.H. ed. 1977: *Lightning*. Vol. 1. London and New York: Academic Press.
Norin, J. 1986: Geomorphological effects of lightning. *Zeitschrift für Geomorphologie* 30, pp. 141–50.

limiting angles (of slopes) Describe the upper and lower angles at which distinct processes or forms may occur either in a given locality or under particular environmental conditions. The upper (maximum) limiting angles for continuous regolith and vegetation cover are commonly regarded as being in the range of 40 to 45° in western

Lightning strikes the earth's surface on average 100,000 times each day. This spectacular example was photographed over the new volcanic island of Surtsey, Iceland.

Europe: in Papua-New Guinea on mud-stones under rain forest the limiting angle is in the range 70–80°. Lower limiting angles for the occurrence of landslides have been quoted for a few areas: in thin regolith under temperate climates this may be in the range of 18 to 40°, but in periglacial environments the angles are much lower and in the range of 1 to 8°. Further examples are given by Young (1972, p. 165). MJS

Reading and Reference
Young, A. 1972: *Slopes*. Edinburgh: Oliver & Boyd.

limiting factors Those factors in eco-systems which are in short supply and can thus inhibit efficient and productive ecological development. The concept of limiting factors was introduced in the 1840s by the German chemist, Justus von Liebig, who found that the yield of a crop could be increased only by supplying the plants with more of the nutrient which was present in the smallest quantities. ASG

Reading
Blackman, F. 1905: Optimal and limiting factors. *Annals of botany* 19, pp. 281–95.
Park, C.C. 1980: *Ecology and environmental management*. Folkestone: Dawson. Pp. 94–9.

limnology The study of freshwater ponds and lakes. This may involve physical, chemical and biological studies. The fact that the water in lakes is essentially standing makes them distinctive aquatic environments. Physical studies include one of thermal stratification and overturning or circulation patterns. Biological communities may be varied and zoned according to light and thermal conditions, while lake chemistry studies may involve nutrient cycling and effects of pollution. Though lakes are geologically temporary features, they are important and complex developing environments. JL

Reading
Hutchinson, G.E. 1957 and 1967: *A treatise on limnology*. Vols I and II. New York: Wiley.

line squall See SQUALL LINE.

lineament A large-scale linear feature on the landsurface, such as a trough or ridge, that is the product of the structural geology of a region.

liquid limit The maximum amount of water an unconsolidated sediment or material can hold before it becomes a turbid liquid.

lithification The process of the formation of a consolidated rock from originally unconsolidated sediments through cementation or other diagenetic processes.

lithology The macroscopic physical characteristics of a rock.

lithosol Surficial deposits which do not exhibit soil horizons.

lithosphere The earth's crust and a portion of the upper MANTLE, which together constitute a layer of strength, relative to the more easily deformable ASTHENOSPHERE below. On the basis of worldwide heat flow measurements, it has been estimated that the lithosphere varies in thickness from only a few kilometres along the crest of mid-ocean ridges where, according to the PLATE TECTONICS model, new lithosphere is being created, to over 300 km beneath some continental areas. Oceanic lithosphere capped by continental crust tends to be thinner but more dense than continental lithosphere, which is capped by continental crust.   MAS

Reading
Pollack, H.N. and Chapman, D.S. 1977: On the regional variations of heat flow, geotherms and lithospheric thickness. *Tectonophysics* 38, pp. 279–96.
Walcott, R.I. 1970: Flexural rigidity, thickness and viscosity of the lithosphere. *Journal of geophysical research* 75, pp. 3941–54.
Wilson, J.T. ed. 1976: *Continents adrift and continents aground.* San Francisco: W.H. Freeman.

litter The remains of dead vegetation material, especially tree leaves which are present on the ground surface. They are broken down into essential nutrients by a wide range of DECOMPOSER organisms (bacteria, fungi) and other associates, among which are earthworms, springtails, mites and millipedes. The rate of breakdown and, conversely, the degree of accumulation of litter varies with climate. The amount of litter present in tropical rain forests, where breakdown is fast, may be only 20 kg ha^{-1}, or less than 1 per cent of the above-ground BIOMASS; whereas in the cold, dry climate of high-latitude boreal forests, litter accumulation is substantial due to slow rates of breakdown, reaching 300 kg ha^{-1}, or *c.*30 per cent of the above-ground biomass (Rodin and Basilevich 1967).   DW

Reading
Rodin, L.E. and Basilevich, N.I. 1967: *Production and mineral cycling in terrestrial vegetation.* Edinburgh: Oliver & Boyd.

Little Climatic Optimum A phase in early medieval times (*c.*AD 750–1200) when conditions were relatively clement in Europe and North America, allowing settlement in inhospitable parts of Greenland, reducing the problems of ice on the coast of Iceland, and allowing widespread cultivation of the vine in Britain.   ASG

Reading
Lamb, H.H. 1982: *Climate, history, and the modern world.* London: Methuen.

Little Ice Age See NEOGLACIAL.

load, stream The total mass of material transported by a stream. The units employed vary according to the time-base considered. Short-term loads may be expressed in kg^{-1} or t day^{-1}, whereas annual loads are expressed in t year^{-1}. The total includes both organic and inorganic material and comprises three major components: first, material carried in solution, secondly, material transported in suspension and, thirdly, material moving on the bed of the stream as bed load. The magnitude of the load and the relative importance of the three load components may vary markedly in both time and space. (See also BED LOAD; DISSOLVED LOAD; SUSPENDED LOAD.)   DEW

load structures Irregular contortions found in fine-grained deposits where sands have been deposited on water-saturated hydroplastic silts or muds. Differences in density, compaction and pore-fluid pressures cause lobes of sand to sink into the underlying silts and/or tongues of mud to rise up into the sand horizons. The resulting load structures exhibit contorted and deformed laminae, often folded, festooned or detached.   JM

local climate The climate of a small area or region, for instance an urban area, is distinguished as a local climate or a *mesoclimate*. The distinguishing factors between the local climate and the more general regional climate (*macroclimate*) are usually caused by topography, such as in valleys producing a frost hollow; or by man-made structures such as buildings in cities; or on coasts producing sea breezes. Air and surface temperatures are most commonly used as the distinguishing meteorological parameters, but local wind velocities can also be combined with local variations in precipitation, humidity and cloudiness.

JET

Reading
Oke, T.R. 1987: *Boundary layer climates*. 2nd edn. London: Routledge.

local winds Those winds which differ from the general winds expected from the pressure pattern due to topographical or urban or other effects. Four main types of local wind have been identified: (1) those winds intensified by topographical features such as a narrow mountain gap or urban canyon; (2) winds that blow along the pressure gradient such as land and sea breezes, mountain and valley winds, and on larger scales föhn, chinook, bora and mistral winds; (3) winds associated with vertical instability such as those accompanying thunderstorms; (4) strong winds due to flow over a level surface with a strong pressure gradient, such as uninterrupted flow over a level surface with a strong pressure gradient, such as sirocco and blizzard.

JET

lodgement till See TILL.

loess The original German word *Löss* was simply a name for a particular form of loose, crumbly earth. In due course, definitions became more constricted, and that of Flint (1957, p. 181) has received wide currency: 'a sediment, commonly nonstratified and commonly unconsolidated, composed predominantly of silt-sized materials, ordinarily with accessory clay and sand, and deposited primarily by wind'. This definition involves a mechanism of formation that is not universally acceptable, though a hundred years ago Ferdinand von Richthofen, after visits to Tibet and Central Asia, had championed the aeolian cause. Some earlier workers, such as Lyell, had envisaged a fluvial origin, while later workers have proposed non-aeolian mechanisms of formation (e.g. the cosmic origin ideas of Keilhack; the *in situ* formation ideas of Berg; see Smalley 1975 for a selection of earlier papers on loess). The other prime argument about loess formation concerns the mechanism whereby silt-sized quartz material is generated. Some workers stress very strongly the importance of glacial

grinding (e.g. Smalley and Vita-Finzi 1968), and the very widespread development of loess deposits around the borders of Pleistocene ice sheets lends some support to this view. These workers have tended to doubt the existence of a mechanism to produce material of appropriate grain size in quartz which is characteristic of the other environment from which loess might be derived – deserts. Processes such as salt weathering may produce silt-sized material in deserts, and the existence of frequent dust storms shows that silt-sized material is present in desert areas and available for wind transportation (Goudie *et al.* 1979).

Whatever their origin, loess deposits are undoubtedly of great importance, partly because of their great areal extent (Mississippi valley, Patagonia, New Zealand, Tunisia, Negev, Central Europe, Soviet Central Asia, China, Pakistan) but also because of their importance as a record of Pleistocene climatic fluctuations. The fossil soils and faunal and floral remains in thick, dated loess sections provide an environmental record that is equalled only by that preserved in the deep sea cores (Kukla 1977). Loess has also been an important influence on human settlement.   ASG

Reading and References
Flint, R.F. 1957: *Glacial and Pleistocene geology*. New York: Wiley.
Goudie, A.S., Cooke, R.U. and Doornkamp, J.C. 1979: The formation of silt from quartz dune sand by salt-weathering processes in deserts. *Journal of arid environments* 2, pp. 105–12.
Kukla, G.J. 1977: Pleistocene land–sea correlations: I. Europe. *Earth science reviews* 13, pp. 307–74.
Pye, K. 1987: *Aeolian dust and dust deposits*. London: Academic Press.
Smalley, I.J. 1975: *Loess lithology and genesis*. Stroudsburg, Pa.: Dowden, Hutchinson & Ross.
— and Vita-Finzi, I.J. 1968: The formation of fine particles in sandy deserts and the nature of 'desert' loess. *Journal of sedimentary petrology* 38, pp. 766–74.

logan stone Any large boulder that is so balanced that it readily rocks.

long profile, river The graph representing the relation between altitude and distance along the course of the river. The profile is usually concave upwards, is graded to a local or regional BASE LEVEL, and may be punctuated by KNICKPOINTS where the river cuts through former valley floors or river terraces. The profile may be plotted for the bed of the river channel or for the bankfull or channel capacity stage where analysis is to be related to contemporary processes, but it will be plotted for the floodplain or valley floor when related to valley development. The long profile of an entire river or valley may be approximated by one of several equations but in detail over short distances the river long profile is punctuated by pools and riffles.   KJG

Reading
Richards, K.S. 1982: *Rivers: form and process in alluvial channels.* London and New York: Methuen. Pp. 222–51.

longshore drift Longshore drift is the transport of beach material along the coast. There are two main processes involved. Beach drifting is caused by the oblique upward transport of material by the swash of short, little refracted WAVES, and its return straight down the swash slope by the BACKWASH, thus moving it alongshore. The process takes place only in the swash zone. Longshore currents in the surfzone, generated by waves approaching the coast at an angle, also move material alongshore. The transport rate depends upon the wave energy and angle of approach mainly. Longshore transport with steep waves is usually at a maximum along the submarine bar crest. The process is of great importance in accounting for long-term coastal erosion and deposition.   CAMK

Reading
Hardisty, J. 1990: *Beaches: form and process.* London: Unwin Hyman.
Komar, P.D. 1971: The mechanism of sand transport on beaches. *Journal of geophysical research* 76.3, pp. 713–21.
— and Inmam, D.L. 1970: Longshore sand transport on beaches. *Journal of geophysical research* 75, pp. 5914–27.

lopolith An igneous intrusion similar to a laccolith but saucer-shaped on both its upper and lower surfaces.

löss See LOESS.

louderback A lava flow on the surface of the dip slope of a faulted block, the presence of which proves that the topography is the product of faulting rather than erosion.

low flow analysis An analysis of the frequency, magnitude and persistence of low discharge and its relationship with climatic and catchment characteristics.   AMG

Reading
Institute of Hydrology 1980: *Low flow studies.* Wallingford, England: Institute of Hydrology.

Task Committee on Low Flows 1981: Characteristics of low flows. *American Society of Civil Engineers: Journal of the Hydraulics Division* 106, pp. 717–32.

lunettes Dunes formed as arcuate mounds on the lee side of deflated lake basins, lagoon or river segments. They may be composed either of normal quartz dune sand, or of clay aggregates (see CLAY DUNES), and are a feature of semi-arid areas (Coque 1979). In the Lake Mungo area of Australia lunettes have received considerable attention because of the evidence preserved within them of the remains of early man, and because of the indications they give of palaeoclimatic conditions (Bowler 1973).   ASG

References
Bowler, J.M. 1973: Clay dunes: their occurrence, formation and environmental significance. *Earth science reviews* 9, pp. 315–38.
Coque, R. 1979: Sur la place du vent dans l'érosion en milieu aride: l'exemple des lunettes (bourrelets éoliens) de la Tunisie. *Mediterranée* 1 and 2, pp. 15–22.

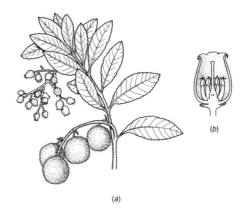

(b)

(a)

The strawberry tree, Arbutus unedo: *(a) plant showing leathery leaves and swollen fruits which are red in colour; (b) cross-section of a flower.*

Lusitanian flora A term used to describe a group of plants with their distributional centre in south-west Europe, particularly the Iberian peninsula (from Lusitania, a province of the Roman Empire mainly corresponding to present-day Portugal). Where such species occur outside this core region, they are known as the lusitanian element in the flora of the area concerned. Examples from the southwest of the British Isles include the strawberry tree (*Arbutus unedo*, see diagram), the pale heath violet (*Viola lactea*), the great butterwort (*Pinguicula grandiflora*) and the Cornish heath (*Erica vagans*). These plants probably spread from Spain and Portugal up the Atlantic seaboard of Europe during postglacial times, but were subsequently cut off by the rising sea levels.   PAS

lynchet A terrace on a hillside, generally held to be man-made and produced by ploughing. Lynchets are widespread in southern England and northern France.

lysimeter An instrument for assessing evapotranspiration losses from a vegetated soil column. Lysimeters may be used to assess either actual or potential evapotranspiration losses and the estimates are derived using a WATER BALANCE approach. A column of soil and vegetation is placed in a container and replaced in the soil so that the vegetation and soil conditions are as similar as possible to their surroundings. The container should be as large as possible to allow free growth of the vegetation and to reduce the significance of boundary effects. There are two main types of lysimeter, the drainage type and the weighing type, although some weighing lysimeters also allow drainage. Input of water to the lysimeter is assessed using rain gauges, output is measured as drainage from the base of the container enclosing the soil column and changes in soil moisture storage are estimated by repeatedly weighing the soil column. These measurements allow the estimation of losses of water through evapotranspiration. If estimates of potential evapotranspiration are required the lysimeter and a surrounding area are irrigated to ensure that the soil moisture is maintained at field capacity.

There is no standard size for a lysimeter. Larger instruments are less influenced by boundary effects but present problems if soil moisture changes are to be determined accurately by changes in weight of the soil column. Occasionally, special environmental circumstances allow the construction of very large drainage lysimeters. For example, large lysimeters with a Sitka spruce cover were employed by both Law (1957) and Calder (1976) where an impermeable subsoil allowed the construction of lysimeters by isolating a slope plot with an impermeable wall penetrating the soil and collecting the drainage from the soil above the impermeable layer. (See also POTENTIAL EVAPORATION.)   AMG

Reading and References

Calder, I.R. 1976: The measurement of water losses from a forested area using a 'natural' lysimeter. *Journal of hydrology* 30, pp. 311–25.

Kovacs, G. 1976: The use of lysimeters in the hydrological investigation of the unsaturated zone. *Hydrological sciences bulletin* 21, pp. 499–516.

Law, F. 1957: Measurement of rainfall, interception and evaporation losses in a plantation of Sitka spruce trees. *Publications of the International Association of Scientific Hydrology* 44, pp. 397–411.

Reyenga, W., Dunin, F.X., Bautovich, B.C., Rath, C.R. and Hulse, L.B. 1988: A weighing lysimeter in a regenerating eucalyptus forest: design, construction and performance. *Hydrological processes* 2, pp. 301–14.

M

maar An old volcanic crater, especially in the Eifel region of Germany. A pond or lake formed in such a depression.

machair A term commonly applied to the landform/vegetation systems of many dune pasture areas of parts of the highlands and islands of Scotland. To the native Gaelic speaker it 'describes a strip of flat, low-lying sandy soil extending along the seaboard of the Hebrides'. The essentials of machair have been summarized by Ritchie (1976) as:

1 a base of blown sand which has a significant percentage of shell-derived materials;
2 lime-rich soils with pH values normally >7.0;
3 a level or low-angle, smooth surface at a mature stage of geomorphological evolution;
4 a sandy, grassland-type vegetation with long dune grasses and other key dune species having been eliminated;
5 biotic interference, such as is caused by heavy grazing, sporadic cultivation, trampling and sometimes artificial drainage, a detectable influence within the recent historical period; and
6 an oceanic location with a moist, cool climatic regime.   ASG

Reference
Ritchie, W. 1976: The meaning and definition of Machair. *Transactions of the Botanical Society of Edinburgh* 42, 431–40.

macrofossils Animal or plant fossil remains visible with the naked eye but which usually require microscopic examination for identification. The commonest macrofossils are those of the genus *Sphagnum*, which occur in and may comprise the bulk of many PEAT deposits; other common macrofossils include seeds and fruits (often abundant in lake sediments), insect remains and molluscs. Many can be identified down to species level. The study of all three groups has been well developed in Quaternary palaeoecology to demonstrate local vegetational changes and hydroseral development, and to investigate wider phenomena such as climatic change (insects, molluscs and peat macrofossils) and marine transgressions (molluscs).   KEB

Reading
Barber, K.E. 1981: *Peat stratigraphy and climatic change*. Rotterdam: Balkema.
Birks, H.J.B. and Birks, H.H. 1980: *Quaternary palaeoecology*. London: Edward Arnold.
Godwin, H. 1975: *History of the British flora*. 2nd edn. Cambridge: Cambridge University Press.

macrometeorology The study of weather systems of large scale, up to and including the scale of the earth itself. It studies the largest in the classification of micro-, meso-, and macroscales of atmospheric motion. The lower limit of the macroscale is variously defined in the range a hundred to a few thousand kilometres. If the lower value is used hurricanes, cyclones and anti-cyclones are included in macroscale, but these are often classed separately as synoptic scale systems. Macroscale systems of larger scale include waves in the westerlies (ROSSBY WAVES), MONSOONS, the southern oscillation, the quasibiennial oscillation and, the largest motion system of all, the mean global flow pattern or GENERAL CIRCULATION.

Waves in the westerlies are vast meanders of the basically westerly flow in the upper troposphere and lower stratosphere, usually numbering about four around a latitude circle (see ROSSBY WAVES).

Monsoonal circulations, on the scale of the continents, develop in response to the differing behaviour of ocean and land in the annual variation of solar input. The processes involved in monsoons are similar to those associated with land and sea breeze circulations, but are on much larger time and space scales. In summer, air over a continent is warmed much more than that over an ocean, so that a thermally induced low pressure area over the land leads to a convergent, cyclonically rotating flow at low levels. In winter the situation is reversed, giving a high over the cold continent and winds in the reverse direction. The principal monsoon circulation is that associated with the land mass of Asia.

The southern oscillation is a fluctuation of the intertropical general circulation, and in particular that part in the Indian and Pacific Ocean regions, also called a Walker circulation. In this there is an exchange of air between the south-east Pacific subtropical high and the Indonesian equatorial low. The circulation is driven by temperature differences between the two areas – the relatively cool south-east Pacific and the warm western Pacific/Indian Ocean region. The complex climatological relationships between these two areas (and others) was first introduced by Sir Gilbert Walker who found that when pressure is high over the Pacific Ocean, it tends to be low in the Indian Ocean; rainfall amount varies in the opposite direction to pressure. The southern oscillation is by no means regular in time, and in this regard the word 'oscillation' is somewhat misleading.

The quasi-biennial oscillation (QBO) is a major reversal of wind direction in the stratosphere with a period of between 22 and 29 months, with westerly winds for roughly one year and easterly winds for the following year. The oscillation has its largest amplitude near the 25 km level, but disappears at the tropopause. The phase of the oscillation varies with height, with the wind direction changes first appearing at about 30 km and propagating downward at a rate of about 1 km per month. The oscillation has its largest amplitude above the equator, decreasing poleward and becoming very small at about 30° latitude. There is a small temperature oscillation associated with the oscillation in wind, of amplitude of about 2°C near 25 km height above the equator. Observations and theory provide evidence that the energy for the QBO is provided by vertically propagating tropospheric waves.   KJW

Reading
Atkinson, B.W. 1981: *Meso-scale atmospheric circulations.* London and New York: Academic Press.
Holton, J.R. 1972: *An introduction to dynamic meteorology.* New York and London: Academic Press.
Lockwood, J.G. 1979: *Causes of climate.* London: Edward Arnold.
Riehl, H. 1979: *Climate and weather in the tropics.* London and New York: Academic Press.

maelstrom A powerful tidal current or whirlpool.

magma Fused, molten rock material found beneath the earth's crust from which igneous rocks are formed. Magma may contain gases and some solid mineral particles.

magnetic storm A high level of magnetic disturbance produced by particles of solar origin, causing rapid field variations over the earth. Such storms disturb the ionosphere causing anomalous radio propagation and adversely affecting cable telegraphy. During storms, aurora, arcs and rays of coloured light appearing in the sky are visible much further towards the equator than their usual position.   KJW

Reading
Scientific American 1979: *The physics of everyday phenomena.* San Francisco: W.H. Freeman.

magnitude and frequency effects Discrete events in the natural world can be characterized by their magnitude and frequency of occurrence. Generally speaking, large events of the same process occur seldom and small events occur often. When the magnitude of the event is plotted against its frequency the resulting curve usually shows an exponential decline, as in the case of earthquakes, storms and floods. For example, large damaging earthquakes are relatively rare for any given place on the surface of the earth, while small imperceptible ones are an everyday occurrence. There are some exceptions, however. Small rockfalls, for example, occur with modest frequency while intermediate-sized ones occur most often (Gardner 1977).

In the most common applications the systematic decline of frequency with increasing magnitude lends itself well to statistical modelling. Research on the subject is especially common in FLOOD FREQUENCY analysis, where four types of probability distributions have seen application: lognormal, Gumbel Type I, Gumbel

Type III extreme value (a logarithmic transformation of the Gumbel Type I), and Pearson Type III (see Chow 1964). Special graph paper is available for use with some of the distributions that allows straight-line plotting of untransformed data (Craver 1980). The choice of which of these extremal frequency functions to use in a particular application has little theoretical support and usually depends on convenience or goodness of fit between the data and the selected distribution.

The declining exponential function describing the relationship between the magnitude and frequency of floods shows some important identifiable values, especially if the annual flood series (a series consisting of the largest flood of each year of record) is considered. On a Gumbel Type I distribution, for example, the arithmetic mean is at 2.33 years, so that the discharge of the mean annual flood is the magnitude that corresponds to the 2.33 year return interval. Similarly, the median annual flood is the one with a magnitude corresponding to the 2.00 year return interval. The most probable annual flood is the one with a return interval of 1.58 years, but when records shorter than one year are included in the analysis the most probable annual flood has the expected value for a return interval of 1.00 year.

Magnitude and frequency are sometimes considered on the basis of the concept of recurrence interval or return interval which is defined as the amount of time that an event of a given magnitude is expected to be equalled or exceeded. Therefore a flood with a 100 year recurrence interval is one that has a magnitude that is expected to be equalled or exceeded only once in a century. The recurrence interval (T) of a given event is defined as:

$$T = (n + 1)/m$$

where n is the number of years of record and m is the rank of the event magnitude, with the largest event having $m = 1$. A plot of recurrence intervals versus associated magnitudes usually produces a group of points that approximates a straight line on semi-logarithmic paper. A predictive function may be fitted to the data numerically, but in many cases a simple graphic line fit is adequate on account of one or two outlying points which unduly affect the numerical fit. The definition of recurrence interval is as sensitive to length of record as it is to the

rank of a given event, so it is sometimes difficult to interpret in those cases where the record is short.

Recurrence interval is directly related to the probability of occurrence or exceedence in any given year as follows:

$$q = 1 - (1 - (1/T)) \exp n$$

where q is the annual probability, T is the recurrence interval of an event of a given magnitude, and n is the number of years considered. Thus in one year's time ($n = 1$) the probability of experiencing the 100-year flood ($T = 100$) is 0.01 or 1 per cent.

The effects of hydrological events such as floods are different for events with different magnitudes and frequencies. For example, in humid regions in stream channels with no entrenchment the channel cross-section is usually adjusted so that the flood with the recurrence interval of 1.50 in the annual flood series fills the channel but does not overflow.

An important consideration in geomorphic studies is the understanding of which events accomplish the most work. If work is defined as sediment being transported out of the basin, in humid regions 90 per cent of the work is performed by processes operating at least once every five to ten years (Wolman and Miller 1960). This implies that although large, rare events appear to be impressive in their impacts it is the more frequent events that are the most important in the long run because they do intermediate amounts of work but they do that work often. This generalization is not consistent from one climatic region to another: Neff (1967) found that in arid regions only 40 per cent of the work (defined as sediment transport) was performed by the events occurring with less than a ten-year recurrence interval. In arid regions, therefore, the less common events are responsible for most of the work.

Different processes may have different relationships between magnitude, frequency, and amount of work. In his study of rockfalls Gardner (1977) found that the most work was performed by the largest event during his observation period, even though the event occurred only once. Landslides demonstrate still other relationships (Wolman and Gerson 1978).

In a larger scale of analysis the relative importance of large, infrequent events and smaller, more frequent ones has been a long-standing debate in the science of geomorphology (Pitty 1982). Two hundred years ago landscape scientists thought that many of the features of the surface of the earth were the product of catastrophic events such as the biblical flood. The concept of UNIFORMITARIANISM provided an alternative view that specified frequent operation of small-scale events over a long time as an explanation of the modern landscape. These two opposing views have merged to a certain degree so that the landscape is now viewed as the product of a long series of events of relatively low magnitude with some evidence of catastrophic events superimposed. Processes in the past are considered to have operated in the same fashion as we observe now, but at some times with different intensities (Pitty 1982).   WLG

References
Chow, V.T. 1964: Frequency analysis. In V.T. Chow ed., *Handbook of applied hydrology*. New York: McGraw-Hill.
Craver, J.S. 1980: *Graph paper from your copier*. Tucson, Arizona: HP Books.
Dunne, T. and Leopold, L.B. 1978: *Water in environmental planning*. San Francisco: W.H. Freeman.
Gardner, J.S. 1977: *Physical geography*. London: Harper's College Press.
Neff, E.L. 1967: Discharge frequency compared to long-term sediment yields. *Publications of the International Association of Scientific Hydrology* 75, pp. 236–42.
Pitty, A.F. 1982: *The nature of geomorphology*. London and New York: Methuen.
Wolman, M.G. and Gerson, R. 1978: Relative scales of time and effectiveness of climate in watershed geomorphology. *Earth surface processes* 3, pp. 189–208.
Wolman, M.G. and Miller, J.P. 1960: Magnitude and frequency of forces in geomorphic processes. *Journal of geology* 68, pp. 54–74.

mallee Eucalyptus scrub characteristic of the semi-arid areas of Australia.

mammilated surface The surface of a rock outcrop that has been smoothed and rounded by erosional processes.

man, evolution How long has our genus and its surviving species (*Homo sapiens sapiens*) been around? The evolutionary tree of mankind is a matter for research and interpretation since the quantity of the fossil record is still small, although its dating has been much improved by POTASSIUM ARGON and FISSION TRACK methods. One interpretation of the lineage has the evolution of two kinds of hominids (*Homo* and *Australopithecus*) during the Tertiary, with bipedalism established certainly by 3.7 million years BP. The locale of most finds is in East, Central and Southern Africa and from here we have evidence that *Australopithecus* became extinct between three and two million years ago. The main line is called *Homo habilis* in its early stages and *Homo erectus* from $\sim$ 1.5 million years ago. Anatomically distinct *H. sapiens* appeared in Africa around 120 thousand years ago and modern *H. sapiens sapiens* around 75,000 years ago. There is controversy about whether *H. sapiens neanderthalensis* is ancestral to modern man or whether they share a common ancestor and the Neanderthals were a short-lived parallel group to *H. sapiens sapiens*. The environmental relationships of early man are generally thought to have been passive: i.e. he was a minor factor in the environment, even after the control of fire which was certainly achieved by the time of *H. erectus* and quite likely before, although whether at 1.4 million years ago, men were fire-creators rather than fire-collectors is unknown. Other interpretations of the human lineage are possible and the whole picture is changing rapidly with the discovery and publication of new fossils and improved dating. (See also DARWINISM; EVOLUTION.)   IGS

Reading
Butzer, K.W. 1977: Environment, culture and human evolution. *American scientist* 65, pp. 572–84.
Gowlett, J.A.J., Harris, J.W.K., Walton, D.A. and Wood, B.A. 1981: Early archaeological sites, hominid remains and traces of fire from Chesowanja, Kenya. *Nature* 294, pp. 125–9.
Leakey, R. and Lewin, R. 1977: *Origins*. London: Macdonald & Jane's.

man–environment relations: alternatives Mankind tends to regard the environment as a set of resources and a receptacle for wastes. During the 1960s the rate of growth of throughput of materials and energy that had been characteristic since the nineteenth century seemed certain, if pursued, to lead to resource depletion and environmental pathologies. Two types of reaction to this situation may be identified. They are described here as the two extremities of a spectrum; the reader will realize that intermediate positions are possible and indeed common. The first reaction can be labelled the environmentalist position and is a form of environmental determinism which suggests that the industrial mode of life is in danger of outrunning the carrying capacity of the earth and that human societies with developed economies must all consume less, especially less energy, and increase the inventory of goods at the expense of throughput. Wealth must needs be transferred directly to less developed countries to help them develop better nutritional and occupational statuses, though without tying themselves to world energy prices or markets. The driving force behind most problems is seen as human population numbers, so this position is characterized by a neo-Malthusian attitude to numbers. Indeed, some writers see the cessation of population growth in both developed and less developed countries as a necessary precondition for improvement. Some facets of this view (especially those about wealth transfer) appeal to advocates of radical politics, others do not (especially those dealing with population control, which are seen as overt attempts by those in power to control the numbers of the currently powerless).

At the other extreme is the economic-technical position, which holds that abundant wealth for everybody can be just around the corner, as large numbers of people are the ultimate resource. But in any case, if people are well off their family size will decrease, as seems to have happened in the West. At the core of this position is the supply of abundant and cheap energy to fuel the various industrial processes and industrialized agriculture which are inevitable, and desirable, concomitants. The actual energy mix may vary from nation to nation but long-term sustainability means that either solar radiation has to be trapped and made available (e.g. electricity or liquid hydrogen) on an industrial scale, or that nuclear power must be the pivot of the strategy. Because uranium supplies are both finite and spatially restricted, breeder reactors which produce more fuel than they consume are thought to be an essential component of the structure in the medium term. In the longer term power from nuclear fusion is seen as plentiful (seawater is a likely source) and quite possibly cheap, though no reactor beyond the laboratory prototype has yet been built. Both these provisions have drawbacks: the environmentalist position has many attractions for the already rich but fewer for the poor; the economic-technical position appears to be a venture of faith towards a land of milk and honey, without making sure that manna will be available. Since the

nation-states of the world agree about so little it is scarcely surprising that few have consciously adopted either alternative as national policy: things grow as best they can from year to year or government to government. This is usually called 'free enterprise' or 'allowing market forces to work'. Some widespread moves towards reducing waste residuals in the environment, increasing energy efficiency of machines, and re-using materials such as metals and paper, have been made.   IGS

Reading
Beckerman, W. 1974: *In defence of economic growth*. London: Cape.
Bookchin, M. 1982: *The ecology of freedom*. Palo Alto, Ca.: Cheshire Books.
Capra, F. 1983: *The turning point*. London: Fontana; New York: Simon & Schuster.
Merchant, C. 1980: *The death of nature*. San Francisco: Harper & Row.
Ophuls, W. 1977: *Ecology and the politics of scarcity*. San Francisco: W.H. Freeman.
Simon, J.L. 1981: *The ultimate resource*. Oxford: Martin Robertson; Princeton, NJ: Princeton University Press.

man–environment relations: ideas
Probably no human society has ever existed which did not regard its environment in a metaphysical way, i.e. try to understand or to manipulate it by means of a set of abstract constructs. Until the emergence of literacy most such ideational structures went unrecorded, though we have occasional glimpses from the ideas of preliterate societies recorded upon contact with Europeans: 'Every part of this soil is sacred in the estimation of my people. Even the rocks . . . thrill with memories of stirring events connected with the lives of my people,' said Chief Seattle in 1853. The environmental ideas of the west have, however, been preserved in literary form and their descent from classical antiquity to the nineteenth century has been traced in a monumental volume by Clarence Glacken (1967). His main thesis is that certain sets of ideas about such relationships show an evolutionary descent, that there has been a continuity of substance, though with adaptations of form. The first set of ideas starts with the essentially religious view of a designed earth – that it, and we, have a purpose and that the earth is made just so, so as to be part of that purpose. This essentially teleological view did not survive the onslaught of Darwinian hypotheses of EVOLUTION intact, but was nevertheless carried forward by the evolutionists in their

concept of a coherent earth where everything had a place into which it had fitted – though by natural selection rather than by divine providence. Such a set of ideas forms an obvious seed-bed for the science of ECOLOGY, with its emphasis on wholes and on the nature of the linkages between the various components of ECOSYSTEMS. These notions go even further in the Lovelock hypothesis that physical features of the earth, such as the composition of the atmosphere, are a consequence of life rather than life adapting to a given chemistry. From a humanistic angle Roszak argues that the planet as a whole can feed back to any components (including mankind) ideas about necessary changes in behaviour.

Secondly, a major set of thoughts concerns environmental influences upon human behaviour. Airs, waters and places affected, according to some ancient Greeks, the bodily humours and then the outlooks and actions of men. During the Enlightenment, this argument become the focus of the antagonism between those who foresaw the perfectibility of man, with nature no hindrance to this Utopian development, and those who saw that disaster, famine and disease were more likely perpetual concomitants of the human species. Chief among the latter was Thomas Malthus, for whom nature exhibited niggardly characteristics. From Malthus' pessimism to the environmental determinism of the nineteenth and twentieth centuries is quite a short step intellectually, though the scale of determinists' examples was in general smaller than the global disaster Malthus saw as being averted only by restraint on the part of every individual.

Global breakdown because the carrying capacity of the earth is being exceeded has been a theme of the environmentalist movement since the 1960s. That the sustained yield levels of the earth are finite and that we must adjust our numbers and our resource consumption to them would have been familiar ideas to Malthus and very possibly to a number of thinkers before him.

Thirdly, there is a set of ideas dealing with the 'second world': man's world within the world of nature, typified by heroic efforts to clear forests, drain marshes and generally improve things. The medieval Cistercian monk had divine sanction for this work: 'laborare est orare', for the Creation was as yet unfinished. That this second world might in some cases be more of a dirge than a psalm was pointed out forcefully by George P. Marsh in his 1864 book (see also MRCFE) where he chronicled, for example, the degradation of Mediterranean ecosystems by millenia of cultivation, burning and grazing. More recent developments in this intellectual tradition have been the concept of ENVIRONMENTAL IMPACT analysis as a means of controlling the unwise or irreversible creation of second worlds, and the development by biologists such as Waddington and philosophers like Jantsch of sophisticated ideas about the co-evolutionary interaction of man and other biota in co-creating a different but progressively better, more stable world in which the duality of man and nature – characteristic of Western rather than Eastern philosophy and religion – is creatively resolved.

Contemporary frameworks for the study of man and environment can perhaps be identified in three main strains conveniently labelled the three Es: ecology, economics and ethology. The ecological view is concerned with man's place in the webs of the ecosphere: as manipulator of flows of energy and matter, as a significant part of the BIOMASS at regional and global scales, and as a reducer of global net primary productivity. This view is sometimes called human ecology, though others argue that ecology makes sense only as a unity, not as a disparate set of component parts. By contrast, economics is a more obviously anthropocentric framework, concerned with the provision of mechanisms for the distribution of resources which are by nature unevenly distributed and which are never in sufficient supply for the plenitude of all. Economics is sometimes claimed to be a value-free discipline but its older name of political economy seems to point to a greater reality. Ethology (behavioural science) cannot avoid considering value systems since these determine much of human behaviour, in, for example, the cognizance taken of exposure to NATURAL HAZARDS or other forms of risk. At one value-laden extreme of the ethological spectrum is ethics, which prescribes how men ought to behave with respect to their use of the environment and its resources. In the twentieth century, the predominant worldview (*Weltanschauung*) has been one of linear exponential growth of resource use and material consumption in a largely mechanistic world. This is held by some students of man–environment relations to be incompatible with the nature of the ECOSPHERE; others aver that human ingenuity and technology will transcend any such limits to growth (see MAN-ENVIRONMENT RELATIONS: ALTERNATIVES). IGS

Reading and References
†Attfield, R. 1983: *The ethics of environmental concern.* Oxford: Basil Blackwell; New York: Columbia University Press.
Glacken, C.J. 1967: *Traces on the Rhodian shore.* Berkeley and Los Angeles: University of California Press.
†Jantsch, E. 1980: *The self-organizing universe.* New York and Oxford: Pergamon.
†Lovelock, J.E. 1979: *Gaia. A new look at life on earth.* Oxford: Oxford University Press.
Marsh, G.P. 1864: *Man and nature.* New York: Scribner. (Reprint 1965, D. Lowenthal, ed. Cambridge, Mass.: Belknap Press.)
†Roszak, T. 1979: *Person planet.* London: Gollancz.
†Sandbach, F. 1980: *Environment, ideology and policy.* Oxford: Basil Blackwell: New York: Allanheld Osmun.

mangrove Plant communities dominated by the genus *Rhizophora, Bruguieria* and *Avicennia* which colonize tidal flats, estuaries and muddy coasts in tropical and subtropical areas. Communities of mangroves, termed *mangals*, play an important role on many tropical coasts. They are highly productive ecosystems which are capable of exporting energy and materials to adjacent communities, they support a diverse heterotrophic food chain, act as nurseries in the life cycles of some organisms, and offer some protection against coast erosion and storm surge attack. At the present time, like many types of wetland, they are under severe anthropogenic pressures. (See p. 318). ASG

Reading
Tomlinson, P.B. 1986: *The botany of mangroves.* Cambridge: Cambridge University Press.

manned earth resources satellites

Satellites carrying a human crew plus photographic and other REMOTE SENSING devices for the production of images of the earth's surface. The US manned satellites of Mercury, Gemini and Apollo are known for their participation in the 'space race' while the later US manned satellites of Skylab and Space Shuttle were experimental space stations. All were designed and operated by the National Aeronautics and Space Administration (NASA). A complementary series of manned earth resources satellites were launched by the CIS but as their image products are not readily available to physical geographers they will not be discussed here.

The first successful photographs of the earth from space were taken in 1961 from a Mercury satellite on its fourth mission. The interest provoked by these photographs led to colour photographs being taken of areas of exposed geology during the satellite's later missions. Following the photographic success of the Mercury missions the two-man crews of the Gemini series of satellites were asked to obtain photographs of 'interesting' areas on the earth's surface.

The prop roots of the mangrove, Rhizophora mangle *lining a creek in Colombia. Mangroves, which play an important role in trapping sediment, are a widespread feature of coastlines in low latitudes.*

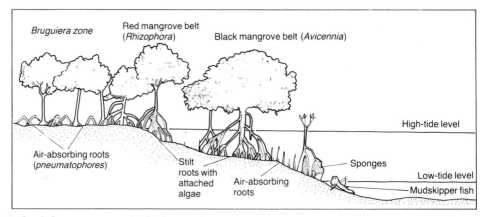

1. *A typical mangrove community showing zonation.*
Source: *P. Furley and W. Newey 1983:* Geography of the biosphere. *London: Butterworths. Figure 13.13.*

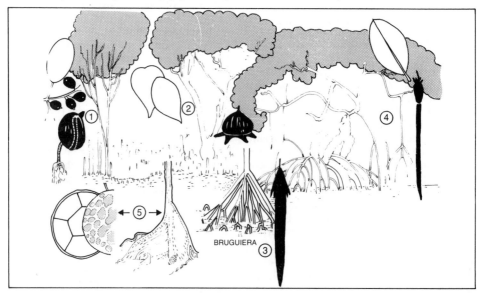

2. *Mangrove of different genera, with details of fruits, foliage and roots:* 1. Avicennia; 2. Sonneratia; 3. Bruguiera; 4. Rhizophora; 5. Xylocarpus.

On the third Gemini mission in 1965 photographs were taken of the south-west United States. The fourth Gemini mission was much longer and enabled over 100 photographs to be taken. The fifth Gemini mission suffered from a lack of power that prevented optimal alignment for ground photography. Nevertheless 1275 photographs were taken of many parts of the world. The sixth and seventh Gemini missions returned 310 photographs between them. During 1966 many more colour photographs were taken as part of a further five Gemini missions, increasing the total number of photographs taken from a Gemini satellite to over 2400.

By the time that the Apollo series of satellites was launched an experiment had been formalized. The aims were first to obtain automatic colour photography from Apollo 6 in 1967, second to obtain handheld colour photography from Apollos 7 and 9 for areas of the earth for which aerial photography and ground data were also available, and third to obtain multiband photography using a multi-camera array from Apollo 9. The first two objectives were fulfilled and over 600 colour photographs were taken. The third objective was also successful and provided the stimulus for NASA to push ahead with the development of the satellite Landsat.

The space station Skylab was first launched by NASA in 1973 into a near-circular orbit 435 km above the earth's surface. It circled the earth every 93 min and passed over each point on its surface between 50° north and 50° south once every five days. Skylab was designed for a number of experiments, one of which was concerned with remote sensing and was named the Earth Resources Experiment Package (EREP). This comprised three types of aerial camera, a 13-channel multispectral scanner, and a number of non-imaging remote sensors. Of greatest interest to physical geographers was the S190A multi-camera array aerial camera which provided good quality photographs with a spatial resolution of 30–80 m in six wavebands and the S190B earth terrain aerial camera which provided good quality photographs with a spatial resolution of 20 m in one or three wavebands.

The Space Shuttle is a series of NASA spacecraft designed to shuttle backwards and forwards between earth and space. They started flying in April 1981 and are likely to continue well into the next century. Each Space Shuttle contains two rocket boosters, one liquid propellant tank and an orbiter vehicle. The orbiter vehicle carries astronauts who take photographs of the earth's surface on a regular basis and two remote sensing packages named OSTA and Spacelab.

The OSTA package is a collection of remote sensing experiments and is named after its NASA designers, the former Office of Space and Terrestrial Applications. The package has included three environmentally useful sensors, a SIDEWAYS LOOKING AIRBORNE RADAR (SAR type), a pair of television cameras and an optical imager. Spacelab, which is built and operated by the European Space Agency, holds a number of sensors, including an aerial camera and a sideways looking airborne radar (SAR type).   PJC

Reading
Barrett, E.C. and Curtis, L.F. 1992: *Introduction to environmental remote sensing.* 3rd edn. London and New York: Chapman & Hall.

Lowman, P.D. 1980: The evolution of geological space photography. In B.S. Siegal and A.R. Gillespie eds, *Remote sensing in geology.* New York: Wiley. Pp. 91–115.

Manning equation In 1891 the Irish engineer Robert Manning summarized uniform flow data in the popular, widely used FLOW EQUATION:

$$v = (k R^{\frac{2}{3}} s^{\frac{1}{2}})/n.$$

Here v is velocity ($m^3\ s^{-1}$), R is the HYDRAULIC RADIUS (mean depth (m) in wide channels), and s is the slope ($m\ m^{-1}$). The Manning ROUGHNESS coefficient n is a general measure of channel resistance which is numerically constant regardless of the measurement units being used. The coefficient k therefore accommodates the variation of measurement system, and is 1 for SI units and 1.49 for Imperial units. In straight natural streams, n ranges from 0.03 for smooth sections to 0.10 for rocky or heavily vegetated sections (Chow 1959, pp. 98–123; Barnes 1967). It is related to the coefficient C in the CHÉZY EQUATION by:

Values of Manning's roughness coefficient for various types of natural channel

Channel type	Normal value	Range
Small channels (width <30 m)		
Low-gradient streams		
Unvegetated straight channels at bankfull stage	0.030	0.025–0.033
Unvegetated winding channels with some pools and shallows	0.040	0.033–0.045
Winding vegetated channels with stones on bed	0.050	0.045–0.060
Sluggish vegetated channels with deep pools	0.070	0.050–0.080
Heavily vegetated channels with deep pools	0.100	0.075–0.150
Mountain streams (with steep unvegetated banks)		
Few boulders on channel bed	0.040	0.030–0.050
Abundant cobbles and large boulders on channel bed	0.050	0.040–0.070
Large channels (width >30 m)		
Regular channel lacking boulders or vegetation	–	0.025–0.060
Irregular channel	–	0.035–0.100

Source: Based on data in Chow, V.T. ed. 1964: *Handbook of applied hydrology.* New York: MacGraw-Hill.

$$C = R^{1/6}/n.$$

KSR

Reading and References
Barnes, H.H. 1967: Roughness characteristics of natural channels. *US Geological Survey water supply paper* 1849.

Chow, V.T. 1959: *Open channel hydraulics*. London: McGraw-Hill.

Manning, R. 1891: On the flow of water in open channels and pipes. *Transactions of the Institution of Civil Engineers of Ireland* 20, pp. 161–207.

mantle The zone within the earth's interior extending from 25 to 70 km below the surface to a depth of 2900 km and lying between the partially molten core and the thin surface crust. The uppermost rigid section of the mantle forms the lower part of the LITHOSPHERE. Extending a few hundred kilometres below this region is the ASTHENOSPHERE, a layer in which the magnesium and iron-rich silicate rocks of the mantle are probably partially molten. Most of the convection within the mantle, viewed by some as a crucial mechanism of PLATE TECTONICS, is generally thought to occur in this zone. MAS

Reading
Davies, P.A. and Runcorn, S.K. eds 1980: *Mechanisms of continental drift and plate tectonics*. London: Academic Press.

Ringwood, A.E. 1975: *Composition and petrology of the earth's mantle*. New York: McGraw-Hill.

Wyllie, P.J. 1976: The earth's mantle. In J.T. Wilson, *Continents adrift and continents aground*. San Francisco: W.H. Freeman. Pp. 46–57.

mantle plume A convectional flow of hot rock that rises through the MANTLE to the base of the LITHOSPHERE and which gives rise to a HOT SPOT on the surface. In the oceans, mantle plumes create lines of large volcanoes (such as the Hawaiian chain) if the overlying plate is moving with respect to the plume. On the continents, mantle plumes have probably been responsible for generating voluminous and extensive accumulations of BASALT flows, such as those of the Karoo (South Africa) and the Deccan (India), many of which are located at PASSIVE MARGINS and were thus originally close to sites of continental breakup. They are also probably capable of causing surface uplift over areas in excess of 1000 km across, both by the direct effects of heating of the lithosphere and by the associated thickening of the crust through the addition of enormous quantities of igneous material. MAS

Reading
Summerfield, M.A. 1991: *Global geomorphology*. London and New York: Longman Scientific and Technical and Wiley.

White, R.S. and McKenzie, D. 1989: Magmatism at rift zones: The generation of volcanic continental margins and flood basalts. *Journal of geophysical research* 94, pp. 7685–729.

maquis Scrub vegetation of evergreen shrubs characteristic of the western Mediterranean, and broadly equivalent to chaparral. Many areas of maquis probably represent human-induced degradation of the natural forest vegetation.

margalitic Soil A horizons which are dark coloured with a high base status, Ca and Mg being the predominant exchangeable cations.

marginal channel A meltwater stream flowing at the margin of a glacier.

marine pollution The pollution of the oceans and seas. At first sight, as Jickells *et al.* (1991, p. 313) point out, two contradictory thoughts may cross our minds on this issue:

The first is the observation of ocean explorers, such as Thor Heyerdahl, of lumps of tar, flotsam and jetsam, and other products of human society thousands of kilometres from inhabited land. An alternative, vaguer feeling is that given the vastness of the oceans (more than 1,000 billion billion litres of water!), how can man have significantly polluted them?

What is the answer to this conundrum? Jickells *et al.* (p. 330) draw a clear distinction between the open oceans and regional seas and in part come up with an answer:

The physical and chemical environment of the open oceans has not been greatly affected by events over the past 300 years, principally because of their large diluting capacity . . . Material that floats and is therefore not diluted, such as tar balls and litter, can be shown to have increased in amount and to have changed character over the past 300 years.

In contrast to the open oceans, regional seas in close proximity to large concentrations of population show evidence of increasing concentrations of various substances that are almost certainly linked to human activities. Thus the partially enclosed North Sea and Baltic show increases in phosphate concentrations as a result of discharges from sewage and agriculture.

Likewise, it is clear that pollution in the open ocean is, as yet, of limited biological significance. GESAMP (1990), an authoritative review of the state of the marine environment for the United Nations Environment Programme, reported (p. 1):

> The open sea is still relatively clean. Low levels of lead, synthetic organic compounds and artificial radionuclides, though widely detectable, are biologically insignificant. Oil slicks and litter are common along sea lanes, but are, at present, of minor consequence to communities of organisms living in open-ocean waters.

On coastal waters it reported (p. 1):

> The rate of introduction of nutrients, chiefly nitrates but sometimes also phosphates, is increasing, and areas of entrophication are expanding, along with enhanced frequency and scale of unusual plankton blooms and excessive seaweed growth. The two major sources of nutrients to coastal waters are sewage disposal and agricultural run-off from fertilizer-treated fields and from intensive stock raising.

Attention is also drawn to the presence of synthetic organic compounds – chlorinated hydrocarbons, which build up in the fatty tissue of top predators such as seals which dwell in costal waters. Levels of contamination are decreasing in northern temperate areas but rising in tropical and subtropical areas due to continued use of chlorinated pesticides there. ASG

References
GESAMP (IMO.FAO/UNESCO/WMO/WHO/IAEA/UN/ENEP Joint Group of Experts on the Scientific Aspects of Marine Pollution) 1990: The state of the marine environment. *UNEP regional seas reports studies* 115, Nairobi.

Jickells, T.D., Carpenter, R. and Liss, P.S. 1991: Marine environment. In Turner, B.L., Clark, W.C., Kates, R.W., Richards, J.F., Mathews, J.T. and Meyer, W.B. eds, *The earth as transformed by human action.* Cambridge: Cambridge University Press. Pp. 313–34.

Markov process A statistical process in which the probability of an event, in a sequence of random events, is influenced by the outcome of the preceding event. Any system in which an outcome depends in part on the previous outcome may be described as exhibiting a Markov property.

Any system in which successive events are partially dependent is said to have a 'memory' of the preceding event. The methodology is formalized using a transition matrix in which entries represent the probabilities of transition from one state to another. When the matrix is powered the probabilities represent transitions from one state to another in two steps. Repeated powering will develop a matrix in which all rows are the same. States may also be defined which are absorptive, when the realization of the process terminates, or reflexive, when a process enters a state and then returns to the previous state. The utility of the process depends on the length of the record on which the probabilities are established, whether the series exhibits the Markov property, is stationary and whether the transition probabilities are invariant with time. DB

Reading
Harbaugh, J.W. and Bonham-Carter, G. 1970: *Computer simulation in geology.* New York: Wiley.

Scheidegger, A.E. 1966: Effect of map scale on stream orders. *Bulletin of the International Association of Scientific Hydrology* 11, pp. 56–61.

Thornes, J.B. and Brunsden, D. 1977: *Geomorphology and time.* London: Methuen.

mass balance Changes in the mass of a sedimentary landform over time. The concept is fundamental to the understanding of a glacier where a zero mass balance describes a situation where accumulation balances ABLATION, a positive mass balance describes an increase in glacier mass and a negative mass balance a reduction in glacier mass. The concept can also be applied to other sedimentary landforms, for example a beach, talus or sand dune. DES

mass movement types The modes of hillslope failure. Various methods have been used to classify mass movements including reference to parent material, the causes and mechanics, slope and landslide geometry, shape of failure surface, post-failure debris distribution, soil moisture and physical properties. The most generally accepted method is the complex variable classification of Varnes (1958). The most common general groupings are linked to the processes of fall, slide, flow and creep.

Falls of rock or soil are the free movement downslope of slope-forming materials. Failures may be in circular, plane, wedge, toppling or settling modes. Slides of rock, mud and soil take place by

Classification and characteristics of the major types of mass movement

Primary mechanism		Mass movement type	Materials in motion	Moisture content	Type of strain and nature of movement	Rate of movement
Lateral component predominant	Creep	Rock creep	Rock (especially readily deformable types such as shales and clays)	Low	Slow plastic deformation of rock, or soil producing a variety of forms including cambering, valley bulging and outcrop bedding curvature	Very slow to extremely slow
		Continuous creep	Soil	Low		
	Flow	Dry flow	Sand or silt	Very low	Funnelled flow down steep slopes of non-cohesive sediments	Rapid to extremely rapid
		Solifluction	Soil	High	Widespread flow of saturated soil over low to moderate angle slopes	Very slow to extremely slow
		Gelifluction	Soil	High	Widespread flow of seasonally saturated soil over permanently frozen subsoil	Very slow to extremely slow
		Mud flow	>80% clay-sized	Extremely high	Confined elongated flow	Slow
		Slow earthflow	>80% sand-sized	Low	Confined elongated flow	Slow
		Rapid earthflow	Soil containing sensitive clays	Very high	Rapid collapse and lateral spreading of soil following disturbance, often by an initial slide	Very rapid
		Debris flow	Mixture of fine and coarse debris (20–80% of particles coarser than sand-sized)	High	Flow usually focused into pre-existing drainage lines	Very rapid
		Debris (rock) avalanche (sturzstrom)	Rock debris, in some cases with ice and snow	Low	Catastrophic low friction movement of up to several kilometres, usually precipitated by a major rockfall and capable of overriding significant topographical features	Extremely rapid
		Snow avalanche	Snow and ice, in some cases with rock debris	Low	Catastrophic low friction movement precipitated by fall or slide	Extremely rapid
		Slush avalanche	Water-saturated snow	Extremely high	Flow along existing drainage lines	Very rapid

Cont'd

Primary mechanism		Mass movement type	Materials in motion	Moisture content	Type of strain and nature of movement	Rate of movement
Lateral component predominant	Slide — Translational	Rock slide	Unfractured rock mass	Low	Shallow slide approximately parallel to ground surface of coherent rock mass along single fracture	Very slow to extremely rapid
		Rock block slide	Fractured rock	Low	Slide approximately parallel to ground surface of fractured rock	Moderate
		Debris/earth slide	Rock debris or soil	Low to moderate	Shallow slide of deformed masses of soil	Very slow to rapid
		Debris/earth block slide	Rock debris or soil	Low to moderate	Shallow slide of largely undeformed masses of soil	Slow
	Slide — Rotational	Rock slump	Rock	Low	Rotational movement along concave failure plane	Extremely slow to moderate
		Debris/earth slump	Rock debris or soil	Moderate	Rotational movement along concave failure plane	Slow
	Heave	Soil creep	Soil	Low	Widespread incremental downslope movement of soil or rock particles	Extremely slow
		Talus creep	Rock debris	Low		Extremely slow
	Fall	Rock fall	Detached rock joint blocks	Low	Fall of individual blocks from vertical faces	Extremely rapid
		Debris/earth fall (topple)	Detached cohesive units of soil	Low	Toppling of cohesive units of soil from near-vertical faces such as river banks	Very rapid
Vertical component predominant	Subsidence	Cavity collapse	Rock or soil	Low	Collapse of rock or soil into underground cavities such as limestone caves or lava tubes	Very rapid
		Settlement	Soil	Low	Lowering of surface due to ground compaction usually resulting from withdrawal of ground water	Slow

Source: Summerfield, M.A. 1991: *Global geomorphology*. London and New York: Longman Scientific and Technical and Wiley. Based largely on Varnes, D.J. 1978: in R.L. Schuster and R.J. Krizek eds, *Landslide analysis and control*. Transportation Research Board Special Report 176. National Academy of Sciences, Washington, DC, 11–33.

movement on a sliding surface which may be circular, non-circular or planar in form. Circular features may be single, multiple or successive in character. Non-circular failures are usually controlled by lithology or rock structures such as bedding planes which impart a translational component to the slide. A graben form at the head or retrogressive failure may also be evident. Planar failures are usually shallow and translational including debris slides and mudslides, but may be of immense size such as the major rock slides along discontinuities in mountain regions.

All slides may disintegrate upon failure to develop into an 'avalanche' or 'flow'. These involve complex processes such as fluidization, liquefaction, cohesionless grain flow, remoulding and possibly air lubrication. Flows represent a transitional set of processes lying between streamflow or mass transport and mass movement. The term debris flow is used for the mass movement of a wet mixture of granular slides, clay, water and air under the influence of gravity with intergranular shear distributed evenly through the mass. They usually occur in three modes, as simple flows on hillslides, as valley-con-

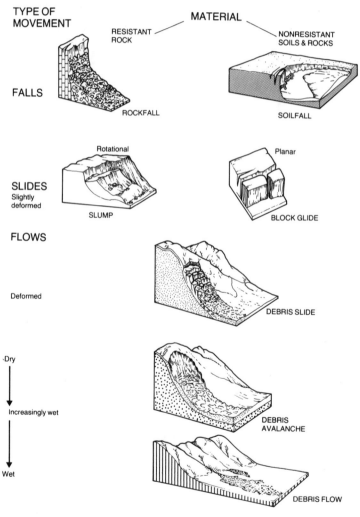

Mass wasting types.
Source: *M.J. Selby ·1982:* Hillslope materials and processes. *Oxford: Oxford University Press. Figure 6.1.*

fined flows debouching onto fans or as catastrophic flows which effectively destroy everything in their path and override topography. This is a convenient division based on scale but the processes are probably similar. Creep includes the continuous, gravity or mass creep of hillslopes affected by the force of gravity in which low-continued stresses deform the slope materials at a fairly constant rate but to a considerable depth. The forms which result are gravity slides or *sackung*, cambering and valley bulging (also produced by periglacial conditions which accelerate the process), outcrop curvature and botanic deformation. Creep is more commonly regarded as a seasonal process in which surface materials are continuously affected by the expansion and contraction of the soil and by soil heave caused by changes of temperature, moisture, freezing, crystal growth or chemical changes. Varieties occur as soils, scree or talus and intense activity develops into faster solifluction processes under periglacial conditions (gelifluction, congelifluction, congeliturbation).   DB

Reading and References
Brunsden, D. 1979: Mass movements. In C.E. Embleton and J.B. Thornes, eds, *Process in geomorphology*. London: Edward Arnold.
Sharpe, C.F.S. 1938: *Landslides and related phenomena*. New York: Columbia University Press.
Terzaghi, K. 1960: Mechanism of landslides. *Bulletin of the Geological Society of America Berkey volume*, pp. 83–122.
Varnes, D.J. 1958: Landslide types and processes. In E.B. Eckel ed., *Landslides and engineering practice*. Highway Research Board special report 29, pp. 20–47.
Zaruba, Q. and Mench, V. 1969: *Landslides and their control*. Prague: Academic and Elsevier.

mass strength (of rock). A measure of the resistance to erosion and instability of an entire rock mass inclusive of its discontinuities, contained water and weathering products. For geomorphic purposes mass strength has been assessed quantitatively by taking into account the following parameters: (1) strength of intact rock; (2) state of weathering of the rock; (3) spacing of joints, bedding planes, foliations or other discontinuities within the rock mass; (4) orientation of discontinuities with respect to a cut slope; (5) width of the discontinuities; (6) lateral or vertical continuity of the discontinuities; (7) infill of the discontinuities; and (8) movement of water within or out of the rock mass.   MJS

Reading
Dackombe, R.V. and Gardiner, V. 1983: *Geomorphological field manual*. London: Allen & Unwin.
Selby, M.J. 1980: A rock mass strength classification for geomorphic purposes: with tests from Antarctica and New Zealand. *Zeitschrift für Geomorphologie NF* 24, pp. 31–51.

massive rock The rock mass as a whole, including discontinuities (see MASS STRENGTH).

mattoral The Chilean equivalent of the evergreen, xerophilous, woody plants of Mediterranean environments. (See also CHAPARRAL and MAQUIS).

Reading
Miller, P.C. 1981: *Resource use by chaparral and matorral*. Berlin: Springer Verlag.

maturity The stage of development of a river system or landscape at which processes are most efficient and vigorous and are tending towards maximum development. In landscape evolution, the period of maximum diversity of form.

Maunder minimum A proposed period from about 1650 to 1700 when there was a relative absence of sunspots, and which may be associated with a period of especially inclement weather during the Little Ice Age.   ASG

Reading
Eddy, J.A. 1976: The Maunder minimum. *Science* 192, pp. 1189–202.

mean sea level The average level of the sea surface, relative to a permanent land-based benchmark, taken over a period long enough to eliminate the effects of waves and tides. Daily means are often computed, but the most commonly used values are monthly and annual means. The arithmetic mean of all the individual hourly levels observed during the period gives a good approximation to the mean sea level. Mean tide level, the level equidistant between tidal low water and high water is a less adequate estimate. Present technology is unable to separate vertical land movements from global changes of sea level.

According to the best estimates of local vertical crustal movements, global sea level is increasing by between 0.10 and 0.15 m per century. Increases may be due to melting of polar ice and to thermal expansion of the oceans, but the relative importance of these factors is not known. In many circumpolar regions, including the northern Baltic Sea, vertical land movements due to isostatic adjustment after removal of a glacial ice burden result in apparent local mean sea level falling by as much as 1.0 m per century.

The mean sea-level surface, undisturbed by meteorological or ocean dynamics, would correspond to the geoid, traditionally used as a datum level for geodetic surveys. In practice, mean sea levels differ from the geoid by up to 1.0 m because of variations of mean air pressure, winds, and the distribution of ocean currents.   DTP

Reading
Pugh, D.T. 1987: *Tides, surges and mean sea level.* Chichester: Wiley.

meandering The sinuous winding of a river, as in the River Menderes in southwest Turkey. Various more restricted definitions have been applied, e.g. that a river may be arbitrarily considered as meandering if it flows in a single channel that is more than one and a half times the direct downvalley length of the reach in question. Other definitions have suggested that meandering must involve the development of regular bends of repeated geometry; these bends may have characteristic amplitude, wavelength and curvature, and they may be approximated by geometrical curves such as circular arcs or so-called sine-

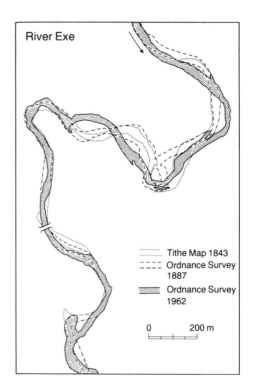

Meanders on the River Exe in Devon showing changes reconstructed from historical sources.
Source: *J.M. Hooke and R.J.P. Kain 1982:* Historical change in the physical environment: a guide to sources and techniques. *London: Butterworth.*

Large meanders developed in the floodplain of the White River, South Dakota, USA.

generated curves, or recently by migration-rate models. Irregular and compound bend patterns are also possible.

The term is also applied to analogous flow patterns (as in the Gulf Stream or atmospheric jet streams) and to the dynamic processes by which such planforms may develop. For rivers, these include outer-bank erosion and inner-bank deposition on channel bends, and these may be related to the hydraulics of flow and sediment flux in curving channels.

Different kinds of meandering may be distinguished: these include free (developing in alluvial materials without hindrance), forced or confined (where barriers such as bedrock valley sides affect meander generation) and valley (rather than just channel) meandering. Meanders in bedrock may be incised, either as they develop (ingrown) or vertically intrenched. (See also CHANNELS, TYPES OF RIVER/STREAM.)   JL

Reading
Collinson, J.D. and Lewin, J. eds 1983: *Modern and ancient fluvial systems*. International Association of Sedimentologists special publication 6. Oxford: Blackwell Scientific.

Gregory, K.J. ed. 1977: *River channel changes*. Chichester: Wiley.

Richards, K. 1982: *Rivers*. London and New York: Methuen.

Mediterranean climate Characterized by hot, dry, sunny summers and a winter rainy season. In KÖPPEN'S CLIMATIC CLASSIFICATION this is the Cs climate, while in Thornthwaite's (1948) it is designated as a subhumid mesothermal climate. Circulation in summer is dominated by expansions of the subtropical high-pressure cells and in winter by travelling mid-latitude depressions (Perry 1981).

Typical monthly mean temperatures are between 21–27°C in summer and 4–13°C in winter with annual rainfall totals between 38 and 76 cm. A number of distinctive LOCAL WINDS are associated with Mediterranean climates, e.g. mistral, bora and Santa Ana (Meteorological Office 1964).

This climatic type is found around the Mediterranean Sea, in southern California, the central Chilean coast, the South African coast near Cape Town and in Western Australia.   AHP

References
Meteorological Office 1964: *Weather in the Mediterranean*. 2 vols. London: HMSO.

Perry, A. 1981: Mediterranean climate – a synoptic reappraisal. *Progress in physical geography* 5, pp. 105–13.

Thornthwaite, C.W. 1948: An approach to a rational classification of climate. *Geographical review* 38, pp. 55–94.

megashear A term introduced by S.W. Carey (1976) to describe a strike-slip fault with a large lateral displacement greatly exceeding the thickness of the crust. In a strike-slip fault a section of the crust moves horizontally with respect to an adjoining section, as on the San Andreas fault system. Megashears form a component of Carey's model of global TECTONICS based on the assumption of an expanding earth.   MAS

Reading and Reference
Carey, S.W. 1976: *The expanding earth*. Amsterdam: Elsevier.

†Neev, D., Hall, J.K. and Saul, J.M. 1982: The Plesium megashear system, across Africa and associated lineament swarms. *Journal of geophysical research* 87, pp. 1015–30.

megathermal climate A climate in which no month has a mean temperature below 18°C. Such conditions are found in the humid tropics and subtropics.

mekgacha Dry or fossil valley systems from the Kalahari of southern Africa. They snake through a largely featureless terrain where at present surface run-off is of very limited occurrence. The origins of mekgacha have been variously ascribed to fluvial activity during wetter episodes, ephemeral flow during high rainfall events, river capture, and groundwater sapping along ancient lineaments. They may be of considerable antiquity.   ASG

Reading
Shaw, P.A., Thomas, D.S.G. and Nash, D.J. 1992: Late Quaternary fluvial activity in the dry valleys (mekgacha) of the Middle and Southern Kalahari, southern Africa. *Journal of Quaternary science* 7, pp. 273–81.

meltout till See TILL.

meltwater Water produced by the melting of snow or ice. In polar periglacial areas the melting of the winter snow cover and ice in the river channels occurs in the spring and early summer. The resulting flood may last from a few days to a few weeks and account for up to 90 per cent of the total discharge. Sediment loads of such floods are commonly suppressed by the still frozen state of the ground and river channel.

Glacier meltwater flow peaks somewhat later in the season and is less sudden in its build up and decline, but sediment loads are high. Meltwater in a glacier is derived mainly from summer melting of the glacier surface, but a small though significant amount is derived from bottom melting of the glacier by geothermal heat and by frictional heating associated with glacier flow (so long as the glacier base is at the pressure melting point).

On glaciers which are below the pressure melting point the meltwater is normally restricted to surface flow. Where the ice is at the pressure melting point, meltwater penetrates the ice via MOULINS until it flows at the glacier bed. It can be distinguished from basally derived meltwater by its lower solute content (Collins 1979). Two hydrological systems exist at the glacier bed, that of the conduits and that of the intervening areas where a film of water exists between glacier ice and bedrock. In certain bedrock depressions water can be trapped to form subglacial lakes up to many kilometres across (Oswald and Robin 1973).   DES

Reading and References
†Arnborg, L., Walker, H.J. and Peippo, J. 1966: Water discharge in the Colville River, Alaska, 1962. *Geografiska annaler* 48A, pp. 195–210.

Collins, D.N. 1979: Quantitative determination of the subglacial hydrology of two alpine glaciers. *Journal of glaciology* 23.89, pp. 347–62.

†Lliboutry, L. 1983: Modifications to the theory of intraglacial water-ways for the case of subglacial ones. *Journal of glaciology* 29.102, pp. 216–26.

Oswald, G.K.A. and Robin, G. de Q. 1973: Lakes beneath the Antarctic ice sheet. *Nature* 245.5423, pp. 251–4.

†Shreve, R.L. 1972: Movement of water in glaciers. *Journal of glaciology* 11.62, pp. 205–14.

†Vivian, R. and Zumstein, J. 1973: Hydrologie sous-glaciaire au glacier d'Argentière (Mont Blanc, France). *Symposium on the hydrology of glaciers, Cambridge, 9–13 September 1969.* International Association of Scientific Hydrology 95, pp. 53–64.

Mercalli scale A scale between 1 and 12 for measuring the intensity of earthquakes based on the amount of structural damage they cause. (See EARTHQUAKE.)

mere A lake, especially in Cheshire and East Anglia where meres are developed in glacial outwash deposits.

meridional circulation Usually defined as the average, over all longitudes, of the flow in the meridional plane. Though very much weaker than the zonal (i.e. along latitudes) component, the convergence-divergence of the meridional component defines the main climatic zones. Spatial unrepresentivity of observations makes the arithmetic unreliable compared to the values of wind speeds of 0.1 to 1 m s^{-1} observed. Annual averages confirm the picture of a direct (upward moving air warmer than downward moving air) tropical cell, bounded on the equatorial side by the INTERTROPICAL CONVERGENCE ZONE (ITCZ) and by an indirect cell (upward moving air cooler than downward moving air) between 30° and 60° latitude. There is some suggestion of a second direct cell between 60° and 90°. Downward velocities bringing dry air towards the surface at 30° (and possibly 90°) coincide with the desert belts at these latitudes. Fluxes of sensible, gravitational potential and latent energy in the meridional circulation are each much greater than their sum. Fluxes of ANGULAR MOMENTUM at high and low levels are each large compared to their sum. Thus the circulation acts largely to convert one type of thermal energy into another and to adjust the vertical distribution of zonal wind. To some extent the circulation is an arithmetic artefact, in that the actual flow takes place in specific geographical areas (such as Indonesia) often on occasions of intense tropical cumulonimbus (see CLOUDS) rather than being spread uniformly in space and time.   JSAG

Reading
Lorenz, E.N. 1967: *The nature and theory of the general circulation of the atmosphere.* Geneva: World Meteorological Organization.

merokarst A little-used term first introduced by Cvijić (1925) to describe non-classical or imperfectly developed limestone scenery. It usually refers to karst regions which comprise very detrital (hence impure) limestones, which retard the development of classic karst features such as KARREN.   PAB

Reference
Cvijić, J. 1925: Le mérokarst. *Comptes rendus de l'Académie des Sciences* 180, pp. 757–8.

mesa A steep-sided plateau of rock, often in horizontally bedded rocks, surrounded by a plain.

mesometeorology In the analysis of the environment, in whole or in part, an appreciation of scale is vital. In meteorology this appreciation is primarily manifest in a recognition of entities of air motion that have significantly different characteristic horizontal sizes. Thus we may visualize the general, global atmospheric circulation to comprise other smaller circulations of different sizes, all interrelated in various

A mesa created by erosion in the Monument Valley Tribal Park, USA. Such features are a good example of one form of slope evolution – parallel retreat.

ways. It is not unreasonable to use the analogy of a car engine, the whole being the 'general circulation' and all the various components being the 'smaller circulations'. Analysis of the energy content of the atmosphere suggests a crude three-tier hierarchy of circulations based upon their characteristic horizontal dimensions (see table). The middle tier of this hierarchy is called the mesoscale, and includes circulations with horizontal sizes of the order of 20 km to 250 km and lifetimes of a few to a few tens of hours. Mesometeorology is primarily concerned with understanding these circulations.

Mesoscale atmospheric circulations may be conveniently categorized into those that are topographically induced and those that are intrinsically products of the free atmosphere. Each of these categories may be divided into two parts. Thus topographically induced systems may result from mechanical and thermal processes and the

Mesometeorology: three scales of atmospheric motion

	Macro	Meso	Local
Characteristic dimension (km)	>483	16–160	<8
	Synoptic	Meso	Micro
Period (h)	> 48	1–48	< 1
Wavelength (km)	>500	20–500	<20

Source: Atkinson 1981.

free atmosphere systems may result from convective and non-convective processes.

Mechanically forced circulations include LEE WAVES, downslope winds (see BORA, FÖHN) and circulations within wakes. Lee waves are just that – waves in the airflow to the lee of the hill that triggers them. Their typical wavelength is about 5–15 km and their crest-to-trough amplitude is about 0.5 km. They are ubiquitous, occurring to the lee of hills as small as the Chilterns and as large as the Rockies. They result primarily from vertical oscillations induced in a stable airstream by the hill. A wind with strong vertical shear and a direction almost perpendicular to the main axis of the hill is also favourable. Strong winds down the leeward slope of the hills frequently accompany lee waves. Circulations in wakes usually take the form of vortices, frequently in trains known as vortex streets. These streets develop behind an obstacle, usually an island, when the flow in the wake region does not mix with that in the surrounding region.

Thermally induced circulations include SEA/LAND BREEZES and slope and VALLEY WINDS. Both are the result of a thermally direct, diurnally varying, vertical circulation. In the case of the sea/land breeze, the land–sea thermal contrast generates low-level flow from the sea by day and low-level flow from the land by night. In hilly areas the daytime mechanism generates upslope and up-valley airflow near the surface, overlain by a 'return-current' aloft that flows from the hills to the plains. At night the reverse occurs, resulting in katabatic flows downslope and down-valley.

Moving gravity waves are non-convective, free-atmosphere systems. Little is currently known about this type of wave but we do know that they have long wavelengths, up to 500 km, and that they are frequently associated with strong vertical shear of the horizontal wind. Convective, free-atmosphere circulations comprise severe local storms (see THUNDERSTORM), shallow cellular circulations and circulations in CYCLONES. Severe local storms have been studied for decades yet still remain a challenge. It now appears that in addition to ATMOSPHERIC INSTABILITY a strong vertical shear of horizontal wind is vital for long-lived (several hours), large (up to 100 km across) storms to exist.

Shallow cellular circulations were first fully documented only in the satellite era. The cells are usually hexagonal, may be open (i.e. cloud-free area surrounded by cloud) or closed (i.e. cloud area surrounded by a hexagonal ribbon of clear air), have diameters of about 30 km and depths of about 2 km. Their full explanation is as yet unknown. Circulations within cyclones are also recently observed features. Radar and autographic rain gauge records have revealed systems about 50 km across, with life times of a few hours, that form, move and dissipate within synoptic-scale frontal areas. At present they are thought to result from CONVECTION which is released in layers of potentially unstable air over the frontal surfaces.

Much remains to be discovered and explained within the realms of mesometeorology. The sub-discipline is taking an increasingly important role within meteorology as a whole. The next decade holds exciting prospects for both observational and theoretical studies of mesoscale atmospheric circulations.   BWA

Reading
Atkinson, B.W. 1981: *Meso-scale atmospheric circulations.* London and New York: Academic Press.

mesophyte A plant which flourishes under conditions which are neither very wet nor very dry.

mesoscale cellular convection Over millions of square kilometres of oceans' heat transfer to the atmosphere takes the form of mesoscale cellular convection (MCC). The convection is visible in the particular cloud forms produced and satellite imagery first noted this form of convection about 1970. The convection is shallow (*c.*2 km deep) and forms distinctive horizontal patterns. In planform the convection takes two main forms: parallel lines of clouds, known as cloud streets; and hexagonal cells. The former is essentially two-dimensional, the latter three-dimensional convection.

The cloud streets may extend for a few hundred kilometres and they lie a few kilometres apart. The hexagonal cells are typically 30 km across and are of two types: closed cells and open cells. The former have uplift in their centre (hence are cloudy and 'closed') and compensatory subsidence on their sides, giving clear air. The open cells have a reversed configuration of vertical motion: uplift on the 'walls' and subsidence in the middle. These different forms are related to vertical stability and vertical shear of horizontal winds, particularly the latter. The details of these relationships are not yet fully clear.

Both types of MCC tend to occur over warm ocean surfaces adjacent to cold land surfaces, such as the Sea of Japan and the Atlantic east of Greenland. In such areas very cold air frequently flows over warm ocean surfaces thus creating the instability required to drive this kind of convection.
                                        BWA

mesosphere A portion of the earth's atmosphere. Either that between the stratosphere (at about 40 km above the surface) and the thermosphere (80 km) or that between the ionosphere (400 km) and the exosphere (1000 km).

mesothermal climate Characterized by moderate temperatures, with the mean temperature of the coldest month between $-3°C$ and $+18°C$. Climates of this type are found mainly between latitudes 30° and 45° but may extend up to latitude 60° on the windward side of continents.

mesotrophic See NUTRIENT STATUS.

Messinian salinity crisis An event that took place around 6.5 to 5.0 million years ago, at the end of the Miocene. The growth of the Antarctic Ice Sheet at that time contributed to a marked sea-level reduction so that the western TETHYS OCEAN lost its connection with the Atlantic. Tectonic activity may have been a contributory factor. Isolation of this ancient Mediterranean from the world ocean caused it to become highly saline, so that enormous gypsum and halite deposits were precipitated. Complete evaporation of this brine is thought to have occurred between 5.0 and 5.5 million years ago. The drying up of so large a body of water over a brief geological period would have had major consequences, facilitating, for example, the incision of rivers and the movements of fauna and flora. ASG

Reading
Adams, C.G., Benson, R.H., Kidd, R.B., Ryan, W.B.F. and Wright, R.C. 1977: The Messinian salinity crisis and evidence of late Miocene eustatic changes in the world ocean. *Nature* 269, pp. 383–6

metamorphism The processes by which the composition, structure and texture of consolidated rocks are significantly altered through the action of heat and pressure greater than that produced normally by burial, or by the introduction of additional minerals as a result of a marked change in the thermodynamic environment, e.g. limestone may be converted to marble, mudstone to slate, and granite to gneiss.

The intensity of metamorphism is referred to as its grade, and high grade metamorphic rocks include gneiss, granulite, blueschist, amphibolites and eclogites.

There are three classes of metamorphism: regional metamorphism which occurs in orogenic belts; contact metamorphism around the boundaries of igneous intrusions: and dislocation metamorphism resulting from friction along fault or thrust planes. ASG

metapedogenesis The modification of soil properties by human agency (Yaalon and Yaron 1966). Some of the major influences of man on the main factors of soil formation are demonstrated in the table, modified after Bidwell and Hole (1965). ASG

References
Bidwell, O.W. and Hole, F.D. 1965: Man as a factor of soil formation. *Soil science* 99, pp. 65–72.
Yaalon, D.H. and Yaron, B. 1966: Framework for man-made soil changes – an outline of metapedogenesis. *Soil science* 102, pp. 272–7.

metasomatism The processes by which a rock is physically or chemically altered,

partially or wholly, as a result of the introduction of new materials. Contact metamorphism is the alteration of rocks adjacent to an igneous intrusion by metasomatism and high temperatures.

meteoric water/groundwater Water that is derived by precipitation in any form from the atmosphere. Originally defined by Meinzer (1923) and contrasted with JUVENILE WATER, meteoric groundwater is water with an atmospheric or surface origin rather

Metapedogenesis: suggested effects of the influence of man on five classic factors of soil formation

PARENT MATERIAL
Beneficial: adding mineral fertilizers, accumulating shells and bones, accumulating ash locally, removing excess amounts of substances such as salts.
Detrimental: removing through harvest more plant and animal nutrients than are replaced, adding materials in amounts toxic to plants or animals, altering soil constituents in a way to depress plant growth.

TOPOGRAPHY
Beneficial: checking erosion through surface roughening, land forming and structure building; raising land level by accumulation of material; land levelling.
Detrimental: causing subsidence by drainage of wetlands and by mining; accelerating erosion; excavating.

CLIMATE
Beneficial: adding water by irrigation; rainmaking by seeding clouds; removing water by drainage; diverting winds, etc.
Detrimental: subjecting soil to excessive insolation, to extended frost action, to wind, etc.

ORGANISMS
Beneficial: introducing and controlling populations of plants and animals; adding organic matter including 'nightsoil'; loosening soil by ploughing to admit more oxygen; fallowing; removing pathogenic organisms as by controlled burning.
Detrimental: removing plants and animals; reducing organic matter content of soil through burning, ploughing, over-grazing, harvesting, etc; adding or fostering pathogenic organisms; adding radioactive substances.

TIME
Beneficial: rejuvenating the soil through adding of fresh parent material or through exposure of local parent material by soil erosion; reclaiming land from under water.
Detrimental: degrading the soil by accelerated removal of nutrients from soil and vegetation cover, burying soil under solid fill or water.

than with a source in the interior of the earth. (See also CONNATE WATER.)　PWW

Reference
Meinzer, O.E. 1923: Outline of groundwater hydrology. *US Geological Survey water supply paper 494.*

meteorological satellites Artificial, earth-orbiting devices designed solely or largely for meteorological observation, supplemented in some cases by meteorological data collection and or relay functions. They should be distinguished from spacecraft (deep space probes) designed to investigate other bodies in the solar system, particularly other planets and their moons although, in several cases, these have investigated atmospheric conditions by similar means to those deployed on meteorological satellites themselves. Observations from meteorological satellites have commonly been made by sensor systems designed to provide cloud imagery, radiation budget data, and vertical profiles through the atmosphere. Although American Vanguard and Explorer satellites orbited in 1959 and 1960 were equipped for earth imaging, this facet of their missions was secondary, and the cloud pictures returned were low in resolution, and therefore indistinct. The first purpose-built meteorological satellite was the American Tiros-I, launched on 1 April 1960. This, a 'television and infrared observation satellite', carried a vidicon camera for visible (reflected) light imaging, and an infrared radiometer to provide target temperature data in the invisible thermal region of the electromagnetic spectrum. The data were transmitted to earth by radio, a procedure followed by all subsequent meteorological satellites. As time passed the value of such satellites was confirmed; in particular it was recognized that they provided more uniform and more complete views of global weather than were available from conventional ('*in situ*') sources. Additionally, their views of weather were new, and supplementary to those obtained from within the atmosphere or from the interface with its parent planet.

Many new satellite configurations, systems and sensors were developed during the 1960s and early 1970s. Russian, European, Japanese and Indian satellites have joined those of the USA to provide an almost overwhelming flood of data useful not only in day-to-day meteorological operations (especially weather forecasting), but also meteorological and climatological research. The last is expanding rapidly now that satellite data sets have begun to achieve climatologically suitable lengths. Key activities in the meteorological satellite arena, organized for the most part chronologically, have included the following:

1　Early testing and development of low-altitude, polar-orbiting satellites by the USA (Tiros series) and the former USSR (some of the Cosmos series), early to mid 1960s.
2　Inauguration of the first fully operational global weather satellite system (Essa series) by the USA in 1966. This consisted of two polar-orbiting satellites, one giving direct read-out data for local use, the other tape-recording data for global archiving. The USSR effectively followed suit with its Meteor series, inaugurated in 1969.
3　Very active testing and research of new sensors and modes of operation, impressively effected through the US Nimbus satellite series from 1964 to the mid 1980s.
4　Implementation of improved operational polar-orbiting satellite systems for civilian use by the American NOAA series inaugurated in 1970, and subsequently by an advanced NOAA series in 1979, capitalizing on experience with earlier civilian satellites, and military satellites of the US Defense Meteorological Satellite Program (DMSP series).
5　Early testing, in the late 1960s, of US 'geostationary' satellites of the Applications Technology Satellite (ATS) series (deployed at 35,400 km above the equator, where they appear to be stationary above a preselected point on the surface of the earth).
6　Inauguration of the first operational geostationary satellites by the USA in 1975 (the GOES series), leading to a complete global encirclement by geostationary satellites in low and middle latitudes for the First GARP Global Experiment (FGGE) year from 1 December 1978 to 30 November 1979. The FGGE coverage necessitated five geostationary orbiters, three American (GOES), one European (Meteosat-1) and one Japanese (GMS (or 'Himawari')-1).

7 The implementation in the early to mid 1980s of geostationary satellites for operational coverage of the European/ African, and Indian/Indian Ocean sectors of the globe by Eumetsat (on behalf of a consortium of European meteorological offices) and India respectively.

The continued internationalization of meteorological satellite activities indicated by (7) above confirms the key role which such satellites are now adjudged to play in meteorology and climatology both globally and regionally. A wide range of satellite products and derived products are now available for weather and climate research. Increasingly *in situ* observing networks are being planned and operated in full conjunction with meteorological satellites, which are now recognized as indispensable parts of any effective broad-scale atmospheric monitoring system.   ECB

Reading
American Society of Photogrammetry 1983: Meteorology satellites. In *Manual of remote sensing*. 2nd edn. Pp. 651–78.

Anderson, R.K. and Veltischev, N.F. 1973: *The use of satellite pictures in weather analysis and forecasting*. WMO technical note 124. Geneva: World Meteorological Organisation.

Brimacombe, C.A. 1981: *Atlas of Meteosat imagery*. ESA SP-1030. Paris: European Space Agency.

National Space Science Data Center periodically: *Reports on active and planned spacecraft and experiments*. NSSDC/WDC-A-R & S. Greenbelt, Md: Goddard Space Flight Center.

NASA 1982: *Meteorological satellites: past, present and future*. NASA conference publication 2227. Washington DC: NASA.

meteorology The science concerned with understanding atmospheres. In recent years since satellites have probed the atmospheres of planets other than earth, it has become acceptable to use the term 'atmospheric science' and to apply the term to any atmosphere, not just that on earth.
                                                   BWA

microclimate The climate in which plants and animals live. Whereas regional climate is largely determined by latitude, global weather patterns and continental/ maritime differences, and local climate is influenced by altitude and exposure, microclimate is inextricably linked to the processes taking place at the surface very close by. The absorption of solar energy by plants, animals, soil and rocks, its conversion to heat in the air or soil, and its use in evaporating water are important factors determining microclimate. Of equal impor-

tance are the influences of WIND speed and turbulence in mixing the air and hence in diffusing local surface effects. Very close to surfaces windspeed is reduced by friction and shelter, and in such cases strong heating or cooling at the surface may produce microclimates much more extreme than in more exposed regions. For example, the daily range of temperature in the air a few centimetres above desert sand may be 50°C larger than the range measured at a height of one or two metres. Microclimates for plants and animals may be much more extreme than implied by standard meteorological data. Animals (including man) can select appropriate microclimates for comfort and survival, and several useful schemes have been developed to relate possible microclimates to known physiological responses, thereby establishing the microclimate space in which an animal can exist.
                                                   MHU

Reading
Jones, H.G. 1983: *Plants and microclimate*. Cambridge and New York: Cambridge University Press.

Rosenberg, N.J. 1983: *Microclimate: the biological environment*. New York: Wiley.

microcracks Can occur in rocks of all kinds. They are generally short (a few centimetres long) and narrow (less than one millimetre) but they are important because they provide one of the main means by which rocks break down. Both physical and chemical weathering can take place within them as they provide a much enlarged surface area for chemical action to take place and the crack tip is an important stress concentrator in brittle fracture. A general classification of such cracks has been put forward by Farran and Thenoz (1965).

microfissures: less than 1 μm in width and about the length of a crystal;
microfractures: about 0.1 mm or less wide;
macrofractures: greater than 0.1 mm wide and may be several metres long.
                                                   WBW

Reading
Farran, J. and Thenoz, B. 1965: L'alterabilité des roches, ses fractures, sa prevision. *Ann. Inst. Tech. Batiments Travaux Publiques* no. 25.

Whalley, W.B., Douglas, G.R. and McGreevy, J.P. 1982: Crack propagation and associated weathering in igneous rocks. *Zeitschrift für Geomorphologie* 26, pp. 33–54.

micro-erosion meter An instrument used to measure accurately the erosion or weathering of a rock from direct measurement on the rock surface. It comprises a spring-loaded probe connected to an engineer's dial gauge. In order to measure exactly the same area each time, the instrument is mounted on three legs which are placed onto studs drilled in the rock surface. Measurements have been achieved to the nearest 0.00001 mm.   PAB

Reading
Goudie, A.S. ed. 1990: *Geomorphological techniques*. London: Unwin Hyman.

micrometeorology The science concerned with the study of physical phenomena taking place on a microscale in the atmosphere close to the surface. The development of micrometeorology has involved not only the examination and analysis of highly accurate observations made in the air layers adjacent to surfaces but also detailed study of processes at surfaces which determine the properties of air in such layers. A principal feature of air near the ground is its turbulence, the complex mixing which is produced by eddies with sizes ranging from microscopic to tens of metres. The turbulence is generated by friction between moving air and the surface, and may be modified by temperature differences between air and ground. Understanding of the structure of turbulence has been a major theme of micrometeorology, of fundamental importance and with practical applications such as the dispersal of spores, and the diffusion of air pollution. Exchange between the surface and the atmosphere leads to vertical gradients of temperature, windspeed and gas concentrations. Much progress in micrometeorology has been made by the detailed analysis of mean gradients (over periods of about 30 min) at sites where there is extensive uniform horizontal ground cover. Based on such analysis, rates of exchange at the surface can be deduced from measurements made entirely in the atmosphere. For example, rates of evaporation or of carbon dioxide uptake by growing crops can be studied in relation to diurnal changes in the weather. Gradient methods in micrometeorology have a sound physical basis but must rely on empirical corrections to allow for effects of temperature gradients on turbulence. Recent developments in instrumentation and computers suggest that, in future, methods of measuring turbulent transfer more directly

by sensing the motion and composition of individual eddies – an eddy correlation approach – will become more readily available for applied research.

The surface processes most directly affecting the structure of the lower atmosphere begin with the absorption of radiant energy at the ground. Micrometeorology is concerned with the interception and fate of that energy, with the way it is used to heat vegetation, soil and water, and with the conversion of some of the energy to latent heat when water is evaporated (see EVAPORATION). Heat stored in the soil may be of great importance as an energy source at night, and the understanding of factors determining the temperature structure of the lower atmosphere at night is a challenging problem for micrometeorologists.

Although this topic can be pursued as a fundamental aspect of meteorological physics, much of micrometeorology has practical applications. Examples are particularly common in AGROMETEOROLOGY, including the study of forest meteorology. Particular progress has been made in understanding atmospheric and surface controls of evaporation, diffusion of pollution from point and area sources, uptake of air pollutants by plants, and the spread of animal and plant disease.   MHU

Reading
Arya, S.P. 1988: *Introduction to micrometeorology*. San Diego: Academic Press.

Munn, R.E. 1966: *Descriptive micrometeorology*. New York and London: Academic Press.

Oke, T.R. 1978: *Boundary layer climates*. London: Routledge.

de Vries, D.A. and Afgan, N.H. eds 1975: *Heat and mass transfer in the biosphere*. Washington: Scripta Press.

mid-ocean ridge Large linear arches on the sea floor which mark the lines of volcanic activity along which basaltic rocks are added to the sea floor as it separates. Volcanic activity along the ridges may produce sea-mounts, guyots and islands.

migration In a wide sense, any movement of animals or plants from one region or habitat to another (see DISPERSAL); more specifically, the seasonal or cyclical movement of animals between two or more areas, a phenomenon common in mammals, birds, fishes and insects. The classic example is probably that of the Arctic tern, which migrates each year from its summer breeding grounds in the Arctic circle right across the world to winter on the coast of Antarctica.

Exactly how and why animals carry out these great feats of migration still remain largely unanswered questions, although it is clear that no one theory is universally applicable throughout the animal kingdom. Navigation appears to be possible through the use of the sun as a compass, a similar use of the stars, the use of the earth's magnetic field and the Coriolis effect, the recognition of the lie of the land and distinctive landmarks, and the use of smell.   PAS

Reading
Baker, R.R. 1978: *The evolutionary ecology of animal migration*. London: Hodder & Stoughton.
Dunbar, R. 1984: How animals know which way to go. *New scientist* 101, 12 January, pp. 26–30.

Milankovitch hypothesis One of the most significant models of earth history, which is seen by some (e.g. Imbrie and Imbrie 1979) as the key to understanding the causes of the climatic fluctuations of the PLEISTOCENE. The hypothesis is based on the fact that the position and configuration of the earth as a planet in relation to the sun is prone to change, thereby affecting the receipt of insolation at the earth's surface. There are three such changes which have been identified, all three occurring in a cyclic manner: changes in the eccentricity of the earth's orbit (a 96,000 year cycle), the precession of the equinoxes (with a periodicity of 21,000 years), and changes in the obliquity of the ecliptic (the angle between the plane of the earth's orbit and the plane of its rotational equator). This last has a periodicity of about 40,000 years.

The earth's orbit around the sun is not a perfect circle but an ellipse. If the orbit were a perfect circle then the summer and winter parts of the year would be equal in their length. With greater eccentricity there will be a greater difference in the length of the seasons. Over a period of about 96,000 years the earth's orbit can stretch by departing much further from a circle and then reverting to almost true circularity.

The precession of the equinoxes simply means that the time of year at which the earth is nearest the sun (perihelion) varies. The reason is that the earth wobbles like a top and swivels its axis round. At the moment the perihelion comes in January. In 10,500 years it will occur in July.

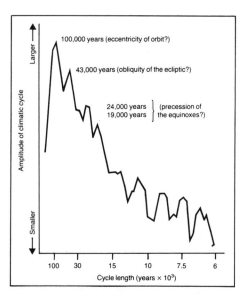

The third cyclic perturbation, change in the obliquity of the ecliptic, involves the variability of the tilt of the axis about which the earth rotates. The values vary from 21°39' to 24°36'. This movement has been likened to the roll of a ship. The greater the tilt, the more pronounced is the difference between winter and summer (Calder 1974).

Appreciation of the possible significance of these three astronomical fluctuations of the earth goes back to at least 1842, when J.F. Adhemar made the suggestion that climate might be affected by them. His views were developed by Croll in the 1860s and by Milankovitch in the 1920s (Mitchell 1965).

The major attraction of these ideas is that while the extent of temperature change caused by them may well only be of the order of 1 or 2°C, the periodicity of these fluctuations seems to be largely comparable with the periodicity of the ice advances and retreats of the Pleistocene. Recent isotopic dating has shown that the record of sea-level changes preserved in the coral terraces of Barbados and elsewhere, and the record of heating and cooling in deep-sea cores, correlates well with theoretical insolation curves based on those of Milankovitch (Mesollela *et al*. 1969).

The variations in the earth's orbit have recently been seen as 'the pacemaker of the ice ages' (Imbrie and Imbrie 1979) for detailed statistical analysis of ocean cores shows that they possess statistically significant wavelike fluctuations with amplitudes of the order of around 100,000 years, 43,000 years and 19–24,000 years (see diagram). The most important of these cycles is the longest one, corresponding to variations in eccentricity (Hays *et al.* 1976). This applies back to 900,000 years ago, but probably not further (Pisias and Moore 1981).

Thus there is some substantial evidence to suggest that astronomical theories may be valid as an explanation of the longer scale of environmental changes (Goreau 1980).

The Croll–Milankovitch model produces a near cyclical series of events which is too long to be relevant to most postglacial fluctuations of climate and too short to throw much light on the spacing of the major ice ages. In addition, the model advocates the development of glaciation in high latitudes by insolation variations, whereas from a glacial mass budget viewpoint, an increase in precipitation over the present minimal levels now received in polar areas may be more important (Andrews 1975). Finally, the computed variations in insolation resulting from this model are never more than a very few per cent so that it is likely that even if this mechanism can initiate change, other factors would be necessary to intensify it.

They may also help to explain the very marked expansion of lakes that took place in tropical and subtropical areas around 9000 BP. Recent analyses of theoretical insolation levels by Kutzbach (1981) indicate that radiation receipts in July at that time were larger than now (by about 7 per cent) and that this led to an intensification of monsoonal circulation and associated precipitation.   ASG

Reading and References
Andrews, J.T. 1975: *Glacial systems: an approach to glaciers and their environments*. North Scituate, Mass.: Duxbury.
Berger, A., Imbrie, J., Hays, J., Kukla, G. and Saltzman, B. eds 1984: *Milankovitch and climate*. 2 vols. Dordrecht: Reidel.
Calder, N. 1974: *The weather machine*. London: BBC.
Goreau, T. 1980: Frequency sensitivity of the deep-sea climatic record. *Nature* 287, pp. 620–2.
Hays, J.D., Imbrie, J. and Shackleton, N.J. 1976: Variations in the earth's orbit: pacemaker of the Ice Ages. *Science* 194, pp. 1121–32.
Imbrie, J. and Imbrie, K.P. 1979: *Ice ages: solving the mystery*. London: Macmillan.
Kutzbach, J.E. 1981: Monsoon climate of the early Holocene: climate experiment with the earth's orbital parameters for 9000 years ago. *Science* 214, pp. 59–61.
Mesollela, K.J., Matthews, R.K., Broecker, W.S. and Thurber, D.L. 1969: The astronomical theory of climatic change: Barbados data. *Journal of geology* 77, pp. 250–74.

Mitchell, J.M. 1965: Theoretical paleoclimatology. In H.E. Wright and D.G. Frey eds, *The Quaternary of the USA*. Princeton, NJ: Princeton University Press. Pp. 881–901.
Pisias, N.G. and Moore, T.C. 1981: The evolution of Pleistocene climate: a time-series approach. *Earth and planetary science letters* 52, pp. 450–8.

mima mound A small, rounded hillock generally 1.0 to 2.0 m in height and 10 m in diameter.

minimum acceptable flow This concept dates from the Water Resources Act 1963 which requires the UK River Authorities to determine and to keep under review minimum acceptable flows for all the main rivers in their areas. Such a flow is very difficult to determine because it necessarily depends upon local circumstances but it should take account of the character of the inland water and its surroundings and should be 'not less than the minimum which in the opinion of the river authority is needed for safeguarding the public health and for meeting (in respect of both quantity and quality of water) the requirements of existing lawful uses of the inland water, whether for agriculture, industry, water supply and other purposes, and the requirements of land drainage, navigation and fisheries . . .' (Water Resources Act 1963, section 19).   AMG

minimum variance theory Depends upon the concept that once an equilibrium condition has been reached there will subsequently be minimal change. A system tends to react to an imposed stress in order to minimize the disturbance, or to restore or to maintain the previous conditions. This theory has been used in fluvial geomorphology and has been employed to explain the interrelations of variables in HYDRAULIC GEOMETRY, the planimetric geometry of MEANDERS and the way in which variables adjust to accommodate change of STREAM POWER.   KJG

Reading
Williams, G.P. 1978: Hydraulic geometry of river cross-sections: theory of minimum variance. *US Geological Survey professional paper* 1029.

mirage An optical illusion produced by refraction in the lower atmosphere as a result of differential heating of the air. Objects beyond the horizon may become visible.

mire A marsh or bog or any area of muddy ground.

misfit See UNDERFIT STREAM.

mist A suspension in the air of very small water drops or wet hygroscopic particles, causing a reduction in the horizontal visibility at the earth's surface. In meteorology the term is used when visibility is so reduced but is still 1000 m or more. Water contents of mists are typically a very small fraction of 1 g m^{-3} and average drop radii are typically less than 1 μm. Owing to relatively high concentrations of solute within the drops, mists may persist with relative humidities as low as 80 per cent.

KJW

mistral A cold dry north or north-west wind affecting the Rhone valley, particularly to the south of Valence. It is typically strong and squally and most violent in winter and spring. A depression over the Tyrrhenian Sea or Gulf of Genoa with high pressure advancing from the west towards Spain provides the necessary synoptic gradient. Market gardens and orchards require protection from the mistral by windbreaks. Topographic channelling causes local strengthening of windspeeds in the Rhone valley (Barsch 1965), while marked diurnal variations are common. In the area of maximum frequency of mistral in the Rhone delta an average of over 100 days per year are recorded.

AHP

Reading and References
Barsch, D. 1965: Les arbres et le vent dans la vallée meridionale du Rhone. *Revue de géographie de Lyons* 40, pp. 35–45.

†Boyer, F., Orieux, A. and Powger, E. 1970: *Le Mistral en Provence occidentale.* Monographs de la météorologie 79. Paris: Nationale.

mixing corrosion The increased degree of solutional corrosion which occurs when two saturated karst waters of different composition mix. (See also TROMBE'S CURVES.)

mixing models Models used to explain or predict temporal variations in the solute concentrations of streamflow by taking account of the mixing of water from different sources or the mixing of water within a store. Gregory and Walling (1976) report several studies where simple mass balance models, based on the mixing of individual run-off components, have been developed. Johnson *et al.* (1969) also describe a model involving the mixing of incoming precipitation with water stored within the basin which in turn supplies the streamflow output.

DEW

References
Gregory, K.J. and Walling, D.E. 1976: *Drainage basin form and process.* London: Edward Arnold. Pp. 222–3.

Johnson, N.M., Likens, G.E., Bormann, F.H., Fisher, D.W. and Pierce, R.S. 1969: A working model for the variation in stream water chemistry at the Hubbard Brook Experimental Forest, New Hampshire. *Water resources research* 5, pp. 1353–63.

mobile belt A linear crustal zone characterized by tectonic activity. The term is applied to both contemporary zones of OROGENY, where mountain ranges are actively being formed, and to ancient zones of intense crustal activity indicated by the effects of metamorphism, granite emplacement and faulting. In the latter sense mobile belts are contrasted with the stable crustal regions of continents or CRATONS.

MAS

Reading
Spencer, E.W. 1977: *Introduction to the structure of the earth.* 2nd edn. New York: McGraw-Hill.

Windley, B.F. 1984: *The evolving continents* 2nd edn. London and New York: Wiley.

moder One of the three main forms of organic matter in soils. See also MOR and MULL. Between the neutral or slightly acid conditions in which mull develops and the very acid conditions of mor, there is an intergrade known as moder. Although much of the organic fraction is well decomposed and incorporated into the soil's mineral profile, the binding between the two remains weak. There is therefore no strong structural development, and a thin layer of litter and fermented material accumulates at the surface.

modulus of elasticity A spring is a mechanical model for behaviour of an elastic material under stress. Such a material responds immediately to an applied load and its change of dimensions is directly proportional to the applied stress. When the load is removed the entire strain (i.e. deformation) is recoverable. The magnitude of the stress (σ) required to produce a given strain (ε) is a characteristic of the material and is known as Young's modulus (E) of elasticity (a modulus is a constant factor):

$$E = \sigma/\varepsilon$$

This relationship between stress and strain is termed Hooke's Law and a perfectly elastic substance is said to be Hookean. The greater the value of E for a material, the less will be the deformation produced by a given value of stress, and the stronger will be the material. Strong rocks are near to being Hookean solids and have values of E in the range 50–100 GPa.   MJS

Reading
Selby, M.J. 1993: *Hillslope materials and processes*. 2nd edn. Oxford: Oxford University Press. Ch. 4.

mogote (haystack hill) A large residual limestone hill which is a remnant of limestone solution and erosional processes. It is roughly circular in shape, with steep sides which terminate abruptly in a flat alluvial plain. Mogotes are usually found in large numbers and are common in south China (where they inspired many classical landscape pictures), North Vietnam, France (Massif Central), the Philippines, Yugoslavia and Java. In each area they have different local names. (See also COCKPIT KARST; KEGELKARST.)   PAB

Reading
Klimaszewski, M. 1964: The karst relief of the Kuelin area (South China). *Geographia Polonica* 1, pp. 187–212.

Mohr–Coulomb equation A widely used relationship to describe the strength of soils in terms of COHESION and FRICTION of materials:

$$s = c + \sigma \tan \phi$$

where s is the shear strength at failure (sometimes denoted by T), c is the cohesion component and ϕ the friction of the soil; σ (sometimes σ_n) is the normal stress. Either cohesive or frictional components may be absent but a plot of normal

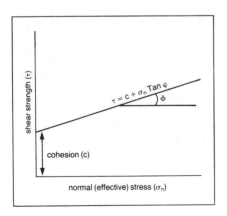

stress and shear stress for a soil with both is of the form shown below. The strength parameters c and ϕ are usually obtained by SHEAR BOX or TRIAXIAL APPARATUS tests. The equation is modified if pore water pressures are taken into account to give an EFFECTIVE STRESS analysis.   WBW

Reading
Mitchell, J.K. 1976: *Fundamentals of soil behaviour*. New York: Wiley.
Statham, I. 1977: *Earth surface sediment transport*. Oxford: Clarendon Press.
Whalley, W.B. 1976: *Properties of materials and geomorphological explanation*. Oxford: Oxford University Press.

Mohorovičić discontinuity Sometimes referred to as the 'moho', lies between the crust of the earth and the underlying MANTLE. Occurring at around 30–40 km below the continents and at about 10 km beneath the oceans, it is a zone where seismic waves are significantly modified.   ASG

moisture index Relates water loss through potential evapotranspiration (PE) to water gain through precipitation (P), as a basis for climatic classification:

$$MI = \frac{100(P - PE)}{PE}$$

The moisture index (MI) falls to -100 when there is no precipitation, and exceeds $+100$ when precipitation greatly exceeds potential evapotranspirational losses.   ASG

molard A term from the French Alps for 'conical mounds of broken slide rock deposited along the typically lobate margins of rock avalanche spoil debris – a debris cone' (Cassie *et al.* 1988). They may be up to 20 m high.   ASG

Reference
Cassie, J.W., Van Gassen, W. and Coudess, D.M. 1988: Laboratory analogue of the formation of molards, cones of rock-avalanche debris. *Geology* 16, pp. 735–78.

molasse A term applied to any thick succession of continental deposits consisting in part of sandstones and conglomerates which were formed as a result of mountain building. The molasse facies is the main diagnostic feature of orogeny (mountain building).

momentum budget Except for the small effects of tidal friction, the absolute

ANGULAR MOMENTUM of the earth and atmosphere combined must remain constant in time; and because the annual average rotation rate of the earth is observed to remain almost constant, the atmosphere alone must also on average conserve its angular momentum. Thus it is possible to consider the fluxes of angular momentum in an angular momentum

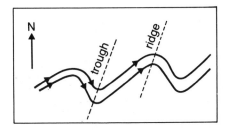

Schematic streamlines for a positive eddy momentum flux.

budget. Such a momentum budget is applied to the time-averaged flow of the atmosphere, that is the GENERAL CIRCULATION.

Angular momentum per unit mass of a westerly current is the product of angular velocity with the distance about the axis of rotation. In the tropics, where surface winds are easterly, the torque, or twisting moment, about the earth's axis exerted by the earth on the atmosphere is such that the atmosphere gains angular momentum from the earth; while in the middle latitudes, where surface winds are westerly, the atmosphere gives up angular momentum to the earth. Therefore there must be a net poleward flux of angular momentum in the atmosphere between the tropics and middle latitudes, otherwise surface friction would decelerate both the easterlies and westerlies. This poleward flux of angular momentum must increase with latitude in the regions of the easterlies and decrease with latitude in the westerlies. In the northern hemisphere it is observed to reach a maximum at about 30° latitude. Poleward of the mid-latitude westerlies, the surface area of the earth is relatively small, so that angular momentum exchange is relatively unimportant. If the total angular momentum of the atmosphere is to remain constant the exchanges between atmosphere and earth in the easterlies and westerlies must be equal but of opposite sense.

In low latitudes the poleward momentum transport is effected partly by the mean meridional flow and partly by eddies. In middle latitudes the mean meridional flow is much too weak to effect a significant flux so that the poleward transport is accomplished mainly by eddies. For eddies to effect a meridional transport of angular momentum they must be asymmetric, with the axes of troughs and ridges tilting from south-west to north-east as shown in the diagram. In such a trough–ridge system zonal flow is larger in those portions where the meridional flow is poleward, so that a poleward angular momentum flux results. Observations show that eddies account for most of the poleward transport except within about 10° latitude of the equator.

The torque exerted by the earth on the atmosphere is due partly to surface friction and partly to pressure differences across mountains. For instance, in the mid-latitude westerlies, observations show that surface pressures on the western slopes of mountains tend to exceed those on the eastern slopes at the same heights. In mid-latitudes this mountain pressure torque is estimated to be about as large as the torque due to surface friction.   KJW

Reading
Holton, J.R. 1992: *An introduction to dynamic meteorology*. New York and London: Academic Press.

monadnock Any residual hill or mountain which is isolated on a flat plain produced by erosion. It is a product of the late stages of the Davisian cycle of erosion, and rises above a peneplain.

monoclimax A theory of vegetation requiring that all the SERES (community sequences) in an area converge on a uniform and stable plant community, the composition of which depends solely on regional climate. Given sufficient time, it is argued, the processes of SUCCESSION overcome any major effects on vegetation of differences in other environmental factors, such as topography and soils.

It is now generally replaced by other theories of CLIMAX VEGETATION. Many vegetational terms had their origin in monoclimax theory. (See also POLYCLIMAX.)   JAM

Reading
Clements, F.E. 1916: *Plant succession*. Publication 242. Washington: Carnegie Institute.

Matthews, J.A. 1979: Refutation of convergence in a vegetation succession. *Naturwissenschaften* 66, pp. 47–9.

Walker, D. 1970: Direction and rate in some British post-glacial hydroseres. In D. Walker and R.G. West eds, *Studies in the vegetation history of the British Isles*. Cambridge: Cambridge University Press. Pp. 117–39.

monocline A zone of steeply dipping strata in an area of horizontally bedded rocks. A zone of rocks which dip steeply to great depth.

monoglaciation The term applied to the Great Ice Age by those who believe only one major glacial advance occurred.

monophylesis The origin of a taxonomic group at one point in space and time, by EVOLUTION.

monsoon Derived from the Arabic word *mausim*, meaning season, the term originally referred to the winds of the Arabian Sea, which blow for about six months from the north-east and for six months from the south-west. The term is now used for other markedly seasonal winds, e.g. the stratospheric monsoon. The characteristics of the monsoon climate are to be found mainly in the Indian subcontinent, where over much of the region annual changes may conveniently be divided as follows:

1 The season of the north-east monsoon: (*a*) January and February, winter season; (*b*) March to May, hot weather season.
2 The season of the south-west monsoon: (*a*) June to September, season of general rains; (*b*) October to December, post-monsoonal season.

Over much of India the bulk of the rainfall comes in the south-west monsoon season.   JGL

Reading
Chang, C.P. 1987: *Monsoon meteorology*. Oxford: Oxford University Press.

Ramage, C.S. 1971: *Monsoon meteorology*. London and New York: Academic Press.

monsoon forest A lowland and montane division of TROPICAL FOREST characterized by seasonal climates with pressure and wind reversal. Monsoon forests occur when the total rainfall during the monsoon is sufficient to maintain forest, but where there is also a marked dry season. The resulting trees may be predominantly evergreen, depending upon the intensity of the wet season. They are, however, more characteristically semi-deciduous, particularly in the upper canopy, as a result of the seasonal water deficiency. The structure is more open than lowland evergreen forest, with greater light penetration to the ground and denser undergrowth, with locally abundant bamboos and lianas.   PAF

Reading
Golley, F.B., Lieth, H. and Werger, M.J.A. eds 1982: *Tropical rain forest ecosystems*. Amsterdam: Elsevier.

Longman, K.A. and Jenik, N. 1987: *Tropical forest and its environment* 2nd edn. London: Longman.

monumented sections River channel cross-sections which are precisely surveyed in relation to at least two fixed points and which can be resurveyed on a number of occasions to indicate the amount of channel change. Such sections were advocated as part of the Vigil Network of REPRESENTATIVE AND EXPERIMENTAL BASINS.   KJG

mor Raw humus which is not admixed with mineral material.

moraine A distinct landform fashioned by the direct action of a glacier. In the past the term was also used to describe glacial sediments, e.g. ground moraine, but now it is accepted that *moraine* should refer to the landforms and the word *till* to the sediment.

One group of moraines exists on the surfaces of glaciers and includes *lateral* moraines, which form through the accumulation of valley-side material on either side of the glaciers, and *medial* moraines which form from the junction of lateral moraines as two glaciers meet. In the ABLATION areas of glaciers in particular such moraines can form prominent upstanding ridges where the debris has protected the underlying ice from melting. The material in lateral and medial moraines is characteristically angular rockfall debris and undergoes minimum modification during transport.

A second group of moraines occurs at the edge of existing glaciers or in areas formerly covered by glaciers. One classification scheme, based on glacier dynamics and position with regard to a glacier is shown in the diagram. Subglacial forms constructed beneath moving glacier ice include uniform till sheets, as for example occur widely in Iowa and Illinois (Kemmis 1981), streamlined and transverse features. Glacial flutes are small-scale examples of ridges streamlined parallel to the direction of ice flow and

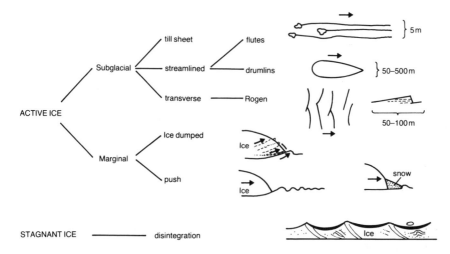

Classification of moraines.
Source: *Chorley et al. 1984. Figure 17.23.*

are generally a few tens or hundreds of metres long and up to 2 m high. They build up in the lee of boulders which become lodged on the bed and thereby create a cavity or low pressure zone in their lee. Drumlins are larger forms streamlined parallel to the direction of ice flow (see DRUMLIN). There is a whole family of ridges, commonly termed Rogen moraines after a lake in Sweden where they abound, which have been constructed transverse to ice flow in a subglacial position. Individual ridges are 5–20 m high and 100–3000 m

long and built of basal till. Although their origin is unclear it seems that they result from some form of overthrusting in a zone of compressing glacier flow, induced perhaps by the shape or nature of the topography or a longitudinal change in basal ice conditions.

Active glaciers which end on land build up moraines at the ice margin. The size of the moraine depends on the period that the margin lies in the same location and also on the amount of rock debris transported to the edge by the glacier. The common 'ice-

Classification of the major types of moraine

Parallel to ice flow	Transverse to ice flow	Lacking consistent orientation
Subglacial forms with streamlining	*Subglacial forms*	*Subglacial forms*
Fluted and drumlinized ground moraine	Rogen or ribbed moraine	Low-relief ground moraine
Drumlins and drumlinoid ridges	De Geer or washboard moraine	Hummocky ground moraine
Crag-and-tail ridges	Subglacial thrust moraine	
	Sublacustrine moraine	
Ice-pressed forms	*Ice-pressed forms*	*Ice-pressed forms*
Longitudinal squeezed ridges	Minor transverse squeezed ridges	Random or rectilinear squeezed ridges
Ice marginal forms	*Ice front forms*	*Ice surface forms*
Lateral and medial moraines	End moraines	Disintegration moraines
Some interlobate and kame moraines	Push moraines	
	Ice thrust/shear moraines	
	Some kame and delta moraines	

Source: Modified from Sugden, D.E. and John, B.S. 1976: *Glaciers and landscape.* London: Edward Arnold. Table 12.1, p. 236. After Prest, V.K. 1968: *Geological survey papers Canada* 67–57.

dumped' moraine is typically a complex landform reflecting many different processes of debris accumulation, including slumping or flow off the glacier surface, and lodgement or deformation of basal till. Another common moraine is associated with deformation of sediments by pushing. Glaciers tend to advance in winter when ablation rates at the snout are low and this advance may physically push sediments into a ridge up to 2 m high. Sometimes it may push a frontal snow bank or sheet of lake ice which itself deforms sediments (Birnie 1977). If a glacier is in overall retreat these annual advances are marked by a succession of small ridges whose spacing can be directly related to the amount of annual ablation. At a larger scale push moraines may involve much larger areas of sediments in front of a glacier, as for example, can occur in the case of a SURGING GLACIER.

Disintegration moraine is the complex remains of a process whereby debris-bearing ice stagnates and melts *in situ*. An irregular landscape of hummocks and kettles, with differing combinations of subglacial and slumped surface debris is the end product. Such moraines form best where compressing flow brings large quantities of debris up to the ice surface.

Moraines have been widely used to delimit the former extent of glaciers. When they are dated unambiguously a relatively sophisticated reconstruction of past climate can be made. Problems have arisen in recent years, however, because many moraines are found to have been built up by successive glacier advances over long time spans.   DES

Reading and References
Birnie, R.V. 1977: A snow-bank push mechanism for the formation of some 'annual' moraine ridges. *Journal of glaciology* 18.78, pp. 77–85.

†Chorley, R.J., Schumm, S.A. and Sugden, D.E. 1984: *Geomorphology*. London: Methuen.

Kemmis, T.J. 1981: Importance of the regelation process to certain properties of basal tills deposited by the Laurentide ice sheet in Iowa and Illinois. *Annals of glaciology* 2, pp. 147–52.

†Moran, S.R., Clayton, L., Hooke, R. Le B., Fenton, M.M. and Andriashek, L.D. 1981: Glacier-bed landforms of the prairie region of North America. *Journal of glaciology* 25.93, pp. 457–76.

morphogenetic regions Those regions in which it is claimed that certain geomorphological processes result from a particular set of climatic conditions, thereby giving distinctive regional landscapes.

morphological mapping A means of mapping landforms. In the strict sense morphological maps display only the shape of the ground with breaks of slope and gradients indicated. Such maps have been found useful by, for example, South African engineers. In a wider sense morphological mapping has been used to map landforms in terms of their origin. This form of morphogenetic mapping, popular in central Europe, has found wide applications in terms of engineering geomorphology, resource surveys and pure research.   DES

Reading
Brunsden, D., Doornkamp, J.C. and Jones, D.K.C. 1979: The Bahrain surface materials resources survey. *Geographical journal* 145, pp. 1–35.

Crofts, R.S. 1974: Detailed geomorphological mapping and land evaluation in Highland Scotland. *Institute of British Geographers special publication* 7, pp. 231–51.

morphometry The quantitative description of forms; in physical geography it refers to the earth's surface (strictly geomorphometry) but in other sciences it is an approach which can relate to fossils or crystals, for example.

Evans (1972, 1990) has distinguished *general geomorphometry* which is based upon an analysis of the entire landsurface as a continuous, rough surface described by the attributes at a sample of points or from arbitrary areas, and *specific geomorphometry* which relates to specific landforms and to the measurement of their size, shape and relationships. In both approaches definitions should be made to allow relationships to process indices. Morphometry is necessary to characterize areas and landforms quantitatively, to allow areas studied by different scientists to be compared easily, to demonstrate how aspects of the landsurface are inter-related and to provide parameters which can relate to processes in relationships from which processes may be estimated where only morphometric parameters are available.

In general geomorphometry, a specific part of the landsurface could be described by an equation but this would have so many terms and be so complex that it has not been used except for specific areas of landforms. Therefore altitude at a point and the derivatives of slope and curvature have often been used as the basis for general geomorphometry and Evans (1990) and others have identified five fundamental attributes which are altitude, gradient, aspect, profile convexity and plan convexity. Profile convexity is the rate of change along a line of maximum gradient and plan convexity is the rate of change of aspect along a contour. These five attributes relate to a point or small area and systems of general geomorphometry which have been put forward are based upon analysis of spatial patterns of some or all of the attributes. Variations between methods which have been suggested depend upon the relative significance accorded to the five attributes but several schemes have been devised because of the relevance to traffic-ability, drainage, suitability for different types of land use, and susceptibility to erosion hazard.

Specific geomorphometry has been used as a more realistic way of simplifying the task of earth surface description and morphometric methods have been devised for the description of coral atolls, karst depressions, glacial cirques, sand dunes, lake basins and many other landforms. Because the morphometry of drainage basins has attracted much attention mor-phometry has sometimes been associated mainly with drainage networks and drainage basin morphometry. The earliest develop-ments in drainage basin morphometry were based upon stream ORDER and HORTON'S LAWS and these could provide the basis for comparisons between areas but have been less valuable because of the statistical nature of these so-called 'laws'. The morphometry of drainage basins has focused (Gregory and Walling 1976) upon the area, length, shape and relief attributes and these have parallels at the level of the drainage NETWORK when DRAINAGE DENSITY is an important mea-sure of relative length. A wide range of morphometric measures has been devised for drainage basins and these have usually been defined as either ratio measures such as ratio of maximum width to breadth giving an index of drainage basin shape, or as measures which depend upon compar-ison with an ideal shape and drainage basins have been compared with a circle or with a lemniscate (pear-shaped) loop for example. (See also CIRQUES; DRAINAGE DENSITY; DUNES; ORDER, STREAM.)   KJG

Reading and References
†Chorley, R.J. ed. 1972: *Spatial analysis in geomorphology.* London: Methuen.
†— and Kennedy, B.A. 1971: *Physical geography: a systems approach.* London: Prentice-Hall.
Evans, I.S. 1972: General geomorphometry, derivatives of altitude and description statistics. In R.J. Chorley ed.,
Spatial analysis in geomorphology. London and New York: Methuen. Pp. 17–90.
— 1990: General geomorphometry. In A. Goudie ed., *Geomorphological techniques.* London: Unwin Hyman. Pp. 31–7.
†Goudie, A. ed. 1990: *Geomorphological techniques.* London: Unwin Hyman.
Gregory, K.J. and Walling, D.E. 1976: *Drainage basin form and process.* London : Edward Arnold. Pp. 37–60.

morphotectonics (or tectonic geomor-phology) The study of the interaction of tectonics and geomorphology.   ASG

Reading
Ollier, C.D. 1991: Morphotectonic and structural geomor-phology. *Zeitschrift für Geomorphologie Supplementband* 82, pp. 1–161.

mosaic-cycle concept An ecological concept associated with forest growth. Natural forests exhibit a cyclic alternation between an optimal phase consisting of trees roughly the same height and age (usually of a single species) which then deteriorates into a phase of decay. This is succeeded by a phase of rejuvenation which in time becomes an optimal phase. Stages with other tree species (some of them pioneer species) are often interpolated between the phases of decay and rejuvena-tion. A primary forest is thus a mosaic of areas in which the same cyclic succession of growth and decay is going on but in which the cycles are out of step.   ASG

Reading
Remmert, H. ed. 1991: *The mosaic-cycle concept of ecosystems.* Berlin: Springer Verlag.

mottled zone The portion of a soil zone or weathering profile immediately beneath a ferricrete or silcrete horizon, in which bleached kaolinitic material occurs with patches of iron staining.

moulin A vertical cylindrical shaft by which surface meltwater flows into a glacier. Moulins tend to form at lines of structural weakness in the glacier and are usually 0.5 to 1.0 m in diameter and up to 25 to 30 m deep.   DES

Reading
Stenborg, T. 1969: Studies of the internal drainage of glaciers. Geografiska annaler 51A. 1–2, pp. 13–41.

mountain meteorology Mountains ex-ert an influence on the atmosphere both *mechanically*: by blocking the airflow, deflecting it over and around the barrier, and through frictional drag; and *thermo-dynamically*: by acting as a direct, elevated

Clouds associated with Everest and Lhotse in Nepal. Mountains generate many distinctive meteorological phenomena of which such clouds are but one example.

heat source, as an indirect heat source through latent heat release in clouds formed over the mountains, and as a moisture sink through precipitation. The scales of mountain effects on the atmospheric circulation include: the planetary wave scale, with upper-air low-pressure troughs located over eastern North America and eastern Asia related, respectively, to the Rocky Mountains and the Tibetan Plateau upwind; the regional-synoptic scale, with the modification of frontal systems as they move across major mountain ranges and the formation of lee cyclones; the mesoscale of mountain-induced lee wave clouds and fall winds (föhn, bora); and the local scale of mountain/valley and slope winds systems resulting from topoclimatic contrasts in diurnal heating patterns. The characteristics of weather and climate in mountain areas are most closely related to the last three categories of meteorological phenomena.

A mountain climate can be considered to exist whenever the relief creates an altitudinal zonation of climatic elements (temperature, precipitation) sufficient to change the local vegetation characteristics.

Exceptions to this criterion may occur, however, where vegetation is absent on hyperarid subtropical or polar mountains. The effect of altitude causes a temperature decrease (environmental lapse rate) of about $5-6°C$ km^{-1}, on average, although a temperature increase with height often occurs in mountain valleys and basins, with wintertime and/or nocturnal temperature inversions. There is also a general altitudinal decrease of water vapour content, a decrease of pressure (approximately 100 mb km^{-1} in the lower troposphere), and an increase of incoming solar radiation (about $5-15$ per cent km^{-1} under cloudless skies). Orography redistributes and in many cases augments the precipitation that would otherwise have occurred through cyclonic or convective processes. The altitudinal enhancement mostly occurs as a result of increased amounts rather than greater frequency of precipitation. On windward mountain slopes, the zone of maximum precipitation, in a climatic sense, typically occurs at low elevations in equatorial zones, about $700-1200$ m in the tropical (trade wind) zones, and at higher levels (up to 3000 m and above) in mid-latitudes. In the lee of many mountain ranges, with respect to the prevailing wind direction, there is a reduction in average precipitation giving rise to a so-called 'rain-shadow'.   RGB

Reading

Barry, R.G. 1992: *Mountain weather and climate*. 2nd edn. London: Routledge.

Browning, K.A. and Hill, F.F. 1981: Orographic rain. *Weather* 36, pp. 326–9.

Smith, R.B. 1979: The influence of mountains on the atmosphere. *Advances in geophysics* 21, pp. 87–230.

Yoshino, M.M. 1975: *Climate in a small area: an introduction to local meteorology*. Tokyo: University of Tokyo Press.

mountain/valley wind A local wind system produced in mountainous regions as a result of temperature differences. The circulation is best developed in summer, when the skies are clear and large-scale motions are weak, and in deep, straight valleys with a north–south axis. During the day the air above the slopes and floor of the valley is heated to a temperature well above that over the centre of the valley. Shallow upslope (ANABATIC) flow results, and is compensated for by air sinking in the valley centre. If the ascending air is moist enough convective clouds may form along the valley ridges. Superimposed on this cross-valley flow is a VALLEY WIND blowing up-valley at low levels from the adjacent plains.

At night the valley surface and the overlying air cool by the emission of infrared radiation, causing the air to flow downslope under the influence of gravity. The convergence of these slope winds near the valley centre produces both a weak ascending motion and a low-level down-valley or mountain wind which flows out of the mountains onto the adjacent plains. At higher elevations a counter flow occurs from the plains to the valleys.   WDS

Reading
Atkinson, B.W. 1981: *Meso-scale atmospheric circulations*. London: Academic Press.

Oke, T.R. 1987: *Boundary layer climates*. 2nd edn London: Routledge.

mountains Substantial elevations of the earth's crust above sea level, which result in localized disruptions to climate, drainage, soils, plants and animals. Increases in altitude tend to repeat the bioclimatic patterns associated with a move towards higher latitudes, although cloudiness, day length and seasonal variations differ from the latitudinal progression. Temperature drops at a rate of approximately 0.5°C 100 m^{-1} and rainfall is often heaviest where moisture-laden winds are forced to rise (over 1500 m in the tropics). Vertical ZONATION of plants and animals is most clearly illustrated where mountains rise from tropical forest to tundra environments.   PAF

Reading
Gerrard, A.J. 1990: *Mountain environments*. London: Belhaven Press.

Moore, D.M. ed. 1983: *Green planet: the story of plant life on earth*. Cambridge and New York: Cambridge University Press.

Price, L.W. 1981: *Mountains and man: a study of process and environment*. Berkeley: University of California Press.

MRCFE A common abbreviation for the volume edited by W.L. Thomas *Man's role in changing the face of the earth*, the proceedings of a USA-based international symposium published in 1956 to honour the diplomat and early environmentalist writer George Perkins Marsh (1801–1882). For many years, this book was the central carrier of the torch of geography as the study of the human habitat: it contained both historical generality and empirical example and was global in scope. As time went on, however, its deficiencies became apparent: technological changes made some of the material badly out of date and some environmentally significant processes had not been foreseen at all, such as the impact of persistent pesticides. Also, the criticism was levelled that there was no theory nor underlying principle of organization: in the end it was held to be mostly anecdotal.

The resurgence of interest in environmental concerns has brought a number of studies in the same tradition, though often with more restricted ambitions. Works by A. Goudie (1990) and A.M. Mannion (1991) employ, respectively, a biogeo-sphere-systematic and a chronological approach to the question of human impact on the 'natural' environment; Simmons (1989) arrays his material around human access to energy sources. The US involvement has come to the forefront again in the establishment of a George Perkins Marsh Institute at Clark University (Worcester, MA) and scholars from that institution have gathered together another large volume (Turner 1990) dealing with human impact. This collected study deals mainly with the past 300 years and is set in a frame of systematic studies of, for example, urban growth and population growth on a world scale. Coverage is systematic, with a few omissions, yet the book lacks any theoretical perspective aimed especially at such work: the lack of such studies presents a challenge to workers in the field, be they geographers or from other fields. The seminal nature of MRCFE, now nearly 40 years old, is confirmed by all such subsequent work and scholarly desiderata.   IGS

References
Goudie, A. 1990: *The human impact on the natural environment*. 3rd edn. Oxford: Blackwell.

Mannion, A.M. 1991: *Global environmental change. A natural and cultural history*. London: Longman.

Simmons, I.G. 1989: *Changing the face of the earth. Culture, environment, history*. Oxford: Basil Blackwell.

Thomas, W.L. ed. 1956: *Man's role in changing the face of the earth*. Chicago: University of Chicago Press.

Turner, B.L. ed. 1990: *The earth as transformed by human action. Global and regional changes in the biosphere in the last 300 years*. Cambridge: Cambridge University Press.

mud lumps Small-scale landforms found from the Mississippi delta region of the USA. Rapid forward growth of distributary channels deposits deltaic sand, mud and organic sediment on top of unstable prodelta clay. This causes loading which in turn causes diapiric intrusions of plastic clays through the overlying sands. Updoming or extrusion occurs, producing the mud lumps.   ASG

Reading
Morgan, J.P., Coleman, J.M. and Gagliano, S.M. 1968:
Mudlumps: diapiric structures in Mississippi delta sediments. *Memoir of the American Association of Petroleum Geologists* 8, pp. 145–61.

mud volcano A mount built up of mud carried to the surface by geysers or gap eruptions in volcanically active regions.

mull Humus admixed with mineral material in the surface horizons of the soil zone.

multispectral scanner An optical remote sensor which is used to derive images of the earth's surface. It possesses three advantages over AERIAL PHOTOGRAPHY. First, it has a very fine radiometric resolution in narrow and simultaneously recorded wavebands. Secondly, these wavebands span a relatively large range of ELECTROMAGNETIC RADIATION from ultraviolet wavelengths to thermal infrared wavelengths and, thirdly, the data can be stored in digital form for DIGITAL IMAGE PROCESSING.

A multispectral scanner measures the radiance of the earth's surface along a scan line perpendicular to the line of aircraft flight. As the aircraft moves forward repeated measurement of radiance enables a two-dimensional image to be built up. These scanners comprise a collecting section, a detecting section and a recording section (see diagram). A telescope directs radiation onto the rotating mirror. The rotating mirror reflects the radiation into the optics. The optics focus the radiation into a narrow beam. The dichroic (doubly refracting) grid splits the radiation into its reflected and emitted components. The reflected radiation is further divided and the emitted radiation goes to the thermal infrared detectors. A prism is placed in the path of the reflected radiation, to divide it into its spectral components. The radiation detectors sense the reflected and emitted radiation. The reflected radiation is usually detected by silicon photodiodes that are placed in their correct geometric position behind the prism. The emitted (thermal infrared) radiation is usually detected by photon detectors, which are held in a cooling flask. After detection the signal is amplified by the preamplifier and is passed in electronic form to the control box. The electronic control console has three components: a signal processor to format the data as required for the recorders, an amplifier to boost the signal level even further and a power distribution unit to balance the signal strength in each waveband. The type of recorder depends upon the make and model of the multispectral scanner. The majority of multispectral scanners have a monitor to

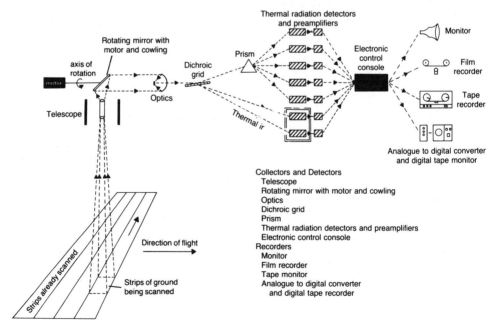

enable the operator to observe the data as they are recorded. Simple multispectral scanners also tend to have analogue recorders; either a film recorder, where the electrical impulses are recorded directly onto film, or an analogue tape recorder, where the electrical impulses are stored on magnetic tape. The analogue tape recorder records the electronic signal in the aircraft. The signals can either be fed into a film recorder to produce images, or can be digitized for later digital image processing. Current scanners use analogue to digital converters to produce a digital output which is recorded by a digital tape recorder.

The two most popular applications for these images are the mapping of vegetation and soil, often as a prerequisite to studies of water quality.   PJC

Reading
Curran, P.J. 1985: *Principles of remote sensing.* Harlow: Longman Scientific and Technical.

Lowe, D.S. 1976: Nonphotographic optical sensors. In J. Lintz and D.S. Simonett eds, *Remote sensing of environment.* Reading, Mass. and London: Addison-Wesley.

murram Laterite or ferricrete in East Africa.

muskeg A Canadian-Indian term for waterlogged depressions in the subarctic zone of Canada and Alaska. There are some 500,000 square miles of such marsh in Canada alone. These depressions are largely filled with peat and characterized by *Sphagnum* moss. It is a region of marshy depressions with scattered lakes, stagnant mosquito-infested pools and slow meandering streams.   AP

Reading
Radforth, N.W. 1969: Environmental and structural differentials in peatland development. In E.C. Dapples and M.E. Hophins, eds, *Environments of coal deposition.* Geological Society of America special paper 114, pp. 87–104.

— and Branner, C.O. eds 1977: *Muskeg and the northern environment in Canada.* Toronto: University of Toronto Press.

Muskingum method A method of FLOOD ROUTING which assumes that along any channel reach the difference between the inflow hydrograph (I) to the reach and the outflow hydrograph (O) from the reach is equal to the stored or depleted water in a specified time interval. Two simultaneous equations can be solved, namely the water balance equation which expresses change in storage:

$$\Delta S = I - O$$

and an equation for storage:

$$S = K\,[xI + (1 - x)\,O]$$

where x is a dimensionless constant for the channel reach and K is a storage constant which is obtained from hydrographs of I and O at each end of the reach.   KJG

Reading
Lawler, E.A. 1964: Flood routing. In V.T. Chow, *Handbook of applied hydrology.* New York: McGraw-Hill. Sect. 25-II.

mutation A change in the structure or amount of DNA in the chromosomes in the cells of an organism, or the resulting change in the organism's characteristics. If a mutation occurs in the gametes (reproductive cells) it is inherited; if it occurs elsewhere (in somatic or non-reproductive cells) it is not. Inherited mutations caused by a change in the structure of the DNA molecule are known as gene mutations; those produced by a change in the amount of DNA are known as chromosomal mutations. These errors in the coding of inherited information occur at a low frequency, apparently spontaneously.

Most mutations are deleterious because of the long period of testing (by natural selection) the genome (the package of genes within the gamete or germ-cell) has undergone. They will be eliminated. The extremely rare beneficial mutation will be incorporated into the genome by the process of natural selection.   PHA

mutualism An interaction that benefits the species involved. The most widespread mutualisms are between plants and animals, to use the animals to improve the efficiency of plants' reproduction and to provide food for the animals in return. For example, birds are attracted to fruits and eat them. The fruits pass through the gut and a portion is defecated shortly afterward. Meanwhile the bird has flown from the site of the parent plant and the seed is left in a supply of fertilizer, ready to germinate.

ASG

N

naled Another word for aufeis or ICING.

nanism (or microsomia) The condition of being dwarfed, often implying stunted, and refers to both plants and animals. Small size is implicit in the expression nanophyllous (small-leaved), or nanoplankton (the smallest plankton), or nanophanerophytes (shrubs under 2 m in height) although in SI terminology the prefix nano- strictly signifies a unit $\times 10^{-9}$. Artificial breeding of dwarf animals and plants is sometimes referred to as nanization.   PAF

nappe A mass of rock which is thrust over other rocks by thrust faulting or a recumbent fold or both.

natural hazard Any extreme event or condition in the natural environment causing harm to people or property. Contemporary theory distinguishes between extreme events in nature (the purely biophysical aspects) and the hazard or risk aspect affecting people. The cause of natural hazards (and in extreme cases the associated disasters) is understood as the process of interaction between people seeking a livelihood by the use of the earth, and the natural processes of the biosphere. In this sense hazards are negative resources, a function of the culture and technology of any human society. Interactions between the human use system and natural events and conditions produce both hazards and resources.

Natural hazards may be classified by principal causal agent. These hazards may cause death, injury and damage or loss of physical property through the disruption of productive economic activities. The amount of loss is affected by the extent to which a society has adjusted to the hazards that exist in its territory and experience. For example, blizzards and heavy snowfalls tend to cause much more damage and disruption in the relatively mild winters of the UK than much more severe conditions in Ontario. Earthquakes of similar magnitude on the Richter scale cause more severe damage and loss of life where buildings are poorly constructed, as shown by recent examples in Italy, Turkey and Pakistan. Tropical cyclones can cause many deaths where there is little possibility for evacuation in advance of the tidal waves, as in Bangladesh, and high property damage where extensive construction of houses and hotels has taken place, as in Florida and parts of the Caribbean.

Natural hazards have been traditionally regarded as acts of God. Efforts have been made to control extreme events in nature wherever technology permits, as in the construction of flood control dams, avalanche and rockfall barriers, sea walls and groynes, and chemical pesticides. Recently, more ecological approaches have been developed in which changes in the human use system are made to make society less vulnerable to natural extremes.

Despite these efforts, there is some evidence that the toll of death and damage from natural hazards (and disasters) is rising. This may be attributed in large part to the growth of human population and the world economy. There is also some indication, however, that new development is occurring disproportionately in areas of

Classification of natural hazards by principal causal agent

Geophysical		Biological	
Climatic and meteorological	Geological and geomorphical	Floral	Faunal
Snow and ice	Avalanches	Fungal diseases, e.g.	Bacterial and viral
Droughts	Earthquakes	athlete's foot, dutch	diseases, e.g.
Floods	Erosion	elm disease, wheat	influenza, malaria,
Frosts	(including soil erosion	stem rust	smallpox, rabies
Hail	and shore and beach	Infestations, e.g. weeds,	Infestations, e.g.
Heatwaves	erosion)	phrectophytes, water	rabbits, termites, locusts
Tropical cyclones	Landslides	hyacinth	Venomous animal bites
Lightning strikes and	Shifting sand	Hay fever	
fires	Tsunami	Poisonous plants	
Tornadoes	Volcanic eruptions		

Source: Burton and Kates 1964.

higher risk such as floodplains, steep slopes, earthquake zones, low-lying coastal areas and drought-prone regions. The magnitude and frequency of natural disasters is an indication of the extent to which the present pattern of development is unsustainable. In response to these developments, the International Decade for Natural Disaster Reduction has been established to work over the decade of the 1990s to find better ways of mitigating natural disasters.   IB

Reading
Burton, I. and Kates, R.W. 1964: The perception of natural hazards in resource management. *Natural resources journal* 3, pp. 412–21.
— and White, G.F. 1993: *The environment as hazard.* New York: Guilford Press.
Garcia, R.V. 1981: *Drought and man: the 1972 case history. Vol. 1: Nature pleads not guilty.* Oxford: Pergamon Press.
Hewitt, K. ed. 1983: *Interpretations of calamity: from the viewpoint of ecology.* London: Allen & Unwin.
Kirby, A. ed. 1990: *Nothing to fear: risks and hazards in American society.* Tucson: University of Arizona Press.
O'Riordan, T. 1986: Coping with environmental hazards. In R.W. Kates and I. Burton eds, *Geography, resources, and environment.* Vol. 2. Chicago: University of Chicago Press. Pp. 272–309.
Palm, R.I. 1990: *Natural hazards: an integrative framework for research and planning.* Baltimore: Johns Hopkins University Press.
Smith, K. 1992: *Environmental hazards.* London: Routledge.
US National Research Council 1987: *Confronting natural disasters: an international decade for natural hazard reduction.* Washington, DC: National Academy Press.

natural selection See DARWINISM; EVOLUTION.

natural vegetation A general term for the total sum of plants in an area, grouped by communities but not as part of a taxonomic system. 'Natural' signifies the sum total of inheritance or genotype, but the term natural vegetation is also associated with environmental factors which encourage or constrain plant growth after an equilibrium between plants and their surroundings has been established. The larger groupings of plants illustrate a ZONATION which has a combined biological and environmental basis. Perhaps the most common usage is to denote the plant cover of any area prior to its modification by humans.   PAF

Reading
Walter, H. 1973: *The vegetation of the earth.* London: English Universities Press.

navigation, effects on river channels The use of river channels for navigation by ships and barges requires certain characteristics of the channel and results in constructed and unintentional impacts on channel form and process. Successful navigation of river channels requires that consistent depths of flow be maintained sufficient to support the river traffic, that the velocity of flow be swift enough to prevent clogging of the channel by sediment but slow enough to permit safe operation of vessels, and that the discharge of the stream be consistent over periods as long as possible.

Most rivers that serve as transportation routes in developed countries have been substantially modified by engineering works to attempt to improve navigability. Meandering streams are frequently shortened by excavating a channel across the narrow necks separating meanders resulting in shorter, steeper channels and more rapid flows at consistent levels sufficient for vessels of deep draught. Dredging prevents excessive build-ups of sediment that reduce depths.

Unintentional impacts from navigation include bank erosion that results from constant pounding by waves generated as wakes from passing vessels. Dredging operations may result in increases in channel depths, but may also result in increased bank or channel erosion as the river adjusts to a new configuration. Debris from the dredging operations (spoil) presents serious disposal problems.

Langbein (1962) proposed a measure which he referred to as specific tractive force which related the horsepower of a vessel to the power required to maintain design speed. This value was then compared to the depth–velocity product to define the minimum specific tractive force required to sustain upstream transport. Using this approach rivers may be assessed for their utility as water transportation routes.   WLG

Reference
Langbein, W.B. 1962: *Hydraulics of river channels as related to navigability*. US Geological Survey water supply paper 1539-W.

neap tide See TIDES.

nebkha A small accumulation of wind-blown sand that collects in the lee of a clump of vegetation or other obstacle.

neck A narrow isthmus or channel. A mass of lava which has solidified in the pipe or vent of a volcano.

needle ice A small-scale heave phenomenon produced by freezing and associated ice segregation at or just beneath the ground surface. Cooling at the ground surface results in ice crystals which grow upwards in the direction of heat loss. The needles, which can range in length from a few mm to several cm, may lift small pebbles or soil particles. The growth of needle ice is usually associated with diurnal freezing and thawing. It is widespread and particularly common in alpine locations in midlatitudes where the frequency of freeze–thaw cycles is at its greatest. Wet, silty, frost-susceptible soils are the sites of most intense needle ice activity. Needle ice frequently occurs in orientated stripes, and both wind direction and sun have been suggested as explanations for the pattern; it is not clear whether orientated needle ice patterns are primarily a shadow effect developed by thawing or a freezing effect.

Thawing and collapse of needle ice is thought significant for frost sorting, frost creep, the differential downslope movement of fine and coarse material, and the origin of certain micro-patterned ground forms. The importance of needle ice as a disruptive agent has probably been underestimated, especially in exposing soil to wind and water in periglacial regions. In other areas it may be responsible for damage to plant materials when freezing causes vertical mechanical stress within the root zone.   HMF

Reading
Lawler, D.M. 1988: Environmental limits of needle ice: a global survey. *Arctic and alpine research* 20, pp. 137–59.
Mackay, J.R. and Mathews, W.H. 1974: Needle ice striped ground. *Arctic and alpine research* 6, pp. 79–84.
Washburn, A.L. 1979: *Geocryology: a survey of periglacial processes and environments*. New York: Wiley. Esp. pp. 91–3.

negative feedback See SYSTEMS.

nehrung A sand or shingle spit which separates a haff from the open sea. A bar which isolates an estuary or lagoon from the sea.

nekton Free-swimming aquatic plants and animals, especially strong swimmers such as fish.

neocatastrophism A term introduced into geoscience in the mid-twentieth century by palaeontologists concerned with sudden and massive extinctions of life forms, such as that which afflicted the great mammals at the end of the Pleistocene. It has been extended into geomorphology by those dealing with rapid outputs from, and rapid inputs to, interfluve systems (Dury 1980). Much modern geomorphology is concerned with events of great magnitude and low frequency, and problems have been encountered with accommodating such an approach within the context of UNIFORMITARIANISM. ASG

Reference
Dury, G.H. 1980: Neocatastrophism? A further look. *Progress in physical geography* 4, pp. 391–413.

neo-Darwinism An evolutionary theory which combines DARWINISM with modern genetics. It regards the gene pool of a population as the fundamental unit in evolution, and takes into account larger mutations as well as the small heritable variations of Darwin. ASG

Reading
Berry, R.J. 1982: *Neo-Darwinism*. London: Edward Arnold.

neoglacial A small-scale glacial advance that occurred in the Holocene, after the time of maximum HYPSITHERMAL glacier shrinkage (Denton and Porter 1970). Fluctuations appear to have been frequent and to have shown sparse temporal correlation between different areas (Grove 1979), though the latest advance, the so-called Little Ice Age, was widespread between *c*.AD 1550 and 1850. ASG

Reading and References
Denton, G.H. and Porter, S.C. 1970: Neo-glaciation. *Scientific American* 222, pp. 101–10.
Grove, J.M. 1979: The glacial history of the Holocene. *Progress in physical geography* 3, pp. 1–54.
Grove, J.M. 1988: *The Little Ice Age*. London: Routledge.

neotectonics The study of the processes and effects of movements of the earth's crust that have occurred during the Late Cainozoic (Neogene). Some investigators use the term in a more restricted temporal sense to refer to post-Miocene or even just Quaternary movements, while others regard neotectonics as involving all tectonic activity which has been instrumental in forming present-day topography. Neotectonic activity which has been directly monitored by geodetic relevelling or other measurement techniques during the present century is commonly referred to as recent crustal movements.

Several lines of evidence have been employed to establish the nature of neotectonic activity, depending on the size of area and time period being considered. Regional subsidence and uplift over several millions of years are largely investigated by the usual methods of structural geology. As the temporal and spatial scale contracts, geomorphological and sedimentological data become more important. For instance, coastal movements can be monitored by raised or down-warped shorelines, while inland the mapping of fluvial features and erosion surfaces in conjunction with detailed study (and especially dating) of associated deposits can provide valuable information. Rapid uplift in mountainous areas may be indicated by geomorphological evidence of the onset of glaciation, while both horizontal and vertical movements along faults can in many cases be related, respectively, to offset drainage and knick-points.

Maximum rates of uplift estimated from geomorphological and other evidence, or measured directly by relevelling, vary from several orders of magnitude over the earth's surface. Average rates of uplift over several millions of years rarely exceed 300 mm ka^{-1}. However, rates of postglacial isostatic uplift (see ISOSTASY) may exceed 20 m ka^{-1}, while contemporary crustal movements in currently highly active tectonic zones average up to 10 m ka^{-1}, or more. Such high short-term rates are clearly not sustained for more than a very limited period in geological terms. MAS

Reading
Fairbridge, R.W. ed. 1981: Neotectonics. *Zeitschrift für Geomorphologie supplementband* 40.
Vita-Finzi, C. 1986: *Recent earth movements – an introduction to neotectonics*. London: Academic Press.
Vyskocil, P., Green, R. and Maelzer, H. eds 1981: *Recent crustal movements, 1979*. Amsterdam: Elsevier.
—, Wasset, A.M. and Green, R. eds 1983: *Recent crustal movements, 1982*. Amsterdam: Elsevier.

nephanalysis The term used to cover the analysis and interpretation of spatially organized cloud data. Coined in pre-satellite days, the term originally applied to 'the study of synoptic charts on which only clouds and weather are plotted' (Berry *et al.* 1945). The observations plotted were of cloud type, cloud amount, precipitation, weather, cloud ceilings and cloud-top heights. With the advent of METEOROLOGI-CAL SATELLITES it soon became clear that the contents of the satellite imagery were so rich and complex that many users would prefer to be provided with simpler cloud charts instead. Such charts are known as *satellite nephanalyses*. The earliest type was designed for use in the US Weather Bureau (see Godshall 1968). Similar manual schemes were implemented in the late 1960s and early 1970s in other major meteorological centres. Proposals for improved satellite nephanalysis, designed in part to standardize results from different analysts, were made for visible imagery by Harris and Barrett (1975), and for infrared imagery by Barrett and Harris (1977). Detailed objective (computer-based) schemes have been proposed by Shenk and Holub (1972), but the problem of automatic cloud identification is not straightforward. Therefore today's quasi-operational objective satellite nephanalyses are the '3-D nephanalysis' of the US Weather Bureau (mapping broad categories of cloud top heights: see Decotiis and Coulan 1971), and a bispectral 'cloud type' procedure applied to Meteosat image data by the European Space Operations Centre. The USAF purports to combine satellite data with conventional data in a more comprehensive nephanalysis, but its procedures have been subject to variation and change. ECB

Reading and References
Barrett, E.C. and Harris, R. 1977: Satellite infra-red nephanalyses. *Meteorological magazine* 106, pp. 11–26.
Berry, F.A., Bollay, E. and Beers, N.R. 1945: *Handbook of meteorology*. New York: McGraw-Hill.
Decotiis, A.G. and Coulan, E.F. 1971: Cloud formation in three spatial dimensions using infra-red thermal imagery and vertical temperature profile data. In *Proceedings of the Seventh International Symposium of Remote Sensing of the Environment*. Ann Arbor, Michigan.
†European Space Agency 1977: *Meteosat meteorological users' handbook*. Meteorological Information Extraction Centre, Darmstadt: European Space Agency.
Godshall, F.A. 1968: Intertropical convergence zone and mean cloud amount in the tropical Pacific Ocean. *Monthly weather review* 96, pp. 172–5.
Harris, R. and Barrett, E.C. 1975: An improved satellite nephanalysis. *Meteorological magazine* 104, pp. 9–16.
Shenk, W.E. and Holub, R.J. 1972: A multispectral cloud type identification method using Nimbus 3 HRIR measurements. In *Proceedings of the Conference on Atmospheric Radiation, Fort Collins, Colorado*. Boston: American Meteorological Society. Pp. 152–4.

nephoscope An instrument for measuring the height, direction of movement and velocity of clouds from a point on the ground.

neptunism The belief that a large proportion of the earth's rocks are precipitates laid down in some chaotic fluid, a theory that was devised and popularized by the German geologist A.G. Werner in the late eighteenth century, and imported into Britain in the early nineteenth century by R. Jameson. It contrasts with plutonism. ASG

Reading
Davies, G.L. 1969: *The earth in decay*. London: Macdonald.

neritic Pertaining to the part of the seas and oceans above the continental shelf.

ness A promontory or headland, especially in Scotland, but also in eastern and southern England.

net primary productivity (NPP) The net augmentation of green plant material per unit area per unit time on land, and of blue-green and other algae, phytoplankton, and higher plants in water bodies; or, the amount of stored cell energy produced by PHOTOSYNTHESIS. It may be expressed by the equation:

$$NPP = \text{gross production} - \text{respiration}$$

(see BIOLOGICAL PRODUCTIVITY), and is measured in dry weight $g\ m^{-2}\ day^{-1}$ or year, dry weight $t\ ha^{-1}\ year^{-1}$, or in assimilated carbon equivalents, or energy equivalents. Although photosynthetic energy fixation is its main determinant, NPP may also be constrained by limiting factors which restrict growth, in particular cold and/or drought, and nutrient availability inadequacies.

On land, NPP may be estimated by means of the harvest method, in which all parts of living plants (including roots if possible) are cut and weighed at the end of a set period of time, and this is especially useful for crops, grasslands and forest plantings. For natural forests and woodlands, some form of forest dimension analysis is normally adopted. The NPP of plankton communities is more often ascertained by the technique of measuring uptake rates of CO_2 labelled with radiocarbon, ^{14}C, which provide accurate values over one or two days.

For land plant communities adequate quantities of energy, water, CO_2 and soil nutrients are required for optimum rates of NPP. Most land plants need very large amounts of water to survive and prosper: e.g. a beech forest in southern England may take in 25,000 to 30,000 kg of water $ha^{-1}\ day^{-1}$ in summer. The majority of this goes in transpiration, so as to keep open the stomata through which CO_2 is received: but some is utilized in a range of metabolic reactions including photosynthesis, and a good deal is settled in cell structures themselves. In dry climates there is an almost linear relationship between water availability and net primary productivity, but this breaks down in more humid areas, for there is a point beyond which an increase in water availability has no effect at all on NPP. Values of NPP are also augmented substantially in areas of high temperatures, especially in the energy-rich tropics. CO_2 is normally present at a level of $c.0.03$ per cent of atmospheric gases: and slight variations in this are known to modify rates of NPP fairly quickly. Most plants demand a large range of nutrients, and shortages in any one may also influence the

NPP: in Australia, for example, deficiencies in molybdenum may lead to the establishment of a heath vegetation, instead of the more productive mallee, or eucalyptus forest. In oceanic communities, the three main restrictions to NPP are, first, the inability of incident solar energy to penetrate very far into the water, so effectively limiting the major productive layer to depths of 30–120 m; secondly, the tendency for a large number of plankton to sink beneath this zone; and, thirdly, a marked deficiency in some nutrients, especially nitrogen and phosphorus.

As might be expected, bearing in mind the low BIOMASS of autotropic organisms in oceans compared to that on land, the mean NPP of the former is low (147 g m^{-2} year^{-1} cf. 780 g m^{-2} year^{-1} on land: Whittaker 1975). In more detail, NPP in oceans can range from c.2 g m^{-2} year^{-1} in Arctic waters under the ice cap, to almost 5000 in some coral reefs, mangrove swamps, and tropical estuaries. Most open oceans, which are poor in nutrients, have values which are equivalent to semi-desert on land, namely 40–200 g m^{-2} year^{-1}. Greater net primary productivities are found in regions of upwelling, and on continental shelves, both of which are nutrient-rich (400–600 g m^{-2} year^{-1}), and when these spatially coincide they can reach 1000 g m^{-2} year^{-1}. Similar rates occur in many temperate-latitude estuaries. A good deal of variation is also found in freshwater systems: OLIGO-TROPHIC streams normally have very low NPPs, while those of EUTROPHIC cattail marshes in Minnesota can reach 2500 g m^{-2} year^{-1}, with 5600 g m^{-2} year^{-1} attained in swamps artificially enriched with sewage in California (Woodwell 1970).

On land the lowest NPPs are located in deserts, semi-deserts and tundras, at 0–250 g m^{-2} year^{-1}. Systems whose growth is restricted by cold and/or drought (boreal forest, semi-desert shrublands, tropical savannahs, steppe grasslands) give values of 250–1000 g m^{-2} year^{-1}; temperate-latitude forests, in which NPP is restrained by seasonal cold, lie within 1000–2000 g m^{-2} year^{-1}; and tropical rain forest and some of the most nutrient-rich marshland may attain 2000–3000 g m^{-2} year^{-1}, and occasionally as high as 5000 g m^{-2} year^{-1} in the case of the tropical rain forest. This latter system, though it covers only c.40 per cent of the earth's surface, may account for 25 per cent of the new land

biomass each year. Ranges of NPP in agricultural systems also vary widely, from c.250–500 g m^{-2} year^{-1} in the case of many tropical subsistence crops, and for most non-intensive temperate-latitude farmland, to 750–1500 g m^{-2} year^{-1} under modern intensive, and frequently irrigated agriculture (see Leith and Aselmann 1983). Sugar cane is perhaps worthy of special mention, since it is one of the most productive of crops: its mean NPP is c.1725 g m^{-2} year^{-1}, but under favourable circumstances this can be augmented to 6700 g m^{-2} year^{-1}, and occasionally even higher.   DW

Reading and References
Leith, H. and Aselmann, I. 1983: Comparing the primary productivity of natural and managed vegetation: an example from Germany. In W. Holzner, M.J.A. Werger and I. Ikusima eds, *Man's impact on vegetation*. The Hague, Boston and London: W. Junk. Pp. 25–40.

†Whittaker, R.H. 1975: *Communities and ecosystems*. 2nd edn. New York: Macmillan.

Woodwell, G.M. 1970: The energy cycle of the biosphere. *Scientific American* 223, pp. 64–74.

net radiation The resultant flux of the solar and terrestrial radiation through a horizontal surface. The downwards (positive) flux of radiation consists of shortwave solar radiation plus infrared atmospheric counter-radiation. The upward (negative) flux consists of reflected shortwave radiation and infrared radiation from the ground surface. The net radiation is considered positive if the flux downwards exceeds that upwards, and in this case will add energy to the surface. Net radiation is also known as radiation balance. Typically, it is positive during the day and negative at night.   JGL

network The structure composed of links and nodes which are the junctions of at least three links. Networks can be identified for any linear form or process, and Haggett and Chorley (1969) recognized two fundamental types of networks, open and closed. Most emphasis has been placed upon drainage networks where network delimitation and density are important considerations (see DRAINAGE DENSITY; DRAINAGE NETWORK). A drainage network has *exterior links* (headwater streams) which are initiated from *sources* and *interior links* which are segments between two *junctions* and *nodes*. stream ORDER is a technique which can be applied to the network and *magnitude* is a more consistent measure of basin size than is stream order and a network of magnitude

n consists of n exterior links, $(n - 1)$ interior links and $(2n - 1)$ links in total. A drainage network has two important features, its density and its structure. Structure can be elucidated by reference to topology which is a branch of pure mathematics. It has been assumed that network topology and geometry develop at random if no constraints are imposed by geology, topography or climate. For a network with n sources there is a range of unique network arrangements that can be created from the exterior links and these are topologically distinct channel networks (TDCNs, see Shreve 1966). Study of the random drainage model has shown that of the TDCNs which occur, some are more likely than others, that links can be subdivided as *cis links* where the bounding tributaries are from the same side and *trans links* where the tributaries at each end of the link are from opposite sides, and that some parameters derived from analyses of network topology may relate to hydrological and sedimentological processes. Although much work on networks has been morphological in character (e.g. Smart 1978) and more definitions have been made (Abrahams and Flint 1983), progress is also being made in establishing how network characteristics relate to travel times of floodwater and to HYDROGRAPHS and to their characteristics.   KJG

Reading and References
Abrahams, A.D. and Flint, J.J. 1983: Geological controls on the topological properties of some trellis channel networks. *Bulletin of the Geological Society of America* 94, pp. 80–91.
Haggett, P. and Chorley, R.J. 1969: *Network analysis in geography*. London: Edward Arnold.
†Richards, K.S. 1982: *Rivers: form and process in alluvial channels*. London and New York: Methuen.
Shreve, R.L. 1966: Statistical law of stream numbers. *Journal of geology*. 74, pp. 17–37.
Smart, J.D. 1978: The analysis of drainage network composition. *Earth surface processes* 3, pp. 129–70.

neutron probe An instrument for determining soil moisture content. It consists of a radioactive source of fast (or high energy) neutrons, a slow neutron detector and a counter unit. The method is based on the principle that fast neutrons emitted into the soil collide with the nuclei of atoms in the soil, notably the hydrogen nuclei of soil water, and as a result lose energy and slow down. A proportion of the resulting cloud of slow neutrons is scattered towards the probe where it is sensed by the slow neutron detector and translated into an estimate of soil moisture content using the mean count rate displayed on the counter unit and a soil moisture calibration curve. The neutron probe is introduced into the soil using a permanently sited access tube and so repeated measurements of soil moisture content can be made at the same site without destruction of the site.   AMG

Reading
Bell, J.P. 1973: *Neutron probe practice*. Institute of Hydrology report 19.
Schmugge, T.J., Jackson, T.J. and McKim, H.L. 1980: Survey of methods for soil moisture determination. *Water resources research* 16, pp. 961–79.

nevé Another, less widely used, word for FIRN.

niche An abstract concept relating to the environmental conditions that are necessary for an organism to maintain a viable population, and the amounts of each of the resources that it requires to do so. A niche is not something that can be seen: it is a characteristic of an organism or species. Habitats, by contrast, are actual places which may contain a whole range of niches.   ASG

Reading
Vandermeer, J.H. 1972: Niche theory. *Annual review of ecology and systematics* 3, pp. 107–32.

nimbostratus See CLOUDS.

nimbus See CLOUDS.

nitrification and denitrification Two essential components of the nitrogen cycle. The former involves the capture of nitrogen from the air, a process which is carried out by bacteria which live symbiotically in association with leguminous plants. They fix nitrogen in the form of highly soluble nitrates, which are taken up by plants, which convert them into useful organic compounds such as proteins. These in turn may be consumed by animals. The process of denitrification takes place when bacteria and fungi decompose plant or animal waste and convert it back again into nitrogen gases.   ASG

nitrogen cycle A vital cycle in the biosphere involving the fixation of nitrogen from the atmosphere by micro-organisms, ionization by lightning and by industrial processes to form nitrogenous fertilizers, followed by the utilization of ammonium and nitrate by plants and animals in amino acids and proteins. Plant and animal wastes, and dead organisms, return the nitrogen to the soil, where it takes place in a number of important reactions with micro-organisms. Nitrogen's unusually large number of valence states, enabling it to form compounds with hydrogen and oxygen (NH_3: ammonia, NO_2: nitrite ion, NO_3: nitrate ion etc.) explains its important role.   KEB

Reading
Bradbury, I.K. 1991: *The biosphere.* London: Belhaven Press.

Royal Society 1984: *The nitrogen cycle of the United Kingdom.* London: Royal Society.

Trudgill, S.T. 1988: *Soil and vegetation systems.* 2nd edn. Oxford: Oxford University Press.

nivation The localized erosion of a hillside by frost action, mass wasting and the sheet flow or rill work of meltwater at the edges of, and beneath, lingering snow patches. The term was introduced by Matthes (1900). The main effect of nivation is to produce nivation hollows, which, as they grow in depth, trap more snow and thereby enhance the process of deepening. Given adequate time and suitable conditions a nivation hollow may evolve into a cirque. Topographic and climatic controls strongly influence the distribution and orientation of nivation hollows. The most favoured locations are on hillsides protected from the sun and with an ample supply of drifted snow. In mid-latitudes these factors favour a north-eastern orientation in the northern hemisphere and south-eastern orientation in the southern hemipsphere.   DES

Reaing and Reference
†Embleton, C. and King, C.A.M. 1975: *Periglacial geomorphology.* London: Edward Arnold.

Matthes, F.E. 1900: Glacial sculpture of the Bighorn Mountains, Wyoming. *US Geological Survey 21st annual report 2*, pp. 173–90.

Thorn, C.E. 1988: Nivation: a geomorphic chimera. In M.J. Clark ed., *Advances in periglacial geomorphology.* Chichester: Wiley. Pp. 3–31.

†Washburn, A.L. 1979: *Geocryology.* London: Edward Arnold.

nivation cirque/corrie A corrie formed by nivation processes. They are generally thought to be formed without glaciers being present, although there is some dispute about this. They may be formed from nivation hollows but this term is also used for depressions which are inclined more horizontally than vertically.   WBW

Reading
Thorn, C. E. 1988: Nivation: a geomorphic chimera. In M.J. Clark ed., *Advances in periglacial geomorphology.* Chichester: Wiley. Pp. 3–31.

nivometric coefficient An index of snowfall efficacy, being the ratio of snowfall (water equivalent) to total annual precipitation. A coefficient of 1 implies precipitation entirely of snow.

Reading
Tricart, J. 1969: *Geomorphology of cold environments.* London: Macmillan.

noctilucent clouds Are found near a height of 80 km where there is a minimum in the temperature. They are so extremely tenuous that they can be seen only against the light scattered by air molecules on summer nights, between latitudes of 50° and 70°, at least one hour after the sun has set at the ground. While the cloud particles are likely to be ice, its origin, and the mechanism for the formation of the clouds is under considerable debate.   JSAG

Reading
Ludlum, F.H. 1980: *Clouds and storms.* University Park, Pa.: Pennsylvania State Press.

nonconformity An angular unconformity or other discontinuity between strata where the older rocks are of plutonic origin.

non-linear system In nature, most systems are non-linear; they are viewed for simplicity as being linear in operation over a restricted range of action, e.g. extension of a spiral spring (Hooke's Law). More complicated systems can be modelled by approximating or simplifying to the linear case. True non-linear systems, however, do not have an easily derived solution by formula, although they can be computed. Non-linear dynamic systems may give rise to responses which are described as chaotic; where the system does not settle down to a fixed equilibrium condition or value. The earliest 'natural' chaotic systems studied were simple meteorological (Edward Lorenz) and the logistic equation in ecology (Robert May). For a steady state condition a 'control parameter' gives a single value (e.g. population in the logistic equation) but at a further value of the control parameter two solutions may be given and at still higher values four, eight etc. and so increasingly rapidly up to the chaotic regime. This is related to the Feigenbaum number.   WBW

Reading
Gleick, J. 1987: *Chaos*. Harmondsworth: Sphere/Cardinal/ Penguin.

normal cycle See CYCLE OF EROSION.

normal fault A fault with the fault-plane inclined towards the side which has been downthrown.

normal stress The stress or load applied to the surface of an object (as either compression or tension). It is perpendicular to the SHEAR STRESSES which act parallel to the surface. A normal stress produces a strain or deformation in the material. WBW

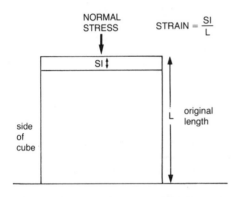

notch Landform that develops at the base of a cliff, platform or reef flat, especially in limestone and on tropical coasts. Deep narrow notches are characteristic of areas with a low tidal range. Their positions are related to lithological and structural controls, tidal characteristics and sea-level history. In general, the higher the amplitude of the waves and the higher the tidal range, the greater is the difference in elevation between the notch roof and the floor. Notches in the humid tropics may be 1–5 m in depth. Although mechanical action of waves may contribute to their development, most investigators now believe that chemical or biochemical corrosion, or biological boring and grazing activities, are important.   ASG

Reading
Woodroffe, C.D., Stoddart, D.R., Harmon, R.S. and Spencer, T. 1983: Coastal morphology and Late Quaternary history, Cayman Islands, West Indies. *Quaternary research* 19, pp. 64–84.

nubbins Small lumps of earth produced by heaving owing to the growth of needle ice.

nuclear waste (radwaste) Waste produced during the operation of nuclear facilities and as a result of decommissioning. Much comes from nuclear power stations, but other sources include hospitals and research institutions. Radwaste can be classified according to its volume, level of activity (high, medium, low) and its form (liquid, solid or gas).

High-volume low-activity solid wastes result from mining and uranium ore processing, from reactor operations, from final plant dismantling (decommissioning) and from soiled clothing, etc. Generally speaking, low-activity wastes are characterized by radionuclides with short half-lives. Commonly, this type of waste is buried in designated shallow trenches, but in the past sea disposal has been used to remove much of this waste.

The disposal of *low-activity liquid waste* from nuclear power plants and fuel reprocessing factories depends upon the siting of the works. Those with a coastal location, or on a large river or lake, remove sufficient radionuclides from liquid streams by distillation or floc precipitation to produce effluents of 'acceptable' purity prior to discharge into the adjacent water body.

Medium-volume medium-activity wastes are produced by both reactor operation and fuel reprocessing, e.g. ion exchange resins, sludges and precipitates, and may include some plutonium-contaminated material. *Solid low-volume high-activity waste* comprises mainly fuel element cladding and solidified material from reprocessing. *High-activity liquid waste* is produced entirely in fuel reprocessing operations.

Disposal of high-activity wastes currently presents problems because of their potential as biological hazards. Not only are they highly active, but they also contain some very long-lived activity. The optional sequence of events in the management of high-level waste is:

(*a*) Storage of fuel elements in ponds for months to several years.

(*b*) Storage of highly active liquor produced in reprocessing fuel for not more than two decades.

(c)   Solidification into borosilicate glass, after which the glass blocks will be artificially cooled, for between ten and twenty years.

(c)   The encapsulation of the blocks and their emplacement in a final deep geological repository.   ASG

Reading
Mounfield, P. 1991: *World nuclear power*. London: Routledge.

nuclear winter A severe deterioration of climate that might take place as a result of multiple nuclear explosions. They might generate so much fire and wind that large quantities of smoke and dust would be emitted into the atmosphere, thereby causing darkness and great cold, the latter resulting from backscattering of incoming solar radiation caused by the reflectance of the fine soot. If an exchange of several thousand megatons took place, ambient land temperatures might be reduced to between −15 and −25°C (Turco *et al.* 1983).   ASG

Reference
Turco, R.P., Toon, O.B., Ackerman, T.P., Pollack, J.B. and Sagan, C. 1983: Nuclear winter: global consequences of multiple nuclear explosions. *Science* 222, pp. 1283–92.

nudation See SUCCESSION, PLANT.

nuée ardente (glowing cloud) A cloud of super-heated gas-charged ash produced by certain acidic volcanic eruptions (e.g. the eruption of Mt Pelée on Martinique in 1902 and 1903). The deposits produced by nuées ardentes are termed ash-flow tuffs, welded tuffs or ignimbrites.   ASG

numerical modelling A method for obtaining particular solutions or deductions for a model which is expressed in mathematical or logical form, and for which general mathematical solutions are not appropriate and/or not available.

Numerical models are commonly although not necessarily implemented on digital computers, although many methods in use predate their development. A numerical solution is often the only one available for any but the simplest model, but has the disadvantage that it lacks the generality of an analytical solution. In numerical modelling all parameters must be given definite values, and all variables assigned initial values. A model run is then a single realization of the model constrained

by these particular values. A large number of trials is therefore needed satisfactorily to explore all the possibilities inherent in any model.

Models vary considerably in style and complexity. In 'black box' models either the whole of a system or parts of it are considered solely in terms of empirical relationships between the input and output of the system. Most models in use in physical geography contain at least elements of this type. The commonest source of material for this type of model is a regression equation based on field observations. As understanding advances the black-box components within the total system become less significant as more components are based on established scientific principles. The level of empirical relationships in a useful forecasting model is partly constrained by our state of knowledge and partly by the level of detail which it is appropriate to represent. Such practical models are commonly developed drawing on the methods of systems analysis.

A stochastic element is often found in numerical models. For example, rainfall inputs to a hydrological model may be drawn at random from a known distribution in order to generate a probability distribution of high and low flows for hydraulic engineering design. Stochastic elements are usually included either to represent a model input of which a direct model is not needed; or to cover variability at scales below the level of resolution of the model. In the example above a stochastic rainfall model may well be more appropriate and economic as an input than an independent model of the general atmospheric circulation! As another example, microtopography might be included within a model of long-term hillslope evolution as a random variation in process rate. The microtopography is below the general level of resolution of the model but could cause some of the observed variability in the forms of neighbouring hillslope profiles apparently subject to identical processes.

Numerical models require an underlying formulation in mathematical terms, and the range of models reflects the range of mathematical possibilities, which is immense. Perhaps the most common type of numerical model in physical geography is a solution to one or a family of differential equations, which are ultimately based on the CONTINUITY EQUATION with the substitution of suitable EQUATIONS OF MOTION or other expressions for the rates of the relevant processes of material or energy transport. In some cases particular classes of solution are sought, e.g. EQUILIBRIUM or KINEMATIC WAVE solutions. In all cases the numerical model can only be run when boundary and initial conditions are specified. Initial conditions must specify the relationships which inputs and outputs must satisfy where they enter/leave the system of interest.

An example of differential equation type is the model of hillslope evolution based on continuity of downslope sediment transport. Ignoring wind-blown dust, the continuity equation is:

$$\partial Q/dx + \partial z/dt = 0$$

where Q is the rate of downslope sediment transport at distance x from the divide, and z is the elevation at x and time t.

For hillslope processes such as soil creep or soil EROSION which are largely transport-limited removal, the sediment transport may be expressed in the form $Q = f(x) \times s^n$ for some function of slope distance on slope gradient s. Where $f(x)$ is constant an analytical solution is available (Culling 1963) but otherwise numerical modelling is the best method. In this case any initial conditions may be used to describe the original slope profile form. The simplest boundary conditions are a fixed divide ($Q = 0$ at $x = 0$) and basal removal at a fixed base level ($z = 0$ at $x = x_1$), although others may be used. Runs of this model may be carried out on small micro-computers to follow long-term profile development. MJK

Reading and Reference
†Carson, M.A. and Kirkby, M.J. 1972: *Hillslope form and process.* Cambridge and New York: Cambridge University Press.
Culling, W.E.H. 1963: Soil creep and the development of hillside slopes. *Journal of geology* 71, pp. 127–61.
†Thomas, R.W. and Huggett, R.J. 1980: *Modelling in geography: a mathematical approach.* London: Harper & Row.

nunatak An Inuit-derived word describing a mountain completely surrounded by glacier ice, normally an ICE CAP or ICE SHEET. The nunatak hypothesis is the idea that plant and animal (but especially the former) communities have been isolated as refugia on nunataks. Sometimes these refugia may be as mountains, as with the normal usage of nunatak, but the nunatak hypothesis may also relate to much larger areas of isolated, ice-free land. WBW

Reference
Gjaerveroll, O. 1963: Survival of plants on nunataks in Norway during the Pleistocene glaciation. In A. Löve and D. Löve eds, *North Atlantic biota and their history.* Oxford: Oxford University Press. Pp. 261–83.

nutrient status A collective term usually applied to soils, peatlands and lakes whose nutrient status may be EUTROPHIC (rich in nutrients), *oligotrophic* (poor) or *mesotrophic* (transitional). Mires are also referred to as OMBROTROPHIC (rain-feeding) or *rheotrophic* (flow-feeding). In this context the use of the suffix 'trophic' is not to be confused with the trophic levels of animal communities feeding off each other and off the autotrophs or primary producers, the plants.

The nutrient status of soils is clearly discussed by Trudgill (1988) who considers in turn the inputs and outputs of the soil–vegetation system and illustrates the general principles with a number of diagrammatic models of soils of low and moderate nutrient status. Soil nutrient status is basically determined by the rock weathering input and the atmospheric input, on the one hand, and the leaching output on the other hand, moderated by biological recycling and storage of nutrients. In a hot, wet climate the potential for vegetative cycling and for weathering will be high but so will the potential for leaching – hence the concern over wholesale clearing of tropical rain forest. The opposite extreme, of a dry cold climate, will limit cycling and leaching, and therefore reduce chemical weathering to a large degree although physical fragmentation of rock may be important. Nutrient status of the soil is therefore low. As Trudgill (1988) goes on to note, there is a real need for integrated studies of the factors leading to low or high nutrient status, for although many studies of important cations, such as calcium, have been made, they have been at different scales, for different purposes and under different climatic regimes, making comparisons difficult. In a humid temperate context one is considering different orders of magnitude in $CaCO_3$ content (e.g. in river water, values of 1–20 ppm for non-limestone areas; 150–200 ppm for chalky areas) and similarly distinct differences in soil biochemistry and vegetative cycling. There are many complicating factors, especially with regard to the relative uptake, storage and release of calcium and other major mineral nutrients (magnesium, potassium, iron, aluminium, sodium, phosphorus and silica).

The nutrient status of lakes, and the way in which conditions have changed over time, have been extensively studied and is a topic of growing importance and scientific effort (Oldfield 1977, Birks and Birks 1980). At the beginning of the Holocene, 10,000 years ago, many lakes, especially those in glaciated lowlands, were eutrophic by virtue of mineral soil erosion inputs. Rising organic productivity and soil stabilization during the early to mid-Holocene changed this situation to one of mesotrophic/oligotrophic status, until man began affecting the soils within the catchment leading to greater nutrient inputs, culminating in some cases in gross 'cultural eutrophication' through sewage and other domestic wastes. The question of whether a lake, in its natural state, has eutrophic or oligotrophic status and what changes will occur in the absence of man, is a complex one. Some lakes in areas rich in nutrients appear to have remained eutrophic throughout more or less the whole Holocene, such as the meres of the Shropshire–Cheshire till plain described by Reynolds (1979), whereas some lakes in the English Lake District and north-west Scotland have been shown (by diatom stratigraphy of their sediments) to have been oligotrophic throughout almost the whole Holocene (Battarbee 1984). The usefulness of lake sediments, with their record of diatoms, chemicals, magnetism, pollen and other indicators, to indicate change within the water body and the catchment, has been amply demonstrated by a number of studies (Birks and Birks 1980; Battarbee 1984).

In the final stages of the terrestrialization of a lake basin *Sphagnam* bog mosses often invade to change the course of the hydrosere from the woodland climax postulated by Tansley and others earlier this century to what we now know as the almost inevitable endpoint of the hydrosere, at least in the British Isles: raised bog (Walker 1970). The nutrient status of these communities is discussed under PEAT-LANDS. The actual levels of pH and calcium ions used to differentiate the various grades of peatland have been reviewed by Ratcliffe (1964). KEB

References

Battarbee, R.W. 1984: Diatom analysis and the acidification of lakes. *Philosophical transactions of the Royal Society* B, 305, pp. 451–77.

Birks, H.J.B. and Birks, H.H. 1980: *Quaternary palaeoecology*. London: Edward Arnold.

Oldfield, F. 1977: Lakes and their drainage basins as units of sediment-based ecological study. *Progress in physical geography*, 1, pp. 460–504.

Ratcliffe, D.A. 1964: Mires and bogs. In J.H. Burnett, ed., *The vegetation of Scotland*. Edinburgh: Oliver and Boyd.

Reynolds, C.S. 1979: The limnology of the eutrophic meres of the Shropshire–Cheshire plain. *Field studies* 5, pp. 93–173.

Trudgill, S.T. 1988: *Soil and vegetation systems*. 2nd edn. Oxford: Oxford University Press.

Walker, D. 1970: Direction and rate in some British post-glacial hydroseres. In D. Walker and R.G. West, eds, *Studies in the vegetational history of the British Isles*. Cambridge: Cambridge University Press.

O

oasis An area within a desert region where there is sufficient water to sustain animal and plant life throughout the year.

obsequent stream A stream or river which is the tributary of a subsequent stream and flows in a direction opposite to the regional dip of the landsurface.

obsidian hydration dating (OHD) Used by archaeologists and geologists to date events ranging in age from a few hundred years to several million years. The principle is based on the dating of obsidian (volcanic glass), which when a fresh surface is formed and exposed to the atmosphere, the diffusion of ambient water will proceed to the formation of a hydration layer. The hydration layers are firmly adherent to the parent glass and resistant to chemical dissolution. The thickness of the hydration layer, which varies from 1 μm to more than 50 μm, depends on the time of exposure. OHD requires that a measurement of hydration thickness or the depth of penetration of water into obsidian be measured and a rate of hydration be known. The measurement can be made optically or by using particle accelerators.

AP

Reading
Trembour, F. and Friedman, I. 1984: The present status of obsidian hydration dating. In W.C. Mahaney, ed., *Quaternary dating methods: developments in palaeontology and stratigraphy 7*. Amsterdam: Elsevier. Pp. 141–51.

occlusion A complex frontal zone associated with the later stages of the life cycle of an extra-tropical cyclone. The name is derived from the associated occluding or uplifting of the warm sector air from the earth's surface. Cold fronts tend to travel more quickly than warm fronts, so the area of warm air between narrows. Eventually

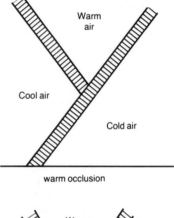

warm occlusion

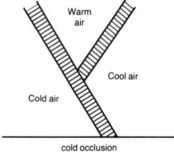

cold occlusion

the warm air is entirely aloft with the occluded front marking the surface juxtaposition of the two cool or cold air masses. The detailed structure of the occlusion will be determined by the temperature difference of these two air masses. Where the cold air behind the cold front is cooler (warmer) than that ahead of the original warm front it is known as a cold (warm) occlusion (see figure). In practice the temperature difference may be quite small and the occlusion difficult to classify. PS

occult deposition The wet deposition of acidic pollutants, particularly SO_2 and NO_x, onto surfaces by the impaction of fog and cloud droplets. Patterns of deposition are influenced by climatic factors which encourage fog and mist and by local variations in wind direction and intensity. Concentration of pollutants in occult deposition can be considerably higher (up to × 20) than those in wet deposition by rainfall. The term is particularly used in connection with acid deposition in urban environments where it precipitates and concentrates pollutants on surfaces protected from rainwash. The process is important in, for example, coastal areas of low rainfall but high relative humidity and fog frequency and deposition can be increased by land-use changes, for example, afforestation, which increase surface area.   BJS

Reference
Building Effects Review Group 1989: *The effects of acid deposition on buildings and building materials in the United Kingdom.* London: HMSO.

ocean The general name for large bodies of saltwater making up around 70 per cent of the earth's surface. Open oceans or oceanic zones are those parts deeper than 200 m whereas shallow coastal waters or neritic zones lie over continental shelves and are usually less than 200 m deep. There is only one ocean basin; the geographical subdivisions are made for convenience, because they are all interconnected. Shallow waters are more affected by changes in temperature, salinity, sedimentation and water movements. They are reached by sunlight and are richer in nutrients, and hence in plant and animal life than the deeper more constant oceans. In general, the nearer the land the higher the NET PRIMARY PRODUCTIVITY.   PAF

Reading
Hedgpeth, J.W. 1957: *Classification of marine environments.* Geological Society of America memoir 671. Pp. 17–28.

oceanography The study or description of the oceans encompassing the sea floor, the physics and chemistry of the seas and all aspects of marine biology.

Reading
Pickard, G.L. and Emery, W.J. 1990: *Descriptive physical oceanography: an introduction.* 5th edn. Oxford: Pergamon.

ogive (also known as Forbes bands) Alternating bands of light and dark ice that extend across the surface of some glaciers below ice falls. They are arcuate in response to the normal pattern of ice flow across a glacier. The combined width of a dark and light band corresponds to the distance the glacier moves in a year. The dark ice corresponds to the ice that traverses the ice fall in summer and is thus exposed to melting, while the light ice reflects the incorporation of snow as the ice traverses the ice falls in winter.   DES

Reading
Paterson, W.S.B. 1981: *The physics of glaciers.* Oxford and New York: Pergamon.

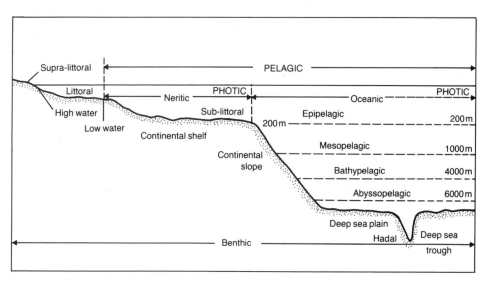

Ocean: Classification of marine biome–zones and divisions.
Source: *P. Furley and W. Newey 1984:* The geography of the biosphere. *London: Butterworth. Figure 13.1.*

oil-shale A shale which contains sufficient quantities of hydrocarbons to yield oil or petroleum gas when distilled.

okta Measurement of the amount of CLOUD cover is one of the two standard scales used by surface meteorological observers worldwide (the other scale being tenths). The observer reports the number of eighths or oktas of the celestial dome which is covered by clouds. Total cloud cover and layer cloud amount are reported in this fashion. Care must be taken to give equal weight to all areas of the sky especially in the case of *cumuliform* clouds when cloud sides as well as cloud bases may be viewed. AH-S

oligotrophic See NUTRIENT STATUS.

ombrotrophic Description of plants or plant communities which are associated with a rain-fed substrate which is poor in nutrients.

omnivore See DIVERSIVORE.

onion-weathering Exfoliation. The destruction of a rock or outcrop through the peeling off of the surface layers.

ontogeny (ontogenesis) The sequence of development during the whole life history of an organism. The term is also applied to the life history of lakes and other systems. ASG

oolite A sedimentary rock, usually calcareous but sometimes dolomitic or siliceous, which is composed of concentrically layered spheres – ooliths – which have formed by accretion on the surface of a grain. The Jurassic rocks of southern England contain the Greater and the Inferior Oolite, and orginally obtained their name because of the supposed resemblance of their fabric to fish roe.

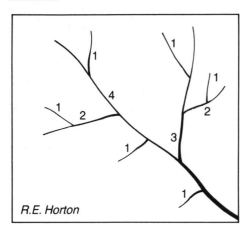

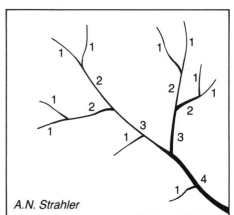

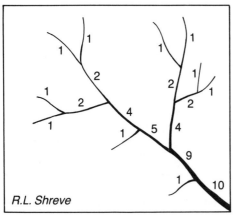

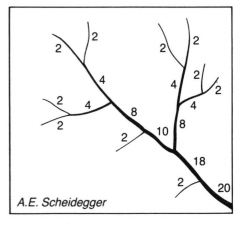

Four methods of stream and segment ordering.

ooze Fine-grained organic-rich sediments on the floors of lakes and oceans.

open channel flow See FLOW REGIMES.

open system See SYSTEMS.

opisometer An instrument for measuring distances on a map.

order, stream Identification of the links in a stream NETWORK is by the process of stream ordering. At least eleven methods of stream ordering have been devised. The first method to be adopted was proposed by R.E. Horton and led to the development of HORTON'S LAWS. In Horton ordering, all unbranched streams were designated first order, two first combined to make a second order and so on, and then the highest order streams were projected to the headwaters. Subsequently Strahler (1952) suggested that the second stage should not be effected and Strahler ordering became widely used. Later developments have endeavoured to propose ordering methods which are mathematically consistent and which allow all components of the network to be differentiated according to their position. Thus in the Horton and Strahler systems of ordering the order of a stream of order *n* is unaffected by the entry of a tributary of order less than *n*. Later ordering systems such as those proposed by Shreve (1967) and by Scheidegger (1965) have overcome this fundamental objection. The objective of all ordering systems is to be able to describe a link in a drainage network anywhere in the world in an unambiguous manner, and also to provide an ordering system that can readily provide an indication of the discharge from a network.   KJG

Reading and References
†Gardiner, V. 1976: *Drainage basin morphometry*. British Geomorphological Research Group technical bulletin 14. Norwich: Geo Abstracts.

†Gregory, K.J. and Walling, D.E. 1976: *Drainage basin form and process*. London: Edward Arnold.

Horton, R.E. 1945: Erosional development of streams and their drainage basins: hydrophysical approach to quantitative morphology. *Bulletin of the Geological Society of America* 56, pp. 275–370.

Scheidegger, A.E. 1965: The algebra of stream-order numbers. *US Geological Society professional paper* 525B, pp. 187–9.

Shreve, R.L. 1967: Infinite topologically random channel networks. *Journal of geology* 75, pp. 178–86.

Strahler, A.N. 1952: Hypsometric (area-attitude) analysis of erosional topography. *Bulletin of the Geological Society of America* 63, pp. 923–38.

organic debris dam Accumulation of organic material which forms when a piece of large woody material falls into a stream and becomes lodged, thereby providing a framework upon which leaves, twigs, tufa etc. can accumulate to give a partially watertight structure.   ASG

Reading
Bilby, R.E. and Likens, G.E. 1980: Importance of organic debris dams in the structure and function of stream ecosystems. *Ecology* 61, pp. 1107–13.

organic weathering The disintegration or destruction of a rock by living organisms or organic processes. It is a much neglected cause of weathering.

orientated lakes A term used in permafrost geomorphology to refer to a THAW LAKE which possesses a preferred long axis orientation. Such lakes occur widely on the Arctic coastal plains of Alaska, Siberia and Canada. Elliptical, oval and D-shaped forms are most common. It is generally agreed that these lakes develop by thermokarst subsidence of ice-rich permafrost terrain. Orientated lakes occur in non-permafrost regions also (see PAN) implying that the presence of permafrost merely facilitates the orientation-forming mechanism. The latter is not clearly understood. In most instances the long axis orientation is at right angles to prevailing winds but the relationship between wind-induced currents and accelerated thermal and mechanical abrasion at the two ends of the lake is problematic.   HMF

Reading
French, H.M. 1976: *The periglacial environment*. London and New York: Longman.

Washburn, A.L. 1979: *Geocryology: a survey of periglacial processes and environments*. New York: Wiley.

orocline A structural arc which has formed by horizontal displacement subsequent to the development of the main structural features of an area.

orogens Total masses of rock deformed during an orogeny (mountain-building episode).

orogeny The event, or mechanism, of the construction of characteristically linear or arcuate mountain chains formed on continents. Such mountain chains, of which the Andes and the Himalayas are contemporary examples, are known as orogenic belts or simply orogens. Investigation of the results

A view to the Kali Gandaki gorge in Nepal. This is one of the steepest places on earth and is created by a combination of mountain building (orogeny) and erosion.

of recent orogeny shows it to produce crustal thickening, deformation and associated volcanic activity, although in some cases, such as the Himalayas, the latter is relatively less important. Many earlier theories of orogeny emphasized the accumulation of a thick wedge of sediments forming a geosyncline before uplift, but these ideas have now been largely subsumed within the more comprehensive model of PLATE TECTONICS. Within this scheme orogeny is of two fundamental types. Where a continent margin, such as that of western South America, is under-ridden by oceanic LITHOSPHERE being reabsorbed or subducted into the earth's interior, largely thermal effects generate uplift and volcanic activity. In contrast, mainly mechanical effects are responsible for the Himalayan orogeny, which is associated with the collision of the northward moving Indian continent and the Eurasian continent.

MAS

Reading
Dennis, J.G. ed. 1982: *Orogeny*. Stroudsburg, Pa.: Hutchinson Ross.
Hsü, K.J. ed. 1982: *Mountain building processes*. London: Academic Press.
Miyashiro, A., Aki, K. and Cecal Sengor, A.M. 1982: *Orogeny*. Chichester: Wiley.

orographic precipitation Precipitation caused by the forced ascent of air over high ground. Uplift of air leads to cooling which, if the air is moist, may lead to condensation and eventually precipitation. The warm sector of an intense extra-tropical cyclone is the synoptic situation which demonstrates the orographic effect most clearly. Even where rain of convectional or cyclonic origins is falling, the orographic influence can still be seen in larger and sometimes longer precipitation events over the hills. The extra uplift will ensure that the precipitation processes in the clouds operate more effectively.

PS

osage-type underfits See UNDERFIT STREAM.

osmosis The movement of a solvent through a membrane from a dilute solution to a more concentrated solution, the membrane being permeable with respect to the solute but not the solvent.

outlet glacier A type of glacier which radiates out from an ICE DOME and often occupies significant depressions. Within an ice dome they can frequently be distinguished by a zone of high-velocity ice termed an ICE STREAM.

outlier An isolated hill lying beyond the scarp slope of a cuesta. A rock outcrop that is surrounded by rocks that are of an older age.

outwash Comprises stratified GLACIO-FLUVIAL sands and gravels deposited at or beyond the ice margin. Outwash usually forms fan, valley bottom (valley train) or plain (SANDUR) deposits, often hundreds of metres thick, built up by aggrading braided or anastomosing meltwater channels which migrate laterally across the outwash surface. Periodic high energy flood events are marked by high rates of transport of sediment reworked from older outwash and till. Where outwash has accumulated on the glacier margin itself, differential ice melt may produce pitted outwash. Close to the glacier outwash is often steeply graded, comprising coarse-grained, imbricated, non-cohesive sediments deposited rapidly during unidirectional, high flow regime conditions, both within channels and on longitudinal bars. Bar deposits exhibit crude horizontal bedding truncated by erosional contacts and scour-and-fill structures representing successive discrete flood events. The bar surfaces are often characterized by SAND LENSES, SILT DRAPES, TRANSVERSE RIBS and coarse gravel lags. CLASTS are typically poorly sorted, subangular to subrounded, and comprise heterogeneous lithologies. Farther from the glacier outwash is more gently graded, with finer-grained, more cohesive, sandy facies types forming transverse and linguoid bars characterized by dune and ripple forms and by bar avalanche face sediments. These distal sediments often exhibit planar and trough cross-bedding, and ripple and ripple drift CROSS-LAMINATION; LOAD STRUCTURES

may also be present. Clasts tend to be better sorted and more rounded, while fewer, more resistant lithologies are represented in both the clast and heavy mineral populations. Coarsening or fining upward sequences may reflect local flood events or periods of channel scour and infill, or longer term periods of ice advance or recession.

JM

Reading
Boothroyd, J.C. and Ashly, G.M. 1975: Process, bar morphology and sedimentary structures on braided outwash fans, northeastern gulf of Alaska. In A.V. Jopling and B.C. McDonald eds. *Glaciofluvial and glaciolacustrine sedimentation.* Society of Economic Paleontologists and Mineralogists special publication 23. Pp. 193–222.
Church, M. 1972: Baffin Island sandurs: a study of Arctic fluvial processes. *Geological Survey of Canada Bulletin* 216.
— and Gilbert, R. 1975: Proglacial fluvial and lacustrine environments. In A.V. Jopling and B.C. McDonald eds, *Glaciofluvial and glaciolacustrine sedimentation.* Society of Economic Paleontologists and Mineralogists special publication 23. Pp. 22–100.
Rust, B.R. 1978: Depositional models for braided alluvium. In A.D. Miall ed., *Fluvial sedimentology.* Canadian Society of Petroleum Geologists memoir 5. Pp. 605–25.

outwash terrace An outwash deposit that has been incised by meltwater to form a terrace. Incision of the outwash deposit is a response by meltwater streams to an increase in channel slope and/or an increase in the discharge/sediment load balance such that a period of proglacial aggradation is followed by a period of increased meltwater flow capacity and stream degradation. This change is likely to occur during glacial retreat when an overall decrease in meltwater and sediment supply occurs. A stepped sequence of outwash terraces may relate to successive periods of ice advance and retreat, or possibly to a more complex response to a single glacial event. Many outwash terraces can be traced up-valley to an associated terminal moraine; the terrace surfaces often exhibit traces of former braided channel networks and are pitted with kettle holes.

JM

overbank deposit Flood sediment laid down by a river beyond its normal-flow channel, generally fine materials deposited from suspension in floodwaters, but it is possible for sheets of coarser bed material to be deposited overbank as well. Finer material may be deposited considerable distances from the channel. JL

overdeepening, glacial Often regarded as a prime characteristic of glacial erosion, overdeepening refers to the long profile of glacial troughs which tend to have a 'down-at-heel' profile with a steep gradient near or at their heads and a gentler slope, sometimes a reverse slope, towards their mouths (Linton 1963). They are 'overdeepened' only when compared to river-long profiles and indeed from a glacial viewpoint river valleys can be regarded as underdeepened! One explanation of the 'overdeepened' glacial long profile is that, unlike most river valleys, a glacier discharge is greatest at the midway EQUILIBRIUM line and decreases towards the snout. One might expect most erosion at the point of maximum discharge. Overdeepening at a different scale involves the excavation of rock basins by glaciers. This is feasible where rock conditions are favourable because glaciers can flow up a bed slope so long as the overall ice surface gradient is in the contrary direction.   DES

Reference
Linton, D.L. 1963: The forms of glacial erosion. *Transactions of the Institute of British Geographers* 33, pp. 1–28.

overflow channels Rock channels incised into cols and interfluves by lake overflows. Such channels were widely recognized in Britain until the 1960s and attributed to the overflows of ice-dammed lakes during deglaciation. However, the majority have now been reinterpreted as subglacial meltwater channels (Sissons 1960, 1961).

Reference
Sissons, J.B. 1960, 1961: Some aspects of glacial drainage channels in Britain. Parts I and II. *Scottish Geographical Magazine* 76, pp. 131–46, 77, pp. 15–36.

overgrazing, hydrological effects of See LAND USE, HYDROLOGICAL EFFECTS OF.

overland flow A visible flow of water over the ground surface, however produced.

Surface flow may be generated through a number of HILLSLOPE FLOW PROCESSES, including excess of rainfall intensity over infiltration capacity; excess of rainfall amount over soil storage capacity and the seepage of return flow. Where storms are brief or local, flow may run overland for some distance and then infiltrate. Flow velocities are typically of 3–150 mm s^{-1}, and flow depths normally of a few millimetres. Flow rarely consists of a uniform sheet of water, and commonly

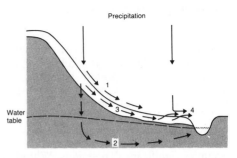

Possible paths of water moving downhill: path 1 is Horton overland flow; path 2 is groundwater flow; path 3 is shallow, subsurface stormflow; path 4 is saturation, overland flow, composed of direct precipitation on the saturated area plus infiltrated water that returns to the ground surface. The unshaded zone indicates highly permeable topsoil, and the shaded zone represents less permeable subsoil or rock.
Source: *T. Dunne and L.B. Leopold 1978:* Water in environmental planning. *San Francisco: Freeman. Figure 9–1.*

concentrates in threads of higher velocity. Overland flow is very important for the removal and transport of debris in SOIL EROSION by water.   MJK

Reading
Emmett, W.W. 1978: Overland flow. In M.J. Kirkby ed., *Hillslope hydrology.* Chichester: Wiley. Pp. 145–76.

overthrust A thrust fault with a fault-plane dipping at a low angle and considerable horizontal displacement.

overtopping The process by which coastal barrier crests are built up as swash flows, of insufficient magnitude to reach across the crest and cause overwash, terminate on top of the crest. Sediment carried in overtopping flows is deposited at the swash limit and hence increments the barrier crest height. There should be a relationship between barrier crest height and the return period of storms generating the swash flows. Any increase in storm return period would be associated with a reduction in barrier crest height as increasing wave height would generate flows capable of forming overwash down the back of the barrier eroding material from the crest in the process. Therefore, overtopping is part of a continuum of cross-barrier flow types, merging into overwash as the magnitude of the flow increases (Orford and Carter 1982). The balance between overtopping and OVERWASHING controls the instability and migration of gravel barriers, whereas on sand barriers aeolian deposition is a more effective control on barrier crestal elevation than wave-generated overtopping.   JO

Reference
Orford, J.D. and Carter, R.W.G. 1982: The structure and origins of recent sandy gravel overtopping and overwashing features at Carnsore Point, southeast Ireland. *Journal of sedimentary petrology* 52, pp. 265–78.

overwashing The process by which storm-generated swash flows transport beach face sediment over the top of the beach ridge and deposit it on the back slope. The definition of a coastal barrier is that it should exhibit some form of landward-dipping back slope; therefore overwash should be recognized as a major contributor to the creation and renewal of the back slope of a barrier. Overwash is regarded as a dominant process by which sand barrier islands are generated (with aeolian deposition as a co-dominant process; Leatherman 1976). Transport is often constrained sufficiently to cut through the crest via an overwash throat. Back slope deposition often occurs in the form of discrete washover fans, but washover can occur along a broad front parallel to the barrier crest. This has been described as the result of sluicing overwash (Orford and Carter 1982) and leaves no obvious sign of overwash throats. Note that 'overwash' is the process and 'washover' the sedimentary result. Fan volume generally correlates with the height of the breaking wave (controlled for sediment size). Some washover fans may exhibit a longshore periodicity. Spatially periodic overwash has been related to the presence of transverse or edge waves (pre-storm or during the storm) influencing the morphology of the beach (Orford and Carter 1984). In particular, high-level periodic cusps on the beach (formed by edge wave interaction) channel and spatially constrain overwash, hence imparting the same periodicity to washover fans. The reflective nature of gravel dominated barriers tends to increase the likelihood of periodic fans, whereas the dissipative nature of low angle sand barriers mitigates against periodic overwash flows. Continual overwashing erodes a barrier's seaward side and extends the landward side. Over time, a barrier appears to be rolling onshore as sediment on the backslope is buried by subsequent overwash and eventually exhumed on the seaward barrier slope as the form of the barrier gradually retreats over its washover foundation. Overwash works towards a reduction of barrier height. The counter process which builds up the barrier crest is known as barrier OVERTOPPING.

JO

References
Leatherman, S.P. 1976: Barrier island dynamics: overwash processes and eolian transport. *Proceedings of the 15th Conference of Coastal Engineers, American Society of Civil Engineers* 3, pp. 1958–74.
Orford, J.D. and Carter, R.W.G. 1982: The structure and origins of recent sandy gravel overtopping and overwashing features at Carnsore Point, southeast Ireland. *Journal of sedimentary petrology* 52, pp. 265–78.
— 1984: Mechanisms to account for the longshore spacing of overwash throats on a coarse clastic barrier in southeast Ireland. *Marine geology* 56, pp. 207–26.

oxbow A lake, usually curved in plan, occupying a CUT-OFF channel reach that has been abandoned. The term may be applied also to an extremely curved active channel meander with only a narrow neck between adjacent reaches or even to the land within such a reach. The term derives from the U-shaped piece of wood fitted around the neck of a harnessed ox. Lakes of this type may become plugged with sediment where they adjoin the channel and then progressively fill in.   JL

oxidation In general terms the loss of electrons from an atom but specifically the loss of oxygen from, or addition of hydrogen to, a substance. It is an important weathering mechanism, and the oxidation of iron compounds in rocks and soil can cause reddening to occur.

ozone A form of oxygen, but whereas the molecules of ordinary oxygen each contain two atoms, the ozone molecule has three. The ozone layer is a relatively high ozone concentration zone which occurs at a height of 16–18 km in polar latitudes and at about 25 km over the equator. This layer helps to control the temperature gradient of the atmosphere and also through absorption controls the amount of ultraviolet (UV-B) radiation reaching the ground.

Ozone is constantly created and destroyed through natural chemical reactions. However, human actions are increasing the concentrations of certain substances that may accelerate the rate of ozone destruction in the stratosphere: oxides of nitrogen, hydrogen, bromide and chlorine. The oxides of nitrogen might be generated by high-flying supersonic aircraft emissions, while the offending chlorine comes from such sources as the chlorofluorocarbons

(CFCs) and carbon tetrachloride (CCl_4). In recent years it has become apparent that the concentrations of ozone in the stratosphere have declined, most notably over the Antarctic, where an 'ozone hole' has been detected by a combination of ground monitoring and satellite observations. The reasons for the development of this zone of depletion in the south polar area include the presence of a well-established polar vortex and the extreme cold which generate ice particles that play a role in the crucial chemical reactions. However, stratospheric ozone depletion is increasingly being recognized in other geographical regions. The most obvious cause for concern is that this depletion will reduce the effectiveness of the ozone layer as a filter for incoming UV-B radiation.

At lower levels in the atmosphere ozone levels may, paradoxically, increase as a result of anthropogenic pollutants. Photochemical reactions can produce ozone. High ozone concentrations can adversely affect human health and damage vegetation.   ASG

Reading
Elsom, D. 1992: *Atmospheric pollution: a global problem.* 2nd edn. Oxford: Blackwell.
Wayne, R.P. 1991: *Chemistry of atmospheres.* Oxford: Clarendon Press.

P

pacific type coast A longitudinal coast where folded belts trend parallel to the coastline.

pack ice See SEA ICE.

padang A vegetation type, developed on poor soils in South-East Asia, consisting of grass and shrubs.

palaeobotany The study of ancient or fossil plants and plant communities.

palaeochannels River or stream channels which no longer convey discharge as part of the contemporary fluvial system. Remnants of palaeochannels may be preserved on floodplains and have clear morphological expression or they may be evident only in sections in sediments because they have been infilled. The size, shape, location and sedimentary infill of palaeochannels may be employed to contrast them with contemporary channels and many palaeochannels have been abandoned because RIVER METAMORPHOSIS has led to the development of a new fluvial system.
KJG

Reading
Collinson, J. and Lewin, J. eds 1983: *Modern and ancient fluvial systems*. International Association of Sedimentologists special publication 6. Oxford: Blackwell Scientific.

palaeoclimatology The study of climate prior to the period of instrumental measurements, few of which pre-date the nineteenth century. Indeed, instrumental records only span a tiny fraction (less than 10^{-7}) of the earth's climatic history and so provide a record that is both an inadequate perspective on climatic variation and a very limited view of the evolution of climates. The foundation of palaeoclimatology is the use of climate-dependent proxy data. The principal sources of proxy data are ice cores, ocean cores, various types of terrestrial sediment (e.g. glacial deposits, periglacial features, loess, relict sand dunes, speleothems, tufas etc.), biological evidence (e.g. tree rings, pollen, plant microfossils, diatoms, insect fossils) and historical records (e.g. writings, paintings, tax returns, phenological records).
ASG

Reading
Bradley, R.S. 1985: *Quaternary paleoclimatology: methods of palaeoclimatic reconstruction*. Boston: Allen & Unwin.

palaeoecology The term widely used for the research field that involves the study of fossils in order to infer past ecological processes, past biological environments and hence past biogeographical patterns. Increasingly these inferences are becoming quantified as we learn more about present plant-environment and animal-environment relationships and fossils become more accurate indicators of past ecological conditions. Fossil plant remains commonly studied are pollen and spores, diatoms, phytoliths and the so-called 'MACRO-FOSSIL' remains of plants, seeds, fruits, leaves and wood. Fossil animal remains commonly studied are cladocera, molluscs, coleoptera (beetles), and skeletal fragments. Using the language of biology and geology, palaeoecologists describe the changing fossil composition of the sedimentary sequence. From these descriptions they infer changing biotic compositions and hence changing environments, and deduce changes in environmental variables such as climate, soil and human influence.
RHS

Reading
Birks, H.J.B. and Birks, H.H. 1980: *Quaternary palaeoecology*. London: Edward Arnold.
— and West, R.G. eds 1973: *Quaternary plant ecology*. Oxford: Blackwell Scientific.

palaeogeography The geography of a former time, especially a specific geological epoch.

palaeohydrology The science of the waters of the earth, their composition, distribution and movement on ancient landscapes from the occurrence of the first rainfall to the beginning of continuous hydrological records (Gregory 1983). The advance of palaeohydrology has been possible with greater understanding of contemporary HYDROLOGY so that morphological evidence, including that from PALAEOCHANNELS, sedimentological evidence including characteristics of PALAEOMAGNETISM and information from organic deposits especially by palynology and diatom analysis, and knowledge of the mechanisms of the hydrological cycle, can be employed to make quantitative estimates of hydrological conditions in the past. Most emphasis has been placed upon Quaternary palaeohydrology but it has also been possible to indicate the palaeohydrological features of geological periods (Schumm 1977). The advance of palaeohydrological investigation depends upon refinement of the relationships between parameters such as precipitation, run-off and temperature and also upon a closer liaison between palaeohydrology and PALAEOCLIMATOLOGY and PALAEOECOLOGY. KJG

Reading and References
Gregory, K.J. ed. 1983: *Background to palaeohydrology.* Chichester: Wiley.
Schumm, S.A. 1977: *The fluvial system.* Chichester: Wiley.
†Starkel, L. and Thornes, J.B. 1981: *Palaeohydrology of river basins.* British Geomorphological Research Group technical bulletin 28.
Starkel, L., Gregory, K.J. and Thornes, J.B. 1991: *Temperate palaeohydrology: fluvial processes in the temperate zone during the last 15,000 years.* Chichester: Wiley.

palaeomagnetism The intensity, direction and polarity of the earth's magnetic field throughout geological time. Palaeomagnetic studies are possible because certain iron-rich rocks containing magnetic minerals become more or less permanently magnetized at the time they are formed. This can occur when a rock cools (thermal remanent magnetism), when an iron mineral is chemically altered to another form (chemical remanent magnetism) or when magnetic particles are deposited in calm water (detrital remanent magnetism). Systematic deviations in the magnetic orientation of rocks of different ages have enabled the previous latitudes of continents to be determined, while periodic polarity reversals of the magnetic field now provide both a record of the creation of new oceanic LITHOSPHERE and a basis for dating suitable deposits. MAS

Reading
Cox, A. ed. 1973: *Plate tectonics and geomagnetic reversals.* San Francisco: W.H. Freeman.
Glen, W. 1982: *The road to Jaramillo.* Stanford, Cal.: Stanford University Press.
Kennett, J.P. ed. 1980: *Magnetic stratigraphy of sediments.* Stroudsburg, Pa.: Dowden, Hutchinson and Ross.

palaeosol A soil of an environment of the past, formed either by burial under later geological materials or because of a change in the climate or topographical conditions of soil formation. They are identifiable by any evidence that indicates the presence of a former landsurface that has undergone some form of alteration in response to *in situ* surface processes. ASG

Reading
Fenwick, I. 1985: Palaeosols – problems of recognition and interpretation. In J. Boardman ed., *Soils and Quaternary landscape evolution.* Chichester: Wiley. Pp. 3–21.

pali ridge A sharply pointed ridge between two stream valleys on deeply dissected volcanic domes, especially in Hawaii.

pallid zone The bleached portion of a soil zone or weathering profile containing a duricrust, and lying beneath the mottled zone.

palsa A peat mound associated with the development of an ice lens. Genetically the feature is similar to a PINGO, but is restricted to peat bogs. DES

Reading
Seppälä, M. 1972: The term 'palsa'. *Zeuschrift für Geomorphologie* NF 16, p. 463.

paludal sediments Deposits of marshes and swamps formed in areas of low and irregular topography, as along the banks of lakes, river floodplains, deltas, etc.

paludification The expansion of a bog caused by the gradual rising of the water table as accumulation of peat impedes water drainage. ASG

palynology See POLLEN ANALYSIS.

pan A closed depression that can occur in great profusion in arid and semi-arid areas such as the Kalahari, Patagonia, Western Australia, and the High Plains of Texas. Pans may result from such processes as solution and animal activity (buffalo or hog wallows), but the prime cause of their development is deflational activity on surfaces composed of susceptible materials (e.g. shales, fine sandstones and sands, lake beds, etc.). On their lee sides they may possess CLAY DUNES or LUNETTES, and they often possess a distinctive kidney-, clam- or heart-shaped morphology.   ASG

Reading
Goudie, A.S. 1991: Pans. *Progress in physical geography* 15, pp. 221–37.
Le Roux, J.S. 1978: The origin and distribution of pans in the Orange Free State. *South African geographer* 6, pp. 167–76.

panbiogeography See VICARIANCE BIOGEOGRAPHY.

panfan The surface produced when a hill or mountain is completely eroded so that the peripheral fans coalesce, as in the end stage of landscape evolution in an arid region.

Pangaea The name given to a postulated continental landmass which split up to produce most of the present northern hemisphere continents. Also the name applied to the former landmass that comprised all the present continental landmasses.

panplain A flat or almost flat landscape that has been produced by lateral erosion by rivers and lowering of divides and interfluves.

pantanal A type of savannah area along the sides of some Brazilian rivers. The land is seasonally flooded by river water but is very dry for most of the year.

parabolic dunes (also called hairpin dunes) Crescentic sand accumulations in which the horns point away from the direction of dune movement (the opposite of BARCHANS). They may occur as rake-like clusters and can develop from blowouts in transverse ridges. Parabolic dunes occur in areas where vegetation provides some impedence to sand flow, as in some coastal dune fields and on desert margins.   ASG

Reading
McKee, E.D. 1979: A study of global sand seas. *US Geological Survey professional paper* 1052.

parallel retreat The phenomenon of denudation of a landscape by lateral erosion of scarp slopes and hills which maintain their slope angle as erosion progresses. It is one of three classic models of slope evolution, the other two being slope downwearing (as in the Davisian cycle) and slope replacement (as in the model by W. Penck). L.C. King explained the great escarpments and inselbergs of Africa through this process.

parallel roads A series of former lake shorelines resembling roads. The best known example, the parallel roads of Glen Roy in Scotland, were first identified by Agassiz (1840–1) as the shorelines of a series of former glacier-dammed lakes.   DES

Reading and Reference
Agassiz, L. 1840–1: On glaciers and the evidence of their having once existed in Scotland, Ireland, and England. *Proceedings of the Geological Society of London* 3, pp. 327–32.
†Sissons, J.B. 1981: Ice dammed lakes in Glen Roy and vicinity, a summary. In J. Neale and J. Flenley eds, *The Quaternary in Britain*. Oxford: Pergamon. Pp. 174–83.

parameterization The calibration process whereby mathematical models are fitted to particular genuine cases. Model parameters, which often represent physical properties, must be numerically evaluated for the model to be applied. Calibration problems are exemplified by parametric hydrology (Nash 1967), which is concerned with describing the hydrological behaviour of drainage basins by generalized numerical properties. A simple example is the rational formula which relates peak catchment discharge Q to basin area A, the run-off coefficient C (which measures the proportion of storm rainfall converted to run-off), and the mean rainfall intensity I_t during the time of concentration t of the flood, by:

$$Q = C I_t A.$$

The parameters C and t are considered characteristic of a basin, and must be estimated before the rainfall intensity during the period t can be assessed and the flood peak predicted. Optimization of such parameter values is achieved by comparing model predictions (output) with actual observed values for given input conditions. Optimization criteria are required, such as the minimization of the sum of squares of deviations between predicted and observed system output values. However, several parallel criteria may be desirable in matching model output to the physical behaviour of the real world (e.g. the height and timing of modelled flood hydrograph peaks). Complex models may require simultaneous optimization of interdependent parameters, and sophisticated parameterization procedures are then necessary (Dawdy and O'Donnell 1965). Alternatively, component sub-models may be parameterized in sequence, with the output from one forming the output to the next. KSR

References
Dawdy, D.R. and O'Donnell, T.R. 1965: Mathematical models of catchment behaviour. *Proceedings of the American Society of Civil Engineers, journal of the Hydraulics Division* 91, pp. 123–37.
Nash, J.E. 1967: The role of parametric hydrology. *Journal of the Institution of Water Engineers* 21, pp. 435–74.

parasite An organism which exists in a somewhat imbalanced SYMBIOSIS with another, in which, although it is able to extract energy (food) from its host for its own preservation it is unlikely to develop this ability to the point where the host is killed. Some human parasites are, however, exceptions to this general rule: and there are others (e.g. the larvae of some Diptera and Hymenoptera). Since parasites do not require their own energy-production systems, they are often structurally simple: for example, mistletoe, which lives through tapping the phloem tissue of certain deciduous trees, often has no chlorophyll. Parasites are usually much smaller than their hosts; and they may co-exist with them externally or internally. DW

parna A word coined by Butler (1956) for aeolian clay deposits found in Australia. Parna deposits occur either as discrete dunes or as thin, discontinuous, widespread sheets. They contain some 'companion sand', in addition to the clay pellets which make up the greater portion of the material. They may be derived from the deflation of material from unvegetated, saline lake floors, or from other soil or alluvial surfaces. Parna is in effect a loessic clay. (See also CLAY DUNE; LUNETTE.)

Reading and Reference
Butler, B.E. 1956: Parna – an aeolian clay. *Australian journal of science* 18, pp. 145–51.

†— 1974: A contribution to the better specification of parna and other aeolian clays in Australia. *Zeitschrift für Geomorphologie* 20, pp. 106–16.

partial area model Forecasts the saturated area around streams and channel heads as a basis for flood hydrograph prediction. This type of model is in direct contrast to the HORTON OVERLAND FLOW MODEL. THROUGHFLOW is generally routed downslope to estimate the areas where SATURATION DEFICIT is zero. This DYNAMIC SOURCE AREA is then assumed to generate OVERLAND FLOW which provides the most rapidly responding part of the stream hydrograph. MJK

Reading
Kirkby, M.J. ed. 1978: *Hillslope hydrology.* Chichester: Wiley. Esp. chs 6–9.

partial duration series A series of events analysed in flood frequency analysis and consisting of all flood peak discharges above a specified threshold discharge. The series is particularly useful when it is needed to know the frequency with which a particular flood discharge is exceeded. (See also FLOOD FREQUENCY.) KJG

particle form Has been used as a composite term to cover several properties of the morphology of sedimentary particles. Thus, in a functional expression:

$$F = f(Sh, A, R, T, F, Sp)$$

where Sh denotes the shape of the particle, A its angularity and R its roundness, T is the surface texture and Sp is the sphericity. Thus a complete description of the form would entail observations of these component parts. However, the ease with which this may be done depends upon factors such as the size of the particles to be examined. Particle size itself does not come into the expression but it is relatively easy to measure axial ratio lengths on gravel size particles, much less so with sands. Both shape and sphericity of particles can be determined by measurement of a, b and c axes. Shape has been defined as 'a measure of the relation between the three axial dimensions of an object'; this is related to, but not the same as, sphericity, 'a measure of the approach of a particle to the shape of a sphere'. Angularity and roundness are at different ends of a continuum and are usually measured by comparing the particle with a 'standard' chart. This method has problems with operator variance and it is likely that computer-based methods using Fourier analysis will supersede it. Analysis of surface texture has been a much-used technique on its own with the advent of the scanning electron microscope, but it too is largely based on qualitative assessments and comparisons. WBW

Reading
Bull, P.A. 1981: Environmental reconstruction by electron microscopy. *Progress in physical geography* 5, pp. 368–97.

Clark, M.W. 1981: Quantitative shape analysis: a review. *Mathematical geology* 13, pp. 303–20.

Goudie, A.S. ed. 1990: *Geomorphological techniques*. 2nd edn. London: Unwin Hyman.

Whalley, W.B. 1972: The description and measurement of sedimentary particles and the concept of form. *Journal of sedimentary petrology* 42, pp. 961–5.

particle size Can refer either to an individual particle or to a mass of particles. It is useful for characterizing or describing one aspect of a particle's morphology, form being another important component. Various names have been given to size ranges, e.g. clay, silt, sand, etc. but these are useful as rough, ordinal scale descriptors only and the actual size range covered by each name varies from scale to scale. For particles that can be measured by picking up an individual and measuring lengths with a rule or calipers, three orthogonal axes give a good measure of size. For smaller particles, especially those which can only be seen easily with a microscope or hand lens, it is more usual to measure a single (long) axis. Because the size of particles may range from microscopic aerosols (less than 1 μm) through clay-size (less than 2 μm), silt, sand and cobbles to even larger blocks, no single measurement technique can be applied. This means that the definition of size varies according to the technique used. This restriction applies even to the normal range of particles seen in a soil.

For instance, if the long *a* axes of cobbles (above 60 mm) are measured with calipers but smaller particles are measured by sieving, the sizes obtained by these methods are not strictly comparable. This difficulty of making measurements on a continuous scale also occurs when sieving becomes difficult and precipitation methods are employed. Sieves give 'nominal diameters' which roughly correspond to the intermediate or *b* axis of a grain. When, as is usual, a sample is taken containing perhaps

Major class intervals used in description of sediment sizes

mm		Size classes	
		Boulder	
256	−8		
64	−6	Cobble	*Gravel*
32	−5		
16	−4	Pebble	
4	−2		
2.83	−1.5	Granule	
2.00	−1.0		
1.41	−0.5	Very coarse sand	
1.00	0.0		
0.71	−0.5	Coarse sand	
0.50	1.0		
0.35	1.5	Medium sand	*Sand*
0.25	2.0		
0.177	2.5	Fine sand	
0.125	3.0		
0.088	3.5	Very fine sand	
0.0625	4.0		
0.031	5.0	Coarse silt	
0.0156	6.0		*Silt*
0.0078	7.0	Fine silt	
0.0039	8.0		
		Clay	*Clay*

Source: Based on Pettijohn, F.J., Potter, P.E. and Siever, R. 1972: *Sand and sandstone*. New York: Springer-Verlag. Table 3–2, p. 71.

many thousands of grains, size fractions are determined by sieving through progressively finer sieves. The weights held on each sieve are recorded and expressed as a 'percentage passing' or 'percentage finer' than a given mesh. Such data may be graphed as histograms or as a cumulative size distribution. Although millimetre and micrometre (μm) sizes can be recorded, it is often convenient to transform the sizes into the ϕ scale by:

$$\phi = -\log_2 d$$

where d is the particle size in millimetres. This transform takes into account the generally log-normal size distribution of most sediments. The usual statistical treatments of distributions can be applied to the data obtained by sieving, for example. Thus mean, SKEWNESS, SORTING and KURTOSIS of a size range can be derived by various methods (most often graphically). Such size parameters can be used to compare sediments, to determine their possible origin, or to use in some other way, e.g. to suggest their geotechnical behaviour.   WBW

Reading
Briggs, D. 1977: *Sediments*. London: Butterworth.

Goudie, A.S. ed. 1990: *Geomorphological techniques*. 2nd edn. London: Unwin Hyman.

Syvitski, J.P.M. ed. 1991: *Principles, methods and applications of particle size analysis*. Cambridge: Cambridge University Press.

passive margin The margin of a continent formed as a result of the breakup of a large continental mass, or supercontinent. Passive margins are so called since, in comparison with the active continental margins which coincide with zones of plate convergence, they are characterized by low levels of tectonic activity once continental rupture has been completed. The fragmentation of the supercontinent of Pangaea, which began about 180 Ma ago, created the many new passive continental margins which form a substantial proportion of the perimeters of the present-day continents. There are two types of passive margin; rifted margins are formed by divergent plate motion, whereas sheared margins are produced where the movement between two adjacent continental blocks has been essentially transform.   MAS

Reading
Summerfield, M.A. 1991: *Global geomorphology*. London and New York: Longman Scientific and Technical and Wiley.

paternoster lake One of a series of lakes occupying basins in a glacial trough. The name is taken from the term for rosary prayer-beads.

patterned ground A general term for the more or less symmetrical forms, such as circles, polygons, nets and stripes, that are characteristic of, but not necessarily confined to, soils subject to intense frost action. Description is usually based upon geometric form and the degree of uniformity in grain size (i.e. sorting). The most common type of patterned ground is the non-sorted circle, sometimes termed earth hummock or mud-boil. The processes responsible for the formation of patterned ground are not clear; some forms may be polygenetic and others may be combination products in a continuous system having different processes as end members. In some instances thermal or desiccation cracking is clearly essential. In others cryoturbation, the lateral and vertical displacement of particles associated with repeated freezing and thawing, combined with site-specific factors such as moisture availability, lithology, and slope appear essential.   HMF

Reading
Mackay, J.R. 1980: The origin of hummocks, western Arctic coast, Canada. *Canadian journal of earth sciences* 17, pp. 996–1006.

Shilts, W.W. 1978: Nature and genesis of mudboils, central Keewatin, Canada. *Canadian journal of earth sciences* 15, pp. 1053–68.

Washburn, A.L. 1979: *Geocryology: a survey of periglacial processes and environments*. New York: Wiley. Esp. pp. 119–70.

pavement, limestone A natural surface of bare rock, often, but not necessarily, flat (see illustration on p. 375). The limestone surface may be smoothed or runnelled, being divided up into blocks by opened joints (GRIKES). The origin of limestone pavements is not known for certain. They have been considered to have been produced by solutional processes beneath a soil cover (to form the grikes) and may have been exposed by glacial stripping.   PAB

Reading
Goldie, H. 1973: The limestone pavements of Craven. *Transactions of the Cave Research Group of Great Britain* 15, pp. 175–90.

peak discharge The highest discharge achieved at a stream gauging site within a specific time period, usually during the passage of a storm hydrograph (see

A limestone pavement showing the development of clints and grikes in Carboniferous rocks above Malham Cove, Yorkshire, England. The surface was originally stripped by glaciation and has been moulded by subsequent solutional activity.

HYDROGRAPHS). Peak discharge is a widely used parameter of the flood characteristics of a site and is the usual parameter to be analysed in FLOOD FREQUENCY ANALYSIS. (See also DISCHARGE.)   AMG

Pearson type III distribution A theoretical probability distribution which may be used to fit the flood frequency distribution in FLOOD FREQUENCY ANA-LYSIS. There are several distributions available and the log-Pearson type III is a further form which may be used.   KJG

peat, peatlands The waterlogged partially decomposed remains of mire plants. They may build up to depths in excess of 10 m and, as individual mires, may cover many hectares, while as collections of mires they may transform whole landscapes; for example, western Ireland, large parts of Finland and Canada. Their impact on the landscape of cool humid regions was, until recent times, comparable to the impact of rain forest in the tropics. Man's efforts at drainage and peat-cutting for fuel and for horticultural or industrial purposes during the past few centuries have tamed many formerly wild areas for agriculture (the Fens, the Somerset Levels) or for recreation (the Norfolk Broads).

Peat will form either directly on a mineral soil surface, or as the end point of the lake infill succession, wherever the basic conditions of waterlogging – leading to anaerobic conditions, and a drastic reduction in decomposition by micro-organisms – is maintained for most of the year. Such conditions may be provided by large, flat estuarine clay areas; by flat-floored valleys with incompetent streams; by the enclosed hollows of a glaciated landscape to which normal drainage has been disrupted; and as the final stage of the sediment build-up of a lake basin or cut-off channel of a river. In wet climates, such as that of north-west Scotland, peat formation may begin, and be maintained, on sloping sites.

The peat itself may be formed from the remains of many plants – sedges and grasses, herbs and dwarf shrubs, trees and mosses – and the often remarkable state of preservation means that these MACRO-FOSSILS can often be easily identified, many to species level. The phytosociology of past mire communities can thus be reconstructed (Rybniček 1973), a unique attribute of mires compared with other plant communities. The plants inhabiting the mire will be determined mainly by its NUTRIENT STATUS and, together with various hydrological parameters, these factors give us the two basic kinds of peatland: fen and bog. There are, of course, various transitional types of peatland, and types peculiar to extreme climatic

Peat is a widespread feature in the Highlands and Islands of Scotland and is here being cut for fuel in Skye. Peat results from the accumulation of organic matter in areas of moisture surplus.

conditions, such as the palsa-mires of the Arctic (Moore and Bellamy 1974), and there have been many classification and zonation schemes proposed (Moore 1984). Many of these are confusing and unnecessarily complex, as easily happens when different workers attempt to classify what is essentially a continuum of vegetation types sharing the basic factor of peat accumulation. The simplest division is perhaps into fen and bog, EUTROPHIC and OLIGOTROPHIC in their nutrient status. Fens tend to be dominated by sedges and grasses (e.g. *Cladium mariscus, Carex paniculata* and *Phragmites australis*) and may be wooded with birch, willow and alder to form fen carr woodland. They are *rheotrophic* and the richness of the plant community depends upon the minerals in the inflowing water. The fen community is often zoned according to the amount of nutrients each area receives. Thus there is 'extreme-rich fen', with pH levels of 7–8 and above and its own characteristic plant communities, including a wide variety of mosses due to water derived from base-rich rocks or till, and 'poor fen' with pH levels of 4–6 or so, an impoverished plant community with sedges (e.g. *Carex rostrata, C. lasiocarpa*) and grasses (e.g. *Molinia caerulea*) due to water draining from base-poor rocks such as granite or from gravels and sand. This grades into valley bog, wherein the various sphagna are more important, along with ericaceous species such as

Calluna vulgaris (common heather) and *Erica tetralix* (cross-leaved heath). Sphagna, of which there are about forty species in Europe, are present in almost all mires, with species such as *Sphagnum squarrosum* common in base-rich situations, *S. fimbriatum* in fen carr and *S. papillosum, S. magellanium, S. cuspidatu* and *S. capillifolium* characteristic of bog habitats. Beyond valley bogs (rheotrophic–mesotrophic/oligotrophic communities) there are bogs at the ombrotrophic–oligotrophic end of the scale. While fens and valley bogs are medium-scale ecosystems, measured in tens and hundreds of metres, ombrotrophic bogs are measured in kilometres and tens of kilometres. They are also, since they are carrying their own water table upwards in the peat, generally much deeper than fens. Ombrotrophic bogs are dominated by the *Sphagnum–Eriophorum–Calluna* association, although there may be a number of other species present (e.g. *Rhynchospora alba, Narthecium ossifragum* etc). The conditions and rate of growth and the end point of such plant communities are not wholly understood (Barber 1985) but their importance can hardly be overestimated. In the British Isles they represent the 'climax' of the hydrosere. As Walker (1970) showed, this is raised-bog and not the oakwood community postulated by Tansley (1939) and others. They represent one of our most untouched and unmodified ecosystems so the protection of our remaining peatlands must have a high priority in any conservation programme.   KEB

Reading and References

Barber, K.E. 1985: Peat stratigraphy and climatic change: some speculations. In M.J. Tooley and G.M. Sheail eds, *The climatic scene: essays in honour of Gordon Manley*. London: Allen & Unwin.

Godwin, H. 1981: *The archives of the peat bogs*. Cambridge: Cambridge University Press.

Ivanov, K.E. 1981: *Water movement in mirelands*. London and New York: Academic Press.

Larsen, J.A. 1982: *Ecology of the northern lowland bogs and conifer forests*. New York and London: Academic Press.

Moore, P.D. ed. 1984: *European mires*. London: Academic Press.

— and Bellamy, D.J. 1974: *Peatlands*. London: Elek.

Rybníček, K. 1973: A comparison of the present and past mire communities of central Europe. In H.J.B. Birks and R.G. West eds, *Quaternary plant ecology*. Oxford: Blackwell Scientific.

Tansley, A.G. 1939: *The British Isles and their vegetation*. Cambridge: Cambridge University Press.

Walker, D. 1970: Direction and rate in some British postglacial hydroseres. In D. Walker and R.G. West eds, *Studies in the vegetational history of the British Isles*. Cambridge: Cambridge University Press.

pedalfer A well-drained soil which has had most of its soluble minerals leached from it. They occur in more humid situations than PEDOCALS.

pediment A term applied by G.K. Gilbert (1880) to alluvial fans flanking the mountains near Lake Bonneville in Utah. Since then it has changed its meaning and is now defined (Adams 1975) as follows: a smooth planoconcave upward erosion surface, typically sloping down from the foot of a highland area and graded to either a local or more general base level. It is an element of a pedimont belt, which may include depositional elements such as fans and playas. The pediment, as defined, excludes such depositional components, although an alluvial cover is frequently present. It is broadly synonymous with the French term, *glacis*. Coalescing pediments create pediplains.

Pediments were initially recognized as being of wide extent in the American southwest. McGee (1897) proposed that they were planed off by sheetfloods, whereas others have invoked the role of lateral planation by rivers or the backwearing of mountain fronts (Cooke and Warren 1973, p. 188 ff). Most pediments have low angle surfaces (generally less than 10°) and the junction between them and the hill mass behind is often marked by an abrupt change of angle. For this to happen a sharp contrast is needed between the nature of sound rock and the weathering debris produced. In particular it is necessary for the products of weathering to be relatively fine-grained and to have a limited size range, so that they can be transported across the low angle pediment slope, thereby permitting development or maintenance of an abrupt break of slope. ASG

References
Adams, G. ed. 1975: *Planation surfaces*. Stroudsburg, Pa.: Dowden, Hutchinson & Ross.

Cooke, R.U. and Warren, A. 1973: *Geomorphology in deserts*. London: Batsford.

Gilbert, G.K. 1880: Contributions to the history of Lake Bonneville. *US Geological Survey annual report* 2, pp. 167–200.

McGee, W.J. 1897: Sheetflood erosion. *Bulletin of the Geological Society of America* 8, pp. 87–112.

pedocal A poorly drained soil which has no soluble salts leached from it. Free calcium occurs in the profile, in the form of concretions, veins, nodules, or layers.

pedon A term used in soil science, and defined as the smallest volume that can be called soil. Its lateral area ranges from 1 to 10 m^2 and is large enough to permit the study of the nature of any horizon present, for a horizon may be variable in thickness and even discontinuous.

A rock cut pediment in granitic desert terrain near Springbok in Namaqualand, South Africa.

peds Soil aggregates: their strength, size and shape give the soil its structure. (See also SOIL STRUCTURE.)

pelagic Refers to that component of an aquatic ecosystem which excludes its margins and substrate. The pelagic environment corresponds to that of the main part of the body of water (Barnes 1980).

The pelagic environment varies in depth. In deep water (e.g. of oceans), light is able to penetrate only a limited distance below the water surface. The lighted or euphotic zone is the site of abundant organic productivity. Here, phytoplanktonic primary producers are consumed by zooplankton (see PLANKTON) and small fish. Beneath the euphotic zone are the mesopelagic and bathypelagic zones, mainly populated by predatory nekton (see NEKTON). These zones benefit from downward-moving biological detritus and are visited by migratory organisms. In the bathypelagic zone, organisms are relatively infrequent. Numerous species emit their own light in order to obtain their prey (Isaacs 1977).

RLJ

Reading and References
Barnes, R.S.K. 1980: The unity and diversity of aquatic systems. In R.S.K. Barnes and K.H. Mann eds, *Fundamentals of aquatic ecosystems*. Oxford: Blackwell Scientific. Pp. 5–23.
Isaacs, J.D. 1977: The nature of oceanic life. In H.W. Menard ed., *Ocean science (Readings from Scientific American)*. San Francisco: W.H. Freeman. Pp. 189–201.
†Moss, B. 1980: *Ecology of fresh waters*. Oxford: Blackwell Scientific.

Penck and Brückner model Developed during the first decade of the twentieth century by A. Penck and E. Brückner, working primarily in Bavaria, to provide a framework for understanding the PLEIS-TOCENE history of the Alps. Four main glacial phases were initially identified – Günz, Mindel, Riss and Würm – together with various interglacials, including the GREAT INTERGLACIAL (see diagram). A more complex Pleistocene history has now been determined on the basis of deep sea core studies, but the Penck and Brückner model has been immensely influential and extended uncritically worldwide. ASG

Reference
Penck, A. and Brückner, E. 1909: *Die Alpen im Eiszeitalter*. Leipzig: C.H. Tauchnitz.

peneplain See CYCLE OF EROSION; EROSION SURFACE.

peninsula A headland or promontory surrounded by water but connected to the mainland by a neck or isthmus.

penitent rocks A residual landscape produced when deeply weathered, slightly dipping schistose rocks are exhumed by erosion. Only the most resistant beds crop out.

Penman formula See EVAPOTRANSPIRATION.

per cent silt clay The percentage of a sediment sample in the silt–clay range that is finer than 0.06 mm or 4 ϕ. There are slight differences in the exact numerical definition that different authorities have proposed, but in analysis this fraction of a sediment sample is below the practical limit for sieving: the grains are hardly visible to the naked eye and may be cohesive. (See also WEIGHTED MEAN PERCENTAGE SILT CLAY.) JL

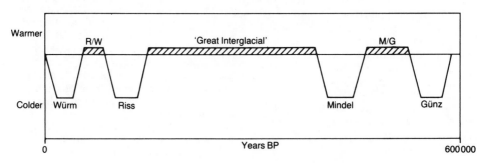

Penck and Brückner model
Source: *A.S. Goudie 1984: The nature of the environment. Oxford and New York: Basil Blackwell.*

Reading
Schumm, S.A. 1977: *The fluvial system*. New York: Wiley.

perched block A boulder or block of rock that is balanced on another rock or outcrop having been deposited there by ice.

Reading
Patterson, E.A. 1984: A mathematic model for perched block formation. *Journal of glaciology* 30, pp. 296–301.

perched groundwater An isolated body of unconfined GROUNDWATER suspended by a discontinuous relatively impervious layer above the main saturated zone and separated from it by unsaturated rock. Groundwater that is perched also has a perched WATER TABLE.   PWW

percolation The process of essentially vertical water movement downwards through soil or rock in the unsaturated (or vadose) zone. Percolation water is the water that has passed through the soil or rock by this process. It may be measured by a PERCOLATION GAUGE.   PWW

percolation gauge An instrument for measuring the quantity of water that passes vertically through soil or rock by the process of PERCOLATION. Such gauges may be established beneath lysimeters to catch the excess water after losses due to evapotranspiration and sometimes they are installed in caves to measure the response at percolation input points to rainfall at the surface.   PWW

percoline A path or seepage line along which moisture flow becomes concentrated, is particularly well developed where soils are relatively deep, and usually presents a dendritic pattern tributary to surface stream channels. After water from precipitation infiltrates it may flow throughout the soil profile as matrix flow, but once it becomes more concentrated – but before a definite stream channel is produced – there will be a percoline which may be indicated by a broad linear depression on the surface and this may be inter-related with SUFFOSION and the feature at the head of seasonally occupied channels such as DAMBOS in tropical landscapes.   KJG

Reading
Bunting, B.T. 1961: The role of seepage moisture in soil formation, slope development and stream initiation. *American journal of science* 259, pp. 503–18.

pereletok A Russian term for a layer of ground between the active layer and the permafrost below which remains frozen for one or several years and then thaws. Use of this term is not recommended since it presupposes that pereletok is not permafrost, although the definition assigns a sufficient duration of time for it to be considered as such. In addition there is little physical difference between pereletok on the one hand and permafrost of only a few years duration on the other. Material identified by this term ought to be referred to either as 'seasonally frozen ground' if it lasts less than one year or 'permafrost' if it exists for the duration of one complete thawing season and then through the next winter.   HMF

Reading
Brown, R.J.E. and Kupsch, W.O. 1974: *Permafrost terminology*. Ottawa: National Research Council of Canada publication 14274.

perennial stream A stream which flows all year. A dynamic drainage network also includes INTERMITTENT STREAMS and EPHEMERAL STREAMS but there should always be flow in a perennial stream channel. For much of the time this flow may be in the form of BASE FLOW or DELAYED FLOW except when QUICKFLOW occurs after rainstorms.   KJG

pericline A dome produced by folding. An ANTICLINE which pitches at both ends.

perigee The point in its orbit at which the moon is closest to the earth.

periglacial A term first used by Walery von Lozinski in 1909 to describe frost weathering conditions in the Carpathian Mountains of Central Europe. The concept of a 'periglacial zone' subsequently developed referring to the climatic and geomorphic conditions of areas peripheral to Pleistocene ice sheets and glaciers. Theoretically, it was a tundra region extending as far south as the treeline. Modern usage refers to a wide range of cold non-glacial conditions regardless of their proximity to glaciers, either in time or space. Periglacial environments exist not only in high latitudes and tundra regions but also below the treeline and in high altitude (alpine) regions of temperate latitudes.

Approximately 20 per cent of the earth's landsurface currently experiences periglacial

Classification of periglacial climates

Polar lowlands	Mean temperature of coldest month <−3°C. Zone is characterized by ice caps, bare rock surfaces and tundra vegetation.
Subpolar lowlands	Mean temperature of coldest month <−3°C and of warmest month >10°C. Taiga type of vegetation. The 10°C isotherm for warmest month roughly coincides with treeline in northern hemisphere.
Mid-latitude lowlands	Mean temperature of coldest month is <−3°C but mean temperature >10°C for at least four months per year.
Highlands	Climate influenced by altitude as well as latitude. Considerable variability over short distance depending on aspect. Diurnal temperature ranges tend to be large.

Source: Based on the classification presented by Washburn, A.L. 1979: *Geocryology*. London: Edward Arnold. Pp. 7–8.

conditions in the form of either intense frost action or the presence of permafrost, or both. There are all gradations between environments in which frost processes dominate, and where a whole or major part of the landscape is the result of such processes, and those in which frost action processes are subservient to others. Complicating factors are that certain lithologies are more prone to frost action than others, and no perfect correlation exists between areas of intense frost action and areas underlain by permafrost.

Unique periglacial processes are the formation of PERMAFROST, the development of thermal contraction cracks, the thawing of permafrost (THERMOKARST), and the formation of wedge and injection ice. Other processes, not necessarily restricted to periglacial regions, are important on account of their high magnitude or frequency. These include ice segregation, seasonal frost action, frost (i.e. cryogenic) weathering, and rapid mass movements.

The most distinctive periglacial landforms are those associated with permafrost. The most widespread are tundra polygons, formed by thermal contraction cracking. They divide the ground surface into polygonal nets 20–30 m in dimension. Ice-cored hills, or pingos, are a less widespread but equally classic periglacial landform; they form when water moves to the freezing plane under hydraulic or hydrostatic pressure. Other aggradational landforms, such as palsas and peat plateaux, are associated with ice segregation. Ground ice slumps, thaw lakes, and irregular hummocky topography with enclosed depressions (thermokarst) result from the melt of ice-rich permafrost.

Many periglacial phenomena form by frost wedging and the cryogenic weathering of exposed bedrock. Frost wedging is associated with the freezing and expansion of water which penetrates joints and bedding planes. The details of cryogenic weathering are poorly understood. Coarse angular rock debris (blockfields), upthrust bedrock blocks, talus (scree) slopes, and certain types of patterned ground are usually attributed to frost action. Angular bedrock masses (tors) may stand out above the debris-covered surfaces, reflecting more resistant bedrock. Flat erosional surfaces, termed cryoplanation terraces, are sometimes associated with tors but, equally, can occur quite independently.

The overall flattening of landscape and smoothing of slopes, thought typical of many periglacial regions, is generally attributed to mass wasting. Agents of transport include frost creep and solifluction. NIVATION, a combination of frost wedging, solifluction and sheetwash operating beneath and downslope of snowbanks, is often regarded as important. In areas dominated by extreme nival regimes and underlain by unconsolidated sediments, fluvial activity can be a significant landscape modifier.   HMF

Reading
Clark, M.J. 1988: *Advances in periglacial geomorphology*. Chichester: Wiley.

French, H.M. 1976: *The periglacial environment*. London and New York: Longman.

Lozinski, W. von 1909: Über die mechansche Verwitterung der Sandsteine im gemässigten Klima. *Bulletin International de l'Academie des Sciences et des Lettres de Cracovie, Classe des Sciences Mathématiques et Naturelles* 1, pp. 1–25.

Washburn, A.L. 1979: *Geocryology: a survey of periglacial processes and environments*. New York: Wiley.

perihelion The point in its orbit about the sun that a planet or comet is closest to the sun.

permafrost The thermal condition in soil and rock where temperatures below 0°C persist over at least two consecutive winters and the intervening summer. Permafrost is defined purely as a thermal condition: moisture, in the form of water and/or ice may or may not be present. The term was first introduced by S.W. Muller in 1945, as a shortened form of permanently frozen ground. Permafrost is not necessarily synonymous with 'frozen ground', however, since earth materials may be below 0°C in temperature but essentially unfrozen on account of depressed freezing points due to mineralized groundwaters or other causes. One solution is to differentiate between 'cryotic' (i.e. below 0°C) and 'non cryotic' (i.e. above 0°C) ground, and to subdivide the former into 'unfrozen', 'partially frozen' and 'frozen', depending upon the amount of unfrozen water present. Equally, permafrost is not 'permanently frozen ground', since changes in climate and terrain may cause it to degrade.

The upper boundary of permafrost is known as the permafrost table, and the near-surface layer which is subject to seasonal thaw is called the active layer. The depth at which annual temperature fluctuations are minimized is termed the depth of zero annual amplitude; this usually varies between 10 and 20 m depending upon climate and terrain factors such as amplitude of annual surface temperature variation, snow cover, and effective thermal diffusivity of the soil and rock. In polar regions, permafrost temperatures may be as low as −15°C at the depth of zero annual amplitude.

Permafrost underlies approximately 20–25 per cent of the earth's landsurface. It is widespread in Siberia, northern Canada, Alaska and China (see diagram on p. 382). Zones of continuous, discontinuous and sporadic (isolated) permafrost are generally recognized, together with alpine, intermontane and subsea (offshore) permafrost. Thicknesses range from a few metres at the southern limits to approximately 500 m in parts of northern Canada and Siberia where it may be relic.   HMF

Reading
Brown, R.J.E. and Kupsch, W.O. 1974: *Permafrost terminology*. National Research Council of Canada publication 14274.

French, H.M. 1982: *The Roger J.E. Brown memorial volume. Proceedings of the Fourth Canadian Permafrost Conference.* National Research Council of Canada.

Muller, S.W. 1947: *Permafrost or permanently frozen ground and related engineering problems.* Ann Arbor, Mich.: J.W. Edwards.

National Academy of Sciences 1973: *North American contribution. Permafrost. Second international conference.* Yakutsk, USSR, 13–28 July 1973.
— 1983 *Permafrost. Fourth international conference.* Vol. I. Washington DC, National Academy Press.

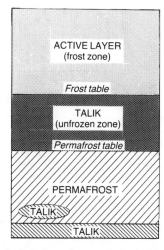

1. Terminology of some features associated with permafrost. Source: *R.U. Cooke and J.C. Doornkamp 1974: Geomorphology in environmental management. Oxford; Clarendon Press. Figure 9.1.*

permeability A measure of the capacity of a rock or soil to transmit fluids. It is often termed INTRINSIC PERMEABILITY or specific permeability to distinguish it from hydraulic CONDUCTIVITY which is sometimes (but now less frequently) called the coefficient of permeability. Permeability (or PERVIOUSNESS) depends upon the physical characteristics of the rock, whereas hydraulic conductivity also takes account of the physical characteristics of the fluid.   PWW

Reading
Freeze, R.A. and Cherry, J.A. 1979: *Groundwater.* Englewood Cliffs. NJ: Prentice-Hall.

permeameter An instrument for measuring the saturated hydraulic CONDUCTIVITY or K of soils or other sediments. Two types of apparatus are in common use. One is a constant head permeameter, the other a falling head permeameter. With reference to the figure, in the constant head case:

$$K = \frac{QL}{Ah}$$

and in the falling head case:

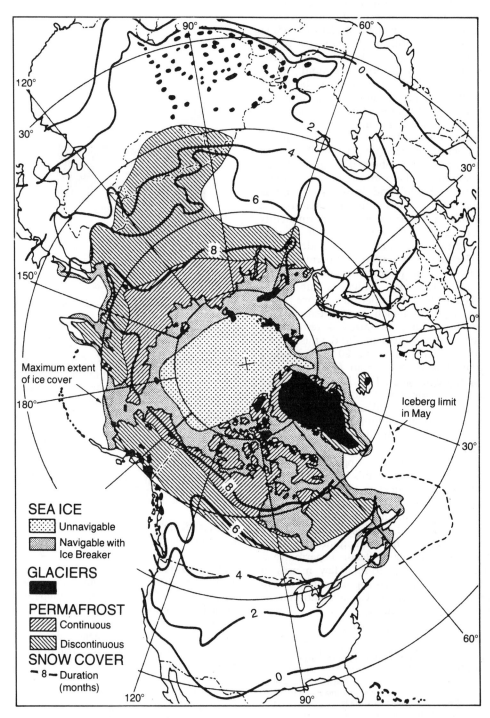

Maximum extent
of ice cover

Iceberg limit
in May

SEA ICE
Unnavigable
Navigable with
Ice Breaker
GLACIERS
PERMAFROST
Continuous
Discontinuous
SNOW COVER
– 8 – Duration
(months)

2. *Snow, ice and permafrost in the northern hemisphere.*
Source: *H.J. Walker 1983:* Mega-geomorphology. *Oxford: Clarendon Press. Figure 3.2.*

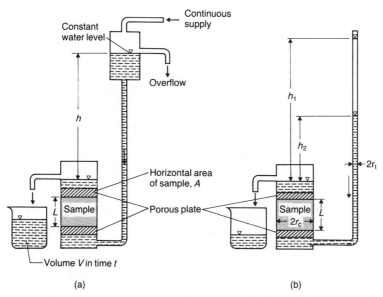

Permeameters for measuring hydraulic conductivity of geological samples. (a) *Constant head;* (b) *falling head.*
Source: *D.K. Todd 1980:* Groundwater hydrology. *Chichester and New York: Wiley. Figure 3.4, p. 73.*

$$K = \frac{aL}{aT}\ln\frac{h_0}{h_1}$$

PWW

Reading and Reference
†Freeze, R.A. and Cherry, J.A. 1979: *Groundwater*. Englewood Cliffs, NJ: Prentice-Hall.

†McGreal, W.A. 1981: Permeability and infiltration capacity. In A. Goudie ed., *Geomorphological techniques*. London: Allen & Unwin. Pp. 94–6.

Todd, D.K. 1980: *Groundwater hydrology*. 2nd edn. New York and Chichester: Wiley.

persistence The continued existence of an ecosystem without significant fluctuations in overall species composition or in the relative number of individuals in different species. Persistence represents one concept of STABILITY and has also been referred to as constancy or 'no-oscillation' stability. This aspect of stability does not take into account the ability of a system to adjust to disturbance.

The linkage of the term persistence to the notion of constancy of ecosystem species populations is unfortunate. Although major fluctuations in population size are thought to increase the risk of extinction during episodes of low population, there is considerable evidence that oscillations of population numbers may enhance the survival of many species by producing episodes of saturation and scarcity with respect to predation.   ARH

Reading
Dunbar, M.J. 1973: Stability and fragility in Arctic ecosystems. *Arctic* 26, pp. 179–85.

perturbation Variation, usually of small amplitude, about some well-defined, usually uninteresting, basic state. It is the conceptual equivalent of gently shaking an unknown package before embarking on more irreversible probing. Perturbation analysis allows all possible states similar to the initial one to be followed at least for a short time. In contrast, numerical integration (see NUMERICAL MODELLING) allows a few possible states to be followed for a longer time. The term is also used to denote various types of weather disturbance such as the waves of the trades and equatorial easterlies.   JSAG

Reading
Haurwitz, B. 1941: *Dynamic meteorology*. New York and London: McGraw-Hill.

perviousness A property of soils and rocks indicating its capacity for transmitting fluids. It is equivalent to PERMEABILITY. Impervious rocks behave hydrologically like an AQUIFUGE or aquiclude.   PWW

pesticides Poisonous chemicals used by man to regulate or eliminate plant and animal pests. The term biocide is sometimes used in the same context but, as Ware (1983) points out, is without a precise scientific meaning and is often used in a rather emotive manner to describe the lethal effects of chemicals on living organisms.

Pesticides are mainly synthetic organic chemicals which possess varying levels of toxicity. They may be selective or non-selective and are aimed at 'target' organisms which are economically or socially undesirable (Cremlyn 1978).

Among elementary forms of life, viruses and bacteria are responsible for numerous plant and animal diseases. Of lower plants, algae which bloom in bodies of water and fungi parasitic on living tissue are pests. Fungi are also the cause of many diseases of flora and fauna. The 30,000 or so higher plants which are pests because they shade crops and/or diminish their supply of moisture and nutrients are termed weeds. Animal pests include some Nematoda (roundworms), Arthropoda (insects), Arachnida (spiders, mites, ticks), Mollusca (slugs and snails), birds and rodents.

There are three main categories of pesticide – herbicides, insecticides and fungicides – in the production of which four major groups of chemicals are employed. Herbicides may operate on either a systemic (penetrative) or non-systemic (contact) basis. Phenoxyaliphatic acids (for example, 2,4-D and 2,4,5-T) are important herbicides, the collapse, wilting and death of target weeds under their application being due to amended phosphate and nucleic acid metabolism and cell division.

Among insecticides, the best known, DDT, is an organochlorine (or chlorinated hydrocarbon) which causes death by upsetting sodium and potassium equilibrium in the nervous system, thereby altering its normal functions. Organophosphates (for example, parathion) have largely supplanted organochlorines as insecticides. Additionally, these are very toxic to vertebrates as they also inhibit normal functioning of the nervous system.

Fungicides need to kill the parasite but not its host. Their usual function is to halt the germination of fungal spores. Inorganic compounds of sulphur, copper and mercury are long-established fungicides, but there is an increasing number of organic fungicides such as dithocarbamates. After metabolism these influence amino-acid activity in disease cells and cause their elimination.

During the first millennium BC, sulphur was used to fumigate Greek houses. Around 150 years ago, sulphur had been joined by arsenic and phosphorus as pesticides. The impetus for much pesticide use came during the Second World War. Some pesticides (for example, DDT) were then used in the control of human disease carriers such as mosquitoes. At this time, research into nerve gas led to the discovery of the insecticidal property of organophosphates. Subsequent use of pesticides was often enthusiastic and without regard to their possible effects upon non-target organisms and their environment (see BIOLOGICAL MAGNIFICATION).

Pests can also become resistant to pesticides. This occurs because a pesticide disrupts a single genetically controlled process in the metabolism of the pest (Ware 1983). This resistance is often a sudden occurrence, and takes place either by the natural selection of hardy individuals which form the basis of subsequent populations, or by mutation to form resistant genotypes.   RLJ

References

Cremlyn, R. 1978: *Pesticides*. Chichester and New York: Wiley.

Ware, G. 1983: *Pesticides: theory and application*. 2nd edn. New York and Oxford: W.H. Freeman.

pF The $\log_{10}$ of the negative head of water (in centimetres) indicating the strength of soil moisture SUCTION at a site. Soil moisture suction is often termed the matric or capillary potential and it results from the attraction of solid surfaces (the soil matrix) for water and also of the water molecules for each other, so that water is held in the soil against gravity by adsorptive and CAPILLARY FORCES. These forces lower the potential energy of the soil water and the degree of this reduction in potential energy in comparison with free water at the same elevation and under the same air pressure may be indicated by a negative head of water. The negative head is measured by TENSIOMETERS and because the negative head may increase rapidly with quite small changes in soil moisture content, it is usually expressed in relation to a $\log_{10}$ base as a pF value. The graph of pF plotted against soil moisture content (by volume) for a soil sample that has been progressively drained of water is called the soil moisture characteristic curve or pF curve. AMG

Reading
Schofield, R.K. 1935: The pF of water in the soil. *Transactions of the Third International Congress on Soil Science* 2, pp. 37–48.
Smedema, L.K. and Rycroft, D.W. 1983: *Land drainage.* London: Batsford.

p-form Smoothed, apparently plastically sculptured forms caused by the action of glaciers. They comprise *grooves* which may be slightly sinuous with soft flowing outlines and sometimes display an overhanging lip, crescentic-shaped depressions or SICHEL-WANNEN 1–10 m in length and 5–6 m wide, with horns pointing down glacier, and POTHOLES and a variety of shallower forms, ranging in size from depressions a few centimetres across to giant potholes 15–20 m deep and 16 m in diameter. Good descriptions are given by Dahl (1965).

Although Gjessing (1965) argued in favour of erosion by saturated till, it may be that the grooved features can be explained by normal glacial abrasion and the remaining forms by high velocity subglacial meltwater action. DES

References
Dahl, R. 1965: Plastically sculptured detail forms on rock surfaces in northerm Nordland, Norway. *Geografisker annaler* 47, pp. 83–140.
Gjessing, J. 1965: One 'plastic scouring' and 'subglacial erosion'. *Norsk Geografisk Tiddskrift* 20, pp. 1–37.

pH The measure of the acidity or alkalinity of a substance measured as the number of hydrogen ions present in one litre of the substance and reported as a figure on a scale the centre of which is 7, representing neutrality. Acid substances have a pH of less than 7 and alkaline substances have a higher pH.

phacolith A lens-shaped igneous intrusion usually situated beneath an anticlinal fold or in the base of a syncline.

phenology The scientific study of the timing of recurring natural phenomena in the life cycle of plants and animals in nature. While all natural phenomena may be included (e.g. the timing of animal migrations, flowering and fruiting of plants, harvest, ripening and seed-time), phenological observations are often restricted to the time at which certain plants come into leaf and flower each year, and to the date of the first and last appearance of birds and animals. Phenological observations have been used as a source of proxy climate change, in order to describe temperature patterns during pre-instrumental times, based on long-term observations of the growth and maturity of cultivated plants, for example wine harvest records (Pfister, 1992). AP

Reading and Reference
Jefree, E.P. 1960: Some long-term meanings from the Phenological Reports (1891–1948) of the Royal Meteorological Society. *Quarterly Journal of the Royal Meteorological Society*, 86, pp. 95–103.
Pfister, C. 1992: Monthly temperature and precipitation in central Europe 1525–1979: quantifying documentary evidence on weather and its effects. In R.S. Bradley and P.D. Jones eds, *Climate since AD 1500*. London: Routledge. Pp. 118–43.

phoresy The transport by one animal of another of a different species to a new feeding site.

phosphate rock An indurated sedimentary deposit which is rich in apatite, a calcium phosphate. Much of it results from the interaction of guano (seabird excrement) and the calcium carbonate of reef sands and elevated limestones. ASG

Reading
Stoddart, D. and Scoffin, T. 1983: Phosphate rock on coral reef islands. In A.S. Goudie and K. Pye eds, *Chemical sediments and geomorphology*. London: Academic Press. Ch. 12.

photic zone The surface zone of a lake, sea or ocean above the maximum depth to which sunlight penetrates.

photochemical smog See SMOG.

photogrammetry The science and technology of obtaining reliable measurements by means of photography. The measurement of aerial photographs to provide details of area, distance and height are skills readily practised by physical geographers. The task of the photogrammetrist is usually considered to be the measurement of aerial photographs both with sufficient accuracy in two dimensions to enable features to be plotted with respect to a national or international grid coordinate system, and with sufficient accuracy in three dimensions to enable them to be located in relation to their height above sea level. PJC

Reading
American Society of Photogrammetry 1981: *Manual of photogrammetry*. 4th edn. Falls Church, Virginia: American Society of Photogrammetry.

Paine, D.P. 1981: *Aerial photography and image interpretation for resource management*. New York: Wiley.

photosynthesis A chemical process, which takes place in the cellular structures of green plants, blue-green algae, phytoplankton and certain other organisms, and which transforms received solar energy into chemically stored foodstuffs, through the conversion of carbon dioxide and water into carbohydrates with the simultaneous release of oxygen, as indicated in the following simplified formula:

$$6CO_2 + 6H_2O \rightarrow C_6H_{12}O_6 + 6O_2 + 673kcal$$

the carbohydrate in this case being glucose. The radiant energy reacts initially with the green pigment chlorophyll, with the result that one electron of this complex molecule is raised above its normal energy level for some 10^{-7} to 10^{-8} s, so triggering other chemical changes (Hutchinson 1970). The 673 kcal noted above represents a fairly large store of energy, which is then used for plant growth and respiration. Photosynthesis is solely a daytime phenomenon. It is thus the basis of the flow of energy through the TROPHIC LEVELS of ECOSYSTEMS and indeed the fundamental process that enables life itself to occur on this planet, with the exception of a few species (mainly bacteria) which obtain their energy from chemicals. DW

Reading and Reference
Gregory, R.P.F. 1989: *Photosynthesis*. Glasgow: Blackie.
Hutchinson, G.E. 1970: The biosphere. *Scientific American* 223, pp. 44–53.

phreatic divide An underground watershed. The WATER TABLE or upper surface of the permanently saturated zone usually has an undulating topography that depends upon variations in hydraulic CONDUCTIVITY in the bedrock and on the location of zones of RECHARGE and discharge. Water flows down the hydraulic gradient (see DARCY'S LAW) away from high points on the water table, which are therefore phreatic divides, i.e. the watersheds of GROUNDWATER basins. But the pattern of groundwater divides need not mirror the surface pattern of topographic watersheds. Unlike the watersheds of surface catchment areas which are permanent, phreatic divides can shift as the water table

rises, lowers or changes its configuration because of localized recharge. (See also GROUNDWATER.) PWW

phreatophytes Plants that have developed root systems with the capability to penetrate great depths in order to draw water directly from groundwater reservoirs. The name is derived from Greek meaning 'well plant'. In arid and semi-arid regions they make up almost all the plants in riparian habitats, where they affect the geomorphological processes of the nearby stream by causing increased hydraulic roughness and concomitant sedimentation. Phreatophytes have important economic impacts on the hydrological cycle because they transpire moisture from the groundwater table which might otherwise be used for pumped irrigation water (Horton and Campbell 1974). WLG

Reference
Horton, J.S. and Campbell, C.J. 1974: *Management of phreatophyte and riparian vegetation for maximum multiple use values*. US Department of Agriculture Forest Service research paper RM-117.

phylogenesis The origin of a taxonomic group by evolution.

physical meteorology That part of meteorology or the atmospheric sciences which deals with the physical properties of the atmosphere and the processes occurring therein. Usually included are atmospheric chemistry, electricity, radiation, thermodynamics, optics and acoustics, cloud and precipitation physics, AEROSOL physics, and physical climatology. Because all of these fields of study interact in one way or another with atmospheric motions, it is somewhat artificial to distinguish between physical meteorology and DYNAMICAL METEOROLOGY. This is especially true in the present day when so much research activity in the atmospheric sciences deals with the numerical modelling of physical processes and their interaction with the atmospheric circulation.

Of the various sub-branches of physical meteorology atmospheric thermodynamics is perhaps the most basic and the most closely related to dynamical meteorology. Usually included under this heading are the gas laws (see EQUATION OF STATE), the HYDROSTATIC EQUATION, the first and second laws of thermodynamics, latent heats and water vapour in the air, ADIA-

BATIC processes, static stability (see VERTI-CAL STABILITY/INSTABILITY), and entropy. Cloud and precipitation physics (see CLOUD MICROPHYSICS) includes not only studies of the formation and growth of cloud droplets, raindrops and ice crystals but also studies involving the processes whereby CLOUDS may be modified to increase or decrease PRECIPITATION, to suppress HAIL, LIGHTNING, and HURRI-CANE winds, and to dissipate FOG.

The sub-branches of physical meteorology are not autonomous. For example, a study of the effect of AEROSOLS on the ALBEDO of clouds would involve aerosol physics, radiative transfer, cloud physics, and possibly atmospheric chemistry. Modern research in physical meteorology may also use REMOTE SENSING, either from SATELLITES or from the ground using radiometers, radar and lidar.   WDS

Reading
Wallace, J.M. and Hobbs, P.V. 1977: *Atmospheric science*. New York: Academic Press.

physiography A word that has obscure origins, although in common currency in eighteenth-century Scandinavia, and in regular usage in the English-speaking world in the nineteenth-century (Stoddart 1975). It was defined by Dana in 1863:

Physiography, which begins where geology ends – that is, with the adult or finished earth – and treats (1) of the earth's final surface arrangements (as to its features, climates, magnetism, life, etc.), and (2) its systems of physical movements and changes (as atmospheric and oceanic currents, and other secular variations in heat, moisture, magnetism, etc.).

One of the most notable exponents of physiography was T.H. Huxley, who published a highly successful text, *Physiography*, in 1877. In the USA W.M. Davis preferred the term to GEOMORPHOLOGY, but he used it without the catholicity of meaning that it had for Huxley.   ASG

References
Dana, J.D. 1863: *Manual of geology: treating on the principles of the science*. Philadelphia: Bliss.
Huxley, T.H. 1877: *Physiography: an introduction to the study of nature*. London: Macmillan.
Stoddart, D.R. 1975: 'That Victorian science': Huxley's *Physiography* and its impact on geography. *Transactions of the Institute of British Geographers*, 66, pp. 17–40.

phytogeography The scientific study of the distribution of plants on the earth.

Although traceable as a theme to the 'father of botany', Theophrastus (*c.*370–287 BC), the first comprehensive studies in plant geography were the *Essai sur la géographie des plantes* of Humboldt and Bonpland (1805, German edition 1807) and the *Géographie botanique raisonée* of Alphonse de Candolle (1855). Used narrowly, the term tends to be confined to the study of geographical or spatial distribution of plant species, genera and families over the surface of the earth; more generally, it is often used, particularly by geographers, to include many aspects of plant ecology and plant biology as well.

PAS

Reading
Moore, D.M. ed. 1982: *Green planet: the story of plant life on earth*. Cambridge: Cambridge University Press.
Stott, P.A. 1981: *Historical plant geography*. London: Allen & Unwin.

phytogeomorphology A concept that 'reflects those sensitive landform–vegetation relationships that are visibly dominant on the landscape' (Howard and Mitchell 1985, p. 5). A major component of its study is the use of REMOTE SENSING as most satellite images of the landsurface reflect terrain and vegetation types which have become the basis for the interpretation of many less clearly visible phenomena, including soils.   ASG

Reference
Howard, J.A. and Mitchell, C.W. 1985: *Phytogeomorphology*. New York: Wiley–Interscience.

phytokarst Features produced by the weathering and erosive action of plants and animals on limestone rocks. It is also called phytokarren or biokarst. The identification of erosive (boring and digestive) action or chemical weathering (chelation) on rocks is very problematic. While phytokarst can form spectacular features (Bull and Laverty 1982), it usually produces random spongework forms (Folk *et al.* 1973). The term BIOKARST is perferred to phytokarst (Viles 1984).   PAB

References
Bull, P.A. and Laverty, M. 1982: Observations on phytokarst. *Zeitschrift für Geomorphologie* 26, pp. 437–57.
Folk, R.L., Roberts, H.H. and Moore, C.H. 1973: Black phytokarst from Hell, Cayman Islands. *Bulletin of the Geological Society of America* 84, pp. 2351–60.
Viles, H. 1984: A review of biokarst. *Progress in physical geography* 8, pp. 523–43.

phytosociology In its widest sense, the study of plants as social or gregarious organisms and thus the study of plant communities; more specifically, the floristic description, classification and naming of community types and the study of their distribution and chief ecological characteristics. The most fully developed approach to phytosociology is that of the Zürich–Montpellier school developed by Braun–Blanquet around 1913, now much refined and widely applied throughout the world. The basic unit of classification in this approach is the ASSOCIATION, which is a recurrent grouping of plant species recorded by means of sampling units termed *relevés*. In recent years, the application of quantitative methods and of computers has revolutionized the study of phytosociology. PAS

Reading
Chapman, S.B. ed. 1976: *Methods in plant ecology*. Oxford: Blackwell Scientific.
Harrison, C.M. 1971: Recent approaches to the description and analysis of vegetation. *Transactions of the Institute of British Geographers* 52, pp. 113–27.
Randall, R.E. 1978: *Theories and techniques in vegetation analysis*. Oxford: Oxford University Press.
Shimwell, D.W. 1971: *The description and classification of vegetation*. London: Sidgwick & Jackson.

piedmont The area of foothills at the edge of a range of mountains.

piedmont glacier A glacier which spreads out into a piedmont lobe as it debouches onto a lowland.

piezometer An instrument for measuring pressure head at a point within a saturated porous medium. A piezometer consists of a tube of greater than capillary cross-section, placed in the soil or rock so that water may only enter the tube at a fixed level (usually at the base of the tube through a porous pot). Water enters the tube and the water level rises until the head of water in the tube balances the water pressure at the entry point. The depth of water in the piezometer is known as the pressure or piezometric head. AMG

Reading
Curtis, L.F. and Trudgill, S. 1974: The measurement of soil moisture. *British Geomorphological Research Group technical bulletin* 13. Norwich: Geo Abstracts.

piezometric surface An imaginary surface to which water levels rise in wells tapping confined aquifers. Confined aquifers (rock formations which contain and can yield significant quantities of water) occur where the aquifer contains GROUNDWATER which is confined at a pressure greater than air pressure by overlying, relatively impermeable rock formations. A well which penetrates a confined aquifer will experience water levels higher than the junction between the aquifer and the overlying rock formations and the piezometric surface is the surface which passes through these well water levels. If the piezometric surface rises above the ground surface a well located at that point will produce flowing water at the surface and is described as an artesian well. AMG

Reading
Walton, W.C. 1970: *Groundwater resource evaluation*. New York: McGraw-Hill Kogakusha International.

pingo An ice-cored hill which is typically conical in shape and can only grow and persist in permafrost. The term is of Inuit origin but has been widely adopted elsewhere. In the former USSR the equivalent term is 'bulgannyakh', of Yakut origin.

Pingos form through the freezing of water which moves under a pressure gradient to the site of the pingo. If water moves from a distant elevated source, the pingo is hydraulic (i.e. open system) in nature. If water moves under pressure arising from local permafrost aggradation and associated pore water expulsion, the pingo is hydrostatic in nature (i.e. closed system). The greatest concentration, about 1450, and some of the largest in the world, occur in the Mackenzie Delta region of Canada where they commonly form in drained lake basins. HMF

Reading
Mackay, J.R. 1979: Pingos of the Tuktoyaktuk Peninsula area, North-west Territories. *Géographie physique et quaternaire* 23, pp. 3–61.

pipes Subsurface channels up to several metres in diameter caused by deflocculation of clay particles in fine-grained, highly permeable soils. Pipes are commonly found in arid or semi-arid regions, less commonly elsewhere. They are usually formed where the soils contain significant amounts of swelling clays such as montmorillonite, illite or bentonite, and where there is a low water table with a steep hydraulic gradient in the near-surface environment. Locally, pipes are usually found in steep slopes and on gully and arroyo sides (Heede 1971). The pipes carry sediment as well as water, and if erosion continues for a long enough period the conduit may enlarge so much that the roof collapses, forming a gully. Pipes are economically important because they are a sign of deteriorating soil conditions and represent accelerated erosion.   WLG

Reading and Reference
Heede, B. 1971: *Characteristics and processes of soil piping in gullies.* US Department of Agriculture Forest Service research paper RM-68.

†Jones, J.A.A. 1981: *The nature of soil piping: a review of research.* British Geomorphological Research Group monograph 3. Norwich: Geo Books.

pipkrakes See NEEDLE ICE.

pisoliths Spherical rock particles of around 5–6 mm in diameter which are formed by the gradual accretion of material around a nucleus. Laterites and calcretes often display pisolithic textures.   ASG

pitometer (or Pitot tube) A means of estimating flow velocity by measuring the pressure of the fluid as it hits an immersed rounded body. It was invented by Henri Pitot in 1732.   AMG

planation surface See EROSION SURFACE.

plane bed Applied to the deformable bed of a fluid flow on which there are no organized BEDFORMS and where the relief is that of individual grains. Such quasi-flat beds can occur when sediment is hardly moving under close-to-threshold conditions of fluid shear (lower stage), and under high shear and high transport rate conditions (upper stage). The latter may be a transitional stage which develops when the Froude number approaches unity and dune forms are destroyed before upper regime bedforms develop.   JL

Reading
Allen, J.R.L. 1970: *Physical processes of sedimentation.* London: Allen & Unwin.

planèze A triangular wedge of resistant lava, tapering towards a point near the cone, which protects the underlying materials from erosion.

planimeter An instrument for the measurement of areas on maps and plans, which is less time consuming and more accurate

than counting squares or estimation but which is less efficient than digitizing methods associated with a mainframe or microcomputer.   KJG

plankton Small freshwater and marine organisms, a substantial number of which are microscopic. While some plankton possess limited mobility, many are inactive. The movement of plankton mainly depends upon the motion of the water in which they are suspended.

There are three major planktonic categories: phytoplankton, zooplankton and bacterioplankton. Phytoplankton (algae) account for the bulk of primary production in aquatic ecosystems, and their BIOLOGICAL PRODUCTIVITY is conventionally high (Barnes 1980). Zooplankton, which include mature and/or larval representatives of numerous important animal groups (e.g. Protozoa, Crustacea and Mollusca), may be herbivores, carnivores or omnivores, either filtering or seizing living planktonic or detrital organic matter (Parsons *et al.* 1977; Parsons, 1980). Bacterioplankton (for instance, *Bacillus* and *Nitrosomonas*) are mainly decomposers. Some are able to perform photosynthesis and chemosynthesis, thereby contributing to primary production (Fogg, 1980).

The distribution and productivity of plankton vary in both space and time. This is due to a combination of environmental factors, among which nutrient availability and climate are of considerable importance.   RLJ

Reading and References
The three following references appear in R.S.K. Barnes and K.H. Mann eds 1980: *Fundamentals of aquatic systems.* Oxford: Blackwell Scientific.

Barnes, R.S.K. 1980: The unity and diversity of aquatic systems. Pp. 5–23.

Fogg, G.E. 1980: Phytoplanktonic primary production. Pp. 24–45.

Parsons, T.R. 1980: Zooplanktonic production. Pp. 26–66.

†Moss, B. 1980: *Ecology of fresh waters.* Oxford: Blackwell Scientific.

Parsons, T.R., Takahashi, M. and Hargrave, B. 1977: *Biological oceanographic processes.* 2nd edn. Oxford and New York: Pergamon Press.

plastic limit The water content of an unconsolidated material when it is at the point of transition from a plastic solid to a rigid mass.

plasticity The behaviour under stress of weak materials such as moist clays and weak rocks. Such weak materials do not deform

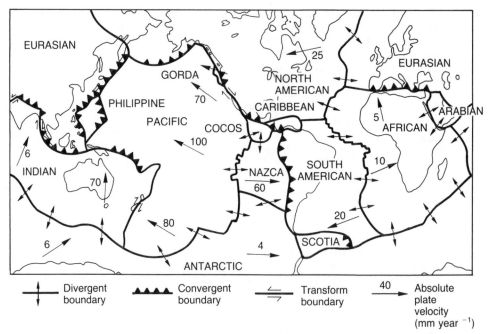

Plate tectonics: *Map of the major lithospheric plates. The various types of plate boundary are shown and the estimated current rates (mm year⁻¹) and directions of plate movements are indicated.*
Source: *M.A. Summerfield 1991: Global geomorphology. London and New York: Longman Scientific and Technical and Wiley.*

under very low magnitudes of stress, but above a critical magnitude, called a yield stress, they deform at a continuous rate if the level of stress is constant: materials exhibiting such behaviours are said to be plastic substances.   MJS

Reading
Selby, M.J. 1993: *Hillslope materials and processe.* 2nd edn. Oxford: Oxford University Press. Ch. 4.

plate tectonics A theory of global TEC-TONICS that holds that the LITHOSPHERE forming the earth's surface is divided into eight major and several minor internally rigid plates which are in motion with respect to each other and the underlying ASTHENO-SPHERE. CONTINENTAL DRIFT is a consequence of plate motion, and earthquakes, volcanoes and mountain building are concentrated in the vicinity of, although are not entirely confined to, plate boundaries.

There are three main types of plate boundary. Divergent boundaries, which are mostly located along the extensive ridge system of the ocean basins, represent the sites of sea-floor spreading where new lithosphere is created and two plates move away from each other. Convergent plate

boundaries occur when two plates move towards each other. This leads to the subduction (reabsorption into the sublitho-spheric mantle) of one of the plates as it plunges down under the leading edge of the other. Sites of subduction are marked by deep ocean trenches and are associated with intense seismicity and volcanic activity. Plate subduction involving only oceanic lithosphere leads to ISLAND ARC formation, but where subduction occurs along the margin of a continent a mountain belt develops (such as the Andes). If two converging plates are capped by continental crust the two continental masses will eventually collide and subduction will be halted. A complex and extensive zone of crustal deformation results, as exemplified by the Himalayas and the Tibetan Plateau which have been created as a result of the collision of India and Eurasia. The third major category of plate interaction occurs along a transform boundary where two plates slip horizontally past each other, such as along the San Andreas Fault System in California. This type of boundary is characterized by numerous earthquakes but low levels of volcanic activity.

Although convection currents in the mantle are clearly involved in plate motion, it does not seem that they are the main driving force. The most important mechanism is probably the pull exerted on the rest of a plate by those parts being actively subducted. Although the plate tectonics model has revolutionized our understanding of the oceans, research over the past decade has emphasized that it provides only a generalized understanding of the morphology and structure of the continents.

MAS

Reading
Cox, A. and Hart, R.B. 1986: *Plate tectonics: how it works*. Palo Alto: Blackwell Scientific.

†Kearey, P. and Vine, F.J. 1990: *Global tectonics*. Oxford: Blackwell Scientific.

Molnar, P. 1988: Continental tectonics in the aftermath of plate tectonics. *Nature* 335, pp. 131–7.

†Summerfield, M.A. 1991: *Global geomorphology*. London and New York: Longman Scientific and Technical and Wiley.

plateau An extensive area of relatively flat land in an area of high relief.

plateau basalt An extensive flow or flows of basalt rock which, owing to erosion of the surrounding less-resistant rocks, forms an upstanding plateau (e.g. the Deccan of India).

playa A closed depression in an arid or semi-arid region that is periodically inundated by surface run-off, or the salt flat within such a closed basin. It is derived from the Spanish word for a 'beach' and is thus probably incorrectly used by English-speaking geomorphologists. In other parts of the world playas are given names such as *chott* and *kavir*.

Playfair's law In his book of 1802 J. Playfair suggested that every river will flow in a valley proportional to the size of the river and that where rivers join their levels will be accordant. This law of accordant tributary junctions came to be known as Playfair's law.

KJG

Reading
Kennedy, B. 1984: On Playfair's law of accordant tributary junctions. *Earth surface processes and landforms* 9, pp. 153–73.

Playfair, J. 1802; *Illustrations of the Huttonian theory of the earth*. Edinburgh: William Creech.

Pleistocene The Pleistocene is the first epoch of the Quaternary, preceded by the Pliocene and succeeded by the Holocene. The Pleistocene was composed of alternations of great cold (glacials, stadials) with stages of relatively greater warmth (interglacials, interstadials), during which worldwide sea levels fluctuated in response to the formation and melting of ice sheets. In glaciated regions, these eustatic changes were accompanied by isostatic depression under the weight of ice cover during glaciations and recovery during the interglacial phases.

The classic interpretation of the history of the Pleistocene, especially in the northern hemisphere, has been based on the study of the extent and character of these alternations of glacial and interglacial deposits on land. However, there is a marked degree of controversy over the number of glaciations, stadials, interglacials and interstadials. This is due to the problem of definition of these events, and also a lack of agreement with regard to correlations of events between different areas.

Until recent times, Pleistocene events were recorded in a chronology based on the location of evidence in geological or archaeological strata. Evidence was dated by correlation with known successions based on the typology, stratigraphy or prehistoric cultures. Increase in the use of new dating techniques and in deep sea core evidence have transformed Pleistocene studies. The traditional view from terrestrial studies indicated that four, five or possibly six glacials existed during the Pleistocene. Indications from ocean cores are that there have been no less than seventeen glacial cycles in the past 1.6 million years.

However, stratigraphic terminology is still to a great extent understandardized, and often no clear distinction is made as to whether the classification system is: lithostratigraphic (based on rock or sediment classification), biostratigraphic (based on the occurrence of fossil fauna and/or flora), chronostratigraphic or a combination of these. A major point of contention in attempts to construct a framework for dating subdivisions of the Pleistocene is the location of the Pleistocene/Pliocene boundary and, as a corollary, the duration of the Pleistocene itself. Some authors suggest a short time-scale (600,000 years); some favour a medium time-scale, e.g. on faunal grounds the Pleistocene/Pliocene boundary has been placed at about 1.6 million years and thus conveniently

During humid pluvial times in the Pleistocene, a high lake occupied the present Salar of Uyuni in Bolivia, and deposited tufa on the basin sides. Such tufas can sometimes be dated isotopically. They are largely composed of calcium carbonate and result from organic and inorganic processes of precipitation.

coincides with a major geomagnetic reversal (the top of the Olduvai event). Others accept the long time-scale (up to 3 million years) on the basis of some major climatic deterioration, namely the marked appearance of mid-latitude, as opposed to polar, glaciers. The variations appear in the main to be due to differing forms of evidence, their interpretation and geographical location, especially marine versus terrestrial evidence. The boundary between the Pleistocene and the Holocene (oxygen stage 1) is arbitrary but is generally regarded as having occurred near 10,000 BP.   AP

Reading
Goudie, A.S. 1992: *Environmental change.* 3rd edn. Oxford: Clarendon Press.

plinian eruption An explosive volcanic eruption which is frequently so violent that the volcanic cone is destroyed.

plinthite A hardpan or soil crust, especially one which forms a capping on unconsolidated sediments and retards their erosion. It is essentially synonymous with *laterite*.

plots, erosion/run-off Plots to determine erosion and run-off are bounded plots of land of known area, slope steepness,

slope length and soil type from which both run-off and soil loss are monitored (Morgan 1979). The plots are edged with a material such as sheet metal which prevents lateral movement of water and sediment and thus ensures that run-off and sediment can only be transported from the plot to collecting vessels at the downslope end. Erosion plots may vary in size but dimensions of 22 m long by 1.8 m wide are frequently used. Analysis of erosion plot records in the USA led to the development of the universal soil loss equation which provides an estimate of average annual soil loss for small areas from indices of rainfall erosion potential, soil erodibility, slope length and steepness, crop practice and conservation practice factors (Wischmeier and Smith 1962).   AMG

Reading and References
Morgan, R.P.C. 1979: *Soil erosion.* London: Longman.
†Task Committee on Preparation of Sedimentation Manual 1970: Ch. IV Sediment sources and sediment yield. *American Society of Civil Engineers' journal of the Hydraulics Division* 96, pp. 1283–329.
Wischmeier, W.H. and Smith, D.D. 1962: Soil loss estimation as a tool in soil and water management planning. *International Association of Scientific Hydrology publication* 59, pp. 148–59.

ploughing block A type of frost creep and/or gelifluction deposit consisting of isolated, commonly boulder-size stones which leave a linear depression upslope and form a low ridge downslope. The depressions and ridges are formed as the stones move downslope under gravity. They are described most frequently from alpine and hilly regions in mid-latitudes which experience seasonal frost. Measurements in the Sudeten Mountains, Poland, indicate that, on slopes between 16 and 34° in angle, blocks move 2.5–9.0 times faster than the soil on which they lie. In northern England maximum movement is 8 cm per year, most movement occurring in winter.   HMF

Reading
Tufnell, L. 1972: Ploughing blocks with special reference to north-west England. *Biuletyn Peryglacjalny* 21, pp. 237–70.

Washburn, A.L. 1979: *Geocryology, a survey of periglacial processes and environments*. New York: Wiley.

plucking A process of glacial erosion describing the removal of discrete blocks of bedrock. It is commonly contrasted with the other main form of glacial erosion, ABRASION, which describes the process of rock wear. Plucking results from failure of the rock along joint planes and reflects two processes. The first is wedging by the pressure of over-riding rock particles. The second is the freezing of blocks to over-riding glacier ice in response to temperature fluctuations at the ice–rock interface as a result of pressure variations.   DES

Reading
Addison, K. 1981: The contribution of discontinuous rock-mass failure to glacier erosion. *Annals of glaciology* 2, pp. 3–10.

Chorley, R.J., Schumm, S.A. and Sugden, D.E. 1984: *Geomorphology*. London: Methuen.

Röthlisberger, H. and Iken, A. 1981: Plucking as an effect of water pressure variations at the glacier bed. *Annals of glaciology* 2, pp. 57–62.

pluton A mass of rock which has solidified underground from intrusions of magma. Plutons have variable shapes, sizes and relationships with the country rock (the invaded rock) surrounding them. Batholiths, dykes, laccoliths, lopoliths, sills and stocks are the main forms.

plutonic Refers to rock material that has formed at depth (e.g. igneous rocks such as granite) where cooling and crystallization have occurred slowly.

plutonism A term, at first used in derision, to describe the ideas of James Hutton in the late eighteenth century. In 1785 Hutton discovered that in Glen Tilt, Perthshire, granite veins were breaking and

displacing local rocks and he postulated that granites were formed by the solidification of molten material intruded into the crust from the earth's hot interior. He invoked a similar igneous origin for basalt, and claimed that all sills, dykes and mineral veins had been filled with molten material rising from deep inside the earth. In contrast to the Neptunists he regarded many igneous rocks as younger than the surrounding strata. He relied heavily upon the power of subterranean heat to form rocks and to raise continents.   ASG

Reading
Davies, G.L. 1969: *The earth in decay*. London: Macdonald.

pluvial Time of greater moisture availability, caused by increased precipitation and/or reduced evapotranspiration levels. Pluvials caused many lake levels in the arid and seasonally humid tropics to be high at various times in the Pleistocene and early Holocene (hence pluvials may also be called lacustrals), helped to recharge groundwater, and caused river systems to be integrated. Pluvials used to be equated in a simple temporal manner with glacials, but this point of view is no longer acceptable.   ASG

Reading
Goudie, A.S. 1992: *Environmental change*. 3rd edn. Oxford: Clarendon Press. Ch. 3.

Street, F.A. 1981: Tropical palaeoenvironments. *Progress in physical geography* 5, pp. 157–85.

pluviometric coefficient The ratio between the mean rainfall total of a particular month and the hypothetical amount equivalent to each month's rainfall were the total to be equally distributed throughout the year.

pneumatolysis See HYDROTHERMAL ALTERATION.

podzol A soil which is characterized by an upper horizon from which aluminium and iron oxides and hydroxides have been leached and a lower horizon where these have accumulated illuvially.

poikilothermy The state of being cold blooded, i.e. possessing a body temperature that is regulated by the environment and not by bodily processes.

point bar deposits Sediments laid down on the inside of a meander bend or 'point', largely by LATERAL ACCRETION. Individual attached BARS commonly form low arcuate ridges or scrolls. Units of accretion are added as the meander loop develops, eventually making up a complex of ridges separated by depressions or swales. The layout of the individual scroll bars may reveal the growth pattern of the point bar as a whole. In some environments, however, this pattern is either irregular or not even apparent at all on the inside of meander bends. The extent of point bar development depends on the amount of blanketing OVERBANK DEPOSIT and on the variable nature of the point bar accretion in conditions with contrasting sediments and river regimes.　　　　　　　　　　JL

Reading
Reading, H.G. ed. 1986: *Sedimentary environments and facies* 2nd edn. Oxford: Blackwell Scientific.

polar front The front separating air of polar origin from that originating within the subtropics. In winter it can often be traced as a band of cloud over thousands of kilometres between 40 and 50° latitude especially over the oceans. Extra-tropical cyclones may be initiated along the strong thermal gradient of the front. Bjerknes (1921) based his theory of frontal evolution upon the presence of this front. In summer the front is more variable in its location and the temperature gradient is weaker.　　PS

Reference
Bjerknes, V. 1921: On the dynamics of the circular vortex with applications to the atmosphere and atmospheric vortex and wave motions. *Geofysiske Publikationer* 2.4

polar meteorology See ANTARCTIC METEOROLOGY; ARCTIC METEOROLOGY.

polder A low-lying area of land that has been reclaimed from the sea or a lake by artificial means and is kept free of water by pumping.

polje An extensive depression feature in karst, closed on all sides, mostly with an even floor, a steep border in places and a clear angle between the polje bottom and the slope. It has underground drainage and can be dry all year round, have ephemeral streams within it, or be inundated continually. The very diverse nature of poljes prevents a definition based on genesis; they are truly polygenetic features.　　PAB

Reading
Gams, I. 1977: Towards a terminology of the polje. *Proceedings of the Seventh International Congress of Speleology, Sheffield.* Pp. 201–2.

pollen analysis (or palynology) Is concerned with seed-bearing plants (Phanerograms) which produce pollen in the male gamete (anther) or with spores which are asexual reproductive cells of cryptogram plants. Primarily, it is the study of fossil pollen grain and spore assemblages, which have been isolated from their sedimented deposit in the recent past or as far back as the Palaeozoic era. Pollen analysis is the most widely adopted and perhaps the most successful technique used in the reconstruction of palaeoenvironments, especially the Quaternary. The identification of pollen and spores enables a picture of the past vegetational history to be made by adding the time dimension. Pollen and spores may normally be preserved in three major site types: lakes, peat and soils. The reconstruction of former vegetation by means of pollen analysis provides a picture of past landscape and environment. In many cases, changes in the vegetation of an area through pollen analysis may lead to strong inferences about the former climate of an area. However, not all changes in vegetation are due to change in climate. For example, fire, insect infestation, plant successional changes, anthropogenic interference, and factors leading to the accumulation and preservation of the material, often make the interpretation of the pollen and spore record complex.

Pollen grains range in size from 5 to 200 μm. Each grain consists of three concentric layers: living cell, intine and exine. As the name implies, the living cell, making up the centre of the grain, is the portion that germinates, effecting fertilization of the female part of the flower. The intine, the layer immediately surrounding the living cell, is composed of cellulose and other elements including protein. Neither the living cell nor intine is preserved in fossil pollen. Pollen is protected by a chemically resistant outer layer, the exine, which is made from sporopollenin, an inert and resistant natural organic compound. Thus, pollen and spore grains may be found in deposits in which other types of fossils have been diagenetically destroyed (usually under anaerobic conditions), although this casing can be destroyed by oxidation processes.

It is the morphology and chemical properties of exine that lie at the foundation of pollen analysis. Pollen and spore grains of many plant families are different morphologically and can be recognized by their distinct shape, size, sculpture and apertures. In some genera (e.g. *Plantago*) it is possible to identify to species level; in other cases identification can only be made to generic (e.g. *Ulmus*) or family (e.g. *Chenopodiacea*). The amount of pollen produced varies from species to species, being dependent on the pollination mechanism.

Since sporopollenin is chemically resistant, chemical methods can be used to separate pollen and spore grains from other components of sediment, after which the fossil grains are mounted onto microscope slides. Examination is undertaken using high-power microscopy ($\times 400$–1000). Identification is made by comparison with prepared reference slide material of living taxa whose specific identity is known with certainty. With pre-Quaternary pollen and spore assemblages, where relationships of fossils to living plants are uncertain, names founded on morphology are needed for fossil taxa.

The study of fossil pollen and spores is being used to elucidate problems of past plant distributions and palaeoenvironments within many scientific fields. Several areas have been examined and documented within geomorphology through the use of this technique with perhaps the major application in the reconstruction of interglacial and interstadial environments. As a result, better understanding of individual interglacial and interstadial environments has been achieved.   AP

Reading
Faegri, K. and Iversen, J. 1990: *Textbook of pollen analysis*. 4th edn revised by K. Faegri, P.E. Kalan and K. Krzywinski. Chichester: Wiley.
Moore, P.D., Webb, J.A. and Collinson, M.E. 1991: *Pollen analysis*. 2nd edn. Oxford: Blackwell.

pollution A condition which ensues when environmental attributes become inimical to the normal existence of living organisms. A contaminant is a substance foreign to an environment and capable of pollution within it. A contaminant has a source from which it is dispersed, usually by means of an atmospheric or aquatic pathway. During this process it may be rendered harmless by transformation or dilution. If this does not occur, the contaminant becomes a pollutant which has a target (Holdgate 1979; Newson 1992). As Mellanby (1972) states, while there are numerous instances of natural

pollution (volcanic emission which becomes toxic and inhibits the development of vegetation in the vicinity, for example), that resulting from human activity is the more significant.

Mature organisms are better able to cope with harmful effects than are young ones. However, as Bailey *et al.* (1978) point out, a substance may only need to reach a concentration of 1 ppm to become a pollutant, the presence of which could ultimately lead to the death of an organism. As pollutants are earth materials, they comprise part of a finite quantity. Thus their components may be changed from one state or position to another but not obliterated (Jørgensen and Johnsen 1981). When change is possible, dilution in air or water (of pesticides, heavy metals and toxic gases, for example), or degradation on land (of garbage and sewage, for instance) is normally involved. Some pollutants, though (certain nuclear wastes and lethal chemicals, for example), are so hazardous and/or of low degradability that they must be sealed and interred rather than released to the environment.

Pollution can occur at a variety of scales and in numerous circumstances. Atmospheric pollutants, for example, may give rise to serious local conditions, and can also be circulated widely. Smog is a localized atmospheric condition formed by the combination of pollutants (such as carbon monoxide and sulphur compounds) and fog. More widespread effects on the atmosphere can be brought about by CO_2, together with sulphur and nitrogen oxides (see ACID PRECIPITATION; GREENHOUSE EFFECT). In a similar vein, an increase in particulate concentration (by dusts, for example) lessens atmospheric transparency and affects the reflectivity of solar radiation.

Aquatic pollution can result from the addition of harmful substances such as acids or hydrocarbons. However, the gradual build-up of essential elements in freshwater subsequent to their application as terrestrial agricultural fertilizers (EUTROPHICATION), may also pollute. In the terrestrial environment, the major pollutant by volume is urban–industrial refuse which, if treated, is either stored, or reduced – usually by BIODEGRADATION and burning.

Pollution is a significant and developing environmental problem. None the less, specific instances involving potentially harmful substances and circumstances often lead to disagreement. As Barbour (1983) notes, fact is frequently obscured by conjecture, while there are sometimes political and socioeconomic undertones to particular cases. Some such factors may be exemplified with reference to noise pollution, much of which (up to the 120 decibel limit when human pain is felt) is a rather subjective experience (see AESTHETIC DEGRADATION).

Pollution control strategies vary. Control at source has been preferred in the United States and on mainland Europe, while in the United Kingdom it has been customary to manage the pathways of pollutants (Newson 1992). The law of England and Wales sets out few exact criteria pertaining to pollution control across the entire territory. Examples of legislation are the Control of Pollution Act 1974 which governs terrestrial waste disposal, and the Clean Air Acts (1956 and 1968) relating to air pollution. Enforcement of the controls has been largely entrusted to bodies (County Councils and Water Authorities for example) with local and regional responsibility for the environment (Macrory 1990).   RLJ

Reading and References
Bailey, R.A., Clark, H.M., Ferris, J.P., Krause, S. and Strong, R.L. 1978: *Chemistry of the environment*. New York and London: Academic Press.
Barbour, A.K. 1983: The control of industrial pollution. In R.M. Harrison ed., *Pollution: causes, effects and control*. London: Royal Society of Chemistry. Pp. 1–18.
†Holdgate, M.W. 1979: *A perspective of environmental pollution*. Cambridge: Cambridge University Press.
Jørgensen, S.E. and Johnsen, I. 1981: *Principles of environmental science and technology*. Amsterdam and New York: Elsevier.
Macrory, R. 1990: The legal control of pollution. In R.M. Harrison ed., *Pollution: causes, effects and control*. 2nd edn. Cambridge: Royal Society of Chemistry. Pp. 277–96.
Mellanby, K. 1972: *The biology of pollution*. London: Edward Arnold.
†Newson, M. 1992: The geography of pollution. In M. Newson ed., *Managing the human impact on the natural environment: patterns and processes*. London and New York: Belhaven Press. Pp. 14–36.

polyclimax A theory of vegetation that allows the coexistence of a number of stable plant communities in an area. According to polyclimax theory, all the SERES (community sequences) in an area do not converge to identify in a MONOCLIMAX, but SUCCESSION produces a partial convergence to a mosaic of different stable communities in different HABITATS. In polyclimax theory, all climax types are of equal rank rather than subordinate to the climatic climax as is required by monoclimax theory. (See also CLIMAX VEGETATION.)   JAM

polygonal karst A limestone landscape entirely pitted with depressions which are smooth rimmed and soil covered, producing a crude polygonal network when viewed from the air. The term was introduced by Williams (1971) when reporting the features from New Guinea.
                                                            PAB

Reference
Williams, P.W. 1971: Illustrating morphometric analysis of karst with examples from New Guinea. *Zeitschrift für Geomorphologie* 15, pp. 40–61.

polymorphism Existing in more than one physical form. Applied to minerals which have the same chemical composition but different physical characteristics, to animals that undergo major physical alterations during individual life cycles and to animal and plant species which, while interbreeding, have markedly different physical forms.

polynya A pool of open water within pack ice or an ice floe.

polypedon A collection of small columns which run through the soil zone.

polytopy Loosely used, a term to describe the occurrence of any organism in two or more completely separate geographical areas; more specifically, a term referring to the process by which an organism may evolve two or more times, quite independently, in differing geographical localities. If the polytopic populations have developed at different times they are also termed polychronic in origin. The term is primarily employed by plant geographers and, with our present understanding of genetics and evolution, such an explanation for a disjunct distribution would only be accepted in very exceptional circumstances. A classic example of polytopy is afforded by the separate, but closely related, sand dune ecotypes of *Hieracium umbellatum* found along the coasts of Sweden. (See also DISJUNCT DISTRIBUTION.)   PAS

Reading
Stott, P. 1981: *Historical plant geography*. London: Allen & Unwin.

ponor A Serbo-Croatian term for a SWALLOW HOLE, often of rather deep type, developed by solutional processes acting upon carbonate rocks. It is a type of KARST feature.

pool and riffle The pool and riffle sequence is a large-scale bedform characteristic of streams with gravelly, heterogeneous bed material. Pools are closed hollows scoured in the bed and commonly floored by relatively fine gravel and sand, while riffles are topographical highs representing accumulations of coarser pebbles and cobbles. These features are created by the pattern of scour and deposition at bankfull discharge, when bed velocity is higher in the pool than in the adjacent riffle, and the coarse sediment then in motion is removed from the pools and deposited on the riffles. At low discharges the flow adjusts to the bed topography, and the water surface slope is flat over the deep, sluggish pools and steep over riffles where flow is shallow and rapid. Fine sediment mobile at this flow stage is removed from the riffles and deposited in pools. The pool–riffle sequence repeats with a mean wavelength of five to seven times the mean channel width, suggesting initial control by a large-scale turbulent eddy scaled to the channel size. Once formed, riffles and pools are fairly stable morphological features although individual sediment grains move through the sequence from riffle to riffle. In regular meander patterns, riffles tend to occur at inflection points and pools at bends, and the meander wavelength is twice the pool–riffle wavelength. This suggests that the pool–riffle feature is a fundamental bedform common to 'straight' and meandering rivers.   KSR

population dynamics The study of changes in population size. This involves a consideration of those factors which might give rise to population growth, and those which might lead to its decline. Growth may be achieved by an increase in the rates of natality over those of mortality, and/or by immigration; and decline normally results from an excess of mortality over natality, and/or emigration. The structure of a population may also influence its dynamics: for example, a population with a relatively high percentage of females of reproductive age is clearly in a favourable position for growth, whereas one without such a representation may not be.

The earliest theoretical studies of population dynamics date back to Malthus (1798),

and the first mathematical representation of population growth, characterized as it is by a sigmoid curve, to P.F. Verhulst in Paris in 1838. The essentials of the latter are as follows. A population will establish itself slowly in a new environment and then, once it is adjusted ecologically and competitively to it, will grow rapidly within its determined NICHE. This explosive phase of growth has been noted widely among both plants and animals, and is associated with r-selection (see r- and K-SELECTION). Subsequently, as it nears the BIOTIC POTENTIAL for that niche and area, environmental resistance will flatten the growth curve until it reaches an equilibrium, which is equivalent to the CARRYING CAPACITY of the area for that species. The process may be expressed by:

$$\frac{\Delta N}{\Delta t} = \gamma N \frac{(K - N)}{K},$$

in which N refers to the numbers of a given population, t to a given period of time, γ to the rate of increase (as determined by the difference between specific birth and death rates for the population), and K is a constant which relates to the upper limit of population growth (in other words, the carrying capacity). Close to the top of the growth curve, most species will be associated with K-selection.

Once at equilibrium level populations may adopt several strategies in order to maintain themselves. In laboratories most species will keep as close to equilibrium as possible, but under field conditions many of the larger organisms will display slight cyclical oscillations around it. These are DENSITY-DEPENDENT, in the sense that when population densities become low enough for a real and permanent decline to become a possibility, in-built compensatory mechanisms (usually in the form of increased birth rates) set in to restore the balance; similar though reverse responses occur when densities become too high. The amplitude of such oscillations is normally greatest when organisms are small (e.g. reaching $\times 40,000$ the minimum in the case of locusts but only $c. \times 2$ for most birds); and, for larger animals, the periodicity customarily fits into the framework of a 4–5 year (mice, voles, foxes, lemming species on both sides of the Atlantic) or a 9–10 year (lynx, snowshoe hare) cycle. Should environmental circumstances change and organisms fail to adapt sufficiently, populations may begin to decline in numbers in a density-independent manner, seriously enough for EXTINCTION to threaten, though if the change is only temporary (e.g. an exceptionally cold winter, which may affect birds which eat freshwater organisms; or a single chemical or heat pollution event in water), such populations may still recover.   DW

Reading and Reference
Malthus, T.R. 1798: *An essay of the principle of population as it affects the future improvement of society.* London. (Various modern editions.)

pore ice A type of ground ice occurring in the pores of soils and rocks. It is sometimes referred to as cement ice. On melting, pore ice does not yield water in excess of the pore volume, in contrast to SEGREGATED ICE. In terms of the total global ground-ice volume, pore ice probably constitutes the most important ground-ice type, primarily because of its ubiquitous distribution.   HMF

Reading
Mackay, J.R. 1972: The world of underground ice. *Annals of the Association of American Geographers* 62, pp. 1–22.

pore water pressure The pressure exerted by water in the pores of a soil or other sediment. Pressure is positive when below the WATER TABLE and negative when above it. Negative pore water pressure in soil is referred to as soil moisture tension (or suction). Pore water pressure can be measured by a tensiometer connected to a mercury manometer, vacuum gauge or pressure transducer (Burt, 1978).   PWW

Reading
Burt, T.P. 1978: An automatic fluid-scanning switch tensiometer system. *British Geomorphological Research Group technical bulletin* 21.

porosity A property of a rock or soil concerned with the extent to which it contains voids or INTERSTICES. It is usually defined as a ratio of the aggregate volume of voids to the total volume of the rock or soil, and is expressed as a percentage. A distinction is sometimes made between primary porosity, arising from intergranular interstices at the time of deposition of the rock, and secondary porosity, arising from later jointing or corrosion. All interstices whether primary or secondary are included in the estimation of the aggregate volume of voids for the purpose of measuring porosity.   PWW

Reading
Davis, S.N. 1969: Porosity and permeability of natural materials. In R.J.M. De Wiest ed., *Flow through porous media.* New York: Academic Press. Pp. 54–89.

positive feedback See SYSTEMS.

postclimax See POTENTIAL CLIMAX.

postglacial See HOLOCENE.

potamology The scientific study of rivers.

potassium argon (K/Ar) dating An isotopic dating technique which utilizes unaltered potassium-rich minerals of volcanic origin in basalts, obsidians, and the like. It is particularly useful for materials more than 50,000 years old.   ASG

Reading
Miller, G.H. 1990: Miscellaneous dating methods. In A.S. Goudie ed., *Geomorphological techniques*, 2nd edn. London: Unwin Hyman. Pp. 405–7.

potential climax The adjacent CLIMAX VEGETATION, which will replace the climax of an area if environmental conditions change, or the theoretical vegetation of an area, given the absence of man and sufficient time under present environmental conditions. It was defined in the first sense as part of MONOCLIMAX theory, though it can also be applied to modern concepts of climax vegetation.

In the second sense it is synonymous with the potential natural vegetation of A.W. Küchler (1967) and is similar to the theoretical climatic climax vegetation of S.R. Eyre (1963). This concept is necessary because of the universality of change in vegetation and the number of plant communities that have been substituted for the original vegetation, directly or indirectly, by the activities of man. (See also CLISERE; ZONATION.)   JAM

Reading and References
Eyre, S.R. 1963: *Vegetation and soils: a world picture.* London: Edward Arnold.
Küchler, A.W. 1967: *Vegetation mapping.* New York: Ronald Press.
†Oosting, H.J. 1956: *The study of plant communities.* San Francisco: W.H. Freeman.

potential energy The energy change when a system is reduced to some standard state, usually applied to gravitational potential energy, with mean sea level defining the reference state. The concept has very wide applicability in physics and, more particularly, in the physics of the natural environment. It has been used very frequently within meteorology for a greater understanding of the energetics and dynamics of atmospheric motion. For example, using:

$$v^2 = 2gh$$

where v is air velocity, g is acceleration due to gravity and h is the height above a specified datum, a parcel of air in the upper TROPOSPHERE has potential energy to a speed of 450 m s^{-1}. Natural parcels do not acquire such speeds because they are unable to fall freely, needing to push other air out of the way. This notion eventually defines the available potential energy and gives:

$$v^2 \simeq 2gh\delta T/T$$

where T is the temperature difference, as they pass, of two parcels being exchanged between two levels in the atmosphere. Hence a more realistic magnitude of air speed is 30 m s^{-1}.   JSAG

potential evaporation The rate of water loss from a surface when water supply to the surface is sufficient to meet the evaporative demand. Since evaporation from a water surface will always be at the potential rate, potential evaporation is often called potential evapotranspiration and is the rate of water loss from a surface other than a water surface, through evaporation and transpiration processes and when these processes are not limited by a water deficiency. In order to ensure comparability of estimates from different areas, Penman (1956) defined potential water loss from a vegetated surface (which he called potential transpiration) as evaporation from an extended surface of 'fresh green crop of about the same colour as green, completely shading the ground, of fairly uniform height and never short of water'.   AMG

Reading and Reference
Allen, R.G. 1986: A Penman for all seasons. *American Society of Civil Engineers, Journal of the Irrigation and Drainage Division* 112, pp. 348–68.
†Morton, F.I. 1983: Operational estimates of areal evapotranspiration and their significance to the science and practice of hydrology. *Journal of hydrology* 66, pp. 1–76.
Penman, H.L. 1956: Estimating evaporation. *Transactions of the American Geophysical Union* 37, pp. 43–6.
Ward, R.C. and Robinson, M. 1989: *Principles of hydrology*. 3rd edn. London: McGraw-Hill.

potential evapotranspiration See EVAPOTRANSPIRATION.

potential temperature (θ) The temperature an air parcel would possess if it were moved from its level to a level with a pressure of 1000 mb dry adiabatically (a rate of 9.8 K km^{-1}). If ascending or descending air parcels are subject only to these ADIABATIC changes their potential temperature will remain constant. In fact, the motion of air within the atmosphere is often close to adiabatic and the θ value of a given sample of air is conserved and acts as a kind of 'label'. Thus when an unsaturated parcel of air ascends or descends it will do so dry adiabatically and will move up or down an imaginary surface of constant potential temperature. If it ascends it will cool at 9.8 K km^{-1} and will warm at this rate in descent – so long as no CONDENSATION occurs.

Because it is a more conservative property than dry-bulb temperature it is often used to highlight frontal changes in vertical cross-section of the atmosphere. Lines of constant θ appear on THERMODYNAMIC DIAGRAMS.   RR

pot-hole A deep, circular hole in the rocky bed of a river which has formed by abrasion by pebbles caught in eddies (see plate on p. 400). Any vertical shaft in limestone.

Pre-Boreal See BLYTT–SERNANDER MODEL.

precession of the equinoxes The change in the relative positions of the equator and the ecliptic which causes the north pole to move in a circle over a cycle of 26,000 years. (See also MILANKOVITCH HYPOTHESIS.)

precipitation The deposition of water from the atmosphere in solid or liquid form. It covers a wide range of particle sizes and shapes such as RAIN, SNOW, HAIL and DEW. In most parts of the world rain is the only significant contribution to annual precipitation totals and the terms are frequently used synonymously. In polar regions and at high altitudes snow will be the dominant type of precipitation. The processes whereby water vapour is converted into precipitation are explained in CLOUD MICROPHYSICS.   PS

Reading
Sumner, G. 1988: *Precipitation: process and analysis*. Chichester: Wiley.

Pot-holes caused by fluvial abrasion eventually lead to the lowering of bedrock channels.

predation The killing of one free-living animal by another for food. Technically this may involve the total removal of a species, or several species, from an environment by a predator, though in mature and/or complex communities, and in natural circumstances, it is unlikely that this would ever happen, for the predator would then have eliminated a potential food resource; moreover, most such communities possess a large number of prey species for each predator, so that the demands on any one are never too heavy. Also most predators prefer a range of different animals in their diet.

However, there is little doubt that continued predation modifies the patterns of COMPETITION in an area and often, in consequence, the local distribution of species. Through reducing their population densities predators tend to lower the competition pressure from prey species in similar NICHES; and this may result in two competitors surviving where, without predation, only one would. Further, it is

possible that balanced predation may actually increase species diversities in many communities. On the Pacific Coast of North America, Paine (1966) has noted, in experiments on rock shores, that the removal of a major predator, the starfish *Pisaster ochraceus*, caused a diminution both in the number of species present in those communities (from 15 to 8), and in their functional variety (from acorn barnacles, limpets, chitons, dog whelks and mussels to predominantly barnacles and mussels, the population of the latter growing explosively). If the idea of HERBIVORES as plant predators is accepted similar consequences can be seen to attend the cessation of balanced grazing: in southern England, meadows closely grazed by sheep may have *c.*20 component species, while those which are taken out of grazing quickly establish a dominance structure in which several of the ground plants are shaded out, and not replaced.

It is considered by some authorities (e.g. MacArthur 1972) that both the number of predators per unit area, and predation pressure generally, increase towards the tropics, and in conditions of reduced physiological stress, both on the major landmasses and in the oceans. But this is not the case in tropical oceanic islands, many of which lack locally evolved predators; and in these the introduction of alien predator species has, unlike the patterns of 'natural' predation, frequently given rise to immense ecological disruption. In the West Indies the planned arrival of the mongoose, which was designed to reduce the number of snakes, also resulted in the total removal of many native birds; and the European dog, cat and rat, together and separately, have wrought similar ecological havoc in Pacific island systems. Recently, human beings themselves have directly augmented rates of predation everywhere, often in an unbalanced way.   DW

References
MacArthur, R.H. 1972: *Geographical ecology: patterns in the distribution of species.* New York: Harper & Row.

Paine, R.T. 1966: Food web complexity and species diversity. *American naturalist* 100, pp. 65–75.

predator–prey relationships The population and energy balances between predators and prey. An intimate relationship exists between these two groups of organisms, for while the former can easily reduce the population of the latter, they are themselves vulnerable to decline and possible EXTINCTION through starvation, should prey become too few. Accordingly, balanced predator–prey interactions depend in large measure on the effective control of the population size of, and by, both sets of participants. They are also important in the EVOLUTION of new species forms by natural selection, selection favouring the efficient predator and the elusive prey: in the latter case the development of a wide variety of cryptic and mimetic coloration in many species.

Animals which may be listed as predators include both 'true predators' and insect parasitoids (often incorrectly termed insect parasites). The latter are extremely numerous, and account for about 10 per cent of the approximately one million known insect species. Most belong to the Diptera (flies) and Hymenoptera (ants, bees, wasps) families, e.g. there are huge numbers of different species of parasitoid wasps, ranging in size from free-living forms to microscopic egg parasitoids. Most, too, are host-specific, i.e. they seek out one host species alone. Unlike true predators only the females of insect parasitoids look for hosts, and then usually only to lay eggs in or on them. The larvae which subsequently emerge feed from the host either internally or externally, but an effective energy balance between them is maintained until the larvae approach maturity, at which point the host's vital organs are eaten, and the host is killed. In this way insect parasitoids also differ from PARASITES, which tend to ensure that the host's life is secured. Relationships between the populations of both host and insect parasitoid are fairly simple in the sense that host mortality depends solely on the ability of females of the prey species, at one particular stage alone in their life cycle, to search out a host; and the reproductive rate of insect parasitoids may be seen to relate clearly to the number of hosts which are colonized. Although some insect parasitoids control the size of their host populations quite closely, this is not always the case.

In contrast, most true predators have more diffuse interactions with their prey, and this is especially the case for vertebrate predators, many of which are not prey-specific, having a wide range of preferred foods. Males, females, and often their young as well, all must search for prey throughout the year, unless they hibernate for part of it; and this is undertaken with different degrees of efficiency. Moreover, the reproductive rates of some true predators are more finely determined by the demographic characteristics of their own populations rather than by the numbers of prey, though this is not an invariable situation.

The population dynamics of the predator–prey dependency are by now fairly well known, and have been reviewed by Hassall (1976). For simple relationships the basic form is expressed by:

$$N_{t+1} = \lambda N_t f(N_t, P_t) \text{ and}$$

$$P_{t+1} = N_t[1 - f(N_t, P_t)]$$

where N_t, N_{t+1} and P_t, P_{t+1}, are respectively the prey and predator populations in successive generations, and λ is the finite net rate of increase of the prey (birth rate minus death rate). The predation rate at time t is an unspecified function of prey and predator sizes. Nicholson (1933) considered that three further assumptions could be made about predator–prey interactions: first, that predators would search randomly for their prey without being influenced by

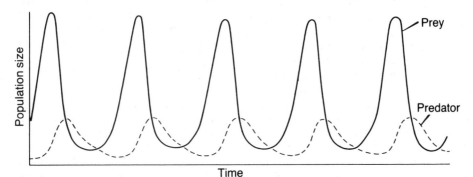

the distribution and density of either the prey or of other potential predators; secondly, that predators' food requirements would be unlimited; and thirdly, that the area in which the search for prey is conducted (a in the equation below) is likely to be constant for a given predator population. With this in mind, and referring back to the first equation:

$$f(N_t, P_t) \text{then} = \exp(-aP_t)$$

and the model changes to:

$$N_{t+1} = \lambda N_t \exp(-aP_t) \text{ and}$$

$$P_{t+1} = N_t[1 - \exp(-aP_t)]$$

This suggests that, first, for each predator–prey system an equilibrium level of population will exist; and, secondly, that this will be inherently unstable, for any movement away from the overall balance will lead quickly to oscillations in both populations, with that of the predator lagging behind that of the prey. In theory such oscillations may increase in size until one of the populations becomes extinct; or, they may subsequently stabilize. Stabilization can ensue, in a predetermined way, from the initial relative numerical balance between predators and prey, in which case it may be termed 'neutral stability' (see diagram); or it may be derived more actively ('actively induced stability') from a density-dependent factor (see POPULATION DYNAMICS) which operates especially on the prey, from resource limitations on the prey (May 1972), or from particular homeostatic predator responses to increases in prey populations, which may be functionally or numerically based.

Of these several possibilities, unstable predator–prey oscillations have been observed in laboratory experiments involving small animals such as protozoa and mites, in simple relationships which become self-annihilating. When laboratory conditions were made more complex stable oscillations resulted and, at least in some cases, the amplitude of these declined over time: they were convergent. This latter event seems to arise from evolutionary response in both predators and prey, in which the prey became more resistant to PREDATION, and the predators in turn began to seek out the prey less assiduously. In the instance of the house fly (*Musca domestica*) and wasp parasite (*Nasonia vitripennis*) populations, these responses are known to be capable of developing in the laboratory in as few as twenty generations (Pimentel *et al.* 1963).

Although it is likely that much of the observed stability in natural biological systems is due to a long-term but similar pattern of co-evolution between predators and prey, there is very little direct evidence of this thus far. Nor has anyone yet discovered a flawless example of a maintained predator–prey oscillatory cycle in field populations. For a long time it was thought that the lynx (*Lynx canadensis*)–snowshoe hare (*Lepus americanus*) predator-prey relationship in Canada, which is centred around a 9–10 year cycle of population growth and decline (see POPULATION DYNAMICS), provided one, but this view was negated by the realization that, in those parts of Canada which were not inhabited by the lynx (especially Anticosti Island, in the Gulf of St Lawrence), snowshoe hares retained that cycle of their own accord (Keith 1963). However, some good field evidence of the general effect of predators on field populations of prey is available. Where predators have been removed entirely from large areas of land, e.g. through shooting, the numbers of prey have frequently been substantially augmented, even to a point where they could no longer be supported by the resources of the environment; the prey population then collapsed. A classic instance of this type of instability has been recorded for the Kaibab plateau of Arizona, in which deer populations soared then crashed after their main predators (wolves and cougar) had been eliminated. Conversely, the presence of predators often encourages both stability and variety among prey populations (see PREDATION): and some of the stability at least may be achieved by the periodic numerical and functional changes in predator feeding habits which are known to occur. Thus, if prey populations become too large, predators may consume more of them; and if they decline too much, predators may switch to an alternative source of food until they recover. Such switching may eventuate most frequently between prey which are almost equally preferred in a diet, and it may be most common in animals that feed in flocks (Murdoch and Oaten 1975) but it has been noted too in relatively solitary predators, such as the English tawny owl (Southern 1970). Curiously, controlled field experiments which have sought to clarify predator–prey population interactions have not yet produced standardized results; and clearly these can be influenced by many environmental factors which are difficult to quantify. But it is likely that the most stable natural predator–prey systems are those in which several species are present, particularly in respect of the prey; those in which safe refuge areas for prey may be found; and those in which the predators select a large number of prey individuals which are no longer of reproductive age.   DW

Reading and References
†Hassall, M.P. 1976: Arthropod predator–prey systems. In R.M. May ed., *Theoretical ecology, principles and applications.* Oxford: Blackwell Scientific.

Keith, L.B. 1963: *Wildlife ten-year cycle.* Madison, Wisconsin: University of Wisconsin Press.

May, R.M. 1972: Limit cycles in predator–prey communities. *Science* 177, pp. 900–4.

Murdoch, W.W. and Oaten, A. 1975: Predation and population stability. *Advances in ecological research* 9, pp. 1–131.

Nicholson, A.J. 1933: The balance of animal populations. *Journal of animal ecology* 2, pp. 131–78.

Pimentel, D., Nagel, W.P. and Madden, J.L. 1963: Space–time structure of the environment and the survival of parasite–host systems. *American naturalist* 97, pp. 141–67.

Southern, H.N. 1970: The natural control of a population of tawny owls (*Strix aluco*). *London journal of zoology* 162, pp. 197–285.

pressure, air The force per unit horizontal area exerted at any given level in the atmosphere by the weight of the air above that level. At sea level the average air pressure is 14.7 lb in^{-2}, 760 mm Hg, 29.92 in Hg, or 1013.25 mbar (or hectopascals). The air pressure decreases most rapidly with height near sea level where the air is most dense. It decreases by about 50 per cent for every 5 km of ascent.   WDS

pressure melting point The temperature at which a liquid becomes solid at a particular pressure. The concept is fundamental to glacial geomorphology since glacier ice may exist at the pressure melting point. Since the temperature at which water freezes diminishes under additional pressure, by a rate of 1°C for every 1400 kPa (140 bar), the melting point at depth will be below 0°C. For example, water was discovered beneath 2164 m of ice at Byrd Station in West Antarctica at a pressure melting point of −1.6°C.   DES

Reading
Paterson, W.S.B. 1981: *The physics of glaciers.* London and New York: Pergamon.

pressure release The process whereby large sheets of rock become detached from a rock mass owing to the continuing relaxation of the pressure within the mass which built up before it was exhumed. For example, the erosion of overlying sedimentary strata from above a granite intrusion may cause pressure release to open joints in the granite when it is exposed.

prevailing wind This term is really an abbreviation for 'prevailing wind direction' and means the wind direction most frequently observed during a given period. The periods most frequently used are days, months, seasons and years.   BWA

primarrumpf An upwarped dome which though still undergoing uplift is being eroded at an equal rate.

prisere A primary successional sequence of plant communities. It is a series of communities that results from the processes of SUCCESSION on newly formed landsurfaces, e.g. emergent land at the coast, lava flows and recently deglaciated terrain. It is distinguished from a SUBSERE by the initial conditions, namely the absence of remnants of previous communities or soil.

Priseres, together with subseres, are commonly classified according to initial environmental conditions. Thus dry sites give rise to xeroseres, which include lithoseres (on rock) and psammoseres (on sand), whereas hydroseres are characteristic of wet sites and include haloseres (saline or alkaline conditions) and oxyseres (acidic conditions). (See also SERE.)   JAM

Reading
Matthews, J.A. 1979: The vegetation of the Storbrean gletschervorfeld, Jotunheimen, Norway. I: Approaches involving ordination and general conclusions. *Journal of biogeography* 6, pp. 133–67.

Olson, J.S. 1958: Rates of succession and soil changes on southern Lake Michigan sand dunes. *Botanical gazette* 119, pp. 125–70.

probable maximum precipitation The rainfall depth, for a particular size of catchment, that approaches the upper limit that the present climate can produce. Some hydrologists think of it only in terms of flood flows, and would define it as the magnitude of rainfall over a particular catchment that will yield flood flow with virtually no risk of being exceeded. The latter is, of course, the probable maximum flood. These definitions imply that there is a physical upper limit to the amount of rainfall in a given climatic zone. This is so because physical restrictions on the joint occurrence of the various rain-forming meteorological mechanisms impose an upper limit on the rainfall magnitude.

Probable maximum precipitation magnitudes may be estimated by the technique of maximizing the storm rainfalls which occur over a given catchment. The starting-point is the identification of storms which have occurred over the study catchment or over catchments which are geographically and climatically similar. These storms are optimized to produce the maximum rainfall by a careful adjustment of the rain-forming mechanisms within the limits of climatic possibility. The most important rainfall-limiting mechanism is the flow of moisture into the storm, and this usually sets an upper limit to the storm precipitation. For a flood to approach the probable maximum value, a number of conditions must be satisfied:

1 The duration of the storm must exceed the time of concentration of the catchment.
2 The storm must cover the whole of the catchment.
3 Over the time specified in (1) and the area specified in (2) the storm must approach the probable maximum intensity.

The intensity of rainfall tends to decrease as the area covered by a storm increases. Storm size and duration are therefore important when considering the possibility of a flood. A small storm may produce a relatively large flood in a small catchment, but little change in the run-off from a large area. In particular, thunderstorms produce deep floods over small areas but have little effect on the flow of large rivers, which are only affected by widespread rains.

Because atmospheric moisture contents are highest in the tropics, probable maximum precipitation is also highest in this region. Large-scale storms are most highly organized in the subtropics and it is here that the highest rainfalls are recorded. If the storm rainfall is enhanced by relief, it can reach very high values and daily rainfalls of up to 175 cm have been reported. Probable maximum precipitation values are therefore highest in the humid regions of the tropics and subtropics.   JGL

Reading
Bruce, J.P. and Clarke, R.H. 1966: *Introduction to hydrometeorology*. Oxford: Pergamon.
Rodda, J.C. 1970: Rainfall excesses in the United Kingdom. *Transactions of the Institute of British Geographers* 49, pp. 49–60.
Schwarz, F.K. 1963: *Probable maximum precipitation in the Hawaiian Islands*. Hydrometeorological report 39 Washington: US Weather Bureau.
Tucker, G.B. 1960: Some meteorological factors affecting dam design and construction. *Weather* 15, pp. 3–13.
Wienser, C.J. 1970: *Hydrometeorology*. London: Chapman & Hall.

process–response system See SYSTEMS.

profile See SOIL PROFILE.

proglacial See ENGLACIAL.

proglacial lake A lake impounded in a depression in front of a glacier. During glacial periods such lakes were well developed in front of the southern margins of the Laurentide and Eurasian ice sheets. The proglacial Lake Agassiz in the area north-west of Lake Superior had an area larger than that of the Great Lakes today and was over 1000 km long. Similar lakes were impounded in the lower valleys of the northward-flowing Asian rivers such as the Ob'.   DES

progradation The extension of a shoreline into the sea through sedimentation.

protalus rampart A narrow ridge, usually a metre or so high and tens of metres long in front of a mountain rock face, composed of rock fragments. It may look like a small moraine, with which they have sometimes been confused, but whereas the former are the result of glacier action a protalus rampart is generally considered to be formed by rock debris sliding over a snowpatch. They have also been called 'winter talus ridges' and 'nival ridges' but protalus rampart (Bryan 1934) is now the accepted name.   WBW

Reading and Reference
Ballantyne, C.K. and Kirkbride, M.P. 1986: Characteristics and significance of some Lateglacial protalus ramparts in upland Britain. *Earth surface processes and landforms* 11, pp. 659–71.
Bryan, K. 1934: Geomorphic processes at high altitudes. *Geographical review* 24, pp. 655–6.
Butler, D.R. 1986: Winter-talus ridges, nivation ridges and pro-talus ramparts. *Journal of glaciology* 32, pp. 543.

protected ecosystems and landscapes The records of history show that there seems to have always been a cultural desire to exempt some parts of nature from contemporary processes, both natural and human-induced. Among most human groups, some species or some places have been so highly valued that they were protected from change. The custody might result in an economic product but the general set of values surrounding the motivation was (and is) different from that of, for example, a managed ocean fish stock. Attention might focus on a particular species of plant or animal, on an assemblage of taxa together with their non-living environment (i.e. an ECOSYSTEM), on a large area in its environmental entirety, or on the visual qualities of a particular landscape. Many historical examples can be traced and the categories remain valid.

Among the species whose perpetuation is currently sought is the Asian tiger, valued little for its ecosystemic predator role but chiefly now for its aesthemic qualities. In East and Central Africa, the assemblage of mammals and birds of the savannah grasslands is subject to conservation measures, partly because of the scientific interest the ecosystems excite, but mainly as a source of income from tourism. Large portions of particular environments may contribute to the maintenance of *in situ* biodiversity (as with reserves in tropical moist forests) or to scientific endeavour, as with Antarctica. The maintenance of beloved scenery is at the heart of the designation of the National Parks of England and Wales and those of Japan.

The successful protection of such phenomena requires that a number of conditions be fulfilled. The general cultural context must be favourable: societies must on the whole feel that protection of various kinds (and probably to varying degrees, depending upon other circumstances) has cultural sanction even if not specific ethical imperatives. Governments must respond by enacting appropriate legislation and then being willing to enforce it. Since so many ecosystems and environments cross international boundaries, then transnational agreements, conventions and treaties become highly important. But the mere passing of laws is not enough. Few protected areas will retain the desired species or stay in the demanded state without management, since they are very often (with the obvious exception of Antarctica) surrounded by more intensely manipulated systems. So expertise in ecosystem, landscape and people management is always required.

This skill must first of all recognize the purpose of management: whether we are discussing a 0.1 ha plot with the northernmost occurrence of a rare flower in one country, the whole of an offshore island and its surrounding seas or indeed a whole continent, it is essential to have an agreed and defined set of goals, with time spans included. With that decided, then management methods can be attempted which recognize that ecosystem behaviour cannot be predicted with the same kind of accuracy as determinate systems in physics or chemistry. These methods are immensely variable: in the case of animal populations they might run all the way from suggesting complete protection of a species using all the machinery of modern science, technology and the state to keep people away from it, to the decision to cull the animal regularly to raise income from saleable parts: attitudes to the African elephant have encompassed such a spectrum of opinion.

Given that the motivation is so often inspired by ethical considerations, a good deal of thought has gone into trying to decide whether protected areas (and especially those which occupy areas little affected by human presence) can be shown more objectively to be necessary to the continued functioning of planetary biogeosystems. Using classic succession theory, E.P. Odum (1971), for example, tried to suggest that a balance between 'mature' ecosystems (i.e. the very wild and indeed the wildernesses) and the 'immature' (agroecosystems) and the inert (volcanoes and cities) had to be maintained. A similar idea might be implicit in some of the Gaia models now popular. But short of a world land-use authority, it is difficult to see how any such ideas, even if thought to be valid could be put into place. Nevertheless, given the high value which most cultures will put upon the protection of parts at least of nature, provided that day-to-day strains are not too great, continued attention to the importance of environmental protection seems at the very least prudent.   IGS

Reading
Engel, J.R. and Engel, J.R. eds 1990: *Ethics of environment and development.* London: Belhaven Press.

IUCN/UNEP/WWF 1991: *Caring for the earth*. Gland, Switzerland: IUCN.

Myers, N. ed. 1985: *The Gaia atlas of planetary management*. London and Sydney: Pan Books.

Odum, E.P. 1971: *Fundamentals of ecology*. 3rd edn. Philadelphia: Saunders.

Simmons, I.G. 1991: *Earth, air and water. Resources and environment in the late 20th century*. London: Edward Arnold. Ch. 7.

Wolf, E.C. 1988: Avoiding a mass extinction of species. In L.R. Brown, ed., *State of the world 1988*. Washington, DC: World Resources Institute. Pp. 101–17.

proximal trough A depression around a steep rock body caused by the increased velocity that can occur when a moving element (e.g. ice, water or wind) flows round an obstruction.   ASG

Reading
Lassila, M. 1986: Proximal troughs and ice movements in Gotland, southern Sweden. *Zeitschrift für Geomorphologie* 30, pp. 129–40.

psammosere See PRISERE.

pseudokarst Landforms produced in non-carbonate rocks which are morphologically similar to those normally associated with karst rocks. These non-calcareous rocks can produce features by solutional processes similar to those types of reaction found in limestone (karren features) or by processes entirely different from these (lava caves).   PAB

Reading
Warwick, G.T. 1976: Geomorphology and caves. In T.D. Ford and C.H.D. Cullingford eds, *The science of speleology*.

London: Academic Press. Pp. 61–126.

pumice A highly porous fine-grained volcanic rock produced when numerous gas bubbles are trapped within the lava when it solidifies.

puna A cold desert, especially one at high altitude as in the Andes.

punctuated aggradational cycles A stratigraphic model which states that most stratigraphic accumulation occurs episodically as thin (1–5 m thick) shallowing upward cycles separated by sharply defined non-depositional surfaces. They are created by geologically instantaneous basin-wide relative base-level rises (punctuation events), with deposition occurring during the intervening periods of base-level stability. Glacial eustatic changes driven by orbital perturbations (see MILANKOVITCH HYPOTHESIS) may be a preferred mechanism.   ASG

Reading
Goodwin, P.W. and Anderson, E.J. 1985: Punctuated aggradational cycles: a general hypothesis of episodic stratigraphic accumulation. *Journal of geology* 93, pp. 515–33.

push moraine See MORAINE.

pyroclastic Refers to fragmental rock products (e.g. ash, volcanic bombs, ignimbrite, etc.) ejected by volcanic explosions.

Q

quasi-equilibrium The 'apparent' balance between opposing forces and resistances. In geomorphology it is applied often to the 'concept' of an apparent balance between the rate of supply, temporary storage and removal of material from a dynamic depositional landform such as a scree slope, alluvial fan, beach or dune. The geometry of the landform or depositional store is dependent on the nature of the balance, the length of time over which it is maintained and the overall volumes involved. In many cases the volume of material passing through the system may be much greater than that in the store at any one time.

The idea may be applied to a surface of transportation such as a hillslope or pediment where there is an apparently close relationship between the geometry of the debris mantle and the form of the surface over which it moves. This implies that there is a negative feedback between the balance of input–storage–output processes of the mantle and the adjustment of the bedrock surface. Thus a decrease in storage thickness may increase weathering rates on the bedrock to restore the mantle to its previous condition. In this way the balance of equilibrium conditions of the depositional layer may partially control overall landform development.

A major weakness of the concept is that the relevant time-scales, over which budgetary or geometrical relationships are achieved or maintained, are unknown. No evaluations have been made of the relaxation times involved and it is not possible, at present, to understand fully the effects of environmental change or the status of landscape relicts.   DB

Quaternary The second period of the Cainozoic Era. It followed the Tertiary and comprises the PLEISTOCENE and HOLOCENE series. On faunal grounds the base of the Quaternary is placed at about 1.6 million years ago (Haq *et al.* 1977). The Quaternary was characterized by widespread glaciation (Bowen 1978).   ASG

References
Bowen, D.Q. 1978: *Quaternary geology.* Oxford: Pergamon.
Haq, B.U., Berggren, W.A. and Van Couvering, J.A. 1977: Corrected age of the Pliocene/Pleistocene boundary. *Nature* 269, pp. 483–8.

Quaternary ecology A number of diverse research fields involved in studying ecological events and processes that have occurred during the most recent geological interval, the Quaternary, which started some 1.6 to 1.7 million years ago. On the basis of information from these fields the Quaternary is divided into the PLEISTOCENE, a period when an extensive but highly variable ice cover occupied much of what is now the temperate northern hemisphere, and the HOLOCENE, a period when the ice cover largely disappeared and when trees and shrubs migrated northwards in Europe and North America to form the present plant assemblages, human population densities increased dramatically, and several species of plants and animals were domesticated. All research into the Quaternary period of the earth's history involves studying objects laid down in sediments as part of a stratigraphic sequence. Inferences are based on the principle of UNIFORMITARIANISM, in which the present is seen as the key to the past. There are three groups of objects that have been widely used in Quaternary ecological studies. Each group has given rise to specific research fields that overlap the more familiar academic discipline boundaries. Sediments are studied by Quaternary stratigraphers who make inferences concerning how they were deposited, where the components of the sediments came from, and what they represent in terms of past geomorphic processes and hence climatic regimes. Many stratigraphers are specifically interested in the magnetic record of sediments (PALAEOMAGNETISM), or the chemical record, or buried soils (PALAEOSOLS). Biotic remains, fossils in the strictest sense of the word, are studied by palaeoecologists who infer in the first instance what the conditions for life were like when a particular organism lived. Most palaeoecologists study specific organisms (e.g. molluscs, ostracods, diatoms) or specific types of remains (e.g. pollen and spores, macroremains of plants, bones). Cultural remains are studied by archaeologists to reconstruct past human economics, including the behavioural characteristics of the people. Archaeologists include those involved in specific cultural groups of specific periods of time or those involved with particular aspects of the economy, such as diet (ethnobotany) or the group's use of animals (ethnozoology). All these objects, physical, biological and cultural, are used by Quaternary ecologists such as palaeoclimatologists to interpret past climates, by palaeolimnologists to interpret past lacustrine conditions, or by oceanographers to interpret past oceanic conditions.

In the past decade our knowledge of the Quaternary Period has increased considerably especially in the following ways:

1 A more sophisticated calibration of the modern object–environmental relationship, allowing us to become more quantitative in our approach to the past.

2 A more sophisticated view of the Quaternary, especially as we recognize

that plant and animal distributions are dynamic and modern assemblages very recent.

3 An increased awareness of conditions in the tropical areas.

4 The use of new objects, e.g. the magnetic record of sediments, and the use of new devices to handle the large volumes of statistical data involved.

RHS

Reading
The journals *Quaternary research* and *Quaternary science reviews* contain inter-disciplinary material on the period.
Flint, R.F. 1971: *Glacial and Quaternary geology*. New York: Wiley.
West, R.G. 1977: *Pleistocene geology and biology*. 2nd edn. London: Longman.
Wright, H.E. Jr and Frey, D.G. eds 1965: *The Quaternary of the United States*. Princeton, NJ: Princeton University Press.

quick clay Water-saturated clay which has insufficient cohesion to prevent heavy objects from sinking into its surface.

quickflow The part of the stream hydrograph which lies above an arbitrary cut-off line drawn on the hydrograph, representing the most rapidly responding hydrological processes and parts of the catchment. The division between quickflow and DELAYED FLOW is usually made by a line which rises from the start of the hydrograph rise at a gradient of $0.55 \ 1 \ s^{-1} \ km^{-2} \ h^{-1}$ until the line meets the falling limb of the hydrograph.

MJK

Reading
Hewlett, J.D. 1961: *Soil moisture as a source of base flow from steep mountain watersheds*. South-eastern Forest Experimental Station paper 132. US Department of Agriculture.

quicksand Water-saturated sand which is semi-liquid and cannot bear the weight of heavy objects.

R

r- and K-selection A theory that natural selection may favour either individuals with high reproductive rates and rapid development (r-selection) or individuals with low reproductive rates and better competitive ability (K-selection), depending on environmental conditions. These two kinds of selection were described by Dobzhansky (1950), although the terms K-selection and r-selection were developed by MacArthur and Wilson (1967). K refers to carrying capacity and r to the maximum intrinsic rate of natural increase (r max). The terms are derived from the logistic growth equation of POPULATION DYNAMICS.

Plant and animal species that exploit resources which do not fluctuate, or which are predictable, can maintain relatively constant population sizes which are at or near the CARRYING CAPACITY of the environment. Under these conditions competition is intense and there is little advantage in being capable of rapid population growth, since the opportunity for such growth rarely exists. Instead, selection favours individuals that maximize the utilization of resources and the production of relatively few offspring which have a high probability of survival. These K-selected species often develop slowly and have larger body size and long life spans (Pianka 1970).

In contrast, disturbed sites or habitats with fluctuating environments, where resources may be temporarily abundant, favour individuals which develop rapidly, have early reproduction and produce a maximum number of offspring. These traits lead to high productivity and are advantageous in the rapid exploitation of resources. These r-selected species are sometimes called opportunistic or colonizing species; they often have a high dispersal ability in order rapidly to reach new areas of disturbance.

An alternative classification of species' adaptive strategies has been proposed by Grime (1979) who suggests that natural selection can favour three strategies, competitors (c-selected) which exploit habitats which have little disturbance and low stress in the sense of shortages of resources such as light, water or nutrients; stress-tolerators (s-selected) which exploit habitats of high stress but low disturbance; and ruderals (r-selected) which exploit habitats of low stress but high disturbance. Grime's model differs from the r–K-selection theory in emphasizing a third strategy, stress tolerance, which is favoured in highly unproductive habitats which often experience extreme stress, for example, Arctic-alpine or arid areas. Grime has linked his concept of three major strategies with that of r- and K-selection by suggesting that the ruderal and stress tolerant strategies correspond respectively to r- and K-selection, and that the competitive strategy occupies an intermediate position on the r–K continuum.

It must be emphasized that attempts to classify organisms with respect to life history strategies can be misleading if they are interpreted too rigidly. Natural selection does not act to produce a maximum value of r or of K. Instead, all species must make a series of trade-offs in the allocation of resources to maintenance, growth and reproduction. It is generally accepted that most organisms fit in to a continuum between the extremes of r- and K-selection and may exhibit life-cycle characteristics which contain elements of both categories.

Several studies have underlined the complexity of life history strategies. For example, Solbrig and Simpson (1974) demonstrated that a dandelion population in a highly disturbed lawn had a high reproductive effort whereas a nearby population in an undisturbed early successional site set fewer seeds but was a better competitor. In this case populations of the same species differ in their position on the r–K continuum as a result of genetic variation. Many organisms inhabiting variable environments have dynamic reproductive strategies which vary over time (Nichols et al. 1976). These species can alter their reproductive effort substantially in response to changes in environmental conditions. Other characteristics which are associated with r- and K-strategies, such as life span and time at first reproduction, can also fluctuate. Consequently, populations of such organisms can shift between relative r- and K-positions over time.   ARH

Reading and References
Dobzhansky, T. 1950: Evolution in the tropics. *American scientist* 38, pp. 209–21.
†Grime, J.P. 1979: *Plant strategies and vegetation processes.* London and New York: Wiley.
MacArthur, R.H. and Wilson, E.O. 1967: *The theory of island biogeography.* Princeton, NJ: Princeton University Press.
Nichols, J.D., Conley, W., Batt, B. and Tipton, A.R. 1976: Temporally dynamic reproductive strategies and the concept of r- and K-selection. *American naturalist* 110, pp. 995–1005.
†Pianka, E.R. 1970: On r- and K-selection. *American naturalist* 104, pp. 592–7.
Solbrig, O.T. and Simpson, B.B. 1974: Components of regulation of a population of dandelions in Michigan. *Journal of ecology* 62, pp. 473–86.
†Southwood, T.R.E. 1976: Bionomic strategies and population parameters. In R.M. May ed., *Theoretical ecology: principles and applications.* Oxford: Blackwell Scientific. Pp. 26–48.

radiation Any object not at a temperature of absolute zero ($-273°C$) transmits energy to its surroundings by radiation, that is, by energy in the form of electromagnetic waves travelling with the speed of light and requiring no intervening medium. This radiation is characterized by its wavelength, of which there is a wide spectrum extending from the very short X-rays through the ultraviolet and visible to infrared, microwaves and radio waves.

A valuable theoretical concept in radiation studies is that of the black body, which is one that absorbs all the radiation falling on it and which emits, at any temperature, the maximum amount of radiant energy. The term arises from the relation between darkness of colour and the proportion of

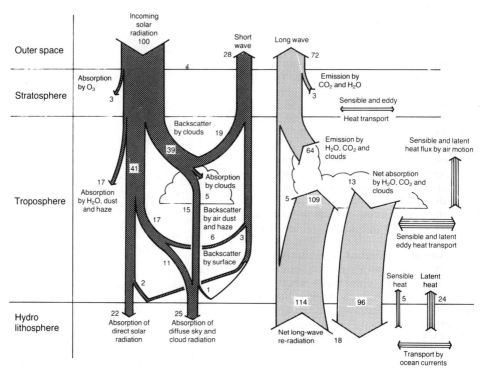

Schematic diagram showing the interactions that radiation undergoes in the atmosphere.
Source: *T.G. Lockwood 1979: Causes of climate. London: Edward Arnold.*

visible light absorbed, since a body which appears white scatters most of the visible light falling on it. For a perfect all-wave black body, the intensity of radiation emitted and the wavelength distribution depend only on the absolute temperature, and in this case several simple laws apply. The Stefan–Boltzmann law states that the amount of energy (F) emitted in unit time from a unit area of a black body is proportional to the fourth power of its absolute temperature (T), i.e.:

$$F = \sigma T^4$$

where σ is Stefan's constant (5.6697×10^{-12} $W\,cm^{-2}K^{-4}$). The higher the temperature of an object the more radiation it will emit.

A black body does not radiate the same amount of energy at all wavelengths for any temperature. At a given temperature, the energy radiation reaches a maximum at some particular wavelength and then decreases for longer or shorter wavelengths. The Wien displacement law states that this wavelength of maximum energy (λ_{max}) is inversely proportional to the absolute temperature, i.e.:

$$\lambda_{max} = \frac{\propto}{T}$$

where $\propto$ is a constant (0.2897 cmK if λ is in centimetres). As the temperature of an object increases, the wavelength of maximum energy decreases, passing from the infrared for objects at room temperature to the visible wavelengths for extremely hot objects.

If the sun is assumed to be a black body, an estimate of its effective radiating temperature may be obtained from the Stefan–Boltzmann law, which suggests an effective surface temperature of 5750 K. For the sun, the wavelength of maximum emission is near 0.5 μm ($1\ \mu m = 10^{-6}$m), which is in the visible portion of the electromagnetic spectrum, and almost 99 per cent of the sun's radiation is contained in so-called short wavelengths from 0.15 to 4.0 μm. Observations show that 9 per cent of this short-wave radiation is in the ultraviolet (less than 0.4 μm), 45 per cent in the visible (0.4–0.6 μm), and 46 per cent in the infrared (greater than 0.74 μm).

The surface of the earth, when heated by the absorption of solar radiation, becomes a source of long-wave radiation. The average temperature of the earth's surface is about 285 K (12°C), and therefore most of the radiation is emitted in the infrared spectral range from 4 to 50 μm, with a peak near 10 μm, as indicated by the Wien displacement law. This radiation may be referred to as long-wave, infrared, terrestrial or thermal radiation.   JGL

Reading
Budyko, M.I. 1974: *Climate and life.* New York: Academic Press.
Houghton, J.T. 1977: *The physics of atmospheres.* Cambridge: Cambridge University Press.
Lockwood, J.G. 1979: *Causes of climate.* London: Edward Arnold.
Sellers, W.D. 1965: *Physical climatology.* Chicago: University of Chicago Press.

radiative forcing A change or perturbation which is imposed upon the climate system and modifies the radiative balance. The causes of such a change include changes in solar radiation input, cloud cover and character, ice, greenhouse gases, volcanic activity etc. Quantification of the effects of radiative forcing is one of the main goals of many climatic models.   ASG

radiocarbon dating See CARBON DATING.

radioisotopes The isotopes of some elements which undergo radioactive decay and alter either to a different element or a stable isotope of the same element.

radiosonde An instrument for measuring the pressure, temperature and humidity of the air at heights of greater than 10 m above the ground. The instrument is attached to a balloon which rises through the TROPOSPHERE, usually bursting near the TROPOPAUSE. The meteorological data are transmitted to a ground station as the balloon ascends. Upper air winds are established by tracking the whole package by radar.   BWA

radon gas A colourless, odourless gas (radon-222) about eight times denser than air. It is derived largely from uranium, which is present in rocks such as granite. It poses an environmental problem when it accumulates in houses, for it may lead to cancer formation. Some areas of Britain are regarded as being at high risk (especially portions of Devon and Cornwall), but the threat can be reduced by appropriate building techniques and ventilation procedures.   ASG

Reading
Clarke, R. and O'Riordan, M. 1990: Rumours of radon. *Science and public affairs* 5, pp. 23–36.

rain Precipitation in the form of liquid water drops. Drop sizes vary up to a maximum of about 0.5 cm in diameter. The smallest diameter of a raindrop is sometimes defined rather arbitrarily as being 0.02 cm. Drops in the range of 0.02–0.05 cm are classed as drizzle and are fairly common on windward locations in temperate latitudes.   PS

rain day In Britain, a climatological period of 24 hours from 09.00 UT within which at least 0.2 mm of precipitation is recorded. The average number of rain days ranges from about 160 in south-eastern England to over 250 in the highlands of Scotland and the Outer Hebrides, where more than 300 rain days may be recorded in wet years. In some countries a different base is taken as the definition of a rain day, e.g. 0.01 inches in the USA.   PS

rain factor A measure of the relationship between temperature and precipitation designed to provide an indication of the climatic aridity of a region. The formula used is:

$$\text{rainfall factor} = \frac{\text{mean annual precipitation in mm}}{\text{mean annual temperature in } °C}$$

It is clearly inappropriate for polar deserts where mean annual temperatures are below freezing. Elsewhere it gives an impression of the dryness of an area, although the seasonal distribution of rainfall will affect what plants can be grown. For example, Hobart in Tasmania and Kaduna in northern Nigeria have a similar rain factor but the former site has rain each month and the latter only for six months.   PS

rain gauge An instrument used to measure rainfall amounts. Gauge characteristics differ in detail between countries, but basically it consists of a funnel and storage system shielded from the free air to prevent evaporation. In the UK the standard gauge made of copper has a collection funnel of 12.7 cm diameter placed on a cylinder 30.5 cm above the ground surface. The water is stored in a glass bottle held within the copper cylinder. Any overflow from the bottle is collected in another container to prevent loss during heavy storms. The rainfall is normally read daily (at 0.900 GMT) by emptying the contents of the collecting vessel into a cylindrical flask graduated to allow a direct reading of the rainfall total. Some gauges are read only weekly or even monthly in remote areas.

Rain gauges do not catch all the rain falling upon the surface because they present an obstruction to the airflow. In strong winds the actual catch may be depleted by up to 50 per cent of the true catch or even more during snowfall. The height of the rain gauge funnel above the ground will influence catch. A taller cylinder will generate more turbulence and reduce the catch compared with an identical gauge close to the surface. Other errors may be caused by splashing, evaporation or even observer mistakes. The total obtained by a rain gauge is therefore only an approximation to the true fall. Where a large proportion of the annual precipitation falls as snow, as in Canada and parts of the former USSR, separate snow gauges are used as standard rain gauges are unsuitable. Because of the different types and sizes of gauge used, comparison of rainfall totals across international frontiers is difficult.

Some gauges make a continuous record of precipitation and its time of occurrence, allowing the calculation of RAINFALL INTENSITY. This is achieved by collecting the water in a tipping bucket, on a weighing system or by a float which measures water level. Each type offers some advantages and disadvantages.   PS

Reading
Sumner, G. 1988: *Precipitation: process and analysis.* Chichester: Wiley.

rain gauge, tilting siphon One type of autographic or recording rain gauge where the pen records on a rotating chart and is connected to a float in a precipitation collecting chamber which is emptied when full by a tilting siphon mechanism. The collecting chamber is precisely balanced against a counter weight, so that when the chamber is full it is sufficiently heavy to overcome the weight of the counter weight and so tilts, causing the contained water to siphon away. The pen records on the rotating chart the rate at which the

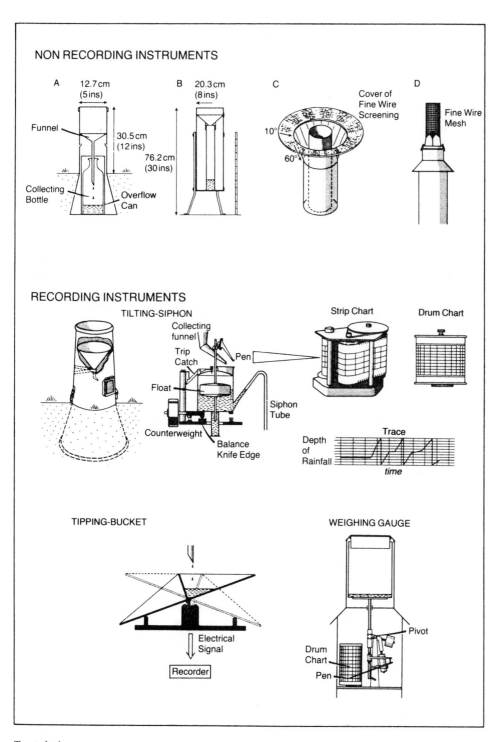

Types of rain gauge.
Source: *K.J. Gregory and D.E. Walling 1976:* Drainage basin form and process. *London: Edward Arnold.*

collecting chamber fills with water and this indicates rainfall intensity as well as amount. The collecting chamber empties rapidly so that the minimum of information on rainfall characteristics is lost.   AMG

Reading
Meteorological Office 1969: *Observer's handbook*. London: HMSO.

rain shadow An area experiencing relatively low rainfall because of its position on the leeward side of a hill or mountain range. Uplift and precipitation over the hills decreases the water content of the air, and this, coupled with the descent of air down the leeward slope, reduces the capacity of the air to produce rain. A good example in the UK is the Moray Firth area to the lee of the Scottish Highlands, where rainfall totals are relatively low.   PS

rain splash See RAINDROP IMPACT EROSION.

rainbow An optical effect consisting of an arc of the spectral colours. It is formed by the passage of sunlight through raindrops. As the beam of light enters the raindrop it is first refracted, then internally reflected from the far side before being refracted again as it leaves the drop. Some of the light may be reflected twice to produce a double rainbow effect. White light from the sun is composed of the colours of the spectrum which are separated in the refraction process because of their slightly different wavelengths. The degree of coloration of the rainbow depends upon the size of the drop and the intensity of the sunlight.   PS

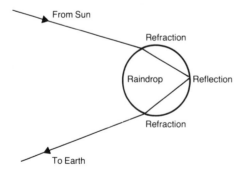

raindrop impact erosion Occurs when energy is released as individual raindrops fall on an exposed soil surface. Soil particles are momentarily lifted from the surface and fall downslope (Mosley 1973). The amount of available energy depends on the rainfall intensity, drop size distribution, and the terminal velocity of the falling drops, but in a typical tropical rainstorm about 20 million $J\,ha^{-1}\,h^{-1}$ is available. Raindrop impact erosion is most effective on slopes of about 33–45°, and is most effective at the beginning of a rainfall when soil particles are not yet bound together by moisture. Overland flow protects soil particles completely when the depth of flow equals or exceeds about three drop diameters.   WLG

Reference
Mosley, M.P. 1973: Rainsplash and the convexity of badland divides. *Zeitschrift für Geomorphologie*, supp. band 18, pp. 10–25.

rainfall The total water equivalent of all forms of precipitation or condensation from the atmosphere received and measured in a RAIN GAUGE. Values are normally expressed as daily, monthly or annual totals. On a world scale, mean annual rainfall amounts are very variable from almost zero in the driest deserts to above 10,000 mm in a few parts of India and Hawaii.   PS

rainfall intensity The rate at which rainfall passes a horizontal surface. Usually this will be the ground where a recording RAIN GAUGE can indicate the value, but it is possible to determine the rates beneath cloud-base by raindrop impaction onto specially designed aircraft equipment or by radar. Rates are normally expressed in $mm\,h^{-1}$ averaged over the hour so are not necessarily representative of short period bursts when much higher rates are possible. For the UK the annual mean rainfall intensity is only $1.25\,mm\,h^{-1}$. In the tropics much higher values are recorded with annual means above $15\,mm\,h^{-1}$.   PS

rainfall run-off That part of the hydrological cycle which connects rainfall and channel flow. When rainfall exceeds the infiltration capacity of the surface materials water begins to collect in small surface depressions. Eventually these surface depressions fill and the water begins to move downslope. The entire overland process is termed Hortonian flow after R.E. Horton who first outlined the process. (See also OVERLAND FLOW; RATIONAL FORMULA.)   WLG

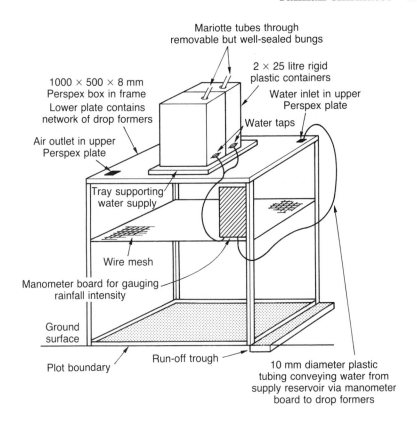

Mariotte tubes through
removable but well-sealed bungs

2 × 25 litre rigid
plastic containers

1000 × 500 × 8 mm
Perspex box in frame
Lower plate contains
network of drop formers

Water inlet in upper
Perspex plate

Air outlet in upper
Perspex plate

Water taps

Tray supporting
water supply

Wire mesh

Manometer board for gauging
rainfall intensity

Ground
surface

Plot boundary

Run-off trough

10 mm diameter plastic
tubing conveying water from
supply reservoir via manometer
board to drop formers

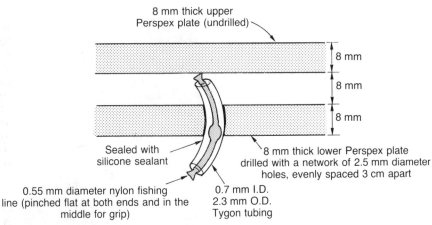

8 mm thick upper
Perspex plate (undrilled)

8 mm

8 mm

8 mm

Sealed with
silicone sealant

8 mm thick lower Perspex plate
drilled with a network of 2.5 mm diameter
holes, evenly spaced 3 cm apart

0.55 mm diameter nylon fishing
line (pinched flat at both ends and in the
middle for grip)

0.7 mm I.D.
2.3 mm O.D.
Tygon tubing

Rainfall simulator: A 'drip-type' simulator and its drop former design.
Source: *T.A.S. Bowyer-Bower and T.P. Burt 1989: Rainfall simulators for investigating soil response to rainfall. Soil technology 2. Figure 3.*

rainfall simulator A rainfall simulator is a mechanical device which produces artificial precipitation over a limited surface area under highly controlled conditions, usually with the objective of measuring run-off, infiltration, or sediment yield. The device usually takes the form of either a sediment tank or a mobile sprinkler. The sediment tank is a large box with an open top containing sediment to be subjected to testing. Vertical pipes with sprinkler heads are arranged around the perimeter of the tank and provide the simulated rainfall by spraying. Mobile rainfall simulators usually consist of a wheel-mounted frame to support pumps, pipes with sprayer heads, and water tanks (see diagram). The device is towed to a field location where natural surfaces are subjected to the artificial precipitation generated by the device. The advantage of rainfall simulators lies in the ability to control rates and intensity of precipitation. The disadvantage of rainfall simulators lies in the difficulty of obtaining an even distribution of water over the test surface. WLG

Reading
Schumm, S.A. 1973: Geomorphic thresholds and complex response of drainage systems. In M. Morisawa ed., *Fluvial geomorphology*. Binghamton, NY: State University of New York Press.

raised beach An emerged shoreline represented by stranded beach deposits, marine shell beds, and wave cut platforms backed by former SEA CLIFFS. During the first decades of the twentieth century it was believed that many raised beaches were the result of eustatic changes brought about by changes in the volume of water stored in ice caps during glaciations and deglaciations, though it was also appreciated that in areas where the earth's crust had been weighed down by the presence of an ice mass that ISOSTASY would have been an important process. In reality, though both these processes are important, there is a wide range of factors that can cause sea-level changes. Raised beaches, which can occur at heights of as much as tens or hundreds of metres above current sea levels, may be warped, so that care needs to be exercised in correlating on the basis of height alone.

ASG

Reading
Guilcher, A. 1969: Pleistocene and Holocene sea-level changes. *Earth science reviews* 5, pp. 69–97.
Rose, J. 1990: Raised shorelines. In A.S. Goudie ed. *Geomorphological techniques*. 2nd edn. London: Unwin Hyman.

randkluft The gap between the back wall of a cirque and the glacial ice that fills the cirque.

random-walk networks Drainage networks can be simulated by a random-walk STOCHASTIC MODEL in which the network evolves on a regular grid, some cells being randomly selected to contain a stream source. The direction of exit from the current cell is then chosen by reference to random number tables. The model may be purely random with equal probabilities ($P = 0.25$) of a move in each cardinal direction (Leopold and Langbein 1962), or biased with, for example, $P = 0.25$ for left and right moves, $P = 0.5$ 'downhill' and $P = 0.0$ 'uphill', to simulate a regional slope. Spatial variation of bias can be allowed to simulate different topographical influences. Constraints to disallow triple junctions, reversals of direction and closed loops are necessary. These simulated networks commonly obey the laws of drainage network composition. KSR

Reading
Leopold, L.B. and Langbein, W.B. 1962: The concept of entropy in landscape evolution. *US Geological Survey professional paper* 500-A.

ranker The name given to a soil which has undergone limited development and has started to have some organic accumulation. Such soils occur on young geomorphological surfaces (e.g. recently deposited alluvium or dune sand).

rapids A steep section in a river channel where the velocity of flow increases and there is extreme turbulence.

Rapids developed on the River Tees above High Force in the north of England. Here, a hard dolerite outcrop caps carboniferous limestone.

rated section A method of obtaining a continuous record of discharge for a river section by continuously recording water stage or level and establishing a relationship between stage and discharge to rate the section. The discharge measurements required to establish this relationship are commonly made using the VELOCITY AREA METHOD. It is important to obtain a stable rating relationship and a section with a bedrock control or artificially stabilized bed and banks is frequently employed. The records obtained are generally less accurate than those provided by WEIRS and FLUMES, particularly for low flows. (See also DISCHARGE.) DEW

rating curve A term frequently used to describe a relationship between discharge and water stage or between suspended sediment and solute transport and water discharge, which can be used to estimate values of the former variable from measurements of the latter. In the case of suspended sediment and solute transport, values of either concentration (mg l^{-1}) or material discharge (kg s^{-1}) may be plotted against water discharge and logarithmic axes are commonly employed. The characteristics of these plots, including slope, intercept, and degree of scatter, have frequently been used to characterize the sediment and solute response of a drainage basin. DEW

Reading
Gregory, K.J. and Walling, D.E. 1976: *Drainage basin form and process*. London: Edward Arnold. Pp. 215–25.

rational formula A simple and long used formula to estimate run-off (Q) in terms of a coefficient (C) depending upon the character of the surface of the drainage area, a measure of precipitation (I), and an index of the basin area (A) in the form:

$$Q = CIA.$$

KJG

Raunkiaer's life forms The life form of a plant is its gross structure. The best-known classification of life forms is that of C. Raunkiaer (1934), who arranged the life forms of plant species into a series primarily based on the position of the perennating buds. He suggested that this reflected an adaptation to climate, arguing that plants' environments were formerly more uniformly hot and moist than at present, that the most primitive life form is the one dominating tropical environments, and that the more highly evolved forms are found particularly in areas with colder climates. The main categories he distinguished are as follows:

Phanerophytes, with the perennating buds on aerial shoots – the most primitive form, in Raunkiaer's view. Evergreen and deciduous

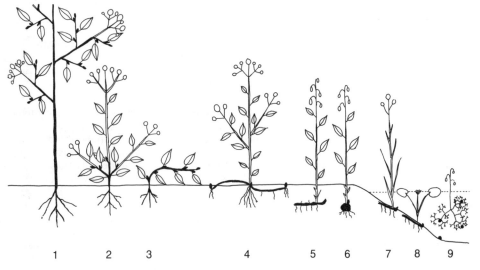

Raunkiaer's life forms: (1) Phanerophytes; (2–3) Chamaephytes; (4) Hemicryptophytes; (5–9) Cryptophytes. The parts of the plants that die during the unfavourable season are shown unshaded, the persistent portions and perennating buds are black.

phanerophytes are distinguished, and each of these categories can be subdivided:

Nanophanerophytes, 2 m in height (shrubs)
Microphanerophytes, 2–8 m
Mesophanerophytes, 8–30 m
Megaphanerophytes, over 30 m (the larger forest trees).

Chamaephytes, with the perennating buds close to ground level. This category includes forms in which aerial shoots die away as the unfavourable season (winter or the dry period) approaches, plants in which the vegetative shoot grows along the ground, and cushion plants.
Hemicryptophytes, where the perennating buds are at ground level, almost all the above-ground material dying with the advance of unfavourable conditions. The group includes plants with stolons and rosette plants.
Cryptophytes, where the perennating buds are below the ground surface or submerged in water. PHA

Reference
Raunkiaer, C. 1934: *The life forms of plants and statistical plant geography.* Translated by H. Gilbert-Carter, A. Fausboll and A.G. Tansley. Oxford: Clarendon Press.

reach A length of channel, as applied to a coastal inlet, the arm of a lake and river channels. The term may be used in more specialized senses, as for a relatively straight section of a navigation waterway, or alternatively for a short length of channel for which discharge or other hydraulic conditions are approximately uniform. JL

reaction time The time which elapses between the application of a change to, or a constraint upon, an earth system and the beginning of adjustment of the system. Reaction time is therefore the period of time after the modification and before RELAXATION TIME begins. KJG

Reading
Graf, W.L. 1977: The rate law in fluvial geomorphology. *American journal of science* 277, pp. 178–91.

recession limb of hydrograph, recession curve The recession limb of the stream discharge hydrograph is the portion after the peak discharge and represents the time after precipitation has ceased when discharge gradually falls and is unaffected by rainfall (for diagram see HYDROGRAPHS). By analysing several hydrographs

for the same gauging station it is possible to derive an average recession curve which may be expressed by a mathematical function such as the exponential form:

$$Q_t = Q_o e^{-at}$$

where Q_t is discharge at time t, Q_o is the initial discharge, a is a constant, t is the time interval and e is the base of natural logarithms. KJG

recharge The process by which water is absorbed, transmitted and ultimately added to the zone of saturation, or by which soil moisture is replaced after drainage and evapotranspiration losses. Areas of particular importance for the replenishment of GROUNDWATER are known as recharge zones. PWW

recumbent fold A fold that is so overturned that the strata lie horizontally.

recurrence interval or return period (T_r) is the expected frequency of occurrence in years of a discharge of a particular magnitude. For a series of flood discharges, which may be obtained as ANNUAL SERIES, ANNUAL EXCEEDANCE SERIES, PARTIAL DURATION SERIES, ranked from the largest to the smallest the formula usually used to calculate recurrence interval is:

$$T_r = \frac{N + 1}{m}$$

where N is the total number of items in the series and m is the rank in the array numbering from the largest (1) downwards. KJG

red beds Sediments, soils or sedimentary rocks which possess red coloration because of the presence of finely divided ferric oxides, chiefly haematite. Most represent iron-stained or cemented clastic sediments. In the continental literature equivalent terms for red beds are *couche rouge*, *rotschicht* and *capa roja* (Pye 1983). A distinction may be drawn between *in situ red beds* (formed in place, without transportation, by processes of direct chemical precipitation or by weathering, soil formation and diagenesis), and *detrital red beds* (formed by erosion, transportation and redeposition of existing red soils or sediments). Only sediments and soils having hues redder than 5YR on the Munsell soil colour chart should be described as red beds. Recent work suggests that if a suitable source of iron and oxidizing conditions exist, reddening can occur very rapidly after deposition of a sediment.   ASG

Reading and Reference
Pye, K. 1983: Red beds. In A.S. Goudie and K. Pye eds, *Chemical sediments and geomorphology*. London: Academic Press.
†Turner, P. 1980: *Continental red beds*. Amsterdam: Elsevier.

redox potential Reduction is the gain of an electron in a chemical reaction; the opposite is known as oxidation. Oxidation is most common in weathering processes. As some elements exist in several oxidation states, e.g. ferrous, Fe(II) is oxidized to the ferric state, Fe(III). The stability of any state depends upon the ease with which oxidation or reduction can occur, i.e. electron transfer. This is often dependent on pH. Redox (reduction–oxidation) potential is a measure of this ease of transfer and is symbolized by Eh. A plot of redox potential (measured in volts) versus pH is a useful tool in determining the stability fields of products of reactions, especially for geochemical environments.   WBW

Reading
Yatsu, E. 1988: *The nature of weathering*. Tokyo: Sozosha.

reef A rocky construction found at or near sea level, formed mainly from biogenically produced carbonates. The term bioherm is also used. Several forms of marine organisms are capable of precipitating calcium and other carbonates in skeletal or non-skeletal forms and it is the accumulation of this material that gives rise to the characteristic form of reefs. Coral reefs, whose form is usually dominated by coral skeletons, are the most common form of reef, but other organisms may also form or contribute to the development of reefs, e.g. algae, vermetids and serpulids. Reefs vary greatly in size from individual clusters of organisms (called patch reefs or micro-atolls) to large-scale rock masses. Large coral reefs exhibit a range of facies controlled by the positions of the corals relative to exposure. Four major forms of large-scale coral reefs have been identified, i.e. fringing reefs, barrier reefs, atolls and table reefs (Stoddart 1969).   HV

Reference
Stoddart, D.R. 1969: Ecology and morphology of recent coral reefs. *Biology reviews* 44, pp. 433–98.

reflective beach Has a steep profile and bars are rarely present. There is no surfzone circulation alongshore. The term implies that a considerable proportion of the incoming wave energy is reflected from the steep face of the beach berm. The other extreme is the dissipative beach with a gently sloping gradient and bars which absorb and dissipate much of the incoming wave energy. The reflective beach is usually the result of the action of constructive waves which build up the foreshore and create a steep-fronted berm. BEACH cusps and rapid alongshore changes in foreshore grain size are common as a reflective beach is developing, but not when it is fully formed.   CAMK

Reading
Davies, J.L. 1980: *Geographical variation in coastal development*. 2nd edn. London: Longman. Ch. 7.
Wright, L.D., Chappell, J., Thom, B.G., Bradshaw, M.P. and Cowell, P. 1979: Morphodynamics of reflective and dissipative beach and inshore systems; south-east Australia. *Marine geology* 32.1–2, pp. 105–40.

refraction, wave Wave refraction is the bending of the wave front as water depth changes when it is less than half the wave length. The wave velocity decelerates as water depth decreases. The wave front, therefore, bends to become more nearly

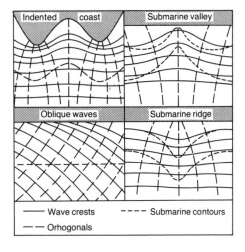

Examples of wave refraction showing wave crests and orthogonals relative to submarine contours.
Source: *C.A.M. King 1980: Physical geography. Oxford: Basil Blackwell.*

The refraction of waves in a bay at Rhossili, Gower, South Wales. Wave refraction is one of the most important controls of bay beach geometry.

parallel to the bottom contours. Orthogonals, which are lines normal to the wave crests and between which energy is constant in deep water, converge on headlands and diverge in bays. Wave energy is concentrated on the headlands and dissipated in bays. Wave refraction diagrams can be drawn manually or by computer where the offshore relief, wave approach direction and wave period are known. These show zones of wave energy concentration and dissipation by constructing the orthogonals or wave rays. CAMK

Reading
Authur, R.S., Munk, W.H. and Isaacs, J.D. 1952: The direct construction of wave rays. *American Geophysical Union transactions* 33, pp. 955–65.

Harrison, W. and Wilson, W.W. 1964: *Development of a method for numerical calculation of wave refraction.* Technical memoir 6. Washington, DC: Coastal Engineering Research Centre.

refugia Localities or habitats in which formerly widespread organisms survive in small restricted populations which are either disjunct or endemic in their distribution. Such localities usually possess some distinctive microclimatic, geomorphological, ecological or historic characteristic which accounts for the survival and persistence of the relict organism at that site. A classic example of such a refugium is the Upper Teesdale area of Yorkshire and Co. Durham in north-east England, which harbours survivals from the open-habitat flora that existed just after the last Ice Age (e.g. the arctic-alpine species, *Betula nana*). PAS

Reading
Holmquist, C. 1962: The relict concept. *Oikos* 13, pp. 262–92.

Pennington, W.A. 1974: *The history of British vegetation.* 2nd edn. London: English Universities Press.

reg A stony desert or any desert region where the surface consists of sheets of gravel.

regelation The refreezing of meltwater. This commonly occurs at a glacier surface to form superimposed ice and at the base of a sliding glacier to form regelation ice. In the latter case ice at the base of a warm-based glacier melts on the upstream side of an obstacle where pressure is locally high and refreezes on the downstream side where pressure is lower. DES

Reading
Paterson, W.S.B. 1981: *The physics of glaciers.* Oxford and New York: Pergamon.

regeneration complex A term coined by Scandinavian botanists in the early twentieth century to describe the small-scale mosaic pattern of vegetation on raised-bog plateaux surfaces which was thought to be the mechanism whereby such OMBRO-TROPHIC bogs grew or regenerated themselves in an autogenic cyclic fashion. The mosaic of species – *Sphagna* (bog mosses) *Calluna* (heather), *Eriophorum* (cotton-grass) etc. – also reflects the microtropography of the mire surface, a pattern of hummocks and hollows, which may contain open water. Each species, because of its tolerance to waterlogging and acidity etc., has a micro-habitat range within this topography; *Calluna* and *Eriophorum vaginatum* generally grow on hummocks with *Sphagnum capillifolium* (formerly *S. rubellum*); the hollows and pools are inhabited by *S. cuspidatum* and other aquatic sphagna, and the hummock sides and flattish 'lawn' areas by *S. papillosum, S. magellanicum* and *Eriophorum angustifolium* etc. Of all the plant associations of an untouched raised bog it is this regeneration complex which is the most vigorous in growth and peat accumulation. Osvald, the Swedish bog ecologist who popularized the term in the 1920s and 1930s after his work on the Komosse bogs, influenced the British ecologists Tansley and Godwin, and the idea of the regeneration complex as a cyclic succession emerged during this period. The basic idea was that as a hummock became higher and drier than the surrounding hollows, its growth would slow down and it would become moribund. The hollows, accumulating peat rapidly, would then grow up to overtop and drown the former hummock which would then in turn become a hollow or pool. This hummock–hollow cycle would be repeated all over the growing bog surface and was seen as the normal growth process over centuries and millenia. Lack of stratigraphic evidence, as well as doubts over climatic influences, did not prevent this theory becoming widely accepted in the ecological literature. Specific testing of the theory, from peat sections of the past two millenia, was carried out by Barber (1981) who rejected the theory in favour of close climatic control of the hydrology, and therefore of the floral composition and peat stratigraphy, of raised bogs.   KEB

Reading and Reference
†Barber, K.E. 1981: *Peat stratigraphy and climatic change: a palaeoecological test of the theory of cyclic peat bog regeneration.* Rotterdam and Salem, New Hampshire: Balkema.
†Godwin, H. 1981: *The archives of the peat bogs*, Cambridge: Cambridge University Press.

regime The average annual variations of climatic or hydrological variables. Seasonal fluctuations of river discharge, represented by the mean monthly flows, are river regimes which vary regionally in relation to precipitation, temperature, evapotranspiration and drainage basin characteristics (Beckinsale 1969). Using the analogy of average climatic statistics, equilibrium channel morphology is also referred to as the regime state, which is constant over a period of years, is adjusted to prevailing hydrological and sedimentological influences, and is predicted using regime theory (Blench 1969). The term is also applied to flow regimes, which are defined by the Froude and Reynolds numbers and associated bed-form types.   KSR

References
Beckinsale, R.P. 1969: River regimes. In R.J. Chorley ed., *Water, earth and man*. London and New York: Methuen.
Blench, T. 1969: *Mobile bed fluviology*. Edmonton: University of Alberta Press.

regime theory The empirical science concerned with the design of stable channel shapes for artificial channels, such as canals and irrigation ditches, formed in natural sediments and carrying a sediment-laden water flow (Blench 1969; Richards 1982, pp. 286–94). It is possible to design stable channels in which the perimeter sediments are at the threshold of motion, but sediment transport must be allowed in larger artificial channels supplied from rivers with heavy sediment loads. Essentially the objective of regime theory is to find the combination of channel width, depth and meander wavelength which allows the water discharge to carry the input sediment load down the required slope without excessive silting or erosion, thereby maintaining stability over a period of years. The methods tend to be empirical because of the indeterminate nature of loose-boundary HYDRAULICS, and distinct sets of regime equations predicting channel width, depth and flow velocity for given discharges and slopes are defined for different sedimentological conditions (e.g. sandbed and gravel-bed channels). However, approximation is often necessary and a degree of self-adjustment is allowed between silting and eroding limits; this was the case in the evolution of the Indo-Gangetic canals whose construction stimulated the development of regime theory in the 1890s. The stable canal relationships of regime theory are very similar to the HYDRAULIC GEOMETRY relations for natural alluvial river channels, which also have self-created morphologies which allow continuity of sediment transport.   KSR

References
Blench, T. 1969: *Mobile bed fluviology.* Edmonton: University of Alberta Press.
Richards, K.S. 1982: *Rivers: form and process in alluvial channels.* London and New York: Methuen.

regolith A term coined by Merrill in 1897 (pp. 299–300) to describe the 'superficial and unconsolidated portion of the earth's crust . . . the entire mantle of unconsolidated material, whatever its nature or origin'. Some later workers have tended to use it as a synonym for weathering products, but such a narrow usage is incorrect (Gale, 1992).   ASG

References
Gale, S.J. 1992: Regolith: the mantle of unconsolidated material at the Earth's surface. *Quaternary research* 37, pp. 261–2.
Merrill, G.P. 1897: *A treatise on rocks, rock weathering and soils.* New York: Macmillan.

regosol Any weakly developed soil.

regur Black, clayey soils which develop in tropical regions of high rainfall and temperatures, notably the north-west Deccan of India.

rejuvenation The renewal of former activity, as of a fault which is reactivated by new tectonic movement or especially a stream whose erosional activity is redeveloped by uplift, a base-level fall or possibly a change in stream sediment load or discharge. The term may be applied to whole landscapes whose reduced relief may be redeveloped as a result of such changes, giving valley-in-valley forms.

The term especially derives from the interpretation of landforms developing in a cyclical fashion from youth to old age, with rejuvenation representing the redevelopment of 'young' surface forms in landscapes otherwise well advanced in a cycle of development. Some of the landforms that were so interpreted more than half a century ago might now be explained in alternative

ways, with emphasis placed on equilibrium adjustments between forms, environments and processes.   JL

relative age The age of an event or feature expressed not in terms of time units, such as years, but in relation to other phenomena (cf. absolute dating).

relative humidity See HUMIDITY.

relaxation time The time taken by any system to adjust morphologically to a change in energy input: the adjustment occurs gradually, over a definite lapse of time, and the system concerned passes through a number of transitory states before attaining a new equilibrium condition. The term was introduced generally into geomorphological use by Chorley and Kennedy (1971) but a fuller and better defined employment of the concept is given by Allen (1974).   BAK

Reading and References
†Allen, J.R.L. 1974: Reaction, relaxation and lag in natural sedimentary systems: general principles, examples and lessons. *Earth science reviews* 10, pp. 263–342.
†Chorley, R.J. and Kennedy, B.A. 1971: *Physical geography: a systems approach.* London: Prentice-Hall.

remanié A glacier which is not directly connected to a snow field but receives ice from avalanches.

remote sensing At its broadest the term refers to the observation and measurement of an object without touching it, and involves the use of force fields, acoustic energy and ELECTROMAGNETIC RADIATION. Within physical geography it involves the use of electromagnetic radiation sensors to record images of the environment which can be used to yield useful information. The four most important sensors are the AERIAL CAMERA, MULTISPECTRAL SCANNER, SIDEWAYS LOOKING AIRBORNE RADAR and THERMAL INFRARED LINE-SCANNER. The two most important platforms are the aircraft and satellite.

The 1960s were formative years for remote sensing during which visual interpretation of black and white aerial photography paralleled research into the use of data from the new aircraft and satellite borne sensors. Remote sensing, especially of the non-photographic type, grew rapidly after the launch of the Landsat 1 satellite in 1972. Today remote sensing is regarded as a powerful and useful technique in physical geography.   PJC

Reading
Barrett, E.C. and Curtis, L.F. 1992: *Introduction to environmental remote sensing.* 3rd edn. London and New York: Chapman & Hall.

Cracknell, A.P. and Hayes, L.W.B. 1991: *Introduction to remote sensing.* London: Taylor & Francis.

Curran, P.J. 1987: Remote sensing methodologies and geography. *International journal of remote sensing* 8, pp. 1255–75.

Lillesand, R.M. and Kiefer, R.W. 1987: *Remote sensing and image interpretation.* 2nd edn. New York: Wiley.

rendzina Dark, organic-rich soil horizons developed upon unconsolidated calcareous materials in areas of chalk and limestone bedrock.

representative and experimental basins Small drainage basins where the representative basins are chosen to be representative of hydrological regions and should experience minimum change during monitoring and where the experimental basins are subjected to deliberate modification so that the impact of the modification on drainage basin dynamics may be identified. The establishment of a large number of representative and experimental basins was one of the major achievements of the International Hydrological Decade (IHD, see HYDROLOGY).

Studies in representative basins involve monitoring hydrological processes to improve understanding of these processes and their inter-relationships. If plentiful data can be built up from catchments which are representative of a particular hydrological region these data should help to improve estimation of the magnitude of hydrological processes, particularly discharge characteristics, in ungauged catchments in the same region. Benchmark catchments (Toebes and Ouryvaev 1970) have been identified as a special type of representative basin which are in their natural state and, therefore, permit the observation of hydrological processes without the effects of man. The Vigil Network also forms a network of representative basins which was originally conceived within the USA (Leopold 1962; Slaymaker and Chorley 1964). In the case of the Vigil Network, the basins are not protected from artificial change but the aim is to monitor a large number of hydrological and geomorphological variables using simple and inexpensive techniques so that

landscape and process change may be identified over long periods of time.

Work in experimental basins involves not only observation of hydrological processes but also deliberate modification of the basin so that the impact of cultural changes on hydrological processes may be measured. Experimental basins are often subjected to a calibration period for several years before modification. This allows the magnitude and variability of processes before and after the change to be observed and compared. If more than one basin is available, where all the basins are virtually identical in size, physical character and vegetation cover, either a paired or a multiple watershed experiment can be carried out. In this type of experiment a long calibration period is desirable but not essential. One or more control basins can be monitored without receiving modification and the remaining basin or basins can be subjected to one or more experimental treatments, preferably using replicate basins for each treatment. In all the basins the same processes are observed so that differences between the control and the treated basins may be identified.   AMG

Reading and References
†IAHS 1980: *The influence of man on the hydrological regime with special reference to representative and experimental basins.* Helsinki Symposium. International Association of Hydrological Sciences publication 130.

†IASH 1965: *Representative and experimental areas.* Budapest Symposium. International Association of Scientific Hydrology publication 66, vols 1 and 2.

Leopold, L.B. 1962: The vigil network. *International Association of Scientific Hydrology bulletin* 7, pp. 5–9.

Slaymaker, H.O. and Chorley, R.J. 1964: The Vigil Network system. *Journal of hydrology* 2, pp. 19–24.

Swank, W.T. and Crossley, D.A. 1987: *Forest hydrology and ecology at Coweeta. Ecological studies,* Vol. 66. Berlin: Springer-Verlag.

Toebes, C. and Ouryvaev, V. eds 1970: *Representative and experimental basins: An international guide for research and practice.* UNESCO studies and reports in hydrology 4. Paris: UNESCO.

resequent stream A stream following the original direction of drainage, but developed at a later stage. This might apply to streams on the back slope of a cuesta of *resistant* rock which did not outcrop on the originally exposed landsurface which guided initial stream development. The term is not now commonly used.   JL

reservoir rocks See GROUNDWATER.

reservoir, storage effects of Reservoirs may be constructed to regulate river flows for supply purposes downstream (regulating reservoirs), to provide water supply directly from direct supply reservoirs or to control floods downstream in which case the reservoir is maintained at less than capacity so that flood discharges may accumulate in the reservoir (flood control reservoirs). Immediately downstream of the dam, scour of the channel bed and banks may occur because the water released is comparatively free of sediment and further downstream there can be major changes of CHANNEL CAPACITY or of channel planform as a response to the changes in FLOOD FREQUENCY and to the altered sediment transport. There will also be changes in the river ecology as a response to changes in flow and to the alterations of aquatic habitats.   KJG

Reading
Eschner, T., Hadley, R.F. and Crowley, K.D. 1983: Hydrologic and morphologic changes in channels of the Platte River Basin in Colorado, Wyoming, and Nebraska: A historical perspective. *US Geological Survey professional paper* 1277-A.
Petts, G.E. 1979: Complex response of river channel morphology to reservoir construction. *Progress in physical geography* 3, pp. 329–62.
Williams, G.P. and Wolman, M.G. 1984: Downstream effects of dams on alluvial rivers. *US Geological Survey professional paper* 1286.

residence time A concept employed in studies of chemical weathering and solute generation in drainage basins to describe the length of time between the input of water as precipitation and its output as run-off. The residence time is seen as a major control on the magnitude of solute uptake by water moving through the drainage basin system and therefore on solute concentrations in streamflow, since a significant period of time may be required for the water to reach chemical equilibrium with the soil and rock.   DEW

residual strength (or residual shear strength) The strength of a soil (usually a clay soil) which is below the peak SHEAR STRENGTH. Large shear strains (deformations) produce a reorientation of the clay mineral particles in the FABRIC and allow a shear plane to develop. The shear strength along this plane is less than the peak strength. The residual strength can be determined in a SHEAR BOX (or a modified version of this called a ring shear) apparatus for soil testing.   WBW

respiration For land plants the chemically reverse process of PHOTOSYNTHESIS or, in other words, the release of energy from food molecules in cells, which is then used in metabolism. Plant 'waste' products of carbon dioxide and water are also transferred back into the atmosphere. The reactions may be represented, in simplified form, by:

$$2(C_3H_6O_3) + 6O_2 \rightarrow 6CO_2 + 6H_2O$$
$$+ \text{heat}$$

The heat goes to form part of the long-wave energy component of the earth's atmosphere, and is unavailable thereafter for use in biological systems. The amount of the plant's available energy store which is used in respiration varies widely, but it is often in excess of 50 per cent of that used in growth, and may be very much more than this (Odum 1971; see also NET PRIMARY PRODUCTIVITY). Unlike photosynthesis, respiration is a day-long not just a daytime phenomenon. For higher land animals respiration is a broadly similar process after atmospheric oxygen has been taken in through breathing.   DW

Reference
Odum, E.P. 1971: *Fundamentals of ecology.* 3rd edn. Philadelphia: W.B. Saunders.

response analysis The representation of observed tidal variations in terms of the frequency-dependent amplitude and phase responses to input or forcing functions, usually the gravitational potential due to the moon and sun, and the radiational meteorological forcing.   DTP

resultant wind The vectorial average of all wind directions and speeds for a given level at a given place for a specified period, such as one month. It is obtained by resolving each wind observation into components from north and east, summing over the given period, obtaining averages and reconverting the average components into a single vector.   BWA

resurgence An emergence point of underground water. The term is often used to describe a karst SPRING, the headwaters of which are initially on the surface but are lost underground by disappearing down a stream-sink (or SWALLOW HOLE). A distinction is sometimes made between this kind of spring and that with no known surface headwaters, which is called an exsurgence. PWW

Reading
Bögli, A. 1980: *Karst hydrology and physical speleology.* Berlin: Springer-Verlag.
Smith, D.I., Atkinson, T.C. and Drew, D.P. 1976: The hydrology of limestone terrains. In T.D. Ford and C.H.D. Cullingford eds, *The science of speleology.* London: Academic Press. Pp. 179–212.

retention curve Or soil moisture retention curve is that obtained by plotting soil moisture content against soil moisture suction. The curve will be of a different form according to whether the soil is drying or wetting. This is due to the phenomenon of capillary hysteresis. PWW

Reading
Childs, E.C. 1969: *An introduction to the physical basis of soil water phenomena.* New York: Wiley.
Ward, R.C. and Robinson, M. 1990: *Principles of hydrology.* 3rd edn. Maidenhead: McGraw-Hill.

retention forces Are those responsible for holding water in the pores of a rock or soil against the force of gravity. The forces involved are capillarity (surface tension), adsorption and osmosis. PWW

Reading
Ward, R.C. and Robinson, M. 1990: *Principles of hydrology.* 3rd edn. Maidenhead: McGraw-Hill.

retrogradation The destruction of a beach profile by large breakers resulting in retreat of the shoreline.

return period The average time between events such as the flooding of a particular level. This information may also be expressed as the level which has a particular return period of flooding, for example, a hundred years. The inverse of the return period is the statistical probability of an event occurring in any individual year. DTP

reversed fault A fault whose fault-plane is inclined towards the upthrown side.

reversed polarity The condition occurring when the earth's magnetic poles change their polarities, the positive becoming the negative and vice versa. Generally applied when the positive magnetic pole lies in the southern hemisphere.

reversing dune A dune that tends to grow upwards but migrate only a limited distance because seasonal shifts in direction of the dominant wind cause it to move alternately in nearly opposite directions.

Reynolds number (R_e) A dimensionless ratio defining the state of fluid motion as laminar or turbulent according to the relative magnitude of inertial and viscous forces. In general:

$$R_e = \rho_f v L / \mu = v L / \nu$$

where ρ_f is fluid density, v is velocity, L is a characteristic length, μ is the dynamic viscosity and ν is the kinematic viscosity ($\nu = \mu/\rho_f$), which is a measure of molecular interference between adjacent fluid layers. The Reynolds number for channel flow is defined by using the hydraulic radius R or the mean depth as the characteristic length, and the mean flow velocity. Flow is turbulent if $R_e > 750$; laminar flow ($R_e < 500$) is rare in rivers and only occurs in shallow overland flow depths. Grain Reynolds numbers can be defined using grain diameter as the length, and grain fall velocity in still water as the velocity. If the grain Reynolds number is <0.1, flow around the falling grain is streamlined and laminar, and the fall velocity varies as the square of the grain diameter (Stokes law). This is the case for silt and clay particle sizes; grain Reynolds numbers are larger for sand particles, and turbulence and flow separation occur around the falling grain as inertial forces dominate. For such grains the fall velocity varies as the square root of grain diameter (Richards 1982, pp. 76–9). KSR

Reference
Richards, K.S. 1982: *Rivers: form and process in alluvial channels.* London and New York: Methuen.

rheidity The capacity of some solid materials to flow under certain conditions.

rheology The study of the deformation and flow of matter. In particular, rheology is concerned with whether a substance behaves as an elastic solid or effectively as a 'viscous' solid or in a manner intermediate between these two. The type of deformation experienced by a particular material is influenced by the duration of the applied stress. This is expressed by the property of rheidity which depends on the relationship between the resistance to viscous flow (viscosity) and the resistance to elastic deformation (rigidity) of a substance and is expressed in units of time. For ice a deforming stress must be applied for only a few weeks before it begins to flow but for subcrustal material the time required is several thousand years.   MAS

Reading
Hager, B.H. and O'Connell, R.J. 1980: Rheology, plate motions and mantle convection. In P.A. Davies and S.K. Runcorn eds, *Mechanisms of continental drift and plate tectonics*. London: Academic Press.
Mörner, N.-A. ed. 1980: *Earth rheology, isostasy and eustasy*. Chichester: Wiley.
Ramberg, H. 1981: *Gravity, deformation and the earth's crust*. 2nd edn. London: Academic Press.

rheotrophic See NUTRIENT STATUS.

rhexistasy See BIOSTASY.

rhizome The underground stem of some plants. The rootstock.

rhizosphere The portion of the soil zone which immediately surrounds the root systems of plants.

rhodoliths Free-living massive and branching spheroidal growths of calcareous red algae which occur in two different settings: (a) shallow lagoonal, reef-flat and back-reef environments, commonly in tidal channels and sea-grass beds, and (b) moderately deep water on fore-reef terraces or shelf edges.   ASG

Reading
Scoffin, T.P., Stoddart, D.R., Tudhope, A.W. and Woodroffe, C. 1985: Rhodoliths and coralliths of Muri Lagoon, Rarotonga, Cook Islands. *Coral reefs* 4, pp. 71–80.

rhourd A pyramid-shaped sand dune, formed by the intersection of other dunes.

ria An inlet of the sea formed by the flooding of river valleys, either by the rising of the sea during the FLANDRIAN TRANS-GRESSION or as a consequence of sinking of the land. They contrast with fjords which are drowned glacial valleys. They are a feature of Pembrokeshire, the south-west peninsula of England, Brittany and Galicia (Spain).   ASG

Richter denudation slope A straight rock-slope unit with an angle of inclination which is at the maximum angle for stability (usually 32–36°) of its thin talus cover. Such slopes are relatively common in the Transantarctic Mountains, and have been recognized in the Cape Mountains of southern Africa and the European Alps. They form as rock fragments fall from the

cliff at the crest of the Richter slope. If the newly fallen material just covers a little of the base of the cliff, the next fall will be over the new talus and hence the base of the cliff will then be higher, so the cliff will extend upwards by a series of minute steps at the angle of the talus. Eventually the free face will be eliminated and a smooth rock slope of uniform inclination will underlie a talus sheet. The talus may subsequently be removed.   MJS

Reading
Selby, M.J. 1993: *Hillslope materials and processes*. 2nd edn. Oxford: Oxford University Press.

Richter scale A measure of the intensity of an earthquake as determined by seismic recorders. A value of 7 and above denotes a major earthquake.

riedel shears A special form of shear fracture which forms roughly perpendicular to the main direction of shear strain or movement. Although they are mainly found in brittle materials they can occur in muds below the plastic limit.   WBW

riegel A step in the rock floor of a glacial valley.

riffle See POOL AND RIFFLE.

rift valley A valley or linear trough formed by subsidence or downthrusting in areas of continental crust in plate interiors where tensional stresses predominate in the lithosphere. The classic interpretation of the morphology is that of a graben, with the rift floor being seen as a downthrown block bounded by normal faults, which create steep bounding escarpments. Seismic data suggest, however, that in many cases this is an oversimplification. The structure of many rifts appears to be asymmetric, with much of the downthrow occurring along a major boundary listric fault on one side only. These major faults tend to be discontinuous and may alternate along the rift, separated by transfer faults. Such rifts have a half-graben structure (see diagram).   ASG

Reading
Summerfield, M.A. 1991: *Global geomorphology*. Harlow: Longman.

rill A small (maximum of a few centi-metres) channel that changes location with every run-off and that can be obliterated by

A salt lake, Lake Natron, in the floor of the Great Rift Valley in Tanzania. Note the steep west wall of the Rift and the Bast Mountains rising beyond. Rift valleys provide one of the most potent indicators of the importance of tectonic processes.

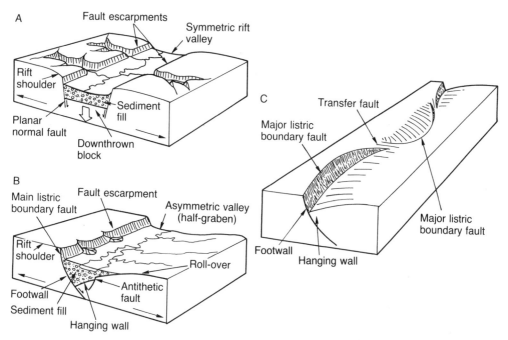

Rift valley: Schematic representation of contrasting rift structures: (A) classic symmetric graben structure with a downthrown block bounded by normal faults; (B) asymmetric, half-graben structure. In both cases the number of faults in real rifts and the complexity of their structure are much greater than indicated here. (C) Highly schematic illustration of alternating half-graben along a rift valley. Source: Summerfield 1991. Figs 4.9 and 4.10.

ploughing. Rills, formed by the merging of sheet wash into channel flow, may join to form larger permanent gullies. Rills are conduits for water and sediment transport: water may flow at velocities greater than 30 cm s^{-1}, and sediment concentrations may be 300,000–500,000 ppm.   WLG

rime ice Ice formed by the accumulation of supercooled water droplets when they strike a cold object and freeze on impact. Rime ice builds up most rapidly in cool humid conditions on surfaces exposed to the wind. On mountain peaks it can accumulate as large cauliflower-shaped excrescences.   DES

Reading
Koerner, R.M. 1961: Glaciological observations in Trinity Peninsula, Graham Land, Antarctica. *Journal of glaciology* 3. 30, pp. 1063–74.

ring complex 'A petrologically variable but structurally distinctive group of hypabyssal or subvolcanic igneous intrusions that include ring dykes, partial ring dykes and cone sheets. Outcrop patterns are arcuate, annular, polygonal and elliptical with varying diameters ranging from less than 1 to 30 km or greater. The majority of ring complexes represent the eroded roots of volcanoes and their calderas' (Bowden 1985, p. 17).   ASG

Reference
Bowden, P. 1985: The geochemistry and mineralization of alkaline ring complexes in Africa (a review). *Journal of African earth sciences* 3, pp. 17–39.

ring-dyke A funnel-shaped or cylindrical intrusion of igneous rock usually surrounded by an older intrusive mass. The dyke appears as a ring of rocks when viewed from the air.

rip current A narrow, fast current flowing seaward through the breaker zone. The term, first introduced by F.P. Shepard in 1936, replaces undertow. Rip currents are fed by longshore currents in the surfzone. Their velocity ranges from about 1 m s^{-1} to over 5 m s^{-1} in severe storms. Channels 1–3 m deep can be scoured in the breaker zone by rip currents. Their spacing depends on variation of wave height alongshore, which may arise through wave refraction or may be related to the width of the surfzone. They may be associated with cuspate shoreline features and edge waves.   CAMK

Reading
Bowen, A.J. 1969: Rip currents. *Journal of geophysical research* 74, pp. 5467–78.
Dalrymple, R.A. 1975: A mechanism for rip current generation on an open coast. *Journal of geophysical research*. 80, pp. 3485–7.
Gruszczyński, M., Rudowski, S., Semil J., Slomiński J., and Zrobek, J. 1983: Rip currents as a geological tool. *Sedimentology* 40, pp. 217–36.

ripple See CURRENT RIPPLES.

rising limb The portion of the stream hydrograph for a storm event between the beginning of the rise in discharge and the peak of the hydrograph. (For diagram see HYDROGRAPH.)

river basin planning (integrated basin planning) The process of managing water resources within the drainage basin in a manner which optimizes water use throughout the basin and minimizes deleterious effects for water, river channels and land use. Although the idea of an integrated approach to river basin development is long established, experience of piecemeal development which has induced subsequent feedback effects in other parts of the drainage basin have fostered the movement towards a more integrated view. This movement has also been encouraged both by the need to design multiple purpose water resource projects that will, for example, supply water and power and reduce floods, and by the use of the drainage basin as the most appropriate unit for analysis in relation to planning. The United Nations first advocated integrated river basin development in 1958 and its report was reissued in 1970.   KJG

Reading and Reference
†Saha, S.K. and Barrow, C.J. 1981: *River basin planning: theory and practice.* Chichester: Wiley.
United Nations 1970: *Integrated river basin development: report of panel of experts.* New York: Department of Economic and Social Affairs.

river classification See CHANNELS, TYPES OF RIVER/STREAM.

river discharge see DISCHARGE.

river metamorphosis Refers to the change of river channel morphology that can occur when changes of discharge and sediment exceed a THRESHOLD condition. Channel changes which occur can be from a multi-thread to a single thread channel and S.A. Schumm (1969) introduced the term

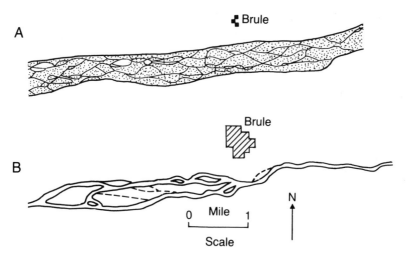

River metamorphosis: The South Platte River at Brule, Nebraska. The channel according to surveys made in 1897 is shown in A and the same channel in 1959 in B is based on aerial photographs.
Source: S.A. Schumm 1971: Channel adjustment and river metamorphosis. In H.W. Shen ed. River Mechanics. *Volume 1. Fort Collins: Water Resources Publications. Pp. 5.1–5.22.*

River metamorphosis: Potential channel adjustments

| Potential adjustments of | Fluvial landform | | |
	River channel cross-section	River channel pattern	Drainage network
Size	INCREASE OR DECREASE OF RIVER CHANNEL CAPACITY Erosion of bed and banks can produce a larger channel which maintains the same shape. Sedimentation can produce a smaller channel which maintains the same shape.	INCREASE OR DECREASE OF SIZE OF PATTERN Increase or decrease of meander wavelength while preserving the same planform shape.	INCREASE OR DECREASE OF NETWORK EXTENT AND DENSITY Extension of channels or shrinkage of perennial, intermittent and ephemeral streams.
Shape	ADJUSTMENT OF SHAPE Width/depth ratio may be increased or decreased.	ALTERATION OF SHAPE OF PATTERN A change from regular to irregular meanders.	DRAINAGE PATTERN CHANGE IN SHAPE Inclusion of new stream channels after deforestation.
Composition	CHANGE IN CHANNEL SEDIMENTS Alteration of grain size of sediments in bed and banks possibly accompanied by development of berms and bars.	PLANFORM METAMORPHOSIS Change from single to multi-thread channel or converse.	NETWORK COMPOSITION CHANGE The replacement of channels with no definite stream channel (dambos in West Africa) by a clearly defined channel.

General changes are indicated in capitals and examples given in lower case letters.

river metamorphosis to signify the range of changes that may arise. Metamorphosis may involve changes of size, shape or composition of aspects of river channel morphology as shown in the table. It is important to be able to predict the degrees of freedom which a river system possesses: width, depth, slope, velocity and plan shape. To predict the character of river metamorphosis it is necessary to suggest what changes will occur, where they will obtain, and when the changes will begin and end. KJG

Reading and References
†Gregory, K.J. 1981: River channels. In K.J. Gregory and D.E. Walling eds, *Man and environmental processes*. London: Butterworth. Pp. 123–43.

Schumm, S.A. 1969: River metamorphosis: *Proceedings of the American Society of Civil Engineers, Hydraulic Division 95*, pp. 251–73.

†— 1977: *The fluvial system*. Chichester: Wiley.

river regime See REGIME.

riverscape The landscape character of the river and the adjacent area. When considering developments which may affect rivers or in the design of river management schemes it is often important to consider the aesthetic quality of the riverscape. Methods have therefore been devised to express the aesthetics of riverscapes in quantitative terms and two broad approaches have been used. One approach has been based upon the components which combine to make the aesthetic character of a particular area of riverscape; values for the physical and chemical character of the river, for biological character, and for human use and interest were used by Leopold and Marchand (1968). A further approach, sometimes developed as an extension of the one based upon components, establishes the extent to which an area of riverscape is unique. Uniqueness was defined by Leopold (1969) in order to decide which of several localities was appropriate for the construction of a new dam, the decision subsequently being taken to site a dam in a valley that was similar to other valleys rather than in a valley which was unique. (See also LANDSCAPE EVALUATION.) KJG

Reading and References
†Dunne, T. and Leopold, L.B. 1978: *Water in environmental planning*. San Francisco: W.H. Freeman.

Leopold, L.B. 1969: Quantitative comparison of some aesthetic factors among rivers. *US Geological Survey circular* 620.

— and Marchand, M.O. 1968: On the quantitative inventory of riverscape. *Water resources research* 4, pp. 709–17.

robber economy The exploitation and destruction of natural resources which cannot be readily replaced. It was a concept introduced by the German geographer Friedrich in 1904 for the depletion of land, wild plants and animals, and was adopted by Jean Brunhes in his *Human Geography* (1920), where he called it 'destructive exploitation'. ASG

References
Brunhes, J. 1920: *Human geography*. London: Harrap.

Friedrich, E. 1904: Wesen und geographische Verbreitung der 'Raubwirtschaft'. *Petermann's Mitteilungen* 50, pp. 68–79, 92–5.

roche moutonnée An asymmetric rock bump with one side ice-moulded and the other side steepened and often cliffed, generally recognized as the hallmark of glacial erosion. The term was introduced by H.B. de Saussure in 1787 in recognition of the similarity of the rocks to the rippled appearance of wavy wigs styled *moutonnées* at the time. The morphology of roches moutonnées seems to reflect the contrast between ABRASION on the smoothed up-ice side and PLUCKING on the lee side. DES

Reading
Chorley, R.J., Schumm, S.A. and Sugden, D.E. 1984: *Geomorphology*. London: Methuen. Ch. 17.

Sugden, D.E., Glasser, N. and Clapperton, C.M. 1992: Evolution of large roches moutonnées. *Geografiska Annaler* 74A, pp. 253–64.

rock drumlin A rock hill streamlined by the passage of over-running glacier ice.

rock flour The fine debris created by ABRASION beneath a glacier. The material, generally less than 0.2 mm in diameter, may be flushed away to give the characteristic brown and blue-green colour of glacial meltwater streams and lakes. (See GLACIER MILK.) DES

Reading
Haldorsen, S. 1981: Grain-size distribution of subglacial till and its relation to glacial crushing and abrasion. *Boreas* 87, pp. 1003–15.

rock glacier A term which has been used to describe a feature, comparatively common in many alpine areas, which looks like a glacier (with apparent flow structures, etc.) but is composed of rock debris. Some workers use the term to imply that such a feature has a specific mode of origin. There are two main theories: that the rock debris has ice mixed in the spaces between the rock (the interstitial ice model) or that the debris is a thick covering on a thin, probably decaying true glacier (the glacier ice model). There is much disagreement as to which model is correct. One major feature is that rock glaciers usually exhibit slow movement, often less than a metre per year, i.e. at least an order of magnitude less than most true ice glaciers.   WBW

Reading
Giardino, J.R., Shroder, J.F. and Vitek, J.D. 1987: *Rock glaciers*. Boston: Allen & Unwin.

Martin, H.E. and Whalley, W.B. 1987: Rock glaciers a review: Part I. *Progress in physical geography* 11, pp. 260–86.

Whalley, W.B. and Martin, H.E. 1992: Rock glaciers: Part II. Models and mechanisms. *Progress in physical geography* 16, pp. 127–86.

rock mass strength An important concept in geomorphology, developed by

Selby (1980) and others in an attempt to gain a quantitative measure of the resistance of a rock mass to erosion. It involves giving a rank of importance to a range of different rock parameters and then summing them to come up with a total rating of strength.
ASG

Reference
Selby, M.J. 1980: A rock mass strength classification for geomorphic purposes: with tests from Antarctica and New Zealand. *Zeitschrift für Geomorphologie* 24, pp. 31–51.

rock quality indices Measures used to relate the numerical intensity of fractures in a rock to the quality of the unweathered rock. One measure is the relationship between the compressional wave velocity measured *in situ* in MASSIVE ROCK and on a core of the intact rock (see INTACT STRENGTH). The fewer the joints, the nearer is the ratio to unity. Rock quality

Classification of rock mass strength

Variable	Weighting (%)	Very strong	Strong	Moderate	Weak	Very weak
Intact rock strength (Schmdit hammer rebound value)	20	100–60 $r = 20$	60–50 $r = 18$	50–40 $r = 14$	40–35 $r = 10$	35–10 $r = 5$
Weathering	10	Unweathered $r = 10$	Slightly weathered $r = 9$	Moderately weathered $r = 7$	Highly weathered $r = 5$	Completely weathered $r = 3$
Joint spacing	30	>3 m $r = 30$	3–1 m $r = 28$	1–0.3 m $r = 21$	300-50 mm $r = 15$	<50 mm $r = 8$
Joint orientations	20	Very favourable. Steep dips into slope, cross joints interlock $r = 20$	Favourable. Moderate dips into slope $r = 18$	Fair. Horizontal dips or nearly vertical dips (hard rocks only) $r = 14$	Unfavourable. Moderate dips out of slope $r = 9$	Very unfavourable. Steep dips out of slope $r = 5$
Joint width	7	<0.1 mm $r = 7$	0.1–1 mm $r = 6$	1–5mm $r = 5$	5–20 mm $r = 4$	>20 mm $r = 2$
Joint continuity and infill	7	None, continuous $r = 7$	Few, continuous $r = 6$	Continuous, no infill $r554$	Continuous, thin infill $r = 4$	Continuous, thick infill $r = 1$
Groundwater outflow	6	None $r = 6$	Trace $r = 5$	Slight <40 ml s^{-1} m^{-2} $r = 4$	Moderate 40–200 ml s^{-1} m^{-2} $r = 3$	Great >200 ml s^{-1} m^{-2} $r = 1$
Total rating		100–91	90–71	70–51	50–26	<26

Source: Modified from Selby 1980. Table 6, pp. 44–5.

designation is the relationship between intact cored rock length to the total length drilled and the 'fracture index' is the frequency of fractures occurring within a rock unit. WBW

Reading
Bell, F.G. 1992: *Engineering properties of soils and rocks*. 3rd edn. Oxford: Butterworth Heinemann.

rock step See RIEGEL.

roddon A sinuous, silty ridge that snakes about above the general level of the peat Fenland of East Anglia (England). Roddons represent ancient river systems that may initially have flowed between levées above the general level of the surrounding land or which have subsequently become relatively elevated as a consequence of peat wastage. They are favoured sites for settlement.

ASG

Reading
Fowler, G. 1932: Old river beds in the Fenlands. *Geographical journal* 79, pp. 210–12.
Godwin, H. 1938: The origin of roddons. *Geographical journal* 91, pp. 241–50.

rollability A property related to the angle of a slope down which a given sedimentary particle will roll. The concept was introduced by Winkelmolen to account for the ease with which particles (usually of sand size) can be rolled in unidirectional fluid flow. It is indexed by the time it takes for grains to travel down the inside of a revolving cylinder inclined at a slight angle. The time taken is a measure of the rollability potential. Usually, a few grammes of the sediment is used in a test. The measure is said to correlate with the influence of grain form on the particle settling velocity. Rollability can be considered as amalgamating several aspects of PARTICLE FORM. WBW

Reading
Goudie, A.S. ed. 1990: *Geomorphological techniques*, 2nd edn. London: Unwin Hyman.
Winkelmolen, A.M. 1971: Rollability, a functional shape property of sand grains. *Journal of sedimentary petrology* 41, pp. 703–14.

Rossby waves Wave motions in the atmosphere of planetary scale. They take the form of vast meanders of airflow around a hemisphere and are most clearly identifiable at upper levels. They owe their existence to the variation with the latitude (ϕ) of Coriolis parameter ($2\Omega \sin \phi$, where Ω

is the angular velocity of rotation of the earth).

In non-divergent, large-scale motion absolute VORTICITY, ζ_a, can be considered constant, given by:

$$\frac{\mathrm{d}}{\mathrm{d}t}(\zeta_a) = \frac{\mathrm{d}}{\mathrm{d}t}(\zeta_r + 2\Omega \sin \phi) = 0$$

where ζ_r is the relative vorticity. In the northern hemisphere if a uniform westerly current with zero initial relative vorticity is displaced poleward, then, as latitude increases, the relative vorticity must become negative (i.e. anti-cyclonic) so that the air turns southward. After moving towards the equator of its original latitude its relative vorticity becomes positive (cyclonic) so that it turns northward. The current thus oscillates about its original latitude giving a series of waves called long waves or Rossby waves (named after Carl-Gustav Rossby, a Swedish-American meteorologist) usually numbering between two and six around a latitude circle.

Rossby wave theory predicts that in a basic westerly current and for a particular number of troughs and ridges around a latitude circle, there is a critical flow speed for which the waves are steady and stationary. If the flow is less than this critical speed the waves drift westward, while if it is greater they drift eastward.

Rossby waves are forced by three principal mechanisms: by orographic forcing resulting from a basic westerly flow impinging on a mountain range (especially the Rockies or the Andes); by thermal forcing due to differential heating of the oceans and continents; and by interaction with smaller scale disturbances such as extra-tropical cyclones.

Rossby waves have also been identified in the oceans. KJW

Reading
Gill, A.E. 1982: *Atmosphere–ocean dynamics*. London: Academic Press.
Houghton, J. 1986: *The physics of atmospheres*. Cambridge: Cambridge University Press.

rotational failure The name given to failure, usually in clays although also in weak rocks, where the shape of the slip surface (akin to a shear plane), which forms the boundary between the stable ground and the mass which has moved, is curved. At the top it frequently forms a step, where tension cracks develop in the early stages of failure. At the base there is usually a bulge. If the mass is rather mobile, as is frequently the case with QUICK CLAY, then there may be spreading at the foot, perhaps to give a secondary earthflow. In some cases the production of one rotational failure may give rise to a second failure.   WBW

Reading
Selby, M.J. 1993: *Hillslope materials and processes*. 2nd edn. Oxford and New York: Oxford University Press.

rotor streaming A condition of unsteady flow over a mountain in which lee eddies are generated, then blow away. It probably involves a mechanism similar to that influencing the vortex streets noted by and named after von Karman.   JSAG

Reading
Atkinson, B.W. 1981: *Meso scale atmospheric circulations*. London and New York: Academic Press.

roughness A composite term embracing all the frictional or retarding influences causing energy loss in fluid motion, and opposing the downslope gravitational driving force which maintains the flow. It is measured by roughness or friction coefficients such as Manning's n, the Chézy C, or the Darcy–Weisbach f (see CHÉZY EQUATION; MANNING EQUATION). Roughness includes the flow resistance due to grain roughness effects. For example, in the STRICKLER EQUATION, Manning's n is shown to increase with the median grain diameter of river bed material. Flow resistance is also affected, however, by the depth of water drowning a given size of grain, and by the spatial arrangement of the coarser bed particles. Form resistance, due to the size, shape and arrangement of sand bedforms such as ripples, dunes and antidunes, also influences roughness. On a larger scale roughness is affected by bends, and channel bars and islands, which force changes of flow direction and result in energy losses. These systematic influences on roughness are augmented by the random effects of overhanging bank vegetation, bed vegetation, tree roots, boulders, etc. It is normally assumed that these roughness sources are additive, and an overall roughness coefficient for a channel is the sum of component indices for the contributing roughness factors. Since these component indices are difficult to quantify, roughness coefficients are often identified from standard sets of photographs of different channel types (Barnes 1967).   KSR

Reference
Barnes, H.H. 1967: Roughness characteristics of natural channels. *US Geological Survey water supply paper* 1849.

roundness Tending towards rounded edges, describing the degree of abrasion of a clastic fragment as shown by the sharpness of its edges and corners, independent of shape. Spherical particles are perfectly rounded, but well-rounded objects (such as an egg), need not be spherical (Waddell 1933).   ASG

Reference
Waddell, H. 1933: Sphericity and roundness of quartz particles. *Journal of geology* 41, pp. 310–31.

routing, flood See FLOOD ROUTING.

r-selection See *r-* AND *K*-SELECTION.

ruderal vegetation Plants which grow in waste land, on or among rubbish and debris. The term was used originally in the sense of stone waste but has been extended to include roadsides, verges and old fields. The plant succession is usually rapid, with profusely seeded and rapidly growing annuals being replaced by hardy perennials, particularly grasses. Naturalized plants are often found in the early seral stages because of the relative absence of competition, but in time they tend to be eliminated in the struggle with native plants.   PAF

run-off A portion of part of the hydrological cycle connecting precipitation and channel flow. Sometimes referred to as storm run-off, direct run-off or quickflow, it occurs when the infiltration capacity of the soil surface is exceeded, and the subsurface can no longer absorb moisture at the rate at which it is being supplied. The precipitation collects in surface depressions and is briefly stored, but when these depressions are filled, the water begins to flow downslope. This filling and overflow process followed by downslope movement is referred to as Hortonian flow, after R.E. Horton who first described it (see HORTON OVERLAND FLOW MODEL).

As run-off proceeds down a slope, its overall form and process may change. As it begins its journey to the channel it may move in a wide shallow sheet as sheetflow. When depths increase, the water may begin to coalesce into small ephemeral channels called rills. As the water proceeds downslope, the channel flow in rills may coalesce into gullies.

The amount of run-off is affected by land use, vegetation and the porosity of the surface. Vegetation intercepts rainfall and delays its journey to the surface, and also allows more moisture to be absorbed at the ground level. Vegetation provides litter on the surface that acts like a blotter to hold moisture in place instead of allowing it to run downhill. Porosity of the soil surface determines how much moisture will move into the subsurface and therefore be lost to the run-off process. Both vegetation and surface porosity are strongly affected by land use, with rural land uses bringing about wholesale substitutions for the natural vegetation to which the run-off

processes, soils and slopes were originally adjusted. Vegetation changes are therefore usually accompanied by readjustments in surfaces and forms through the run-off process. In urban areas the most significant changes are in the installation of impervious surfaces that permit larger than natural amounts of precipitation to become part of the run-off, resulting in flooding problems.   WLG

Reading
Dunne, T. and Leopold, L.B. 1978: *Water in environmental planning*. San Francisco: W.H. Freeman.

ruware A low, dome-shaped exposure of bedrock projecting from a cover of alluvium or weathered bedrock. It is either an incipient or a relict INSELBERG.

S

sabkha (sebkha) An Arabic word that refers either to coastal salt flats, such as those that fringe parts of the Arabian Gulf, or to inland salt basins (PLAYAS).

salars Basins of inland drainage in a desert which are only occasionally inundated with saline water. They are also known as *playas*, *sabkhas* or *chotts*.

salcrete A light-coloured surface crust of halite-cemented beach sand caused by the concentration by evaporation of swash or spray blown onshore by breaking waves (Yasso 1966).   ASG

Reference
Yasso, W.E. 1966: Heavy mineral concentration and sastrugi-like deflation furrows in a beach salcrete at Rockaway Point, NY. *Journal of sedimentary petrology* 36, pp. 836–8.

salinity problems The problems caused by the presence of salts in the waters and soils of arid, semi-arid and coastal areas. In dry areas salts tend to accumulate because of high rates of evaporation and limited leaching. Sources of salts include atmospheric inputs, rivers, groundwater seepage, the weathering of rocks and old evaporites, and the sea. Salinity is a normal characteristic of areas of water deficit, but man has increased its extent, especially by the introduction of irrigation schemes. In coastal areas salinity problems are created by seawater incursion brought about by over-pumping. In certain parts of the world salinization has resulted from vegetation clearance (Peck 1978). Man-induced salinity (salinization) is not a new problem, and has been interpeted as a source of decline in agricultural yields 4000 years ago in Mesopotamia (Jacobsen and Adams 1958). Among other consequences of salinity are changes in soil structure, reduced plant growth, accelerated salt weathering of stone and construction materials, and reduced potability of water supplies. Salinity problems can be reduced by improved irrigation technology, better drainage, the addition of gypsum to soils, and freshwater injection into groundwater.   ASG

Reading and Reference
Jacobsen, T. and Adams, R.M. 1958: Salt and silt in ancient Mesopotamian agriculture. *Science* 128, pp. 1251–8.

Peck, A.J. 1978: Salinization of non-irrigated soils and associated streams: a review. *Australian journal of soil research* 16, pp. 157–68.

Thomas, D.S.G. and Middleton, N.J. 1993: Salinization: new perspectives on a major desertification issue. *Journal of arid environments* 24, pp. 95–105.

†Worthington, E.B. ed. 1977: *Arid land irrigation in developing countries: environmental problems and effects.* Oxford: Pergamon.

salt marsh Salt marshes are vegetated mud-flats in the high intertidal zone found commonly on many low-lying coasts in a wide range of temperate environments. On tropical coastlines salt marshes tend to be replaced by MANGROVE swamps, although sometimes they are found together. Salt marshes, which support a range of halophytic vegetation, grade seawards into mud- or sand-flats. Salt marsh plants themselves play an important role in trapping sediment and in building up the marsh surface. In turn, the development of the marsh encourages a succession of plants from early colonizers such as *Salicornia* spp. and *Spartina* spp. to plants that are less tolerant of frequent inundation by seawater. Salt marshes vary greatly throughout the world in both ecology and geomorphology, but are often characterized by intricate creek systems and salt pan development.   HV

Reading
Adam, P. 1990: *Saltmarsh ecology*. Cambridge: Cambridge University Press.

Salt weathering is a major cause of rock decay and of building disintegration, particularly in arid areas. This wall in Ras al Khaimah in the United Arab Emirates is being undermined by the crystallization of salts from upwardly migrating groundwater (the 'wick effect'). A hole large enough for a substantial man to crawl through has developed.

Allen, J.R.L. and Pye, K. eds 1992: *Saltmarshes: morpho-dynamics, conservation and engineering significance.* Cambridge: Cambridge University Press.

salt weathering The breakdown of rock by HALOCLASTY; it is caused primarily by physical changes produced by salt crystallization, salt hydration, or the thermal expansion of salts. Among the most effective salts are sodium sulphate, sodium carbonate and magnesium sulphate. Salt weathering has been recognized as an important process in desert, coastal, polar and urban areas, and is a serious hazard to concrete structures in saline regions.   ASG

Reading
Cooke, R.U. 1981: Salt weathering in deserts. *Proceedings of the Geologists' Association of London* 92, pp. 1–16.
Goudie, A.S. 1985: Salt weathering. *Research papers series, School of Geography, University of Oxford* 32.

saltation Has two main meanings in physical geography. The first refers to the hopping motion of sand grains transported by a fluid (water or air). The grains are ejected from the bed in a near-vertical trajectory by lift forces, accelerate in the flow direction when affected by fluid drag, then fall to the bed again on a path inclined at $6°–12°$ which is the result of gravitational and drag forces. Sand is only two to three times more dense than water, so the inertia of the rising grain only carries it to a height of about three grain diameters. The viscosity of water allows the grain to settle gently back to the bed. Saltation in air, however, involves trajectories up to 2–3 m high, especially after bouncing impacts on rock or pebble surfaces. The return impact also splashes other grains into the air, allowing a mechanical chain reaction of accelerated transport once motion has been initiated by fluid forces.

The second usage occurs in biogeography and refers to a theory which postulates the origin of major new taxonomic groups by the occurrence of single massive mutations.   KSR

salt-dome A rounded hill produced by the upward doming of rock strata as a result of the diapiric movement of a halite bed or other evaporite deposit.   ASG

Reading
Goudie, A.S. 1989: Salt tectonics and geomorphology. *Progress in physical geography* 13, 597–605.

salt-flat A near horizontal stretch of salt crust representing the bed of a former salt lake.

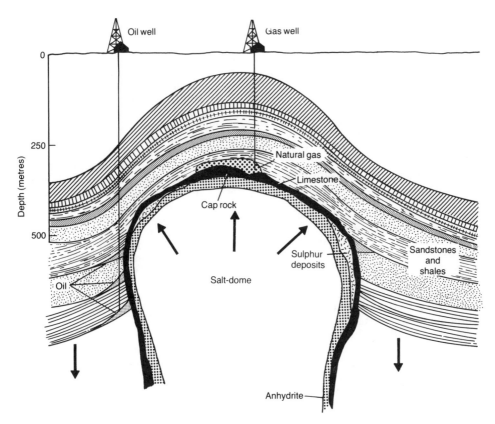

Salt-dome. If the dome pierces the overlying rocks it will be attacked by solutional processes and may also flow as a 'salt glacier'.

saltwater intrusion See GHYBEN-HERZBERG PRINCIPLE.

sand Unconsolidated material composed of mineral particles ranging in diameter from about 63 μm to 2.0 mm.

sand banks Form significant depositional features in many coastal regions and on continental shelves, and are often fed and capped by sand waves. Two main types have been identified (Dyer, 1986). Linear sand banks or ridges occur in shallow tidal seas where sand is present and current velocities exceed about 0.5 m s^{-1}. They can be up to 80 km long, and typically 1–3 km wide and tens of metres high. Headland (or banner) banks develop in association with promontories, again where current strengths exceed about 0.5 m s^{-1}. They are only a few kilometres in length and have an elongated pear-shaped form, the broad end being directed towards the top of the headland.   ASG

Reference
Dyer, K.R. 1986: *Coastal and estuarine sediment dynamics.* Chichester: Wiley.

sand lens A discontinuous layer of sand in a sedimentary sequence representing the remnant of a former channel infill or overbank deposit comprising dunes, ripples, plane-bedded sands or sand sheets, or channel margin sediments.   JM

sand rose A circular histogram depicting the amount of sand potentially moved by winds from various compass directions at a given topographical locality.

sand volcano A small mount of sand with a smaller conical depression at the apex. Dimensions range from less than 2.5 cm in diameter up to 5 cm, and heights rarely exceed 3 cm. They are a surface expression of sediment compaction and dewatering. As the sediment settles, interstitial water may be expelled as small springs, at the mouth of which sand and mud particles may be deposited in a cone.

ASG

Reading
Picard, M.D. and High, L.R. 1973: *Sedimentary structures of ephemeral streams*. Amsterdam; Elsevier Scientific. Pp. 139–41.

sandbed channels Are common in semi-arid environments. They have bed material in the 0.0625–2 mm sand size-range which is mobile even during low sub-bankfull flows, and is transported by SALTATION, or in suspension at high flow stages. During transport the sand travels in bedforms (ripples, dunes, antidunes) which change systematically with the FLOW RE-GIME, strongly influencing the bed ROUGHNESS, and which may be preserved as sedimentary structures in the sand river deposits. Sandbed rivers have characteristically high rates of bedload transport, are migratory on their floodplains, and are often morphologically unstable with multiple, changing, braided channels and sand bars.   KSR

sandstone An indurated sedimentary rock composed of cemented particles of sand, with a range of grain sizes between 0.0625 mm and 2 mm. They can be subdivided into various types on the basis of grain size and mineral composition. Types with a limited matrix content (less than 15 per cent) are called *arenites*, whereas those with a greater matrix content are termed *wackes*. Within the arenites, distinction must be made between *quartz* (less than 5 per cent feldspar or rock fragments), *lithic* (more than 25 per cent rock fragments, excluding feldspar), *arkose* (more than 25 per cent feldspar) and *volcanic* (more than 50 per cent volcanic fragments) subtypes. Sandstones cover very approximately the same area of the continents as do granites and carbonates, but have been much less the scene for the development of any particular geomorphological approach than the other two rock types. None the less, sandstone landscapes do have some distinctive geomorphological features that have been the subject of a review by Young and Young (1992).   ASG

Reference
Young, R. and Young, A. 1992: *Sandstone landforms*. Berlin: Springer-Verlag.

sandstorm An atmospheric phenomenon occuring when strong winds entrain particles of dust and sand and transport them in the atmosphere.

sandur (pl. sandar; Icelandic) An extensive plain of glaciofluvial sands and gravels

Sandstone terrain with characteristic ruiniform structures in the Cederberg Mountains near Clanwilliam, Cape Province, South Africa.

deposited in front of an ice margin by a system of braided or anastomosing meltwater streams which migrate across the sandur surface. The whole sandur is rarely flooded except during jökulhlaup events. A valley sandur is confined between valley walls. Pitted sandur forms on an ice margin that melts out to produce kettle holes. For morphological and sedimentary characteristics, see OUTWASH.   JM

Reading
Bluck, B.J. 1974: Structure and directional properties of some valley sandur deposits in Southern Iceland. *Sedimentology* 21, pp. 533–54.
Hjulström, F. 1952: The geomorphology of the alluvial outwash plains (sandurs) of Iceland, and the mechanics of braided rivers. *International Geographical Union, seventeenth congress proceedings.* Washington: DC. Pp. 337–42.
Krigström, A. 1962: Geomorphological studies of sandur plains and their braided rivers in Iceland. *Geografiska annaler* 44A, pp. 328–45.
Price, R.J. 1969: Moraines, sandar, kames and eskers near Breidamerkurjökull, Iceland. *Transactions of the Institute of British Geographers* 46, pp. 17–43.

sapping The undermining of the base of a cliff, with subsequent failure of the cliff face. It can be accomplished by a variety of processes, including lateral erosion by streams, wave action and groundwater outflow. This last is often called 'spring sapping', and has been held responsible for the development of the steep heads of DRY VALLEYS, and for the formation of certain canyons (Higgins 1984).   ASG

Reference
Higgins, C.G. 1984: Piping and sapping: development of landforms by groundwater outflow. In R.G. La Fleur ed., *Groundwater as a geomorphic agent.* Boston, Mass.: Allen & Unwin. Pp. 18–58.

saprolite Weathered or partially weathered bedrock which is *in situ.*

sapropel Amorphous organic compounds which collect in various water basins: lakes, lagoons, shallow marine basins and estuaries are termed sapropels. The sapropel is formed by predominantly anaerobically decomposing remains of phyto- and zooplankton. These are richer in fatty and protein substances than is peat. The decomposition and putrefaction of the organic content leads to the formation of various hydrocarbons, which are believed to be the basis of the origin of petroleum and natural gas compounds which form after compression under accumulated sediment. The progressive accumulation of sapropel is governed largely by rapid multiplication of the organisms responsible for it.   AP

Reading
Pettijohn, F.J. 1984: *Sedimentary rock.* 3rd edn. Delhi: CBS Publishers.
Rossignol-Strick, M. 1985: Mediterranean Quaternary sapropels, an immediate response of the African monsoon to variation of insolation. *Palaeogeography, palaeoclimatology, palaeoecology* 49, pp. 237–63.

saprophyte An organism, usually a plant, which obtains nutrients from dead or dying organisms. Most fungi are saprophytes.

sarn A rocky causeway exposed only at low tide. It possibly consists of a fossil beach ridge, or even a moraine, submerged by rising sea level.

sarsen A block of silica-cemented sandstone, breccia or conglomerate found in many parts of southern England, notably on the margins of the London and Hampshire basins. Sometimes called 'grey wethers' or 'pudding stone', sarsens are thought to be the result of weathering processes in a surface or near-surface environment under conditions of Tertiary warmth. Man has often employed such blocks to make monuments, such as Avebury stone circle or Windsor Castle. Sarsens are in effect a fossil silcrete duricrust, and comparable deposits in the Paris Basin are called *meulières.*   ASG

Reading
Summerfield, M.A. and Goudie, A.S. 1980: The sarsens of southern England: their palaeoenvironmental interpretation with reference to other silcretes. In D.K.C. Jones ed., *The shaping of southern England.* Institute of British Geographers special publication 11, pp. 71–100.

sastrugi Furrows and ridges in the surface of ice and snow accumulations through the action of wind.

satellite meteorology Meteorology which depends largely, or completely, on data generated by METEROLOGICAL SATELLITES. As such it is a relatively young branch of the parent science, but one which grew very rapidly after the launching of the first specialized meteorological satellite in 1960. With meteorological satellite platforms and sensor systems still undergoing active development as part of the present day rise of environmental remote sensing, satellite meteorology continues to increase in both its scope and sophistication. Each meterological satellite or satellite system is supported by ground facilities which receive satellite data for dissemination to the user community. There, initial preprocessing (e.g. removal of extraneous data contents) and processing (e.g. geographical rectification) are carried out, and ranges of satellite products are prepared to meet the needs of the primary user community. These include:

1 Manual and basic products, including unenhanced and enhanced visible and infrared imagery; facsimile maps (e.g. nephanalyses, and snow and ice charts); and alphanumeric outputs (e.g. satellite weather bulletins, moisture analyses and plume winds).
2 Man–machine combined products, including cloud motion vectors and precipitation fields.
3 Computer–derived image products, including period minimum brightness composites, and cloud field analyses (cloud types, cloud top heights, etc.).
4 Computer–derived digital products, e.g. vertical temperature and moisture profiles, and sea surface temperature data.
5 Archival products, e.g. magnetic tapes of raw and/or processed data, and photographic images, for use in atmospheric research.

These products form a very important part of the complete data pool for use in weather forecasting. Some (data (e.g. cloud images) may be used *qualitatively* in the analysis of synoptic situations, and forecasting for synoptic, subsynoptic, or mesoscale regions, either on the ground, or even aloft. Others are used *quantitatively*, e.g. satellite-derived vertical profiles, and satellite winds: such data may be added to the available conventional (*in situ*) observations to provide improved data arrays for numerical (computer-based) forecasting procedures. The satellite inputs are particularly important for tropical and polar regions, and some of the more remote continental areas in middle latitudes, i.e. those regions traditionally least well observed by surface weather stations. Satellite-derived quantitative data are also important in vertical profiling of the atmosphere in these regions because of the sparseness of their upper-air weather observatories – but satellite soundings are of value almost everywhere because they penetrate the atmosphere downwards, whereas radio-sondes and similar conventional devices inspect it from the bottom upwards, usually to the tropopause, and rarely far beyond it.

Equally important and varied uses are made of satellite data in *meteorological research*. On the global scale evaluations of earth/atmosphere radiation, and related energy budgets have benefited greatly from satellite radiation data. Since satellite sensor systems orbit above the top of the earth's atmosphere a number of budget components which were previously amenable only to estimation can now be measured. These include the earth/atmosphere albedo, long-wave radiative fluxes towards space, and the net radiation budget for the globe, or any selected area, and their attendant columns of the atmosphere. Studies of spatial and temporal changes of these and other quantities are yielding vital insights into the behaviour of the atmosphere, involving both long-distance interactions ('teleconnections') and vertical inter-relationships among its different layers. Satellite imagery has revolutionized our knowledge and understanding of many regions, and types of weather systems, ranging in the first case from tropical latitudes to polar regions, and in the second from synoptic down to mesoscale features. Virtually no region or type of weather system has not been elucidated in some way. The First GARP Global Experiment (FGGE) in 1979 depended heavily on satellite data, and is likely to be the forerunner of other large and complex meteorological research projects in the future.

As the runs of data from meteorological satellites have lengthened, so increasing attention has been paid to *climatological analysis* of these types of data sets. Particular advantages have accrued in studies of the climatologies of cloud distributions, synoptic weather systems (e.g. the intertropical cloud band and associated features, hurricanes, jet streams, and mid-latitude depressions), and the structure and behaviour of the upper atmosphere. As the data sets lengthen further, so they may be expected to support an increasing number and range of studies of climatic change. It is also likely that microwave data will feature increasingly in both meteorological and climatological satellite applications.   ECB

Reading
Anderson, R.K. and Veltischev, N.F. 1973: *The use of satellite pictures in weather analysis and forecasting.* WMO technical note 124. Geneva: World Meteorological Organization.

Barrett, E.C. and Curtis, L.F. 1992: *Introduction to environmental remote sensing*. 3rd edn. London: Chapman & Hall.

— and Martin, D.W. 1981: *The use of satellites in rainfall monitoring*. London: Academic Press.

Cracknell, A.P. 1981: *Remote sensing in meteorology, oceanography and hydrology*. Chichester: Ellis Horwood; New York: Wiley.

Kondratiev, Ya. 1983: *Satellite climatology*. Leningrad: Hydrometeozdat.

NASA 1982: *Meteorological satellites: past, present and future*. NASA conference publication 2227. Washington, DC: NASA.

Tanczer, T., Gotz, G. and Major, G. 1981: First FGGE results from satellites. *Advances in space research* 1.4. Oxford: Pergamon.

saturated wedge and zone Consists of a layer of saturated soil at the base of a hillslope, thickening downslope and in some areas reaching the soil surface to define an area of surface saturation.

Saturation develops in response to THROUGHFLOW collecting from the hillside above. Its total volume generally increases downslope, whereas its rate of flow may decrease if the slope profile is concave or if water is backed up from the stream. The magnitude of this effect may be seen for conditions of steady net rainfall at intensity I, falling in a collecting area of a per unit contour width. The total outflow along 1 m of contour is then $Q = Ia$. This flow may also be expressed in terms of the depth h of saturated flow in the soil and the slope (or strictly the total potential) gradient S, in the form $Q = Sf(h)$ where f is a function which increases with h and with soil permeability. Equating these expressions for the flow, the saturated wedge has thickness h given by

$$f(h) = Ia/S$$

The effect of slope profile concavity (decreasing S) and of a large collecting area or converging flow (large a) may readily be seen. It is also evident that saturated zones/wedges only persist seasonally in areas where net rainfall (rainfall minus evapotranspiration) remains positive seasonally. Saturated wedges rise to the soil surface where h is greater than the soil water storage, giving rise to a DYNAMIC SOURCE AREA of surface saturation.   MJK

Reading
Kirkby, M.J. 1978: *Hillslope hydrology*, Chichester: Wiley. Esp. chs 7 and 9.

saturation coefficient or degree of saturation (S) is the water content of a rock after free saturation, expressed as a percentage of the maximum water content. The free saturation is the water which can be absorbed into the pore spaces of a sample when it is just immersed in water, while the maximum water content is the amount of water absorbed when it is forced in under vacuum. The saturation coefficient is important in the testing of the mechanical breakdown of rocks.   WBW

saturation deficit The depth of water required to bring saturation up to the soil surface and initiate overland flow. Since THROUGHFLOW is not necessarily connected to the groundwater table at depth, a zero saturation deficit need not imply complete saturation of the soil profile. The saturation deficit is important in establishing areas of current overland flow production, and gives directly the amount of storm rainfall needed to initiate overland flow. The deficit is also important as one of the controls on evaporation, particularly from unvegetated soil surfaces.   MJK

saturation overland flow See OVERLAND FLOW.

savannah A grassland region of the tropics and subtropics. The word is probably of American Indian origin, signifying treeless grasslands. Savannahs today are taken to mean plant formations dominated by grasses and grass-like species (graminoids) with herbaceous non-grass species (forbs), often possessing a light to dense scattering of trees (see illustration on p. 442). Woody savannahs prevail where grasses are at least co-dominant, grading to savannah woodland where trees assume dominance. Seasonal climates impose water stresses during the dry period, which inhibit tree growth and encourage drought-resistant plants. Burning, both natural and man-induced, also favours a herbaceous formation with xeromorphic characteristics.   PAF

Reading
Bourlière, F. ed. 1983: *Tropical savannas*. Amsterdam: Elsevier.

Huntley, B.J. and Walker, B.H. 1982: *Ecology of tropical savanna*. Berlin: Springer-Verlag.

Werner, P.A. 1991: *Savanna ecology and management: Australian perspectives and intercontinental comparisons*. Oxford: Blackwell Scientific.

scabland An area where, as a result of erosion, there is no soil cover and the landsurface consists of rock and rock fragments.

A palm savannah bordering the River Uruguay in Argentina. Savannahs show a wide range of forms from relatively open grasslands to dense woodland.

scar A cliff or very steep slope or a rocky outcrop.

sciophyte A plant that lives in a well-shaded environment.

sclerophyllous Refers to species of evergreens that, like olive and cork oak, have adapted to the lengthy seasonal drought experienced in regions with a 'Mediterranean type' of climate, by producing tough, leathery leaves to cut down moisture loss caused by transpiration.

scoria A volcanic rock or slag consisting of angular rock fragments and numerous voids which were originally filled with volcanic gases.

scour and fill The processes of cutting and subsequent filling of fluvial channels.

scree An accumulation of primarily angular clasts which lies at an angle of around 36° beneath an exposed free face or cliff. The prime cause of deposition is rock fall, but other processes, such as debris flows, may contribute to their development. The largest clasts occur at the base of the scree. ASG

Reading
Statham, I. 1973: Scree slope development under conditions of surface particle movement. *Transactions of the Institute of British Geographers* 59, pp. 41–53.

S-curve A method of extending a unit hydrograph. It represents the surface run-off hydrograph caused by an effective rainfall of intensity I/T mm h^{-1} applied indefinitely where T is the duration of the effective rainfall (see UNIT HYDROGRAPH).

sea cliffs Steep slopes that border ocean coasts. They are ubiquitous and occur along approximately 80 per cent of the ocean coasts of the earth (Emery and Kuhn 1982). They can be classified into three main types according to their stage of development: *active cliffs*, where bedrock is exposed by continuous retreat under the influence of both marine and subaerial forces; *inactive*, where their bases are mantled by talus, and where there is some vegetation cover; and *former*, where the influences of marine erosion have disappeared, so that subaerial erosion rounds the crests and provides material for stream deposition beyond their bases. Among the important factors in their development are the nature of the landward topography, the structure and lithology of the materials in which they have been eroded, the nature of marine erosion, and the power of subaerial processes (rockfalls, mass movements, frost weathering, etc.). Many cliff profiles display complex forms brought about by long continued changes in sea level, climate, and the balance between subaerial and marine processes (Steers 1962). ASG

References
Emery, K.O. and Kuhn, G.G. 1982: Sea cliffs: processes, profiles, classification. *Bulletin of the Geological Society of America* 93, pp. 644–53.

Steers, J.A. 1962: Coastal cliffs: report of a symposium. *Geographical Journal* 128, pp. 303–20.

sea ice Ice which forms on the sea surface when water temperatures fall to around −1.9°C. Such conditions apply to considerable areas of the Arctic and Antarctic (see diagram). In polar latitudes the equilibrium thickness of sea ice is around 3 m and any surface melting is equalled by bottom freezing. When the sea ice is attached to the land it is known as *fast* ice and when floating free under the influence of currents and wind it is called *pack* ice. The latter consists of ICE FLOES centred on relatively resistant ice, leads of open water which may freeze over rapidly in winter, and irregular fractured and crumpled ice-forming features known as *pressure ridges* or *keels* (see illustration on p. 444). DES

Reading and Reference
†Lewis, E.L. and Weeks, W.F. 1971: Sea ice: some polar contrasts. In G. Deacon ed., *Symposium on Antarctic ice and water masses, Tokyo, September 1970.* Cambridge: Scientific Committee for Antarctic Research Pp. 23–31.

†Nansen, F. 1897: *Farthest north.* 2 vols. London: Constable.

†Sugden, D.E. 1982: *Arctic and Antarctic.* Oxford: Basil Blackwell; Totowa, NJ: Barnes & Noble.

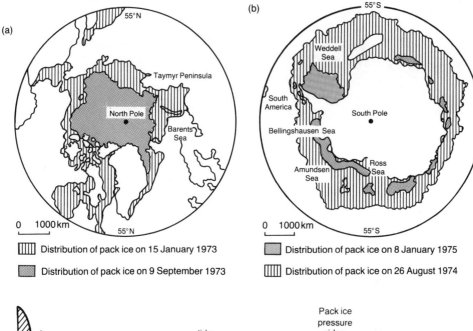

Distribution of pack ice on 15 January 1973

Distribution of pack ice on 9 September 1973

Distribution of pack ice on 8 January 1975

Distribution of pack ice on 26 August 1974

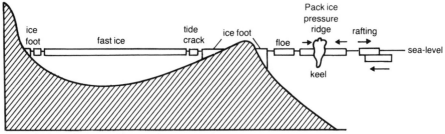

Sea-ice: 1. (a) *Distribution in the Arctic on 15 January and 9 September 1973, mapped from passive microwave images.* (b) *Distribution in the Antarctic on 26 August 1974 and 8 January 1975, mapped from passive microwave images. The area of each map is identical.*
Source: *Sugden 1982. Figure 6.12. (After Zwally & Gloersen 1977.)*

2. A morphological and environmental classification of sea ice. The shaded area is land.
Source: *D. Sugden 1982: Arctic and Antarctic. Oxford: Basil Blackwell. Figure 6.7.*

Sea ice photographed on the Imperial Trans-Antarctic Expedition 1914–16.

Zwally, H.J. and Gloersen, P. 1977: Passive microwave images of the Polar Regions and research applications. *Polar record* 18, 116, pp. 431–50.

sea/land breeze The sea and land breezes form part of a diurnally varying, vertical circulation of air that is ultimately induced by temperature differences between land and sea. Given a morning of near calm conditions and clear skies, a land area heats up more rapidly than does the adjacent water body. Consequently, as a result of turbulent transfer and convection of heat, the air over the land becomes warmer than that at the same height over the sea. In turn, this means that the land air is less dense than the sea air and thus, by the HYDROSTATIC EQUATION, the vertical gradient of pressure is less over land than over sea. Assuming that pressure was initially uniform over land and sea, the difference in vertical gradients must result in a higher pressure over the land at a height H (say) than at the same height over the sea. Consequently, at height H, air moves from land to sea. In turn, this disturbs the hydrostatic equilibrium of the air columns over both land and sea and also generates a horizontal pressure gradient from sea to land near the surface. The air motion resulting from this gradient force is the sea breeze. At the landward and seaward extremities of this breeze the air rises and sinks respectively. Consequently a vertical circulation is completed.

The land breeze results from a reversal of these processes when the land is cooler than the sea.   BWA

Reading
Atkinson, B.W. 1981: *Meso-scale atmospheric circulations*. London and New York: Academic Press. Esp. ch. 5.

sea level The mean surface elevation of the sea. There are, however, two basic problems with this definition. First, the sea surface is not level, being affected by many short-term influences such as wind, waves or tides, and also showing considerable variation in terms of the shape of the geoid. Secondly, over what time period should we calculate a mean value? For sea level is always changing in response to tidal variations, fluctuations in pressure and temperature, upwelling, river discharges, and so forth. If all these influences are

Factors in sea-level change

Eustatic (worldwide)	Local
Glacio-eustasy	Glacio-isostasy
Infilling of basins	Hydro-isostasy
Orogenic-eustasy	Erosional and depositional isostasy
Decantation	Compaction of sediments (autocompaction)
Transfer from lakes to oceans	Orogeny
Expansion or contraction of water volume	Epeirogeny
because of temperature change	Ice–water gravitational attraction
Juvenile water	
Geoidal changes	

Source: Goudie 1992. Table 6.1.

excluded, then progressive (or secular) changes in sea level can be observed. These may be caused by global factors (eustatic) or more local effects (tectonic and isostatic) (see table).

Sea levels have changed repeatedly during the history of the earth (see, for example, EUSTASY; FLANDRIAN TRANSGRESSION etc.). At the present time there is great interest in whether or not sea level is rising because of global warming, and debate about the future speed of likely sea-level rise. Because of the problems of disengaging local from eustatic factors there is some controversy about the current rate of global

sea-level rise (see Pirazzoli 1989), which probably ranges from a rate of about 4 cm to 20 cm over the past hundred years. It is possible that over the next hundred years the rate of rise will be rather greater, possibly at around 60 cm per hundred years (Houghton *et al.* 1990).   ASG

Reading and References
Goudie, A.S. 1992: *Environmental change.* Oxford: Oxford University Press.

Houghton, J.T., Jenkins, G.J. and Ephraums, J.J. eds 1990: *Climate change: the IPCC scientific assessment.* Cambridge: Cambridge University Press.

Pirazzoli, P.A. 1989: Present and near-future sea level changes. *Palaeography, palaeoclimatology, palaeoecology* 75, pp. 241–58.

The south-west coast of Turkey shows clear evidence of sea-level change. Apart from major drowned valleys, there are also examples of archaeological sites being submerged, as illustrated by this Lycian tomb.

sea mount A mountain or other area of high relief on the sea floor which does not reach the surface. Flat-topped seamounts are called GUYOTS.

secondary depression A region of low pressure which forms within the circulation of a pre-existing depression. It sometimes occurs at the 'triple point' where the occlusion and warm and cold fronts meet, or along the individual fronts as warm or cold front waves. The development of the secondary is often preceded by an increase in the spacing of the isobars locally and once formed, the secondary can even absorb the parent low.

Secondary depressions also form occasionally in the unstable airstreams as polar lows.   RR

secondary flows Those currents in moving fluids which have a velocity component transverse to the local axis of the primary, or main flow direction. In river channels they are associated with the longitudinal vortex known as HELICAL FLOW. Secondary flows in straight channels are probably caused by non-uniform distribution of boundary shear stress and by turbulence generated at the base of the channel banks. Two circulatory cells occur side by side, with flow alternately *converging* at the surface then downwelling, and *diverging* at the surface where upwelling occurs in mid-channel, at locations spaced apart at five to seven times the channel width. Downwelling intensifies the local bed shear stress exerted by the flow and encourages scour, while upwelling results in a low shear stress and deposition. This secondary circulation flow pattern thus relates to the POOL AND RIFFLE sequence. In curved channels, secondary flows result from skewing of the main flow towards one bank, and the consequent transverse HYDRAULIC GRADIENT then drives a transverse current which crosses the channel at the bed after plunging near the outer bank and assisting bed scour and undermining of the bank. Secondary flows represent an important control of the spatial distribution of erosion and deposition within a river channel, and of channel pattern change.

KSR

sediment, fluvial Has been defined as particles derived from rock or biological material that are, or have been, transported by water. It has provided the focus for numerous studies by physical geographers, although a dichotomy exists between studies of contemporary *transport* and studies of the *deposits* associated with contemporary and past fluvial activity. The movement of fluvial sediment also has important practical implications for the

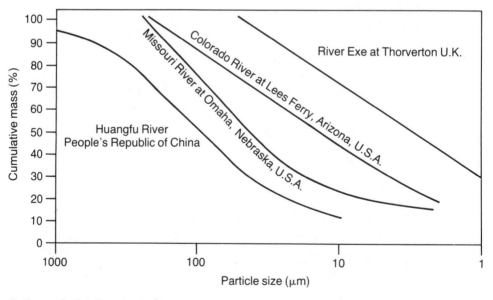

Sediment, fluvial: Examples of particle size distribution.

problems of channel management, river intake and irrigation canal maintenance, reservoir and harbour siltation, debris dumping and floodplain accretion that stem directly from its deposition. Furthermore there is an increasing awareness of the important role of fine sediment in the transport of contaminants such as heavy metals and pesticides by rivers.

Information on the properties of fluvial sediment is of significance in all these contexts, and particle size must rank as the most important since it exerts a major control on entrainment, transport and deposition processes. The size of particulate material transported by rivers ranges between fine clay and colloidal particles of less than 0.5 μm in diameter to large boulders moved during flood events. Documentary evidence points to boulders weighing up to 7.5 t being transported by the Lynmouth Flood, which occurred in North Devon, UK, in 1953. A useful distinction may be made between material moving as BED LOAD and that transported as SUSPENDED LOAD. The latter generally involves particles <0.2 mm in diameter and the diagram provides several examples of typical particle size distributions of suspended sediment from rivers in different areas of the world. Although such particle size analyses commonly relate to the size of the discrete sediment particles, it is important to recognize that many of these will be transported within larger aggregates. In the case of suspended sediment, the particle size characteristics are largely governed by the nature of the source material. With the larger particles comprising the bed load, the boundary shear stress or force available to initiate movement is the dominant control, although the range of sizes available may also be important. The degree of sorting will vary according to the precise character of the fluvial environment and the transport distances involved.

Measurements of particle shape and roundness have also been employed to demonstrate downstream changes in bed material character.

Fluvial sediment transported as bed load consists almost entirely of inorganic material and this will closely resemble the parent rock in terms of mineral composition. The finer sediment transported in suspension may, however, incorporate a considerable proportion of organic material. Its mineralogy may differ considerably from that of the parent rock due to the chemical weathering processes involved in the disintegration of the rock and the selectivity of the detachment and transport processes. This selectivity will result in the suspended sediment showing considerable enrichment in clay-sized particles and inorganic matter, when compared to the source material. Organic matter contents are typically of the order of 10 per cent, although values as high as 40 per cent have been encountered in some rivers. Enrichment in fine material has important implications for sediment-associated transport of contaminants, because clay-sized particles exhibit considerably greater specific surface areas (typical values are 200 m^2 g^{-1} for clay, 40 m^2 g^{-1} for silt and 0.5 mm^2 g^{-1} for sand) and cation exchange capacities.

Information on the average chemical composition of inorganic suspended sediment transported by world rivers and its relation to that of surficial rocks, abstracted from Martin and Meybeck (1979) and Meybeck (1981) is listed below.

These data indicate that in general suspended sediment is enriched in Al, Fe and Ti with respect to parent rock, while Na, Ca and Mg are strongly depleted. This tendency is more marked for tropical rivers than for rivers in temperate and Arctic regions because of the greater efficacy of chemical weathering in tropical areas.

Average chemical composition of inorganic suspended sediment and surficial rocks

	Concentration (mg g^{-1})							
	Al	Ca	Fe	K	Mg	Na	Si	Ti
Tropical rivers	114	7.5	62	18	9.6	5.1	264	7.3
Temperate and Arctic rivers	72	36	45	23	12	8.6	293	4.9
World rivers	90	25	52	21	11	7.1	281	5.8
Surficial rocks	70	45	36	24	16	14	275	3.8

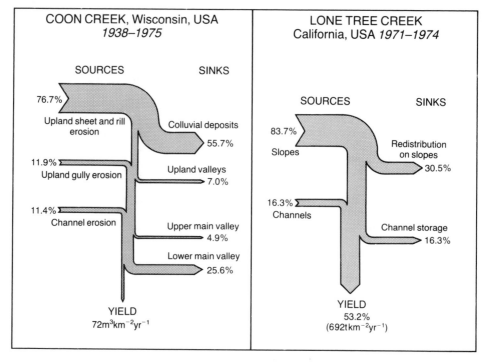

Several workers have attempted to estimate the total annual sediment load transported from the landsurface of the globe to the oceans, although in the absence of data on bed load transport these values relate to suspended sediment. The estimate of Milliman and Meade in 1983 pointed to a mean annual tranport to the ocean of 13.5×10^9 t year^{-1}. This is equivalent to a sediment yield of approximately 135 t km^{-2} year^{-1} from the landsurface of the globe and is 3.6 times greater than the total dissolved load transport to the oceans of 3.72×10^9 t year^{-1} suggested by Meybeck (1979).

Existing estimates of the relative importance of bed load and suspended load in the total transport of fluvial sediment to the oceans are extremely tentative, but they place the bed load contribution at about 10 per cent of the suspended load. This proportion varies markedly for individual rivers. In Arctic streams on Baffin Island measurements suggest that coarse material or bed load constitutes 80–95 per cent of the total transport of fluvial sediment, whereas data from the Volga river in the former USSR indicate that 98–99.7 per cent of the sediment transported is composed of fine material in suspension.

Studies of fluvial sediment deposits have been largely concerned with the coarse fraction which moves relatively slowly through a river system and which may come to rest temporarily or permanently in depositional sinks or stores. These deposits are commonly classified on a genetic basis and they include, for example, point, longitudinal and marginal bars, lag deposits, splays and alluvial fans. Whereas most channel deposits are composed of relatively coarse material (bed load), finer material may be deposited by rivers on floodplains and in lakes or reservoirs, and where fine sediment is transported downslope by unconcentrated surface run-off this may frequently be deposited before reaching the stream channel.

The dichotomy between studies of fluvial sediment transport and of the associated deposits can usefully be bridged by considering the *sediment budget* of a drainage basin. With this concept the transport of sediment out of the basin is seen as the result of the various processes involved in mobilizing sediment within the basin and of the deposition and storage of the sediment within the basin. Only a small proportion of the sediment mobilized may be transported out of the basin and some material may be deposited in temporary storage to be remobilized on a subsequent occasion, for example, during a higher magnitude event. The deposits are therefore treated as an integral part of the conveyance system. Two examples of sediment budgets are illustrates. The first, for Coon Creek, Wisconsin, USA, is based on the work of Trimble (1981) and illustrates a case where only 6.7 per cent of the fluvial sediment mobilized was transported out of the basin. More than two-thirds of the sediment mobilized by upland sheet and rill erosion were deposited as colluvium before reaching the channel network, and more than 80 per cent of the sediment moving through the channel network were incorporated in valley deposits. In the case of Lone Tree Creek, California, USA investigated by Lehre (1982) more than half of the sediment mobilized was transported out of the basin, but the sinks associated with slope and channel deposition still represent a very significant component of the budget. DEW

Reading and References
Abrahams, A.D. and Marston, R.A. eds. 1993: Drainage basin sediment budgets. *Physical Geography* 14, pp. 221–320.

†American Society of Civil Engineers 1975: *Sedimentation engineering*. New York: American Society of Civil Engineers.

Lehre, A.K. 1982: Sediment budget of a small coast range drainage basin in North-Central California. In F.J. Swanson, R.J. Janda, T. Dunne and D.N. Swanson eds, *Sediment budgets and routing in forested drainage basins*. US Forest Service general technical report PNW-141. Pp. 67–77.

Martin, J.M. and Meybeck, M. 1979: Elemental mass-balance of material carried by major world rivers. *Marine chemistry* 7, pp. 173–206.

Meybeck, M. 1979: Concentration des eaux fluviales en elements majeurs et apports en solution aux océans. *Revue de géologie dynamique et de géographice physique* 21, pp. 215–46.

— 1981: Pathways of major elements from land to ocean through rivers. In *River inputs to ocean systems*. UNEP/UNESCO report.

Milliman, J.D. and Meade, R.H. 1983: World-wide delivery of river sediment to the oceans. *Journal of geology* 91, pp. 1–21.

†Richards, K. 1982: *Rivers: form and process in alluvial channels*. London: Methuen.

†Statham, I. 1977: *Earth surface sediment transport*. Oxford: Clarendon Press.

Trimble, S.W. 1981: Changes in sediment storage in the Coon Creek Basin, Driftless Area, Wisconsin, 1853 to 1975. *Science* 214, pp. 181–3.

†UNESCO 1982: *Sedimentation problems in river basins*. Paris; UNESCO.

sediment yield The total mass of particulate material reaching the outlet of a drainage basin. Values of sediment yield are commonly evaluated on an annual basis ($t\ year^{-1}$) and may also be expressed as specific sediment yields or yields per unit area ($t\ km^{-2}\ year^{-1}$). The total sediment yield comprises material transported both as SUSPENDED LOAD and BED LOAD and separate measurements of the two components will generally be necessary. However, where the sediment load of a river is deposited in a lake or reservoir, it may be possible to estimate the total yield directly by monitoring the volume of deposited sediment. The sediment yield from a drainage basin will commonly represent only a small proportion of the gross erosion within the basin. Much of the eroded material will be deposited before reaching the outlet of the basin and the ratio of sediment yield to gross erosion is termed the sediment delivery ratio.

The magnitude of the sediment yield from a drainage basin will reflect control by several factors including climate, topography, lithology, vegetation cover and land

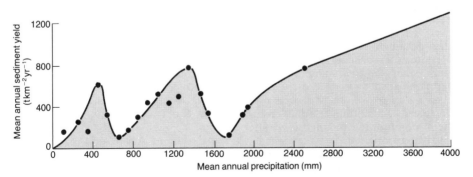

Sediment yield: *The relationship between mean annual sediment yield and mean annual precipitation.*
Source: *D.E. Walling and A.H.A. Kleo 1979: Sediment yields in areas of low precipitation: a global view. Proceedings of the Canberra symposium on the Hydrology of Areas of Low Precipitation. IAHS Publication 128, pp. 479–93.*

use. Maxiumum values in excess of 20,000 t km^{-2} year^{-1} have been recorded in the severely eroded loess areas of the Middle Yellow River basin in China and in the high rainfall areas of South Island, New Zealand. DEW

Reading
Laronne, J.B. and Mosley, M.P. eds 1982: *Erosion and sediment yield*. Benchmark papers in geology 63. Stroudsburg, Pa: Hutchinson Ross.

Walling, D.E. and Webb, B.W. 1983: Patterns of sediment yield. In K.J. Greogory ed., *Background to palaeohydrology*. Chichester: Wiley.

sedimentary rock Rock composed of the fragments and particles of older rocks which have been eroded and the debris deposited by wind or water often as distinct strata. Some sedimentary rocks may be of organic origin.

seeding of clouds See WEATHER MODIFICATION.

segregated ice Ice formed by the migration of pore water to the freezing plane where it forms into discrete lenses, layers or seams ranging in thickness from hairline to greater than 10 m. Segregated ice commonly occurs in alternating layers of ice and soil. Ice structure tends to be parallel to the freezing surface (i.e. usually dominantly horizontal) with air bubbles tending to be elongated and aligned normal to the horizontal layering. When dealing with massive icy bodies, as in the Mackenzie Delta, or within pingos, it is sometimes difficult to differentiate segregated ice from injection ice. HMF

Reading
Mackay, J.R. 1972: The world of underground ice. *Annals of the Association of American Geographers* 62, pp. 1–22.

seiche The oscillation of a body of water at its natural period. Coastal measurements of sea level often show seiches with amplitudes of a few centimetres and periods of a few minutes due to oscillations of the local harbour, estuary or bay, superimposed on the normal tidal changes. DTP

seif A linear or longitudinal dune, groups of them commonly making a parallel straight pattern orientated in the direction of the prevailing wind or winds. They are characterized by their considerable length, often more than 20 km (Lancaster 1982). Around 30 per cent of all aeolian depositional surfaces appear to be composed of linear dunes, but this varies widely from one desert to another, from 85 per cent of the area for the south-western Kalahari to only 1.5 per cent of the Ala Shan. The dunes are typically 0.5–3.0 km apart, 0.2–1.5 km wide, and 5–200 m high. The relationship of seifs to wind regimes is not entirely clear. On the one hand there are those who believe that they form parallel to the direction of the dominant or prevailing winds in which roll vortices occur, while on the other there are those who believe that they form parallel to the resultant or vector sum direction of winds from two or more directions (be it diurnally or seasonally). ASG

Reference
Lancaster, I.N. 1982: Linear dunes. *Progress in physical geography* 6, pp. 475–504.

seismicity The intensity and frequency of earthquakes in an area. Earthquake intensity is measured on the logarithmic Richter scale, the largest shocks having a magnitude of a little over 8.5. For each decrease in unit magnitude the frequency of earthquakes increases by a factor of between 8 and 10. About twenty-five shocks a year with a magnitude of 7 or more are registered, but the total annual number of earthquakes of all magnitudes exceeds one million. Most seismic activity is located along plate boundaries (see PLATE TECTONICS) and is particularly concentrated in the circum-Pacific belt, which accounts for about 80 per cent of global seismicity. MAS

Reading
Bolt, B.A. 1978: *Earthquakes: a primer*. San Francisco: W.H. Freeman.

Gubbins, D. 1990: *Seismology and plate tectonics*. Cambridge: Cambridge University Press.

Wyss, M., ed. 1979: Earthquake prediction and seismicity patterns. *Pure and applied geophysics* 117, pp. 1079–315.

self-mulching A process whereby swelling and shrinking in soils, resulting either from alternate wetting and drying or from freezing and thawing, gives rise to a surface layer, composed of well-aggregated granules or fine blocks, which does not crust.

selva See TROPICAL FOREST.

semi-desert A semi-arid region.

sensible temperature Used in the context of thermal comfort, it is the indoor temperature that would produce the same sense of comfort or discomfort to a lightly clothed person, as the actual outdoor weather environment. It is dependent upon windspeed, humidity and the radiation balance as well as the air temperature.

Sensible heat is the same as enthalpy which represents the total heat or total energy content of a substance per unit mass. In the atmosphere a change of sensible heat of a mass of gas is the heat gained or lost by the gas in an exchange at constant pressure. The transport of sensible heat horizontally in the atmosphere is of fundamental importance to the general circulation of the atmosphere as heat is transferred from the tropics towards the poles. JET

sensitivity Of a clay soil is the ratio of undisturbed, undrained strength to the remoulded, undrained strength when tested at the same moisture content and (dry) density. A reduction in strength on remoulding is usually seen for most normally consolidated soils (see CONSOLIDATION) but it is rarely greater than 10 per cent and for over-consolidated soils it is usually near zero. However, for certain types of clay (quick-clays) the value may be 100 or more. This catastrophic decrease in strength can be caused in the field by an earthquake, for example, and has led to some very large slope failures, in Canada and southern Scandinavia in particular. Loesses may also exhibit a certain amount of sensitivity.
WBW

Reading
Maerz, N.H. and Smalley, I.J. 1985: The nature and properties of very sensitive clays: a descriptive bibliography. *University of Waterloo, Ontario, Bibliography* 12.

serac Pinnacles and cuboid masses of ice associated with rapid glacier flow, as for example found on ice falls and SURGING GLACIERS.

seral community A development or successional plant or animal community. In a community sequence or SERE, all the communities not at the stable terminal (CLIMAX) stage are seral. Seral communities tend to be dominated by opportunistic species adapted for rapid dispersal and growth, often small and capable of completing their life cycle comparatively rapidly (*r*-selection). They may also be short-lived communities, of low diversity, small biomass, high net production, with short food chains and poorly developed homeostatic

mechanisms. According to E.P. Odum, many of their features are also characteristic of agricultural plant communities which are similarly unstable ecologically. (See also SUCCESSION.) JAM

Reading and Reference
†Drury, W.H. and Nisbet, I.C.T. 1973: Succession. *Journal of the Arnold Arboretum* 54, pp. 331–68.
Odum, E.P. 1969: The strategy of ecosystem development. *Science* 164. pp. 262–70.

serclimax A plant community forming a relatively stable stage in a SERE. It is a long-persisting SERAL COMMUNITY that gives at least an impression of permanence, as particular species or environmental factors may stabilize a community well before the climax stage. (See also ALLELOPATHY; CLIMAX VEGETATION; SUCCESSION.) JAM

Reading
Connell, J.H. and Slatyer, R.O. 1977: Mechanisms of succession in natural communities and their role in community stability and organization. *American naturalist* 111, pp. 1119–44.
Moravec, J. 1969: Succession of plant communities and soil development. *Folia geobotanica et phytotaxonomica* 4, pp. 133–64.

sere A sequence of communities, usually plant communities, at a particular site. It is a series of stages that follows from the process of SUCCESSION. Each sere is made up of SERAL COMMUNITIES and may eventually terminate in a stable community. Different seres in a landscape have for long been viewed as at least partially convergent; thus the later stages in a dry environment (xerosere) are supposed to become increasingly similar to the later stages in a wet environment (hydrosere). Many factors, both natural and anthropogenic, may arrest, disturb or deflect such a simple pattern of development. JAM

Reading
Daubenmire, R. 1968: *Plant communities: a textbook of plant synecology*. New York and London: Harper & Row.
Matthews, J.A. 1979: A study of the variability of some successional and climax plant assemblage-types using multiple discriminant analysis. *Journal of ecology* 67, pp. 255–71.

serir A REG. A desert with a surface mantled by sheets of pebbles.

seston 'Includes all of the mineral particles and non-living organic matter that is suspended in the flows, whether derived from allochthonous (tributary, etc., external) or autochthonous (living, internal) sources' in rivers (Petts 1985, p. 88). ASG

Reference
Petts, G.E. 1985: *Impounded rivers*. Chichester: Wiley.

Seventh Approximation for Soil Classification (Comprehensive Soil Classification System) A hierarchical system with a complex terminology developed by the US Department of Agriculture. At the broadest scale are ten main soil orders (see table 1). The name of each order is based on syllables that are intended to convey the major attributes of its class. *Enti*sols, for example, are soils that either have not existed long enough to develop

Table 1 *US Department of Agriculture soil classification: The Seventh Approximation (1975)*

Order	Sub-order	Characteristics/environment
Alfisols (soils with an argillic horizon and moderate to high base content)	Aqualfs	With gleying features
	Boralfs	Others in cold climates
	Udalfs	Others in humid climates (including most leached brown soils)
	Ustalfs	Others in sub-humid climates
	Xeralfs	Others in sub-arid climates
Aridisols (desert and semi-desert soils)	Argids	With argillic horizon (i.e. zone of clay accumulation)
	Orthids	Other soils of dry areas
Entisols (immature, usually azonal soils)	Aquents	With gleying features
	Arents	Artificially disturbed
	Fluvents	On alluvial deposits
	Psamments	Sandy or loamy sand textures
	Orthents	Other entisols
Histosols	Fibrists	Plant remains very little decomposed
	Folists	Freely draining histosols
	Hemists	Plant remains not recognizable because of decomposition. Found in depressions
	Saprists	Plant remains totally decomposed (black)
Inceptisols (moderately developed soils, not in other orders)	Andepts	On volcanic ash
	Aquepts	With gleying features
	Plaggepts	With a man-made surface horizon
	Tropepts	In tropical climates
	Umbrepts	With an umbric epipedon (i.e. dark-coloured surface horizon of low base status); hills and mountains
	Ochrepts	Other inceptisols (including most brown earths) of mid–high latitudes
Mollisols (soils with a dark A horizon and high base status, e.g. chernozems, rendzinas)	Albolls	With argillic and albic horizons
	Aquolls	With gleying features
	Rendolls	On highly calcareous materials
	Borolls	Others in cold climates
	Udolls	Others in humid climates
	Ustolls	Others in sub-humid climates
	Xerolls	Others in sub-arid climates
Oxisols (soils with an oxic horizon or with plinthite near surface)	Aquox	With gleying features
	Humox	With a humose A horizon
	Torrox	Oxisols of arid climates
	Orthox	Others in equatorial climates
	Ustox	Others in sub-humid climates
Spodosols (soils with accumulation of free sesqui-oxides and/or organic carbon, e.g. podzols)	Aquods	With gleying features
	Ferrods	With much iron in spodic horizon
	Humods	With little iron in spodic horizon
	Orthods	With both iron and humus accumulation
Ultisols (soils with an argillic horizon, but low base content)	Aquults	With gleying features
	Humults	With a humose A horizon
	Udults	Others in humid climates
	Ustults	Others in sub-humid climates
	Xerults	Others in sub-arid climates
Vertisols (cracking clay soils with turbulence in profile)	Torrerts	Usually dry (cracks open for 300 days per year)
	Uderts	Usually moist (cracks open and close several times per year)
	Ustets	Cracks remain open 90 days per year (in monsoon climates)
	Xererts	Cracks remain open 60 days per year

Table 2 *Formative elements in names of sub-orders of the Seventh Approximation*

Formative element	Meaning
alb	Presence of albic horizon (a bleached eluvial horizon)
and	Ando-like (i.e. volcanic ash materials)
aqu	Characteristics associated with wetness
ar	Mixed or cultivated horizon
arg	Presence of argillic horizon (a horizon with illuvial clay)
bor	Of cool climates
ferr	Presence of iron
fibr	Fibrous
fluv	Floodplain
fol	Presence of leaves
hem	Presence of well-decomposed organic matter
hum	Presence of horizon of organic enrichment
ochr	Presence of ochric epipedon (a light-coloured surface horizon)
orth	The common ones
plagg	Presence of a plaggen epipedon (a man-made surface 50 cm thick)
psamm	Sandy texture
rend	Rendzina-like
sapr	Presence of totally humified organic matter
torr	Usually dry
trop	Continually warm
ud	Of humid climates
umbr	Presence of umbrid epipedon (a dark-coloured surface horizon)
ust	Of dry climates, usually hot in summer
xer	With annual dry season

mature horizonation (i.e. they are rec*ent*) or they lie on parent materials, such as quartz dune sand, that does not readily evolve into horizons. To construct class names at the sub-order level there are two formative elements, the first indicating the characteristics of the soil or its environment (such as *aqu*, indicating wetness) (see table 2), and the second being a suffix derived from the name of the order. Thus the sub-order *Aquox*, is an oxisol with gleying features indicative of wetness.   ASG

Reading
Soil survey staff 1975: *Soil taxonomy: a basic system of soil classification for making and interpreting soil surveys.* US Department of Agriculture handbook 436.

shakehole A roughly circular depression in the landscape in which water drains into an underground limestone cave system. The term is used synonymously with DO-LINE and swallowhole, but should be restricted in usage to a depression formed by the collapse of underlying limestone strata (Warwick 1976). In practice, shakeholes, etc. are produced by a combination of solution and collapse processes. The term derives from Derbyshire, England, but is now used extensively for any circular depression in limestone regions.   PAB

Reference
Warwick, G.T. 1976: Geomorphology and caves. In C.H.D. Cullingford and T.D. Ford eds, *The science of speleology*. London: Academic Press. Pp. 61–126.

shale A compacted sedimentary rock composed of fine-grained particles usually clay-sized. Shales are characteristically fissile in the plane of their bedding.

sharp-crested weir A particular type of flow gauging WEIR, sometimes referred to as a thin plate weir, in which the crest is formed by a sharpened metal plate and which is widely used for measuring the discharge of small streams. In practice the 'sharp' crest is produced by bevelling the edge of a thickness of *c.* 2 mm. With reference to the shape of the notch or crest, through which the water flows, sharp-crested weirs may be classified as rectangular, triangular, trapezoidal or of other geometric form. Calibration formulae are published for most configurations. The basic formula for a rectangular sharp-crested weir is:

$$Q = KbH$$

where Q is the discharge, b is the width of the weir crest, H is the head of water above the crest, and K is the weir or discharge coefficient, which depends on the dimensions of the installation. (See also DISCHARGE; WEIR). DEW

Reading
Ackers, P., White, W.R., Perkins, J.A. and Harrison, A.J.M. 1978: *Weirs and flumes for flow measurement.* Chichester: Wiley,
Kulin, G. and Compton, P.R. 1975: *A guide to methods and standards for the measurement of water flow.* US National Bureau of Standards special publication 421.

shear box An apparatus used to measure the shear strength of a soil. In its simplest form it consists of a split box into which the soil to be tested is placed. A normal stress is applied to the top of the sample via weight and, while one half of the box is moved parallel to the base to provide the shear stress, the other half resists the movement, the magnitude of the resistance being measured by a 'proving ring'. In this form the test is strictly called the direct shear test as it differs from simple shear which is an angular, rather than a linear, displacement. At least three tests are performed with different values of normal stress; a plot of the resisting stress corresponding to the normal stress gives values of COHESION and FRICTION which can be used in the MOHR–COULOMB EQUATION. The device can also be used to measure the RESIDUAL STRENGTH of the soil. Different sizes of box are required – the larger the size of grain under test the larger the size of the box; thus 30 cm square boxes may be required for testing gravelly soils. Although mainly a laboratory test, some versions are for *in situ* field determinations. The TRIAXIAL APPARATUS does a similar job but has advantages over the direct shear box. WBW

shear strength A measure of the ability of a material to resist shear stress. This is an important parameter in determining the engineering and geomorphic properties of materials. The shear strength of a soil is controlled by components of the MOHR–COULOMB EQUATION. Soils have a maximum strength value (peak strength) usually determined by a triaxial test or SHEAR BOX test. Various types of test conditions may be used to determine the shear strength. These relate to consolidation and drainage conditions. Particularly important are the 'undrained test' where no PORE WATER PRESSURE dissipation is allowed during shearing and the 'drained test' where such drainage is allowed. Generally speaking, the undrained test, which is done with a high strain rate, gives a lower shear strength value. WBW

Reading
Atkinson, J.H. and Bransby, P.L. 1978: *The mechanics of soils.* London: McGraw-Hill.
Whalley, W.B. 1976: *Properties of materials in geomorphological explanation.* Oxford: Oxford University Press.

shear stresses Two perpendicular loads or stresses (force per unit area) applied parallel (tangential) to the surface of a body. They are themselves perpendicular to the NORMAL STRESS. The diagram illustrates their orientation with respect to a cube of material. Shear stresses produce angular deformation (shear strain) in the body. WBW

Reading
Statham, I. 1977: *Earth surface sediment transport.* Oxford: Clarendon Press.

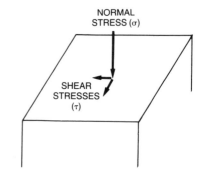

sheet erosion The process of erosion of a broad area by a sheet flood, occurring especially in arid regions.

sheetflow Unchannelled flow of water over the soil surface. OVERLAND FLOW on all but the smoothest surfaces, and on all natural hillsides, breaks into threads of high velocity separated by areas of slower and shallower flow. Sediment entrainment by the flow occurs only within the more rapid threads. In the sheetflow areas between, sediment detached or dislodged by rainsplash will be transported farther than in the absence of flow, so that sheetflow has a small influence on sediment transport even though it cannot initiate erosion. MJK

Reading
Kirkby, M.J. and Morgan, R.P.C. 1980: *Soil erosion.* Chichester: Wiley. Esp. chs 4–6.

sheeting The formation of joints in a massive rock such that the outer layers of the rock separate in shells or spalls and exfoliate (i.e. peel away from the parent rock mass). The shells are large, with minimum areas of several square metres and thicknesses of tens of centimetres. They commonly form, and may be responsible for, the shape of rounded outcrops and boulders. Successive exfoliation of concentric rock sheets results in the maintenance of rounded forms of some tors and bornhardts.

Sheeting developed on rock outcrops in the Namib Desert in South-West Africa (Namibia). The rounded form of inselbergs may in part result from the development of dilatation joints associated with this process.

Sheeting commonly occurs in massive rocks which have once been deeply buried and have subsequently been brought to the ground surface by removal of overburden or overlying ice. The uncovered rock undergoes release of confining stresses and expands outwards towards the ground surface and bounding joints. Extension fractures form perpendicular to the minimum stress direction at the time of failure: this occurs because the original maximum stress direction is controlled by the overburden; as the overburden is eroded the maximum stress converts to being the minimum stress and sheeting develops nearly parallel to the unloading surface. Well-known examples occur in the granites of Yosemite Valley, California and curved forms in sandstones of the Colorado Plateau are described by Bradley (1963).

MJS

Reading and Reference
†Bradley, W.C. 1963: Large-scale exfoliation in massive sandstones of the Colorado Plateau. *Bulletin of the Geological Society of America*, 74, pp. 519–27.

shelf The continental shelf. The sea floor lying between the coast and the steeper slope down to the deep ocean.

shell pavements Accumulation of shell valves which occur in all places where mixtures of sand and shell are subjected to selective erosional winnowing, leaving a superficial lag deposit of the coarse shell material. Pavements of this type are common in coastal situations where, for example, fine sediment can be moved by tidal currents or by aeolian processes above high-water mark. ASG

Reading
Carter, R.W.G. 1976: Formation, maintenance, and geomorphological significance of an aeolian shell pavement. *Journal of sedimentary petrology* 46, pp. 418–29.

shield A continental area of exposed Precambrian rocks within a CRATON and bordered by a platform area covered by post-Precambrian sedimentary strata. Shields are highly stable low-lying areas which have experienced little deformation or volcanic activity since the Precambrian. Examples include the Canadian Shield of North America and the Baltic Shield in northern Europe. MAS

Reading
Spencer, E.W. 1977: *Introduction to the structure of the earth.* 2nd edn. New York: McGraw-Hill.

Windley, B.F. 1984: *The evolving continents* 2nd edn. London and New York: Wiley.

shillow Loose rock fragments on limestone bedrock slopes, the equivalent of the clitter of granite slopes. ASG

Reading
Frank, R. 1991: Shillow–a neglected Holocene regolith on linestone. *Quaternary newsletter* 65, pp. 1–7.

shoal Area of shallow water in a lake or sea. A sand bank which lies just beneath the surface of a lake or sea.

shore The area of land immediately adjacent to a body of water.

shore platforms Intertidal rock surfaces of low slope angle. The term is preferred to 'wave cut platform' because processes other than mechanical wave action can play a role in their formation, not least weathering and bio-erosion. The form of the platforms depends on the nature of the main processes operative upon them, the nature of the rocks and their structures, on tidal

A shore platform developed on Lias mudstones near Charmouth, Dorset, southern England.

characteristics, on their age, and on their history.   ASG

Reading
Trenhaile, A.S. 1980: Shore platforms: a neglected coastal feature. *Progress in physical geography.* 4, pp. 1–23.

sial A term introduced by Suess to describe that part of the earth's crust with a granitic-type composition dominated by minerals rich in silicon (Si) and aluminium (Al). It is contrasted with the term SIMA. Sial forms at least the upper part of the crust of the continents and has a mean density of about 2700 kg m^{-3} and a silica content of between 65 and 75 per cent.   MAS

sichelwannen Bow-shaped furrows, with arms generally pointing in the direction of flow of the glacier that created them. They are of the order of 1–2 m long, relatively shallow, and tend to occur in large localized assemblages on glaciated surfaces. (See also P-FORM.)   ASG

Reading
Allen, J.R.L. 1984: *Sedimentary structures.* Amsterdam: Elsevier. Pp. 264–6.

sideways looking airborne radar (SLAR) A microwave remote sensor used to derive images of the earth's surface.

The SLAR senses the terrain to the side of an aircraft's track. It does this by pulsing out long, up to radio, wavelengths of ELECTROMAGNETIC RADIATION and then recording, first, the strength of the pulse return to the aircraft, to detect objects, and, secondly, the time it takes for the pulse to return, to give the range of objects from the aircraft. The name radar is the acronym of these functions of radio detection and ranging. As these pulses are emitted at right angles to the aircraft track, the movement of the aircraft enables pulse lines to be built up to form an image.

Like a MULTISPECTRAL SCANNER or THERMAL INFRARED LINESCANNER, the SLAR possesses collectors, detectors and recorders and, in addition, a transmitter and antenna (see diagram).

The transmitter produces pulses of microwave energy which are timed by a synchronizer and standardized to a known power by a modulator. For a fraction of a second the transmit/receive switch is switched to transmit, as the transmitter releases a microwave pulse from the antenna. The transmit/receive switch then returns to its original position and the antenna continues to receive pulses that have been backscattered from the earth's surface. These pulses are converted to a

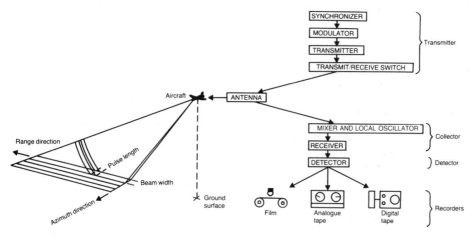

Sideways looking airborne radar

form suitable for amplification and further processing by a mixer and local oscillator before being passed to a receiver. The receiver amplifies the signal before passing it to the detector which produces an electronic signal suitable for recording on to photographic film or analogue or digital tape. To improve the spatial resolution of this sensor a larger antenna can be synthesized electronically. Such an SLAR, which is termed a synthetic aperture radar (SAR), is the primary SLAR for environmental research. The SAR is carried by aircraft and several UNMANNED EARTH RESOURCES SATELLITES, most notably ERS-1 and JERS-1.

Four characteristics of SLAR imagery determine its fields of application. They are its relatively high cost, its rapid rate of data acquisition (as it is unhindered by cloud or nightfall) and its sensitivity to both surface roughness and surface moisture content. SLAR was used initially for geological exploration as the likely financial returns were high, the areas to be covered were large and surface roughness and moisture content often varied between the areas of interest.

Today SLAR imagery is used in geomorphology, the mapping of soil moisture and vegetation, the estimation of forest biophysical properties and the location of oil pollution and sea ice.   PJC

Reading
Trevett, J.W. 1986: *Imaging radar for resources surveys.* London: Chapman & Hall.

Ulaby, F.T., Moore, R.K. and Fung, A.K. 1981: *Microwave remote sensing, active and passive. Volume 1: Fundamentals and radiometry.* Reading, Mass. and London: Addison-Wesley.

sieve deposits Occur on alluvial fans when the sediment load of the flood is deficient in fine-grained sediment. A highly permeable older deposit causes the flow to diminish rapidly as infiltration of water occurs, and as a result a clast-supported gravel lobe is deposited.

silcrete A highly siliceous indurated material formed at, or near, the earth's surface through the silicification of bedrock, weathering products or other deposits by low temperature physicochemical processes. Silcrete of Cainozoic age is particularly well developed in areas of inland Australia, in southern Africa and in northwest Europe. It may attain a thickness in excess of 5 m, and through its resistance to weathering and erosion it plays an important role in armouring erosion surfaces. It forms in areas of minimal local relief under both semi-arid and humid climatic regimes.   MAS

Reading
Langford-Smith, T. ed. 1978: *Silcrete in Australia.* Armidale, NSW: Department of Geography, University of New England.

Summerfield, M.A. 1983: Silcrete. In A.S. Goudie and K. Pye eds, *Chemical sediments and geomorphology.* London: Academic Press. Pp. 59–91.

sill A tabular sheet of igneous rock injected along the bedding planes of sedimentary or volcanic formations.

silt An unconsolidated material composed of particles ranging in size from *c.* 2.0 μm to *c.* 63μm. A soil containing more than about 80 per cent material in this size range.

silt drape A thin veneer of silt that is deposited as fallout from suspension.

siltation The accumulation of fine sediment (strictly speaking of silt) in a body of water. Applied technically to the settling out of fine particles in water and more generally to the filling or choking of lakes, reservoirs or water courses.   JL

sima A term introduced by Suess to describe that part of the earth's crust with a basaltic-type composition dominated by minerals rich in silicon (Si) and magnesium (Mg). It is contrasted with the term SIAL. It forms the crust of the ocean basins and the lower portion of the crust of the continents. Sima has a higher mean density than sial $(2800-3400$ kg m$^{-3})$ and a lower silica content (less than 55 per cent).   MAS

simulated basin Various component systems of drainage basins may be modelled using computer SIMULATION methods. For example RANDOM-WALK NETWORKS simulate the evolution and structure of the drainage pattern. However, simulation is most advanced in the analysis and prediction of drainage basin response to precipitation input. Mathematical moisture accounting procedures route this input through a series of conceptual stores forming a cascade of linked sub-systems. This routine is achieved by mathematical expressions governing transfers of moisture. For example, the infiltration rate of the soil surface is modelled by a negative exponential function of time during a rainstorm, and this is used to allocate rainfall to surface run-off and soil moisture storage.   KSR

simulation Hypothesis-testing in physical geography often makes use of the vicarious experiment procedure of simulation in which genuine phenomena are represented by scale, analogue or mathematical models. Scale models are small-scale hardware replications such as flumes, wave tanks or kaolin glaciers. Analogue models involve equivalent systems; for example, an electrical potential model of groundwater is feasible because of the

mathematical equivalence of the laws governing current flow in a circuit and flow in a porous medium. Computer-based mathematical simulation models (Thornes and Brunsden 1977, pp. 157–71), which may be deterministic or probabilistic, are often used to test alternative hypotheses by comparing model outputs under different assumptions with actual behaviour. Links between system components are represented by mathematical relations, whose constants are adjusted by PARAMETER-IZATION procedures until the best fit with reality is achieved. (See also RANDOM-WALK NETWORKS; SIMULATED BASIN.)

KSR

Reference
Thornes, J.B. and Brunsden, D.B. 1977; *Geomorphology and time.* London and New York: Methuen.

singing sands When in motion certain dune sands generate clearly audible sounds that have been variously reported as roaring, booming, squeaking, musical and singing (Curzon, 1923). A unique combination of granulometric properties appears to be responsible, including a high sorting value, uniform grain size and a high degree of roundness of grains (Van Rooyen and Verster, 1983).   ASG

References
Curzon, G.N. 1923: The singing sands. In *Tales of travel.* London: Hodder & Stoughton. Ch. 11.

Haff, P.K. 1986: Booming dunes. *American scientist* 74, pp. 376–81.

Van Rooyen, T.H. and Verster, E. 1983 Granulometric properties of the roaring sands in the south-eastern Kalahari. *Journal of arid environments* 6, pp. 215–22.

sinkhole A roughly circular depression in the landscape into which water drains and collects. Specifically, it is a depression in limestone terrain, often connecting with an underground cave system through which the water drains. It is used synonymously with SHAKEHOLE and DOLINE and was originally an American term.   PAB

Reading
Beck, B.F. and Wilson, W.L. 1987: *Karst hydrology: engineering and environmental applications.* Rotterdam: Balkema.

sinter A precipitate of silica or calcium carbonate associated with geysers and hot springs.

sinuosity The degree of wandering or winding, applied especially to river channels. It may be defined as the ratio of actual channel distance between identified points compared to the straight or down-valley distance.   JL

siphon A vertical or inverted U-shaped portion of a subterranean stream channel in which the water is in hydrostatic equilibrium.

skewness A statistical measure which is widely used for PARTICLE SIZE analysis. It measures the degree of asymmetry of a statistical distribution as well as whether the distribution has an asymmetrical tail to the left or right.   ASG

Reading
Folk, R.L. 1974: *Petrology of sedimentary rocks*. Austin, Texas: Hemphill.

slab failure A term usually used of strong rocks (although it can occur in muds and weak rocks) where the TRANSLATIONAL SLIDE is along discontinuities (cracks, joints, etc.) which dip outwards from the face. Failure is largely controlled by the FRICTION between the blocks so that the angle of the discontinuities needs to be high enough to allow most of the frictional strength to be exceeded but less than the cliff slope angle. The final 'trigger' which causes failure may be caused by ice wedging, where the blocks moved are small, or by (cleft) water pressure which changes the EFFECTIVE STRESS conditions. Quantitative assessment of cliff stability can be made with various ROCK QUALITY INDICES. (See also TOPPLING FAILURE.)
                                                   WBW

Reading
Attwell, P.B. and Farmer, I.W. 1976: *Principles of engineering geology*. London: Chapman & Hall; New York: Wiley.
Selby, M.J. 1993: *Hillslope materials and processes*. 2nd edn. Oxford: Oxford University Press.

slack A depression or hollow in an area of sand dunes or mud banks.

slaking The disintegration of a loosely consolidated material on the introduction of water or exposure to the atmosphere.

slickenside A polished or scratched rock surface produced by the friction during faulting.

slide A landslide. The landform produced by mass movement under the influence of gravity.

slip A landslide or mudslide. A fault.

slip face The steeply sloping portion of some sand dunes on their downwind side where oversteepening causes sand to cascade downwards and rest as its maximum angle of repose.

slip off slope The more gently sloping bank of a river on the inside of a meander.

slope A word with two applications in physical geography:

1 In a general sense it is used to refer to the angle which any part of the earth's surface makes with a horizontal datum. Synonyms for this usage include: inclination, declivity, and gradient.
2 In geomorphology 'slope' refers to any geometric element of the earth's solid surface whether that element is above or below sea level. Slope elements thus form entire landscapes. In a more restricted use the term is often applied to escarpments and valley sides, and thus excludes floodplains, terrace surfaces and other nearly horizontal elements. To avoid confusion the more explicit word 'hillslope' is in common use.

Hillslopes are regarded as three-dimensional forms produced by weathering and erosion with basal elements which may be either depositional or erosional in origin. The development of hillslopes is consequently the principal result of denudation and the study of such features is a major part of geomorphology.   MJS

Reading
Carson, M.A. and Kirkby, M.J. 1972: *Hillslope form and process*. Cambridge: Cambridge University Press.
Finlayson, B. and Statham, I. 1980: *Hillslope analysis*. London: Butterworth.
Selby, M.J. 1993: *Hillslope materials and processes*. 2nd edn. Oxford: Oxford University Press.
Young, A. 1972: *Slopes*. Edinburgh: Oliver & Boyd.

slope replacement A model of slope evolution formulated by the German geomorphologist, Walther Penck, in which the maximum slope angle decreases through time as a result of replacement from below by gentler slopes, causing the majority of the slope profile to become occupied by a concavity.

slope wind See ANABATIC FLOWS; KATA-BATIC FLOWS.

smog A term originally used to describe a combination of smoke and FOG but now used for any visibly polluted air. Dr Harold Antoine Des Voeux first used the word in 1911 to describe a series of pollution episodes in Glasgow, Scotland, during the autumn of 1909. Photochemical smog forms when hydrocarbons, originating from vaporized gasoline and other petroleum products, combine with nitrogen oxide molecules, also emitted by combustion engines, and water vapour in the presence of ultraviolet sunlight.   WDS

Reading
Lewis, H.R. 1965: *With every breath you take.* New York: Crown.

snout, glacial The terminus of a glacier.

snow Solid precipitation composed of single ice crystals or aggregates known as snowflakes. Ice crystals are most frequent when temperatures are much below freezing and the moisture content of the air is small. As temperatures increase towards 0°C, the ice crystals grow and cluster into flakes.

Graphic Symbol	Examples			Symbol	Type of Particle
⬡				F1	Plate
✳				F2	Stellar crystal
▭				F3	Column
→				F4	Needle
⊕				F5	Spatial dendrite
⊟				F6	Capped column
⌒				F7	Irregular crystal
△				F8	Graupel
△				F9	Ice pellet
▲				F0	Hail

The principal types of snow crystals.
Source: *L.W. Price 1981:* Mountains and man. *California: University of California Press. After La Chapelle 1969.*

Snow is difficult to measure accurately as it tends to block standard rain gauges or be blown out.   PS

snow line The altitudinal limit on land separating areas in which fallen snow disappears in summer from areas in which snow remains throughout the year. The altitudinal distribution over the globe is the same as for the FIRN LINE but unlike the latter it is not restricted to glaciers.   DES

snow patch An isolated area of snow which may last throughout the summer and initiate processes associated with NIVATION.

snowblitz theory A popular and extreme version of the view that glaciations may start rapidly as a result of positive feedback processes related to the increased albedo of a high latitude continent covered with persistent snow. A few years of excess snow could modify atmospheric circulation patterns and enhance snow accumulation (Lamb and Woodroffe 1970).

Developing the concept for Britain, Calder (1974, p. 118) wrote:

> In the snowblitz the ice sheet comes out of the sky and grows, not sideways, but from the bottom upwards. Like airborne troops, invading snowflakes seize whole counties in a single winter. The fact that they have come to stay does not become apparent, though, until the following summer. Then the snow that piled up on the meadows fails to melt completely. Instead it lies through the summer and autumn, reflecting the sunshine. It chills the air and guarantees more snow next winter. Thereafter, as fast as the snow can fall, the ice sheet gradually grows thicker over a huge area.

The theory is far from being established for Arctic Canada, let alone the lush meadows of Britain. Yet evidence of rapid ice sheet build-up as a result of sudden changes in atmospheric and oceanic circulation is emerging. (See also ICE AGE.)   DES

Reading and References
Calder, N. 1974: *The weather machine and the threat of ice.* London: BBC publications.
†Denton, G.H. and Hughes, T.J. 1983: Milankovitch theory of ice ages: hypothesis of ice-sheet linkage between regional insolation and global climate. *Quaternary research* 20, pp. 125–44.
†Imbrie, J. and Imbrie, K.P. 1979: *Ice ages.* London: Macmillan.

†Ives, J.D., Andrews, J.T. and Barry, R.G. 1975: Growth and decay of the Laurentide ice sheet and comparisons with Fenno-Scandinavia. *Die Naturwissenschalt* 62, pp. 118–25.

Lamb, H.H. and Woodroffe, A. 1970: Atmospheric circulation during the last Ice Age. *Quarternary research* 1.1, pp. 29–58.

snowmelt The part of run-off that is generated by melting of a snowpack on the ground surface. Portions of the snowpack that do not melt may remain on the surface long enough to be compressed by subsequent snowfalls into glacial ice. It may not melt over a period of many years and thus become a semi-permanent snow body, or its mass may be lost through sublimation and deflation.

The quality of the snowpack refers to the potential amount of run-off that may be generated by snowmelt and is measured as the weight of ice divided by the total weight of a unit volume of the snowpack. When the snowpack nears the time for melting, its quality is usually in the 0.90 to 1.00 range (US Army Corps of Engineers 1960).

                                                            WLG

Reference
US Army Corps of Engineers 1960: *Runoff from snowmelt.* US Army Corps of Engineers engineering manual 1110–2–1406.

soil The material composed of mineral particles and organic remains that overlies the bedrock and supports the growth of rooted plants.

soil classification See SEVENTH APPROXIMATION FOR SOIL CLASSIFICATION.

soil erosion The natural process of removal of top soil by water and wind. It is a process whose rates may be magnified by man (accelerated erosion). On a global scale the fastest rates occur in zones with highly seasonal precipitation, as in monsoonal, Mediterranean and semi-arid climates (Walling and Kleo 1979). There is a long history of the study of accelerated soil erosion (see e.g. Marsh 1864; Bennett 1938), and in spite of the introduction of soil conservation measures such as those discussed by Hudson (1971), it continues to be a serious environmental problem (Carter 1977; Pimentel 1976). In the USA soil erosion on agricultural land operates at a rate of about 30 t ha^{-1} year^{-1}. Water run-off delivers around four billion tonnes of soil to the rivers of the forty-eight contiguous states, and three-quarters of this comes from agricultural land. Another billion tonnes of soil are eroded by the wind, a process which created the Dust Bowl of the 1930s. In addition to the soil erosion caused by deforestation and agriculture, urbanization, fire, war and mining are often significant in accelerating erosion of the soil.   ASG

Reading and References
Bennett, H.H. 1938: *Soil conservation.* New York: McGraw-Hill.

Carter, L.J. 1977: Soil erosion: the problem still persists despite the billions spent on it. *Science* 196, pp. 409–11.

Hudson, N. 1971: *Soil conservation.* London: Batsford.

Marsh, G.P. 1864: *Man and nature.* New York: Schribner.

†Morgan, R.P.C. 1986: *Soil erosion and conservation.* Harlow: Longman.

Pimentel, D. 1976: Land degradation: effects on food and energy resources. *Science* 194, pp. 149–55.

Walling, D. and Kleo, A.H.A. 1979: *Sediment yields of rivers in areas of low precipitation: a global view.* International Association of Scientific Hydrology publication 128, pp. 479–93.

soil moisture deficit 'Soil moisture deficits are considered to have been set up when evapotranspiration exceeds precipitation and vegetation has to draw on reserves of moisture in the soil to satisfy transpiration requirements' (Grindley 1967).

The evaluation of soil moisture deficit is essential to the estimation of irrigation need since it provides an estimate of the degree to which soil moisture content has dropped below field capacity. Field capacity is the soil moisture condition when excess water has drained out of a saturated or near-saturated soil and it is thought to be the soil moisture condition which will promote maximum plant growth, with transpiration occurring at the potential rate (i.e. transpiration is not limited by moisture availability). Any reduction of soil moisture content below field capacity will create a soil moisture deficit which may be removed by irrigation.

Soil moisture deficits can be estimated using field instruments or evaporation estimation equations combined with evaluation of the soil WATER BALANCE. LYSIMETERS and NEUTRON PROBES allow changes in soil moisture storage to be directly observed so that the deficit may be calculated but other instruments, including evaporation pans and atmometers, as well as the majority of evaporation equations, provide estimates of open water evaporation (Eo) or potential evapotranspiration (PEt), which may be actual evapotranspiration (Et) and a soil water budget for the site.

In calculating Et from PEt or Eo, it is necessary to take into account the degree to which a decrease in soil moisture content will reduce the Et rate below the PEt rate. This is a controversial topic which is reviewed by Baier (1968) but the method adopted by the UK Meteorological Office (Grindley 1967) employs the concept of root constants and related drying curves described by Penman (1949). Grindley (1967) explains the way in which the difference between PEt and precipitation in consecutive time periods can be partitioned to calculate Et and soil moisture surplus or soil moisture deficit.   AMG

Reading and References
Baier, W. 1968: Relationship between soil moisture, actual and potential evapotranspiration. In *Soil moisture*. Proceedings of the Hydrology Symposium 6, University of Saskatchewan, 15–16 November 1967. Ottawa: Queen's Printer. Pp. 155–91.

†Calder, I.R., Harding, R.J. and Rosier, P.T.W. 1983: An objective assessment of soil moisture deficit models. *Journal of hydrology* 60, pp. 329–55.

Grindley, J. 1967: The estimation of soil moisture deficits. *Meteorological magazine* 96, pp. 97–108.

†Ministry of Agriculture, Fisheries and Food 1967: *Potential transpiration*. Technical bulletin 16. London: HMSO.

Penman, H.L. 1949: The dependence of transpiration on weather and soil conditions. *Journal of soil science* 1, pp. 74–89.

soil profile The full sequence through the soil zone from the surface down to the unaltered bedrock.

soil structure The grouping of aggregates within a soil. Individual structures have a variety of forms (see diagram) and may range in size from tiny granules to large blocks. Among the controls of soil structure are the presence of clay, humus and soluble silts. Soil aggregates are sometimes referred to as PEDS.   ASG

soil texture The character of the soil imparted by the proportions of sand, silt and clay within a sample. Two examples of soil textural classification schemes are shown in the diagram.   ASG

soil-landscape systems Examine how soils are organized within the landscape. The interaction between soils and topography can be treated at several levels of scale, but one important concept is that of the CATENA. The diagram shows the hypothetical nine-unit landsurface model developed by Dalrymple *et al.* (1968), which shows the relationship between slope position and dominant soil and landforming processes.   ASG

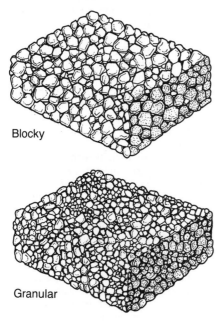

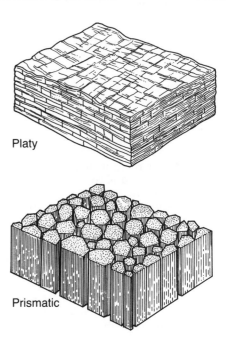

Soil structure: Some common forms.

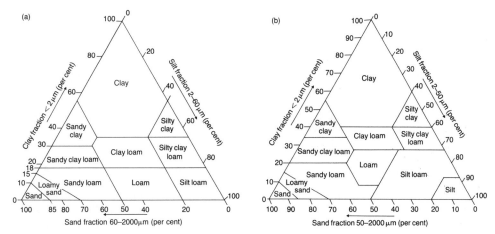

(a)

100 — 0

Clay fraction <2μm (per cent)

Silt fraction 2-60μm (per cent)

Clay

Sandy clay

Silty clay

Clay loam

Silty clay loam

Sandy clay loam

Loamy sand | Sandy loam

Loam

Silt loam

Sand

Sand fraction 60-2000μm (per cent)

(b)

100 — 0

Clay fraction <2μm (per cent)

Silt fraction 2-50μm (per cent)

Clay

Sandy clay

Silty clay

Clay loam

Silty clay loam

Sandy clay loam

Loam

Sandy loam

Silt loam

Loamy sand

Silt

Sand

Sand fraction 50-2000μm (per cent)

Soil texture: *Triangular classification based on the limits laid down by:* (a) *Soil Survey of England and Wales*; (b) *US Department of Agriculture.*

INTERFLUVE 1

SEEPAGE SLOPE 2

CONVEX CREEP SLOPE 3

4 FALL FACE

⊗ INDICATES MOVEMENT IN A DOWNVALLEY DIRECTION

TRANSPORTATIONAL MIDSLOPE 5

COLLUVIAL FOOTSLOPE 6

ALLUVIAL TOESLOPE 7

CHANNEL WALL 8

CHANNEL BED 9

ARROWS INDICATE DIRECTION AND RELATIVE INTENSITY OF MOVEMENT OF WEATHERED ROCK AND SOIL MATERIALS BY DOMINANT GEOMORPHIC PROCESSES

APPROX. LOWER LIMIT OF SOIL FORMATION

1 Pedogenic processes associated with vertical subsurface soil water movement.

2 Mechanical and chemical elluviation by lateral subsurface water movement.

3 Soil creep; terracette formation.

4 Fall; slide; chemical and physical weathering.

5 Transportation of material by mass movement (flow, slide, slump, creep); terracette formation; surface and subsurface water action.

6 Redeposition of material by mass movement and some surface wash; fan formation, transportation of material; creep; subsurface water action.

7 Alluvial deposition; processes from subsurface water movement.

8 Corrasion, slumping, fall.

9 Transportation of material downvalley by surface water action; periodic aggradation and corrasion.

1   2   3   4   5   6   7   8   9

Soil-landscape systems: *The hypothetical nine-unit landsurface model.*
Source: *Dalrymple et al. 1968.*

Reading and Reference
Dalrymple, J.B., Blong, R.J. and Conacher, A.J. 1968: A hypothetical nine unit landsurface model. *Zeitschrift für Geomorphologie* 12, pp. 60–76.
†Gerrard, A.J. 1992: *Soil geomorphology*. London: Chapman and Hall.

solar constant The rate at which solar radiation is received outside the earth's atmosphere on a surface normal to the incident radiation, and at the earth's mean distance from the sun. Its exact value is still a little uncertain but it is nearly 1380 W m^{-2}. Despite its name it is probably slightly variable in time.   BWA

solfatara A small volcanic vent through which acid gases are emitted, usually in areas where violent volcanism has ceased.

Boiling mud in sulphur solfatara fields in northern Iceland is associated with volcanic activity connected with sea-floor spreading along the Atlantic's mid-ocean ridge.

solifluction A term first used by J.G. Andersson in 1906 to describe the 'slow flowing from higher to lower ground of waste saturated with water' which he observed in the Falkland Islands. It has subsequently been applied elsewhere to the slow gravitational downslope movement of water saturated, seasonally thawed materials. In contrast to gelifluction, solifluction does not require permafrost for its occurrence, but modern use of the term does imply the existence of cold climate conditions. It is a form of mass wasting (i.e. viscous flow), faster than soil creep, often in the order of 0.5–5 cm year^{-1}. Features produced by solifluction include uniform sheets of locally derived materials, tongue-shaped lobes, and alternating stripes of coarse and fine sediment. When associated with the active layer (i.e. in permafrost

regions) the term gelifluction should be used.   HMF

Reading and Reference
Andersson, J.G. 1906: Solifluction, a component of subaerial denudation. *Journal of geology* 14, pp. 91–112.
Benedict, J.B. 1970: Downslope soil movement in a Colorado alpine region: rates, processes and climatic significance. *Arctic and alpine research*, 2, pp. 165–226.
Washburn, A.L. 1979: *Geocryology: a survey of periglacial processes and environments*. New York: Wiley.

solonchak A group of soils which are the result of salinization, and occur where there is an accumulation of soluble salts of sodium, calcium, magnesium and potassium in the upper horizon. The anions found are mostly sulphate and chloride. In contrast to SOLONETZ soils, which are highly alkaline, solonchaks, often called white alkali soils, are only slightly alkaline, their pH seldom rising much above pH 8.
       ASG

solonetz An intrazonal group of soils which have surface horizons of varying degrees of friability underlain by dark, hard soil characterized by a columnar structure. The hard layer is usually highly alkaline, with the high pH resulting from the adsorbed sodium and the presence of sodium carbonate. The soil colloids, both inorganic and organic, become dispersed and tend to move slowly down the profile, while the frequently observed dark colour of the surface crust is due to dissolved organic matter. These soils, sometimes known as black alkali soils, occur in semi-arid and subhumid areas.   ASG

solstice The day of maximum or minimum declination of the sun. Either the longest or shortest day of the year.

solum The soil zone above the weathered parent material, in effect the A and B horizons.

solutes All natural waters contain organic and inorganic material in solution or solutes. The oceans constitute about 97 per cent of the hydrosphere and its average chemical composition is therefore essentially that of seawater, with a total solute concentration of approximately 34,558 mg l^{-1}. The solute content of the remaining water associated with the terrestrial phase of the hydrological cycle exhibits considerable spatial and temporal variation and is of greater interest to the physical geographer. It has been widely studied as a means of investigating chemical weathering processes and rates of chemical denudation, evaluating nutrient cycling by vegetation communities, and elucidating the processes and pathways involved in the movement of water through the drainage basin system. The major solutes contained in these waters are Ca^{2+}, Mg^{2+}, Na^+, K^+, Cl^-, HCO_3^-, SO_4^{2-}, NO_3^- and SiO_2.

Precipitation inputs to the landsurface contain significant concentrations of solutes as a result of rain-out and wash-out of atmospheric aerosols. The magnitude and composition of this solute content will vary according to the relative importance of terrestrial and marine aerosols. Highest total solute concentrations are found in coastal areas where marine aerosols dominate and where Na^+, Cl^-, Mg^{2+} and K^+ are the dominant ions. The overall solute content of precipitation declines inland and, at distances in excess of about 100 km from the coast, terrestrial aerosols predominate as solute sources. Here the major ions are Ca^{2+} and SO_4^{2-}. Meybeck (1983) provides estimates of the average solute content of precipitation over coastal and inland zones as shown in table 1.

As it moves through the vegetation canopy and the soil and rock of a drainage basin the solute content of the water will increase and its chemical composition will frequently change. Solutes will be leached and washed from vegetation and solute levels within the soil will be influenced by concentration and precipitation mechanisms, interactions with the soil matrix, release of solutes through chemical weathering, and biotic uptake and release of nutrients. Further evolution of the solute content may occur within the groundwater body.

The solute content of streamflow will therefore reflect the characteristics of the upstream drainage basin, including its geology, topography and vegetation cover, and the pathways and RESIDENCE TIME associated with water movement through the basin. Concentrations will vary through time in response to hydrological conditions and will frequently exhibit a DILUTION EFFECT during storm run-off events.

At the global scale, climate and lithology exert a major influence on the solute content of river water. Solute concentrations commonly exhibit an inverse relationship with mean annual run-off and demonstrate marked contrasts between major rock types. Total solute concentrations in rivers draining basins underlain by sedimentary rocks are on average about five times greater than from basins underlain by crystalline rocks and about 2.5 times greater than from basins underlain by volcanic rocks. Typical ranges of concentrations associated with individual solute species in major world rivers are listed in table 2.

Calcium is the dominant cation and HCO_3^- the dominant anion in nearly all major rivers. A greater degree of variation in solute concentrations is to be found when considering data from small streams.

Measurements of the solute input into a drainage basin and the output in streamflow provide a means of establishing a solute budget for the basin. The net solute yield (t year^{-1}) (output–input) reflects the production of solutes within the basin, which may in turn be related to the products of chemical weathering, the uptake of atmospheric CO_2 by weathering reactions and the mineralization of organic material. On a

Table 1 *Average solute content of precipitation*

	Concentration (mg l^{-1})				
	Ca^{2+}	Mg^{2+}	Na	K	Cl^-	SO_4^{2-}
Coastal	0.29	0.45	3.45	0.17	6.0	1.45
Inland	0.43	0.19	0.37	0.15	0.75	1.73

Source: Meybeck 1983.

Table 2 *Range of concentrations of solutes in major world rivers*

	Concentration (mg l^{-1})							
	Ca^{2+}	Mg^{2+}	K	Na	HCO_3^-	Cl^-	SO_4^{2-}	SiO_2
Minimum	2.0	1.0	0.5	1.0	10	1.0	1.5	2.0
Maximum	55	15	4	40	170	45	65	20

Source: Meybeck 1983.

global basis, approximately 50 per cent of the solutes found in river water represent the products of chemical weathering, but this value will vary markedly between individual catchments. In catchments underlain by resistant crystalline rocks the contribution from chemical weathering may be negligible, whereas in areas of sedimentary rocks this contribution will be dominant.

Interest in the solute content of natural waters has necessitated the development of a wide range of analytical methods, which are documented in a number of laboratory manuals. Many of these methods are now semi-automated and provide a means of dealing with the large numbers of samples produced by automatic samplers or intensive manual sampling programmes. Measurements of specific CONDUCTANCE are widely employed as a simple means of estimating the total solute content of a sample and this parameter may be continuously recorded using simple equipment. Progress has also been made in the development of apparatus for continuous monitoring of individual solute species using specific ion electrodes.   DEW

Reading and Reference
†American Public Health Association 1971: *Standard methods for the examination of water and wastewater.* New York: American Public Health Association.

†Golterman, H.L., Clymo, R.S. and Ohnstad, M.A.M. 1978: *Methods for chemical analysis of fresh waters.* Oxford: Blackwell Scientific.

†Hem, J.D. 1970: *Study and interpretation of the chemical characteristics of natural water.* US Geological Survey water supply paper 1473.

†Likens, G.E., Bormann, F.H., Pierce, R.S., Eaton, J.S. and Johnson, N.M. 1977: *Biogeochemistry of a forested ecosystem.* New York: Springer-Verlag.

Meybeck, M. 1983: Atmospheric inputs and river transport of dissolved substances. In *Dissolved loads of rivers and surface water quantity/quality relationships.* IAHS publication 141. Pp. 173–92.

†Walling, D.E. 1980: Water in the catchment ecosystem. In A.M. Gower ed., *Water quality in catchment ecosystems.* Chichester and New York: Wiley.

solution, limestone The change of limestone from the solid state to the liquid state by combination with water. When water charged with carbon dioxide comes in contact with limestone (either as free CO_2 or as HCO_3^-) it dissolves the rock (CORROSION). When this occurs beneath the water table in the phreatic zone it dissolves bedding planes or joints to produce characteristically oval cave passages. When it occurs in normal stream situations it produces notches at the side of the stream on a line approximating the water level. Limestone solutional processes can also produce many different micro-features (KARREN).   PAB

Reading
Picknett, R.G., Bray, L.G. and Stenner, R.D. 1976: The chemistry of cave water. In T.D. Ford and C.H.D. Cullingford eds, *The science of speleology.* London: Academic Press. Pp. 213–66.

Trudgill, S.T. ed. 1986: *Solute processes.* Chichester: Wiley.

sorting of a particle size distribution is a measure of the standard deviation of the sample. It relates to the way in which material is differentially removed by particular geomorphic agencies, e.g. wind action tends to leave a well-sorted residual which is represented by a low sorting value, i.e. a predominance in a narrow size range roughly around the mean for a log normal distribution. It can be obtained graphically from phi (ϕ) percentiles by:

$$So = \frac{\phi 90 + \phi 80 + \phi 70 - \phi 30 - \phi 20 - \phi 10}{5.3}$$

A method of calculation using moment measures is also available.   WBW

Reading
Briggs, D. 1977: *Sediments.* London: Butterworth.

Tucker, M.E. 1981: *Sedimentary petrology: an introduction.* Oxford: Blackwell Scientific.

source area The area of a catchment which is physically producing OVERLAND FLOW at any time. This area is changing dynamically during and after storms. Its estimation is central to PARTIAL AREA MODELS, and the term is generally used in the context of overland flow produced by subsurface saturation rather than because the INFILTRATION capacity has been exceeded. Before a storm the previous rainfall establishes the pattern of the SATURATED ZONE within a catchment. Storm rainfall is added to this layer of saturated water, and increases the source area during the storm. After the storm THROUGHFLOW gradually diminishes the saturated wedge and the source area declines with it. The source area typically consists of river floodplains, together with a narrow strip along the base of concave hillsides and a larger area of converging flow in streamhead hollows. The total area involved varies from 1 to 3 per cent of catchments under dry conditions up to 10–50 per cent during and immediately after major storms, although there are wide differences between catchments. MJK

Reading
Kirkby, M.J. 1978: *Hillslope hydrology.* Chichester: Wiley.

southern oscillation See MACROME-TEOROLOGY.

speciation See DARWINISM; EVOLUTION.

species–area curve A graph of the relationships between plant or animal numbers and the area of sample plots. In general, the number of species present will increase as area increases within any given community. Eventually, the number of new species found in successively larger plots will become progressively fewer and the species–area curve will flatten and become approximately horizontal. Species–area curves are nevertheless useful guides in the determination of a satisfactory quadrat size for sampling and for comparing the size of the fauna or flora of different islands or various sized landmasses. PAS

Reading
Hopkins, B. 1957: The concept of minimal area. *Journal of ecology* 45, pp. 441–9.
Kershaw, K.A. 1973: *Quantitative and dynamic plant ecology.* 2nd edn. London: Edward Arnold.
Krebs, C.J. 1978: *Ecology: the experimental analysis of distribution and abundance.* 2nd edn. New York: Harper & Row.
Preston, F.W. 1962: The canonical distribution of commonness and rarity. *Ecology* 43, pp. 185–215, 410–32.
Randall, R.E. 1978: *Theories and techniques in vegetation analysis.* Oxford: Oxford University Press.

species–energy theory Suggests that the present-day species richness of plants and animals for largish regions can be explained in terms of available energy. The hypothesis is that subject to water supply and other factors not being limiting, diversity in terrestrial habitats is to a great extent controlled by the amount of solar energy available, declining with increasing latitude in accordance with the poleward reduction in the receipt of solar radiation (Wright 1983). ASG

Reference
Wright, D.H. 1983: Species–energy theory: an extension of the species–area theory. *Oikas* 41, pp. 496–506.

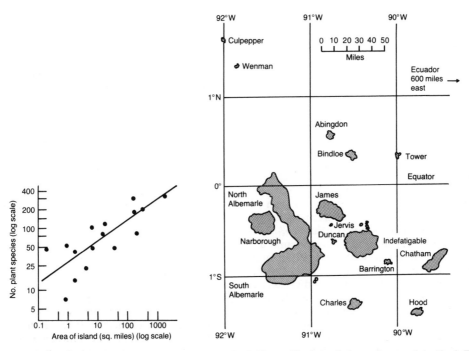

Species–area curve: Number of land-plant species on the Galápagos Islands in relation to the area of the island. The islands range in area from 0.2 to 2249 square miles and contain from 7 to 325 plant species.
Source: *Preston 1962.*

specific conductance See CONDUC-
TANCE, SPECIFIC.

specific retention The volume of water
which a rock or soil retains against the
influence of gravity if it is drained following
saturation. The difference between POR-
OSITY and specific retention is the SPECI-
FIC YIELD. PWW

Reading
Ward, R.C. and Robinson, M. 1990: *Principles of hydrology.*
3rd edn. Maidenhead: McGraw-Hill.

specific yield The volume of water that a
water-bearing rock or soil releases from
storage under the influence of gravity. In an
unconfined AQUIFER, it is expressed as the
volume per unit surface area of aquifer per
unit decline in the WATER TABLE. PWW

Reading
Freeze, R.A. and Cherry, J.A. 1979: *Groundwater.* Englewood Cliffs, NJ: Prentice-Hall.

speleology The scientific study of caves,
their formation and processes. It includes
studies of speleogenesis (cave formation
processes), cave survey, biology, geology
and chemistry. The science used to be
conducted by amateurs or semi-profes-
sionals but, since 1970, in response to the
increasing technicalities of the subject, most
of the pertinent research is carried out by
full-time scientists and researchers.
PAB

Reading
Bögli, A. 1980: *Karst hydrology and physical speleology.* New
York: Springer-Verlag.
Ford, T.D. and Cullingford, C.H.D. eds 1976: *The science
of speleology.* London: Academic Press.

speleothem A general term for depositional features which include stalactites,
stalagmites, columns, flowstone, helictites
and curtains. Speleothems are commonly
calcareous, crystalline deposits but can be
made of a number of different materials;
silica, gypsum, peat and ice have all been
recorded. Calcareous speleothems are
formed when rainwater seeps through
organic rich soils and absorbs carbon
dioxide. On contact with limestone it
dissolves some of the rock, but when it
reaches a cave roof it comes in contact with
air that is not charged with as much carbon
dioxide. The air absorbs some of the carbon
dioxide, the water becomes less aggressive
(indeed, supersaturated) and deposits some
calcium carbonate in the cave. PAB

Reading
Warwick, G.T. 1962: Cave formations and deposits. In
C.H.D. Cullingford ed. *British caving.* London: Routledge
& Kegan Paul. Pp. 83–119.

sphenochasm and sphenopiezm The
former is the triangular gap of oceanic crust
separating two cratonic blocks with fault
margins converging to a point, and is
interpreted as having originated by the
rotation of one of the blocks with respect
to the other (e.g. the Bay of Biscay). By
contrast, the latter is a wedge of crust
caused by the squeezing together of blocks
(e.g. the Pyrenees). ASG

sphericity The degree to which a particle
tends toward the shape of a sphere.

spheroidal weathering Exfoliation,
onion-weathering. The disintegration of a
rock by the peeling of the surface layers
which tends to round boulders and cobbles.

spits Spits are generally linear desposits of
beach material attached at one end to land
and free at the other. There are many
different types, including single, recurved,
looped, hooked, complex and double spits.
They occur where LONGSHORE DRIFT
carries material beyond a change in
orientation of the coast or at river mouths.
They usually have a narrow proximal part
and a broader distal end, where recurves are
common. Spits may be called bay head, mid
bay or bay mouth according to their
position in an embayment. They occur
mainly on indented coasts where abundant
sediment can move alongshore freely.
Nearly all spits have formed since sea level
stabilized about 4000 years ago, and many
are much younger. CAMK

Reading
Schwartz, M.L. ed. 1972: *Spits and bars.* Stroudsburg.
Penn: Dowden, Hutchinson and Ross.
Zenkovich, V.P. 1967: *Processes of coastal development.*
Edinburgh: Oliver & Boyd.

spring line See SPRINGS.

springs The natural concentrated outflow
points of water that has flowed underground. In older scientific literature they
are sometimes called fountains or wells.
Springs are classified in many different
ways depending on the characteristics
considered. According to temperature one
may distinguish thermal (or geothermal)
springs from cold water springs. Water

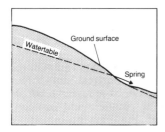

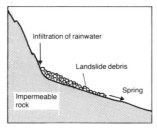

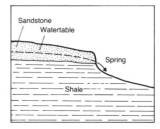

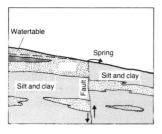

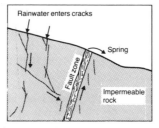

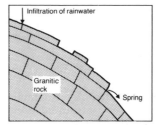

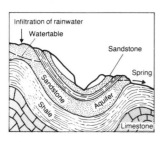

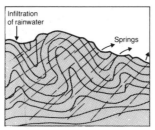

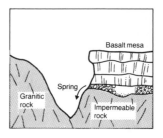

Geological factors in the location of springs.
Source: *S.N. Davis and R.J.M. DeWiest 1966:* Hydrology. *Chichester and New York: Wiley.*

quality criteria have been used to identify mineral and saline springs – often the basis of spas and mineral water industries – and hardwater or karst springs.

Very often more than one of the above characteristics may be combined at a given site, for artesian springs are sometimes also mineral springs and ebbing and flowing springs are usually karst springs.

Consideration of discharge characteristics has led to the recognition of perennial and intermittent springs, ARTESIAN and vauclusian springs, and ebbing and flowing springs with alternating pulsating high and low flows. Locational considerations identify intertidal and submarine springs, overflow and underflow springs.

An unusual but interesting feature is a spring that under certain hydrological conditions reverses its flow direction to become an inflow point rather than a source of outflow. Such features are termed *estavelles* and occur in karst terrains. Under normal circumstances the WATER TABLE rises upstream of the spring, beneath the hill from which the spring water emerges. But sometimes the lowland or enclosed basin (such as a POLJE in karst) is temporarily deeply flooded and the hydraulic gradient is reversed, the water table being higher in the inundated basin than beneath the adjoining hill. Subterranean passages then permit the floodwaters to escape by flowing into the estavelle, which becomes a temporary stream-sink.

Springs are sometimes called RESURGENCES if they represent the reappearance points of known surface headwaters, but if the headwaters are unknown they are termed exsurgences.

Springs often occur in groups, sometimes in clusters and sometimes in lines along the foot of hills. Those in clusters drain the same AQUIFER, with those of lower altitude commonly having the largest and least varying discharge. These are underflow springs. Other members of the group at higher elevations typically have widely varying discharges and can sometimes dry up completely. They are overflow points for the aquifer under high water table conditions. PWW

Reading
Bögli, A. 1980: *Karst hydrology and physical speleology*. Berlin and New York. Springer-Verlag. Ch. 9.

Meinzer, O.E. 1942: Occurrence, origin, and discharge of ground water. In O.E. Meinzer ed., *Hydrology*. New York and London: McGraw-Hill. Pp. 385–477.

Sweeting, M.M. 1972: *Karst landforms*. London: Macmillan. Ch. 11.

Todd, D.K. 1980: *Groundwater hydrology*. 2nd edn. New York and Chichester: Wiley.

squall line A few cumulonimbus storms in a row, organized to produce strong along-line winds and heavy rain. It may be hundreds of kilometres long but only a few wide, so wind speeds increase very rapidly, followed quickly by heavy rain over a large transverse distance, causing widespread damage. Two broad categories of 'tropical' and 'mid-latitude' are not necessarily confined to those regions. Extensive cirrus sheets seen on satellite pictures help diagnosis in regions where data are sparse. JSAG

Reading
Ludlam, F.H. 1980: *Clouds and storms*. Englewood Cliffs, NJ: Pennsylvania State University Press.

stability The ability of an ecosystem to maintain or return to its original condition following a natural or human-induced disturbance. This concept of stability has been widely used by scientists in recent years, but many other meanings have been attached to the term 'stability' (Orians 1975). For example, the term has been used in reference to the constancy or PERSISTENCE of species populations or ecosystems. Two major aspects of ecosystem stability in relation to disturbance have received most attention. Even in this context a confusing variety of terms has been used. The first property, which is often labelled 'resistance', is the ability of a system to remain unaffected by disturbances. This property is referred to as 'inertia' by Orians (1975) and Westman (1978) and as 'resilience' by Holling (1973). The second attribute is usually termed 'resilience' and is the ability of the system to recover to its original state following a disturbance. The more general term 'stability' has also been applied to this property by May (1973) and Holling (1973).

The concepts of resistance and resilience are of considerable interest not only to scientists engaged in basic research but also to environmental planners and managers. Knowledge of the varying ability of ecosystems to resist change, or to recover quickly following disturbance, is of obvious value in planning development projects and assessing potential damage from pollutants.

Resistance and resilience can be measured in a variety of ways. The particular ecosystem charactistics anlysed will depend on the nature of the disturbance and the goals of the research, and may range from a focus on individual species populations to overall system properties such as species diversity, primary production and nutrient losses in drainage water. Resistance to a disturbance can be measured by the magnitude of the system response and by the time delay before a response occurs. These parameters provide an assessment of the relative resistance of an ecosystem to different types of stress or alternatively of different ecosystems of the same stress. For example, a study by Vitousek *et al.* (1981) of nitrate losses from disturbed forest plots revealed a major peak in nitrate losses within six months of disturbance in Indiana maple and oak forests, whereas much smaller losses occurred in hemlock and Douglas-fir forests in Oregon. A pine forest in Indiana exhibited an extended delay in response, with substantial losses beginning almost two years later.

Westman (1978) examines a variety of measurement problems associated with resilience and suggests that four aspects of this component can be evaluated. Elasticity refers to the rapidity of the system's return to its original state; amplitude to the zone from which the system can recover; hysteresis, the extent to which the recovery pathway differs from the pattern of disruption which occurred in response to the disturbance; and malleability, the degree to which the new stable state established after disturbance differs from the original steady state. The amplitude aspect of resilience is of particular interest because it deals with a threshold beyond which the system cannot recover to its initial state. It may be possible to suggest that a system is reaching a threshold by studying the rate of change of various characteristics in relation to the range of intensity of a particular stress. Baker (1973) has studied the amplitude response of saltmarsh vegetation to oil pollution. His data suggest that recovery was good when the vegetation was exposed to not more than four oil spillages. Substantial damage and very slow recovery occurred after 8–12 successive oilings, suggesting the presence of a threshold of recovery.

Ecosystem resistance and resilience may not necessarily be closely linked. The initial degree to which a system is altered by disturbance may in some cases be a poor indicator of the ultimate ability of the system to cope with stress. In the Great Lakes many species of fish including herring, walleye and lake trout withstood fishing pressure for many years without any obvious signs of decline, but all these species experienced sudden collapses of populations to near-extinction levels without any advance warning (Holling 1973).

Much research has been devoted to the study of relationships between stability and other ecosystem properties, particularly species diversity (Goodman 1975, Pimm 1984). The notion that more complex systems involving large numbers of interacting species should be more stable than simple systems with few species is intuitively attractive since there should be more alternative pathways for feedback and ajustment to disturbance in the complex system. MacArthur's (1955) hypothesis that stability was a function of the complexity of feeding linkages between organisms in an ecosystem was therefore rapidly accepted and several lines of evidence were used to support the relationship. This evidence, which is reviewed by Elton (1958), involved data from laboratory experiments with one prey/one predator systems, which revealed the occurrence of large population fluctuations followed by rapid extinction. Emphasis was also placed on the vulnerability of simple agricultural systems to pest outbreaks and the contrast between prominent population oscillations in Arctic tundra and the apparent lack of such fluctuations in the complex and species-rich tropical rain forests. A more critical evaluation of the linkage between stability and diversity in recent years has underlined the weakness of this evidence. For example, instability in the laboratory predator–prey system does not provide a valid analogy to

the real world since even very simple ecosystems contain many different species. Similarly, the instability of crop monocultures can probably be attributed to the absence of coevolution over long time periods of the species involved.

Empirical studies suggest that there is no simple link between diversity and stability (Goodman 1975). The total species diversity of an ecosystem may not be an adequate measure of complexity, which can take a variety of forms in relation to the trophic level involved and the spatial organization of the system. Watt (1968) has proposed that stability at any herbivore or carnivore trophic level increases with the number of competitor species at that level, decreases with the number of competitor species that feed upon it, and decreases with the proportion of the environment containing useful food. The question of the relationship between diversity and stability is further complicated by evidence that some ecosystems contain a single species high in the food web which can influence the system structure (Paine 1969). The stability of such systems would be highly dependent on the effects of disturbance on the 'keystone' species rather than on the overall species diversity of the system. Extensive examination of the diversity–stability hypothesis has been undertaken, using mathematical models (May 1973). Defining stability as the ability of the system to return to equilibrium after stress, May found that complex model systems are less stable than simple ones. In view of the complexities revealed by recent research and the fact that existing evidence is contradictory, many ecologists now regard the equating of stability with diversity as a tentative hypothesis rather than an axiom. The influence of other ecosystem properties on stability has generally been neglected, although it is likely that the resistance and resilience of systems to human-induced stresses can be affected by the spatial organization of the ecosystem and a variety of linkages between biological and abiotic components (Hill 1975).   ARH

Reading and References
Baker, J.M. 1973: Recovery of salt marsh vegetation from successive oil spillages. *Environmental pollution* 4, pp. 223–30.

Elton, C.S. 1958: *The ecology of invasions by animals and plants.* London: Methuen. Ch. 8, pp. 143–53.

†Goodman, D. 1975: The theory of diversity–stability relationships in ecology. *Quaterly review of biology* 50, pp. 237–66.

†Hill, A.R. 1975: Ecosystem stability in relation to stresses caused by human activities. *Canadian geographer* 19, pp. 206–20.

Holling, C.S. 1973: Resilience and stability of ecological systems. *Annual review of ecology and systematics* 4, pp. 1–23.

MacArthur, R.H. 1955: Fluctuations of animal populations and a measure of community stability. *Ecology* 36, pp. 633–6.

May R.M. 1973: *Stability and complexity in model ecosystems.* Princeton, NJ: Princeton University Press.

†Orians, G.H. 1975: Diversity, stability and maturity in natural ecosystems. In W.H. Van Dobben and R.H. Lowe-McConnell eds, *Unifying concepts in ecology*. The Hague: W. Junk. Pp. 139–50.

Paine, R.T. 1969: A note on tropic complexity and community stability. *American naturalist* 103, pp. 91–3.

†Pimm S.L. 1984: The comnplexity and stability of ecosystems. *Nature* 307, pp. 307, pp. 321–6.

Vitousek, P., Reiners, W.A., Melillo, J.M., Grier, C.C. and Gosz, J.R. 1981: Nitrogen cycling and loss following forest perturbation: the components of response. In G.W. Barrett and R. Rosenberg eds, *Stress effects on natural ecosystems*. London and New York: Wiley. Pp. 114–27.

†Watt, K.E.F. 1968: *Ecology and resource management.* New York and London: McGraw-Hill. Ch. 3, pp. 39–50.

†Westman, W.E. 1978: Measuring the inertia and resilience of ecosystems. *Bioscience* 28, pp. 705–10.

stability analysis The procedure for examining the likelihood of failure of a soil or rock slope. The types of analysis and the way they are carried out depend largely upon the nature of the materials to be investigated but they generally require a knowledge of the COHESION and FRICTION properties of the slope material as well as the jointing characteristics if the material is a rock slope. Slope geometry is also crucial, as is a knowledge of the water availability as pore water pressure (in soils) or cleft water pressure (in rocks). Many stability analyses are two-dimensional but digital computing methods can now make three-dimensional analyses relatively easy. In soil mechanics such analyses are used to determine a factor of safety so that a safe design can be produced, but they can be used in a more geomorphological way to help determine the characteristics once a slope has failed.

WBW

Reading
Bell, F.G. 1992: *Engineering properties of soils and rocks* 3rd edn. Oxford: Butterworth-Heinemann.

Lambe, T.W. and Whitman, R.V. 1981: *Soil mechanics.* New York: Wiley.

stable equilibrium A condition of a system in which very limited displacement in any direction is followed by a return to a persistent state or condition. Huggett (1980, figure 1.3) employs the mechanical analogy (from Spanner 1964) of a ball resting in a deep cup: the ball may move from side to side or round and round if the cup is shaken, but will always ultimately return to the bottom of the cup as long as the stable equilibrium condition prevails.

BAK

References
Huggett, R. 1980: *Systems analysis in geography.* Oxford: Clarendon Press. Spanner, D.C. 1964: *Introduction to thermodynamics.* London: Academic Press.

stack A free-standing pinnacle of rock, usually in the sea which represents an outlier of a coastal cliff.

A spectacular stack, the Old Man of Hoy, developed in Old Red Sandstone in the Orkney Islands. Its height is 140 m.

stadial A short cold period with smaller ice volumes than the full glacial stages of an ice age. The warmer intervals between them are called interstadials.

staff gauge An instrument for determining water depth at a site on a river system. It is often located at a stream gauging site to provide a datum for setting and checking continuous stage recorders. The staff gauge consists of a plate or board with painted or engraved elevation divisions, which is firmly fixed in the water at a river cross-section so that water levels can be read by eye from the divisions on the staff gauge at all flow stages, usually to an accuracy of 0.5 cm. Single staff gauges are usually installed to stand vertically and close to one bank of the river, but in some locations more than one staff gauge will be installed to cover different ranges of stage and occasionally the staff gauge will be inclined to lean against the bank so that flow disturbance is minimized

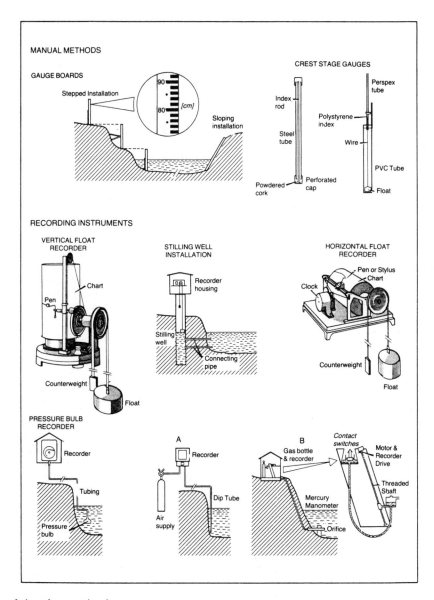

Some techniques for measuring river stage.
Source: *K.J. Gregory and D.E. Walling 1973:* Drainage basin form and process. *London: Edward Arnold.*

and so that the depth scale may be more easily read. (See also DISCHARGE; HYDRO-METRY.)   AMG

stage The term used to describe water depth or the elevation of the water surface at a location on a river system. Instruments which continuously monitor water surface elevation at a gauging station site are called stage recorders. The river stage fluctuates through time in response to precipitation events. In a stable and straight river section a relationship can be established between discharge and stage so that discharge can be estimated for any water stage, and this relationship between stage and discharge is known as the discharge rating curve for that site. (See also DISCHARGE.)   AMG

stagnant ice topography The view that many glacial deposits in formerly glaciated areas are the result of stagnant ice downwasting *in situ* was championed in North America in the 1920s (Cook 1924; Flint 1929), in Scandinavia in the 1940s and 1950s (Mannerfelt 1945; Hoppe 1959) and in Scotland (Sissons 1967). The idea is that the ice, by virtue of shallow surface gradients or by its isolation from the main ice mass, no longer flows actively but melts from the surface downwards, the last remnants being preserved in depressions. The landforms that are formed depend both on the shape of the underlying topography and on the debris characteristics of the glacier. In places, MORAINE landforms of disintegration are dominant and consist of irregular mounds and KETTLES built of varying quantities of basal and supraglacial TILL. The basal till may have reached the glacier surface as a result of COMPRESSING flow before being redistributed through the action of slumping and surface sediment flows as the ice melts away. Such conditions also favour considerable meltwater activity. One fine example of such topography with a relief of 30–45 m occurs on the Canadian prairies and covers areas of thousands of square kilometres (Prest 1983).

In hillier parts of the world the role of valleys in influencing the location of the stagnant remnants of ice seems to favour another association of landforms. Once isolated from the ice sheet such stagnant ice masses are plugged by glaciofluvial deposits associated with meltwater streams from both the ice surface and the surrounding hills and their courses are marked by

KAME TERRACES, KAMES and associated ESKERS and irregular mounds. Mixed with the glaciofluvial deposits may be irregular mounds of till. Such landscapes abound in upland Scandinavia and Scotland.   DES

Reading and References
Cook, J.H. 1924: The disappearance of the last glacial ice-sheet from eastern New York. *Bulletin of the New York State Museum* 251, pp. 158–76.

†Flint, R.F. 1929: The stagnation and dissipation of the last ice sheet. *Geographical review* 19, pp. 256–89.

Hoppe, G. 1959: Glacial morphology and inland ice recession in north Sweden. *Geografiska annaler.* 41, pp. 193–212.

Mannerfelt, C.M. 1945: Några glacialmorfologiska Forme-lement. *Geografiska annaler.* 27, pp. 1–239. (English summary and figure captions.)

†Moran, S.R., Clayton, L., Hooke, R. Le B., Fenton, M.M. and Andriashek, L.D. 1981: Glacier-bed landforms of the prairie region of North America. *Journal of glaciology* 25.93, pp. 457–76.

Prest, V.K. 1983: *Canada's heritage of glacial features.* Geological Survey of Canada miscellaneous report 28.

Sissons, J.B. 1967: *The evolution of Scotland's scenery.* Edinburgh: Oliver & Boyd.

stalagmite, stalactite See SPELEO-THEM.

Stanford watershed model See WATERSHED MODEL.

star dune A pyramidal dune, roughly star-shaped, with three or more radial buttresses or arms extending in various directions from a high central cone. Slipfaces dip in at least three directions. Star dunes develop in areas with complex wind regimes.   ASG

Reading
Lancaster, N. 1989: Star dunes. *Progress in physical geography* 13, pp. 67–9.

steady flow The flow of water in open channels is classified according to its temporal and spatial variability, and steady flow occurs when the water depth, discharge and therefore the velocity are temporally constant. Since open channels have a free surface exposed to the atmosphere the flow responds to rainfall and run-off inputs and is naturally temporally unsteady during storm hydrographs. The rate of change of depth and discharge is often sufficiently slow for a steady flow to be assumed during the time interval under consideration, and this forms the basis for the development and application of simplified FLOW EQUATIONS. Temporally steady flow may be spatially uniform or varied, and varied flow may be gradually varied or rapidly varied.   KSR

steady state The notion that the input, output and properties of a SYSTEM remain constant over time. The concept of steady state is a powerful means of simplification which allows the mathematical modelling of complex natural systems.   DES

steam fog Fog formed when cold air passes over warm water. Both heat and water vapour are added to the air which quickly becomes saturated. Any additional water vapour evaporated from the warm water will rapidly condense forming a swirling, steam-like fog. Steam fog is common over heated swimming pools in winter, above lakes in autumn and early winter mornings, over thermal ponds, such as those in Yellowstone National Park, all the year round, and above open water in polar regions, where it is called arctic sea smoke.   WDS

Reading
Ahrens, C.D. 1982: *Meteorology today.* St. Paul, Minn.: West.

Stefan's method Heat flow problems concerned with permafrost are complex. A method of determining, for instance, the depth of penetration of a 'cold wave' into the ground can be performed by the relatively simple Stefan's formula. This is based on the principle that heat absorbed is equal to heat conducted through a 1 cm^2 area and that the temperature gradient in the frozen zone is linear. It requires knowledge of the thermal gradients, thermal conductivity of frozen soil, ice density and latent heat of fusion of ice.   WBW

stem flow The drainage of intercepted water down the stems of plants. Precipitation may be intercepted by vegetation and subsequently lost through evaporation (INTERCEPTION or interception loss) or it may drip through the vegetation canopy (THROUGHFALL) or it may drain across the leaves and stems of plants or down the branches and trunks of trees as stem flow. Stem flow is a means by which precipitation may reach the ground surface and it may form quite an important route for water, particularly if the structure of the vegetation encourages this form of drainage.   AMG

Reading
Courtney, F.M. 1981: Developments in forest hydrology. *Progress in physical geography* 5, pp. 217–41.

Sopper, W.E. and Lull, H.W. eds 1967: *International symposium on forest hydrology.* Oxford and New York: Pergamon.

step-pool systems Commonly occur in mountain streams with steep gradients and coarse bed materials. They are formed by high-magnitude, low-frequency flood events, and the staircase-like structure tends to be relatively stable for long periods of time during low flows. The steps and pools alternate to produce a characteristic, repetitive sequence, with the steps composed of an accumulation of cobbles and boulders that are transverse to the channel. Finer materials fill the pools. Logs may contribute to the formation of steps.   ASG

Reading
Chin, A. 1989: Step pools in stream channels. *Progress in physical geography* 13, pp. 391–407.
Whittaker, J.G. and Jaeggi, M.N.R. 1982: Origin of step-pool systems in mountain streams. *Proceedings of the American Society of Civil Engineers, Journal of the Hydraulics Division* 108, pp. 758–73.

steppes Mid-latitude grasslands with few trees. The Russian equivalent of the North American prairies and Argentinian pampas.

stick slip The jerky motion by which glaciers slide over bedrock. Sudden slip phases of 1–3 cm displacement are interspersed with longer quiescent phases. Each slip phase is highly localized beneath a glacier and probably relates to the failure of a local bond between part of the glacier and the bed.   DES

Reading
Robin, G. de Q. 1976: Is the basal ice of a temperate glacier at the melting point? *Journal of glaciology* 16.74, pp. 183–96.

stilling well A large diameter tube installed at the edge of a river channel, or

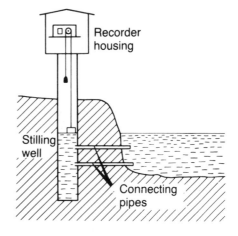

in a river bank, in order to obtain accurate measurements of river level from a still water surface. The tube is connected to the river by intake pipes, in order to ensure that the water level in the well is identical to that in the channel. The diameter of these pipes should be sufficiently small to damp out turbulence or short-term oscillations, but must be large enough to permit instantaneous response to changes in river level. Provision must be made for flushing or clearing the intake pipes. A FLOAT RECORDER is often associated with a stilling well.   DEW

stillstand A period of stability between two phases of tectonic activity in the earth's history. Also a period during which mean sea level is constant.

stochastic models Any mathematical model which represents a stochastic process – a phenomenon whose temporal or spatial sequence is characterized by statistical properties – is a stochastic model. The incorporation in the model of a sequential time or space function of the probability of occurrence is what distinguishes a stochastic model from a purely probabilistic model (such as the probability distribution of floods of different magnitude). The selection of an appropriate model depends on the nature of the process being modelled and the type of data collected to summarize the process numerically (Thornes and Brunsden 1977, pp. 5–7, 70–87). A continuous process is observed continuously through time or space, even if the measured variable only takes a discrete set of values, as in the case of a binary variable denoting presence or absence. For convenience, a continuous process is often represented by data obtained at discrete time or space intervals (usually equal), either as discrete sampled data read at specific points, or as discrete aggregate data which are summed or averaged over a period (e.g. daily rainfall). Another class of stochastic phenomena involves point processes, in which either the frequency of *events* in successive discrete time periods is counted, or the distribution of time *intervals* between events is assessed.

Discrete approximations of continuous time or space series may be modelled using the general linear random model:

$$z_t - \phi_1 z_{t-1} - ... - \phi_p z_{t-p} = e_t - \theta_1 e_{t-1}$$
$$... - \theta_q e_{t-q}$$

where the zs are values of a variable measured at times t, $t-1$, etc., the es are random 'shocks' at these times, and the ϕ_1 and θ_1 are coefficients. In fitting this mixed autoregressive-moving average model, the objective is to minimize its complexity by reducing the 'order' of dependency defined by the lags p and q (Chatfield 1975; Richards 1979). The stochastic point process (Cox and Lewis 1966) may be described by appropriate probability distributions such as the binomial or Poisson distributions, and modelled as a series using the theory of queues.   KSR

References

Chatfield, C. 1975: *The analysis of time series: theory and practice*. London: Chapman & Hall.

Cox, D.R. and Lewis, P.A.W. 1966: *The statistical analysis of series of events*. London: Methuen.

Richards, K.S. 1979: *Stochastic processes in one-dimensional series: an introduction*; Catmog 23. Norwich: Geo Abstracts.

Thornes, J.B. and Brunsden, D. 1977: *Geomorphology and time*. London: Methuen.

stochastic process A statistical phenomenon in which the evolutionary sequence in time and/or space follows probabilistic laws. 'Stochastic', from the Greek word meaning 'guess', implies a chance process which contrasts with a deterministic phenomenon whose future values can be predicted with certainty if the existing values of the controlling variables are known. In a stochastic process exact prediction is impossible because dependence is partly on past conditions, and partly on random influences. Nevertheless, stochastic models can be used to represent the process mathematically, and to provide both efficient forecasts and the distribution of forecast errors. Natural phenomena may be inherently stochastic, but often the randomness apparent in their behaviour reflects the scientist's incomplete understanding, and inaccurate measurement (Mann 1970). In practice, many phenomena display the mixed deterministic-stochastic behaviour typified by climatic and hydrological processes (Yevjevich 1972). For example, daily river flows vary seasonally with a fundamentally deterministic cycle related to annual variation of radiation receipt and evaporation loss. Superimposed on this is the random occurrence of sharp increases of flow caused by individual storm inputs, followed by the gradual decrease of discharge in the flood recession curve, caused by water retention in the drainage basin and consequent slow outflow. Thus the stochastic component of the hydrological process involves both random 'shocks' and a system 'memory', which can be modelled by an autoregressive STOCHASTIC MODEL. Note that the hydrologist treats rainfall as a random input, whereas the meteorologist seeks to explain rainfall deterministically.

KSR

References
Mann, C.J. 1970: Randomness in nature. *Bulletin of the Geological Society of America* 81, pp. 95–104.

Yevjevich, V. 1972: *Stochastic processes in hydrology.* Fort Collins, Col.: Water Resources Publications.

stock A large, irregularly shaped intrusion of igneous rock.

stone line A horizon of gravel-sized rock fragments within a soil profile or accumulation of relatively fine-grained sediments.

stone pavement (or desert pavement) An armour or lag of coarse particles, normally of broken pieces of rock, overlying material which contains a much larger proportion of fine materials (sand, silt, clay, salt etc.). It results either from the horizontal removal of fine particles by wind or sheetwash from a sediment with a mixture of grain sizes, or as a result of vertical sorting processes, such as frost or wetting and drying, causing coarse particles to migrate upwards (Cooke 1970).   ASG

Reference
Cooke, R.U. 1970: Stone pavements in deserts. *Annals of the Association of American Geographers* 60, pp. 560–77.

storage Describes the stores or reservoirs of water included in the hydrological cycle. We can think of surface storage, soil moisture storage and groundwater storage as locations where dynamic reservoirs of water may exist in a drainage basin. In hydrological modelling complex distributions of stored water may be represented as one or more mathematically defined stores or reservoirs, the simplest of which is the linear store. In a linear store water outflow (Q) is directly proportional to water storage (S):

$$S = kQ$$

where k is the storage coefficient.

The storage equation is combined with a continuity equation which states that the difference between inflow (I) to the store and outflow from the store is accommodated by a change in the amount of water in storage:

$$\frac{\partial S}{\partial t} = I - Q$$

AMG

Reading
Kirkby, M.J. 1975: Hydrograph modelling strategies. In R. Peel, M. Chisholm and P. Haggett, eds, *Progress in physical and human geography.* London: Heinemann.

— ed. 1978: *Hillslope hydrology.* Chichester: Wiley.

storm run-off See HYDROGRAPHS; RUN-OFF.

storm surges Changes in sea level generated by extreme weather events. They appear on sea-level records as distortions of the regular tidal patterns, and are most severe in regions of extensive shallow water. When maximum surge levels coincide with maximum high-water levels on spring tides, very high total sea levels result. Low-lying coastal areas are then vulnerable to severe flooding. In tropical regions severe surges are occasionally generated by cyclones, hurricanes or typhoons: the actual levels depend on the intensity of the meteorological disturbance, the speed and direction with which it tracks towards the coast, and the simultaneous tidal levels. Areas at risk include the Indian and Bangladesh coasts of the Bay of Bengal, the south-east coast of the USA and the coast of Japan. Satellite and radar tracking of the weather patterns are used to give advanced warning of imminent flood danger. Extra-tropical surges, generated by meteorological disturbances at higher latitudes, usually extend over hundreds of kilometres, whereas the major effects of tropical surges are confined to within a few tens of kilometres of the point where the hurricane meets the coast. Flood-warning systems for extra-tropical surges must take account of the total response of a region to the weather patterns.   DTP

Reading
Pugh, D.T. 1987: *Tides, surges and mean sea level.* Chichester: Wiley.

stoss The direction from which wind, water or ice moves. The windward side of a sand dune.

strain A measure of the deformation of a body when a load or STRESS is applied. It is usually expressed as a dimensionless value (ratio or percentage) as, for linear strain, it is the change in length divided by the original length. Similarly, areal and volumetric strains can be defined. In the SHEAR BOX and TRIAXIAL APPARATUS, the tests are normally done at a constant deformation rate and are called 'strain controlled' tests.

WBW

strain rate The rate at which a body deforms in response to stress. It is often represented by the symbol $\overset{\circ}{\varepsilon}$. (See also STRAIN.)

strandflat The term was introduced to describe an undulating rocky lowland up to 65 km wide in western Norway (Reusch 1894; Nansen 1922). It is partly submerged, forming an irregular outer belt of skerries, and is backed by a steeply rising coast. Similar features may have been recognized in Iceland, Svalbard, Novaya Zemlya, East and West Greenland, Baffin Island and the Antarctic Peninsula. Nansen's view that the strandflat was cut by freeze–thaw processes adjacent to the shoreline has survived to the present day, but there is an alternative view, namely that it is a marine or subaerial lowland, subsequently modified by glacial action.

DES

Reading and References
†Gjessing, J. 1966: Norway's paleic surface. *Norsk geografisk tidsskrift* 21, pp. 69–132.
†Holtedahl, H. 1960: Mountain, fjord, strandflat: geomorphology and general geology of parts of western Norway. In J.A. Dons ed., *Guide to excursions A6 and C3*. International Geological Congress twenty-first session. Oslo: Norden.
Nansen, F. 1922: *The strandflat and isostasy*. Oslo: Videnskapssel Skapets Skrifter 1.
Reusch, H. 1894: The Norwegian coast-plain. *Journal of geology* 2, pp. 347–9.

stratified scree See GRÈZES LITÉES.

stratigraphy The study of the order and arrangement of geological strata. Lithostratigraphy (rock stratigraphy) is concerned with the organization of strata into units based on their lithological characteristics. Biostratigraphy is concerned with the organization of strata into units based on their fossil content. Chronostratigraphy (time stratigraphy) is concerned with the organization of strata into units based on their age relationships.

ASG

Reading
Bowen, D.Q. 1978: *Quaternary geology*. Oxford: Pergamon.

stratocumulus See CLOUDS.

stratosphere The layer from heights of about 10–30 km immediately above the TROPOSPHERE, in which temperature is nearly independent of height. Heat transfer is dominated by thermal RADIATION which tends to eliminate temperature differences. Emden's theory of radiative equilibrium suggests a stratospheric temperature some $2^{-1/4}$ that of the troposphere which, at 215 K (0.84 times a tropospheric temperature of 255 K), is about right.

JSAG

Reading
Goody, R.M. and Walker, J.C.G. 1972: *Atmospheres*. Englewood Cliffs, NJ: Prentice-Hall.

strato-volcano A composite volcano. A volcano that emits both molten and solid material and builds up a steep-sided cone.

stream ordering See ORDER, STREAM.

stream power A concept that relates fluvial energy to sediment transport. To transport sediment, work (defined as the product of force and distance) must be performed. Power is the rate of doing that work, and stream power per unit length of stream. It is the rate of energy supply at the channel bed that is available to overcome friction and to transport sediment. Stream power (Ω), or the power per unit length of stream, is measured in W (watts) or J s^{-1} (joules per second) and is defined as

$$\Omega = \varrho_w g \, QS$$

where ϱ_w is the density of water, g the acceleration of gravity, Q discharge and S channel gradient.

ASG

streamline A line whose tangent at any point in a fluid is parallel to the instantaneous velocity of the fluid at that point. A map of streamlines gives an instantaneous 'snapshot' of the flow. Alternatively one may think of such a map as being one frame in a moving film of the flow. The streamline pattern changes with time. Only in a steady state flow do the streamlines coincide with the trajectories of the fluid particles.

BWA

strength See INTACT STRENGTH; MASS STRENGTH.

strength equilibrium slopes Are formed on exposed bedrock which has an inclination adjusted to the MASS STRENGTH of the rock. This type of slope is controlled by processes of erosion and by geomorphic resistance operating at a scale of individual joint blocks. A distinction is made between (1) equilibrium slopes, and (2) those which are formed with critical angles for stability dipping out of the slope – these slopes fail by large-scale landsliding along the critical joints and have forms controlled by this process.   MJS

Reading
Selby, M.J. 1982: Controls on the stability and inclinations of hillslopes formed on hard rock. *Earth surface processes and landforms* 7, pp. 449–67.

stress Is produced by a system of forces in equilibrium tending to produce STRAIN in a body. Stress can be produced in tension or compression, by hydrostatic pressure or by

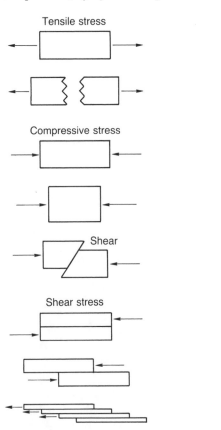

Stress: Schematic illustration of tensile, compressive and shear stresses.
Source: *M.A. Summerfield 1991*: Global geomorphology. *London and New York: Longman Scientific and Technical and Wiley.*

shear stress. The units of stress are (force per unit area) newtons per square metre, $N\,m^{-2}$. In most geomorphological examples, the forces will give values as $kN\,m^{-2}$ or the equivalent kPa where a pascal (Pa) is equal to a $N\,m^{-2}$. Tensile stress is an extensional force which tends to stretch or pull material apart. Compressive stress is a force which tends to compress material and thereby change its shape. Shear stress is a force which deforms a mass of material by one part sliding over another along one or more failure plains.   WBW

Reading
Whalley, W.B. 1976: *Properties of materials and geomorphological explanation.* Oxford: Oxford University Press.

stress ecology Defined by Barrett (1981) as a subdiscipline which attempts to measure and evaluate the impact of natural or foreign perturbations on the structure and function of ecological systems. Perturbations include pesticides, fire, nutrient enrichment and radiation. Stress itself can be defined as a perturbation that is applied to a system by a stressor which is foreign to that system or which may be natural to it but, in the instance concerned, is applied at an excessive level (e.g. phosphorus or water). Stress therefore involves an unfavourable deflection, whereas *subsidy* involves a favourable deflection.   ASG

Reference
Barrett, G.W. 1981: Stress ecology: an integrative approach. In G.W. Barrett and R. Rosenberg eds, *Stress effects on natural ecosystems.* Chichester: Wiley.

striated soil (also called needle ice, striped ground and striated ground) Consists of a miniature pattern characterized by a distinct alignment of the surface soil particles. The orientation of the stripes does not necessarily coincide with slope gradient. Wind direction and the alignment of the early sun's rays may play a role. NEEDLE ICE is the predominant formative process.   ASG

striation Scratches etched onto a rock surface by the passage over it of another rock of equal or greater hardness. Striations are characteristic of erosion by glaciers but may also occur beneath snow patches (Jennings 1978) and on coasts affected by sea ice (Laverdière *et al.* 1981; Hansom 1983). Glacial striations are generally up to a few millimetres in width and rarely more than a metre in length. Larger striations grade into grooves. Striations are best displayed on rock surfaces which face up-ice, mainly because pressure melting in these locations forces the rock tools against the bedrock.   DES

Reading and References
†Embleton, C. and King, C.A.M. 1975: *Glacial geomorphology*. London: Edward Arnold.

Hansom, J.D. 1983: Ice-formed intertidal boulder pavements in the Sub-Antarctic. *Journal of sedimentary petrology* 53. 1, pp. 135–45.

Iverson, N.R. 1991: Morphology of glacial striae: implications for abrasion of glacier beds and fault surfaces. *Bulletin of the Geological Society of America*. 103, pp. 1308–16.

Jennings, J.N. 1978: The geomorphic role of stone movement through snow creep, Mount Twynam, Snowy Mountains, Australia. *Geografiska annaler* 60A, pp. 1–8.

Laverdière, C., Guimont, P. and Dionne, J.C. 1981: Marques d'abrasion glacielles en milieu littoral Hudsonien. Québec Subarctique. *Géographie physique et Quaternaire* 35.2, pp. 269–75.

Strickler equation In 1923 Strickler analysed data from Swiss gravel-bed rivers lacking bed undulations to develop an equation permitting estimation of the ROUGHNESS coefficient n in the MANNING EQUATION from the measured bed material particle size. It can be shown theoretically that:

$$n = 0.0132 \, k_s^{1/6}$$

where k_s is a grain roughness height in millimetres. Strickler shows that if k_s is taken as the median grain diameter D_{50}, this relationship becomes:

$$n = 0.0151 \, D_{50}^{1/6}$$

This can be used to estimate the Manning coefficient if grain roughness is the primary source of flow resistance, in wide, flat-bed gravel-floored channels. It is, however, often preferable to define a friction coefficient in terms of a depth : grain size ratio, since the resistance of a grain is partly dependent on the water depth covering it (Richards 1982, p. 66).   KSR

Reading and References
Richards, K.S. 1982: *Rivers: form and process in aluvial channels*. London and New York: Methuen.

†Strickler, A. 1923: Beitrage zur Frage der Geschwindigheits formel und der Rauhigkeitszahlen für Strome, Kanole und Geschlossene Leitungen. *Mittelungen des Eidgenössischer Amkes für Wasserwittschaft*. Bern.

strike The direction of a horizontal line drawn in the same plane as strata are bedded but at right angles to their dip.

string bog An area of water-logged land characterized by ridges of peat separated by water-filled troughs.

A series of cauliflower-like lacustrine stromatolites from Lake Chew Bahir (Stephanie) in southern Ethiopia.

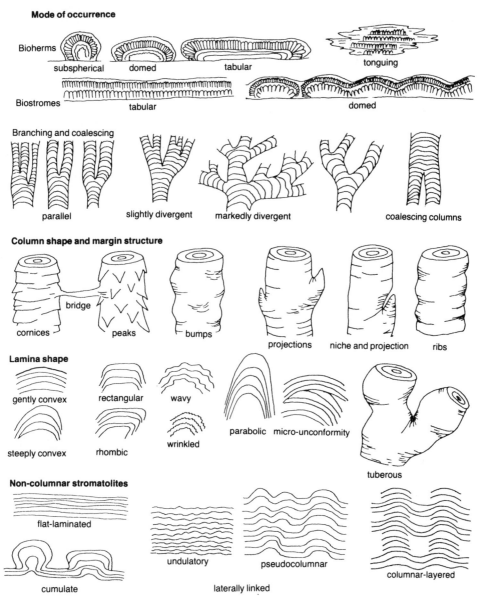

Stromatolite: The main terms used in the description of stromatolite bodies and stromatolitic lamination.
Source: *J.D. Collinson and D.B. Thompson 1982: Sedimentary structures. London: Allen & Unwin. Figure 8.6.*

stromatolite (stromatolith) A term first used in 1908 by E. Kalkowsky to describe some sedimentary structures in the Bunter of North Germany. A currently favoured definition (Walter 1976, p. 1) is that stromatolites are 'organosedimentary structures produced by sediment trapping, binding and/or precipitation as a result of the growth and metabolic activity of micro-organisms, principally cyanophytes'. They can develop in marine, marsh and lacustrine environments and, though they form today where conditions permit, they reached the acme of their development in the Proterozoic (Hofmann 1973). The largest known forms are mounds several hundreds of metres across and several tens of metres high. Gross morphologies vary in the extreme and range from stratiform crustose forms, through nodular and bulbous mounds and spherical oncoids, to long slender columns, erect to inclined, and with various styles of branching (see diagram). ASG

References
Hofmann, H.J. 1973: Stromatolites: characteristics and utility. *Earth science reviews* 9, pp. 339–73.
Kalkowsky, E. 1908: Oolith und Stromatolith im nord-deutschen Buntsandstein. *Zeitschrift Deutsche Geologische Gesellschaftz* 60, pp. 68–125.
Walter, M.R. 1976: Stromatolites. *Developments in sedimentology* 20.

sturzstrom Very large rock avalanches (with volumes >5 Mm^3) initially falling or avalanching from high cliffs may develop low coefficients of internal friction, and therefore travel large horizontal distances (5–30 km) at velocities of 90–350 km h^{-1}. They are the most powerful forms of mass-wasting and may be major, but rare, causes of erosion in very high mountain ranges and on large volcanoes. These very large avalanches were named *sturzstroms* by Hsü (1975). MJS

Reading and Reference
Hsü, K.J. 1975: Catastrophic debris streams (sturzstroms) generated by rockfalls. *Bulletin of the Geological Society of America* 86, pp. 129–40.
Selby, M.J. 1993: *Hillslope materials and processes*. 2nd edn. Oxford: Oxford University Press. Ch. 14.

subaerial Occurring or existing at the landsurface.

suballuvial bench The lower portion of a rock pediment where it is overlain by alluvial sediments.

Sub-Atlantic See BLYTT–SERNANDER MODEL.

Sub-Boreal See BLYTT–SERNANDER MODEL.

subclimax Any plant community related to and closely preceding the true climax community for an area. Usage is normally in the sense of a stable community resembling the climax but prevented from developing towards it by some disturbance or other arresting factor. If the arresting factor is removed, a subclimax is expected to proceed to the climax stage. The term is also used simply for a long-persisting SERAL COMMUNITY that appears to be climax. Many subclimaxes are the result of the activities of man and domesticated animals,

particularly burning and grazing. (See also CLIMAX VEGETATION; DISCLIMAX; MONO-CLIMAX.) JAM

Reading
Eyre, S.R. 1966: *Vegetation and soils: a world picture*. London: Edward Arnold.
Oosting, H.J. 1956: *The study of plant communities*. San Francisco: W.H. Freeman.

subduction zone An area where the rocks making up the sea floor are forced beneath continental rocks at a plate margin and are reincorporated in the magma beneath the earth's crust. Such areas are characterized by intense seismic and volcanic activity.

subglacial The environment beneath a glacier.

Reading
Menzies, J. and Rose, J. eds 1989: Subglacial bedforms – drumlins, rogen moraine and associated subglacial bedforms. *Sedimentary geology* 62, pp. 117–430.

subhumid One of the five basic humidity provinces recognized in Thornthwaite's climate classification (1948). The classes are defined by calculating a PRECIPITATION efficiency (P/E index) which is the sum of twelve monthly values of the ratio of mean precipitation to mean EVAPORATION. Sub-humid falls in the middle of the range so defined, with a P/E value of 32 to 63 and characterizes areas of grassland type vegetation. The savannah regions of Africa which experience an extensive dry season every year fall in this category. RR

Reference
Thornthwaite, C.W. 1948: An approach to a rational classification of climate. *Geographical review*, 38, pp. 55–94.

sublimation The process of direct deposition of atmospheric water vapour on to an ice surface or evaporation from an ice surface. Cirrus clouds are sometimes formed as a result of a direct phase change from water vapour to ice crystals. RR

sublittoral The area of the seas between the intertidal zone and the edge of the continental shelf. Also the deeper parts of a lake in which plants cannot root.

submarine canyon A canyon-like valley form cut into the CONTINENTAL SHELF, continental slope or continental rise. Its resemblance to a subaerial valley extends to the existence of tributary valleys and the presence of knickpoints along its profile. It can extend beyond the continental slope almost as far as the deep ocean floor. Most submarine canyons contain sediments apparently deposited by high density flows initiated by submarine slides known as turbidity currents. These are probably capable of preventing the canyons from filling with sediment, but the initial formation of most submarine canyons is probably attributable to fluvial erosion prior to subsidence.   MAS

Reading
Barnes, N.E., Bouma, A.H. and Normark, W.R. 1985: *Submarine fans and related turbidity systems* New York: Springer-Verlag.
Kennett, J.P. 1982: *Marine geology*. Englewood Cliffs, NJ and London: Prentice Hall.
Shanmugan, G. and Miola, R.J. 1988: Submarine fans: characteristics, models, classification and reservoir potential. *Earth science reviews* 24, pp. 383–428.
Shepard, F.P. and Dill, R.F. 1966: *Submarine canyons and other sea valleys*. Chicago: Rand McNally.
Whitaker, J.D. McD. 1974: Ancient submarine canyons and fan valleys. In R.H. Dott Jr and R.H. Shaver eds, *Modern and ancient geosynclinal sedimentation*. Society of Economic Paleontologists and Mineralogists special publication 19, pp. 106–25.

submerged forest A submerged forest is, as its name implies, an area of forest vegetation that has become submerged. Such features are commonly associated with coastlines that have experienced a recent submergence of a metre or so. Remnants of tree slumps are usually visible at low tide in most cases.   HV

subsequent stream A stream which follows a course determined by the structure of the local bedrock.

subsere A secondary successional sequence of plant communities. It is a series of community stages, the result of SUCCESSION on incompletely bared surfaces, or beginning with a community not truly climax in status (e.g. a SERAL COMMUNITY, SUBCLIMAX or DISCLIMAX). The essential characteristic of any subsere is its initiation from at least the vestige of a previous community at the site. This may involve residual species, seedlings, existing soil with its seed bank, or a complete community from which some controlling factor has been removed. In many instances, therefore, the subsere may be viewed as a manifestation of the recovery process in damaged ecosystems. (See also PRISERE.)
                                                        JAM

Reading
Cairns, J. ed. 1980: *The recovery process in damaged ecosystems*. Ann Arbor: Ann Arbor Science.
Fontaine, R.G. Gomez-Pompa, A. and Ludlow, B. 1978: Secondary successions. In *Tropical forest ecosystems: a state of knowledge report prepared by UNESCO/UNEP/FAO*. Paris: UNESCO. Pp. 216–32.

subsidence Landsurface sinking resulting from such processes as the withdrawal of groundwater, geothermal fluids, oil and gas; the extraction of coal, salt, sulphur and other solids through mining; the hydrocompaction of sediments; oxidation and shrinkage of organic deposits (notably peats); the development of thermokarst in areas underlain by permafrost; and karstic collapse.   ASG

Reading
Johnson, A.I. 1991: Land subsidence. *IAHS publication* 200, pp. 1–690.

succession The complex of processes producing a gradual directional change in the structure and species composition of ecosystems at a particular site. Sudden disruptions, seasonal rhythms, cyclic regeneration and evolutionary change are excluded. In his classic work, Clements (1928) produced the first major synthesis of ideas about succession and recognized seven basic successional processes: nudation (initiation); migration of propagules; ecesis (establishment); competition between species; action of environmental factors on organisms; reaction of organisms on their environment; and stabilization. These categories still provide a valid framework for discussion, although stabilization is best regarded as one possible effect of the other processes.

Clements emphasized reaction as the main driving force of succession and this process remains the basis of one of the most influential mechanistic models of successional change: the so-called facilitation model. However, the importance of this mechanism, whereby the establishment of later colonizers is facilitated by the previous occupation of the site by different species that modify their habitat to such an extent that they bring about their own replacement, remains largely untested. Other causes of succession of known importance in particular instances include life-cycle differences, environmental change, nutrient supply, chemical inhibition (allelopathy), seed banks, and interactions with micro-organisms, herbivores, predators

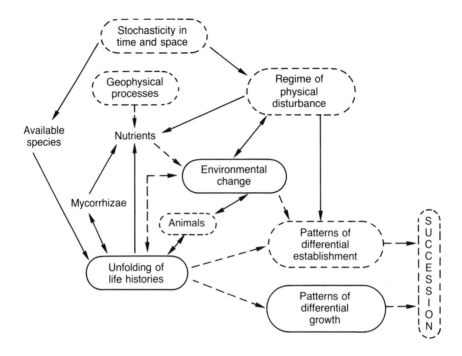

Succession: *Schematic representation of the concept indicating classic (broken arrows) and modern (solid arrows) elements, areas where processes and mechanisms are well understood (solid compartments) and areas where relatively little is known (broken compartments).*

Source: *D.C. West, H.H. Shugart and D.B. Botkin eds 1981:* Forest succession, concepts and application. *New York: Springer. Figure 1.2.*

and pathogens. Modern interpretations also emphasize the interactions of individual species populations, stochasticity, geophysical processes, disturbance regimes and multiple process interactions (see figure). Thus, the controlling processes are both AUTOGENIC (organism or community driven) and ALLOGENIC (driven by external physical environmental factors), and succession may be considered as a geoecological concept (Matthews 1992).

Succession embraces not only plant succession, traditionally terminating in CLIMAX VEGETATION, but also related processes of animal succession and soil development (e.g. Moravec 1969), and it is equally applicable to biotic communities or whole ecosytsems. The effect of succession is often to produce a sequence of communities (a SERE), made up of more or less distinct stages each of which is characterized by different dominant species and/or species with different environmental requirements and reflecting different evolutionary strategies.

Succession may be primary or secondary (see PRISERE and SUBSERE). Primary succession occurs, for example, on recently deglaciated terrain, where herbaceous pioneer species may eventually be replaced by shrubby heath species, and is the cause of the transition from bare ash to forest on volcanic islands such as Krakatoa (Whittaker *et al.* 1989). Secondary succession tends to be more rapid and occurs, for example, following the abandonment of agricultural land or after forest clearance. With increasing human impact on the landscape, secondary succession has become almost ubiquitous. The theory of both primary and secondary succession therefore underpins many ecosystem management programmes and has important applications to the restoration of land, ranging from overgrazed grassland to colliery spoil tips (e.g. Bradshaw and Chadwick 1980; Salzberg *et al.* 1987; Luken 1990).

Continuing controversy over the nature and significance of succession is partly due to the time-scale of the processes involved, which means that evidence is based largely on inference rather than on direct observation or experiment. There are three main ways of circumventing this problem. The first is to employ space-for-time substitution (chronosequences), such as the zonation of communities on terrain of varying age in front of a retreating glacier, an approach that is not without limitations (Pickett 1988). Secondly, on-site fossil records can be investigated where the remains of earlier succesional stages have been preserved, as is the case in the peaty waterlogged habitats characteristic of hydroseres (Walker 1970). Thirdly, successional processes may be simulated or otherwise modelled using numerical techniques (e.g. Shugart 1984; Usher, 1987). Future progress in understanding the complexities of succession will undoubtedly require a concerted effort using all the available approaches.   JAM

Reading and References
Bradshaw, A.D. and Chadwick, M.J. 1980: *The restoration of land: the ecology and reclamation of derelict and degraded land.* Oxford: Blackwell Scientific.

Burrows, C.J. 1990: *Processes of vegetation change.* London: Unwin Hyman.

Clements, F.E. 1928: *Plant succession and indicators: a definitive edition of plant succession and plant indicators.* New York: H.W. Wilson.

Finegan, B. 1984: Forest succession. *Nature* 312, pp. 109–14.

Luken, J.O. 1990: *Directing ecological succession.* London and New York: Chapman & Hall.

Matthews, J.A. 1992: *The ecology of recently deglaciated terrain: a geoecological approach to glacier forelands and primary succession.* Cambridge: Cambridge University Press.

Miles, J. 1987: Vegetation succession: past and present perceptions. In A.J. Gray, M.J. Crawley and P.J. Edwards eds, *Colonization, succession and stability.* Oxford: Blackwell Scientific. Pp. 1–29.

Moravec, J. 1969: Succession of plant communities and soil development. *Folia geobotanica et phytotaxonomica* 4, pp. 133–64.

Pickett, S.T.A. 1988: Space-for-time substitution as an alternative to long-term studies. In G.E. Likens ed., *Long-term studies in ecology: approaches and alternatives.* New York: Springer. Pp. 110–35.

Salzberg, K., Fredriksson, S. and Webber, P.J. eds 1987: Restoration and vegetation succession in circumpolar lands. *Arctic and alpine research* 19, pp. 337–577.

Shugart, H.H. 1984: *A theory of forest dynamics: the ecological implications of forest succession models.* New York: Springer.

Usher, M.B. 1987: Modelling successional processes in ecosystems. In A.J. Gray, M.J. Crawley and P.J. Edwards eds, *Colonization, succession and stability.* Oxford: Blackwell Scientific. Pp. 31–5.

Walker, D. 1970: Direction and rate in some British post-glacial hydroseres. In D. Walker and R.G. West eds, *Studies in the vegetation history of the British Isles.* Cambridge: Cambridge University Press. Pp. 117–39.

Walker, L.R. and Chapin III, F.S. 1987: Interactions amongst processes controlling successional change. *Oikos* 50, pp. 131–5.

Whittaker, R.J., Bush, M.B. and Richards, K. 1989: Plant recolonization and vegetation succession on the Krakatau Islands, Indonesia. *Ecological monographs* 59, pp. 59–123.

suction Describes the energy state of soil moisture under unsaturated conditions. Soil moisture suction is often called the matric potential or capillary potential of the soil moisture and it is described as a suction because it is a negative pressure potential resulting from the capillary and adsorptive forces due to the soil matrix holding moisture in the soil at a pressure less than atmospheric pressure. The magnitude of soil moisture suction can be indicated by a negative head of water (or mercury) which can be determined using a tensiometer and which is often expressed as a pF value. (See also CAPILLARY FORCES.)   AMG

Suess effect The relative change in the $^{14}C/C$ or $^{13}C/C$ ratio of any carbon pool or reservoir caused by the addition of fossil-fuel CO_2 to the atmosphere. Fossil fuels are devoid of ^{14}C because of the radiactive decay of ^{14}C to ^{14}N during long underground storage and are depleted in ^{13}C because of isotopic fractionation eons ago during photosynthesis by the plants that were the precursors of the fossil fuels. Carbon dioxide produced by the combustion of fossil fuels is thus virtually free of ^{14}C and depleted in ^{13}C. The term 'Suess effect' originally referred to the dilution of the $^{14}C/$ C ratio in atmospheric CO_2 by the admixture of fossil-fuel produced CO_2, but the definition has been extended to both the ^{14}C and ^{13}C ratios in any pool or reservoir of the carbon cycle resulting from human disturbances.   ASG

Reading
Keeling, C.D. 1979: The Suess effect: ^{13}carbon – ^{14}carbon interrelations. *Environment international* 2. 6, pp. 229–300.

suffosion An erosional process occurring in areas where limestone bedrock is overlain by unconsolidated superficial materials. The sediments slump down into widened joints and cavities in the bedrock surface, producing an irregular landsurface. It has been likened to an egg-timer effect.

sulphation The reaction between materials containing calcium carbonate and sulphur dioxide in humid atmospheres. The sulphur dioxide is oxidized to sulphur trioxide in a reaction which can be catalysed, for example, by vanadium oxide produced by internal combustion or by iron oxides on the material surface. Sulphur trioxide then reacts with the calcium carbonate to form gypsum and carbon dioxide.

$$CaCo_{3(s)} + SO_{2(g)} + 1/2\,O_{2(g)} + 2H_2O_{(g)} \rightarrow$$
$$CaSO_4.2H_2O_{(s)} + CO_{2(g)}$$

This reaction particularly occurs on calcareous stonework in polluted urban environments which is protected from rainwash. A layer of gypsum can form on these surfaces which can incorporate combustion particles to form a *black crust*. Gypsum from these films may be washed into underlying porous stonework by periodic wetting and can contribute to stone disruption by *salt weathering* mechanisms. BJS

summit plane See EROSION SURFACE.

sunspots Vortex-like disturbances with large associated magnetic fields that afflict the sun. They are relatively dark regions on the disc of the sun, with an inner 'umbra' of effective radiation temperature about 4500 K, and an outer 'penumbra' of somewhat higher temperature. Their frequency is quasi-periodic, with an average period of around 11 years. Various relationships between sunspot activity and climatic fluctuations have been proposed.

supercooling Denotes the presence of a substance in the liquid phase at a temperature which is below its normal freezing point. It is a common occurrence in the atmosphere where cloud droplets often occur in the supercooled state down to $-20°C$ and even down to $-35°C$ in rare cases. RR

superimposed drainage The pattern of a drainage network which developed on a landscape or bedrock which has since been removed by erosion, the network being preserved on the new landsurface. With antecedence it is one of the main reasons why drainage may appear to be unadjusted to present structures.

superimposed ice Formed when water comes into contact with a cold glacier surface and freezes. Under such circumstances it is a process of glacier accumulation and it is common in dry continental climates, such as northern Canada, where

90 per cent of the glacier ice may have formed in this way. Here summer temperatures are high enough to melt the winter snow, but it refreezes when it comes into contact with the glacier surface, which has been chilled by low winter temperatures. DES

Reading
Koerner, R.M. 1970: The mass balance of the Devon Island ice cap, NWT, Canada, *Journal of glaciology* 9. 57, pp. 325–36.

superimposed profile Diagrams made up of a series of topographical profiles taken at regular intervals across an area.

superposition, law of The law which states that the upper strata in sedimentary sequences postdate those which they overlie.

supersaturation A metastable state occurring when a solution contains more of a solute than is necessary to saturate it. The term is usually applied to a solution of limestone and is important in the process of SPELEOTHEM growth in caves.

supraglacial See ENGLACIAL.

surf Surf is formed on coasts as waves initially break on the shore. It consists of complex, transitional forms of waves and develops in the surfzone on beaches, between the breaker and swash zones. The surfzone is also where longshore currents develop.

surface detention The part of precipitation which remains in temporary storage during or immediately after a storm before it moves downslope by OVERLAND FLOW. The majority of surface detention water will form a part of the storm hydrograph but some of the water may infiltrate the soil or may be evaporated after the storm ends. (See also DEPRESSION STORAGE; SURFACE STORAGE.) AMG

surface run-off See RUN-OFF.

surface storage Water stored on the ground surface within a drainage basin. In some circumstances the volume of water stored in this way may be very large so surface storage is of particular importance in drainage basins containing large reservoirs, natural lakes or swamps. Frozen surface water in the form of ice and snow also forms part of surface storage. In relation to the Horton (1933) model of run-off generation, surface storage is considered to be the sum of DEPRESSION STORAGE and SURFACE DETENTION (Chorley 1978).   AMG

References
Chorley, R.J. 1978: The hillslope hydrological cycle. In M.J. Kirkby ed., *Hillslope hydrology*. Chichester: Wiley.
Horton, R.E. 1933: The role of infiltration in the hydrological cycle. *Transactions of the American Geophysical Union* 14, pp. 446–60.

surface tension The force required per unit length to pull the surface of a liquid apart. It is the result of the change in orientation of molecular bonds as the interface with another substance is approached. A molecule beneath the surface of a liquid is attracted in all directions by surrounding molecules within its 'sphere of molecular attraction' and so the resultant force on it is zero. However, a molecule on the surface of the liquid has a resultant force towards the main body of the liquid because there are more molecules within the part of its 'sphere of molecular attraction' within the liquid than there are in the part which falls within the overlying vapour.   AMG

Reading
Baver, L.D., Gardner, W.H. and Gardener, W.R. 1991: *Soil physics*. 5th edn. New York: Wiley.

surge See STORM SURGES.

surging glacier A glacier which flows at a velocity of an order of magnitude higher than normal. Whereas normal glaciers flow at 3–300 m year^{-1}, surging glaciers may flow at velocities of 4–12 km year^{-1}. Some glaciers flow permanently at surging velocities as is the case of the Jakobshavn Isbrae in West Greenland which, nourished by the Greenland ice sheet, flows at 7–12 km year^{-1}. Others experience periodic surges whereby a wave of ice moves downglacier at velocities of 4–7 km year^{-1} and may represent a velocity of 10–100 times higher than the pre-surge velocity. In such cases the jump from normal to surging flow takes place very suddenly. The wave of fast moving ice may plough into formerly stagnant ice near the glacier margin or extend beyond it. The wave is associated with strong COMPRESSING flow in front and extending flow behind and, following its passage, has the effect of lowering the glacier long profile. After a surge, the glacier experiences a quiescent phase, often of several decades, while it builds up to its original profile before surging once more.

One possible explanation of surging behaviour is that there are two modes of basal sliding, one normal and one fast. Budd (1975) has related the discharge of glaciers to glacier velocity and suggested that large outlet glaciers from big ice sheets have sufficient ice supply to maintain the fast mode of sliding permanently. Normal glaciers only have enough ice to maintain normal flow. Periodically surging glaciers occupy an intermediate position and may have enough ice supply for them to be able to cross the threshold from normal to fast flow, but then they cannot provide enough ice to maintain the fast mode of flow and so revert to normal flow. The two modes of flow probably relate to water thicknesses beneath the glacier and the development of cavities (Hutter 1982).   DES

Reading and References
†Many articles on glacier surges are contained in a special theme volume: *Canadian journal of earth sciences* 6.4 (1969), pp. 807–1018.
Budd, W.F. 1975: A first simple model for periodically self-surging glaciers. *Journal of glaciology*. 14, pp. 3–21.
Hutter, K. 1982: Glacier flow. *American scientist* 70, pp. 26–34.
†Kamb, B., Raymond, C.F., Harrison, W.D., Englehardt, H., Echelmeyer, K.A., Humphrey, N., Brugman, M.M., and Pfeffer, T. 1985: Glacier surge mechanism: 1982–83. Surge of variegated glaciers, Alaska, *Science* 227, pp. 469–79.
Sharp, M. 1988: Surging glaciers: behaviour and mechanisms. *Progress in physical geography* 12, pp. 349–70.

suspended load Sediment transported by a river in suspension. The material is carried within the body of flowing water, with its weight supported by the upward component of fluid turbulence. Particles are commonly less than 0.2 mm in diameter and in many rivers the suspended load will be dominated by silt- and clay-sized particles (i.e. <0.062 mm diameter). This fine-grained material is frequently referred to as wash load and is supplied to the river by erosion of the catchment slopes. The coarser particles usually represent suspended bed material and are derived from the channel perimeter. Transport rates are supply controlled and the suspended load is therefore a non-capacity load. Measurements of suspended sediment concentration expressed in mg l^{-1} or kg m^{-3} are required to calculate transport rates. (See also SEDIMENT YIELD.)   DEW

sustained yield Some resources are, in theory, renewable. They can be infinitely recycled through the biosphere and through human societies, either because they are basically unchanged by their use (e.g. water) or because they are self-regenerating (e.g. plants and animals). Sustained yield is a management concept which aims to regulate the system providing the resources so as to maintain the yield at a desired level into the foreseeable future and perhaps longer. In the case of agroecosystems, for example, it means the maintenance of soil structure and nutrient status, the control of weeds, pests and diseases and the selection of biota to respond to, for example, climatic change. In the case of whaling, it means avoiding the rate of cull that undermines the species' ability to reproduce themselves, an attempt that has so far been notably unsuccessful. Calculation of rates of sustained yield requires a thorough empirical knowledge of the ecosystem in question and possibly sophisticated modelling as well. It is made much more complicated when, as in the case of fish populations, for example, the natural condition seems to be one of considerable cyclic fluctuation. The optimum yield for one year, then, may be markedly different from the next and neither science nor politics may be able to cope with this amplitude of coming and going. IGS

Reading
Dasmann, R.F., Milton, J.P. and Freeman, P.H. 1973: *Ecological principles for economic development.* London and New York: Wiley.

suturing The act of splitting. Fissuring of the earth's crust resulting in the emission of lava.

swale A depression in regions of undulating glacial moraine, a trough between beach ridges produced by erosion, or an area of low ground between dune ridges.

swallet A sinkhole. A hole in limestone bedrock which has been produced by solution and through which stream water disappears.

swallow hole (or swallet) A feature through which surface water goes underground in a limestone area. They have a variety of local names (e.g. AVEN, PONOR) and a range of forms, from deep shafts (like Gaping Gill in Yorkshire) to less obvious

zones in a stream bed where discharge is lost as a result of downward percolation. ASG

swamp A type of wetland, dominated by woody plants. By comparison the term 'marsh' tends to be reserved for wetland dominated by graminoids, grasses or herbs. ASG

Reading
Gore, A.J.P. ed. 1983: *Mires: swamp, bog, fen and moor.* Amsterdam: Elsevier. 2 vols.
Williams, M. ed. 1990: *Wetlands: a threatened landscape.* Oxford: Blackwell.

symbiosis The 'living together' of organisms in close association, which occurs when one or both members of two species come to depend on the presence of the other for survival and reproduction. There can be different degrees of attachment in such a relationship. The weakest, in which one species benefits but the other does not (and neither is harmed) is termed COMMENSALISM. When the bonds between the two are stronger, and positive gains flow in both directions, MUTUALISM may be said to have developed. Lichens, which are composite structures of algae and fungi, provide good examples of this. Another instance may be found in the well-known linkages between the nitrogen-fixing bacterial genus *Rhizobium*, representatives of which live within the root nodules of legumes, including alfalfa, peas, beans and clover; and, as they do so, they obtain from the plants the energy which they need to give rise to the N-fixing reaction:

$$2N_2 + 6H_2O \rightarrow 4NH_3 + 3O_2$$

Some of the ammonia which is produced is in turn made available to the legumes for the synthesis of amino-acids (Delwiche 1970). The specific use of legumes in crop-rotation schemes on deficient soils in Nigeria has given rise to a symbiotic fixation of nitrogen of up to 36.4 kg ha^{-1} year^{-1} (Nye and Greenland 1960). Other plant species associated with different symbiotic bacteria, which together are capable of N-fixation, include cycads, gingkoes, alders, sea buckthorn, bog myrtle, *Ceanothus* shrubs, and the tropical *Casuarina* (Watts 1974). The symbiotic relationships between fungi and roots (*mycorrhiza*) are also important for the expeditious transfer of nutrients between soil and plants in environments as different as tropical rain forest and heather moorland in Europe.

When symbiotically associated species come to rely on each other totally, this may be termed *obligate mutualism*. This often involves *evolutionary symbiosis*, over long periods of time (see COEVOLUTION). Probably the best example relates to the evolution of lobeliads in Hawaii which, in the absence of their normal insect pollinators, emerged along with a group of nectar-eating birds, the honey-creepers (family: Drepanididae) (Amadon 1950). A further extreme form of symbiosis involves PARASITES.   DW

References
Amadon, D. 1950: The Hawaiian honey-creepers. *Bulletin of the American Museum of Natural History* 95, pp. 1–262.
Delwiche, C.C. 1970: The nitrogen cycle. *Scientific American* 223, pp. 136–47.
Nye, P.H. and Greenland, D.J. 1960: *The soil under shifting cultivation*. Harpendon: Commonwealth Bureau of Soils.
Watts, D. 1974: Biogeochemical cycles and energy flows in environmental systems. In I.R. Manners and M.W. Mikesell eds., *Perspectives on environment*. Washington, DC: Association of American Geographers.

sympatry Originating in or occupying the same geographical area. The term sympatric is used to describe species or populations with overlapping geographical ranges which are, therefore, not spatially isolated. Where the ranges are adjacent, but not exactly overlapping, the term used is parapatric. Species isolated are said to be allopatric. When brought together artificially, allopatric species often hybridize quite freely. (See also BIOTIC ISOLATION and ISOLATION, ECOLOGICAL.)   PAS

Reading
Stebbins, G.L. 1950: *Variation and evolution in plants*. New York and London: Columbia University Press.
— 1977: *Processes of organic evolution*. 3rd edn. Englewood Cliffs, NJ: Prentice-Hall.

syncline A trough in folded strata.

synecology The ECOLOGY of whole communities (plant, animal or biotic communities) as opposed to individuals or single species. Emphasis is given to the reciprocal relationships between communities and their environments. The unit of study (the community) is at a higher level of organization than the individual organism or the species population but it is a lower level of organization than the ECOSYSTEM or the BIOGEOCOENOSIS. (See also AUTECOLOGY.)   JAM

Reading
Daubenmire, R. 1968: *Plant communities: a textbook of plant synecology*. New York and London: Harper & Row.
Dice, L.R. 1968: *Natural communities*. Ann Arbor: University of Michigan Press.
Whittaker, R.H. 1970: *Communities and ecosystems*. London and Toronto: Collier-Macmillan.

synforms See ANTIFORMS.

synoptic climatology An aspect of climatology concerned with the description of local or regional climates in terms of the properties and motions of the atmosphere rather than with reference to arbitrary time intervals, such as months. There are two stages to a synoptic climatological study: the determination of categories of atmospheric circulation type (often referred to as WEATHER TYPE) and, secondly, the assessment of mean, modal and other statistical parameters of the weather elements in relation to these categories.

Although the first investigations of synoptic patterns were made in the nineteenth century, in many countries longer-standing popular weather lore had already associated cold, heat or precipitation with particular wind directions. Modern synoptic climatology developed during and after the Second World War in response to the needs of military operations to assess likely weather conditions (Barry and Perry 1973). Since then both subjective and objective means of classifying the totality of weather patterns have been widely used.

With the advent of high-speed computers and digitized grid-point data sets of sea-level pressure and geopotential height fields, there has been a transformation in procedures for preparing catalogues of synoptic types since the 1960s (Perry 1983). Two approaches have been widely adopted in developing classifications: (1) a determination of pattern similarity based on correlation methods, (2) the use of one of a range of statistical techniques to extract components of the fields, perhaps combined with a clustering approach to obtain pattern types. New types of data, e.g. satellite imagery (see SATELLITE METEOROLOGY), are now being employed in the classification process and are helping to extend the synoptic approach to the tropics where hitherto it has been little used.

Classification catalogues exist for several areas of the world for the available period of synoptic weather maps (Smithson 1986). In addition to being used in the description and analysis of such persistence and recurrence models as singularities, weather spells and natural seasons, they have found applications in studies of air chemistry, climatic fluctuations and the reconstruction of past climatic states. Synoptic climatology can also serve as an important check on computer-derived numerical climate studies and because the method relates local climate conditions to the atmospheric circulation it provides a realistic basis for much climatological teaching and project work.   AHP

Reading and References
Barry, R.G. and Perry, A.H. 1973: *Synoptic climatology: methods and applications.* London: Methuen.
Perry, A.H. 1983: Growth points in synoptic climatology. *Progress in physical geography* 7, pp. 90–6.
Smithson, P.A. 1986: Dynamic and synoptic climatology. *Progress in physical geography* 12, pp. 119–29.

synoptic meteorology An aspect of meteorolgy concerned with the description of current weather and the forecasting of future weather using synoptic charts which provide a representation of the weather at a particular time over a large geographical area.

Brandes was the first to develop the idea of synoptic weather mapping by comparing meteorological observations made simultaneously over a wide area. It was not until the invention of the electric telegraph that the rapid preparation of a map of weather observation became possible; the first British daily weather map was sold at the Great Exhibition of 1851. The loss of numerous lives at sea encouraged the issue of regular gale warnings by Admiral Fitzroy who was appointed to the first official meteorological post in the UK. From the 1860s onwards national weather services were established in many countries and by the 1890s the first upper-air soundings gave a better understanding of the vertical structure of weather systems (Bergeron 1981). Just after the First World War the Norwegian meteorologists J. Bjerknes and H. Solberg produced a synoptic model of the mid-latitude frontal cyclone, using observations from a dense network of observing stations. A knowledge of such synoptic models, and a combination of experience, skill and judgement, was the mainstay of the weather forecaster up to the 1950s. The current weather situation was analysed on surface maps by drawing isobars and fronts by hand and distinguish-

ing areas of significant weather, while on upper air charts pressure contours and THICKNESS lines were indicated. Examination of the circulation patterns and extrapolation of the movement and development of the weather systems was also part of the forecasting process.

The development of weather prediction by numerical methods using high-speed electronic computers has transformed synoptic meteorology in the past forty years, and the role of the meteorologist today is that of monitoring the output of computer-based forecasts, modifying its products in the light of experience of atmospheric behaviour.

New sources of data, including radar echoes of rainfall intensity, and global satellite cloud pictures, have widened the information available to the forecaster.

Regional meteorological centres, such as the one at Bracknell in the UK, collect coded synoptic observations and transmit regional sets to three world centres which in turn produce global data sets for distribution to national meteorological centres. At the Meteorological Office in Bracknell numerical forecasts are produced twice daily for the following 24 hours and the programme centres continue to give a medium-range forecast for the next 72 hours, as well as a six-day prediction. A further forecast model is run to predict the weather in greater detail over the British Isles for the following 36 hours.   AHP

Reading and Reference
Bergeron, T. 1981: Synoptic meteorology: an historical review. In G.H Liljequist ed., *Weather and weather maps.* Stuttgart: Birkhauser Verlag.
†Hardy, R., Wright, P., Gribbin, J. and Kingston, J. 1982: *The weather book.* London: Michael Joseph.

synthetic unit hydrograph See HYDROGRAPHS.

systems Simply 'sets of interrelated parts' (Huggett 1980, p. 1). They are defined as possessing three basic ingredients: 'elements, states, and relations between elements and states'. There can be both concrete and abstract systems; for example, the hot water system of a house or the set of moral values of a society; and the elements may therefore be real things or concepts, each of which is held to possess a variety of properties (or can be said to exist in a variety of states). The overall state of the whole system is then defined by the character of these properties at a given moment. Because the system is defined as a *set* of parts, it follows that there is some boundary which distinguishes that particular set from all other possible sets; and the boundedness of systems is both an important theoretical attribute and a source of immense practical difficulty.

While it is possible to view some systems as completely isolated from all external influences, which in physical examples implies that no movement of energy or mass can occur across their boundaries, most are defined as either *open* or *closed*. An open system exchanges both energy and mass with its surroundings, whereas a closed system is open only to the transfer of energy. These *inputs* of mass and/or energy are termed *forcing functions* and they are generally of considerable importance in the concrete systems of interest to physical geographers. The *throughput* of energy and/or mass creates the linkages or relations between the system elements, which may adjust in the process, either by *negative* feedback mechanisms (*homeostasis*) so that the system state remains unchanged, or by *positive* feedback, so that a net change in the system state results. The outcome of the transfers is termed the system *output*, which may be energy and/or mass and/or a new system state.

Finally, systems are regarded as hierarchical sets: the whole system at any one level being merely a component or element of some higher-order set and its own elements being, in reality, smaller-scale systems in their own right. To give an example: a drainage basin with a single stream channel may be studied as a first-order drainage system; yet its slopes and stream channel may equally well be viewed as individual systems; and the whole basin readily becomes just one element of a larger drainage network.

From this description of systems it should be clear that the concept may be applied to an infinitely wide range of phenomena. This, indeed, was seen as its chief methodological merit by Von Bertalanffy, who introduced the notion in 1950, with the explicit hope that a focus on such a general feature of the physical and mental environment would encourage the sciences, in particular, to adopt a unified methodology. From this came the idea of GENERAL SYSTEM THEORY. The concept of the open system was introduced into geomorphology

by Strahler (1950) and its merits were widely advocated by Strahler's pupil, Chorley (see e.g. Chorley 1960, 1967; Chorley and Kennedy 1971; Bennett and Chorley 1978). An extremely persuasive and influential use of the concept has been made by yet another of Strahler's pupils, Schumm (1977), and several other substantial works of a more general nature have also appeared (e.g. Huggett 1980; Sugden 1982).

More influential, however, has been the application of the idea of systems in the field of ecology, where the concept of the ecosystem – as a formal statement – goes back to Tansley (1935), although its basics can readily be traced into the far earlier views of Darwin and Haeckel (Stoddart 1967). Studies such as H.T. Odum's monumental analysis of the Silver Springs ecosystem (1957) remain classic examples of the possibilities and limitations of what has come to be known as systems analysis. It seems clear that it was in large measure the apparent success with which the systems concept was used as an analytical device by ecologists in the 1950s which spurred its adoption in mainstream physical geography. The growing realization of the problems posed by identification and isolation of ecosystems as objects of analysis has, similarly, been accompanied by a reduction in the emphasis placed upon their investigation in other fields.

The system concept seems to have made four different kinds of appearance in physical geography. First, it is now widely and loosely used as a source of jargon and terminology: the ideas of various forms of equilibrium are particularly persistent and poorly defined borrowings. Secondly, the idea has been adopted as a pedagogic framework into which the results of earlier studies, or the fruits of different concepts, can be slotted for ease of exposition: Chorley and Kennedy (1971) and Sugden (1982) are examples of the category. Thirdly, the concept may be taken as the basis for a substantive investigation of the workings of some portion of reality, in line with ecological studies such as that of Odum (1957): these inquiries are most commonly directed towards drainage basin hydrology, ecology and climatology and few of them have actually been undertaken by geographers (although human geographers have made more explicit use of systems in this particular way); an exception, in both soils (1975). Finally, the idea of systems has been adopted as a useful framework in which to view questions relating to environmental management and planning: a fairly early, clear and comprehensively explained example is Hamilton *et al.* (1969) and the whole approach has been exhaustively discussed by Bennett and Chorley (1978).

The application of the systems concept to any substantive investigation of the actual or potential workings of portions of the physical environment encounters two principal difficulties. The first is the need to define system boundaries. While arbitrary lines can, of course, be drawn anywhere to define a system of any magnitude, the very nature of the concept is of an inter-related set of elements which is in some real sense functionally or at least morphologically distinguishable from all adjacent sets. In practice, even lakes, islands and river basins – the most clearly definable of the physical units of interest to geographers – do not actually possess sharp and precise boundaries. The resulting difficulties of definition are horribly but understandably akin to those related to the definition of regions: they are equally hard to resolve.

The second problem stems from the need to evaluate and interpret the results of system analyses (see Kennedy, 1979). Langton (1972) points out that the only effective way of comparing different systems is with reference to their success in attaining some preferred or most efficient state of operation or morphology. It is clear that we can use this criterion very readily to deal with engineered, planned or 'control' systems, since it is possible to specify the preferred or 'best' outcome. It is very far from clear that it is justifiable to regard natural systems in such a light. What is the 'goal' of an ecosystem? Or a drainage basin? Or the general circulation of the atmosphere? Could this be the reason why substantive studies of the workings of the natural systems of interest to the physical

geographer are actually rather thin on the ground?

The concept of the system is, in essence, simple and it can be widely applied. The actual applications in physical geography to date have been most numerous in pedagogic and applied areas and there little fresh insight has been gained into the actual workings of the natural environment. The greatest impact seems to have been made in the wholesale introduction of the terminology of systems analysis.   BAK

Reading and References
†Bennett, R.J. and Chorley, R.J. 1978: *Environmental systems: philosophy, analysis and control*. London: Methuen.

Chorley, R.J. 1960: *Geomorphology and general systems theory*. US Geological Survey professional paper 500-B.

— 1967: Models in geomorphology. In R.J. Chorley and P. Haggett eds, *Models in geography*. London: Methuen. Pp. 59–96.

— and Kennedy, B.A. 1971: *Physical geography: a systems approach*. London: Prentice-Hall International.

†Hamilton, H.R. *et al.* 1969: *Systems simulation for regional analysis. An application to river-basin planning*. Cambridge, Mass.: MIT Press.

†Huggett, R.J. 1975: Soil landscape systems: a model of soil genesis. *Geoderma* 13, pp. 1–22.

— 1980: *Systems analysis in geography*. Oxford: Clarendon Press.

Kennedy, B.A. 1979: A naughty world. *Transactions of the Institute of British Geographers* 4, pp. 550–8.

Langton, J. 1972: Potentialities and problems of a systems approach to the study of change in human geography. *Progress in geography* 4, pp. 125–79.

Odum, H.T. 1957: Trophic structure and productivity of Silver Springs, Florida. *Ecological monographs* 27. pp. 55–112.

Schumm, S.A. 1977: *The fluvial system*. London: Wiley.

Stoddart, D.R. 1967: Organism and ecosystem as geographical models. In R.J. Chorley and P. Haggett eds, *Models in geography*. London: Methuen. Pp. 511–48.

Strahler, A.N. 1950: Equilibrium theory of erosional slopes approached by frequency distribution analysis. *American journal of science* 248, pp. 673–96, 800–14.

Sugden, D.E. 1982: *Arctic and Antarctic: a modern geographical synthesis*. Oxford: Basil Blackwell; Totowa. NJ: Barnes & Noble.

Tansley, A.G. 1935: The use and abuse of vegetational concepts and terms. *Ecology* 16, pp. 284–307.

Von Bertalanffy, L. 1950: The theory of open systems in physics and biology. *Science* 111, pp. 23–9.

syzygy One of the two points at which the moon or a planet is aligned with the earth and the sun.

T

tafoni Named after features in Corsica; are cavernous weathering forms. They frequently occur in granitic rocks, but are known from other types, including sandstone. SALT WEATHERING , wind scour, and chemical changes involving case-hardening, contribute to their development. ASG

taiga The most northerly coniferous forest of cold temperate regions. It does not exist as a zone in the southern hemisphere, and generally refers to open woodland lying to the south of TUNDRA and to the north of the dense BOREAL FOREST. It has also been used more broadly to include the entire area covered by coniferous forest of high latitudes and high mountain slopes. More literally (from the Russian), it is characterized by open, rocky landscapes dominated by conifers but with scattered deciduous trees, such as birch and alder, locally dense along river valleys, with a fairly continuous carpet of lichens and heathy shrubs. The area is often poorly drained and peat-filled. PAF

Reading
Larsen, J.A. 1980: *The boreal ecosystem.* New York: Academic Press.

takyr A desert soil with a bare, parquet-like surface, broken up by a network of splits into numerous polygonal aggregates. Takyrs are typical landscape elements of the deserts of Central Asia, but comparable claypans are found in the deserts of Australia, Iran and North Africa. They have no higher vegetation, a crusted surface, occur in the lower parts of piedmont plains, and their formation requires a seasonal flooding of the surface by a thin layer of water, carrying

Cavernous weathering features called tafoni developed in granitic rock in the Atacama Desert. Weathering and abrasional processes may contribute to their development.

suspended clay material and soluble salts (Kovda *et al.* 1979).   ASG

Reference
Kovda, V.A., Samoilova, E.M., Charley, J.L. and Skujinš. J.J. 1979: Soil processes in arid lands. In D.W. Goodhall, R.A. Perry and K.M.W. Howes eds, *Arid-land ecosystems: structure, functioning and management.* Vol. I. Cambridge: Cambridge University Press. Pp. 439–70.

talik A layer of unfrozen ground below the seasonally frozen surface layer and above or within PERMAFROST.

talsand ('valley sand') A widely used concept in north German Quaternary geology. Talsands are large-scale sandy infillings of ice marginal valleys which consist of fluvial beds with a capping of windborne sands.   ASG

Reference
Schwan, J. 1987: Sedimentologic characteristics of a fluvial to aeolian succession in Weichselian Talsand in the Emsland (FRG). *Sedimentary geology* 52, pp. 273–98.

talus A sloping accumulation of rock fragments at the foot of a cliff or rock outcrop. The material forming such a feature.

tank A man-made pond or lake. A natural depression in bare rock that is filled with water throughout the year.

tarn A small mountain lake (northern England). Compare Scottish 'lochain' and Welsh 'llyn'.

taxon cycle A concept introduced by P.J. Darlington in 1943 and so named in 1961 by the island biogeographer E.O. Wilson; it attempts to portray the stages through which an invading species-population passes when colonizing an archipelago or particular islands. It examines the way in which the population might adapt to new habitats and undergoes evolutionary diversification in the process. The cycle has been demonstrated for a number of groups of organisms, including ants in Melanesia and birds in the Solomon Islands. Most cycles so far described exhibit three or four stages, usually involving the expansion of a species without geographical divergence, the fragmentation of the distribution with speciation or subspeciation and a contracting stage, in which species are limited to relict areas, leading to extinction or to a secondary expansion.   PAS

Reading
Pielou, E.C. 1979: *Biogeography.* New York and Chichester: Wiley.

Ricklefs, R.E. and Cox, G.W. 1972: Taxon cycles in the West Indian avifauna. *American naturalist* 106, pp. 195–219.

taxonomy The study, description and classification of variation in organisms, including the causes and consequences of such variation. This is a modern wide definition which makes taxonomy synonymous with the now interchangeable term, systematics. Traditionally, taxonomy was often restricted to the narrower activity of the classification and naming of organisms; in this sense, it was only a part of systematics.

The origins of taxonomy lie in man's need to classify the discontinuities of variations he saw in nature. Scientific taxonomy reached its great flowering in the eighteenth century with the still fundamental contributions of the Swedish biologist, Carolus Linnaeus (1707–1778). These included his *Systema Naturae* (1735 and many later editions), in which he classified all the then known animals, plants and minerals.

Today two major approaches to biological classification are recognized. The first is *phenetic* which expresses the relationships between organisms in terms of their similarities in characters, without taking into account how they came to possess these. The second is *phylogenetic*, or evolutionary, in which some aspects of their evolution are taken into account when making the classification, including their CLADISTIC relationship, which refers to the pathways of ancestry. These pathways are usually expressed in the form of a tree-diagram or cladogram. (See also VICARIANCE BIOGEOGRAPHY.)   PAS

Reading
†Jeffrey, C. 1977: *Biological nomenclature.* 2nd edn. London: Edward Arnold.

†— 1982: *An introduction to plant taxonomy.* 2nd edn. Cambridge: Cambridge University Press.

Ridley, M. 1983: Can classification do without evolution? *New scientist* December, pp. 647–51.

Ross, H.H. 1974: *Biological systematics.* Reading, Mass.: Addison-Wesley.

Stace, C.A. 1980: *Plant taxonomy and biosystematics.* London: Edward Arnold.

tear fault A fault with a vertical fault plane, the blocks either side of which move horizontally. A transcurrent or strike fault.

tectonics The study of the broad structures of the earth's lithosphere and the processes of faulting, folding and warping that form them. Tectonics focuses on structures at the regional scale and above, the investigation of smaller-scale structural features usually being described as structural geology. Tectonics is concerned not only with the determination of the three-dimensional form of geological structures but also their history, origin and relationship to each other. Tectonic landforms are those, such as fault-scarps, produced directly by tectonic mechanisms and those larger features, such as warped erosion surfaces, which owe their general, though not detailed, form to such processes. (See also PLATE TECTONICS.   MAS

Reading
Kearey, P. and Vine, F.J. 1990: *Global tectonics.* Oxford: Blackwell Scientific.

Ollier, C. 1981: *Tectonics and landforms.* London: Longman.

Spencer, E.W. 1977: *Introduction to the structure of the earth.* 2nd edn. New York: McGraw-Hill.

teleconnections Simultaneous atmospheric events in areas remote from each other. Positive and negative correlations with atmospheric and oceanic circulation features and weather elsewhere have been found widely. Early descriptive work in the 1920s by Sir Gilbert Walker led to the discovery of the southern oscillation (see MACROMETEOROLOGY), a zonal circulation of air with sinking in the eastern South Pacific and rising air over the Indonesian Ocean. Work on dynamical explorations of regional-scale patterns resembling steady waves and based on ROSSBY WAVE propagation principles has recently predominated (Perry 1983), and has included investigation of the North Atlantic oscillation with its 'see-saw' in winter temperatures between Greenland and northern Europe (Loewe 1966).   AHP

Reading and References
†Bjerknes, J. 1969: Atmospheric telecommunications from the equatorial Pacific. *Monthly weather review* 97, pp. 162–72.

Glantz, M.H., Katz, R.W. and Nicholl, N. eds 1991: *Teleconnections linking worldwide climate anomalies: scientific basis and societal impact.* Cambridge: Cambridge University Press.

Loewe, F. 1966: The temperature see-saw between western Greenland and Europe. *Weather* 21, pp. 241–6.

Perry, A.H. 1983: Growth points in synoptic climatology. *Progress in physical geography* 7, pp. 90–6.

†Wallace, J.M. and Gutzler, D.S. 1981: Teleconnections in the geopotential height field during the northern hemisphere winter. *Monthly weather review* 109, pp. 784–812.

temperate ice Ice which is at the pressure melting point. In temperate glacier ice, water is present throughout and generally amounts to between 0.1 and 2 per cent of the total volume (Lliboutry 1976). *Temperate* ice is contrasted with *cold* ice which is below the pressure melting point. It is common to classify whole glaciers as temperate or cold, but this is misleading since both types of ice are common in most glaciers. For example, a glacier which consists wholly of temperate ice in summer may have a cold surface layer of ice in winter, while it is also likely that cold patches exist at the bottom as a result of pressure changes around obstacles (Robin 1976). Furthermore many 'cold' glaciers have ice at the pressure melting point at depth.   DES

Reading and References
Lliboutry, L. 1976: Physical processes in temperate glaciers. *Journal of glaciology* 16, pp. 151–8.

†Paterson, W.S.B. 1981: *The physics of glaciers.* London and New York: Pergamon.

Robin, G. de Q. 1976: Is the basal ice of a temperate glacier at the pressure melting point? *Journal of glaciology* 16, pp. 183–96.

temperature To the general public temperature is a confusing parameter. When used to measure how warm or how cold substances feel it is confusing to have different temperature scales, so that, for instance, the temperature at which pure ice melts can be stated as 0° Celsius or 32° Fahrenheit or 273.15° Kelvin. In science the Kelvin scale is used because it avoids negative temperatures and is a more accurate reflection of the energy possessed by the molecules in a substance. Temperature is really a measure of the molecular kinetic energy of a substance, in other words the average speed at which the molecules are moving in gas, or vibrating in a solid. The Celsius and Fahrenheit scales are used for convenience in that they give easier numbers for people to handle for the range of temperatures normally encountered at the earth's surface.   JET

temperature humidity index Originally called the Discomfort Index when it was introduced by E.C. Thom (1959) in the USA, the temperature humidity index (THI) is a simplified form of effective temperature:

$$THI = 0.4\,(T_{DB} + T_{WB}) + 4.8$$

when the dry-bulb and WET-BULB TEMPERATURES are recorded in °C. Windspeeds and solar radiation are not taken into consideration and so the index must be used with caution. The THI is widely used in weather forecasts for the general public in the USA during summer. At a THI below 70 there is no discomfort. When the THI reaches 80 everyone feels uncomfortable.

DGT

Reference
Thom, E.C. 1959: The discomfort index. *Weatherwise* 12, pp. 57–60.

tensiometer An instrument for estimating the matric or capillary potential of soil moisture. A tensiometer consists of a porous pot at the required point in the soil and connected through watertight tubing to a manometer or pressure gauge. The whole instrument is filled with water before the cup is placed in a carefully augured hole, refilled as near to its former condition as possible. Water may pass through the walls of the pot in response to suction forces in the surrounding soil. As water passes into the soil a partial vacuum builds up inside the instrument and this is monitored as a negative head of water by the vacuum gauge or manometer. Water will move into or out of the pot until the soil moisture suction is balanced by the strength of the partial vacuum or the weight of a negative head of water (or, more usually, mercury). Tensiometers are only suitable for use in comparatively moist soil because of problems in keeping the instrument airtight at high suctions.

AMG

Reading
Burt, T.P. 1978: *An automatic fluid-scanning switch tensiometer system.* British Geomorphological Research Group technical bulletin 21. Norwich: Geo Abstracts.
Curtis, L.F. and Trudgill, S. 1974: *The measurement of soil moisture.* British Geomorphological Research Group technical bulletin 13. Norwich: Geo Abstracts.

tepee An overthrust sheet of limestone which appears as an inverted V in a two-dimensional exposure, so named because of its two-dimensional resemblance to the hide dwellings of early American Indians. Tepees are found in tidal areas, around salt lakes, and in CALCRETE, developing as a result of deformation or desiccation and contraction processes related to fluctuations in water levels and in the nature of chemical precipitation.

ASG

Reading
Kendall, C.G. St C. and Warren, J. 1987: A review of the origin and setting of tepees and their associated fabrics. *Sedimentology* 34, pp. 1007–27.
Warren, J.K. 1983: Tepees, modern (Southern Australia) and ancient (Permian – Texas and New Mexico) – a comparison. *Sedimentary geology* 34, pp. 1–19.

tephigram A meteorological thermodynamic chart used for plotting and analysing dry-bulb and dewpoint temperature from a RADIOSONDE ascent, usually from the surface to the lower STRATOSPHERE. The vertical axis is atmospheric PRESSURE decreasing upwards on a logarithmic scale and TEMPERATURES are plotted with reference to the horizontal axis although the isotherms are skewed to run from bottom left to top right of the diagram. Dry adiabats, saturation adiabats and lines of constant humidity mixing ratio are also printed in the chart. The tephigram is basically an aid to forecasting; for example, fog formation or shower development.

RR

tephra The solid material ejected from a volcano which includes dust, ash, cinders and volcanic bombs.

tephrochronology A dating technique based on the principal that different ash falls from a volcanic eruption may be recognizable on the basis of petrology and chemical composition. If the age of each fall is known (either because of historical records or because of isotopic dating) the ash can be used as a marker horizon in otherwise undated sections.

ASG

terlough Depressions, with a sinkhole, which fill with water when the watertable rises. The rise may be associated with tidal effects. They are a feature of parts of western Ireland.

terminal grades The fine fraction due to a glacial comminution which reflects the mineral components of the debris. Dreimanis and Vagners (1971) suggested that once a glacial TILL has been broken down to mineral-sized fragments it experiences relatively little further breakdown, thus meriting the use of the term terminal grades. This grain-size group comprises one of the bimodal grain-sized products of abrasion, the other being rock fragments. Under some circumstances, however, fine

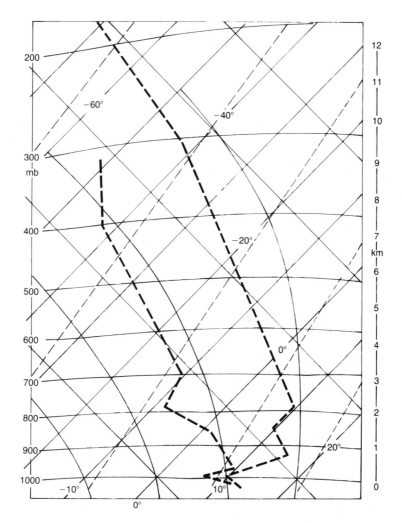

Tephigram plotted at Camborne at 1200Z, 13 May 1979.

material fragments do break down further to finer material (Haldorsen 1981).　　DES

References

Dreimanis, A. and Vagners, U.J. 1971: Bimodal distribution of rock and mineral fragments in basal tills. In R.P. Goldthwait ed. *Till:* A *symposium.* Ohio State University Press. Pp. 237–50.

Haldorsen, S. 1981: Grain-size distribution of subglacial till and its relation to glacial crushing and abrasion. *Boreas* 10, pp. 91–105.

terminal moraine The moraine at the terminus of a glacier. (See also MORAINE.)

terminal velocity That with which a falling body moves relative to a fluid if the resultant force on the body is zero. The terminal velocity V of a sphere, radius r and density σ, falling in a fluid, density ρ, under the influence of gravity, is given by Stoke's law from:

$$V = 2/9(\sigma - \rho)\eta r^2 g$$

where g is the acceleration due to gravity and η is the coefficient of VISCOSITY. WBW

terminations The boundaries in deep sea core sediments that separate pronounced oxygen isotopic maxima from exceptionally pronounced minima. They are in effect rapid deglaciations, and are conventionally numbered by Roman numerals in order of increasing age. The segments bounded by two terminations are called glacial cycles. Nine terminations have occurred in the past 0.7 million years. ASG

Reading
Kukla, G.J. 1977: Pleistocene land-sea correlations. I. Europe. *Earth science reviews* 13, pp. 307–74.

termites Of which there are several thousand species, are insects of the Isoptera order, and about four-fifths of the known species belong to the Termitidae family (Harris 1961). They vary in size according to their species, from the large African Macrotermes, with a length of around 20 mm and a wing span of 90 mm, down to the Middle Eastern Microcerotermes which are only around 6 mm long with a wing span of 12 mm. Major recent taxonomic and ecological surveys include that of Brian (1978), while Lee and Wood (1971) provide a detailed study of the effects of termites on soils, and Goudie (1988) reviews their geomorphological impact.

A grotesquely bulbous termite mound from the sandy plains near Derby in tropical north-west Australia. Termites play a major role in translocating sediment and soil in such environments.

Termites, though 'fierce, sinister and often repulsive' (Maeterlinck 1927), are remarkable for having been adapted to living in highly organized communities for as long as 150–200 million years, and much of their success is due to their development of elaborate architectural, behavioural, morphological and chemical strategies for colony defence. They occur in great numbers: 2.3 million ha^{-1} in Senegal and 9.1 million ha^{-1} in the Ivory Coast (UNESCO/UNEP/FAO/1979). Maeterlinck (1927) regarded them as 'the most tenacious, the most deeply rooted, the most formidable, of all the occupants and conquerors of this globe.' The vast majority of termite species are found in the tropics. ASG

Reading and References
Brian, M.V. ed. 1978: *Production ecology of ants and termites.* Cambridge: Cambridge University Press.

Goudie, A.S. 1988: The geomorphological role of termites and earthworms in the tropics. In H.A. Viles ed., *Biogeomorphology.* Oxford: Blackwell. Pp. 166–92.

Harris, W.V. 1961: *Termites: their recognition and control.* London: Longman.

Lee, K.E. and Wood, T.G. 1971: *Termites and soils.* London and New York: Academic Press.

Maeterlinck, M. 1927: *The life of the white ant.* London: Allen & Unwin.

UNESCO/UNEP/FAO 1979: *Tropical grazing and land ecosystems.* Paris: UNESCO.

terra rossa Red soils developed on the iron-oxide-rich residual material on limestone bedrock, particularly in warm temperate regions.

terrace, pluvial Lake basins may contain pluvial terraces that are landforms consisting of water-worked materials now located at elevations different from modern lake levels. The word 'pluvial' refers to forms and processes derived from a period of increased precipitation, and was first applied to all surface processes by Tylor (1868). The term still maintains this broad perspective, but its general use is most common in referring to lake basins in modern arid regions that were once more moist. Because the upper surfaces of the terraces are the products of the actions on a plane coincidental with a level of the lake waters at some time in the past, the upper surfaces are usually flat or nearly so.

Pluvial terraces may be erosional or depositional. Erosional terraces are produced by wave erosion levelling either rock or pre-existing unconsolidated materials. When the relative level of the lake changes the wave-cut terrace may either be elevated or drowned. Pluvial terraces are more commonly depositional in origin, with sediment supplied by lacustrine processes or by fluvial processes bringing new materials into the lake basin. In this latter process streams draining into the lake build deltas that are graded to a particular lake level, but when that level changes a massive sedimentary accumulation with a nearly flat top is left at the original water elevation.

Abandoned beaches also commonly form pluvial terraces. At a given lake level shoreline erosion dislodges material through wave action, and after a period of transportation along the shore by littoral currents the materials may be deposited in the form of a beach. When lake levels change, old beach lines are left behind as bench-like terraces. The littoral or long-shore currents also distribute materials and deposit them into a variety of bars, hooks and spits. When these features are abandoned as lake levels change, they can also be considered as pluvial terraces, especially if they are positioned along slopes that are near to what was once the perimeter of the lake.

The significance in landscape interpretation of pluvial terraces lies in their role as indicators of past water levels in lake basins. In the basin of ancient Lake Bonneville in the western USA pluvial terraces record scores of levels of a lake that was thirteen times the size of the present one (Flint 1971). In arid and semi-arid regions with closed lake basins (that is, without an outlet) the pluvial terraces and their associated lake levels provide clues to past climatic and hydrological conditions because the lake elevations represent equilibrium levels between input of water by precipitation and streams on one hand and evaporation on the other. In one example Snyder and Langbein (1962) found that a Pleistocene lake existed in a Nevada valley under conditions whereby evaporation was at least 30 per cent less and precipitation at least 66 per cent more than at present. In other cases the lake levels are useful in reconstructing the tectonic history of the lake basin because the terraces represent marker horizons in the progressive displacement of the mountain blocks in relation to the basin blocks.   WLG

References
Flint, R.F. 1971: *Glacial and Quaternary geology.* New York: Wiley.

Snyder, C.T. and Langbein, W.B. 1962: The Pleistocene lake in Spring Valley, Nevada, and its climatic implications. *Journal of geophysical research* 67, pp. 2385–94.
Tylor, A. 1868: On the Amiens gravel. *Geological Society of London quaterly journal* 24, pp. 103–25.

terracette Miniature terrace or ridge extending across a slope, usually normal to the direction of maximum slope. Terracettes are rarely more than 0.5 m wide and deep. Their origin is still a matter for debate. Some may be animal tracks, but as others occur in areas where animals are very rare it would seem that some other mechanisms are involved. They are probably a consequence of *soil* mantle instability on steep slopes.   ASG

terrane 'A mappable structural entity which has a stratigraphic sequence and an igneous, metamorphic and structural history quite distinct from those of adjacent units. Each terrane is separated from its neighbours by a structural break which may take the form of a normal fault, a reverse fault, a wrench fault, or an overthrust' (Barber 1985, p. 116).

During the 1980s the terrane concept became an important one for many earth sciences (Howell 1989). It was developed most notably in the context of the North American Cordillera system where, it was maintained, substantial portions of the Cordillera were 'exotic' blocks and slithers that had 'docked' onto the North American Craton. Subsequently, the importance of terranes has been recognized with respect to areas as diverse as Highland Scotland and the archipelagos of south-east Asia. Terranologists hold that major oceanic belts often consist of 'collages' of fault-bounded crustal and/or lithospheric fragments, of diverse origins and different sizes. However, as Sengòr and Dewey (1991) point out, reactions to the concept range from enthusiastic applause to abusive rejection. They themselves express some cogent doubts, suggesting, for example (p. 6) that 'terranology not only does not go beyond plate tectonics, it takes a backward step; by confusing primary and secondary collage components, it confuses also their genetic implications.' They continue (p. 17) 'the word terrane is a lump term for a number of older and more informative non-genetic (block and sliver) and genetic (fragment, nappe, strike-slip duplex, microcontinent, island arc, suture, etc) terms. Because it is less informative, it is less useful than any of these and also because, historically, the term "terrane" has a number of different meanings it is best avoided. Terrane analysis is neither a new way of looking at orogenic belts, nor a particularly helpful one.'   ASG

Reading and References
Barber, A. 1985: A new concept of mountain building. *Geology today* 1, pp. 116–21.
Howell, D.G. 1989: *Tectonics of suspect terranes*. London: Chapman & Hall.
Sengòr, A.M.C. and Dewey, J.F. 1991: Terranology: vice or virtue? In J.F. Dewey, I.G. Gass, G.B. Curry, N.B.W. Harris and A.M.C. Sengör eds, *Allochthonous terranes*. Cambridge: Cambridge University Press.

terrestrial magnetism The natural magnetism of the earth, also referred to as geomagnetism. The shape of the earth's mainly dipole magnetic field suggests that it is related to circular electrical currents flowing approximately normal to the axis of rotation. These electrical currents may be induced by slow convective movements within the partially molten iron-rich core of the earth, with large-scale eddies producing the regional variations in the main field. Secular changes in the field include the continuous movement of magnetic north, variations in the strength of the field and periodic reversals of polarity.   MAS

Reading
Parkinson, W.D. 1983: *Introduction to geomagnetism*. Edinburgh: Scottish Academic Press: Amsterdam: Elsevier.
Smith, P.J. 1981: The earth as a magnet. In D.G. Smith ed., *The Cambridge encyclopaedia of earth sciences*. Cambridge: Cambridge University Press: New York: Crown Publishers. Pp. 109–23.

territory, animal An area held and defined by an animal or group of animals. Territories may contain resources such as food or mates, or may be a display or rutting stand, and may or may not be fixed in space and/or time. Notable examples of animals exhibiting marked territorial behaviour are tawny owls, which have a fixed exclusive spatial area; and cats, which have fixed exclusive areas in time. Such territorial behaviour must have certain advantages. The concept of 'economic defendability' suggests that an animal should only defend a territory if there is a net benefit, in terms of propagating its genes, in doing so. Single parameters such as food, mates and predation will act together to determine spacing behaviour which will probably always be a compromise moulded by these selective pressures. The consequences of

territorial behaviour may be to limit populations by the establishment of exclusive feeding or breeding rights in an area, the animals excluded then being culled by exposure, starvation and predation. Spacing by territories may also be advantageous to the survival of camouflaged prey.   KEB

Reading
Krebs, C.J. 1985: *Ecology*. 3rd edn. New York and London: Harper & Row.
Krebs, J.R. and Davies, N.B. eds 1984: *Behavioural ecology: an evolutionary approach*. Oxford: Blackwell Scientific.

Tertiary Refers to the first part or period of the CAINOZOIC era, comprising the Palaeocene through to the Pliocene.   ASG

Tethys Ocean An enormous seaway initially formed in the Palaeozoic Era and attaining its maximum development during the Mesozoic, which extended from what is now the Mediterranean eastwards as far as south-east Asia. Beginning about 75 million years ago extensive volcanism, followed by intense deformation and uplift of sediments accumulated in the Tethys Ocean throughout the Mesozoic, led to the formation of the present-day Alpine-Himalayan mountain belt. The opening and closure of the Tethys Ocean (see MESSINIAN SALINITY CRISIS) is currently interpreted in terms of CONTINENTAL DRIFT and the operation of PLATE TECTONICS.   MAS

Reading
Sonnenfield, P. ed. 1981: *Tethys: the ancestral Mediterranean*. Stroudsburg, Pa.: Dowden, Hutchinson & Ross.

thalassostatic A term used to describe river terraces which are produced by aggradation during periods of rising or high sea level and by incision at times of low sea level.

thalweg The line of maximum depth along a river channel. It may also refer to the line of maximum depth along a river valley or in a lake.

thaw lake A shallow, rounded lake occupying a depression resulting from the melting of ground ice (see also THERMO-KARST). Thaw lakes are extremely common on the North American and Siberian Arctic lowlands and are ubiquitous wherever there is a flat lowland with silty alluvium and a high ice content. Most lakes are less than 300 m in diameter and less than 3–4 m deep. Following random exposure of ground ice, a water-filled depression soon develops into a roughly circular lake. Eventually vegetation protects the banks from further thawing and in a matter of 2000–3000 years or so the lake is infilled and the cycle of development complete. Many thaw lakes are elongate in shape with a systematic orientation of the long axis at right angles to the prevailing wind (Carson and Hussey 1963). (See also ORIENTATED LAKE.)   DES

Reading and References
Carson, C.E. and Hussey, K.M. 1963: The oriented lakes of Arctic Alaska: a reply. *Journal of geology* 71, pp. 532–3.

†French, H.M. 1974: *The periglacial environment*. London: Longman.

†Washburn, A.L. 1979: *Geocryology*. London: Edward Arnold.

thermal depression A region of low surface pressure which is generated by strong solar heating on fine days over land in the summer. Thermal depressions can range from the large, seasonal thermal low of the south Asian monsoon to the short-lived lows that sometimes form over England on hot summer days. Their cyclonic inflow tends to be strongest near the surface and to weaken with height since they are shallow features, usually a few kilometres deep.

Fine summer weather in Britain occasionally breaks down in association with an extending thundery thermal depression over France.   RR

thermal efficiency The term was used by C.W. Thornthwaite (1931) in his first classification of climate. The thermal efficiency index (TEI) indicates the plant growth potential of a location and is calculated by summing the 12 monthly values of $(T - 32)/4$, where T = mean monthly temperature in °F. The index ranges from zero on the polar limit of the tundra ('frost climate') to over 127 in the tropics. Six temperature provinces were defined. In this classification Thornthwaite made moisture, in the form of precipitation effectiveness (PEI), the primary factor for a T-E Index of over 31 (taiga/cool temperate boundary). In his second classification of climate, Thornthwaite (1948) used potential evapotranspiration as a measure of thermal efficiency.   DGT

Reading and References
Thornthwaite, C.W. 1931: The climates of North America according to a new classification. *Geographical review* 21, pp. 633–55.

†— 1933: The climates of the earth. *Geographical review* 23, pp. 433–40.

— 1948: An approach toward a rational classification of climate. *Geographical review* 38, pp. 55–94.

thermal equator Variously defined, but most commonly the line which circumscribes the earth and connects all points of highest mean annual temperature for their longitude. Sometimes the seasonal or monthly variation of this line is considered. The annual mean position departs by as much as 20° latitude from the equator. In recent years the term thermal equator or heat equator has also been taken as synonymous with the INTERTROPICAL CONVERGENCE ZONE, the belt along which the thermally driven trade winds of the two hemispheres converge and are forced to rise.   WDS

thermal infrared linescanner An optical remote sensor used to derive thermal images of the earth's surface. Early thermal infrared linescanners have two thermal detectors and record an image onto a photographic film (see diagram), while more recent thermal infrared linescanners are often part of a MULTISPECTRAL SCANNER which records data digitally. With both these systems the physical geographer must choose the waveband to be used, the time of day when the image is to be recorded and the method of approximate calibration. The two thermal infrared wavebands used by thermal infrared linescanners are defined by the two transmitting atmospheric 'windows' located between the wavelengths of 3–5 μm and 8–14 μm. The choice of which of these wavebands to employ depends upon the application. The peak of radiant emission from the earth's surface occurs in the 8–14 μm region and this waveband has proved to be the most popular in physical geography.

Thermal infrared linescanners are used most frequently at night when there is no interference from reflected solar radiation. The usual flying time is just before dawn, when the effects of differential solar heating are minimized. Flights are occasionally made during daylight hours, either because AERIAL PHOTOGRAPHY is to be taken or because it is advantageous to have terrain details enhanced by differential solar heating and shadowing. Owing to the variation in emissivity within a scene and the presence of a thermally variable

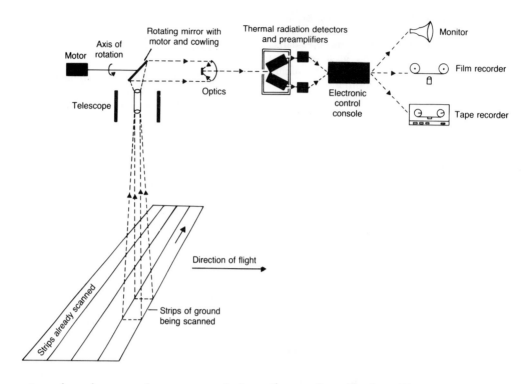

atmosphere between the sensor and the ground, it is not possible to calibrate absolutely the radiant temperature sensed by the detector. Various approximate calibration methods have been employed of which the most accurate involves repeated flights at a range of altitudes. By using the temperature of thermally stable objects of known emissivity as standards, a graph is constructed of temperature versus height. This is extrapolated to the ground surface to give the temperature of enough points to enable calibration of the imagery.

Thermal infrared linescanner data can be presented in image form to be interpreted like an aerial photograph. The interpreter generally uses the images not to map an area but to search for thermal patterns that give a clue to some past, present or future environmental process such as soil movement, frost hollows, water stress in crops, vulcanism or thermal pollution of water.

PJC

Reading
Cracknell, A.P. and Hayes, L.W.B. 1991: *Introduction to remote sensing*. London: Taylor & Francis.
Lillesand, T.M. and Kiefer, R.W. 1987: *Remote sensing and image interpretation*. 2nd edn. New York: Wiley.

thermal pollution The pollution of water by increasing its temperature. Many fauna are affected by temperature so that this environmental impact has some significance. Among the main sources of thermal pollution of stream waters are condenser cooling water released from electricity generating stations, the urban 'heat island effect', reservoir construction, shade removal by deforestation, and changes in the width–depth ratios of channels. The effects of thermal pollution are especially severe at times of low flow.

ASG

Reading
Langford, T.E.L. 1990: *Ecological effects of thermal discharges*. London: Elsevier.
Pluhowski, E.J. 1970: Urbanization and its effects on the temperature of the streams on Long Island, New York. *United States Geological Survey professional paper* 627-D.

thermal wind (v_t) The vector difference between the geostrophic wind at two levels in the atmosphere. It is calculated by subtracting the lower level wind (v_l) from the upper level wind (v_u) and is therefore not a real wind, which would be observed in the intervening layer, but an expression of the shear of the horizontal wind within the layer.

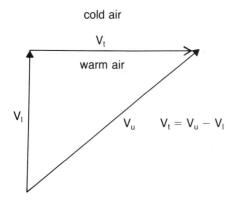

It is termed 'thermal' because its strength and direction are dependent on the thermal pattern of the layer involved. Thus (v_t) is aligned parallel to the layer's mean isotherms with cold air to its left and warm air to its right. Its magnitude or length is proportional to the strength of the layer's mean temperature gradient so that thermal winds are strongest in regions where steep horizontal temperature gradients occur in depth, in the polar front zone for example.

Thermal winds are also therefore parallel to THICKNESS contours and can in fact be calculated precisely with reference to thickness variations across a map.   RR

thermistor A type of semi-conductor resistor which has a high (usually negative) temperature coefficient of resistance. Thus they are often used as temperature sensors or measuring devices. The resistance response to temperature is not linear but can be linearized either in hardware or software in the measuring instrument or subsequent data processing. Despite this disadvantage, it is often preferred to the linear response of the platinum resistance thermometer.   WBW

thermoclasty See INSOLATION WEATHERING.

thermocline A layer of water within a lake or ocean through which the rate of decrease of temperature with depth is much greater than in adjacent layers. It is particularly well developed in tropical oceans where a *permanent* thermocline, up to a few hundred metres in thickness, lies with its upper boundary some 25–150 m below the ocean surface. The temperature gradient may reach $10°C (100\ m)^{-1}$ in the upper part of the layer. Towards midlatitudes the permanent thermocline becomes thicker and less intense with its upper surface as much as 600 m deep; it is entirely absent at latitudes greater than about 60°. The relatively stable stratification of the thermocline layer inhibits interchange between the warm waters of the surface mixed layer and the deeper cold waters forming the main body of the ocean (see GLOBAL OCEAN CIRCULATION).

A *seasonal* thermocline readily develops in both lakes and oceans whose surface temperatures undergo a significant annual variation. Lying close to the surface, its depth and intensity will depend on the amount of summer insolation and extent of turbulent mixing by the wind.   JEA

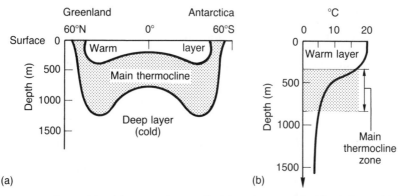

The main temperature characteristics of the oceans: (a) a profile of the main zones of the Atlantic Ocean; (b) the average temperature depth profile for the open ocean in low latitudes.
Source: *Goudie, A. S. 1993*: The Nature of the environment. *3rd ed, Oxford: Blackwell. Figure 2.10.*

Reading
Harvey, J.G. 1976: *Atmosphere and ocean*. London: Artemis Press.

Open University Oceanography Course Team 1989: *Seawater: its composition, properties and behaviour*. Oxford: Pergamon; Milton Keynes: Open University Press.

thermocouple A temperature measuring device. It employs the phenomenon that when a metal wire is interposed in a wire of a different metal then if one junction is heated relative to the other an electromotive force is produced (Seebeck effect). Generally, one junction is used as the sensor, the other being held at 0°C. Various metal combinations produce different emfs and are used for different purposes.   WBW

thermodynamic diagram A chart on which are plotted observations of pressure, temperature and humidity from a given RADIOSONDE ascent. In its simplest form it is a diagram whose vertical axis is pressure and horizontal axis temperature with the former being logarithmic with pressure decreasing upwards and the latter a linear scale with temperature increasing to the right.

Operational thermodynamic diagrams are more complex, however, and are designed so that area on the chart represents energy; the TEPHIGRAM is one example. This is useful when a forester has to assess the likelihood of fog clearance on the basis of the amount of energy required to evaporate it – and whether solar heating at various times of year will be great enough to meet the requirement.

A variety of derived quantities are easily obtained once the temperatures (dry-bulb and dewpoint) have been plotted, for example the POTENTIAL TEMPERATURE and both relative and absolute humidity.

Thermodynamic diagrams are also used in the graphical estimation of the nature and intensity of VERTICAL STABILITY in the atmosphere for the time and place represented.   RR

Reading
Atkinson, B.W. 1968: *The weather business*. London: Aldus Books.

thermodynamic equation Expresses a fundamental relationship in meteorology in which the time rate of change of an air parcel's TEMPERATURE as it moves through the atmosphere is related to both ADIABATIC expansion or compression and to diabatic heating. It is a predictive equation.

In a typical middle latitude disturbance, a cyclone for example, air parcels at middle levels undergo adiabatic temperature changes of about 30 K d^{-1} while the diabatic changes (due to absorption of solar radiation, absorption and emission of infrared radiation, latent heat release, etc) tend to compensate one another and have a net value of about 1 K d^{-1}.   RR

thermograph A meteorological instrument which is housed in a weather screen and provides a continuous record of air temperature. The response of a sensor, for example a bimetallic coil, to fluctuations of air temperature is magnified by a long pen to which it is connected mechanically. The pen traces a line on a daily or weekly strip chart which is fixed to a slowly revolving clockwork-driven drum.   RR

thermokarst Topographical depressions resulting from the thawing of ground ice (Washburn 1979). There are many kinds of thermokarst, including collapsed PINGOS, ground-ice mudslumps, linear and polygonal troughs, THAW LAKES and ALASes. Some thermokarst features result from climatic change, but most relate to minor environmental changes which promote thawing, for example, the shift of a stream channel, natural and man-induced disturbance to tundra vegetation. Thermokarst is sometimes used in a wider sense also to include thermal erosion by flowing water at a temperature above 0°C (French 1974). As such it would include thermo-erosional niches and overhangs and various features associated with slopewash.   DES

Reading and References
†French, H.M. 1974: *The periglacial environment*. London: Longman.

†Washburn, A.L. 1979: *Geocryology*. London: Edward Arnold.

thermoluminescence A technique that has come into use for the dating first of pottery and more recently of Quaternary sediments over time spans of the order of 10^3–10^6 years. In the latter case it employs quartz and dune sand, and is based on the principle that if a sample has been irradiated and subsequently heated, light is emitted as a function of temperature. This is called a glow curve, the intensity of which depends in part on the age of the sample. The technique has great potential but is still in its infancy.   ASG

Reading
Dreimanis, A., Hutt, G., Raukas, A. and Whippey, P.W. 1978: Dating methods of Pleistocene deposits and their problems: I. Thermoluminescence dating. *Geoscience Canada* 5, pp. 55–60.
Wintle, A.G. and Huntley, D.J. 1982: Thermoluminescence dating of sediments. *Quaternary science review* 1, pp. 31–53.

thermopile A radiation detecting device which uses a series of thin wire THERMO-COUPLES to measure radiation incident on the hot junctions of the thermocouples. The cold junctions are shielded and kept at a uniform, measured temperature. This device differs from a bolometer where a blackened platinum foil is heated by the radiation and the increase in resistance is measured.   WBW

thickness The difference in height above mean sea level of two PRESSURE surfaces above a given point. It is proportional to the mean TEMPERATURE of the layer in question so that the larger values indicate relatively warm air and smaller values relatively cold air. Thickness values are obtained from RADIOSONDE observations and are plotted on a base map which commonly depicts the variations for the 1000–500 mb layer. Contours are drawn, say, every 60 m to produce a thickness analysis (which is also effectively an isotherm analysis of the layer involved) which is of importance in synoptic meteorology.   RR

Reading
Atkinson, B.W. 1968: *The weather business*. London: Aldus Books.

Thiessen polygon Defines the horizontal area which is nearer to one rain gauge than to any other in a rain gauge network. The areas of the polygons are used to weight rain-gauge catches when calculating the average areal precipitation. Thiessen polygons are constructed from the perpendicular bisectors of the horizontal projections of straight lines joining adjacent rain gauges. These perpendicular bisectors are extended and joined to leave each rain-gauge site at the centre of a polygon and the area of the polygon is known as its Thiessen weight. The method is widely applied because it is easy to use and makes allowance for the uneven distribution of rain gauges.   AMG

Reading
Damant, C., Austin, G.L., Bellon, A. and Broughton, R.S. 1983: Errors in the Thiessen technique for estimating areal rain amounts using weather radar data. *Journal of hydrology* 62, pp. 81–94.
Thiessen, A.H. 1911: Precipitation averages for large areas. *Monthly weather review* 39, pp. 1082–4.
Wiesner, C.J. 1970: *Hydrometeorology*. London: Chapman & Hall.

tholoid A volcanic cone situated within a large vocanic crater or caldera.

threshold, geomorphological A threshold of landform stability that is exceeded either by intrinsic change of the landform itself, or by a progressive change of an external variable (Schumm 1979). The concept is closely bound up with the view of a landform as a system or part of a system in which there is normally some sort of balance between the morphology and the processes involved.

An *intrinsic* threshold implies that changes can take place within the system without a change in an external variable. An example is a SURGING GLACIER which may exhibit periodic surges although the input of snow remains identical over many decades. In this case there is a build-up of snow and ice to a critical level which causes a sudden change in the process of basal sliding. The sudden transition to a fast mode of flow lowers the ice surface until the glacier reverts to a slow mode of flow and the cycle starts once more. In this particular case the threshold is peculiar to a particular glacier and probably relates to different modes of sliding related to critical basal conditions (Budd 1975). Other normal glaciers may never quite reach the initial threshold allowing fast flow. Other geomorphological examples of intrinsic thresholds involve river channel patterns and river terrace changes (Schumm 1973, 1979).

An *extrinsic* threshold describes an abrupt change which is triggered by a progressive change in an external variable. Well-known examples are the threshold velocities required to set in motion particles of a given size. With a continuous change in velocity the response of river channel bed forms or of aeolian forms will suddenly change. Another example is the progressive depletion of vegetation which may allow the threshold of gully formation and thereby active soil erosion to be crossed suddenly.

The widespread adoption of a systems approach in geomorphology went hand in hand with the concept of some sort of balance between process and form (Chorley and Kennedy 1971). Systems were seen as hunting for an equilibrium and they were analysed on the assumption that they operated in STEADY STATE. Such an approach made it difficult to study landforms which were evolving rapidly over time. Also it was difficult to predict the response of a landform to change without knowledge of its sensitivity or indeed its proximity to a threshold of change, a vital requirement if geomorphology is to be applied successfully. Threshold analysis offers a powerful means of tackling these limitations. It focuses on one of the mechanisms of change and also opens up the prospect of an effective applied geomorphology which can predict and prevent problems before they occur. Viewed in such a light the study of thresholds is a way of developing the potential of a systems approach to geomorphology.

There is currently much interest in the nature of threshold changes and the possible applications of CATASTROPHE THEORY which offers a mathematical means of modelling such discontinuities in a system. There are several different types of discontinuity and the nature of a particular threshold may vary according to the direction of change in a system. Thus the initial water thickness needed to initiate surging glacier velocities may be different from the thickness required to maintain or stop them (Hutter 1982). Another example concerning ephemeral streams is provided by Thornes (1982).   DES

Reading and References
Budd, W.F. 1975: A first simple model for periodically self-surging glaciers. *Journal of glaciology* 14.70, pp. 3–21.

Chorley, R.J. and Kennedy, B.A. 1971: *Physical geography: a systems approach.* Englewood Cliffs, NJ: Prentice-Hall.

Hutter, K. 1982: Glacier flow. *American scientist* 70, pp. 26–34.

Schumm, S.A. 1973: Geomorphic thresholds and complex response of drainage systems. In M. Morisawa ed., *Fluvial geomorphology.* Binghamton, NY: State University of New York Press. Pp. 299–310.

†— 1979: Geomorphic thresholds: the concept and its applications. *Transactions of the Institute of British Geographers* n.s. 4.4, pp. 485–515.

Thornes, J.B. 1982: Structural instability and ephemeral channel behaviour. *Zeitschrift für Geomorphologie*, pp. 233–44.

†— 1983: Evolutionary geomorphology. *Geography* 68, pp. 225–35.

threshold slopes Hillslopes with inclinations controlled by the resistance of their soil cover to a dominant degradational process. Such slopes are recognized within areas of consistent rock and soil types and

erosional processes by nearly uniform inclinations of characteristic hillslope units. It has been postulated that these units are at the maximum inclination, for temporary stability, permitted by the soil strength and pore water pressures within the soil cover.

Three characteristic maximum hillslope angles have been identified for areas prone to erosion by landsliding processes (Carson 1976): (1) a frictional threshold slope angle where the soil is a dry rock rubble and the angle of inclination equals the angle of plane sliding friction of the rubble; (2) a semi-frictional threshold slope angle for cohesionless soils where the water table can rise to the surface and seepage is parallel to it; the slope angle then approximates to half that of the effective angle of plane sliding friction of the soil; and (3) an artesian condition in which water flows out of the soil and the slope angle is less than that of case (2).

The concept of threshold slopes applies primarily to straight slope segments between upper convexities and lower concavities. In areas of rapid uplift and incision of stream channels slope angles may be steeper than threshold angles until landsliding reduces the slope angle to the threshold angle. Threshold slopes will eventually be eliminated by creep, wash and other processes: they are consequently temporary features of a landscape dominated by landsliding.   MJS

Reading and Reference
Carson, M.A. 1976: Mass-wasting, slope development and climate. In E. Derbyshire ed., *Geomorphology and climate.* London: Wiley. Pp. 101–30.

†— and Kirkby, M.J. 1972: *Hillslope form and process.* Cambridge: Cambridge University Press.

†Selby, M.J. 1993: *Hillslope materials and processes.* 2nd edn. Oxford: Oxford University Press.

throughfall The net rainfall below vegetation cover excluding STEM FLOW. Throughfall comprises both precipitation which falls straight through the vegetation canopy and precipitation which has been intercepted by the vegetation but then drips onto the ground.   AMG

Reading
Ford, E.D. and Deans, J.D. 1978: The effects of canopy structure on stemflow, throughfall and interception loss in a young Sitka spruce plantation. *Journal of applied ecology* 15, pp. 905–17.

Sopper, W.E. and Lull, H.W. eds 1967: *International symposium on forest hydrology.* Oxford and New York: Pergamon.

throughflow (or subsurface flow) Down-slope flow within the soil. Where there are well-defined aquifers downward percolation is commonly rapid enough to prevent appreciable throughflow. Where the bedrock is not highly permeable, lateral flow within the soil is, for many sites, the most effective form of downslope flow because the soil is more permeable and more porous than the bedrock. This is particularly so in open-structured soils like those beneath many mature woodlands. Infiltrated water percolates downward until it meets an impeding horizon, where it is diverted laterally as saturated throughflow. Impedance may be due to general saturation below, or to the reduction in permeability with depth which is a normal feature of soils, with the greatest reduction near the base of the 'A' horizon.   MJK

Reading
Kirkby, M.J. ed. 1978: *Hillslope hydrology*. Chichester: Wiley.

throw of a fault The vertical displacement of a fault.

thrust A low-angle reverse fault. Also a fault occurring on the overturned limb of a fold.

thufur A soil hummock found in periglacial environments.

Reading
Schunke, E. and Zoltai, S.C. 1988: Earth hummocks (thufur). In M.J. Clark ed., *Advances in periglacial geomorphology*. Chichester: Wiley. Pp. 231–45.

thunderstorm Storm accompanied by lightning and therefore thunder, rarely heard more than about one minute after the flash. Separation of electrical charge probably demands, at some stage, water droplets going down colliding with ice crystals going up and reinforcing an initial electrostatic field. More comprehensive theories are numerous, elaborate and uncertain. The name is often used for any severe convective storm of middle latitudes. Tropical storms are rarely accompanied by thunder.   JSAG

Reading
Lane, F.W. 1966: *The elements rage*. Newton Abbot: David & Charles. Ludlam, F.H. 1980: *Clouds and storms*. University Park, Pa.: Pennsylvania State University Press.

tidal currents The periodic horizontal motions of the sea, generated by the gravitational attraction of the moon and sun. They are linked hydrodynamically with tidal changes of sea level and have similar spring to neap modulations. Places which have a large TIDAL RANGE also have large tidal currents. Large tidal currents may also occur where tidal ranges are small, for example near amphidromes, or through narrow straits which separate two regions having different tidal regimes. Typically, tidal currents on the continental shelf have speeds of 1 m s⁻¹.   DTP

tidal palaeomorph 'Oversized, often meandering valleys thought to be former tidal channels created when sea level was higher than it is now. Such valleys have underfit rivers in them' (Geyl 1985, p. 1).   ASG

Reference
Geyl, W.F. 1985: Tidal paleomorphs in Eastern Virginia. *Research papers in geography, University of Newcastle NSW*, no. 29.

tidal prism The total amount of water that flows in or out of a coastal inlet with the rise and fall of the tide, excluding any freshwater discharges. For any given period it is the product of the mean and the high- and low-water surface areas of the bays behind the inlet entrance and the TIDAL RANGE in each segment.   ASG

tidal range The vertical distance between tidal low water and high water. It varies between spring and neap tides over a period of 14 days. Ocean tidal ranges are usually less than a metre, but ranges increase as the tides spread onto the shallower continental shelves. Here typical ranges are 2–5 m, but there are many local variations. In exceptional cases where large spring tides excite local hydrodynamic resonance, as in the Minas Basin of the Bay of Fundy, ranges in excess of 15 m can occur. Tidal amplitude is a half of the tidal range.   DTP

tides The regular movements of the oceans and seas, generated by the gravitational attraction of the moon and sun. They are most easily observed as changes in coastal sea levels, but the associated horizontal currents are equally important for mariners. There are also tidal movements of the atmosphere and of the solid earth which are not apparent to the casual observer.

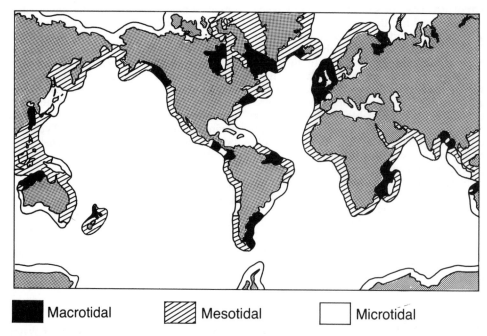

Macrotidal | Mesotidal | Microtidal

Distribution of microtidal (<2 m), mesotidal (2–4 m) and macrotidal (>4 m) ranges.
Source: *J.L. Davies 1964: A morphogenic approach to the world's shorelines.* Zeitschrift für Geomorphologie 8, pp. 127–42.

On average the gravitational attraction between the earth and moon balances the orbital centrifugal force. On the side of the earth nearest to the moon the gravitational force is slightly greater than the centrifugal force, whereas on the opposite side it is weaker. This gives two tidal maxima each day (semi-diurnal tides) as the earth rotates about its axis; however, the times of the maximum lunar tides are later by an average of 52 minutes each day because of the advance of the moon on its orbit. Changes of declination cause daily (diurnal) tides. Longer period tides are generated by varying lunar and solar distances.

Solar tides have average amplitudes which are 46 per cent of the lunar tides, but their maximum values at a site occur at the same times each solar day. Every 14 days, at new and full moon, when the lunar and solar tidal maxima coincide, the combined spring tidal range is large. Between, small neap tide ranges occur when the solar tides tend to cancel the lunar tides.

Tides calculated directly from gravitational theory, the equilibrium tides, are not observed in the ocean because of several additional effects: these include the land boundaries which prevent their westward movement, the deflection of tidal currents

caused by the earth's rotation, the tidal movements of the solid earth, and the natural periods of oscillation of the oceans and shelf seas. If the natural period is near to a period in the tidal forcing, large resonant tides are produced. Because the oceans have natural periods close to 12 hours, the observed tides are predominantly semi-diurnal.

From the oceans the tides spread to the adjacent continental shelf regions. Here they are modified by amplification, by local resonances and by reflections at land boundaries. A reflected tidal wave can interfere with an incoming wave to produce zero tidal range at a distance of a quarter wavelength from the reflecting coast; because of the earth's rotation the tides circulate around an amphidromic point of zero tidal amplitude. Tidal energy is eventually dissipated by the frictional drag of the sea bed in shallow water. Schemes to use tidal power have been developed over many centuries. Early examples of tidal mills are found in East Anglia and along the coast of New England. Large schemes have been proposed for the Bristol Channel and the Bay of Fundy. The first scheme to use modern technology is La Rance, near

Saint Malo in France, which began operating in 1966.   DTP

Reading
Cartwright, D.E. 1977: Oceanic tides. *Reports on Progress of Physics* 40, pp. 665–708.
Pugh, D.T. 1987: *Tides, surges and mean sea level*. Chichester: Wiley.

till A Scottish word, popularly used to describe a coarse, bouldery soil, which was adopted to describe unstratified glacier deposits by A. Geikie (1863). The term is now understood to refer to the sedimentary material deposited directly by the action of a glacier. As such it supplants the former, oversimplistic term boulder clay.

Till covers the landsurface in many former land-bound sectors of mid-latitude ice sheets in the former USSR, northern and north-western Europe, Canada and the northern USA. It has been the focus of much interest among geomorphologists and geologists both because of its importance in the understanding of glacier activity and because of its engineering implications. Till has spawned an enormous literature including several significant symposium volumes (Goldthwait 1971; Legget 1976; *Boreas* 1977).

The table shows a recent classification of till based on the processes of debris release and the position of the debris deposition. Following Lawson (1981) the processes are subdivided into those which are primary and influence the nature of the sediment directly and those which are closely related secondary processes which modify the sediment.

Meltout till forms by the direct release of debris from a body of stagnant, debris-rich ice by melting of the interstitial ice. When it occurs subglacially, the debris is let down onto the underlying bed without modification. Thus it retains a fabric (see TILL FABRIC ANALYSIS) inherited from transport in the ice, with preferentially orientated pebbles. The deposit frequently consists of structureless, pebbly, sandy silt, discontinuous laminae, and bands of sediment which may be deposited over large clasts. These latter characteristics are the direct result of the melting of debris-bearing REGELATION ice. Meltout also occurs at the surface, but here secondary processes of flow and slumping may affect the sediment.

In cold, arid environments, such as Victoria Land in Antarctica, the interstitial

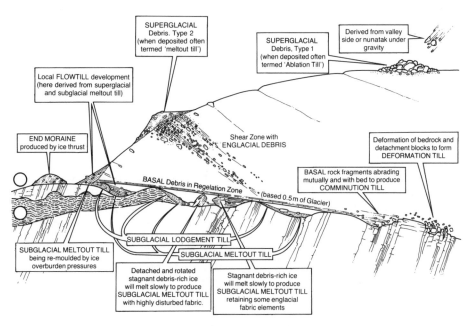

Till: *Generalized relationships of ice and debris in a temperate glacier – vertical scale greatly exaggerated. Two cases of end moraine formation shown: A, glacier piles up ridge as it slides over bedrock; B, glacier rests on thick saturated till and produces 'squeeze-up' moraines.*
Source: *E. Derbyshire, K.J. Gregory and J.R. Hails 1979:* Geomorphological processes. *London: Butterworth. Figure 5.56.*

Classification of till based on the processes of
debris release and position in relation to the glacier

	Primary	Secondary
Subglacial	Meltout	Deformation
	Sublimation	Settling through
	Lodgement	standing water
Supraglacial	Meltout	Sediment flow
	Sublimation	Gravitational
		slumping
		Settling through
		standing water

Source: Lawson 1981.

ice may be lost by sublimation rather than by melting (Shaw 1979). This second primary process produces *sublimation* till, which also retains textural and structural characteristics inherited from glacial transport.

The third primary depositional process is subglacial and produces *lodgement* till. Particles lodge when the frictional resistance with the bed exceeds the drag of the moving ice. Since the tills are deposited beneath the weight of overlying ice, they tend to be characterized by a high degree of compaction, high SHEAR STRENGTH and low porosity. Foliation, slip planes, fine lamination and horizontal joints are also typical. The fabric of the till usually shows a preferred orientation with elongated pebbles parallel to the direction of ice flow. Occasionally, large boulders may have been moulded into minor ROCHE MOUTONNÉE shapes as a result of overriding during lodgement (Sharp 1982).

The secondary processes are intimately associated with till deposition. *Deformation* till has been deformed by glacier movement after its primary deposition. The most common occurrence is beneath glaciers where lodgement occurs. Such a layer of deformation till is massive, relatively poorly consolidated and, when saturated, easily collapses beneath the weight of a person! This is the 'mud' surrounding so many glaciers, and it is commonly up to 70 cm thick. Deformation till and lodgement till are two types of subglacial till which have in the past been called *ground moraine*.

Sediment flows and *gravitational slumps* are the result of surface processes which modify supraglacial debris. Sediment flows occur when fine-grained debris is exposed on the surface of a glacier (see COMPRESSING FLOW). They are frequently called *flow tills*

(Hartshorn 1958; Boulton 1968). The till becomes saturated and flows down the local ice slope. Lawson (1982) studied such processes on the Matanuska Glacier in Alaska and recognized four main types of sediment flow, depending on the proportion of water they contained and the amount of sorting present. On Matanuska Glacier sediment flows account for 95 per cent of till deposition. Gravitational slumps occur when the surface debris contains insufficient fine material to flow. Instead it is stable on ice slopes up to *c*.35° but further steepening causes it to slump down the ice slope.

A further secondary process is settling through a water column. This is the situation common around the Antarctic where basal debris melts out from the bottom of an ICE SHELF or ICEBERG and falls to the sea floor. Anderson *et al.* (1980) distinguish such a glaciomarine deposit from normal basal tills on several criteria. The most diagnostic characteristics are the horizontal randomly orientated pebble fabrics of the dropstones and the distinctive marine microfauna.   DES

Reading and References
Anderson, J.B., Kurtz, D.D., Domack, E.W. and Balshaw, K.M. 1980: Glacial and glacial marine sediments of the Antarctic continental shelf. *Journal of geology* 88.4, pp. 399–414.

†*Boreas* 1977: *A symposium on the genesis of till. Boreas* 6.2.

Boulton, G.S. 1968: Flow tills and related deposits on some Vestspitzbergen glaciers. *Journal of glaciology* 7.51, pp. 391–412.

Geikie, A. 1863: On the phenomena of the glacial drift of Scotland. *Transactions of the Geological Society of Glasgow* 1, pp. 1–190.

†Goldthwait, R.P. ed. 1971: *Till, a symposium.* Ohio State University Press.

Hartshorn, J.H. 1958: Flowtill in south-eastern Massachusetts. *Bulletin of the Geological Society of America* 69, pp. 477–82.

†Lawson, D.E. 1981: Distinguishing characteristics of diamictons at the margin of the Matanuska Glacier, Alaska. *Annals of glaciology* 2, pp. 78–84.

— 1982: Mobilization, movement and deposition of active subaerial sediment flows, Matanuska Glacier, Alaska. *Journal of geology* 90, pp. 279–300.

Legget, R.F. ed. 1976: *Glacial till: an interdisciplinary study.* Royal Society of Canada special publication 12.

Sharp, M.J. 1982: Modification of clasts in lodgement tills by glacial erosion. *Journal of glaciology* 28.100, pp. 475–81.

Shaw, J. 1979: Tills deposited in arid polar environments. *Canadian journal of earth sciences* 14.6, pp. 1239–45.

till fabric analysis Measurement of the direction and dip of elongated stones in glacial till (see FABRIC). Elongated stones in lodgement TILL and subglacial melt-out till tend to be deposited with their long axes parallel to the direction of ice flow.   DES

Reading
Andrews, J.T. 1971: Techniques of till fabric analysis. *British Geomorphological Research Group technical bulletin* 6.
Glen, J.W., Donner, J.J. and West, R.G. 1957: On the mechanism by which stones in till become orientated. *American journal of science* 255, pp. 194–205.
Holmes, C.D. 1941: Till fabric. *Bulletin of the Geological Society of America* 52, pp. 1299–354.

tillite A consolidated sedimentary rock formed by LITHIFICATION of glacial till, especially pre-Pleistocene till.

timberline (or treeline) The upper (altitudinal) or polar (latitudinal) margins of tree growth. Timberlines may be sharply defined or diffuse, responding to increasing climatic constraints through lower temperatures and greater exposure. Generally the trees thin out and become progressively smaller and more stunted, having KRUMMHOLZ characteristics. In the northern hemisphere, the timberline is approximately coincident with the Arctic Circle. On mountains, the timberlines generally decrease in altitude with distance from the equator, where they are found at around 3300–4000 m, but the weight varies greatly with local conditions.   PAF

tjaele A Russian term equivalent to permafrost.

tolerance The ability of organisms to withstand environmental conditions. Plants and animals within a particular environment have limits of tolerance beyond which they cannot exist. These tolerance limits reflect a range between minimum and maximum values for essential materials such as heat, light, water and nutrients, which are necessary for growth and reproduction. The relative degree of tolerance is expressed by a series of terms that utilize the prefix 'steno' meaning narrow and 'eury' meaning wide. For example, stenothermal and eurythermal refer to narrow and wide temperature tolerance respectively.   ARH

Reading
Odum, E.P. 1971: *Fundamentals of ecology.* 3rd edn. Philadelphia and London: W.B. Saunders. Ch. 5, pp. 106–39.

tombolo A bar or spit connecting an island to the mainland or to another island.

toposequence A sequence or grouping of related soils that differ from each other on account of their topographical position. (See also CLINOSEQUENCE and CATENA.)

toppling failure A type of slope failure (usually in rocks) characterized by overturning of columns of rock as they fall from a cliff. The mode of failure is common where bedding planes and joints are inclined to the valley side but dip downwards only to a maximum of around 35°. Beyond this value sliding is more common, in which case SLAB FAILURE results. Triggering of falls may be due to water pressures in the joints or, for small blocks, ice wedging.   WBW

Reading
de Freitas, M.H. and Watters, R.J. 1973: Some examples of toppling failure. *Geotechnique* 23, pp. 495–514.

topset beds Horizontal sedimentary layers laid down on the surface of inclined beds, as in deltaic environments and aeolian sands.

tor 'An exposure of rock *in situ*, upstanding on all sides from the surrounding slopes . . . formed by the differential weathering of a rock bed and the removal of the debris by mass movement' (Pullan 1959, p. 54). This definition is basically the same as that used by Caine (1967), 'residuals of bare bedrock usually crystalline in nature, isolated by freefaces on all sides, the result of differential weathering followed by mass wasting and stripping'. The definition employed by Linton (1955, p. 476) introduced some genetic implications that have not been accepted as desirable: 'a residual mass of bedrock produced below the surface by a phase of profound rock rotting effected by groundwater and guided by joint systems, followed by a phase of mechanical stripping of the incoherent products of chemical action'. Some workers would not recognize a stage of prior deep chemical weathering as being a *sine qua non* for tor development, pointing to the possible role of physical disintegrative processes or the concurrent operation of weathering and stripping processes on rock of variable strength (e.g. Palmer and Radley 1961).

Although Linton's definition has problems, his description of what tors are like is indeed graphic:

Lithologies associated with British tors

Area	Lithology
Dartmoor (Devon)	Granite
Exmoor (Valley of Rocks)	Sandstone
Weald (Kent, Sussex)	Ardingly sandstone (Lower Tunbridge Wells sandstone)
Charnwood Forest (Leics.)	Granite, microdiorite and hornstone
Pennines (Derbyshire)	Millstone and gritstone
Derbyshire	Dolomite
Bridestones (NE Yorks.)	Passage beds of the Corallian (silicified grits and calcareous beds)
Cheviot Hills (Northumberland)	Granite
Stiperstones (Shropshire)	Quartzite
Pembrokeshire (Wales)	Flinty rhyolite
Prescelly Hills (W. Wales)	Dolerite
Cairngorms (Scotland)	Caledonian granite
Ben Loyal (Scotland)	Syenite
Caithness (Scotland)	Arkose of the old red sandstone

Source: Goudie and Piggott 1981.

They rise as conspicuous and often fantastic features from the long swelling skylines of the moor, and dominate its lonely spaces to an extent that seems out of all proportion to their size. Approach one of them more closely and the shape that seemed large and sinister when silhouetted against the sunset sky is revealed as a bare rock mass, surmounted and surrounded by blocks and boulders; rarely will the whole thing be more than a score or so feet high. (Linton 1955, p. 470)

Though he was talking about the granite tors of Dartmoor, south-west England, comparable forms occur on a wide range of rock types elsewhere in Britain.   ASG

References
Caine, N. 1967: The tors of Ben Lomond, Tasmania. *Zeitschrift für Geomorphologie* NF 4, pp. 418–29.
Goudie, A.S. and Piggott, N.R. 1981: Quartzite tors, stone stripes and slopes at the Stiperstones, Shropshire, England. *Biuletyn Peryglacjalny* 28, pp. 47–56.
Linton, D.L. 1955: The problem of tors. *Geographical journal* 121, pp. 470–87.
Palmer, J. and Radley, J. 1961: Gritstone tors of the English Pennines. *Zeitschrift für Geomorphologie* NF 5, pp. 37–51.
Pullan, R.A. 1959: Notes on periglacial phenomena: tors. *Scottish geographical magazine* 75, pp. 51–5.

toreva blocks Large masses of relatively unfractured rock that have slipped down a cliff or mountain side, rotating backwards towards the cliff in doing so.

tornado A violent rotating storm with winds of 100 m s^{-1} circulating round a funnel cloud some 100 m in diameter which includes aerial debris such as doors, bushes and frogs. It is associated with violent cumulonimbus of right-hand parity, identifiable on radar, and is a menace to the mid-western USA. Houses with closed windows may explode, due to sudden imposition of low external pressure.   JSAG

Reading
Lane, F.W. 1966: *The elements rage*. Newton Abbot: David & Charles.

A rare twin funnel tornado photographed near Elkhart, Indiana, USA on 11 April 1965. Tornadoes are a major natural hazard in the mid-west of the USA.

torrent A swift, turbulent flow of water or lava.

tower karst Residual limestone hills rising from a flat plain. They are distinguished from KEGELKARST in that the hills have near vertical slopes and are separated from each other by an alluvial plain or swamp. The extremely steep sides of the hills may be caused by marginal solution or fluvial erosion at the edge of the swampy plains. PAB

Classic tower karst rising above level plains in China. Although lithological controls are important in explaining distribution patterns of such features at the regional scale, some broad climate control is evidenced by their preferential development in lower latitudes.

Reading
McDonald, R.C. 1976: Hillslope base depressions in tower karst topography of Belize. *Zeitschrift für Geomorphologie* NF supplementband 26, pp. 98–103.

tracers Substances introduced into streams or groundwater in order to determine the subterranean course or direction of flow.

tractive force The drag force exerted when a fluid moves over a solid bed. In UNIFORM STEADY FLOW in open channels this force is the effective component of the gravity force acting on the water body in the direction of flow. For a reach of length L, cross-section area A and slope S, this is $\gamma_f ALS$, where γ_f is the unit weight of water. The average value of tractive force per unit of bed area (the mean bed shear stress, τ_0) is:

$$\tau_0 = \gamma_f ALS/PL = \gamma RS \approx \gamma_f dS$$

if the wetted perimeter is P. The simplification follows because the HYDRAULIC RADIUS $R = P/A$, and is approximated by depth d in wide channels. Channel perimeter sediments have a maximum permissible tractive force or shear stress, the critical or threshold shear stress, τ_{0c}. If the flow exerts a stress in excess of this, entrainment of erosion (traction) occurs at a rate dependent on the excess stress (see DU BOYS EQUATION). It is theoretically possible to design channels to carry clear water with no sediment transport by ensuring that the perimeter sediments are everywhere at or below the threshold state: this is the tractive force theory of channel design (Richards 1982, pp. 281–6). KSR

Reference
Richards, K.S. 1982: *Rivers: form and process in alluvial channels*. London and New York: Methuen.

trade winds Winds with an easterly component which blow from the subtropical high pressure areas around 30° of latitude towards the equator. Although only surface winds, they are a major component in the general circulation of the atmosphere as they are the most consistent wind system on earth. Together the north-east and south-east trade winds occupy most of the tropics. BWA

trafficability The capacity of a soil or a particular type of terrain to permit the movement of vehicles.

translational slide Occurs where the failure of the soil or rock is along planes of weakness (such as bedding planes or joints) which are approximately parallel to the ground surface. This term is often used in a broad sense and can include a variety of types of mass movement such as mudflows, debris flows, etc. Solid rock movements such as wedge failures can also be translational. WBW

transpiration Plant perspiration, or the loss of water vapour mainly from the cells of the leaves through pores (stomata) but also from the leaf cuticle and through lenticels of the stem. The cooling effect is secondary to the fundamental role of the transpiration, bringing a stream of water, and dissolved mineral nutrients, from the root hairs through the stem vessels (xylem) to the leaves, which is maintained by the vapour pressure gradient of the transpiring cell surfaces. The velocity of the transpiration stream varies from 1–2 m h^{-1} in coniferous trees to 60 m h^{-1} in some herbs. The main control mechanism on transpiration is the opening and closing of the stomatis, induced by osmotic pressure changes consequent upon water balance changes due to high temperatures or other factors. KEB

Reading
Etherington, J.R. 1982: *Environmental and plant ecology*. 2nd edn. Chichester: Wiley.

transportation slope A type of slope on which at each point the amount of material brought in from upslope is more or less balanced by the amount of material carried away downslope. Thus there is neither a net loss or a net gain of land.

transverse dune An asymmetric sand ridge (leeward slope, steep; windward slope, gentle) in which the long dimension is normal to the dominant wind direction. It has one slipface.

transverse rib A narrow ridge of well-imbricated pebbles and cobbles, which lies transverse to streamflow direction. Transverse ribs normally form a series of regularly spaced ridges, apparently associated with the development of antidune breaking waves. JM

treeline The upper altitudinal limit to which trees can grow, and this depends on such factors as latitude, aspect, exposure and soil type. In arid areas there may be a lower treeline, the position of which is largely controlled by moisture availability, a commodity which tends to become scarcer at lower altitudes.

triaxial apparatus A device used to measure the SHEAR STRENGTH of a soil according to the MOHR–COULOMB EQUATION criterion. Unlike the SHEAR BOX, failure is not along a predetermined line but takes place in a cylinder loaded axially across the ends by a compressive stress. In its simplest form (for cohesive soils only) the curved surfaces of the cylinder are at atmospheric pressure and failure produced by the axial compression gives an 'unconfined' strength. More usually, a surrounding pressure is applied to the sample, tested in a water-filled cell. This surrounding pressure is changed so that three tests give failure at different axial loads. A plot involving the applied load and cell pressure is a Mohr circle construction from which the strength of the soil can be derived. In clays a shear plane can develop in the soil but granular materials tend to bulge and a given STRAIN value is used to indicate failure (often 10 per cent). WBW

Transverse dunes with steep lee side slip faces at Maspalomas, Grand Canary.

Reading
Lambe, T.W. and Whitman, R.V. 1981: *Soil mechanics*. New York: Wiley.

Trombe's curves A graph portraying the relationship between the calcium content of saturated solutions at different temperatures and the pH. As pH falls from alkali to acid conditions Trombe suggests that there is a curvilinear increase in the amount of calcium able to be held in solution. These curves are now superseded by more recent work.   PAB

Reading
Sweeting, M.M. 1972: *Karst landforms*. London: Macmillan.

trophic levels Literally, 'nourishment' or feeding levels within a biological system, which represent stages in the transfer of energy through it: the concept links in with that of the FOOD CHAIN or food web. Thus, in the grazing food chain, all producer organisms (green plants, blue-green and other algae, phytoplankton, etc.) are placed in the first trophic level, which contains the maximum store of energy which is available for use in any given system; the second is comprised of HERBIVORE consumers, the third of CARNIVORE consumers, the fourth of top carnivore consumers, and so on.

The amount of energy present in any trophic level is determined by the constraints set by the first two laws of thermodynamics, the first of which (the law of the conservation of energy) may be expressed by:

$$\Delta E = Q + W$$

where ΔE refers to changes in the internal energy of that level, Q represents the heat given off by it (mainly, the heat associated with RESPIRATION), and W is the work done within it (i.e. the energy retained in cells for growth). The second states that most energy will eventually degrade to heat. The application of the theoretical relationships to real-life situations may be seen by reference to the classic series of experiments conducted at Silver Springs, Florida, by H.T. Odum (1957), a summary of the results of which is given in the diagram. In this small aquatic system the amount of energy (E) fixed, mainly through photosynthesis, in the first tropic level amounted to 20,810 kcal m^{-2} year^{-1}; respiration (Q) was equivalent to the loss of 11,977 kcal m^{-2} year^{-1}, and a small quotient went into a subsidary, DECOMPOSER food chain. Some 8428 kcal m^{-2} year^{-1} were stored at this level (W), this being its NET PRIMARY PRODUCTIVITY or, in other words, the quantity available for passage to the second trophic level, at which point a new 'first law' balance comes into operation. The process is then repeated at subsequent trophic levels, causing a further diminution in the amount of available energy in each, until eventually all the energy within the system has been utilized. Normally, the amount of usable energy passed from one trophic level to another is of the order of 10 per cent of net production, but reference to the diagram shows that this can vary widely even in the same system, and it is known to range generally from 5 per cent to 80 per cent (Kormondy 1976). Marine systems are customarily more efficient at transferring energy between trophic levels than are land systems. Both result in a characteristic 'pyramid' of energy flow.

In view of the above the number of trophic levels which any biological system can support is cleary limited. The usual maximum on land is four or five. But in marine environments, where the ratio of plant to animal BIOMASS is much more balanced, and energy transfer efficiencies are greater, there may be up to seven. Unstable and unpredictable communities tend in general to have fewer trophic levels than do stable ones (May 1975); and so also do polluted ones. (See also BIOLOGICAL PRODUCTIVITY; ECOLOGICAL ENERGETICS.)   DW

References
Kormondy, E.J. 1976: *Concepts of ecology*. 2nd edn. Englewood Cliffs. NJ: Prentice-Hall.

To next trophic level (k cal/m²/yr)					Respiratory loss (k cal/m²/yr)
Exported	Retained				
8	0		21	(top Carnivores)	13
46	21		383	(Carnivores)	316
1555	383		8428	(Herbivores)	6490
405	8428		20810	(Producers)	11977

May, R.M. 1975: *Stability and complexity in model ecosystems*. Princeton, NJ: Princeton University Press.

Odum, H.T. 1957: Trophic structure and productivity of Silver Springs, Florida. *Ecological monographs* 27, pp. 55–112.

tropical cyclones A collective term which refers to the intense cyclonic vortices that are observed principally across tropical oceans and exhibit maximum sustained surface winds of 33 m s^{-1} (64 kn). They are known as hurricanes in the Caribbean, typhoons in the north-west Pacific and cyclones in the Indian Ocean. They are formed from pre-existing disturbances most frequently in the 10°–15° latitude band and are known to be favoured by the following conditions:

1 Strong low-level cyclonic relative VORTICITY.
2 A reasonably large CORIOLIS FORCE in order for an organized circulation to develop.
3 A small difference between the disturbance velocity and the vertical profile of the horizontal wind of the large-scale surrounding current (this small 'ventilation' aids the concentration of heating in a vertical column).
4 Sea-surface temperature warmer than 27°C.
5 An unstable LAPSE RATE from the surface to middle levels.
6 High humidity at mid-tropospheric heights.

Tropical cyclones are characterized by cyclonic inflow (counterclockwise in the northern hemisphere) which is strongest in the lowest 2 km, with inward spiralling cloud bands hundreds of kilometres long and ten kilometres or so wide. These bands converge towards the deep wall cloud which surrounds the eye of the system and is the zone of strongest WINDS (up to 100 m s^{-1} in extreme cases) and heaviest PRECIPITATION (50 cm d^{-1} in vigorous cyclones). The eye is a circular region in which the air subsides, winds decrease and precipitation ceases and it is typically 10–15 km in diameter.

These systems are classically asymmetric in plan view because one flank (right in the northern and left in the southern hemisphere) is characterized by stronger flow as a result of the compounding effect of the large scale steering current when 'added' to the disturbance's wind pattern. The elevated water levels of storm surges, for example, in the Gulf of Mexico and Bay of Bengal, are associated with a cyclone's low central pressure (typically 920–950 mb) and damage is also caused by the high winds and heavy precipitation which commonly occur in tropical cyclones.

The strength of the inflow decreases to a minimum at middle levels (near 6 km) where the ascent is strongest while at the top of the circulation (above 9 km) air spirals out anti-cyclonically (clockwise in the northern hemisphere) to balance roughly the mass flowing in at low levels. The high level outflow is marked by an extensive shield of cirrus which is clearly visible from weather satellites. In general, tropical cyclones are around 650 km in diameter and thus substantially smaller than middle latitude cyclonic disturbances.

For the period 1958–77 an annual average of 54 tropical cyclones were observed in the northern and 24 in the southern hemispheres. Of this total 33 per cent occurred in the north-west Pacific, 17 per cent in the north-east Pacific, 13 per cent in the Australasian area, 11 per cent in the north-west Atlantic, 10 per cent in the South Indian Ocean and 8 per cent in both

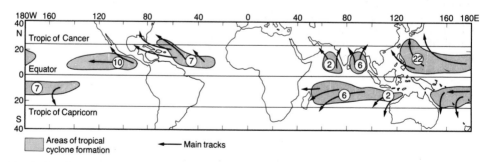

Tropical cyclones: *Distribution, frequency and movement (hurricanes). Encircled numbers indicate approximate number in each area per year.*
Source: *A.S. Goudie 1984:* The nature of the environment. *Oxford and New York: Basil Blackwell.*

the South Indian and Pacific Oceans. The time of maximum frequency coincides with the period of highest sea-surface temperatures: 72 per cent occur from July to October in the northern hemisphere (21 per cent in September) and 68 per cent occur from January to March in the southern hemisphere (25 per cent in January). RR

Reading
Pielke, R.A. 1990 *The hurricane*. London: Routledge.

tropical forest Defined literally, an area lying between the lines of the tropics, with trees as the dominant life form. In practice, similar forests extend outside the tropics. There are three broad groups of tropical forest: evergreen, deciduous and mangrove. Evergreen forests are the most widespread, characterized by leaf-exchange mechanisms and little or no bud protection. They occur in lowlands below around 1000 m, in mountain and cloud formations and in water-saturated areas. Deciduous forests range from areas with predominantly evergreen subcanopy trees to dominantly deciduous trees throughout the vertical structure of the forest. Mangrove forests have a range of adapations to saline conditions and waterlogging. PAF

Reading
Golley, F.B., Leith, H. and Werger, M.J.A. eds 1982: *Tropical rain forest ecosystems*. Amsterdam: Elsevier.

Longman, K.A. and Jenik, N. 1987: *Tropical forest and its environment*, 2nd edn. London: Longman.

Mather, A.S. 1990: *Global forest resources*. London: Belhaven Press.

Whitmore, T.C. 1975: *Tropical rainforests of the Far East*. Oxford: Clarendon Press.

tropical meteorology The study of the atmosphere and its interaction with the earth's surface in tropical latitudes. In the atmospheric sciences the tropics are usually considered to extend poleward from the equator in both hemispheres to the fluctuating boundary between the middle latitude westerlies and the subtropical easterlies. This boundary coincides with the descending branch of the HADLEY CELLS and with the centre of the subtropical highs and usually lies at latitudes between 25° and 40° in both hemispheres, being farthest poleward during the hemispheric summer.

The atmospheric circulation in the tropics is more complex than it was once thought to be, although still simpler and less chaotic than the flow in higher latitudes. The classical theory tacitly assumes that solar

RADIATION acts like the steady flame of a laboratory experiment in heating the tropics. Although this is true when the combined earth–atmosphere system is considered, it is not true for the atmosphere alone. The atmosphere emits more radiation than it absorbs. The resulting radiative cooling is strongest in the upper tropical TROPOSPHERE where it exceeds $1.7°C\ d^{-1}$. This process alone would soon wipe out the temperature difference between low and high latitudes and greatly weaken the atmospheric circulation.

The warmth of the tropical atmosphere is maintained primarily by CONDENSATION. The fuel is water vapour evaporated from the subtropical oceans and drawn into the tropical furnace by the trade winds. Condensation of water vapour not only provides heat for the tropical atmosphere but also, by preserving the equator to pole temperature gradient, sustains the atmosphere's store of POTENTIAL ENERGY available to be converted to kinetic or wind energy through vertical motions.

The energy input into the tropical atmosphere by condensation is far from steady, being associated with PRECIPITATION events that are spatially and temporally erratic and not simple, obvious or reliable. The largest portion of the heat from condensation and precipitation is produced within a few relatively narrow zones of active weather. These are commonly associated either with disturbances along the INTERTROPICAL CONVERGENCE ZONE or with easterly waves developing within the trade wind belt.

Even in the southern hemisphere the belt of high pressure on the poleward side of the trade winds is not continuous but broken up into a series of high-pressure centres. A similar fragmentation of the Hadley cells seems likely. Otherwise, air flowing poleward aloft at all longitudes would conserve its absolute ANGULAR MOMENTUM and produce a much stronger subtropical JET STREAM than is actually observed.

The complexity of the wind field in the tropics is enhanced by the presence of longitudinal temperature differences which produce zonal thermal circulations. The best known of these is the Walker circulation created by air rising over the warm waters of the western equatorial Pacific Ocean and sinking over the cool waters of the eastern Pacific. The cool water is primarily the result of upwelling of subsurface water to replace that driven westward and away from the equator by the trade winds. The Walker circulation, the trade winds, the Hadley cells, and upwelling are all very closely inter-related. Their fluctuations produce a see-sawing effect on the sea-level pressure difference between the eastern and western Pacific. This is known as the southern oscillation. Closely connected with it are the EL NIÑO EFFECT off the coasts of Chile and Peru and significant weather and climate TELECONNECTIONS in middle latitudes.

The statistical significance of the Hadley cells as a major feature of the tropical atmosphere is greatly enhanced by the MONSOON winds of south-east Asia and India. The dry, cold north-east monsoon winds of winter and humid, warm south-west monsoon winds of summer are associated, respectively, with the Hadley cells of the northern and southern hemispheres. The summer rains of this region arise from disturbances that move westward along the intertropical convergence zone.

Although there is a great deal of persistence in tropical weather, it is far from constant. Further, even relatively minor fluctuations in this energy-rich region may be amplified and affect the climate and weather of the rest of the world. WDS

Reading
Riehl, H. 1979: *Climate and weather in the tropics.* London: Academic Press.

tropopause The boundary between the TROPOSPHERE and the STRATOSPHERE, usually revealed by a fairly sharp change in the LAPSE RATE of temperature. The change is in the direction of increased ATMO-SPHERIC STABILITY from regions below to regions above the tropopause. The height of the tropopause is about 20 km in the tropics and 10 km in polar regions, the decline with increasing latitude being step-like, with jet-streams occupying the steep rises. The tropopause is frequently difficult to locate, largely due to its comprising several 'leaves', giving rise to the idea of a multiple tropopause, rather than a single continuous surface. BWA

troposphere The portion of the atmosphere lying between the earth's surface and the TROPOPAUSE. Due to the varying height of the tropopause, tropospheric depths vary, on average, from 10 km to 20 km, the low values being in polar regions, the high in tropical regions. Within the troposphere the mean values of temperature, water vapour content and pressure decrease with height. Horizontal WIND speeds increase with height and VERTICAL MOTION is substantial. As a result, the troposphere is the part of the atmosphere that contains all the WEATHER we experience from day to day. BWA

trottoir (from the French word for a pavement or sidewalk) A narrow organic reef constructed by such organisms as *Lithophyllum tortuosum*, *Vermetidae* and *Serpulidae*. Trottoirs are common in the Mediterranean and in low latitudes and develop in the intertidal zone. ASG

Reading
Tzur, Y., and Safriel, U.N. 1979: Vermetid platforms as indicators of coastal movements. *Israel journal of earth sciences* 27, pp. 124–7.

trough In meteorology this term virtually always relates to PRESSURE. Thus a pressure trough is an elongated area of relatively low pressure: the opposite is a ridge, of relatively high pressure. The words trough and ridge clearly derive from the valleys and ridges familiar to us in the solid earth. Troughs occur on scales ranging from mesoscale to continental, the most familiar being those appearing on the synoptic weather map on television. The very largest troughs, found within the planetary, or ROSSBY WAVES are critical to the formation of extra-tropical CYCLONES. BWA

truncated spur Steepened bluff on the side of a glacial trough in between tributary valleys. It arises through the widening and straightening of a pre-existing sinuous river valley by glacial action. Such features have long been regarded as characteristic of glacial erosion. DES

tsunami Popularly called tidal waves, are sea-surface waves generated by submarine earthquakes and volcanic activity. Physically they propagate as long waves with a speed given by:

$$(\text{water depth} \times \text{gravitational acceleration})^{1/2}$$

In ocean depths they can only be detected by sensitive bottom-pressure measurements. When they reach shallow coastal regions, amplitudes may increase to several metres. Around the Pacific Ocean, which is particularly vulnerable to tsunami, a network of tide gauges is coordinated to give advanced warning of their arrival. Tsunami damage results from flooding, and from wave impacts on coastal structures coupled with the erosion of their foundations.   DTP

Reading
Bernard, E.N. 1991: *Tsunami hazard. A practical guide to Tsunami hazard reduction*. Dordrecht: Kluwer.

tufa A freshwater carbonate deposit which, according to Pentecost (1981: p. 365), is 'a soft, porous, calcareous rock formed in springs, waterfalls and lakes in limestone regions.' The term is often used interchangeably with *travertine*, although some authors consider one as being a special case of the other. Pentecost (1981, p. 365), for example, states that 'travertine is identical in composition to calcareous tufa but is a hard non-porous variety used for building.' *Sinter* or *calc-sinter* is another commonly used term but is usually restricted to inorganically precipitated deposits. Tufas can form significant landforms (terraces, barrages, dams etc.) and may contain much palaeoenvironmental information. Organic processes (e.g. precipitation by blue-green algae) probably play a major role in their development.   ASG

Reading and Reference
Pentecost, A. 1981: The tufa deposits of the Malham District, North Yorkshire. *Field studies* 5, pp. 365–87.

Viles, H.A. and Goudie, A.S. 1990: Tufas, travertines and allied carbonate deposits. *Progress in physical geography* 14, pp. 19–41.

tuff Consolidated clastic material ejected from volcanoes with a predominance of fragments less than 2 mm in diameter.

tundra Vast, level, treeless and marshy regions, usually with permanently frozen subsoil. Originally derived from northern Eurasia, the term has expanded to include all Arctic and Antarctic areas polewards of the TAIGA, and also to similar alpine environments above the TIMBERLINE on mountains. Drier tundra sites are characterized mainly by herbaceous perennials, with occasional trees and scattered heath plants, grasses, lichens and mosses, while cotton grass, hygrophytic sedges and willows are typical of wet sites. Cryptophytic communities develop in snow and ice.   PAF

Reading
Bliss, L.C., Heal, D.W. and Moore, J.T. eds 1981: *Tundra ecosystems: a comparative analysis*. Cambridge and New York: Cambridge University Press.

Ives, J.D. and Barry, R.T. eds 1979: *Arctic and alpine environments*. London and New York: Methuen.

A small tufa dam developed in a creek draining from the limestone Napier Range in north-west Australia.

tunnel valleys Form by subglacial stream action. They tend to have flat floors, steep sides and irregular long profiles, and are a feature of northern Germany and Denmark.

tunnelling, tunnel gully erosion A form of erosion, initiated by subsurface water movement, which often causes surface collapse, leading to open gullying. Water movement through soil cracks eluviates material, thereby leading to the development of tunnels which continue to erode as gullies, following tunnel collapse. It is thus related to PIPES.   ASG

Reading
Lynn, I.H. and Eyles, G.O. 1984: Distribution and severity of tunnel gully erosion in New Zealand. *New Zealand journal of science* 27, pp. 175–86.

turbidity current A density current. A sinking mass of sediment-laden air or water. Their erosive activity is thought to contribute to the formation of some submarine canyons on the continental shelves. Sediments deposited by turbidity currents are known as turbidites.

turbulence A concept very difficult both to define and to explain. In lay terms it is the gustiness experienced in a brisk WIND. Its effects can be seen qualitatively in the behaviour of flows as wide ranging in size as those of cigarette smoke, factory chimney smoke and the material from volcanoes such as Mount Saint Helen, whose dust was visible on SATELLITE imagery stretching virtually the whole way across North America. In more rigorous fashion, Sutton (1955, p. 9) defined turbulence as 'a state of fluid flow in which the instantaneous velocities exhibit irregular and apparently random fluctuations so that in practice only statistical properties can be recognized and subjected to analysis.' Transfer through space of heat, water and momentum by turbulence is a fundamental, and highly problematical, mechanism within the atmosphere.   BWA

Reference
Sutton, O.G. 1955: *Atmospheric turbulence*. London and New York: Methuen.

turbulent flow Involves an aggregate forward flow direction characterized by the eddy motion of small bodies of fluid which follow independent, irregular paths until they mix with their surroundings. If a thin stream of dye is injected into turbulent flow, it diffuses throughout the depth rather than maintaining a coherent thread. Various scales of turbulent eddy occur, from small random eddies associated with bed ROUGHNESS to larger systematic eddies controlled by channel shape (see SECONDARY FLOWS). Turbulent eddies influence grain motion on unconsolidated sedimentary surfaces, and maintain the suspension of sediment particles within the flow. Turbulent flow arises when the flow inertia swamps the effects of fluid viscosity and the REYNOLDS NUMBER exceeds 2000. In LAMINAR FLOW a microscopic (even molecular) scale of interference occurs between adjacent moving layers but in turbulent flow the eddies cause microscopic interference, and water parcels of high momentum are transferred close to the bed. A turbulent BOUNDARY LAYER has a high flow velocity close to the bed, and a profile in which velocity increases as the logarithm of height above the bed.   KSR

turmkarst A term of German origin describing steep-sided residual limestone hills which rise above alluvial or swampy plains. They are also called TOWER KARST.   PAB

typhoon See TROPICAL CYCLONES.

U

ubac The side of a hill or valley that is most shaded from the sun.

unconformity A discontinuity between sedimentary strata which testifies to a temporary interruption in the process of accumulation (see diagram).

underfit stream A stream which is much smaller than expected from the size of its valley. An underfit stream could have occurred as a result of river capture when the beheaded CONSEQUENT STREAM would be smaller than expected. However, Dury (1977) has shown that underfit streams are a widespread occurrence, that they reflect the impact of climatic change, and that the wavelength of VALLEY MEANDERS may be three to ten times greater than the wavelengths of the underfit stream meanders. *A manifestly underfit stream* is an underfit stream which meanders within a more amply meandering valley and an *osage type* of stream has a much smaller pool–riffle

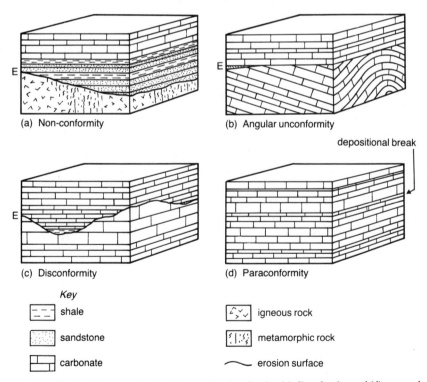

(a) Non-conformity

(b) Angular unconformity

depositional break

(c) Disconformity

(d) Paraconformity

Key

shale

sandstone

carbonate

igneous rock

metamorphic rock

erosion surface

Four types of unconformity: (a) *non-conformity;* (b) *angular unconformity;* (c) *disconformity; and* (d) *paraconformity.*
Source: *J.D. Collinson and D.B. Thompson 1982: Sedimentary structures. London: Allen & Unwin. Figure 2.8.*

spacing than would be expected from the size of the valley meanders. KJG

Reference
Dury, G.H. 1977: Underfit streams: retrospect, perspect and prospect. In K.J. Gregory ed., *River channel changes*. Chichester: Wiley. Pp. 281–93.

underplating A potentially highly important tectonic process caused when magma generated over a mantle plume is accreted to the base of the crust. According to the underplating model, the addition of volcanic rock in this way thickens the crust and the resulting isostatic adjustment leads to the formation of a broad *hot-spot* swell up to 2000 km across and with an increase in surface elevation of up to 2000 m. This has substantial implications for river network evolution. ASG

Reading
Summerfield, M.A. 1991: *Global geomorphology*. London and New York: Longman Scientific and Technical and Wiley.

unequal slopes, law of Unequal angles of hillslope are the result of variations in the rate of slope erosion, which in turn result from differences in geological resistance and from variations in the power of erosional processes. The statement concerning inequality of slopes is more commonly expressed in its reverse form, propounded by G.K. Gilbert (1877) as the 'law of uniform slopes', which states that under uniform conditions of erosion and resistance 'if steep slopes are worn more rapidly than gentle, the tendency is to abolish all differences of slope and produce uniformity' (p. 109). MJS

Reference
Gilbert, G.K. 1877: *Report on the geology of the Henry Mountains*. Washington, DC: Department of the Interior, US Geographical and Geological Survey of the Rocky Mountain Region.

uniclinal Pertaining to a formation of rock strata which dip uniformly in one direction.

uniclinal shifting The process whereby a stream or river flowing in an asymmetric valley in an area of gently dipping rocks migrates down the dip slope of the valley, cutting back the steeper scarp slope.

uniform steady flow Exists when the water depth is equal at every section in a channel reach. Unsteady uniform flow would require the water surface to remain parallel to the channel bed as discharge changes, which is practically impossible. Accordingly, spatially uniform flow is temporally steady. Discharge, and flow depth, width, cross-section area and velocity are all constant from section to section, and the ENERGY GRADE LINE, water surface and bed profile are all parallel. This is rare in natural channels with variable width and POOL AND RIFFLE bedforms.

The CHÉZY EQUATION and MANNING EQUATION define the mean velocity of uniform flow as a function of depth, slope and ROUGHNESS, and are often applied to short natural river reaches where uniform flow can be assumed. If local velocities at every point in the cross-section are constant along a reach, the entire velocity distribution is uniform, the turbulent BOUNDARY LAYER is fully developed, and the logarithmic vertical velocity profile occurs. KSR

Uniformitarianism A practical tenet held by all modern sciences concerning the way in which we should choose between competing explanations of pheneomena. It rests on the principle that the choice should be the simplest explanation which is consistent both with the evidence and with the known or inferred operation of scientific laws. Uniformitarianism is therefore applicable to both historical inference (or 'postdiction') and to prediction of the future outcome of the operations of natural processes (Goodman 1967). It is, in consequence, as Shea (1982, p. 458) has forcibly emphasized, a concept 'with no substantive content – that is, it asserts nothing whatever about nature. Uniformitarianism must be viewed as telling us how to behave as scientists and not as telling nature how it must behave.'

In physical geography Uniformitarianism is usually linked with James Hutton's demonstration (1788) that the simplest explanation for the nature of the earth's surface topography and rock strata was not the invocation of divine intervention at a single moment of creation and then again by the biblical flood, but rather the assumption that processes of erosion, lithification and uplift comparable to those whose operations could be observed or inferred in the modern world, acting over immensely long periods of time, were responsible. Shea (1982) points out that Hutton was not the first to adopt this viewpoint, but as he certainly was the first to provide an extensive working-out of its implications it is reasonable to regard him as the founder of modern earth science. Hutton's views conflicted sharply with those of other natural philosophers, notably Werner and Kirwan, who came to be termed Catastrophists (see Chorley *et al.* 1964). These latter produced interpretations, often incredibly complex, of rocks and relief which derived from an implicit belief that the biblical chronology was sacrosanct and that God had interverned directly to control the mechanisms of earth sculpture.

One of the major sources of confusion which has come to enshroud discussions of Uniformitariansim in earth science in general, and in geomorphology in particular, has derived from a change in the interpretation of 'Catastrophism'. Increasingly, the term has been taken to imply a belief that large, sudden and (to human eyes) 'catastrophic' events have more significance in earth history than the slow and virtually continuous operation of 'normal' processes (see e.g. the extraordinary influential paper by Wolman and Miller 1960). This change in meaning has left Uniformitarianism apparently standing for a view in which the *simplest* explanation is equated with that which requires the slowest and/or most *uniform* rate of operation of processes: this fallacy – termed 'gradualism' (see Hooykaas 1963) – is a complete misinterpretation of the Uniformitarian tenet. Moreover, any careful reading of Hutton, or Playfair (1802), or any edition of Lyell's *Principles of geology* (first published 1830–3), make it abundantly clear that all those early Uniformitarians ascribe a very important role to what would now be termed 'high magnitude, low frequency' events. This tendency becomes particularly marked in the later editions of Lyell's *Principles* (e.g. the 9th, 1853) as his congenital reluctance to believe that 'normal' fluvial processes are *actually* responsible for substantial earth sculpture, leads him to an increasing emphasis on the role of sudden, large and locally 'catastrophic' occurrences. It cannot be emphasized too forcibly that Uniformitarianism does not, as a principle, require any presupposition about the rates of operation of processes, other than those limits apparently fixed by the laws of physics and chemistry.

Nor does the concept involve – as another fallacy proposes – the belief that only processes which can actually be observed in operation may be properly invoked as explanations. In consequence, it equally does not assume that the nature and rates of operation of processes have remained unchanged over time. It is, for example, both apparent and entirely consistent with the Uniformitarian tenet, that the nature and rates of processes on the earth must have been very different from today either before the evolution of land plants, or at the height of one of the Pleistocene glacial advances.

Probably the most instructive example of the application and misapplication (or misconstruction) of the Uniformitarian principle, and of the conflict which can be generated, is the case of J.H. Bretz (1923) and his interpretation of the channelled scabland of the north-west USA as the product of an almost unimaginably large flood. Baker (1981) has provided a fascinating (and sobering) collection and discussion of the major papers in the channelled scabland debate, which should be required reading for all geomorphologists.

The essence of this controversy concerns the most probable origin of the huge complex of deep channels cut through loess and basalt in the Columbia Plateau region of eastern Washington State (including the site of the Grand Coulee Dam). Using the evidence from painstaking field studies, Bretz in a series of papers from 1923 onwards argued that the simplest interpretation of the data called for a great flood or 'debacle', which cut channels over a 40,000 km^2 area. The source of the water for this 'Spokane flood', Bretz found in the site of glacial Lake Missoula: it was suggested that the ice damming the lake had been suddenly breached, releasing some 500 mile^3 of water into the scabland tract. Bretz's explanation was entirely consonant with Uniformitariansim. He observed scabland features which, while huge, were clearly products of running water and all that was required to explain them was, therefore, a way of providing a very large flow of water in a very short period. Glacial lake dam bursts are well-documented occurrences.

The Spokane flood theory was attacked – and very viciously attacked – on the grounds both that the proposed explanation smacked far too much of the Diluvial Catastrophists' interpretations of *all* valleys in terms of the mighty waters of Noah's flood and, in addition, because no flood as large as the one hypothesized had ever been observed. Both lines of argument depend upon fallacious interpretations of the Uniformitarian principle. Their proponents, attempting to provide 'permissible' explanations for the channelled scablands without Bretz's flood, tied themselves in increasingly complicated knots.

Ironically, it was ultimately rather small-scale and geomorphologically undramatic evidence to which the Uniformitarian principle was correctly applied, which led to the vindication of Bretz's earlier hypothesis. The vital evidence consisted of the discovery of giant current ripple marks both within the Grand Coulee area and on the floor of the former Lake Missoula. The hydraulic and hydrodynamic relationships between depth and velocity of water movement and the dimensions of bedforms such as current ripples are, in fact, so well established that the simplest and therefore Uniformitarian explanation for the giant examples was a water body with all the characteristics of Bretz's Spokane flood.

While the details of Bretz's explanation of the channelled scabland have been modified by later studies (there were, it seems, several different dam bursts and floods), in essence his 1923 views have been accepted. Indeed, his 'outrageous' mechanism has since been used to explain the apparently similar 'channelled scablands' seen on the planet Mars (see Baker 1981), as it is a very sound application of the Uniformitarian tenet to assume that direct analogy may be the simplest explanation of apparently directly analogous forms.

Nevertheless, it must be emphasized again that Uniformitarianism is a guiding tenet of science and *not* a rule of nature. As theories about the operation of nature change, so it is possible – and, indeed, inevitable – that one 'Uniformitarian' explanation will come to replace another.   BAK

Reading and References

†Baker, V.R. ed. 1981: *Catastrophic flooding*. Stroudsburg, Pa.: Dowden, Hutchinson & Ross.

Bretz, J.H. 1923: The channelled scablands of the Columbia Plateau. *Journal of geology* 31, pp. 617–49.

Chorley, R.J., Dunn, A.J. and Beckinsale, R.P. 1964: *The history of the study of landforms*. Vol. I. London: Methuen.

Goodman, N. 1967: Uniformity and simplicity. In C.C. Albritton ed., *Uniformity and simplicity*. Geological Society of America special paper 89, pp. 93–9.

Hooykaas, R. 1963: *Natural law and divine miracle: the principle of uniformity in geology, biology and theology*. Leiden: E.J. Brill.

Hutton, J. 1788: Theory of the earth; or an investigation of the laws observable in the composition, dissolution and restoration of land upon the globe. *Transactions of the Royal Society of Edinburgh* I, part II, pp. 209–304.

Lyell, C. 1830–3: *Principles of geology*. 3 vols. London: John Murray.

— 1853: *Principles of geology*. 9th edn. London: John Murray.

Playfair, J. 1802: *Illustrations of the Huttonian theory of the earth*. London: Cadell & Davies.

†Shea, J. 1982: Twelve fallacies of Uniformitariansim. *Geology* 10, pp. 455–60.

Wolman, M.G. and Miller, J.P. 1960: Magnitude and frequency of forces in geomorphic processes. *Journal of geology* 68, pp. 54–74.

unit hydrograph A characteristic or generalized hydrograph for a particular drainage basin. A unit hydrograph of duration t is defined as the hydrograph of direct run-off resulting from a unit depth of effective rainfall generated uniformly in space and time over the basin in unit time. The unit depth was originally one inch but is now usually one centimetre and t is chosen arbitrarily according to the size of the basin and to the response time to major events and can be 1, 6 or 16 hours, for example. The technique was developed by L.K. Sherman in 1932 and it has been used to predict hydrographs for the engineering design of reservoirs, flood detention structures and urban stormwater drainage. Since many streams are still ungauged, the discharge records from all stations in an area can be analysed and synthetic unit hydrographs developed from the relations between unit hydrograph parameters and drainage basin characteristics. The drainage basin characteristics of the basin above an ungauged site can then be used to obtain the synthetic unit hydrograph for that site. To compare drainage basins of different areas dimensionless unit hydrographs can be constructed with the discharge ordinate expressed as the ratio to the peak discharge and the time ordinate expressed as the ratio to the lag time. The instantaneous unit hydrograph is a mathematical abstraction produced when the duration of the effective precipitation becomes infinitesimally small and this is used in the investigation of rainfall–run-off dynamics. Unit hydrograph theory depends upon a number of assumptions including the HORTON OVERLAND FLOW MODEL and, with the advent of the PARTIAL AREA MODEL of run-off formation, it has been necessary to revise the use and analysis of the unit hydrograph.   KJG

Reading and Reference
†Dunne, T. and Leopold, L.B. 1978: *Water in environmental planning.* San Francisco: W.H. Freeman. Pp. 329–50.
†Shaw, E.M. 1983: *Hydrology in practice.* Wokingham: Van Nostrand Rheinhold. Pp. 326–44.
Sherman, L.K. 1932: Stream flow from rainfall by the unitgraph method. *Engineering news record* 108, pp. 501–5.

unit response graph The theoretical quickflow hydrograph produced by 1 inch of effective rainfall and derived from the actual quickflow hydrography by assuming a linear extension such as is carried out in the derivation of simple unit hydrographs (Walling 1971). The derivation of a unit response graph is similar to the derivation of a unit hydrograph but flow separation is based upon the method proposed by Hibbert and Cunningham (1967) (see HYDROGRAPHS) and each unit response graph will vary in shape in relation to the contributing area generating storm run-off.

Classic unit hydrograph theory assumes that the whole catchment contributes to storm run-off and so the form of the hydrograph will reflect the characteristics of the whole catchment and any variation in hydrograph form will result entirely from variations in the time distribution of effective rainfall. Hydrograph separation can be achieved using any consistent technique although Linsley *et al.* (1982) stress the importance of using a separation technique which ensures that the time base of direct or storm run-off remains relatively constant from storm to storm. An hour unit hydrograph will be storm run-off response to a unit of effective rainfall (usually 1 cm or 1 in) falling with even intensity over the entire catchment during a period of *n* hours. Because of the assumed linear relationship between effective rainfall and storm run-off, it is possible to derive standard hydrographs for different rainfall intensities and for different rainfall durations by applying simple transformations to a unit hydrograph or to a derivative of a unit hydrograph such as an *S*-curve or an instantaneous unit hydrograph. The unit hydrograph concept is explained in detail by Linsley *et al.* (1982).

In the case of the unit response graph, the magnitude of the ordinates of the quickflow hydrograph are adjusted so that there is a volume of run-off equivalent to a unit of effective rainfall over the catchment, but unit response graphs to storms of the same duration would not be expected to have the same form because of the influence of the size and shape of the contributing area of the speed with which water can drain from the catchment.   AMG

References
Hibbert, A.R. and Cunningham, G.B. 1967: Streamflow data processing opportunities and application. In W.E. Sopper and H.W. Lull eds, *International symposium on forest hydrology.* Oxford and New York: Pergamon. Pp. 725–36.
Linsley, R.K., Kohler, M.A. and Paulhus, J.L.H. 1982: *Hydrology for engineers.* 3rd edn. New York: McGraw-Hill.
Walling, D.E. 1971: Streamflow from instrumented catchments in south-east Devon. In K.J. Gregory and W.I.D. Ravenhill eds, *Exeter essays in geography.* Exeter: University of Exeter. Pp. 55–81.

unloading The stripping of rock or ice from a landscape and the resulting effects the release of pressure has on the exhumed landsurface.

unmanned earth resources satellites Satellites carrying a range of REMOTE SENSING devices for the production of images of the earth's surface. There are five groups of unmanned earth resources satellites. Groups one and two both record radiation in visible and near visible wavelengths. Group one comprises the Landsat series, which were the first generation of earth resources satellites, and group two comprises the second generation of earth resources satellites and includes SPOT. Group three carries sensors that record thermal wavelengths and includes HCMM; group four carries sensors that record microwave wavelenghs including Seasat, ERS-1 and Radarsat; and group five comprises the 'polar platform' satellites that will be providing physical geographers with a major source of environmental data until well into the next century.

1 The Landsat series
After the success of ‚MANNED EARTH RESOURCES SATELLITES the National Aeronautics and Space Administration (NASA) of the USA and the US Department of Interior developed an experimental earth resources satellite series to evaluate the utility of images collected from an unmanned satellite. The first satellite in the series carried two types of sensor, a four waveband multispectral scanning system (MSS) and three return beam vidicon (RBV) television cameras. When launched

in July 1972 it was called ERTS, the Earth Resources Technology Satellite, a name that it held until January 1975 when it was renamed Landsat. The main advantages offered by the imagery collected from this satellite were: ready availability, low cost, repetitive multispectral coverage and minimal image distortion. Landsat 1 had a high and fast orbit at an altitude of 900 km and a speed of 6.5 km s^{-1}. The orbit was circular, flying within 9° of the north and south poles and sun-synchronous as it kept pace with the sun's westward progress as the earth rotated. To obtain repeat coverage of an area the orbits were moved westwards each day and this enabled an image to be taken of each area of the earth's surface every 18 days. Landsat 1 lasted for almost six years until January 1978 and for part of its life shared the heavens with Landsat 2 which was launched in 1975. Landsats 3, 4 and 5 were launched in 1978, 1982 and 1984. Landsats 6 and 7 are due for launch in 1993 and 1997. For Landsats 4 and 5 the satellite body was changed to increase stability and payload capability and the orbit altitude was lowered to 705 km, thus giving a faster repeat cycle of 16 days and a changed orbit spacing.

The Landsat satellites all carry, or have carried, two sensors: a multispectral scanning system (MSS) and either a thematic mapper (TM) or RBV television cameras. The MSS records four images of a scene, each covering a ground area of 185 × 185 km at a nominal spatial resolution of 79 m. These four images cover green, red, near infrared and infrared wavebands and were identified by the channels they occupied in the satellite's telemetry system, which were 4, 5, 6 and 7 respectively. The MSS sensor has undergone very little change since the launch of the first Landsat. The three important changes are, first, the addition of an extra waveband, known as band 8, to the MSS of Landsat 3. This recorded thermal infrared images but as it failed shortly after launch few images have been used. Secondly, to compensate for the lower orbit altitude of Landsats 4 and 5, the spatial resolution of the MSS images was decreased by 3 m to 82 m and the field of view was increased by 3.41° to 14.93°. Thirdly, the numbering of the MSS wavebands was modified from Landsat 4 onwards. Landsat MSS data were initially used to obtain a synoptic view of a large area of the earth's surface for visual interpretation. Today, owing to the availability of DIGITAL IMAGE PROCESSING, digital Landsat MSS data are frequently used for classifying land cover, estimating characteristics of the earth's surface and for monitoring change.

The thematic mapper (TM) carried by Landsats 4 and 5 records image areas of 185 × 185 km in seven wavebands with a spatial resolution of around 30 m in six of them. The TM is an important sensor for physical geographers and is used extensively throughout the subject.

Return beam vidicom (RBV) television cameras were carried on Landsats 1, 2 and 3. On Landsats 1 and 2, three cameras were used, each filtered into a different waveband, camera 1 into green, camera 2 into red and camera 3 into near infrared. Unfortunately, the RBV on board Landsat 1 returned only 1690 images before it was turned off in August 1972. The RBV on Landsat 2 returned even fewer images and so for Landsats 1 and 2 the MSS images were their primary product. Landsat 3 carried two RBVs and these were both filtered to a broad red to near infrared waveband. Their design was similar to the RBVs carried by Landsats 1 and 2 except for their focal length, which was increased to give a nominal ground resolution of around 30 m and an image area of 98 × 98 km.

2 Second generation of earth resources satellites
Like the Landsat satellites, these will carry optical sensors but they will be linear array multispectral scanners. The satellites so far named include the French satellite Système Probatoire d'Observation de la Terre (SPOT). SPOT 1 and SPOT 2 were launched in 1986 and 1990 and SPOT 3 and 4 are due for launch in 1993 and 1996. The satellites are operated by the Centre National d'Etudes Spatials and have a near-polar, sun-synchronous, 832 km high orbit, which will provide repeat ground coverage every 26 days. It carries two high-resolution visible (HRV) sensors, two tape recorders and telemetry equipment to transmit data to earth. The HRV records an area of 60 × 60 km with a spatial resolution of 10 m in one panchromatic waveband or 20 m in green, red and near infared wavebands. These images can be recorded obliquely to give a stereoscopic view of the terrain, avoid patchy cloud and decrease the revisit time.

Japan has launched two earth resources satellites. The first, in 1990, was the Marine Observation Satellite (MOS-1) and the second, in 1992, was the Japanese Earth Resources Satellite (JERS-1). MOS-1 has an orbit and altitude similar to the Landsat series of satellites and carries three sensors, the most important of which is a linear array multispectral scanner named the Multispectral, Electronic, Self-scanning Radiometer (MESSR). This senses in four wavebands from green to near infrared, has a spatial resolution of 50 m and an image area of 200 × 200 km. JERS-1 carries a stereoscopic linear array sensor and also a SIDEWAYS LOOKING AIRBORNE RADAR (type SAR).

The Indian Remote Sensing (IRS) satellites carry Linear Imaging Self-scanning (LISS) multispectral scanners with design specifications similar to those of the sensors on board Landsat. IRS-la and IRS-lb were launched in 1988 and 1991 with future IRS launches planned for 1993, 1994 and 1996.

3 Satellites carrying thermal sensors
Landsats 3, 4 and 5 and the Heat Capacity Mapping Mission (HCMM) carried thermal sensors with a low spatial resolution. The HCMM satellite was launched in April 1978 and lasted until September 1980. Its orbit was near-circular at an altitude of 620 km. The satellite contained a scanning radiometer which recorded in a visible and near infrared waveband and a thermal infrared waveband. The orbits of the satellite were arranged to ensure that images were obtained of each scene during times of maximum and minimum surface temperature for the determination of thermal inertia. The multispectral scanner had a very wide scan angle of 60° resulting in an image width of 720 km. The spatial resolution decreased from around 0.6 km at the centre of the image to around 1 km at the edge of the image. The data from the HCMM sensors were intended for geological mapping but they have also been used for microclimatology, pollution monitoring and hydrology.

4 Satellites carrying microwave sensors
The first unmanned earth resources satellite to carry a sideways looking airborne radar (type SAR) was Seasat. This was an experimental satellite designed by NASA to establish the utility of microwave sensors for remote sensing of the oceans. Images of

the land were also obtained giving physical geographers a synoptic view of the earth in microwave wavelengths. Seasat had a circular non-sun-synchronous orbit at an altitude of 800 km, sensing the earth's surface from 72° N to 72° S, orbiting the earth 14 times a day and passing over the same area every 152 days. The satellite carried two sensors of potential interest to physical geographers: a multispectral scanner and a sideways looking airborne radar (type SAR). The multispectral scanner recorded in two wavebands, visible at a spatial resolution of 2 km and thermal infrared at a spatial resolution of 4 km. The sideways looking airborne radar (type SAR) produced images of 100 km wide swaths with a nominal spatial resolution of 25 m. These data have been used for many applications such as the monitoring of sea state and the mapping of vegetation, sea ice and urban form.

The first remote sensing satellite to be launched by the European Space Agency (ESA) was the Earth Resources Satellite (ERS-1) in 1991. It was launched into a sun-synchronous orbit at an altitude of around 700 km with a great cycle of three days. It carries several sensors, the most important as far as physical geographers are concerned is the sideways looking airborne radar (type SAR). The Japanese Earth Resources Satellite (JERS-1), launched in 1992, carries a sideways looking airborne radar (type SAR). This sensor records at longer wavelengths than that on board ERS-1 and can be used to sense beneath vegetation canopies.

Canada proposes to launch the satellite Radarsat in 1995. It will carry a sideways looking airborne radar (type SAR). The major application of data collected from the satellite is for mapping ice, especially in the offshore oil drilling areas of northern Canada.

5 Satellites carrying a suite of environmental sensors
A series of major earth resources satellites is being developed. These 'polar platforms' will carry suites of sensors that will enable the simultaneous estimation of fundamental environmental processes on land, in the oceans and in the atmosphere. The first two of these satellites are to be launched in 1998. The US Earth Observing System (EOS) satellite AM-1 will carry four sensors and the ESA Polar Orbit Earth

Observing Mission (POEM) satellite EN-VISAT will carry up to seventeen sensors.

PJC

Reading
Barrett, E.C. and Curtis, L.F. 1992: *Introduction to environmental remote sensing.* 2nd edn. London and New York: Chapman & Hall.
CEOS 1992: *The relevance of satellite missions to the study of the global environment.* Committee on Earth Observation Satellites. London: British National Space Centre.
Cracknell, A.P. and Hayes, C.W.B. 1991: *Introduction to remote sensing.* London: Taylor & Francis.
Curran, P.J. 1985: *Principles of remote sensing.* Harlow: Longman Scientific and Technical.
Mather, P.M. ed. 1992: *TERRA-1 understanding the terrestrial environment.* London: Taylor & Francis.

unstable channels A river or tidal channel that is shifting through erosion and deposition. Some writers restrict the term to those that are shifting rapidly, changing their pattern or adjusting to changed conditions, because many channels quite normally shift their courses without being in a state of imbalance or disequilibrium with environmental controls. By contrast, engineering design commonly aims to achieve stable channels or canals that remain fixed in position and which will not require costly maintenance works. (See also CHANNELS, TYPES OF RIVER/STREAM.)

JL

unstable equilibrium If we had two spheres, one larger than the other, and we placed the smaller one upon the larger one in such a way that, upon letting it go, it remained where we put it, then, in the absence of any disturbing force, the two spheres would be in equilibrium. But we are all aware that it would be extraordinary difficult to achieve the above result and that, even should we succeed, the merest hint of a breath would disturb the equilibrium, sending the smaller sphere increasingly further away from its original position in equilibrium. Thus we had initially a state of unstable equilibrium. In many natural systems, and particularly in fluids, this type of equilibrium may exist. For example, very warm parcels of air may remain near the ground until some small disturbance triggers their release.

BWA

unsteady flow Occurs in an open channel (e.g. a river or canal) when the depth and discharge of water at different points along a reach change through time because of the passage of a flood wave or surge along the channel. Simplified FLOW EQUATIONS cannot be applied to such translatory wave processes. Analysis of changes of flow conditions at a section, or of the shape of the flood wave as it travels downstream, therefore require the application of wave theory or FLOOD ROUTING methods.

KSR

upwelling The vertical movement of deeper water towards the sea surface. It occurs where a divergence of surface currents must be compensated for by vertical flow, e.g. wind-driven offshore currents may be balanced by coastal upwelling. This deeper water is often rich in nutrients, which allow a high productivity of phytoplankton near the surface. As a result, many of the world's most important fisheries are in areas of upwelling. These include the seas off north-west Africa, Oregon and Peru. Every five years or so the Peruvian upwelling is inhibited when the tropical Pacific Ocean responds to a relaxation of the trade winds. The phenomenon, known as the EL NIÑO EFFECT, has serious consequences for the Peruvian fishing industry.

DTP

ural (type) glacier A small glacier developed in the lee of prevailing winds of a mountain or plateau. Snow is deposited in a 'rotor' in the lee (as a large snowdrift). A glacier formed in this way may continue to exist even though the mountain and its Ural glacier may be below the regional snowline. Named after the Ural mountains but other examples exist, e.g. in northern Iceland and in the Colorado Rockies (USA).

WBW

uranium series dating The determination of the age of a substance by measuring the extent of decay of any radioactive isotopes of the uranide elements.

urban hydrology The study of the hydrological cycle and of the water balance within urban areas. Extensive impervious areas mean that surface storage is reduced, infiltration is not possible and evapotranspiration is much less than in rural areas. Impervious areas increase the amount of surface run-off and this is accentuated by the stormwater drainage system which collects water from roads, roofs and other impervious surfaces. Modern stormwater drainage systems are installed separately from foul water drainage systems but in

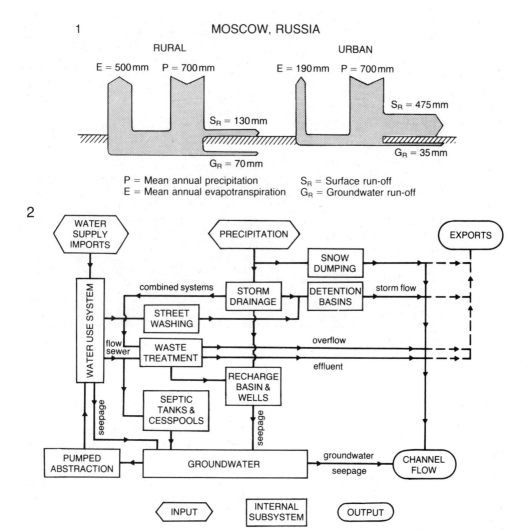

1 MOSCOW, RUSSIA

RURAL

E = 500 mm   P = 700 mm

S_R = 130 mm

G_R = 70 mm

URBAN

E = 190 mm   P = 700 mm

S_R = 475 mm

G_R = 35 mm

P = Mean annual precipitation S_R = Surface run-off
E = Mean annual evapotranspiration G_R = Groundwater run-off

2

WATER SUPPLY IMPORTS

PRECIPITATION

EXPORTS

WATER USE SYSTEM

combined systems

SNOW DUMPING

STORM DRAINAGE

DETENTION BASINS   storm flow

STREET WASHING

flow sewer

WASTE TREATMENT   overflow

effluent

RECHARGE BASIN & WELLS

seepage

SEPTIC TANKS & CESSPOOLS   seepage

PUMPED ABSTRACTION

GROUNDWATER   groundwater   CHANNEL FLOW
seepage

INPUT   INTERNAL SUBSYSTEM   OUTPUT

Source: *Walling 1981.*

the past a single system was often employed in urban areas. Stream discharge from urban areas tends to have higher peak flows and lower base flows than discharge from rural areas and the FLOOD FREQUENCIES along rivers draining urban areas will be significantly changed from the time when the urban area did not exist. Increased flooding has often been observed within and downstream from urban areas as urbanization has occurred and the larger and more frequent floods may have led to increased river channel erosion. Problems of increased frequency and extent of flooding have often been mitigated by engineering works. The area of Moscow, Russia is shown (Lvovich and Chernogaeva 1977) to have a decrease of evapotranspiration of 62 per cent, of groundwater run-off of 50 per cent, and an increase in total run-off of 155 per cent (figure 1). In urban hydrology it is not simply a modification of the rural hydrological cycle but there can also be a series of other components supplying or reducing water (figure 2). Urban areas also generate a characteristic water quality with water temperatures often higher than those of rural areas, with higher solute concentrations reflecting additional sources including pollutants, and suspended sediment concentrations particularly high during building activity but much lower when the urban area is established and the sources of

The extension of urbanization has modified hydrological processes and contributed to such phenomena as accelerated flood generation. The spread of urban areas onto floodplains such as that of the Neches River at Beaumont, Texas, USA, exposes human populations to the flood hazard.

suspended sediments are no longer exposed. (See also LAND USE, HYDROLOGICAL EFFECTS.)   KJG

Reading and References
†Hollis, G.E. 1979: *Man's impact on the hydrological cycle in the United Kingdom*. Norwich: Geo Books.

†Kilber, D.R. ed. 1982: *Urban stormwater hydrology*. Water resources monograph 7. Washington DC: American Geophysical Union.

Lvovich, M.I. and Chernogaeva, G.M. 1977: The water balance of Moscow. *Effects of urbanization and industrialization on the hydrological regime and on water quality*. International Association of Hydrological Sciences publication 123. Pp. 48–51.

Walling, D.E. 1981: Hydrological processes. In K.J. Gregory and D.E. Walling eds, *Man and environmental processes*. London: Butterworth. Pp. 57–81.

urban meteorology The study of atmospheric phenomena attributable to the development of human settlements. It encompasses work on the process involved (physical, chemical and biological), the resulting climate effects, and the application of this knowledge to the planning and operation of urban areas. It is one of the clearest examples of man's role in climate modification.

Urban development disrupts the climatic properties of the surface and the atmosphere. These, in turn, alter the exchanges and budgets of heat, mass and momentum which underlie the climate of any site. Every land clearance, drainage, paving and building project leads to the creation of a new microclimate in its vicinity, and the collection of these diverse, man-affected microclimates is what constitutes the urban climate in the air layer below roof level (henceforward called the urban canopy layer, or UCL, Oke 1987). These very localized effects tend to be merged by turbulence above roof level where they form the urban boundary layer (UBL) which appears like a giant urban plume over and downwind of the city.

A city exerts both roughness and thermal influences on winds. When synoptic winds are strong the greater roughness produces greater turbulence (by 10–20 per cent), increased frictional drag, slower winds (by about 25 per cent), cyclonic turning, and a general tendency towards uplift. In the downwind rural area the near surface flow recovers its original characteristics but urban effects are detectable in the elevated UBL plume for tens of kilometres. The drag may even retard the passage of weather fronts. In windy conditions, flow in the UCL is extremely variable. While some

A demonstration of the microclimatic significance of the urban boundary layer in Vancouver, Canada.

areas are sheltered, others may be experiencing strong across-street vortices, gustiness or jets (especially near tall buildings). When synoptic winds are light or absent, thermal effects associated with the heat island (see below) become evident. The city may generate its own thermal circulation, analogous to SEA/LAND BREEZE, with 'country' breezes converging on the city centre, rising and diverging aloft to form a counter flow. Urban thermal effects can also lead to acceleration near the surface both as a result of the heat pressure field and because thermal turbulence helps transport momentum downwards.

Considering the major changes in the physical environment wrought by urban development, the changes in the energy (heat) budget are surprisingly small. For example, despite all the radiant fluxes being altered (by pollution or changed surface properties) the net radiation in cities is usually within 5 per cent of that of their rural surroundings. It is true that the city's

energy budget is supplemented by heat released by combustion, but though this heat source may be important to climate in some locations, in most places it is minor (Kalma and Newcombe 1976). Usually more important is the fact that the city channels more energy into sensible rather than LATENT HEAT. This is because of the removal of many sources of water for EVAPOTRANSPIRATION. As a result more heat is used to warm the air and ground (including buildings etc.). The relative warmth of the city is called its 'heat island' (Landsberg 1981; Oke 1982). On an annual basis the canopy layer of a large city (10 million inhabitants) is typically 1–3°C warmer than its surrounding countryside. This may seem small, but the magnitude of the heat island varies diurnally (largest near midnight, the smallest in the afternoon) and in response to weather (largest with calm and no cloud). The difference is also related to city size (measured by its population or better by the geometry of its central street

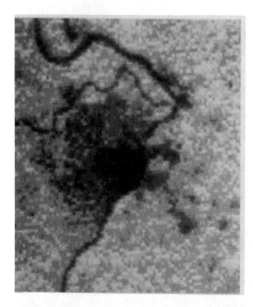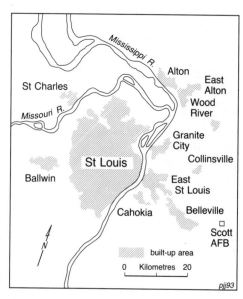

Urban meteorology *The heat island of St Louis given by a computer-enhanced NOAA satellite image of surface infrared emission in the late evening of 28 July 1977. There is excellent correspondence between built-up area and warmth. The surface heat island magnitude of the city is between 4 and 5 °C. The grey scale has 22 steps from 0 °C (white) to 32 °C (black) as follows: 0, 4, 8, 12, 14, 16, then upward in 1 ° increments.*
Source: M. Matson, E.P. McClain, D.F. McGinnis jr and J.A. Pritchard 1978: *Satellite detection of urban heat islands.* Monthly weather review 106, pp. 1725–34.

canyons, Oke 1982). On the most favourable nights in a large city differences of 10°C and more have been recorded. Spatial variation of temperature within the UCL bears a strong relation to land use and building density and there is a sharp gradient at the urban/rural boundary. The city's warmth extends down into the underlying ground and upward into the UBL above. At night the heat island maintains a weak mixed layer above the city (tens to hundreds of metres deep) when rural areas are stable.

The exchange of moisture between the surface and the air is altered by changes in the availability of water and energy, and in the perturbed airflow. Normally values of atmospheric moisture in the daytime UCL are lower than in the country (on account of less evapotranspiration and greater mixing), but the reverse holds at night (because of decreased dewfall and the release of water vapour from combustion). The effects seen in the UCL are also evident in the UBL plume. An exception is provided by high latitude cities in winter where evaporation from frozen surfaces is very small so that humidity is largely governed by vapour from combustion, with the result that the city is more humid by both day and night. At temperatures below −30°C ice fog is a common, and unpleasant, fact of urban life. Above freezing, urban effects on fogs are complex: extra warmth may decrease their frequency but extra condensation nuclei may increase their density and severity. AEROSOL is also responsible for a general increase in daytime haze in the subcloud layer of the UBL, and a deterioration of visibility (Braham 1977).

Urban modification of PRECIPITATION is a subject that has received considerable research study, especially through Project METROMEX in St Louis (Changnon 1981). There seems to be a consensus view that cities enhance precipitation in their downwind areas. These effects seem to be most marked in relation to summer convective rainfall, especially heavy rain, and severe weather (thunder- and hail-storms) rather than frontal precipitation. Annual increases of up to 10 per cent are commonly reported, but the exact role of urban versus non-urban influences is often hard to determine. There is also difficulty in isolating the most important causes. It is possible that the microphysics of urban clouds is altered (e.g. cloud droplet sizes and numbers) and/or that cloud dynamics are changed by the UBL (e.g. strength of uplift, height of mixed layer) leading to more favourable precipitation conditions. The latter changes seem most important in St Louis but much more work is needed.

The field, which began in the early nineteenth century with descriptive studies, is now engaged in the study of meteorological processes and attempts to devise models which link processes and effects. Its most significant deficiencies are in having little knowledge of tropical urban climates and its failure to develop applied science aspects (Page 1970).   TRO

Reading and References
†Braham, R.R. Jr 1977: Overview of urban climate. *Proceedings of the Conference on Metropolitan Physical Environment: USDA Forest Service general technical report NE-25.* Upper Darby, PA. Pp. 1–17.

†Changnon, S.A. Jr ed. 1981: METROMEX; a review and summary. *Meteorological monographs.* 18.40. Boston: American Meteorological Society.

Kalma, J.D. and Newcombe, K.J. 1976: Energy use in two large cities: a comparison of Hong Kong and Sydney, Australia. *Environmental studies* 9, pp. 53–64.

†Landsberg, H.E. 1981: *The urban climate.* New York: Academic Press.

†Oke, T.R. 1987: *Boundary layer climates* 2nd edn. London: Methuen.

— 1982: The energetic basis of the urban heat island. *Quaterly journal of the Royal Meterological Society* 108, pp. 1–24.

Page, J.K. 1970: *The fundamental problems of building climatology considered from the point of view of decision-making by the architect and urban designer.* WMO technical note 109. Geneva: World Meteorological Organization.

urstromtäler An anastomosing pattern of meltwater channels in northern Germany. Individual channels may be hundreds of kilometres long and often more than 100 m deep with irregular long profiles. Some channel patterns are buried by later glacial deposits (Ehlers 1981). Although there is still discussion about their origin, it seems that they were cut primarily by subglacial meltwater erosion.

                 DES

Reference
Ehlers, J. 1981: Some aspects of glacial erosion and deposition in northern Germany. *Annals of ecology* 2, pp. 143–6.

uvala A depression or large hollow in limestone areas produced when several sinkholes coalesce.

V

V notch weir An alternative name for a triangular SHARP-CRESTED WEIR. The triangular form is particularly suited to accurate measurement of small discharges: 90° and 120° notches are most common. The discharge formula for a V notch weir is:

$$Q = K \tan(\theta/2)H^{S/2}$$

where Q is the discharge, θ is the angle of the V notch, H is the head of water above the apex of the V notch, and K is the weir coefficient.   DEW

Reading
British Standards Institution 1981: *Methods of measurement of liquid flow in open channels. Part 4A: thin plate weirs and venturi flumes.* BS 3680. London: British Standards Institution.

valley bulges (valley-bottom bulges) Consist of strata that have bulged up in the base of a valley as a result of erosive processes. They are widespread in the sedimentary rock terrains of the English Midlands, where limestones, sandstones and clays occur in close juxtaposition. The mechanism of formation invoked for those of the Stroud area in Gloucestershire (Ackermann and Cave 1977) is that during the Pleistocene severe erosion and valley incision occurred at a time when permafrost conditions pertained. At the end of the cold period the rocks thawed out and the susceptible clays, silts and sands, highly charged with water, became plastic, and under the weight of the more competent limestones above were extruded through the weakest points of the recently developed valley floors. Cambering of strata would occur on the valley sides.   ASG

Reference
Ackermann, K.J. and Cave, R. 1977: Superficial deposits and structures, including landslip, in the Stroud District, Gloucestershire. *Proceedings of the Geologists' Association of London* 78, pp. 567–86.

valley meanders Meanders which are usually cut in bedrock and which usually have a greater wavelength than that of the contemporary river pattern. Dury (1977) has shown that valley meanders were produced during periods of higher run-off and higher peak discharges before stream shrinkage which led to comtemporary underfit streams. Some writers have suggested that the valley meanders may not indicate stream shrinkage but are rather related to rare high magnitude events, to the contrast between bedrock and fluvial deposits, and to the pattern of stream migration (see Dury 1977 and papers cited therein.)   KJG

Reading and Reference
†Dury, G.H. 1976: Discharge prediction, present and former from channel dimensions. *Journal of hydrology* 30, pp. 219–45.
— 1977: Underfit streams: retrospect, perspect and prospect. In K.J. Gregory ed., *River channel changes.* Chichester: Wiley. Pp. 281–93.

valley wind The up-valley flow which develops during the day, especially in north–south orientated valleys during anticyclonic weather in summer. The flow is induced by strong heating of the valley air, making it much warmer than the air at the

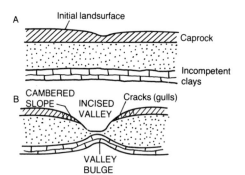

same elevation over the adjacent plain. Valley winds are usually as reliable as the mountain, or down-valley, winds which develop at night.   WDS

Reading
Atkinson, B.W. 1981: *Meso-scale atmospheric circulations.* London: Academic Press.

vallon de gélivation A small valley formed by the widening of bedrock joints by ice action rather than by normal fluvial processes.

valloni The drowned river valleys of a Dalmatian-type coastline.

Van't Hoff's rule The rule which states that when a system is in thermodynamic equilibrium a lowering of temperature will promote an exothermic reaction and a raising of the temperature an endothermic one.

vapour pressure The pressure exerted by the molecules of a given vapour. In meteorology, the vapour in question is usually water vapour and the pressure is the partial pressure, i.e. the contribution by water vapour to the total pressure of the atmosphere. It may be calculated by using a humidity slide rule or tables in conjunction with values of dry-bulb and WET-BULB TEMPERATURE and is expressed in millibars. Water vapour's concentration is highly variable in the atmosphere, so vapour pressure changes substantially in time and space with the highest values (15–20 mb) being found in the humid tropics and the lowest (1–2 mb) across wintertime high latitude continents.   RR

varves Traditionally defined as being sedimentary beds or lamina deposited in a body of still water within the course of one year. The term has normally been applied to thin layers, usually deposited by meltwater streams in a body of water in front of a glacier. A glacial varve normally includes a lower 'summer' layer consisting of relatively coarse-grained sand or silt, produced by rapid ice melt in the warmer months, which grades upwards into a thinner 'winter' layer composed of finer material deposited from suspension in quiet water while the streams feeding the lake are frozen. However, it is becoming increasingly clear that varves may be deposited in a wide range of environments, both lacustrine and marine, and an alternative term, rhythmite, is now widely used.   ASG

Reading
O'Sullivan, P.E. 1983: Annually laminated sediments and the study of Quaternary environmental changes – a review. *Quaternary science review* 1, pp. 245–313.
Schlüchter, Ch. 1979: *Moraines and varves: origins, genesis, classification.* Rotterdam: Balkema.

vasques Wide (up to several decimetres), shallow pools with flat bottoms, which form a network consisting of a tiered, terrace-like series of steps on limestone coastal platforms, especially in aeolianite. The pools are separated from each other by winding, narrow, lobed ridges, 10–200 mm in height, and running continuously for tens of metres. They develop between high and low tide levels, especially in intertropical and Mediterranean climatic regions.   ASG

Reading
Battistini, R. 1983: La morphogénèse des plateformes de corrosion littorale dans les grès calcaires (plateforme supériere et plateforme à vasques) et le problème des vasques, d'après des observations faites à Madagascar. *Revue de géomorphologie dynamique* 30, pp. 81–94.

vauclusian spring See SPRINGS.

veering See WIND.

velocity area method A widely used method of measuring the discharge of a river, in which a series of verticals is used to subdivide the cross-section into a number of segments, the discharge of each segment is determined as the product of average *velocity* and cross-sectional *area*, and the total discharge is calculated as the sum of the values for the individual segments. Verticals are spaced at intervals of no greater than 1/15th of the width. Measurements of mean velocity in the vertical are obtained by using a rotating current meter and these are assumed to be representative of the average velocity in the adjacent segment. (See also DISCHARGE.)   DEW

Reading
British Standards Institute 1964: *Methods of measuring liquid flow in open channels. Part 3: Velocity area methods.* BS 3680. London: British Standards Institution.

velocity profile and measurement The velocity of flow in a river channel will vary both vertically and laterally in response to boundary resistance. Maximum velocities will generally occur near the surface at the centre of the channel. A velocity profile depicts the variation of velocity with depth; the graph

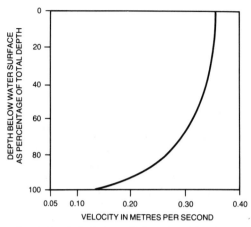

A typical vertical velocity profile for a river channel.

shows a typical example. Approximate values of mean velocity in the vertical can be obtained from measurements at 0.6 depth.

A wide variety of techniques has been employed to measure flow velocity. These include rotating current meters, optical current meters, electromagnetic current meters, tracers, floats, acoustic equipment and devices sensing impact pressure (e.g. pitot tubes) and vane deflection. Some will provide only values of mean velocity or surface velocity, and the rotating current meter is commonly used for measuring point velocities and velocity profile (See also CURRENT METER.)   DEW

Reading
Herschy, R.W. ed. 1978: *Hydrometry, principles and practices.* Chichester: Wiley.

ventifact A stone that has been shaped by the wind, especially in arid and polar areas. Abrasion is achieved by sand, dust or snow, and the stones become shaped (often into three-sided DREIKANTER), and have surface textures that may be polished, pitted or fluted. They have some utility for estimating past and present wind directions.   ASG

Reading
Whitney, M.I. and Dietrich, R.V. 1973: Ventifact sculpture by windblown dust. *Bulletin of the Geological Society of America* 84, pp. 2561–81.

vertical motion (in the atmosphere). The vertical component of air velocity. Persistent net horizontal flow out of a region, called horizontal DIVERGENCE, would result in the depletion of air in the region, so that its density would decrease;

but observed density changes are small, so that vertical motion must result. Such persistent outflow leads to downward motion above a flat surface (where the vertical motion is necessarily zero). In the upper troposphere persistent horizontal divergence usually leads to upward motion because the great stability of the stratosphere above is unfavourable for vertical motions. Persistent horizontal convergence leads to vertical motions of the opposite sign.

Vertical motion in the troposphere is normally upwards in CYCLONES and downwards in ANTI-CYCLONES. Magnitudes of vertical motions in such systems are typically only a few centimetres per second, but, because they persist, large vertical displacements of air are involved. The vertical motion in frontal zones is greater than the average values within depressions, being typically a few tens of centimetres per second.

Vertical motions within cumulus clouds are of the order of a metre per second, but in cumulonimbus cloud they may be as much as several tens of metres per second.   KJW

Reading
Battan, L.J. 1974: *Fundamentals of meteorology.* Englewood Cliffs, NJ: Prentice-Hall.

vertical stability/instability (in the atmosphere) Terms which relate to the response of say, a parcel of air to the imposition of an impulse and whether this will lead the parcel to move a small distance and then return to its original position (stability) or to move away from its original position (instability). The qualification 'vertical' indicates that the concept relates to rising or sinking motion.

The usual way of assessing the nature of the stability is first to plot the TEMPERATURE profile from a RADIOSONDE ascent on a THERMODYNAMIC DIAGRAM. This is looked upon as representing a vertical 'snapshot' of the temperature lapse within the local environment (the environmental LAPSE RATE or ELR) and forms the basis for the graphical evaluation of the air parcel's stability.

If an air bubble is unsaturated and starts to rise it will cool at the dry adiabatic lapse rate (DALR) of 9.8 K km^{-1}. If this is less than the rate within the environment (the ELR) through which it ascends, the bubble will be warmer than its surroundings and

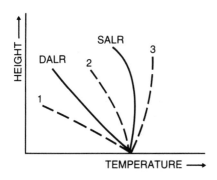

Vertical temperature profiles and stability.

will continue to ascend (case 1 in the diagram). This situation is known as absolute instability since within such an environment the air parcel will always be buoyant. The air close to strongly heated ground in the summertime exhibits such an absolutely unstable lapse rate.

If, however, an air parcel is saturated, the rate at which its temperature decreases as it ascends depends partly on how much moisture is available to act as a source of latent heating to offset the adiabatic cooling. This means that a saturated parcel of air cools much more slowly than a dry parcel and illustrates the important role played by moisture in atmospheric processes of this kind. At low levels the saturated adiabatic lapse rate (SALR) is typically half the dry rate. In nature the most commonly observed ELR is illustrated by case 2 in the diagram which is stable for dry ascent and unstable for saturated ascent. Because the instability depends on whether the bubble is saturated or not, it is known as conditional instability.

Finally case 3 in the diagram illustrates the antithesis of case 1. For both dry and saturated ascent the profile is stable because either parcel is always cooler than its environment. This absolute stability occurs for example within INVERSIONS OF TEMPERATURE such as are observed usually 1–2 km above the surface in anti-cyclones. These stable inversion layers commonly act as 'lids' which dampen convective cloud formed in the unstable layer below.   RR

vesicular Possessing numerous large pores and internal voids.

vicariance biogeography In general terms, the study of groups of plants or animals which are descended from a common ancestor but which are now spatially isolated from each other in disjunct or endemic distributions; more specifically, a recently developed school of biogeography which traces its ancestry through the writings of de Candolle, Croizat and Hennig, focusing on allopatric differentiation (see SYMPATRY) and endemism, linking the study of CLADISTICS and TAXONOMY with that of biogeography.

At the simplest level, many vicariants, or vicarious taxa, form species-pairs, such as the American *Maianthemum canadense* and the Eurasian *Maianthemum bioflium* or *Drosera uniflora* in South America and *Drosera arcturi* in New Zealand. But others have a more complex grouping and represent multiple vicariism, a phenomenon particularly common on isolated oceanic islands. A fine example of such multiple vicariism is provided by the 30 or so species of the palm genus, *Pritchardia*, on the Hawaiian Islands, where the geographical isolation of the taxa is a result of both island separation and the internal topography of the islands themselves. On one island, for example, there are nine different species, each of which grows in a separate valley.

This classic concept of vicariance involves allopatric species which have descended from a common ancestral population and which attained some degree of spatial isolation. Vicariance is essentially a passive condition when contrasted with ADAPTIVE RADIATION, a process involving active selection by different environmental conditions. This traditional form of vicariance is described as horizontal or geographical vicariance. The differences in the isolated pairs of groups will have arisen through the lack of gene exchange, chance differences in genetic composition and, possibly, genetic drift.

Three other forms of vicarious distribution have also been recognized. The first of these is altitudinal vicariance in which the related taxa form lowland/highland pairs. The second is habitat or ecological vicariance where the species-pairs occupy different environmental niches. This is well exemplified by the sea arrow-grass (*Triglochin maritima*) and the marsh arrow-grass (*T. palustris*), which inhabit respectively saltwater and freshwater marshes. Paradoxically, the term has also been used to describe taxa unrelated from the evolutionary standpoint, but which appear to be adapted to the same ecological niche in different locations. Finally, phenological vicariants have been recognized in which the separation is seasonal, with the related taxa flowering or breeding at different times in the year. Such taxa may actually be sympatric.

The proponents of the new school of vicariance biogeography seek a far more comprehensive philosophy and methodology and see themselves in direct and crusading opposition to what they term 'dispersalist' biogeography in which, they argue, the traditional explanations for allopatric species depend on the long-range dispersal of taxa from centres of origin. They claim that this 'misguided' approach is based on the views and writings of Charles Darwin, whose *On the origin of species* (1859) and ideas have been dismissed as 'piffle' by one of the main protagonists of the new school, Leon Croizat. Unfortunately, the whole debate has been marred by totally unnecessary and often gratuitous *ad hominem* abuse, so much so that it has frequently proved very difficult to grasp the thrust of the arguments involved. The contrived opposition of vicariance versus dispersal also seems to many biogeographers somewhat forced, as these are by no means the only processes involved in the development of plant and animal areas.

Two particular developments underpin the growth of the new school, which is most fully documented in the pages of the journal, *Systematic zoology*. The first is traced through the phylogenetic systematics, now normally referred to as cladistics, of the entomologist, Willi Hennig (1966). Cladistics is simply a method of classification, giving rise to graphs (cladograms) of relative affinity. They make no *a priori* assumptions about the nature of the relationships involved. On replacing the taxa represented by cladograms with the localities they inhabit, cladograms of affinities of areas or biological area-cladograms result. These are crucial to the new approach, for, as Nelson (in Nelson and Rosen 1981) is at pains to stress, vicariance biogeography begins by asking the question: 'Is there a cladogram of areas of endemism?'

Secondly, the new biogeography implicitly accepts that the motor of vicariance is continental movement, brought about through sea-floor spreading and plate tectonics, and it is this geological or geophysical model of the earth which has given the concept of vicariance biogeography such a boost since the geologists' crucial synthesis of plate tectonics in the 1960s. Nevertheless, most vicariance biogeographers insist that the sequence of distributional events must be derived from the taxa themselves *before* the geological evidence is introduced and that the biological evidence must stand whether or not it matches the 'fashionable' geology.

Thus, the new biogeography attempts above all to link recent developments in systematics with biogeography. It is first and foremost concerned with the coincidence of pattern, a point clearly emphasized by the essential character of Croizat's original methods in what are seen as founder publications, particularly *Panbiogeography* (1958) and *Space, time, form: the biological synthesis* (1964). His idea was to map distributions of whole varieties of taxa and then to establish concordances of pattern in order to identify ancestral biotas. This then necessitated explanations for the fragmentation of previously more extensive distributions. The phenomenon of fragmentation was vicariance. If a given type of distribution – 'an individual track' – recurs in group after group of organisms, the region delineated by the coincident distributions – 'the generalized track' – becomes statistically and thus geographically significant.

The method is well exemplified by the study of Parenti (1981) on cyprinodontiform fishes. Cladograms of areas are derived from cladograms of taxa and then from the coincidence of areal patterns for different taxa 'a pattern of earth history is suggested.' It should be especially noted that this pattern is independent of geological hypotheses although, it is admitted, a geological model may be found to fit the pattern. Above all, the absence of such a model is not seen as invalidating the biogeography.

Many biogeographers, of course, regret this rejection of independent sources of information and are far from happy with the newest manifestations of the underlying cladism (e.g. Ridley 1983). Moreover, they regard the whole method of vicariance biogeography as inductive, mechanical, unrelated to process, lacking in any true sense of geography beyond that of simple geographical coordinates, and over-encumbered with a new and less than euphonious jargon. Undoubtedly, the case has not been helped by the stridency of some of its proponents. Nevertheless, vicariance is an important feature of plant and animal distributions and, as with dispersal, biogeography would be considerably the poorer if it were not given its due prominence. PAS

Reading and References
Croizat, L. 1958: *Panbiogeography, or an introductory synthesis of zoogeography, phytogeography, and geology.* Caracas: privately published.

— 1964: *Space, time, form: the biological synthesis.* Caracas: privately published.

Hennig, W. 1966: *Phylogenetic systematics.* Urbana: University of Illinois Press.

†Nelson, G. 1978: From Candolle to Croizat: comments on the history of biogeography. *Journal of the history of biology* 11, pp. 269–305.

†— and Platnick, N.I. 1981: *Systematics and biogeography: cladistics and vicariance.* New York: Columbia University Press.

— and Rosen, D.E. eds 1981: *Vicariance biogeography: a critique.* New York: Columbia University Press.

Parenti, L.R. 1981: A phylogenetic and biogeographic analysis of cyprinodontiform fishes (Teleostei, Atherinomorpha). *Bulletin of the American Museum of Natural History* 168, pp. 335–557.

†Patterson, C. 1981: Biogeography: in search of principles. *Nature* 291, pp. 612–13.

Ridley, M. 1983: Can classification do without evolution? *New scientist* 1 December, pp. 647–51.

†Stoddart, D.R. 1977, 1978, 1981, 1983: Biogeography. *Progress in physical geography.* 1, pp. 537–43; 2, pp. 514–28; 5, pp. 575–90; 7, pp. 256–64.

†Stott, P.A. 1984: History of biogeography. In J.A. Taylor ed., *Themes in biogeography.* London: Croom Helm. Pp. 1–24.

vigil network See REPRESENTATIVE AND EXPERIMENTAL BASINS.

virgation The formation of trails of ice crystals falling from a cloud.

viscosity The property of fluids by virtue of which they resist flow. Newton's law of viscous (non-turbulent) flow is given by:

$$F = \eta A \, dv/dx$$

where F is the tangential force between two parallel layers of liquid of area A, a distance dx apart, moving with relative velocity dv, and η is the coefficient of viscosity, dynamic viscosity or just viscosity. It is measured in $N \, s \, m^{-2}$. It should be carefully distinguished from the kinematic viscosity v which is η divided by the fluid density ρ; the units here are in $m^2 \, s^{-1}$. WBW

Reading
Whalley, W.B. 1976: *Properties of materials and geomorphological explanation.* Oxford: Oxford University Press.

VISSR A visible infrared spin scan radiometer (VISSR) is carried by geostationary meteorological satellites for the production of visible imagery with a fine spatial resolution and thermal infrared imagery with a coarse spatial resolution. PJC

Volcanic activity plays a major role in the development of many mountain ranges. These eroded ignimbrites are in the western Cordillera of the Andes in Peru.

Types of volcanic eruptions

Type	Characteristics
Icelandic	Fissure eruptions, releasing free-flowing (fluidal) basaltic magma; quiet, gas-poor; great volumes of lava issued, flowing as sheets over large areas to build up plateaux (Colombia).
Hawaiian	Fissure, caldera and pit crater eruptions; mobile lavas, with some gas; quiet to moderately active eruptions; occasional rapid emission of gas-charged lava produces fire fountains; only minor amounts of ash; builds up lava domes.
Strombolian	Stratocones (summit craters); moderate, rhythmic to nearly continuous explosions, resulting from spasmodic gas escape; clots of lava ejected, producing bombs and scoria; periodic more intense activity with outpourings of lava; light-coloured clouds (mostly steam) reach upward only to moderate heights.
Vulcanian	Stratocones (central vents); associated lavas more viscous; lavas crust over in vent between eruptions, allowing gas build-up below surface; eruptions increase in violence over longer periods of quiet until lava crust is broken up, clearing vent, ejecting bombs, pumice and ash; lava flows from top of flank after main explosive eruption; dark ash-laden clouds, convoluted, cauliflower-shaped, rise to moderate heights more or less vertically, depositing tephra along flanks of volcano. (Note: ultravulcanian eruption has similar characteristics but results when other types, e.g. Hawaiian, become phreatic and produce large steam clouds, carrying fragmental matter.)
Vesuvian	More paroxysmal than Strombolian or Vulcanian types; extremely violent expulsion of gas-charged magma from stratocone vent; eruption occurs after long interval of quiescence of mild activity; vent tends to be emptied to considerable depth; lava ejects in explosive spray (glow above vent), with repeated clouds (cauliflower) that reach great heights and deposit tephra.
Plinian	More violent form of Vesuvian eruption; last major phase is uprush of gas that carries cloud rapidly upward in vertical column for miles; narrow at base but expands outward at upper elevations; cloud generally low in tephra.
Peléean	Results from high-viscosity lavas; delayed explosiveness; conduit of stratovolcano usually blocked by dome or plug; gas (some lava) escapes from lateral (flank) openings or by destruction or uplift of plug; gas, ash and blocks move downslope in one or more blasts as nuées ardentes or glowing avalanches, producing directed deposits.
Katmaian	Variant of a Peléean eruption characterized by massive outpourings of fluidized ash flows; accompanied by widespread explosive tephra; ignimbrites are common end products, also hot springs and fumaroles.

Source: Short, N.M. 1986: Volcanic landforms. In N.M. Short and R.W. Blair eds, *Geomorphology from space – a global overview of regional landforms*. Washington, DC: NASA. Table 3–1, p. 187.

void ratio The ratio of the volume of interstitial voids in a portion of sedimentary rock to the volume of that portion.

volcano An opening or vent through which magma, molten rock, ash or volatiles erupt on to the earth's surface, or the landform produced by the erupted material. Volcanoes tend to be conical in shape but can have a variety of forms, depending on the nature of the erupted material (particularly its viscosity), the character of recent eruptive activity and the extent of post-erupted modification by erosion (see illustration on p. 539). Most volcanoes are concentrated at convergent and divergent plate boundaries (see PLATE TECTONICS) but others located in the interior of plates, are associated with HOT SPOTS.   MAS

Reading
Francis, P.W. 1976: *Volcanoes*. Harmondsworth: Penguin.
Ollier, C.D. 1988: *Volcanoes*. Oxford: Blackwell.
Williams H. and McBirney, A.R. 1979: *Volcanology*. San Francsico: Freeman, Cooper.

vorticity A microscopic measure of rotation in a fluid. It is a vector quantity defined as the curl of the velocity and has dimensions of $(\text{time})^{-1}$. In cartesian coordinates the x, y and z components of vorticity (ζ) are given by:

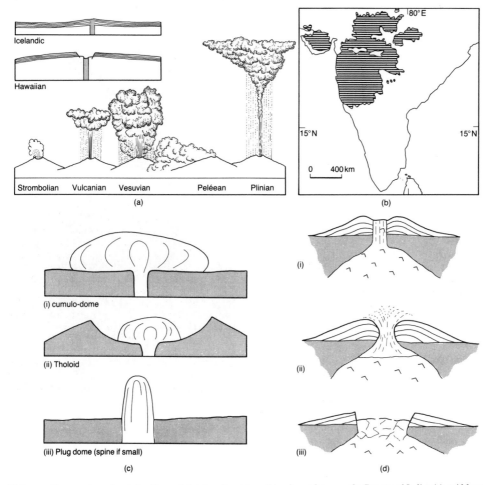

Volcano: *Some major volcanic landforms: (a) types of eruption; (b) a lava plateau – the Deccan of India; (c) acid lava (viscous) extrusion forms; (d) the stages in the formation of a caldera by collapse: (i) initial volcano, (ii) explosion, (iii) collapse.*
Source: *A.S. Goudie 1984:* The nature of the environment. *Oxford and New York: Basil Blackwell.*

$$\zeta_x = \frac{\partial w}{\partial y} - \frac{\partial v}{\partial z}; \zeta_y = \frac{\partial u}{\partial z} - \frac{\partial w}{\partial x}$$

$$\zeta_z = \frac{\partial v}{\partial x} - \frac{\partial u}{\partial y};$$

where u, v and w are the x, y and z components of velocity. The absolute vorticity is given by the curl of the absolute velocity and relative vorticity is given by the curl of the velocity relative to the earth. In the northern hemisphere relative vorticity in a cyclonic sense is positive and in an anti-cyclonic sense is negative. Fluid in solid

rotation with angular velocity ω has vorticity 2ω.

In large-scale motions in the atmosphere the vertical components of absolute and relative vorticity are of chief importance and these terms are often used without the explicit modifier 'vertical component of'. The difference between the vertical component of absolute and relative vorticity is given by the vertical component of vorticity of the earth due to its rotation, being $2\Omega \sin \phi$, where Ω is the angular velocity of the earth and ϕ is latitude.

In adiabatic, frictionless motion a quantity called potential vorticity is conserved, this being expressed by:

$$(\text{absolute vorticity}) \times \frac{\partial \theta}{\partial p} = \text{constant}$$

where θ is potential temperature and p is pressure.   KJW

Reading
Atkinson, B.W. ed. 1981: *Dynamic meteorology: an introductory selection*. London and New York: Methuen.
Gill, A.E. 1982: *Atmosphere–ocean dynamics*. London: Academic Press.
Houghton, J. 1986: *The physics of atmospheres*. Cambridge: Cambridge University Press.

vugh A void or cavity within a rock which can be lined with mineral precipitates.

vulcanism The movement of magma or molten rock and associated volatiles onto or towards the earth's surface. Extrusions of material onto the earth's surface can occur through a vent known as a VOLCANO or through linear openings in the crust, called fissures. Intrusion of magma or molten material into the upper part of the crust can give rise to large masses of igneous rock, such as BATHOLITHS, which may cause domal uplift of the overlying strata. The term vulcanism is also used as a synonym of volcanism and in this sense refers only to extrusion of material onto the earth's surface.   MAS

Reading
Williams, H., and McBirney, A.R. 1979: *Volcanology*. San Francisco: Freeman, Cooper.

Vulcanism is frequently associated with the margins of the major global plates. The Volcano de Fuego in Guatemala is still active and forms a part of the great 'ring of fire' around the Pacific Ocean.

W

wadi An Arabic word, sometimes also spelt *oued*, generally used as a term for ephemeral river channels in desert areas. Wadis may flow only occasionally, and then sometimes discontinuously, along their courses.

A desert river system – wadi – developed in Sahra el-Arabiya, Egypt. In spite of low annual rainfall totals, fluvial activity is a potent force in many desert areas.

Wallace's line A zoogeographical boundary, originally put forward in 1858 by Alfred Russel Wallace, which runs through the middle of the Malay archipelago and which, Wallace argued, marked the meeting of the Asian and Australian faunas. The original Wallace line ran between the islands of Bali and Lombok and then between Borneo and Celebes (Sulawesi), and was based primarily on the distribution of barbets, cockatoos and parrots. Later, in 1910, Wallace moved his line to the east, so that it lay between Celebes and the Moluccas. Many alternative lines have been suggested and, in 1928, the whole island area with its complex distributions was designated by Dickerson, following a suggestion previously made by Wallace himself, a separate zoogeographical region called Wallacea. Plate tectonic studies have now revealed that the 'anomlaous' island of Celebes comprises geological elements of both Gondwanaland and Laurasia, which collided there in the middle Miocene. This may help to explain the unique assemblage of organisms on this and neighbouring islands. (See diagram on p. 544.) PAS

Reading
Whitmore, T.C. ed. 1981: *Wallace's line and plate tectonics.* Oxford: Clarendon Press.

Wallace's realms A division of the world into six zoogeographical regions, defined by their distinctive faunas and proposed by Alfred Russel Wallace in his classic of zoogeography, *The geographical distribution of animals* (1876); also called the Sclater–Wallace system of zoogeographical regions. The names of the six regions were adapted from the continents or were of a classical form, and are normally given as: the Pal(a)earctic, the Nearctic, the Neotropical, the Ethiopian, the Oriental and the Australian (see FAUNAL REALMS for a full description and illustration). Wallace based his system on the earlier work of Philip Lutley Sclater (1858), who had likewise recognized six regions based on his study of the distribution of birds. Wallace chose mammals to verify Sclater's six divisions and, in doing so, he drew on his wealth of personal exploration, which ranged from his

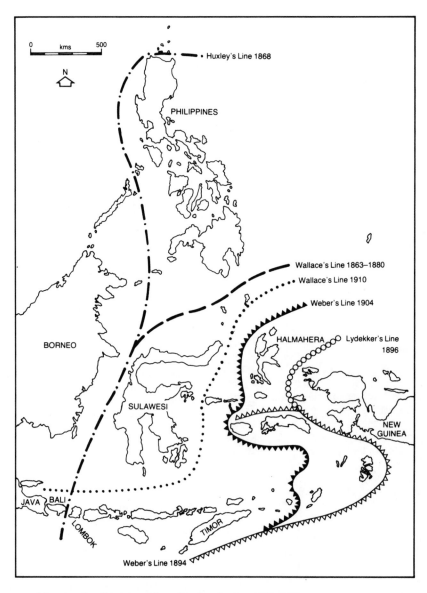

Lines suggested for separating Oriental and Australian faunal regions 1863–1910.

visits to the Amazon and Rio Negro (1848–52) to south-east Asia (1854–62). Though frequently modified and expanded, the Sclater–Wallace regions remain the basis of our understanding of the great 'realms of life'. (See also FLORISTIC REALMS; WALLACE'S LINE.)   PAS

Reading
George, W. 1962: *Animal geography*. London: Heinemann.
— 1964: *Biologist philosopher: a study of the life and writings of Alfred Russel Wallace*. London: Abelard-Schuman.
Sclater, P.L. 1858: On the general distribution of the members of the class *Aves*. *Journal of the Linnean Society of London* 2, pp. 130–45.
Smith, C.H. 1983: A system of world faunal regions. 1. Logical and statistic derivation of the regions. *Journal of biogeography* 10, pp. 455–66.
Udvardy, M.D.F. 1969: *Dynamic zoogeography*. New York: Van Nostrand Reinhold.

waning slopes Depositional hillslope units formed at the base of a talus (scree) slope as weathering and rain wash fine-grained particles from the talus and deposit them as a concave unit which may become progressively flatter (Wood 1942). L.C. King (1957) adopted the basic definitions of slope units, proposed by Wood, and the waning slope is identified as a rock-cut pediment with a veneer of sediment produced by surface wash.   MJS

References
King, L.C. 1957: The uniformitarian nature of hillslopes. *Transactions of the Edinburgh Geological Society* 17, pp. 81–102.
Wood, A. 1942: The development of hillside slopes. *Proceedings of the Geologists' Association* 53, pp. 128–40.

warm front A frontal zone in the atmosphere where, from its direction of movement, cool air is being replaced by rising warm air. As the front approaches, cirrus clouds are gradually replaced by cirrostratus, altostratus and finally nimbostratus clouds. Precipitation usually occurs within a wide belt up to about 400 km ahead of the surface front. As the surface front passes, temperatures and dewpoint increase, winds veer and pressure stops falling. Most fronts exhibit considerable differences from this standard model outlined, partly depending upon the rate of uplift within the warm air. Well-developed warm fronts are relatively rare in the southern hemisphere extra-tropical cyclone belt where sources of warm air for the warm sector are limited. Kinematically and dynamically there is no fundamental difference between cold and warm fronts, despite the views of the Norwegian School of Meteorology which made a sharp distinction between their roles in the precipitation process.   PS

warm ice See TEMPERATE ICE.

warm occlusion See OCCLUSION.

warm sector The area of warm air lying between the warm and cold fronts of an EXTRA-TROPICAL CYCLONE. It eventually disappears from the surface as the cyclone evolves to become occluded. Temperatures in the warm sector are noticeably higher than in the preceding and following air streams. Cloud and precipitation are very variable. With a strong ridge of high pressure there may be clear skies but more frequently the skies are overcast. Heavy rain is likely over upland areas if the warm sector is potentially or conditionally unstable.   PS

warping The ending and deformation of extensive areas of the earth's crust without the formation of folds or faults.

washboard moraine Morainic ridges a few metres high lying transverse to the direction of former ice flow. Also called *cross-valley*, *De Geer* and *ribbed* moraine. There are almost as many theories about transverse moraines as there are researchers, but one common characteristic is that they are often associated with the presence or former presence of lakes. They may be a subaqueous example of a push moraine (perhaps annual). Others may relate to subglacial thrusting as proposed for Rogen moraines. (See also MORAINE.)   DES

Reading
Embleton, C. and King, C.A.M. 1975: *Glacial geomorphology*. London: Edward Arnold.

water balance The water balance or water budget of an area over a period of time represents the way in which precipitation during the time period is partitioned between the processes of evapotranspiration and run-off, taking account of changes in water storage. The water balance summarizes the changes in the components of the HYDROLOGICAL CYCLE during a particular time period and may be expressed for a drainage basin as:

$$P = Q + Et \pm \Delta SS \pm \Delta SMS \pm \Delta AZS \pm \Delta GS \pm DT$$

where all the variables are expressed as a depth of water over the catchment area for the time period studied. P is precipitation, Q is run-off, Et is evapotranspiration losses, ΔSS is change in surface storage, ΔSMS is change in soil moisture storage, ΔAZS is change in aeration zone storage, ΔGS is change in groundwater storage, and DT is deep transfer of water across the watershed.

The water balance can be evaluated by direct field measurements, or a climatic water balance can be calculated by making simplifying assumptions about the role and operation of different stores. A water balance can be calculated for any size of area from small LYSIMETERS up to whole continents and it can be used to isolate the hydrological effects of man's activities, but water balance evaluation is most widely applied at the drainge basin scale. A water

Water balance components in millimetres for world continents

Continent	Precipitation	Evapotranspiration	Run-off
Europe	657	375	282
Asia	696	420	276
Africa	696	582	114
Australia (with islands)	803	534	269
Australia (without islands)	447	420	27
North America	645	403	242
South America	1564	946	618
Antarctica	169	28	141

Source: Baumgartner and Reichel 1975.

balance can also be calculated for any time period, but it is simplest to evaluate for a time period where storage is approximately the same at the start and end of the period. In the UK the WATER YEAR which runs from 1 October to 30 September is often used for water balance calculations because this starts and ends at the period of minimum storage in most years. If the storage can be assumed to be the same at the start and end of the required period the water balance equation simplifies so that precipitation is only partitioned between evapotranspiration and run-off.

Estimation of the water balance of individual basins is often used to check or to estimate the magnitude of a water balance component that is difficult to measure. Evapotranspiration loss has been frequently studied in this way and formed the basis of the Institute of Hydrology's original Plynlimon catchment study, where losses from an afforested catchment and a grassland catchment were estimated using a water balance approach and were compared with potential evapotranspiration rates calculated from observations from automatic weather stations. Water balance studies are also useful in identifying the impact of catchment modifications on hydrological processes. This is well illustrated by Clarke and Newson's analysis (1978) of the impact of the 1975–6 drought in the Institute of Hydrology's research catchments and the implications for the water balance of drainage basins under different vegetation cover.

Calculation of the water balance from climatic data was first attempted by Thornthwaite (1948) in his classification of world climates. The evaluation of a climatic water balance requires observations or estimates of precipitation and potential evapotranspiration (see POTENTIAL EVAPORATION) and a means of correcting potential evapotranspiration to an actual evapotranspiration rate by taking account of the degree to which evapotranspiration is limited by availability of soil moisture. Baier (1968) reviewed various proposals for the relationship between the ratio of actual to potential evapotranspiration and percentage soil moisture availability. The UK Meteorological Office (Grindley 1967) assumes a mix of riparian, short-rooted and long-rooted vegetation. The riparian vegetation is assumed continually to transpire at the potential rate and the short- and long-rooted vegetation are assumed to transpire at the potential rate up to soil moisture deficits of 75 mm and 200 mm respectively, and then at a gradually reducing rate. Thus, the water budget which partitions precipitation between evapotranspiration, soil moisture storage and rainfall excess can be calculated from an area and for consecutive time periods, most frequently producing a monthly water budget.

The evaluation of the water balance provides an essential stage in the estimation of the water resources of an area and so mapping of water balance components (see HYDROLOGICAL MAPS), notably, precipitation, evapotranspiration and run-off have received a great deal of attention. Gurnell (1981) reviews attempts at evapotranspiration mapping and many of these maps fall within a context of water balance mapping, ranging from maps of the water balance of administrative areas within countries, to national maps and continental maps (Doornkamp *et al.* 1980), to maps for the whole world (Baumgartner and Reichel 1975; UNESCO 1978a and b). The table summarizes the water balance of the continents according to Baumgartner and Reichel (1975). Mapping the water balance for such large areas requires the combined use of measurements of precipitation, run-off and climatic variables which allow the estimation of evapotranspiration; the evaluation of climatic water balances and the use of regional relationships between altitude and each of the water balance variables. In this way isoline maps of the water balance components over large areas may be produced.

Water balance information is a useful tool in water resources planning because it can represent both long-term average and extreme conditions at a site. Extreme values of soil moisture deficit and water surplus may give an indication of the drought and flood producing potential of an area and temporal trends in the water balance may reflect climatic change or the influence of catchment modification. The water balance equation and the concept of the hydrological cycle are the foundations of hydrological studies. AMG

References
Baier, W. 1968: Relationship between soil moisture, actual and potential evapotranspiration. In *Soil moisture*. Proceedings of the Hydrology Symposium 6. University of Saskatchewan, 15–16 November 1967. Ottawa: Queen's Printer.

Baumgartner, A. and Reichel, E. 1975: *The world water balance: mean annual global , continental and maritime precipitation, evaporation and runoff*. New York: Elsevier.

Clark, R.T. and Newson, M.D. 1978: Some detailed water balance studies of research catchments. *Proceedings of the Royal Society* series A, 363, pp. 21–42.

Doornkamp, J.C., Gregory, K.J. and Burn, A.S. eds 1980: *Atlas of drought in Britain 1975–1976*. London: Institute of British Geographers.

Grindley, J. 1967: The estimation of soil moisture deficits. *Meteorological magazine* 96, pp. 97–108.

Gurnell, A.M. 1981: Mapping potential evapotranspiration: the smooth interpolation of isolines with a low density station network. *Applied geography* 1, pp. 167–83.

Thornthwaite, C.W. 1948: An approach toward a rational classification of climate. *Geographical review* 38, pp. 55–94.

UNESCO 1978a: *World water balance and water resources of the earth*. Paris: UNESCO.

— 1978b: *Atlas of the world water balance*. Paris. UNESCO.

water mass A body of water having approximately uniform characteristics typically acquired in a source region in contact with the atmosphere. Oceanic water masses are usually identified from their values of temperature and salinity which are conservative properties, although non-conservative properties – for example, the concentrations of dissolved oxygen or of nutrients such as phosphate and silica – can be useful in tracing their movements. Upper water masses are found above the THERMOCLINE where their formation is controlled by the pattern of surface currents

(see CURRENTS, OCEAN). For example, the subtropical GYRES enclose central waters of different types all of which have relatively high temperatures and salinities. The source regions of deeper water masses are predominantly at high latitudes where the thermocline is weak or absent. Most of the Atlantic Ocean basin is occupied by North Atlantic deep water; south of about 30°N Antarctic bottom water is found close to the sea bed and Antarctic intermediate water at depths in the range 500–1500 m. These deep water masses of the Atlantic are characterized by values of temperature (T) and salinity (S) that vary within very narrow limits; such water masses may be designated *water types*. Other water masses, including the central waters mentioned above, have a characteristic range of values of T and S (see table on p. 552). Investigations of the stratification of the oceans and the identification, movement and mixing of water masses is greatly facilitated by the plotting of data values on a graph of T as ordinate against S as abscissa (a T–S diagram) on which lines of constant density (isopycnals) are convex upwards. (See pp. 548–51 for diagrams.) JEA

Reading
Harvey, J.G. 1976: *Atmosphere and ocean*. London: Artemis Press.

Mamayev, O.I. 1975: *Temperature–salinity analysis of world ocean waters*. Amsterdam: Elsevier.

Open University Oceanography Course Team 1989: *Seawater: its composition, properties and behaviour*. Oxford: Pergamon; Milton Keynes: Open University Press.

Sverdrup, H.U., Johnson, M.W. and Fleming, R.H. 1942: *The oceans*. Englewood Cliffs, NJ: Prentice-Hall.

Tolmazin, D. 1985: *Elements of dynamic oceanography*. Boston and London: Allen & Unwin.

water table The surface defined by the level of free standing water in fissures and pores at the top of the saturated zone. It is an equilibrium surface at which fluid pressure in the voids is equal to atmospheric pressure. The equivalent term in continental European literature is the PIEZOMETRIC SURFACE. A potential source of confusion arises because the latter is sometimes used in English literature to describe the water-level elevations in wells tapping a confined artesian AQUIFER. The term *potentiometric surface* is preferred for this (Freeze and Cherry 1979). PWW

Reference
Freeze, R.A. and Cherry, J.A. 1979: *Groundwater*. Englewood Cliffs, NJ: Prentice-Hall.

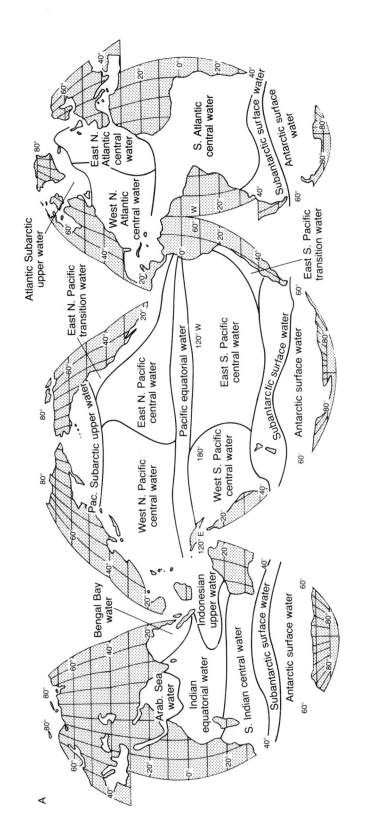

Water mass: *(A) The global distribution of upper water masses.*

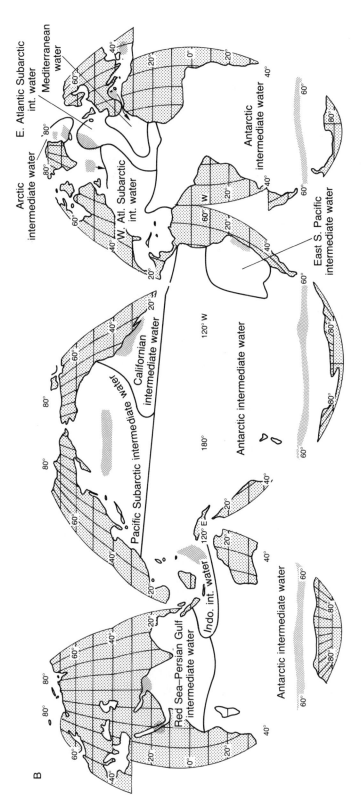

Water mass: (B) The global distribution of intermediate water masses (between 550 and 1500 m depth). The source regions of the water masses are indicated by dark tone. Note that Antarctic intermediate water is by far the most widespread intermediate water mass.

B

Arctic intermediate water

E. Atlantic Subarctic int. water

Mediterranean water

W. Atl. Subarctic int. water

Antarctic intermediate water

East S. Pacific intermediate water

Pacific Subarctic intermediate water

Californian intermediate water

Antarctic intermediate water

Red Sea–Persian Gulf intermediate water

Indo. int. water

Antarctic intermediate water

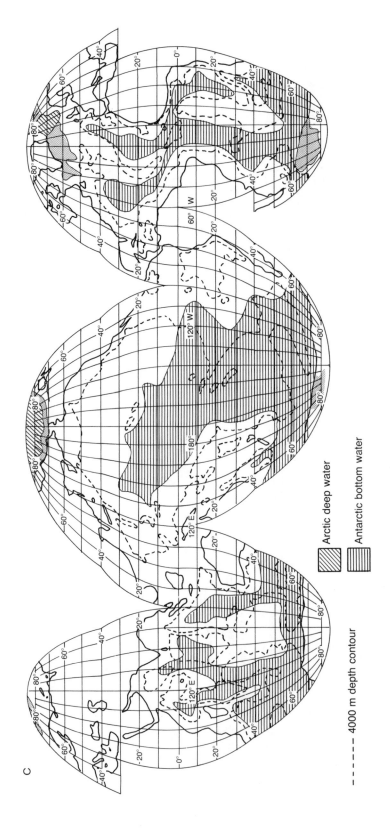

Water mass: (C) The global distribution of deep and bottom water masses (between a depth of about 1500 m and the sea floor). The source regions are shown by dark tone. The fine dashed line is the 4000 m isobath.
Source: Open University Oceanography Course Team 1989: Ocean circulation. Oxford: Pergamon Press, Milton Keynes: Open University Press. Figures 6.12, 6.13 and 6.17.

▨▨▨ Arctic deep water

▥▥▥ Antarctic bottom water

− − − − − 4000 m depth contour

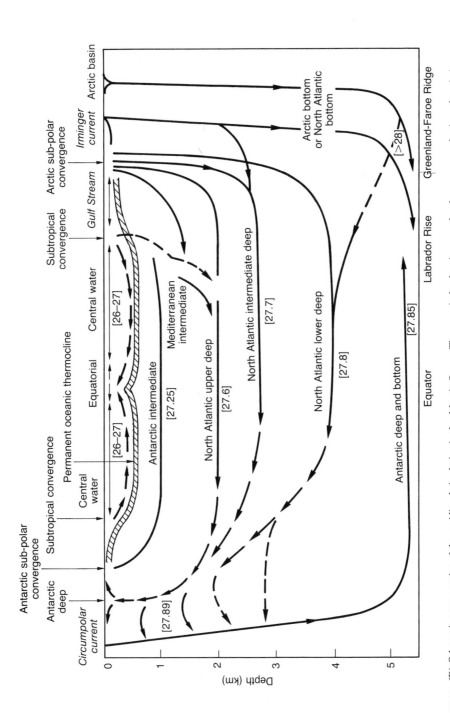

Water mass: *(D) Schematic representation of the meridional circulation in the Atlantic Ocean. The numerical values in square brackets are mean density values (σ_t).*
Source: *Tolmazin 1985. Figure 7.8.*

Major water masses of the world oceans and their T–S characteristics

Location	Atlantic Ocean			Indian Ocean			Pacific Ocean		
	Name	T(°C)	S(%)	Name	T(°C)	S(%)	Name	T(°C)	S(%)
Central waters	North Atlantic	20.0	36.5	Bay of Bengal	25.0	33.8	Western North Pacific	20.0	34.8
	South Atlantic	18.0	35.9	Equatorial	25.0	35.3	Eastern North Pacific	20.0	35.2
				South Indian	16.0	35.6	Equatorial		
							Western South Pacific	20.0	35.7
Intermediate waters	Atlantic subarctic	2.0	34.9	–			Pacific subarctic	5–9	33.5–33.8
	Mediterranean intermediate	11.9	36.5	Red Sea intermediate	23.0	40.0	North Pacific intermediate	4–10	34.0–34.5
	Antarctic intermediate	2.2	33.8	Timor Sea intermediate	12.0	34.6	South Pacific intermediate	9–12	33.9
				Antarctic intermediate	5.2	34.7	Antarctic intermediate	5.0	34.1
Deep and bottom waters	North Atlantic deep and bottom	2.5	34.9	Antarctic deep and bottom	0.6	34.7	Antarctic deep and bottom	1.3	34.7
	Antarctic deep	4.0	35.0						
	Antarctic bottom	−0.4	34.66						

Source: Tolmazin 1985, table 7.1. After Mamayev 1975; Sverdrup et al. 1942.

water year (or hydrological year) runs from 1 October to 30 September in the UK.

waterfall A stream that falls from a height. Waterfalls are often the sites of greatest concentration of energy dissipation along the course of a stream and have generally been regarded as forming where a soft rock is eroded from beneath a harder rock (the caprock model). However, in reality, waterfalls have more diverse forms than this simple model would suggest. It is likely that it is applicable only in areas with gently dipping strata. In addition to such structural control, waterfalls depend for their development on such factors as glacial overdeepening, tectonic changes and base-level change. ASG

Reading
Young, R.W. 1985: Waterfalls: form and process. *Zeitschrift für Geomorphologie*, supplementband 55, pp. 81–95.

watershed The boundary which delimits a drainage basin as the basic hydrological unit. On large-scale topographical maps it is often drawn as a line according to the contour information, but this surface watershed may not correspond with the subsurface boundary of the basin delimited according to the WATER TABLE as the PHREATIC DIVIDE. Sometimes, especially in North America, watershed is used synonymously with CATCHMENT or drainage basin area. KJG

watershed model A physical or mathematical representation of a drainage basin, and the processes operating within it, that provides improved explanation of its behaviour or permits prediction of its response. Many different criteria have been used to classify watershed models, but a simple scheme could distinguish between physical, analogue and mathematical models. A physical model is essentially a scaled-down physical representation of a drainage basin that can be studied in the laboratory. In an analogue model use is made of an analogous system, such as the flow of electricity through a circuit, to represent the watershed.

Mathematical models are the most widely used and may be further subdivided into stochastic and deterministic models. A stochastic model places emphasis on the statistical properties of hydrological data and may employ random variables with probability distributions. In a deterministic model, an attempt is made to represent the actual processes operating by means of physically based equations or simplified relationships, and a distinction may be made between lumped and distribution approaches. In the former case the model does not explicitly take account of spatial variability of inputs or watershed behaviour and is usally represented by a set of ordinary differential equations. A distributed model incorporates spatial variations in inputs and process response and often consists of a set of partial differential equations. Mathematical models have been successfully employed to represent both run-off processes and sediment and solute production from watersheds. DEW

Reading
Clarke, R.T. 1973: *Mathematical models in hydrology*. FAO irrigation and drainage paper 19. Rome: FAO.
Fleming, G. 1975: *Computer simulation techniques in hydrology*. New York: Elsevier.
Haan, C.T., Johnson, H.P. and Brakensick, D.L. eds 1982: *Hydrologic modeling of small watersheds*. St Joseph, Michigan: American Society of Agricultural Engineers.

waterspout A vortex disturbance that forms in the atmosphere over a body of water when the atmosphere is unstable and the air stagnant. Waterspouts are favoured by conditions of high temperature and humidity, and still air, and can cause damage to and even the foundering of large boats. ASG

Reading
Gordon, A.H. 1951: Waterspouts. *Weather* 6, pp. 364–71.

watten Tidal marshland between the mainland and an offshore sand bar.

wave cut platform See SHORE PLATFORMS.

waves Regular oscillations in the water surface of large water bodies. They are produced by the pushing effect of wind on the water surface. Energy is transferred from the wind to the water and eventually dissipated when the waves hit a shoreline. Waves are of great geomorphological importance because of the role they play in coastal sedimentation and erosion processes. In deep water, wave form can be predicted by a variety of equations, and movement of individual water particles takes the form of circular orbits within the wave. On approaching shallow water, wave forms become more complex and there is a

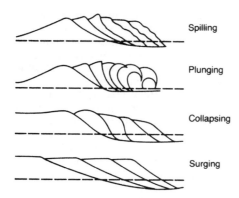

Spilling

Plunging

Collapsing

Surging

Successive profiles of various breaking waves.
Source: J.E. Costa and V.R. Baker 1981: Surficial geology.
Chichester and New York: Wiley. Figure 14.2.

general forward trend to the movement of
water particles.   HV

Reading
Carter, R.W.G. 1988: *Coastal environments.* London: Academic Press.

waxing slopes Hillslope units forming
crests of convex profile at the intersection
of the cliff and the hilltop or plateau surface.
It is assumed by Wood (1942) that these
slope units will increase in significance in
the landscape as cliffs (free faces) and talus
(scree) slopes are eliminated, and that they
will lengthen to become major parts of the
total hillslope as a result of weathering and
soil creep.   MJS

Reference
Wood, A. 1942: The development of hillside slopes.
Proceedings of the Geologists' Association 53, p. 128–40.

weather The overall state of the atmosphere on a time-scale of minutes to
months, with particular emphasis on those
atmospheric phenomena that affect human
activity. Thus sunshine, temperature, rainfall, wind, cloud contribute to weather
whereas air density does not. In contrast
to weather, CLIMATE is concerned with the
long-term behaviour of the atmosphere.
                                               BWA

weather forecasting The science of
predicting the future state of the atmosphere from very short periods of less than
one hour up to seven or even ten days
ahead. In recent years the science has been
revolutionized by the application of modern technology, e.g. by satellites and radar,
and by the development of advanced
mathematical models which can be run

on the most powerful of modern electronic
computers.

In many of the more advanced forecasting
centres operations are carried out as an
automatic routine with many thousands of
observations flowing in continuously from
around the globe. These come from
manned surface and upper air stations,
automatic data buoys in remote areas,
flight level reports from commercial aircraft, and polar orbiting and geostationary
satellites. The data are accepted automati-
caly into the computing system and are first
assimilated from their real geographically
scattered pattern into a regular global
network of so-called gridpoints at typically
fifteen levels from the surface to the
STRATOSPHERE. These points number
around 36,000 at a given level or approximately 500,000 in total. Given that seven
variables are assimilated at each point
(including TEMPERATURE, WIND and HU-
MIDITY), it is apparent that about 3.5×10^7
numbers are stored to define the state of the
atmosphere at the given observation time.

At every point a set of predictive
equations is solved in order to calculate
the values of temperature, wind, humidity,
etc. for a sequence of 'time-steps' to
produce a forecast for, say, 12, 24, 48 and
72 hours ahead. The longer term or
medium range (up to seven days) forecasts
are based on observations taken at the main
synoptic hours of 00 and 12 GMT. The
most powerful computers in existence are
used operationally and are currently capable
of carrying out 5×10^7 instructions per
second; a ten-day global weather forecast
involves around 5×10^{11} operations exe-
cuted over a four-hour period.

Modern weather forecasting by this
means provides a vast range of products
for appraisal by forecasters who are still
required to interpret and to communicate
the information to the user. Predicted fields
include the traditional mean sea-level
pressure charts and upper air charts along
with more modern fields of 'large-scale' and
'convergence scale' precipitation totals on a
regular grid and forecast TEPHIGRAMS. Very
short-term prediction of precipitation
(nowcasting) in the period nought to six
hours ahead is underway in Britain, based
on the linear extrapolation of radar network
observations of precipitation patterns.

Weather forecasting services are provided
for many users, for example offshore oil
developments (sea and swell, deck level

winds), civil aviation (wind, temperature and clear air turbulence along the flight), ship routeing (optimum course for a given journey) and agriculture (e.g. crop spraying operations and weather-related disease outbreak).   RR

Reading
Aktinson, B.W. 1968: *The weather business*. London: Aldus Books.
Browning, K.A. ed. 1982. *Nowcasting*. London: Academic Press.
Gadd, A.J. 1981: Numerical modelling of the atmopsher. In B.W. Atkinson ed., *Dynamical meteorology; an introductory selection*. London and New York: Methuen. Pp. 194–204.
Wickham, P.G. 1970: *The practice of weather forecasting*. London: HMSO.

weather modification For centuries mankind has attempted to alleviate the stresses of the weather by prayers, incantations, cannon fire, and various other means, but the modern era of purposeful weather modification did not begin until the 1950s when scientists in the USA discovered that finely powdered dry ice dispersed by aircraft into thin winter stratus clouds could dissipate these clouds by turning the water vapour into ice crystals which in turn fell out of the clouds as snow.

Since then major efforts have been made by most nations of the world to modify the weather. The goals have been to increase precipitation (rain or snow) to make water for agricultural production, hydroelectric power, or water supplies; to reduce storm damages from hail, high winds and hurricanes; and to reduce or dissipate fog. Fog suppression is now established: chemicals turn cold fogs into snow, a process which removes fogs at many airports around the world, and heat sources such as jet engines literally burn away warm fogs. Fog dispersal systems work, but they are sufficiently expensive to be cost-beneficial only at major commercial and military airports.

Since the early cloud-seeding experiments of the 1950s, there have been major experiments in the USA, the former USSR, Australia, and France to develop techniques for weather modification. Most techniques involved the use of chemicals such as silver iodide, lead iodide, dry ice, or salt to alter the microphysical properties of clouds; this alters the rate of development of cloud droplets so that precipitation falls and may increase the vertical motions inside the clouds. These chemical materials are delivered into clouds in a variety of ways including the use of surface generators, particularly in areas where the air is lifted into clouds over mountains. Aircraft flying in or below the clouds are often used, as are rockets or anti-aircraft guns, a practice used in the former USSR and satellite nations.

An important aspect of weather modification over the past thirty years has been parallel efforts to conduct scientific field experiments, often very expensive multi-year efforts, while at the same time utilizing existing techniques in 'operational projects' where users, usually farmers in distress, wish to have relief from periods of too much hail or too little rainfall. Experimentation in the USA has been extensive, with twelve major field experiments since 1960. Some of these have demonstrated the ability to produce 10–25 per cent increases in snowfall in mountainous areas under certain weather conditions. Efforts to modify warm season convective rainfall have yet to be convincing. Efforts to increase precipitation, either snow in winter, or rain in summer, have been widely pursued particularly in nations that have semi-arid climates such as Australia, Mexico, Argentina, Chile, South Africa, Israel and several North African nations.

Hail suppression experiments in Europe and the USA have been inconclusive; statistical evaluation of long-term operational hail suppression projects in the Great Plains of the USA and South Africa, however, suggest 20 to 40 per cent decreases in hail. Areas in the major mountain chains in the former USSR, Europe, Africa and the USA have been other prime sites for hail suppression projects. The basic aim of most cloud seeding for hail suppression is to fill the upper portion of the storm with silver iodide ice crystals which compete for available water, leading to the production of many small hailstones rather than a few large ones. Considerable research and utilization of hail suppression was conducted in the Soviet Union from the late 1950s. Several operational projects exist using either large rockets or anti-aircraft guns to launch lead iodide or silver iodide into the centre of thunderstorms so as to 'overseed them', leading to decreases in the number and size of hailstones. Their claims are for 60–80 per cent reduction in hail loss.

Limited efforts have been made to modify hurricanes. Overseeding of the storms in the wall of the storm was intended to reduce convection, and in turn to lower the speeds of the damaging winds in the wall of the storm. Experimentation has been basically inconclusive.

Another aspect of the extensive research into planned weather modification, principally in the USA, has involved social and environmental issues with attention to the impacts of weather changes. Large-scale weather changes are shown to have considerable socioeconomic impacts over large areas. In some places there have been religious concerns. A few major storms occurred in areas where cloud seeding was active, leading to lawsuits and court cases. However, inability to establish a connection between the cloud seeding and extreme storm events has led to a decline in claims of damage. Economic studies have shown that increases of precipitation of about 10–25 per cent would produce sizeable benefit–cost ratios, of the order of 10 to 1.

The inability of the atmospheric science community to develop satisfactory weather modification techniques other than those for fog suppression and snowfall enhancement has led researchers to ask more basic atmospheric questions. Many of the early experiments recognized the lack of adequate atmospheric knowledge and used randomization (seed some days not others) in hopes of finding a technology, but evaluation of weather modification projects of this type has been very difficult and inconclusive. There has been speculation about how land use changes (such as changing an earth's colour and thus altering the albedo), or altering ocean flows (such as damming the Bering Straits) might lead to weather and climate modification.   SAC

Reading
Arnett, D. 1980: *Weather modification by cloud seeding.* London: Academic Press.
American Geophysical Union 1983: *Reviews of geophysics and space physics.* Vol. 17.

weather type The categorization of the principal modes or patterns of atmospheric circulation, on a local, regional or hemisphere scale produces a series of classes known as weather types. Each type tends to be associated with a particular type of weather, and weather typing represents the first stage in a SYNOPTIC CLIMATOLOGY study. Objective methods of typing pressure patterns are superseding subjective methods, as computer resources and suitable gridded data sets become more available (Barry 1980; Perry 1983). Among the longest daily weather type catalogues are those of the UK (Lamb 1972) and for central Europe.   AHP

References
Barry, R.G. 1980: Recent advances in synoptic and dynamic climatology. *Progress in physical geography* 4, pp. 88–96.
Lamb, H.H. 1972: *British Isles weather types and a register of the daily sequences of circulation patterns 1861–1971.* Meteorological Office Geophysical memoir 118. London: HMSO.
Perry, A.H. 1983: Growth points in synoptic climatology. *Progress in physical geography* 7, pp. 90–6.

weathering One of the most important of geomorphological and pedological processes, occuring when the rocks and sediments in the top metres of the earth's crust are exposed to physical, chemical and biological conditions much different from those prevailing at the time the rocks were formed. Two main types are recognized (see table). Mechanical weathering involves the breakdown or disintegration of rock without any substantial degree of chemical change taking place in the minerals which make up the rock mass, while chemical weathering involves the decomposition or decay of such minerals. In most parts of the world both types of weathering may operate together though in differing proportions, and one may accelerate the other. For example, the mechanical disintegration of a rock will greatly increase the surface area that is then exposed to chemical attack. The rate at which weathering occurs will depend

Classification of weathering processes

Processes of disintegration (physical or mechanical weathering)
Crystallization processes
salt weathering (by crystallization, hydration and thermal expansion)
frost weathering
Temperature change processes
insolation weathering (heating and cooling)
fire
expansion of dirt in cracks
Wetting and drying (especially of shales)
Pressure and release by erosion of overburden
Organic processes (e.g. root wedging)
Processes of decomposition (chemical weathering)
Hydration and hydrolysis
Oxidation and reduction
Solution and carbonation
Chelation
Biological chemical changes

both on climatic conditions (e.g. the frequency of frost, the amount of water available for chemical reactions, etc.) and rock character (e.g. porosity, the density of jointing, and the susceptibility of the minerals). Consequences of weathering are duricrust formation, the development of karst, and the provision of material for removal by erosive processes.   ASG

Reading
Goudie, A.S. 1981: *Weathering*. In A.S. Goudie ed., *Geomorphological techniques*. London: Allen & Unwin. Pp. 139–55.
Ollier, C.D. 1969: *Weathering*. Edinburgh: Oliver & Boyd.
Yatsu, E. 1988: *The nature of weathering*. Tokyo: Sozosha.

weathering front The limit of the zone of weathering of bedrock beneath the landsurface.

weathering index A quantitative or semi-quantitative indicator of the degree of weathering. Indices of both chemical and physical weathering have been proposed. Chemical indices of the alteration of rocks, W_I, may be expressed as the ratio of unweathered to weathered minerals in a volume of material, or as the ratio of the chemically more mobile to the chemically less mobile species, in the general form:

$$W_I = \frac{\text{Proportion of chemical reactants}}{\text{Proportion of residual products}}$$

or more specifically in a form such as:

$$W_I = \frac{\text{feldspar} + \text{mica} + \text{calcite}}{\text{clay minerals} + \text{quartz}}$$

Physical indices of weathering have been expressed in a number of ways such as: capacity of unweathered material to take up water compared with the capacity of weathered material; the softening effect of water on material; the degree of swelling and slaking which results from water absorption; changes in strength or hardness of minerals; and velocity of ultrasound (seismic) waves through materials.   MJS

Reading
Selby, M.J. 1993: *Hillslope materials and processes*. 2nd edn. Oxford: Oxford University Press. Ch. 8.

weathering potential index This is a measure of the ease with which rocks decay chemically. It is the ratio of the mole percentage of alkaline and alkaline earth metals (Ca, Na, Mg, K) less combined water to the mole percentage total metals (including Fe and Ti) less water. It provides a means of assessing the Goldich weathering stability series in numerical form. However, the computed value of quartz (zero) is not in agreement with its generally stable nature.

WBW

Reference
Wahlstrom, E.E. 1948: Pre-Fountain and recent weathering on Flagstaff Mountain near Boulder, Colorado. *Geological Society of America bulletin* 59, pp. 1173–89.

weathering rinds Oxidation phenomena which stain the parent rock red-yellow when exposed to air or near-surface groundwater for some time. They may extend for more than a millimetre into the rock, whereas DESERT VARNISH is generally much thinner. The thickness of weathering rinds may have some utility for relative dating of surface outcrops (Anderson and Anderson 1981).   ASG

Reference
Anderson, L.W. and Anderson, D.S. 1981: Weathering rinds on quartz arenite clasts as a relative age indicator and the glacial chronology of Mount Timpanogos, Wasatch Range, Utah. *Arctic and alpine research* 13, pp. 25–31.

weathering-limited slopes Hillslopes on which the potential rate at which weathered soil and debris can be removed by erosional processes exceeds the rate at which the material can be produced by weathering. The rate of ground loss is consequently controlled by the rate of weathering, and the form of the hillslope unit is controlled by the relative resistance to weathering of the rock masses on which the slope is formed.   MJS

Reading
Young, A. 1972: *Slopes*. Edinburgh: Oliver & Boyd.

wedge failure The removal of blocks of rock (or, more unusually, clay) where two or three fractures, often joints, intersect at a high angle in a cliff, and dip down towards the valley. The size of blocks from this kind of fall is rarely greater than a few 10 m^3. (See also SLAB FAILURE; TOPPLING FAILURE.)   WBW

weeds See ALIENS.

weighted mean percentage silt clay A method of expressing the PARTICLE SIZE of sediment in the perimeter of a river channel cross-section proposed by Schumm (1960) as a way of accounting for the nature of the particles making up the bed and banks of the stream. The percentage is calculated by the formula:

$$M = [(Sb \times w) + 2(Sc \times d)]/(w + 2d)$$

where M is the weighted mean percentage, Sb is the percentage silt and clay in the banks, w is the channel width, and d is the channel depth. The significance of the measure is that channels with high values of weighted mean percentage silt clay are likely to be rivers that are relatively narrow and deep and carry a relatively high percentage of their sediment load in suspension (Schumm 1977). Channels with low values of weighted mean percentage silt clay are likely to be wide, shallow streams carrying most of their sediment as bedload.
WLG

Reference
Schumm, S.A. 1960: The shape of alluvial channels in relation to sediment type. *US Geological Survey professional paper* 352-B, pp. 17–30.
— 1977: *The fluvial system.* New York: Wiley.

weir A structure built across a river or stream channel in order to measure the flow. It may be built of concrete, metal or

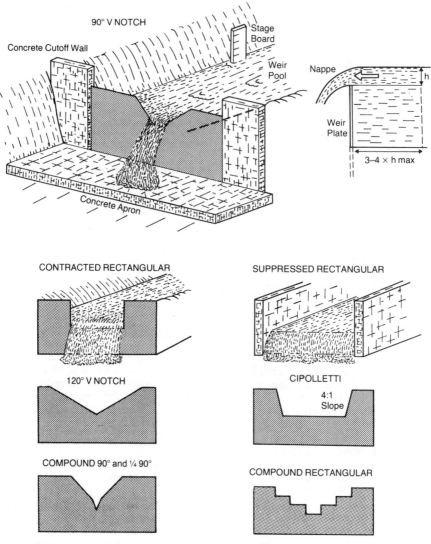

1. *Sharp-crested weirs.*
Source: *K.J. Gregory and D.E. Walling 1973: Drainage basin form and process. London: Edward Arnold.*

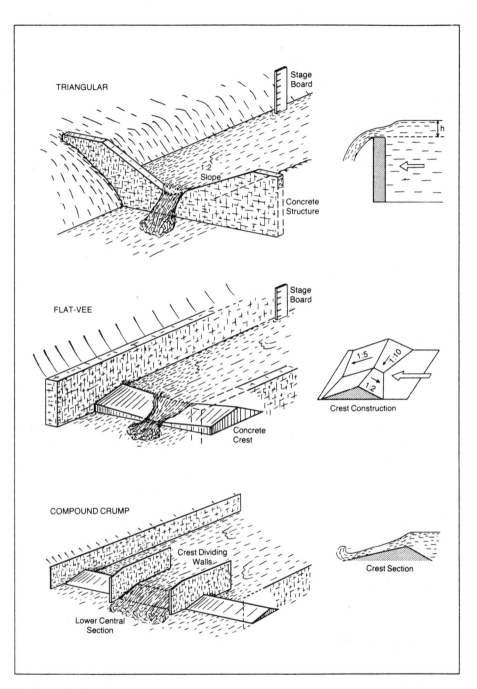

2. Broad-crested weirs.
Source: *K.J. Gregory and D.E. Walling 1973:* Drainage basin form and process. *London: Edward Arnold.*

wood and possesses two major features. First, it ponds back the flow to create a pool with flow velocities. Secondly, it incorporates a crest or notch over or through which the water flows freely from the upstream pool. A unique and stable relationship exists between the depth or head of water above the weir crest and water discharge. Values of discharge may be obtained from measurements of water stage by using published formulae or calibration curves established using other methods of flow measurement. Weirs may be further classified, according to the form of the crest, into SHARP-CRESTED and BROAD-CRESTED WEIRS. (See also DISCHARGE.)   DEW

Reading
Ackers, P., White, W.R., Perkins, J.A. and Harrison, A.J.M. 1978: *Weirs and flumes for flow measurement.* Chichester: Wiley.

wells Vertical dug or bored shafts penetrating to the saturated zone for the purpose of exploiting GROUNDWATER. In some old literature and on old maps, wells may refer to SPRINGS used as water supplies.   PWW

westerlies Belts of winds, generally south-westerly in the northern hemisphere and north-westerly in the southern hemisphere, with average position between about latitudes 35° and 60°. These belts move poleward in the winter and equatorward in summer and are part of the GENERAL CIRCULATION OF THE ATMOSPHERE, being the low-level branch of the FERREL CELL.
                                    KJW

Reading
Lutgens, F.K. and Tarbuck, R.E.J. 1982: *The atmosphere.* 2nd edn. Englewood Cliffs, NJ: Prentice-Hall.

wet-bulb temperature The temperature at which a sample of air will become saturated by evaporating pure water into it at constant pressure. It is measured by a thermometer whose bulb is covered by muslin which is kept constantly wet with distilled water. The lower the humidity is, the stronger the evaporation from the muslin and the stronger the cooling of the thermometer bulb. Wet-bulb temperature is an indicator of absolute humidity and its difference from the dry-bulb is a measure of relative humidity.   RR

wetland Defined by Maltby (1986) as 'a collective term for ecosystems whose formation has been dominated by water,

and whose processes and characteristics are largely controlled by water. A wetland is a place that has been wet enough for a long time to develop specially adapted vegetation and other organisms.' They include areas of marsh, fen, peatland or water, whether natural or artificial, permanent or temporary, with water that is static or flowing, brackish or salt, including marine water the depth of which at low tide does not exceed 6 m.   ASG

Reference
Maltby, E. 1986: *Waterlogged wealth. Why waste the world's wet places?* London: Earthscan.
Williams, M. ed. 1990: *Wetlands: a threatened landscape.* Oxford: Blackwell.

wetland drainage Wetlands are areas that are nearly level with water-logged surfaces and persistently high water tables; they are characterized by marsh or bog vegetation communities. Wetlands are frequently desirable development areas for agricultural or urban uses because of their relatively level surfaces. However, a drainage of their excess moisture usually takes one of four forms. Channels may be constructed on the surface to conduct runoff away from the area, perforated pipes may be buried in the subsurface to drain off excess groundwater, tiles may also be buried within a metre of the surface, and groundwater may be pumped out and conducted elsewhere. These water-reduction strategies produce drier conditions, but the materials left behind compress as their moisture content is reduced, so that surface subsidence may present problems. (See also LAND DRAINAGE.)   WLG

wetted perimeter The perimeter of a river channel which is covered by water at a specific stage of flow. Such a wetted perimeter (p) is used together with the cross-sectional area (a) of the water at the same stage to give the HYDRAULIC RADIUS as a/p.   KJG

wetting front The subsurface limit to which a soil, particularly in a desert region, becomes saturated by infiltrating rainwater.

whirlwind Loosely used term to describe rotating winds of scales up to that of a tornado. They are usually manifestations of intense convection over very small areas. The winds are frequently strong enough to raise surface materials (causing dust devils) and can become extremely hazardous when associated with large fires, such as may result from heavy bombing, or as occur in semi-arid bush areas.   JSAG

width–depth ratio A simple measure of the shape of a river channel cross-section usually obtained as w/d where w is the top width of the cross-section to be characterized and d is derived as an average value by dividing the cross-sectional area (a) by w in the form:

$$d = \frac{a}{w}$$

The ratio is conventionally evaluated for the bankfull stage, and Riley (1976) suggested two alternative measures: namely bed width, and the exponent x in the equation:

$$w_1 = cd^x$$

where w_1 is width at stage d_t.   KJG

Reference
Riley, S.J. 1976: Alternative measures of river channel shape and their significance. *Journal of hydrology* (NZ) 15, pp. 9–16.

wilderness A wild, uncultivated area. Ideally, wilderness areas should never have been subject to human activity which has resulted in manipulation, either deliberate or unconscious, of the ecology of the area (Simmons 1981, p. 62).   ASG

Reference
Simmons, I. 1981: *The ecology of natural resources*, 2nd edn. Oxford: Blackwell.

wind Air in motion relative to the surface of the earth. It is important to note that this definition covers more than the layman's view of simple horizontal airflow. First, the motion may also have a substantial vertical component, and indeed it is this component that is a primary cause of all the WEATHER we experience on this planet. Secondly, it is vital to appreciate that the motion is *relative* to the surface of the earth. Thus, in 'calm' conditions, when air is stationary relative to the surface, it is still moving through space at the same velocity as the surface of the earth. This characteristic has important repercussions on the direction of airflows and upon the momentum budget of the atmosphere.

Wind is specified in terms of speed and direction. Speeds are reported in knots or metres per second, the reading being taken over a few minutes to reduce the effects of very short duration gusts, and direction (which is the direction from which the air flows) is reported to the nearest 10° of the compass. If wind direction varies with time a clockwise shift is called veering and an anticlockwise one backing. At the surface of the earth winds are measured by ANEMOMETERS, with the actual sensor preferably being at a height of 10 m. At higher levels winds are measured by pilot balloons and RADIOSONDES. In both cases the measurements are made of only the horizontal component of the wind. Hence we have no routine direct measurements of vertical winds, largely because they are notoriously difficult to measure at all, let alone accurately. This is so because of the smallness of vertical windspeeds: the speed of vertical winds is usually of the order of cm s^{-1}, whereas that of horizontal winds is of the order of m s^{-1}. This lack of information on vertical motion is most frustrating both for the theoretician and for the practical meteorologist, such as a forecaster. It is precisely the phenomenon that we wish to measure that, up to now, has proved unmeasurable.

The meteorologist's frustration can be more fully understood when we appreciate the fundamental role of airflow in the workings of the atmosphere. Virtually all the weather and climate we experience results from air motion, albeit on many scales. Sunshine at the ground occurs only in the absence of cloud, which results from condensation of water vapour that occurs when air rises to its condensation level. Rainfall results from cloud micro-physics and airflows within clouds. Evaporation requires air motion to remove water molecules. Humidity is partly determined by mixing and the temperature at any place is frequently as much a function of advection of warm/cold air as it is of radiation. Clearly, if we analyse air motion we are investigating the main problem of meteorology. In fact, the aim of dynamical meteorology is to do just that.

Although observations of winds have been, and continue to be, vital to our knowledge of the atmosphere, it is true to say that deeper understanding has emerged from theoretical analyses of airflow. These analyses all rest on Newton's Second Law, which states that if a force is applied to a mass, it will accelerate that mass in the direction of the force. This law can readily be applied to a particle of air in the atmosphere. The value in doing this lies in the acceleration part of the law. Acceleration is the change of velocity with time.

Hence, if we can derive an acceleration of air at a given time from the application of Newton's Law, we can then derive an air velocity at a future time. This has the double advantage not only of giving information about the velocity of the airflow, which has been shown to be fundamental to weather and climate, but also of giving its value at a future time, that is a forecast. The acceleration of the air largely results from the application of three forces: the pressure gradient force, the CORIOLIS FORCE and the FRICTION force. Hence the equation of motion describing the flow in a horizontal direction – say x – is:

$$\frac{\partial u}{\partial t} = -\frac{1}{\rho}\frac{\partial p}{\partial x} + 2\Omega v \sin\phi$$

Acceleration   Pressure   Coriolis force
gradient force

$$+\frac{\partial}{\partial z}\left(\frac{K\partial u}{\partial z}\right)$$

Frictional force

where u is the air velocity in x direction, t is time, ρ is air density, p is air pressure, Ω is the angular velocity of the earth's rotation, v is the air velocity along a direction perpendicular to x, ϕ is latitude, z is the vertical coordinate, and K is the EDDY DIFFUSIVITY of momentum.

In large-scale meteorology it is customary to make the x-direction east–west, the y-direction north–south and the z direction vertical. The equation is thus complemented by similar ones in the y and z directions, but for present purposes it suffices.

The fact that four terms appear in the equation does not mean that all four are important to all airflows at all times. This was appreciated by Jeffreys in the 1920s and by scrutiny of each term he was able to classify winds dynamically. In addition Jeffreys showed that certain types of wind were associated with circulations of certain sizes. This classic paper deserves to be more widely known among non-meteorologists. Jeffreys argues that the pressure gradient force is an important term, being the force responsible for most air motion relative to

the earth's surface. If, of the remaining three terms, the Coriolis force is much larger than both the acceleration and frictional terms, then, in the equation, the pressure gradient term is balanced by the Coriolis term and the resultant wind is known as the geostrophic wind, a term suggested by Sir N. Shaw. Such a wind blows parallel to the isobars, with low pressure to the left and high pressure to the right in the northern hemisphere and the converse in the southern hemisphere. If, secondly, the Coriolis and frictional terms are small in comparison with the acceleration term, then the latter balances the pressure gradient term to give Eulerian winds. If, thirdly, the frictional terms exceed the Coriolis and accelerational terms, then friction must balance the pressure gradient. The wind blows along the pressure gradient and the friction, assumed to act opposite to the flow, is sufficient to prevent the velocity of a particular mass of air from increasing steadily throughout its journey. Jeffreys termed such a wind antitriptic.

Crude as was Jeffreys' classification it provided a firm, rational foundation for an understanding of air motion *per se* and the organization of air flow into recognizably different configurations. Jeffreys himself noted that there would be many cases in which the number of terms in the equations of motion comparable with the pressure gradient term is two or three. Such a case is the wind in the lowest kilometre or so of the atmosphere, where the pressure gradient, Coriolis and frictional forces are in balance. In this boundary layer the windspeed is zero at the ground, but increases in value (and direction veers) with height until the geostrophic condition is reached at the top of the friction layer. A further case occurs when the isobars are curved as opposed to straight. This means that a centripetal force enters the equation and the resultant (friction free) wind is known as the GRADIENT WIND, a flow resulting from a balance of pressure gradient, Coriolis and centripetal forces. Clearly one can further complicate each of these basic types by allowing the relative magnitude of any term to change.

Observations have confirmed the basic validity of Jeffreys' results. The general atmospheric circulation comprises entities ranging in characteristic horizontal dimension from thousands of kilometres (the ROSSBY WAVES) to a few kilometres or even a few hundreds of metres (locally induced flows such as cold air drainage). For the Coriolis term to have a major influence upon airflows, their characteristic horizontal size must be greater than 70 km (at the poles) and 400 km (10° from the equator). Thus, cyclones, anti-cyclones and Rossby waves are manifestations of geostrophic and gradient flow. Systems smaller than the critical size (mesoscale and smaller) are largely unaffected by the Coriolis force (the sea breeze is the notable exception). They fall into Jeffreys' 'antitriptic' category if friction is an important force: sea breezes, mountain breezes exemplify this type. If, however, friction may be ignored but the accelerational terms may not, then these small systems fall into Jeffreys' Eulerian category. Severe local storms, such as thunderstorms and tornadoes, exemplify this type of airflow.

As noted earlier, this fundamental approach to the classification of winds also allows the possibility of calculating future values from equations such as the one above. We now know that certain combinations of forces produce certain types of circulation which, in turn, have certain values for the accelerational term $\partial u / \partial t$. Consequently, if we know (from observations or, in a theoretical treatment, simply by specification) initial distributions of air velocity, and pressure, the estimated acceleration allows us to forecast a future distribution of air velocity. From this a forecaster can infer the weather that is likely to ensue. If the equations are suitably modified and the calculations pushed far into the future on a global scale it is possible to 'create' the whole general atmospheric circulation – or global climate. This is now being done by several research groups throughout the world, with most promising results. (See also GENERAL CIRCULATION MODELLING.)   BWA

Reading
Atkinson, B.W. 1981: *Dynamical meteorology – an introductory selection.* London and New York: Methuen. Esp. ch. 1.
— 1986: *Meso-scale atmospheric circulations.* New York and London: Academic Press.
— 1982: Atmospheric processes. In J.M. Gray and R. Lee eds. *Fresh perspectives in geography.* Special publication 3. Department of Geography. Queen Mary College, University of London.
Barry, R.G. 1992: *Mountain weather and climate,* 2nd edn. London and New York: Routledge.
Barry, R.G. and Chorley R.J. 1993: *Atmosphere, weather and climate.* 6th edn. London and New York: Routledge.

Defant, F. 1951: Local winds. In T.F. Malone ed. *Compendium of meteorology.* American Meteorological Society. pp. 655–72.
Jeffreys, H. 1922: On the dynamics of winds. *Quarterly journal of the Royal Meteorological Society* 48, pp. 29–46.
Meteorological Office 1978: *A course in elementary meteorology.* 2nd edn. London: HMSO.
Yoshino, M.M. ed 1976: *Local wind bora.* Tokyo: University of Tokyo Press.

wind chill An index of the degree of atmospheric cooling experienced by a person. The amount of heat loss or gain can be measured either by the actual heat exchange in watts per square metre of skin exposed or as an equivalent temperature, i.e. the temperature in still air that would correspond to the cooling (or heating) generated by the particular combination of temperature and windspeed. Other indices have been devised to take account of clothing type and thickness.   PS

Reading
Dixon, J.C. and Prior, M.J. 1987: Wind chill indices – a review. *Meteorological magazine* 116, pp. 1–17.

wind rose Illustrates graphically the climatic characteristics of *wind* direction, and also frequently windspeed, at a particular location. Radii are drawn outwards from a small circle, proportional to the frequency of wind from each direction. The percentage frequency of calms is usually indicated inside the circle. Windspeed can be depicted by varying the thickness of the radiating lines. There are many variations of this basic method. A wind rose can be adapted to show the relationship between wind direction and other meteorological variables, such as precipitation or the incidence of thunderstorms.   DGT

Reading
Monkhouse, F.J. and Wilkinson, H.R. 1971: *Maps and diagrams.* London and New York: Methuen. pp. 240–3.

wind shadow The region downwind of an obstacle, shielded from the wind. Hedges are less of an obstacle than walls and hence reduce the strength of lee eddies while still providing shelter. The effect persists a distance downwind about thirty times the height of the obstacle in neutrally stratified conditions, much further for strong INVERSIONS. It may lead to complete stagnation of air in city basins, as in Los Angeles, Rome and London.   JSAG

Reading
Monteith, J.L. 1973: *Principles of environmental physics.* London: Arnold.

windthrow The overturning and damage to trees caused by high-velocity winds. This is probably a major cause of habitat change in woodlands and also affects soils and surface topography because of the translocation of material attached to root balls.

ASG

Reading
Mayer, H. 1989: Windthrow. *Philosophical Transactions of the Royal Society of London* 324B, pp. 267–81.

woolsack A corestone, or large rounded boulder of unweathered rock incorporated within a mass of *in situ* weathered rock. Further stripping may expose it as a small, tor-like feature.

world conservation strategy In 1980, IUCN, UNEP and WWF jointly published a policy document which asked for responses from all national governments. The *Strategy* explains the contribution of the CONSERVATION of living resources to human survival and to sustainable development; identifies the priority conservation issues and the main requirements for dealing with them; and proposes effective methods of implementation. It is thus addressed to governments and their policy advisers, to conservation activists and to development practitioners and planners, and tries to show that the prospects for sustainable development are enhanced by conservation methods. Three main objectives are proposed: (1) the maintenance of essential ecological processes and life-support systems, such as soil regeneration and protection, and the recycling of nutrients; (2) the preservation of genetic diversity found in the world's living organisms; and (3) the sustainable utilization of species and ecosystems, especially fish and other wildlife, forests and grazing lands. The *Strategy* explains that the planet's capacity to support people is being irreversibly reduced in both developed and developing countries, that malnourished people are compelled to destroy the resources that might free them from poverty and starvation, that the cost of energy and other costs of providing goods and services are rising quickly, and that the resource base of major industries is shrinking. It sees the major obstacles as the belief that conservation is a limited sector rather than a process that cuts across all other sectors; the failure thus to integrate conservation with development; a lack of infrastructural capacity to conserve; and failure to deliver conservation-based development to, for example, developing rural areas.

IGS

Reference
IUCN/UNEP/WWF 1980: *World conservation strategy.* Gland: IUCN.

wrench fault A strike-slip fault with a fault plane tending towards vertical.

X

xenolith An isolated block of country rock incorporated within and metamorphosed by an igneous intrusion.

xerosphere The 'dry' part of the globe. Scanty supplies of water result in xeric adaptations by plants and animals. True xerophytes are plants capable of enduring recurrent periods of drought by means of specialized structural and functional adaptations, and are mostly succulent and non-succulent perennials. Xerophyllous refers to leaves adapted to reduce transpiration and to conserve water. Xeromorphic refers to adaptations of form. The xerosphere is characteristically associated with hot and cold deserts but more localized dry locations can be found all over the terrestrial globe, on bare rock and tree-bark surfaces or in porous soil.   PAF

Reading
Furley, P.A. and Newey, W.W. 1983: *Geography of the biosphere*. London: Butterworth.
Walter, H. 1971: *Ecology of tropical and subtropical vegetation*. Edinburgh: Oliver & Boyd.

xerothermic index Measures the effects of drought intensity on plant growth. In the UNESCO/FAO scheme for the mapping of the bioclimates of the Mediterranean zone, in which the importance of aridity is emphasized, a xerothermic index summarizes the duration and severity of the dry season (UNESCO/FAO 1963). To calculate the xerothermic index for a dry month (August) with 3 days of rain, 6 days of mist and dew and a mean atmospheric humidity of 70 per cent:

$$x_m = 31 - \left(3 + \frac{6}{2}\right) \times 0.8 = 20$$

The xerothermic index (x) for the dry season indicates the number of 'biologically' dry days. The bioclimate type is defined in terms of threshold temperatures and the value of the xerothermic index. The classification can be applied to areas outside the Mediterranean.   DGT

Reference
UNESCO/FAO 1963: Bioclimatic map of the Mediterranean zone. Explanatary notes, pp. 11–20.

Y

yardang An aerodynamically shaped desert landform produced by the wind erosion of bedrock or sediments. Yardangs occur in both unconsolidated Pleistocene materials and in hard dolomites and sandstone, and in the vicinity of Tibestsi (central Sahara), Luxor (Egypt), and the Lut (Iran) are sufficiently large to be evident on satellite photographs. They tend to be composed of parallel ridges and troughs (groovy ground) with the shape of an upturned ship's keel.

ASG

Reading
Greeley, R. and Iversen, J.D. 1985: *Wind as a geological process*. Cambridge: Cambridge University Press.
McCauley, J.F., Grudier, M.J. and Breed, C.S. 1977: Yardangs of Peru and other desert regions. *US Geological Survey interagency report: astrogeology 81.*

yazoo A tributary stream that runs parallel to the main river for some distance.

Younger Dryas Between around 11,000 and 10,000 BP there appears to have been a cold phase of climatic deterioration characterized by major ice advance in north-west Europe. This phase has been called the Younger Dryas (termed the Loch Lomond stadial in the British Isles) named after the Alpine plant Mountain Avens (*Dryas octopetula*), the distribution of which was more southerly in north-west Europe. It has traditionally been referred to as a north-west European climatic anomaly; however, indications of this episode of cooler climatic conditions are found eastwards into Russia, Spain and Portugal, North Africa and North America. This has been interpreted as having resulted from precipitation variations due to latitudinal changes in the positions of the polar atmospheric and oceanic fronts in the north Atlantic. Outside the glaciated areas in the Younger Dryas time, periglaciation, with discontinuous permafrost, was both extensive and highly effective.

AP

Reading
Dawson, A.G. 1992: *Ice age earth: late Quaternary geology and climate*. London: Routledge.

Z

zeuge A tabular mass of rock perched on a pinnacle of softer rock as a result of erosion, usually by wind, of the underlying materials. Although common in textbooks describing desert landscapes, zeuges are in reality rather rare.

zibar A type of low sand dune which has limited slip face development. Most observers have noted that they are associated with sands having many coarse grains. They often occur in the corridors between higher dunes.   ASG

Reading
Warren, A. 1971. Dunes in the Teneré Desert. *Geographical journal* 137, pp. 458–61.

zonal circulation Any flow along latitude circles. The term is used in a more specific sense to indicate a high index circulation in mid-latitudes with a strong westerly wind between 35 and 55° and little meridional air mass exchange.

At the surface, pressure systems have a dominantly east–west orientation (see INDEX CYCLE) and in middle latitudes unsettled weather with alternating troughs of low pressure and ridges of high pressure move in quick succession giving changeable conditions on any specified area.   AHP

zonal soil A soil which occurs over a wide area and owes its characteristics to climatic rather than local topographical or geological factors.

zonation In biogeography and ecology, one of the most important patterns of floral and faunal distribution. On a world scale, characteristic groups of organisms occupy different zones of the terrestrial and aquatic surface, which are primarily determined by climate. Each zone forms an idealized latitudinal band around the earth, although this is frequently modified by local factors (BIOMES). There are also approximately concentric zones around MOUNTAINS, consisting of lateral bands of plants and animals typical of the changing environment from base to summit. The term has also been extended to include any ecological unit with spatial dimensions, e.g. the tension or boundary zone between different, competing biota.   PAF

Reading
Furley, P.A. and Newey, W.W. 1983: *Geography of the biosphere*. London: Butterworth.

zoogeography The science of the distribution of animals, linking the subject matter of the discipline of zoology with the viewpoint of geography. Since the work of Hesse in 1924 it is now common to draw a distinction between ecological animal geography and historical animal geography, the first focusing on the relationships between animals and the environment, the second on the definition and interpretation of the spatial distribution of animals over the surface of the earth. This latter subject is sometimes known as chorology or faunistics.

The origins of zoogeography are clearly traceable, along with the first scientific biology, to Aristotle, but it was the great scientific expeditions of the nineteenth century that really laid the foundations of the subject as we know it today. The first important organizer of zoogeographical knowledge was the German biologist and philosopher Ernst Haeckel, who introduced the term ecology into the subject (1866). Early synthesists of historical zoogeography were P.L. Sclater (1858), T.H. Huxley (1868) and J.A. Allen (1871), but the modern 'Father of zoogeography' is usually regarded as Alfred Russel Wallace, who published his now classic work, *The geographical distribution of animals*, in 1876 (see WALLACE'S LINE; WALLACE'S REALMS). Another pioneer, this time in the ecological approach, was K. Möbius (1877), who developed the concept of biotic communities in his paper on the oyster.

Early interpretations of animal distributions were based on the account of Noah's ark given in Genesis. This view was first challenged by Augustinus *c.*AD 400, when he postulated separate creations, after the biblical flood, to account for the unique faunas of islands (see ISLAND BIOGEOGRAPHY). With the discovery of America a further impetus was given to this view of separate creations, e.g. Paracelsus, although many others talked of land-bridges linking the continents. With the evolutionary thinking of Darwin, Wallace, Huxley, Lamarck and many others and with the publication of *On the origin of species by means of natural selection* (1859), the whole framework of thought changed and zoogeography grew in stature as a scientific discipline. The earliest writers in this new mould (e.g. Wallace) still believed in the permanency of the continents, a view soon to be challenged by Wegener's theory of continental drift, developed in the early part of the twentieth century. The essential vindication of this theory in the 1960s with the models of plate tectonics and sea-floor spreading has, in contrast, provided a completely new insight into the causes of animal distributions and has led to a lively and often bitter debate on the relative merits of dispersal and vicariance, the latter term meaning the separate development of faunas from common ancestors on the separating continents. This development has been accompanied by the rise of a new school of zoogeography, called VICARIANCE BIOGEOGRAPHY which, in turn, is linked with the development of phylogenetic systematics and CLADISTICS in systematics (see TAXONOMY). PAS

Reading
George, W. 1962: *Animal geography*. London: Heinemann.
Illies, J. 1974: *Introduction to zoogeography*. London: Macmillan.
Nelson, G. and Rosen, D.E. 1981: *Vicariance biogeography: a critique*. New York: Columbia University Press.
Udvardy, M.D.F. 1969: *Dynamic zoogeography*. New York: Van Nostrand Reinhold.
See also the journal *Systematic zoology*, 1970 onwards.

Index

Note: Alphabetic arrangement is word by word. Subheadings are also arranged alphabetically, word by word, but ignoring words such as 'in', 'on', 'and' and 'by'. Abbreviations appear at the beginning of each letter sequence while acronyms are treated as ordinary words. Page numbers in bold type indicate the main discussion of a topic which is referred to on a number of other pages, but in less detail. Page numbers in italic type refer to illustrations. Readers are advised to look for information under the most specific term possible. For example, information on forests is most easily found by looking up 'forests', or even particular types of forest, rather than simply 'vegetation'. 'See' and 'see also' references provide a guide to related terms. In addition, readers should note the references in the main text to related topics.

abîmes, 1
abiotic components, 1
ablation, 1
  in balance with accumulation, 185
  increase in, 130
  and moraine, 340
ablation zone, 185
  compressing flow, 105
abrasion, 1
  comminution by, 103
  facets produced by, 202
  rock flour creation, 430
  ventifacts creation, 155, 536
absolute age, 1
absolute humidity, 257–8, 560
  and dewpoint, 140
abundance, species, 1
abysses, ocean, 1
abyssobenthic zone, 1
abyssopelagic zone, 1–2, 361
accelerated erosion, 461
acceleration, centripetal, 81–2
  see also **Coriolis force**
accessory minerals, 2
accommodation, 2
accordant junctions, law of
  (Playfair's law), 2, 391
accordant summits, 2
accretion, 2
accumulated departure, 2
accumulated temperature, 2
accumulation zone, 185
  extending flow in, 105
acid precipitation, 2–4, 141
  see also **dry deposition; occult deposition**
acid rocks, 4
  see also **igneous rocks**
acidity see **pH**
acidity profile, 4
aclinic line, 4
actinometer, 4
active channels, 82
active layer, 4, 381
  baydzharakh formation in, 49

cryostatic pressure in, 117
and patterned ground, 374
pereletok below, 379
processes in: cryoturbation, 117, 374; frost creep, 4, **220**; frost heave, 4, **220**; see also **freeze–thaw cycles**
activity ratio, 4, 9
actual evaporation, 191, 193
actual evapotranspiration, **194–5**, 310, 546
actualism, 523
adaptation, 246
  and convergent evolution, 197–8
adaptive radiation, **4–5**
  on islands, 5, 285
  marsupials, 204
adhesion ripples, 5
adiabatic lapse rate, 301, 536–7
adiabatic processes, 5, 504
  condensation produced by, 106
  katabatic winds warmed by, 214, 293
  and potential temperature, 399
  and temperature inversion, 283
adobe, 5
adret, 5
adsorption, 5
advection, 5
  convection sometimes confused with, 111
  and fog formation, 100, 214
  hodograph information on, 255
  and temperature inversion, 283
  see also **convection; winds**
adventitious roots, 5
adventitive cones, 5
aeolianite, 5, **535**
aeration zone (vadose zone), **5–6**, 242–3, **263**, 379
aerial cameras, 6
aerial films, 6
aerial photography, **6–7**
  measurement (photogrammetry), 385
  from satellites, 317–19

multispectral scanners compared with, 346
aerobic processes, 7
aerobiology, 7–8
aerography, 8
aerology, 8
aeronomy, 8
aerosols, 8
  dry deposition on, 3
  haze caused by, 250, 532
  and precipitation solute content, 465
  study of, 7–8
  sulphate, 143
aesthetic degradation, 8
aestivation, 8
affluent streams, 8
after-glow, 8
aftershocks, 8
Agassiz, Lake, 404
age
  absolute, 1
  relative, 422
  see also **time**
ageostrophic motion, 8–9
agglomerate, 9
aggradation
  punctuated cycles, 406
aggregation ratio, 4, 9
aggresivity, 9
agonic line, 9
agricultural land
  bush encroachment, 72
  productivity, 60, 353
  soil erosion, 461
agriculture
  climatology applied to, 9
  energy use, 175
  hydrological effects, 299–300
  instability, 471
  land capability for, 298
  meteorology applied to, **9–10**, 28, 29, 334
  preceded by hunting and gathering, 258
  trees in (agroforestry), 9

agroclimatology, 9
**agroecosystems, sustained yield
in,** 488
agroforestry, 9
agrometeorology, 9–10, 169, 334
 see also **microclimate;
 micrometeorolgy**
Agussiz, Louis, 115
aigulles, 10
air *see* **atmosphere**
air conditioning
 degree days used in
 determining, 130
air masses, 10–11
 baroclinity between, 44
 compressional, 106
 divergence in, 149
air movement *see* **atmospheric
 circulation**
air parcels, 11
 potential energy, 399
 temperature, 399, 504
 vertical stability, 536–7
air pollution, 11, 36, 395
 and acid precipitation, 2–3
 aerosols, 8
 air shed concept in modelling,
 12
 by chlorofluorocarbons, **89–90,**
 367–8
 dry deposition, 156
 haze caused by, 32
 ice fog caused by, 33, **272,** 532
 by ozone, 368
 recirculation, 304
 smog, 395, **460**
 and sulphation, 485–6
 see also particular pollutants
air pressure *see* **pressure**
air sheds, 12
air–sea interaction, 11–12
aircraft contrails, 110
Airy, G.B., 288
aklé dune pattern, 12
alases, 12
albedo, 12–13
 clouds, 13; and
 dimethylsulphides, 143
 and snowblitz theory, 460
 and surface temperature, 251
alcoves, 14
alcrete, 14
alfisols, 14, 452
algae, 14
 calcareous, 112, 114
 planktonic: dimethylsulphide
 production, 143
 proliferation, 14
 rhodoliths, 426
 symbiotic: associated with
 corals, 112, 113
algal bloom, 14
alidade, 14
alien species, 14
 control by natural predators,
 282–3
 exotics compared with, 198
 islands, 14, **286,** 400
alimentation, 14
alkalinity *see* **pH**
allelopathy, 14–15
Allen's rule, 15

Allerød interstadial, 15
allochthonous sediments, 15
allogenic streams, 15
allogenic succession, 15, 40
allometric growth, 15–16
allopatric species, 16, 537, 538
allophane, 16
alluvial fans, 17, 43
 fanglomerate on, 203
 sieve deposits on, 457
alluvial fill, 17–18
alluvial terraces, 498–9
alluvial toeslopes, *463*
alluvium, 18–19
 arroyos in, 35
 carses, 78
 channels in, **16–17,** 148;
 classification, 85, *86, 87,* 88;
 hydraulic geometry, 259
 colluvium differentiated from,
 102
alpha diversity, 19, 55
alpine orogeny, 19
alpine zone, 19
alps, 19
Alps
 aigulles, *10*
 couloirs, 114
 formation, 19, 500; and flysch
 formation, 213
 plants, 32
altimetric frequency curve, 19
altiplanation (cryoplanation), **19,**
 117
 terrace formation, 236, 380
altithermal (hypsithermal) **phase,
 19,** 268, 351
altitude, 268
 alimetric frequency curves in
 determination of, 19
 and animal distribution, 567
 and climate, 344, 345
 and glacier equilibrium line, 185
 measurement, 269
 as morphometric parameter, 342
 and temperature, 301
 and tree growth, 296, 511, 514
 and vegetation, **247–8,** 567
 and vicariance, 537
 see also **mountains**
altocumulus clouds, 99, 100
altostratus clouds, 98, *99,* 100,
 545
aluminium
 bauxite main ore of, 49
 in fragipans, 217
 in laterite, 303
 in sial, 456
alveolar structures, 19
ambient temperature, 19
amensalism, 19
amino acid racemization, 19
amphidromic point, 508
ana-front, 20
anabatic flows, 20, 344
 see also **valley winds**
anabranching channels, 20, 86,
 88
anaclinal rivers, 20
anaerobic organisms, 20
anaerobic processes
 fermentation, 62–3

 and peat formation, 375
analemnas, 20
analogue models, 458
analysis of variance, 275
anamolistic cycle, 20
anaseism, 20
anastomosing channels, 20, 70,
 155
 on beaches, *49*
 on outwash, 365, 533
anchor ice, 20
andosols, 20
Andrews, E.D., 82
andromy, 20
anemograph, 20
anemometer, 20
aneroid barometer, 44–5
angiosperms, 20
angle of dilation, 20–1
angle of initial yield, 21
angle of internal friction, 218
**angle of internal shearing
 resistance,** 21
angle of plane sliding friction,
 21
angle of repose, 21
 constant slopes, 109
 graded slopes, 237
angle of residual shear, 21
angular momentum, 21
 fluxes, 328, 338–9
angular unconformity, 21
angularity, particle, 372
animalia, 295
animals
 adaptation to habitat conditions,
 248
 adaptive radiation, 4–5
 aestivation, 8
 alien species, **14,** 198, **286;**
 predatory, 282–3, 400
 aposematic coloration, 27
 areographic studies of, 34
 bioluminescence, 60
 biomass, 60
 biotic isolation, 63
 biotic potential, 63
 body size, Bergmann's rule on,
 54
 body temperature: maintenance,
 255; regulated by
 environment, 393
 carnivorous, 77, 215, 515
 circadian rhythm, 90
 classification, 294–5, 305
 and climate, 164
 coevolution, 101
 colonizing, **103,** 300
 competition between *see*
 competition
 conservation, 108
 diseases, pesticides for control
 of, 384
 dispersal, 146
 distribution, 203–4, 543–4,
 567–8; endemic, 174;
 vicariant, 537–9; zonal, 567
 diversivorous, 150
 domestication, 152; and energy
 use, 175
 as dominant organisms, 153
 dwarf, 348

estuarine, *188*
exotic, 198
fossil, 311; environmental
  reconstruction from, 369
and geomorphology, 58
grazing, 240, 400
heat conservation, 15
herbivorous, 215, **253**, 400, 515
on islands, 14, **285–6**, 400
life cycle phenology, 385
limb size, Allen's rule on, 15
limestone weathering by, 387
migration, 334–5
mutualism between plants and,
  347
nanism, 348
parasitic, 77, **372**, **384**, 401
phoresy, 385
pigmentation, 236
polymorphism, 396
population dynamics, 105,
  165–6, 397–8, 400–3
predation, 77, 383, **400–3**; and
  coevolution, 101; and pest
  control, 282–3
productivity, 59–60
respiration, 164, 424
seral communities, 451
species diversity, 467
species-area curves, 467
territory, 500
tolerance limits, 511
toxins in: biological
  magnification, 58–9
see also particular types and
  species, e.g. micro-organisms
anisotrophy, 21–2
annual exceedance series, 22
annual series, 22
annular drainage, 22, *155*
Antarctic
  ablation, 1
  air masses, 11
  cold pole, 102
  ice floes, 271–2
  ice sheet, 230, *231*
  ice shelves, 273
  meteorology, 22–4
  ocean currents around, 234, 236
  ozone hole over, 368
  sea ice, 443, *444*
  water masses, 234, 236, *548–50*,
    552
Antarctic Bottom Water, 234,
  236, *550*, 552
Antarctic kingdom, 211
antecedent drainage, 24, 154
antecedent moisture, 24
  and intermittent streams, 280–1
  and overland flow, 264
antecedent platform theory, 24
antecedent precipitation index,
  24
anteconsequent streams, 24
anthropochores, 24
anthropogene, 25
anthropogenic processes, 25
anthropogeomorphology, 25
antibiosis, 25
antibiotics, 63
anticentre, 25
anticlines, 25, *26*, 214, 379

diapir, 140–1
  phacoliths below, 385
anticyclones, 25–6
  and drought, 156
  pressure gradient in, 82
  temperature inversions in, 283
  vertical motion in, 536
antidunes, **26**, 222
antiforms, 26
antipleions, 26
antipodal bulge, 26
antipodes, 26
antitrades, 26
antitriptic winds, 562, 563
apatite, 385
aphanitic texture, 26
aphelion, 26
aphotic zone, 26
aphytic zone, 26
apogee, 27
Apollo satellites, 318
aposematic coloration, 27
Applications Technology
  Satellites, 332
applied climatology, 29
applied geomorphology, 27–8
applied hydraulics, 261
applied hydrology, 265
applied meteorology, 28–31
aquaculture, 31
aquatic macrophytes, 31
aquicludes, 31
aquifers
  artesian, 35
  coastal: saltwater intrusion,
    229–30
  draw down in, 155
  hydraulic conductivity, 106–7
  hydraulic diffusivity, 258
  karst, 242–3
  mapping, 261
  piezometric surface, 388
  pumping: cones of depression
    caused by, 107
  seepage from, 137–8
  specific storage, 142
  specific yield, 468
  springs draining, 469
  see also **groundwater**
aquifuges, 31–2
aquitards, 31
archaeology, 408
Archaeopteryx, *196*, 197
arches, natural, 32
archipelagos, 32
Arctic, 32
  air masses, 11
  cold pole, 102
  glacial maxima, *231*
  ice floes, 271–2
  ice fog, 272
  meteorology, 32–3
  plants, 32
  sea ice, 443
  sediment transport in, 448
  see also **permafrost; tundra**
arctic haze, 32
arctic smoke (frost smoke), 221
Arctic tern, 334
Arctogaea, 204
arenaceous rocks, 34
arenas, 34

arenites, 438
areography, 34
arêtes, 34
argillaceous rocks, 34
argon, 36
arid zone, 34
  brousse tigrée in, 71–2
  pans in, 371
  phreatophytes in, 386
  pipes in, 388–9
  playas in, 391
  pluvial terraces in, 499
  salinity problems, 435
  sheet erosion in, 454
  see also **deserts; drought;**
    semi-arid areas
aridisols, 34, 452
aridity
  rain factor as measurement of,
    412
  xerothermic index of, 565
arkose sandstone, 34
armoured mud balls, 34
armouring, 34
arroyos, 35
artesian aquifers, 35
  draw down in, 155
  mapping, 261
artesian springs, 469
aspect, slope, 35
  and plant growth, 248
associations, plant, 35, 388
asthenosphere, 35, 320
  hydrostatic equilibrium with
    lithosphere, 287–8
astroblemes, 35
asymmetrical folds, 36, *214*
asymmetrical valleys, 36
Atacama Desert, *159*, *493*
Atlantic coastlines, 36
Atlantic Deep Water, 236, 547,
  552
Atlantic Ocean
  currents and circulation, 120,
    234, *235*, 236, *551*
  hurricanes in, *516*, 517
  Tethys Ocean isolated from,
    330–1
  thermocline, *503*
  water masses, 236, 547, *548–51*,
    552
Atlantic Tropical Experiment,
  12
atmometers, 36, **191–2**
atmosphere, 36
  adiabatic processes in *see*
    adiabatic processes
  advection in, 5
  aerosols in *see* **aerosols**
  after-glow in, 8
  angular momentum, 21, 328,
    338–9
  aurora borealis in, 39
  baroclinity, **44**, 290
  boundary layers, **37**, **171–2**, 562;
    urban, 530, 532
  carbon dioxide in *see* **carbon**
    dioxide
  circulation *see* **atmospheric**
    circulation
  composition, 36–7
  convection in, 111, 330

atmosphere (*continued*)
dust in, 8, **160–1**, 296
energetics, 37; *see also* **latent heat; potential energy; thermodynamic diagrams**
fronts in *see* **fronts**
haze in, 32, 160, **250**, 532
human adjustment to, 28–31
humidity *see* **humidity**
hydrostatic equilibrium, 268
ice crystals in, 270, 460; *see also* **clouds**
instability, 37, 536–7
kinetic energy, 225; *see also* **wind**
layers, 37; chemosphere, 89; heterosphere, 253; ionosphere, **284– 5**, 312; mesosphere, 330; *see also* **stratosphere; troposphere**
observation: radiosonde, 411, 496, 504, 505; satellite, 332–3, 439–40
oceans' interaction with, 11–12, 330; Southern Oscillation, **312**, 495, 518
ozone in, 3, 37, 89, 249, **367–8**
physical properties, 386–7; *see also particular properties*
pollution *see* **air pollution**
potential energy, 225, 249, **399**, 517
predictability *see* **weather: forecasting**
pressure *see* **pressure**
radiation interactions in, *410*
refraction in: mirages caused by, 336
solar radiation modified by, 251
supercooling in, 486
temperature structure, *36*
turbulence, 334, 520
vertical motion, 536; and cloud formation, 98, 99–100; lack of data on, 561
vertical stability/instability, 11, 504, 518, **536**
vortex disturbances: and waterspouts, 553
vorticity, 541–2
waves in, **38**, 225–6, 311, 330; and cyclone formation, 200; dishpan simulation, 146; lee waves, 253, **304**, 329; *see also* **Rossby waves**
see also **air masses; meteorology**
atmospheric circulation, 225–6
ageostrophic, 8–9
centripetal acceleration in, 82
and cloud formation, 98
and Coriolis force, 113–14, 562–3
cyclostrophic forces in, 123
divergence, 149
in eddies *see* **eddies**
forecasting *see* **weather: forecasting**
GARP project on, 223
in Hadley cells *see* **Hadley cells**
indices, 90, 276
macroscale, 311–12
Mediterranean climate, 327

meridional, 205, **248–9**, **328**, 339
mesoscale, 328–30; *see also* **katabatic flows; lee waves; sea breezes**
microscale, 334
modelling and simulation, 38, **146, 224–5**
momentum budget, 339
and mountains, 343–4
and potential energy, 399
satellite data on, 440
sensible heat transport, 451
teleconnections, 495
tropical, 517–18
weather determined by, 561
and weather types, 489–90, 556
zonal, 567
see also **cyclones; eddies; Hadley cells; monsoons; Rossby waves; wind**
atollons, 39
atolls, 24, **38**
atterberg limits, 39
aufeis (icings), 274, 348
auge winds, 39
aulacogens, 39
aureole, metamorphic, 39
aurora australis, 39
aurora borealis, 39
Australian kingdom, 211
Australian realm, 204, 543
Australopithecus, 314
autecology, 39
autochthonous sediments, 15, 39
autocorrelation, 40
in hydraulic geometry, 258
autogenic succession, 40
autotrophic organisms, 40, 215
avalanches, 40–1, 324
rock, 41, *302*, 322, 482
snow, 41, 322; remanié glaciers fed by, 422
avens, 41
avulsion, 41
azimuth, 41
azoic period, 41
azonal soils, 41
azotobacters, nitrogen fixation by, 41, 354, 488

backing winds, 561
backshore, 42, *45, 50*
backswamps, 42
backwalls, 42, 90
backwash, 42, 309
backwearing, 42
bacteria, 384
nitrogen-fixing, 41, 354, 488
symbiotic, 488
bacterioplankton, 389
badlands, 42–3, *133*, 153
bajadas, 17, 43
ball lightning, 305
bank erosion, 43
bank storage, 43–4
bankfull capacity, definitions of, 82
bankfull discharge, 44
and pool and riffle formation, 397
banner clouds, 44

barchans and barchanoid ridges, 44, *158*, 222
baroclinic waves, 38, 225
baroclinicity, 44, 290
barograph, 45
barometers, 44–5
barotropic fluids, air masses as, 10
barotropic motion, 45
barrancas, 45
barrier islands, 45
overtopping, 366–7
overwashing, 367
barrier reefs, 45–6
barrier spits, 46
bars, 46
and braiding, 70, *86*
coastal, 44, *50*, 511
and cut-off development, 121
in deltas, 131
longshore, *50*
at meanders, 393–4
migratory, 53
in outwash, 365
see also **barrier islands; current ripples**
barysphere, 46
basal complex, 46
basal ice, 47
basal sapping, 47
in arête and cirque formation, 34, 90
basal sliding, 47, **105**, 272
surging glaciers, 487, 505
basal till, 474
basalt, 47
origins, 393
plateau, 320, 391
sima, 458
springs on, *469*
see also **sima**
base exchange (cation exchange), *47*, 80
base flow, 47, *243*, 262, **263**, 264
depletion curves, 47, **137–8**
dry weather flow as, 157
base level, 47–8
denudation rates influenced by elevation from, 134
fluctuations, 48; and alluvial filling, 18; and cycle of erosion, 122
base saturation, 48
basement complex, 48
basic rocks, 48
basin-and-range terrain, 48
basins *see* **drainage basins**
batholiths, 49, 66, 117, *284*
bathymetry, 49
bathypelagic zone, *361*, 378
bauxite, 49, 303
baydzharakhs, 49
beach rock, 51
beaches, 49
abandoned: on pluvial terraces, 499
backwash, 42
bars, 44, *50*, 511
beach rock on, 51
berms, 42, *45, 49, 50*, **54–5**
cusps, *49*, 121, 367, 419
deflation, 130

dissipative, 147, 419
dunes on, 100–1
edge waves along, 120, 169,
    170, 367
equilibrium, 185
erosion protection, 244
longshore drift along, 309
profiles, 49, *50*; retrogradation,
    425
raised, *232*, 416
recession: Bruun rule, 72
reflective, 419
ridges, 49–51; chenier, 89;
    overwashing, 367; on sarns,
    439; swales between, 488, *see
    also* **berms**
salcrete on, 435
wave refraction, 419–20
zones, 49, *50*; backshore, 42,
    45, *50*; breaker, 428;
    eulittoral, 189; foreshore, *45*,
    50; surf, *50*, 147, 309, 486;
    swash, *50*, 54–5, 309
beaded drainage, 51
beaded eskers, 187
bearing capacity, soil, 51–2
Beaufort scale, 52, 75
Beaumont period, 52
Beaumont (Texas), *530*
bed load, 52–3, *307*
    and channel classification, 85, *87*
    critical erosion velocity for
        entrainment, 116
    equations, 53, 157
    equilibrium, 184
    measurement, 53
    mineral composition, 447
    particle sizes, 447
    relative importance, 448
    sandbed channels, 438
bed roughness *see* **roughness**
bed shear stress, 259
bedding, graded, 237
bedding planes, 53
bedforms, 53–4
    absence from plane bed, 54, 389
    antidunes, *26*, 222
    braided channels between, 70
    current bedding, 117–18
    and flow regime, 212, 222
    pool–riffle sequence, *84*, 239,
        397
    response to flow changes, 297
    sandbed channels, 212, 438
    transverse ribs, 514
    see also **bars; dunes; ripples**
bedrock, 54
    channels in, 16
    colluvium-filled depressions, 103
    cutters, 121
    frost shattering, 64
    slopes, boulder-controlled, 67
Begon, M., 77–8
benches, suballuvial, 482
benchmark catchments, 423
benefit–cost ratio, 114
benioff zone, 54
benthic zone, 54, *361*
berg wind, 54
**Bergeron–Findeisen
    mechanism,** 98–9
berghlaups, 54

Bergmann's rule, 54
bergschrund, 54
bergsturz, 54
berms, 42, *45, 49, 50*, 54–5
Bernouilli's theorem, 55
best units analysis, 55
beta diversity, 55
Bible *see* **Genesis**
biennial oscillation, 312
bifurcation ratio, 55
bigeocenose, 57
biocenose *see* **biocoenosis**
**biochemical oxygen demand
    (BOD),** 55, 56
biocides, 384
bioclastic rocks, 55
bioclimatology, 62
biocoenosis, 55–6, 57
    biotopes, 63–4
    ecotopes, 168
biocontrol, 58
biodegradation, 56, 59
biodiversity, 56, 150
bioengineering, 62
biogeochemical cycles, 56–7
    and deforestation, 130
    geological cycle as, 122
    see also **carbon cycle; nitrogen
        cycle**
biogeocoenosis, 57
biogeography, 57–8
    areography compared with, 34
    dispersalist, 538
    island, *285–6*, 400, 494, 537
    vicariance, 537–9; *see also*
        cladistics
biogeomorphology, 58
bioherms, 58, 419, *481*
biokarst, 58, 387
biological control, 58, 282–3
biological hazards, 349
biological magnification, 58–9
biological productivity, 59–60,
    167, 515
    and downwelling, 153
    estuaries, 188
    rain forest, 184
    see also **biomass; net primary
        productivity**
biological weathering, 88–9
biology
    interaction with engineering, 62
bioluminescence, 60
biomass, 60
    and energy flows, 165
biomes, 60–1
**biometeorological index,
    effective temperature as,**
    103
biometeorology, 61–2
    see also **microclimate**
biosphere, 62
    biogeography as study of, 57
    forests as percentage of, 129
    management for sustainability,
        108
    see also **ecosphere**
biostasy, 62
biostratigraphy, 478
biota, 62
biotechnology, 62–3

biotic communities, 55–6, 57,
    60–1
    biotopes, 63–4
biotic isolation, 63
biotic potential, 63, 397
biotopes, 63–4
bioturbation, 64
birds
    guano from, 244
    migration, 334
birth rates, 397
    and carrying capacity, 77–8
bise (bize) **wind,** 64
black alkali soils, 464
black body radiation, 411–12
black box models, 357
black and white film, 6
blanket bogs, 64
bleaching, coral, 113
blind valleys, 64
blizzards, 64
block faulting, 48, 64
block fields and block streams,
    64, 98
block slides, *302*
blocking anticyclones, 64, 156
blood rain, 64
blow-holes, 64
blowout dunes, 64–5, 158, 159
blueholes, 65
Blytt-Sernander model, 65
BOD, 55, 56
boddens, 65
bogaz, 65
bogs, 65
    blanket, 64
    bursts, 65
    cushion, *66*
    floating, 209
    ombotrophic, 376, 421
    paludification, 370
    raised, 359, **421**
    reclamation as carrs, 77
    string, 480
    vegetation, 376, 421
Bølling interstadial, 65
bolsons, 65
bones, dating of, 19
Bonneville, Lake, 499
bora wind, 65–6
boralfs, 452
boreal climate, 66
boreal forests, 66, 307
bores, tidal, 66, 188
bornhardts, 279
bosses, 66
botryoidal form, 67
bottom-set beds, 67
Bouguer anomaly, 67
boulder clay *see* **till**
boulder trains, 67
boulder zone, 85
boulders
    in block fields/streams, **64** 98
    case hardening, 78
    demoiselle protection, 131
    erratics, 187
    fluvial transport, 447
    logan stones, 308
    size, 373
    slope control, 67
    within weathered rock, 113, 564

boulders (*continued*)
  see also **clasts**
boundary conditions, 67–8
  GCMs, 224–5
  numerical models, 358
boundary layers, 68–9, 261
  atmospheric, **37, 171–2**, 562;
    urban (UBL), 530, 532
  frictional resistance within, 218
  turbulent, **68–9**, 520, 522
bournes, 69
Bowen's ratio, 193, 252
Bowen's reaction series, 69
Boyle's law, 223
brackish water, 69
braided rivers, **69–71**, *155*
  alluvial, 17
  anabranching, **20**, *86*, 88
  anastomosing, 20, 70
  bars in, 46
  and channel classification, *86,
    87*, 88
  floodplains, 210
  lateral accretion, 302
  on outwash, 365, 439
brash ice, 75
brash rock, 71
braunerde, 71
Braunton Barrows, *216*
Brazil, savannah in, 82, 88
break-point bars, 44
breaker zone, *50*
  rip currents in, 428
breakers, *170*, *554*
  on dissipative beaches, 147
breccia, 71, 78, 439
breezes
  on Beaufort scale, 52
  sea and land, 444
Bretz, J.H., 79, 523–4
brittle fracture, 217
broad-crested weirs, 71, *559*
brodel, 71
brousse tigrée, 71–2
brown earth, 71
brown forest soil, 71
Bruckner, E., 378
Bruckner cycle, 72
brunizem, 72
Bruun rule, 72
Bryan, K., 67
Bryce Canyon, *256*
Bubnoff units, 72
Budel, J., 96
Budyko, M.I., 96
buffer solutions, 72
building materials
  acid precipitation effects on, 3
  durability, 160
buoyancy waves *see* **gravity
  waves**
bush encroachment, 72
bushveld, 73
buttes, 73
Buys Ballot's law, 73, 238
  see also **Coriolis force;
    geostrophic wind**
bysmaliths, 73, *284*

CFCs, **89–90**, 367–8
c-selection, 409
caatinga woodland, 74

caballing, 74
Cainozoic, 74, 228
  glaciations, 270, 271
  silcrete, 457
  tectonic activity, 351
  see also **Quaternary; Tertiary**
calc-sinter, 519
calcicoles, 74
calcifuges, 74
calcium
  and plant growth, 74
  in saturated solutions, 515
  in water, 465
calcium carbonate
  in aeolianite, 5
  in beach rock, 51
  in calcrete, 74
  deposition, 468
  in helictites, 252
  reef formation from, 419
  in sinter, 458
  solution *see* **limestone**
  sulphation, 485–6
calcrete, 74, 152
  pisoliths in, 389
  tepees in, 496
Calder, N., 460
calderas, 75, *541*
calibration
  infrared imagery, 502
  mathematical models, 371–2
caliche, 74
Calluna, 421
calms, 75
calories, energy measured in,
  164
calving, 75
cambering, 75, 244, *302*
  and valley bulges, 534
Cambrian, 228
cameras, aerial, 6
Canada
  braided rivers, 70
  muskeg, 347
canal design, regime theory in,
  421–2
canopies, vegetation, 75
canyons, submarine, 482
cap-rock, 76
Cape kingdom, 211
capillary forces, 75–6
capillary potential, 384–5, 485
  measurement, 496
capture, river, 76
carapace, 76
carbon cycle, 76
  and greenhouse effect, 57, 76
  Suess effect in, 485
carbon dating, 77
carbon dioxide
  atmospheric, 36, 37, 77 and
    acid rain, 2; and
    deforestation, 129; and
    greenhouse effect, 77, **240–1**;
    increase in, 76, 77, 129, 241;
    from respiration, 424; Suess
    effect, 485
  in photosynthesis, 386
  in water, 77; limestone solution,
    466
carbon-14, 288, 485
carbonation, 77

Carboniferous, 228
Carey, S.W., 327
carnivores, 77, 215, 515
carrs, 77
carrying capacity, **77–8**, 397
  for humanity, 152, 166–7
  and natural selection, 409
carses, 78
cascading systems *see* **systems**
case hardening, 78
cataclasis, 78
cataclinal rivers, 78
catastrophe theory, 78, 79
  and thresholds, 506
catastrophes, 78
  see also **hazards**
catastrophism, 79, 314, 523
  see also **diluvialism;
    neocatastrophism;
    uniformitarianism**
catchments *see* **drainage basins**
catena concept, 80, 462, *463*
cation exchange (base exchange),
  47, 48, 80
cation-ratio dating, 80
causality, **80–1**
causse, 81
  see also **karst**
caverns, *151*
caves, 81
  avens in, 41
  collapsed passages, 152
  sea: blow-holes in, 64
  shakeholes in, 453
  speleothems in, 252–3, 468
  study of, 468
cavitation, 81
cays, 38
Cederberg Mountains, *438*
Celsius scale, 495
cementation, 5, 78, 140, 276,
  307
centrifugal force, 239
centripetal acceleration, **81–2**,
  238
  see also **Coriolis force**
centripetal drainage, *155*
Cenozoic *see* **Cainozoic**
cerrado, 82
chalk
  bournes on, 69
  clay-with-flints on, 94
  coombes, 103
  karst scenery not developed on,
    293
  soils, 423
chamaephytes, *417*, 418
channelization, 54, **83**, *84*
  see also **land drainage**
channelled scabland debate,
  523–4
channels
  abandoned (palaeochannels),
    369
  active, 82
  alluvial, 16–17; classification,
    85, *86*, *87*, 88; hydraulic
    geometry, 259
  alluvial fill in, 17–18
  anabranching, 20, *86*, 88
  anastomosing, 20, *49*, 70, *155*,
    365

armouring, 34
artificial, 83, *84*, 421–2
avulsion, 41
bank storage, 43–4
bedforms *see* **bars; bedforms**
braided *see* **braided rivers**
capacity, **82–3**; changes in, 83,
    429; and discharge, 44, 82–3,
    152; downstream of
    reservoirs, 424
chutes, 90
classification, 83–8
colks in bed, 102
cross-profiles, 116–17
currents in, 120–1
cut-off, 121, 367
deposition *see* **bars; bedforms**
design: regime theory, 421–2;
    tractive force theory, 513
discharge *see* **discharge**
dissection by, 146, 205
distributary, 148
divagation, 148
energy grade line, **175**, 522
entrenchment: arroyos formed
    by, 35
ephemeral, 183, 543
erosion: banks, 43; by
    cavitation, 81; clearwater, 94;
    by evorsion, 198; and helical/
    secondary flows, **252**, 446; by
    navigation, 350
flood routing along, 210, 347
flow: competence, 104; and
    drainage density, 154;
    equations for, 89, **211–12**,
    474; Froude number, 212,
    222, 260, 389; gradually
    varied, 81, 212, **238–9**, 474;
    helical/secondary, 252, 446;
    measurement, 118, 267, 536;
    minimum acceptable, 336;
    over plane bed, 389; Reynolds
    number, 68, 212, **425**, 520;
    steady, 474; tractive force,
    513; uniform steady, 89, 212,
    218–19, 474, 513, **522**;
    unsteady, 528; velocity
    profile, 535–6
flow regimes, **212**, 222
grade, 237
hydraulic autogeometry, 258
hydraulic force in, 258–9
hydraulic geometry, 258, **259**,
    422; and alluvial filling, 18
hydraulic gradient, 260, 446
hydraulic jump, 260
hydraulic radius, 89, **260**
ice jams in, 272
lateral accretion, 302–3, 393–4
lateral migration, 303
long profile, 308–9
meandering *see* **meanders**
meltwater *see* **meltwater**
monumented sections, 340
morphological changes, 428–30
navigational use, 349–50
overflow, 366
percoline preceding, 379
perennial, 379
perimeter sediments, 513, 557–8
plane bed, 389

pool–riffle sequence in, *84*, 239,
    397, 446
pot-holes in, 399, *400*
reaches, 418; inflow and outflow
    hydrographs, 347
regime state, 421
and reservoir construction, 424
resistance, 83
roughness, 53, **83**, **433**;
    Manning equation for, **319–20**,
    522; Strickler equation for,
    480
sandbed, 212, 222, **438**
secondary currents in, 446
sinuosity, *86*, 87–8, 458–9
stability, 85, *87*, 88
step–pool systems in, 475
storage, 83
subterranean: siphons, 459
thalweg, 500
unstable, 528
wetted perimeter, 560
width and depth: adjustment to
    discharge, 259
width-depth ratio, 561
see also **rills**
chaotic systems, 355
chapada, 88
chaparral, *61*, 88, 320
Charles's law, 223
Charmouth, *456*
chattermarks, 88
cheiorographic coasts, 88
chelates, 89
chelation, 88–9
cheluviation, 89
chemical remanent magnetism,
    370
chemical weathering, 556
    carbonation, 77
    corrosion, 114
    hydration, 258
    hydrolysis, 266
    index of, 557
    in microcracks, 333
    oxidation, 367, 419
    periglacial, 221–2
    potential for, 557
    and streamwater composition,
        466
    sulphation, 485–6
    and suspended load chemical
        composition, 447
    see also **karst; limestone:
        solution**
chemicals, ecotoxicological
    studies of, 168
chemosphere, 89
chenier ridges, 89
chernozem, 89
chert, 89
Chew Bahir, Lake, *480*
Chézy equation, **89**, 319–20, 522
    see also **Manning equation**
China, tower karst in, *513*
chines, 89
chinook, 89, 214
Chitale, S.K., *87*
chlorinated hydrocarbon
    pesticides, 321, 384
chlorofluorocarbons, **89–90**,
    367–8

chlorophyll, 386
Chorley, R.J., 226, 491
chorology, 567
chotts, 90, 93, 391, 435
chronosequences, 90, 485
chronostratigraphy, 478
Church, M.A., 15, 16
chute cut-offs, 121
chutes, 90
circadian rhythm, 90
circles, stone, 374
circulation *see* **atmospheric
    circulation** *and under* **oceans**
circulation indices, 90, 276
cirques, **90–1**
    backwalls, 42
    basal sapping, 47
    bergschrund, 54
    glaciers in, 255
    nivation, 355
    randklufts, 416
cirrocumulus clouds, 98, *99*,
    100
cirrostratus clouds, 98, *99*
cirrus clouds, 13, *99*, 100, 482,
    545
cities *see* **urban areas**
cladistics (phylogenetic
    systematics), 91–3, 538
cladograms, *91*, 92, 538
classification
    organisms, 294–5, 305, 417–18,
        494
    see also under particular topics, e.g.
        rivers
clastic sediments, 93
clasts, 93
    fabric analysis, 202
    in outwash, 365
    in screes, 442
    see also **boulders**
clathrates, 93
clay balls, 34
clay dunes, **93**, 371
    see also **parna deposits**
clay–humus complex, 93
clay-with-flints, 94
claypans, 93
clays
    allophanes in, 16
    cambering over, 75
    cations bonded onto, 47, 80
    cohesion, 102
    colloids, 102
    composition, 93
    consolidation, 108
    contaminant transport, 447
    coordination numbers, 111
    expansion and contraction,
        198–9
    fabric, 202
    flocculation, 209
    flysch deposits, 213
    in fragipans, 217
    gilgai, 230
    gley soils, 233
    gumbo, 244
    kaolin, 291
    lessivage, 305
    minerals in soil: and activity
        ratio, 4; and aggregation ratio,
        9

clays (*continued*)
in parna deposits, 372
particle size, 373
and pipe formation, 388
plasticity, 389–90
proportion in sediment, 378,
557–8
proportion in soils, *463*
quick, 408, 433, 451
residual strength, 424
rocks composed of, 34
rotational failures, 432–3
sensitivity, 451
soil particles coated with, 121
clear water erosion, 94
cleavage
mineral, 94
Clements, F.E., 483
cliffs, *49*
corniches at base, 114
flatiron, 208
free face, 217–18
Richter slopes at base, 426
rock avalanches, 482
sapping, 439
sea, 442
stacks as outliers of, 472
talus at base, 494
toppling failures, *302*, **511**
CLIMAP, 94
climate
and agriculture, 28, 29
and animals, 164, 333, 567
and biome, 61
changes, *see* **climatic change**
classification, 95–6, 295;
moisture index in, 338;
precipitation efficiency in,
482; thermal efficiency in,
501
cyclic changes, 121–2
definition, 94
and deforestation, 129
and denudation rates, 135
and drought perception, 155–6
fluctuations *see* **climatic change**
and humans, **28–31,** 61–2, 97;
urban areas, 530–3; *see also*
**global warming; nuclear
winter**
and landforms, 96–7, 184
local, 307–8; urban areas, 530–3
see also **karren**
microscale variation, 333
modelling, 224–5
modification, 555–6
oceanic influence on, 11–12
radiative forcing, 411
seasonality: and expansive soils,
198
and soil formation, 331
types: Antarctic, 22–4; boreal,
66; continental, 109;
Mediterranean, 327;
megathermal, 327;
mesothermal, 330; monsoon,
340; mountain, 344;
periglacial, 380
variability, 29–30
and vegetation, **246–7,** 333, 339,
417- 18, 567; *see also* climax
vegetation; climatology;

weather *and specific aspects of
climate*
climatic change, 94–5
academic and political response
to, 29–30
CLIMAP study, 94
and climax vegetation, 97–8,
398
cyclic, 121–2
evidence: deep sea cores, 94,
335–6, 497–8; loess deposits,
308; moraines, 342; pluvial
terraces, 499; pollen analysis,
394–5
influences on, *95*
and landform development,
96–7
Little Climatic Optimum, 307
Milankovitch hypothesis, 335–6
pre-instrumental data, 369, 385
and sunspot frequency, 325
Younger Dryas, 566
see also **glacials; global
warming; interglacials**
climatic geomorphology, 96
and equifinality, 184
climatic optimum (altithermal/
hypsithermal), 19, 268
**climato-genetic
geomorphology,** 96–7
climatology, 97
applied, 9, 29
satellite data in, 440
synoptic, 489–90
see also **climate; meteorology**
climatopes, 168
climax vegetation, 35, **97**
in biomes, 61
and climatic change, 97–8, 398
disclimax resulting from
disturbance, 146
monoclimax theory, 339
polyclimax theory, 396
potential, 398
subclimax preceding, 482
climbing dunes, 97, 159
climograms, 97
clines, 97
clinographic curves, 97
clinometer, 97
clinosequences, 97
clints, 97, *375*
see also **karren**
cliseres, 97–8
clitter, 98
closed systems, 491
cloud cover, measurement of,
362
cloud forests, 98
cloud streets, 99, 330
clouds
albedo, 13, 143
analysis and interpretation,
351–2
Antarctic, 24
Arctic, 33
condensation nuclei, **98,** 143
development and form, 98,
99–100
droplet size, 99
and extra-tropical cyclones, 200,
201, 545

ice crystals in, 98–9, 249, 539
lee waves identified by, 304
measurement, 352
microphysics, 98–9
mountains, *344*
precipitation production, **98–9,**
266
seeding, 555, 556
supercooling in, 486
in tropical cyclones, 516
types, 100; banner, 44; cirrus,
99, 100, 482, 545; fractus,
217; noctilucent, 355; *see also*
**cumulonimbus clouds;
cumulus clouds; fog**
urban areas, 533
cluses, 100
coasts
aquifers: saltwater intrusion,
229–30
classification, 100, *101*
downwelling along, 153
erosion, 244, 442, 455–6
fractal dimensions, 217
hydrogeomorphology, 261
landforms: arches, 32; bars, 44,
50; corniches, 114; dunes,
100–1, 215, *216;* geos, 227;
notches, 356; sand banks,
437; sea cliffs, 442; shell
pavements, 455; shore
platforms, 455–6; spits, 468;
strandflats, 478; submerged
forests, 483; vasques, 535; *see
also* **beaches**
longshore drift along, 309
management, 181
pollution, 321
precipitation solute content, 465
progradation, 404
recession: Bruun rule on, 72
storm surges and flooding, 209,
477
tidal ranges, 507, *508*
tropical: mangroves on, 317
types, 100; *101;* Atlantic, 36;
barrier, 45, 366–7;
cheiorographic, 88; dalmatian,
124; Pacific, 36, 369
see also **estuaries; salt marshes**
cobbles, 373
cockpit karst (kegelkarst), 101,
151, 293, 338
coefficient of permeability, 381
coevolution, 101–2, 489
see also **symbiosis**
cohesion *see under* **soils**
cold fronts, 102, 200, *201,* 219,
360
cold ice, 495
cold occlusions, 360
cold pole, 102
cold-blooded animals, 393
cold-hardiness, 247
colks, 102
collision coasts, *101*
colloids, 102
colluvial footslopes, *463*
colluvium, 102–3
badlands in, *43*
in bedrock depressions, 103
dongas in, 153

colonization, 103, 494
see also **aliens; dispersal;**
K-selection
colour film, 6
cols, 102
combes, 103
comfort zone, 103
commensalism, 103
comminution, 104
comminution till, *509*
communities, 104
competition in, 105
consociation in, 108
disclimax, 146
diversity, 149–50; and
predation, 400
dominance in, 152–3
plant *see* **plant communities**
priseres, 403
sequence of *see* **seral**
communities; serclimax;
seres
synecological studies of, 489
see also **ecosystems**
compaction, soil, 104
compensation flow, 104
competence, 104
see also **Hjulström curve**
competition, 104–5
and carrying capacity, 78
and extinction, 199
and natural selection, 409
between plants, 168
and predation, 400
competitive exclusion, 199
complex response, 48, **105**
compressing flow, 105–6, **487**
compression, 106
stress produced by, 479
computers
drainage basin simulation, 458
for GCMs, 224
GIS use, 227
in weather forecasting, 440, 490,
554
conchoidal fractures, 106
concordant coastlines, 36
concordant intrusions, *284*
concordant summits, 106
concretion, 106
condensation, 106
and dew formation, 140
energy release, 301
and fog formation, 214, 475
tropical warmth maintained by,
517
condensation nuclei, 98, 143,
270
condensation trails (contrails),
110
conductance, specific, 106, 466
conduction, 250
conductivity, hydraulic *see*
hydraulic conductivity
cones, alluvial, *17*
cones of depression, 107
congelifluction, 108
congelifraction, 108
congeliturbation, 108
conglomerate, 108
inselbergs in, 279
molasse, 338

sarsens, 439
coniferous forest, *61*, 66, 493
coniology, 296
connate water, 108
consequent streams, *76*, 108
conservation, 108, 181–2
peatlands, 376
protection as, 108, 405–6
soil, 300, 461
and sustainability, 108, **488**, 564
tropical forests, 286
world strategy for, 564
conservation of mass,
equations for, 110
consociation, 108
consolidation, soil, 104, **108**,
451
constant slopes, 109
construction industry, applied
meteorology in, 28–9
consumption, environmental
impact of, 179, 180
continental air masses, 11
continental climates, 95, 109
continental drift, 109
and animal distribution, 204,
538, 568
and floristic realms, 211
and hot spots, 257
mechanism, 390–1
and Tethys Ocean, 500
see also **Gondwanaland**
continental islands, 109, 287
continental shelves, 110, *361*,
455
epeiric seas on, 182
productivity, 353
submarine canyons on, 482–3
tidal ranges, 507, 508
continental slopes, 110, *361*,
482–3
continents
moho depth below, 338
passive margins, 374
continuity equations, 110
in GCMs, 224
for kinematic waves, 294
in numerical models, 357–8
contour lines, 110, 217
contraction hypothesis, 110
contrails, 110
contributing areas, 110–11, 257
control structures *see* **flumes;**
weirs
convection, 111, 250
and cloud formation, 98, 99
in cyclones, 330
mesoscale cellular, 330
whirlwinds caused by, 560–1
convergence, 149
convergent boundaries, 390
convergent evolution, 197
convex creep slopes, *463*
convexity, 342
coombes, 103
Coon Creek, *448*, 449
coordination numbers, 111
coprolite, 111
coquina, 111
coral algal reefs, 111–13, 419
bioherms, 58
blueholes in, 65

formation, 111–12, 122;
antecedent platform theory,
24; Daly's theory, 124; high
wave energy preceeding, 253
lagoons: Daly level, 124
types: atolls, 38; barrier, 45–6
corals, 112
bleaching, 113
core, 113
corestones, 113
Coriolis force, 113–14
and ageostrophic motion, 9
and atmospheric waves, 38
centripetal acceleration similar
to, 82
and cyclostrophic forces, 123
Ferrel's law based on, 205
and geostrophic wind, 229, 562
and gradient wind, 238
and ocean currents, 120, 234
and Rossby wave formation, 432
and tropical cyclones, 516
and wind direction, 113–14,
562–3
see also **geostrophic wind**
Coriolis parameter, 114, 171
cornices, *113*, 114
corniches, 114
corrasion, 114, *463*
correlation *see* **autocorrelation**
correlograms, 40
corries *see* **cirques**
corrosion, 114, 337
see also **Trombe's curves**
cost–benefit ratios, 114
coulées, 114
couloirs, 114
Coulomb equation *see* **Mohr–**
Coulomb equation
coversand, 115
crab-holes, 115
Cracraft, J., 91, 92
CRAE method, 195
crag-and-tail, 115, 341
craters, 115
calderas, 75, *541*
old (maars), 311
cratons, 115
denudation rates, 134
epeirogenic uplift, 182
mobile belts contrasted with,
337
separated by sphenochasms, 468
shields within *see* **shields**
creationism, 115–16, 126
creep, 116, 322, **325**, *463*
cambering caused by, 75, *302*
frost heaving as cause of, **220**,
350, 392–3
crescentic bars, 44
Cretaceous, 228
crevasses, 54, 116
critical erosion velocity, 116
critical load, 116
Croizat, Leon, 538
Croll–Milankovitch model,
335–6
crops
agrometeorological studies, 10
respiration rates, 59
yield: limiting factors, 306

cross-bedding, 116, 302–3, 365
  current bedding as, 117–18
cross-lamination, 116, 118, 365
cross-profiles, river and valley, 117
crumb structure, 117
crust
  composition: sial, 456; sima, 458
  continental shelves part of, 110
  cymatogeny, 123
  hot spots, 257
  isostatic adjustment, 262, 288
  Mohorovičić discontinuity beneath, 338
  sphenochasms and sphenopiezms in, 468
  suturing, 488
  tectonics see faults;
    neotectonics; tectonic activity
  underplating, 522
  warping, 545
  see also lithosphere
cryergic processes see periglacial areas
cryofronts, 218
cryogenic weathering, 380
cryoplanation (altiplanation), 19, 117
  terrace formation, 236, 380
cryostatic pressure, 117
cryoturbation, 117, 374
  see also frost heave
cryovegetation, 117
cryptophytes, 417, 418
cryptovolcanoes, 117
crystallization, mineral, 69
cuestas, 117
  butte temoins on, 73
  dipslopes, 143
  outliers, 365
  resequent streams on, 423
cuirasses, 117
cumecs, 117
cumulative soil profiles, 117
cumuliform clouds, 13, 98
cumulonimbus clouds, 98, 99
  hail from, 249
  in tornadoes, 512
  vertical motion in, 111, 536
cumulonimbus storms, 470
cumulus clouds, 37, 98, 99, 100
  in cloud streets, 99
  vertical motion in, 111, 536
cupolas, 117
current bedding, 117–18
current meters, 118, 267
current ripples, 118, 119
  see also bars
currents
  density, 133
  extreme: probability of extreme sea levels and, 290
  longshore, 309
  nearshore, 118–20
  ocean see under oceans
  rip, 169, 170, 428
  river, 120–1; secondary, 446
  tidal, 121, 312, 507
  turbidity, 483, 520
cushion bog, 66
cusps, beach, 49, 121, 367, 419

cut-offs, 121, 367
cutan, 121
cutters, 121
Cuvier, Baron Georges, 79
cwms see cirques
cycle of erosion, 122
  Darwinian influence on, 126–7
  ergodic assumptions, 186
  planation surface end product of, 186–7
  rejuvenation concept, 422
  river capture integral to, 76
cycles, climatic, 122, 335–6, 486
cyclone (baroclinic) waves, 38, 225
cyclones
  circulation within, 330
  storm surge generation, 477
  vertical motion in, 536
  see also extra-tropical cyclones; tropical cyclones
cyclostrophic forces, 123
cyclothems, 123
cymatogeny, 123

DDT, 59, 384
DNA, 63, 347
dalmatian coasts, 124
Dalton's law, 223
Daly level, 124
dambos, 124
dams, 124
  clear water erosion below, 94
  collapse: and flooding, 209
  construction: landscape evaluation prior to, 430
  design discharge for, 140
  ice, 271; and flooding, 209, 290, 523–4
  organic debris, 363
Dana, J.D., 387
Darcy's law, 106, 124–6, 141, 185, 260
Dartmoor, tors on, 98, 512
Darwin, Charles
  on coral reefs, 45, 80
  on evolution by natural selection, 126, 127, 196–7
  ideas rejected, 127, 538
Darwinism, 92–3, 126–7, 350–1
data collection and analysis, 275
  see also remote sensing; satellites
dating techniques
  amino acid racemization, 19
  cation-ratio, 80
  electron spin resonance, 173
  fission track, 207
  lichenometry, 305
  obsidian hydration, 360
  for planation surfaces, 187
  for Pleistocene, 391–2
  potassium argon, 398
  radioactive decay, 288, 528;
    carbon, 77
  tephrochronology, 496
  thermoluminescence, 504
Davies, J.L., 100
Davis, W.M.
  on base level, 48

climatic geomorphology, 96
cycle of erosion, 76, 122, 126–7
  on graded slopes, 237
  on rejuvenation, 295
  mentioned, 213, 387
dayas, 127
De Geer moraines, 130, 545
dead ice topography, 474
death rates, 78, 397
debâcles, 128
debris
  flows, 322, 324–5
  glaciers composed of, 430–1
  plants on, 433
  slides, 404
  see also mass movements, etc.;
    talus
Deccan, 541
deciduous forests, 61, 128
  alfisols, 14
  tropical, 517
décollement, 128
decomissioning, nuclear plant, 128
decomposition, 56, 128, 307
  denitrification by, 354
  in detrital food chains, 215
deep weathering, 128
deep-sea sediments, climatic evidence in, 94, 335–6, 497–8
deflation, 130, 160
  in pan formation, 371
  see also dunes
deforestation, 128–30
  hydrological impact, 216
deformation, 130, 425–6
  see also flow
deformation till, 509, 510
deglaciation, 130
  deep sea core evidence, 497–8
degradation see erosion
degradation, aesthetic, 8
degree days, 130, 218
  see also accumulated temperature
delayed flow, 130–1, 408
dells, 131
deltas, 18, 131–2
  bottom-set beds, 67
  distributaries, 131, 148, 155, 345
  foreset beds, 215
  topset beds, 511
demoiselles, 131
dendritic drainage, 155
dendroecology, 131–2
denitrification, 354
density, 132
  and baroclinity, 45
  and barotropic motion, 45
density currents, 133
density dependence, 133, 397–8
density slicing, 142
denudation, 133–5
  chronology, 136;
    geomorphology broader than, 227–8; planation surfaces central to, 187
  rates: calculation, 136–7, 148;
    factors determining, 133–5

deoxyribonucleic acid (DNA), 63, 347
depletion curves, 47, **137–8**, 262
deposition
  coastal see **barrier islands; beaches; berms; spits,** etc.
  fluvial, 17–19, 120–1, 365, 448–9, *463*; see also **alluvial fans; bars; beforms; floodplains**
  glacial: and compressing flow, 105; see also **glaciofluvial landforms; moraine; till**
  pluvial terraces, 499
  punctuated cycles, 406
  quasi-equilibrium landforms, 407
  and soil development, 90
  on waning slopes, 544–5
  see also **sediments**
deposition, dry, 3, **156**
depression storage, 138
depressions
  lee, 304
  secondary, 446
  thermal, 501
  see also **extra-tropical cyclones**
depth–area curves, 138
depth–duration curves, 138
deranged drainage, *155*
desalinization (desalination), 138
desert (stone) **pavements,** 130, 230, **477**
desert varnish, 139
desertification, 139–40
deserts, 138–9
  animals: aestivation, 8
  biomass and biological productivity, 60, 353
  distribution, *61*
  insolation weathering, 279
  landforms: dreikanters, 155; dunes, 157–60; gibbers, 230; hamadas, 249; kavirs, 293; mekgacha, 327; oases, 360; pans, 371; pavements, 130, 230, *477*; pediments, *377*; playas, 391; salars, 435; wadis, 543; yardangs, 566; zeugen, 567
  soils, 34; gypcrete, 245; takyrs, 493–4
  temperature, 251, 333
  types: ergs, 186; punas, 406; regs, 420; serirs, 451
  vegetation, 247, 386, 565
  winds: deflation and dust storms caused by, 130, **160**, 308; helical flow, 252; see also **dunes**
desiccation, 140
design discharge, 140
desquamation see **exfoliation**
determinism, environmental, 314–15, 316
deterministic models, 553
detrital food chains, 215
detrital remanent magnetism, 370
developing countries
  deforestation, 129

environmental management, 182
Devensian Glacial, 301
Devonian, 228
dew, 140
  freezing, 254
dewpoint, 140
diabatic processes, 5, 504
diachronous sediment, 140
diaclinal rivers, 140
diagenesis, 140
diamictite, 140
diamond dust, 33
diapirs, 140–1
diastrophism, 141
diatremes, 141
die-back, 141
diffluence, 141
diffusion equation, 141–2
diffusivity, hydraulic, 258
digital image processing, 142, 346
dikakas, 143
dilation, angle of, 20–1
dilation (dilatation), 143
dilution effect, 143
dilution gauging, 143, **145,** 267
diluvialism, 143, 523–4
dimensionless unit hydrographs, 524
dimethylsulphide (DMS), 143
dip, 143
dip slopes, 143, 522
dirt cones, 143
disasters, 143–4, 348–9
discharge
  analysis, 145
  annual exceedance series, 22
  annual series, 22
  and bank storage, 43–4
  bankfull, 44; and pool and riffle formation, 397
  base flow see **base flow**
  and channel capacity, 82–3
  compensation flow, 104
  contributing areas for, 110–11
  definition, 144
  design level, 140
  dominant, 4, 152
  drainage basin, 48–9
  effective, 82
  flashiness, **208**, 300
  flow duration curves, 211
  glaciers, 366, 487
  glaciofluvial streams, 232
  and hydraulic geometry, 259
  and hydraulic jump, 260
  hydrographs see **hydrographs**
  low flow analysis, 309
  measurement, **144–5,** 266, 267; cumecs as unit for, 117; electromagnetic gauging, 145, 267; float recorders for, **208–9,** 267; flumes, 144, **213,** 267; at gauging stations, 224; at rated sections, 417; slope area method, 145, 267; stage recorders, 267, **473–4,** 475–6; stilling wells, 475–6; tracer dilution, 143, **145,** 267; ultrasonic gauging, 145, 267; velocity area method, 144–5,

417, **535;** weirs, 111, **144,** 267, **453–4,** 534, **558–60**
  meltwater streams, 327
  minimum acceptable, 336
  partial duration series, 372
  peak, 375
  random component, 476–7
  rating curve, 417
  recurrence intervals, 140, 209–10, 418
  regimes, 421
  reponse to rainfall: influence of storage, 263–4
  and river currents, 120
  and sediment transport: artificial channels, 421–2
  and solute concentration, 269
  and suspended load, 269
  urban areas, 300, 528–9
  for valley meander formation, 534
disclimax community, 146
Discomfort Index, 495–6
disconformities, 146
discordant intrusions, *284*
dishpan experiments, 146
disintegration moraines, 341, 342
disjunct distributions, 146
dislocation metamorphism, 331
dispersal, plant and animal, 103, **146**
dispersalist biogeography, 538
dissection, 146, 205
dissipative beaches, 147, 419
dissolved load, 147, 307, **465–6**
  concentration: and residence time, 424
  dilution, 143, 277
  and discharge, 269, 417
  ion concentration, 283–4
  mixing models, 337
  not included in dendudation rate calculations, 137
  rating curve, 417
  and residence time, 424
  total dissolved solids, 106, 147–8
dissolved oxygen, 55, 56, **147**
dissolved solids, 147–8, 283–4
distributaries, 131, 148, *155*
  mud lumps in, 345
distribution graphs, 148
distributions
  disjunct, 146
  see also under **animals** and **plants**
diurnal tides, 148
divagation, 148
divergence, 149
divergent boundaries, 390
divergent erosion, 149
diversity, 56, 149–50
  alpha and beta distinguished, 19, **55,** 149
  and area, 467
  concern about loss of, 56
  and ecosystem stability, 471
  and energy availability, 467
  measurement, 149
  and predation, 400, 402

diversivores, 150
doabs, 150
dodo, 285
doldrums, 150–1
dolines (sinkholes/shakeholes),
    151–2, **293, 453,** 458, 488
    coalescence, 533
    in terloughs, 496
dolocrete, 152
dolomite, yardangs in, 566
dome dunes, 152, *158*
domestication, 152
    and energy use, 175
dominant discharge, 44, 152
    see also **channels: capacity**
dominant organisms, 152–3
dominant wind, 153
dongas, 153
dormant volcanoes, 153
double mass analysis, 153
downwelling, 153
draas, 153
drainage, 153–4
    beaded, 51
    deserts, 90, 391, 435, 543
    and geology, 136, 153–4;
        antecedent, 24, 154;
        anteconsequent, 24;
        cataclinal, 78; diaclinal, 140;
        inconsequent, 274;
        obsequent, 360; resequent,
        423; subsequent, 360, 483
    ice-dammed, 209, 271, 290,
        523–4
    patterns *see* **drainage networks**
    see also **land drainage**
drainage basins
    base flow, 47; depletion curves,
        137–8; dry weather flow as,
        157
    base level fluctuations, 48
    boundaries, 553
    characteristics, 154; and
        discharge, 48–9; *see also*
        drainage density
    complex responses, 105
    composition: Horton's laws, 16,
        257
    contributing areas, 110–11
    definition, 153
    discharge, 48–9; *see also*
        discharge
    graded time-scale studies, 238
    hydrographs *see* **hydrographs**
    hydrological cycle *see*
        hydrological cycle
    modelling and simulation,
        371–2, 458, 553
    morphometry, 343
    planning and management, 80,
        428
    representative and experimental,
        199, 423
    residence time in, 424
    sediment budget, 448–9
    sediment yield *see* **sediment
        yield**
    solutes in, 465–6
    source area, 161, 372, 441,
        466–7
    storage in, 138, 263–4, 276–7,
        477, 486–7

    as system, 226, 491
    transfers between, 279–80
    unit hydrographs, 524–5
    water balance, 48, 49, **545–7**
drainage density, 154
drainage networks, 154–5, 353–4
    anastomosing, 20, *49*, 70, *155*
    annular, 22, *155*
    bifurcation ratio, 55
    changes in, 429
    classification, 154, *155*
    endoreic, 174
    ephemeral streams in, 183, 543
    evolution, 153–4; river capture
        in, 76
    insequent, 274, 279
    links in *see* **stream order**
    random-walk simulation, 416
    superimposed, 486
    see also **drainage density**
drains, 153
draw down, *107,* **155**
    coastal aquifers, 229–30
dreikanters, 155
drift potential, 155
dripstone, 155, 468
drought, 155–6
    and desertification, 140
    and vegetation, 141, 247, 442,
        565
    xerothermic index of, 565
    see also **arid zone; deserts**
drumlins, 156, *341*
    rock, 430
dry adiabatic lapse rate, 301,
    536–7
dry bulb temperatures, 52, 496
    and vapour pressure, 535
dry deposition, 3, **156**
    see also **acid rain**
dry valleys, 156–7, *327,* 439
dry weather flow, 157
du Boys equation, 157
ductile fracture, 217
dune pastures (machair), 311
dune sands, cementation of, 5
dunes, 54, 157–60
    absent or undistinct on
        coversand, 115
    aklé pattern, 12
    on barrier beaches, *45*
    blowouts, **64–5,** 159
    in draas, 153
    and flow regime, 212
    foreset beds, 215
    fuljes between, 222
    on lee sides of pans, 371
    singing sands in, 458
    slacks between, 459
    slip faces, 459
    types, *158,* 159; barchan, **44,**
        158, 222; clay, 93; climbing,
        97; coastal, 100–1, 215, *216;*
        dome, **152,** *158;* echo, 164;
        falling, 159, **202;** foredunes,
        215, *216;* impeded, 274;
        lunette, 159, **309;** parabolic,
        158, 159, **371,** *371;* reversing,
        158, **425,** *425;* rhourds, 426;
        seif, 450; star, *158,* 474;
        transverse, *158,* 514; zibars,
        567

    and vegetation, 100, *159*
    see also **dune sands; machair;
        parna deposits**
durability, 160
Durdle Door, *32*
duricrusts, 160, 370
    deep weathering associated with,
        128
    fossil: sarsens, 439
    pisoliths in, 389
    silcrete, 160, 343, **457**
    types: alcrete, 14; calcrete, **74,**
        152, 496; ferricrete, **205,** 343,
        347; laterite, **303,** 347, 392
duripans, 160, 249
dust, 160–1, 296
    air pollution by, 395
    deposition *see* **loess**
    haze caused by, 250
    in rain, 64
Dust Bowl, 461
dust veil index, 161
duststorms, 160, 248
duyodas, 49
dykes, 161, *284,* 428
dynamic equilibrium, 161, 237
dynamic meteorology, 161, 386
    see also **satellite meteorology;
        synoptic meteorology**
dynamic pressure, 55
dynamic source area, 161, 372,
    440
dynamics, 162

Eh (redox potential), 419
earth
    albedo *see* **albedo**
    cooling: contraction hypothesis,
        110
    gravity *see* **gravity**
    heat budget, 250–2
    magnetism *see* **magnetic poles;
        magnetism**
    orbit round sun: ecliptic, 164,
        335, 399; perihelion, 380
    rotation *see* **Coriolis force**
    see also **crust; lithosphere;
        mantle,** *etc.*
earth hummocks, 163
**Earth Observing System
    satellite,** 527
earth pillars, 163
**Earth Resources Experiment
    Package (EREP),** 319
Earth Resources Satellite, 527
earth resources satellites,
    317–19, 525–8
**Earth Resources Technology
    Satellite,** 525–6
earthflows, 322
earthquakes, 163–4
    aftershocks, 8
    anaseism, 20
    anticentre, 25
    benioff zone, 54
    compressional waves, 106
    damage due to, 348
    denudational role, 134
    energy equivalents, 164
    epicentre, 183
    intensity, **163,** 328, 426, 450;
        and frequency, 312, 450

perception of, 144
and soil strength, 451
at transform boundaries, 390
tsunami generation, 518
see also **faults; volcanic
eruptions** etc.
easterly waves, 163
echo dunes, 159, 164
ecliptic, 164
changes in, 335, 399
ecogeographical rules, 164
Allen's, 15
Bergmann's, 54
ecological explosions, 165–6
ecological introductions, 24,
282–3, 286, 400
ecological isolation, 287
**ecological replacement
principle,** 199
ecological transition, 166
ecology, 166–7
biogeography compared with,
57, 58
of communities, 489
concepts see **biological
productivity; population
dynamics; succession,** etc.
Darwinist principles, 127
of individual organisms and
species, 39
landscape, 301
of past environments, 369,
407–8; tree-ring studies in,
131–2
systems ideas in, 491; see also
ecosystems
**economic growth,
environmental
management and,** 182
economics, 316
environmental, 176–7
ecosphere, 166, 167
see also **biosphere**
ecosystems, 167–8, 491
abiotic components, 1
alien species in, 14, 286, 400
aquatic: acid precipitation
effects, 3
biological magnification in, 58–9
biological productivity, 167
carrying capacity, 77–8, 167
damaged: subseres initiated
from, 483
diversity, 56, **149–50;** and area,
467; and energy availability,
467; and stability, 150, 471
energy flows, 164–5, 515; by
photosynthesis, 386
eutrophication, **191,** 353, 359,
376
holism, 167
human impact on see **human
impact**
limiting factors in, 306
management, **181–2,** 484
nutrient flows in, 167
persistence, 383
perturbations in, 167, 479
protection, 405–6
resilience, 167, **470–1**
resistance, 470, 471
species composition, 383

stability, 167, 383, **470–2;** and
diversity, 150, 471
stress, 479
succession in see **succession**
sustained yield, 488
trophic levels see **trophic levels**
types: coral, 112–13; forest,
129- 30; wetland, 560;
wilderness, 561; see also
islands
see also **biogeocoenosis**
ecotones, 168
ecotopes, 57, 63, **168**
see also **biotopes; habitats**
ecotoxicology, 168
ecotypes, 168, 224
edaphic factors, 168–9
see also **soils**
edaphology, 169
edaphotopes, 168
eddies, 169, 334
diffusivity, 169
dishpan simulation, 146
lee, 304
momentum flux, 339
turbulent, 520; see also
turbulence
eddy correlation, 193
edge waves, 169, *170*
and overwashing, 367
and rip currents, 120
edge-lines, 169
Edinburgh, crag-and-tail in,
115
effective discharge, 82
effective rainfall, 169–70, *171,*
442
unit hydrograph from, 524
unit response graph from, 525
effective stress, 170, 459
effective temperature, 103,
495–6
effluent, 63, 171
effluent streams, 171
egres, 66, 171
Ekman spiral, 120, **171–2**
El Niño effect, 12, 172
elastic rebound theory, 172
elastic waves, atmospheric, 38
elasticity, modulus of, 337–8
elbow of capture, 76
Eldredge, N., 91, 92
electrical resistance blocks, 172
electromagnetic gauging, 145,
267
electromagnetic radiation,
172–3, 410–11
digital images, 142
remote sensing, 422, 456–7;
aerial cameras, 6, 7;
multispectral scanners, 346;
satellite, 525–7
electron spin resonance, 173
eluviation, 173, *463*
eluvium, 102, 173
emissivity, 278
endangered species, 173–4
endemic species, 174
and genetic drift, 226
on islands, 285
in refugia, 420

see also **vicariance
biogeography**
endogenetic processes, 174
endoreic drainage, 174
endrumpfs, 174
energy
adjustment to changes in, 422
and biological productivity, 59
black body emission, 410–11
for change of state, 301–2
flows, 174–5; in atmospheric
circulation, 225–6, 248–9;
through food chains, 164–5,
215, 515; in meridional
circulation, 328; in oceanic
circulation, 120; photo-
synthesis as mechanism for,
164, 386; in systems, 491
food, 60
human use of, 175–6, 315
kinetic see **kinetic energy**
loss: in channels, 83, 175
nuclear, 176
parasites' use of, 372
potential see **potential energy**
represented on thermodynamic
diagrams, 504
for respiration, 424
for sediment transport, 478
solar see **solar radiation**
and species diversity, 467
tidal, 508–9
waterfalls, 553
see also **heat; radiation**
energy balance
gradually varied flow, 239
models, 224
between predators and prey, 401
wet surfaces, 251–2
energy grade line, 175
gradually varied flow, 239
uniform steady flow, 522
energy resources
and applied meteorology, 28
engineering
environmental: geomorphology
applied to, 177
interaction with biology, 62
river modification, 83, *84*
unforeseen consequences, 27
englacial debris, *509*
englacial environment, 176
**ENSO (El Nino/Southern
Oscillation),** 12, 172, **312,**
495, 518
entisols, 452–3
entrainment, 176
velocity, 116, 254
entropy, 176
environment
aesthetic degradation, 8
assessment and evaluation, 176,
430
conservation see **conservation**
in cost–benefit analysis, 114
destructive exploitation, 430
diversity: adaptive radiation in
response to, 4–5
global change, 234; see also
climatic change

environment (*continued*)
  human relationship with, 58,
    166–7, **177–81**, **314–17**, 345;
    see also **pollution**
  management, *178*, 180, **181–2**;
    geomorphologists' role, 27;
    systems concept in, 492
  reconstruction: from fossil
    plants, 369; from pollen
    analysis, 394–5; Quaternary,
    407–8
  sustainable use, 108, **488**, 564
environmental determinism,
    314–15, 316
environmental economics,
    176–7
**environmental engineering
    geomorphology**, 177
**environmental impact
    assessment**, *179*, **180–1**, 316
**environmental impact
    statements**, 180–1
environmental lapse rate, 536
**environmental life cycle
    analysis**, 305
environmentalism, 314–15, 316
ENVISAT, 527
enzymes
  biotechnological use, 62, 63
  in decomposition, 128
Eocene, 74, 228
EOS, 527
epeiric seas, 182
epeirogeny, 182–3
ephemeral plants, 183
ephemeral streams, 183, 543
epicentre, 183
epidemics, 165
epilimnion, 183
epipedons, 183
epiphytes, 183
epochs, geological, 183
equations
  continuity, 110, 224, 294, 357–8
  hydrostatic, 224, **268**, 444
  of motion, 183, 562; and
    Coriolis force, 113; in GCMs,
    224; in numerical models,
    358
  of state, **183**, 224
  thermodynamic, 224
equator
  magnetic, 4
  thermal, 501
equatorial forests *see* **tropical
    forests**
equatorial trough, 184
  see also **intertropical
    convergence zone**
equifinality, 184
equilibrium, 184–5
  alluvial channels, 17
  beaches, 185
  conceptual limitations, 506
  dynamic, **161**, 237
  equations of state representing,
    183
  and graded time, 238
  hydrostatic, 268
  minimal change from, 336
  population, **397**, 402
  punctuated, 79, 197

quasi, 407
relaxation time before, 422
stable, 303–4, 472; carrying
  capacity as, 78; *see also*
  stability
thermodynamic: Van't Hoff's
  rule, 535
unstable, 528
equilibrium line, glacier, **185**,
  366
equilibrium shoreline, 185
equilibrium slopes, 479
**equilibrium theory of island
  biogeography**, 285–6
equilibrium tides, 508
equinoxes, precession of, 335,
  399
equipotentials, **185**, 261
eras, geological, 186
ergodic hypothesis, 186
ergs, 186
Erhart, H., 62
Eriophorum, 421
erodibility, 186
erosion, 130, 186
  banks, 43
  base level for, 47–8
  coastal, 244, 442
  cycle of *see* **cycle of erosion**
  demoiselles protected from, 131
  divergent, 149
  fluvial *see under* **channels**
  glacial *see under* **glaciers**
  history: reconstruction, 136
  landforms: badlands, **42–3**, *133*,
    153; inverted relief, 283; sea
    cliffs, 442; shore platforms,
    455–6; stone pavements, 130,
    230, **477**; *see also* **gullies;
    planation surfaces;** *and
    under* glaciers
  measurement, 333–4, 392
  processes: abrasion, 1, 103, 155,
    202, 430; corrasion, 114;
    nivation, 355; raindrop
    impact, 414; sturzstrom, 482;
    suffosion, 485; wind, 130,
    155, 477, 536
  resistance to, 134, 325, 431
  soil *see* **soil erosion**
  unequal, 522
  see also **denudation;
    deposition; weathering**
erosion plots, 392
erosion surfaces *see* **planation
  surfaces**
erosivity, 187
erratics, 187
eruptions, volcanic *see* **volcanic
  eruptions**
escarpments
  gulls on, 244
  retreat: and inselberg formation,
    279
eskers, 187
Essa satellites, 332
estavelles, 469
estuaries, 188–9
  biological productivity, 60, 188
  sapropels in, 439
  tidal bores in, 66
  tidal currents in, 121

estuarine zone, 85
etchplains, 189
etesian wind, 189
ethics, 316
Ethiopian realm, 204, 543
ethology, 316–17
eucalyptus, 314
eugeogenous rock, 189
Eulerian winds, 562, 563
eulittoral zone, 189
euphotic zone, 378
Europe
  environmental impact
    assessment, 180
  environmental management, 182
European Space Agency, 527
eustasy, 48, **189–90**, 445
  and hydroisostasy, 262
  and raised beaches, 416
eutrophication, 191
  fenland, 376
  lakes, 14, 359
  and productivity, 353
Evans, L.S., 342
evaporation, 191–4, *263*
  actual and potential
    distinguished, 191, 399
  and drought, 155
  and effective rainfall, *171*
  energy loss during, 251–2
  factors controlling, 191
  forests, 216
  measurement, 36, **191–3**
  and precipitation efficiency, 482
  solar energy saved up by, 301–2
  total dissolved solids determined
    by, 147–8
  see also **evapotranspiration**
evaporation pans, 192
evaporite, 194
evapotranspiration, **194–6**, *243*,
  247, *263*, 399
  actual and potential
    distinguished, 194–5, 546
  climatic classification based on,
    96
  and deforestation, 129, 130
  deserts, 138
  forests, 216–17
  mapping, 265
  measurement, 195, 310
  in moisture index, 338
  during pluvials, 393
  and soil moisture deficits, 461–2
  urban areas, 531
  in water balance calculations,
    545–7
  see also **evaporation**
Everest, Mount, *344*
evergreen vegetation, 196, 442
  heathlands, 252
  tropical forests, 517
evolution, **196–8**, 386
  adaptive radiation, **4–5**, 285
  and animal distributions, 568
  cladistic view of, 91–3
  convergent, 197–8
  creationism opposed to, 115–16
  Darwinian, 92–3, 126–7
  and extinction, 199
  genetic drift as mechanism for,
    226–7

humans, 314
interrelated species
 (coevolution), **101–2**, 489
marsupials, 204
neo-Darwinian, 350–1
polytopic, 396
and taxomomy, 494
vicariant species, 537–9
see also **natural selection**
evorsion, 198
exaration, 198
Exe, River, *326*
exfoliation, 140, 198, 362, **454–5,**
 468
exhaustion effects, 198
exhumation, 198
exogenetic processes, 198
exotic species, 198
expansive soils, 198–9
experimental catchments, 199,
 423
experimental design, 274–5
experiments, laboratory
 fluvial and hydrological, 297
exosphere, 37
exsurgence, 425
extending flow, 105, 487
externalities, 176
extinction, **199,** 398
 island species, 285, 286
 and predation, 401
 see also **evolution**
extra-tropical cyclones, 199–201
 and Antarctic temperatures, 23
 cloud formation, 98, 100
 formation, 37, 44, **122–3,** 304
 fronts, 102, *201,* 219–20, 293,
 545
 in lee of mountains, 304
 occlusions, *201,* 219, 360
 on polar front, 394
 precipitation, 219–20, 364, 545
 secondary depressions, 446
 storm surge generation, 477
 weather, 122, 123, 200
extreme value analysis, 244
extrusion, volcanic, 201
extrusion flow, 201
exudation basins, 201
eyes (cyclonic), 201, 516
eyots, 201

fabric, till, 202, 510–11
facets, 202
facies, 202
facilitation model, 483
Fahrenheit scale, 495
fall faces, *463*
falling dunes, 159, **202**
falls
 rock avalanches, 41, *302,* 322,
 482
 rockfalls, *302,* 321, 323, *324*
 slab failures, *302,* 459
 toppling failures, *302,* **511,** 511
false colour/near infrared film,
 6
false-bedding, 202
fanglomerate, 203
Fanning, P.C., 71–2
fans, 202
 alluvial, **17,** 43, 203, 457

coalescence, 371
washover, 367
farmers
 ecological transition, 166
 see also **agriculture**
fast ice, 443
fatigue failure, 203
faults, 203
 and earthquakes, 163
 elastic rebound theory of, 172
 landforms: basin-and-range
 terrain, 48; graben, 236;
 horsts, 256; klippes, 295; rift
 valleys, 426, *427;* slickensides,
 459
 and spring location, *469*
 and terranes, 499
 types: block, 48, 64;
 compressional, 106; gravity,
 239; normal, 356; overthrust,
 366; reversed, 203, 425;
 strike-slip, 327; tear, 494;
 thrust, 203, 348, 366, 507;
 wrench, 564
fauna *see* **animals**
faunal realms, 203–4, 543–4
faunistics, 567
feather edges, 205
feedback, 184, 491
 in Gaia concept, 223
 in le Chatelier principle, 303
 and quasi-equilibrium, 407
 in snowblitz theory, 460
feedstocks, 62
feldspar
 in granite, 239
 in greywacke, 241
 in sandstone, 34
felsenmeer *see* **block fields**
fens, 205, 376
feral relief, 205
Ferguson, R.I., 87
fermentation, 62–3
ferrallitization, 205
Ferrel cell, 205
Ferrel's law, 205
ferricrete, 205, 343, 347
fertilizers, water pollution by,
 395
fetch, 205
fiards, 205
field capacity, **205,** 461
field drainage, 205–6, 298
films, aerial, 6
Finger Lakes, 206
fiords, 206
fire
 early human use, 258, 314
 ecosystem modification, 175,
 206, 258; deforestation, 129
 hydrological effects, 206–7
firn, 14, 207, 270, 354
firn line, 185, 460
fish, migration, 20
fish farming, 31
**fishing, ecosystem resistance
 to,** 471
fishscale dune pattern, 12
fission track dating, 207
fissure eruptions, 207
fjords 206
Flandrian transgression, 207–8

and high energy window, 253
 ria formation in, 426
flash floods, 208
flashes, 208
flashiness, 208
 urban areas, 300
flatirons, 208
flint
 in clay-with-flints, 94
float recorders, 208–9, 267
floating bogs, 209
flocculation, 209
flood, Biblical, 143, 523–4
flood frequency analysis,
 209–10, 312– 13
 annual series in, 22
 Gumbel extreme value theory
 in, 244
 partial duration series in, 372
 of peak discharges, 375
 Pearson type III distribution in,
 375
flood geomorphology, 210
flood hydrographs *see*
 hydrographs
flood routing, 210
 Muskingum method, 347
 in unsteady flow, 528
floodplains, 210
 backswamps, 42
 and bankfull capacity, 82
 inundation, 209
 lateral accretion, 302
 levées on, 305
 palaeochannels on, 369
floods, 209
 catastrophic: geomorphic effects,
 79, **523–4**
 coastal, 209; storm-surge
 generated, 477
 flash, 208, 300
 frequency: and magnitude,
 312–13; and reservoir
 construction, 424; urban
 areas, 529; *see also* **flood
 frequency analysis**
 geomorphological study, 210
 human responses to, 209
 on outwash, 365
 overbank deposits by, 210, 365
 prediction, 266
 prevention, 209; design
 discharge for, 140; by
 drainage, 298
 probable maximum, 403–4
 recurrence intervals, 266, 313,
 418, 425
 sheet erosion by, 454
 and takyr formation, 493–4
 tsunamis as cause of, 519
 urban areas, 300, 529, *530*
 see also **floodplains**
floodway zone, 85
floristic realms, 211
flow
 in channels *see under* **channels**
 diffluence, 141
 diffusion equation for, 141–2
 energy *see under* **energy**
 equations for, 211–12; based on
 steady flow, 474; Chézy, **89,**

flow (*continued*)
319–20, 522; diffusion, 141–2; inapplicability, 238, 528
glaciers: compressing and extending, 105–6
groundwater, 124–6, 244
on hillslopes, **253–4**, 277, 366
hydraulic conductivity as measure of, 106
hydraulics as study of, 260–1
hydrograph components *see* **hydrographs**
ice, 272
laminar, 126, 212, **297–8**, 425, 520
lateral, 303
overland *see* **overland flow**
resistance to, 425–6; *see also* **roughness; viscosity**
streamlines, 169, 478
turbulent, 212, 425, 520
velocity *see* **velocity**
see also **discharge; fluxes**
flow duration curves, 211
flow regimes, 212, 222
flow till, *509,* 510
flowmeters, 267
flows, debris, 322, 324–5
fluid mechanics, 212–13, 260–1
see also **flow, equations for**
fluid potential (hydraulic potential), 185, 254, **260**
fluids
boundary layers *see* **boundary layers**
viscosity, 298, 425, 497, **539**
vorticity, 540–1
see also **flow**
flumes, 111, 144, **213,** 267
fluvial hydrogeomorphology, 261
fluvial sediments *see* **bed load; suspended load** *and particular landforms*
fluvioglacial streams, 232–3
fluviokarst, 213
dry valleys, **156–7,** 327, 439
fluxes, 213
divergence, 213
in meridional circulation, 328
flyggbergs, 213
flysch, 213–14
fog, 100, **214**
occult deposition by, 361
suppresion and dispersal, 555
types, 214; frost smoke, 221; ice, 33, **272,** 532; smog, 395, **460;** steam, 475
urban areas, 532
föhn, 54, **214**
folds, 214–15
landforms: décollement, 128; periclines, 379
types, 214–15; antiforms, 26; asymmetrical, 36; isoclines, 287; recumbent, 76, 214–15, 348, 418; synclines, 214, 385, 489; *see also* **anticlines**
foliation, 215
food
collection: by hunter-gatherers, 258

energy: from solar energy, 386; and secondary productivity, 60
production: and domestication, 152; and environmental management, 182
food chains and webs, 215, 515
biological magnification, 58–9
diversivores in, 150
herbivores' role in, 253
stability, 471
Forbes bands (ogives), 361
forcing functions, 491
forecasting *see under* **weather**
foredunes, 159, 215, *216*
foreset beds, 215
foreshore, *45, 50*
forests
decline, 216
ecological structure complex, 129
growth: mosaic-cycle concept, 343
herbivores in, 253
human disturbance, 470
hydrology, 216–17
interglacial, 280
krummholz in, 296
net primary productivity, 353
removal, 128–30, 216
respiration rates, 59
types: cloud, 98; coniferous, *61,* 66, 493; deciduous, 14, *61,* **128** 517; gallery, 223; monsoon, 340; submerged, 483; *see also* **tropical forests**
forked lightning, 305
form lines, 217
form ratio, 217
fossil fuel combustion
carbon dioxide from, 485
fossils
creationists' interpretation, 115–16
dating: by amino acid racemization, 19
evolutionary evidence from, *196,* 197
extinction evidence from, 199
macrofossils, 311
past ecological conditions deduced from, 369
plant: in peat, 375
pollen: analysis of, 394–5
successional stages preserved in, 485
fractal dimension, 217
fractures, 217
conchoidal, 106
fractus cloud, 217
fragipans, 217
frazil ice, 217
free faces, 217–18
freeze–thaw cycles, 4, **218**
landforms: grezes litées, 241; patterned ground, 374; ploughing blocks, 392–3; strandflats, 478; striated soil, 479
and needle ice formation, 350
freeze–thaw processes *see* **congelifluction; frost heave;**

frost shattering; frost weathering *etc.*
freezing, cryostatic pressure caused by, 117
freezing front, 218, 273
freezing index, 218
frequency curve, alimetric, 19
friction, 218–19
and slab failure, 459
and slope stability, 472
soil cohesion not governed by, 102
and soil strength, 338
Friedrich, E., 430
fringing reefs, 45
frontal fog, 214
frontogenesis, 220
fronts, 219–20
Arctic, 33
formation, 220
temperature inversions in, 283
types: anafronts, 20; cold, 102, 200, *201,* 219, 360; katafronts, 293; occluded, *201,* 219, **360,** 360; polar, 225, **394;** warm, *201,* 219, 360, 545
vertical motion, 536
frost, 220
ground, 241
hoar, 254
see also **periglacial areas; permafrost**
frost creep, 220, 350
ploughing block formation, 392–3
frost heaving, 220
see also **cryoturbation**
frost shattering, 218
block field formation, 64
grezes litées formation, 241
frost smoke, 221
frost weathering, 10, 220, **221–2,** 380
frost wedging, 220, 221, 380
see also **ice wedges**
Froude number, 212, **222**
and hydraulic jump, 260
plane bed, 389
fulgurites, 222
fuljes, 222
fumaroles, 222
fungi, 295, 439
symbiotic, **488–9**
fungicides, 384

GCMs, 224–5
GIS, 142, **227**
gabbro, 223
Gaia concept, 223
Galapagos Islands, 14, *467*
gales, 52
gallery forests, 223
Ganges River, 76
GARP, 12, **223,** 332, 440
garrigue, 223
gas laws, 223
GATE, 12
gauging stations
discharge measurement at, 144–5, 224, *473, 474* and flood frequency, 210

staff gauges at, 473
geest, 224
gelifluction, 224, 322
  and frost creep, 220
  ploughing block formation,
    392–3
  see also **solifluction**
Gemini satellites, 317–18
gendarmes, 224
genecology, 224
general circulation *see*
  atmospheric circulation
general circulation models,
  224–5
general system theory, 226, 491
genes, manipulation of, 63
Genesis
  challenges to, 568
  creationists' acceptance, 115
  diluvialism based on, 143, 523–4
genetic drift, 226–7
genetics
  and Darwinism, 127, 350–1
  see also **inheritance**
geo-ecology, 301
geocryology, 227
geodes, 227
geodesy, 227
geographic information
  systems, 142, **227**
geoid, 227, 326
  eustasy, 190
geological cycle, 122
geology
  catastrophism in, 79
  Darwinist principles underlying,
    127
  and drainage *see under* **drainage**
  geomorphology regarded as part
    of, 227
  and spring location, *469*
  time-scale, 227, 228
geomagnetism, 500
geomorphological mapping, 27
geomorphological processes
  catastrophic, **78–9,** 314, 350
  emphasis on, 136, 213, 228
  flux divergence essential to, 213
  magnitude and frequency
    effects, **312- 14,** 350
  morphogenetic regions based
    on, 342
  rates, 523; measurement, 27
  uniformitarian, 314, **522–4**
  see also **deposition; erosion;**
    weathering etc.
geomorphology, 227–8
  applied, 27–8, 177
  climatic, **96,** 184
  climato-genetic, 96–7
  Darwinist principles, 126–7
  denudation chronology as
    original focus of, 136
  in environmental engineering,
    177
  explanation in: equifinality, 184;
    indeterminacy, 274–5
  human influence on, 25, 27
  hydrogeomorphological studies,
    261
  neocatastrophism in, 350
  organisms' role in, 58

physiography as term for, 387
processes *see* **geomorphological**
  processes
purpose, 227–8
system concept in, 226, 491–2,
  505–6
tectonic (morphotectonics), 343
geomorphometry, 342–3
geophysical hazards, 349
geophytes, 229
geos, 227
geostationary satellites, 332–3
geostrophic wind, 171, **229,** 502,
  562
see also **Coriolis force;**
  gradient wind
geosynclines, 229
geothermal heat, geysers
  heated by, 229
Gerlach trough, 229
geysers, 229, 458
ghibli wind, 294
Ghyben–Herzberg principle,
  229–30
gibbers, 230
Gilbert, G.K., 237, 377, 522
gipfelflur, 230
glacial till *see* **till**
glacials and glaciation, 230–1
  Croll–Milankovitch model, 336
  cyclical nature of, 122
  dating, 391–2; by lichenometry,
    305
  and dry valley formation, 157
  evidence of, 270–1
  isostatic effects, 288
  and loess formation, 308
  monoglaciation theory, 340
  Penck and Bruckner model, 378
  and sea level changes, 189–90
  snowblitz theory, 460
  see also **deglaciation;**
    interglacials; interstadials;
    pluvials
glaciation level, 231
glacier evacateur, 231
glacier milk, 232
glacier reservoir, 231
glacier tables, 232
glacier winds, 293
glacierets, 232
glacierization, 232
glaciers, 231–2
  ablation, **1,** 130, 185, 340; *see*
    also **meltwater**
  accumulation, 14, 185; *see also*
    firn
  activity, 185
  altitudinal limit, 231
  basal ice, 47
  bergschrund, 54
  boulder trains carried by, 67
  calving from, 75
  crevasses in, 116
  deposition: erratics, 187; glacial
    flutes, 340, *341*; perched
    blocks, 379; *see also*
    glaciofluvial landforms;
    moraine; till
  diffluence, 141
  discharge, 366, 487
  distribution, 135, *382*

englacial environment, 176
equilibrium line, **185,** 366
erosion: abrasion, 1; basal
  sapping, 34, 90; dilation joints
  caused by, 143; exaration,
  198; and extending flow, 105;
  plucking, 393; rock flour
  created by, 430
erosional landforms: arêtes, 34;
  cirques, **90–1,** 255, 355;
  drumlins, 156; fiords, 206;
  glacial troughs, **365–6,** 374,
  426; hanging valleys, 249;
  horns, 255; knock-and-
  lochan, 295; p-forms, 385;
  roches moutonnées, 430; rock
  drumlins, 430; sichelwannen,
  456; striations, 479–80;
  truncated spurs, 518
ice falls on, 271
iceberg calving from, 273–4
mass balance, 321
meltwater *see* **meltwater**
moraine *see* **moraine**
moulins in, 343
movement, 233–4, 272; basal
  sliding, **47,** 105, 272, 487,
  505; compressing and
  extending flow, **105,** 487;
  creep, 116; extrusion flow,
  201; indicated by ogives, 361;
  and seracs, 451; stick slip,
  475; surging, 342, 451, **487,**
  505
ogives in, 361
proglacial lakes, 404
protective role, belief in, 231
regelation on, 420
retreat, 130; and outwash
  terrace formation, 365
shear stress in, 233–4, 272
snouts, 460
stagnant *see* **stagnant ice**
  topography
stones resting above, 232
superimposed ice, 185, **486**
temperate ice in, 495
types, 231–2; cirque, **90–1,** 255,
  416; outlet, 273, 274, **365;**
  piedmont, 388; remanié, 422;
  rock, 430–1; surging, 342,
  451, **487,** 505; ural, 528
velocity, 487
see also **ice; ice caps; ice**
  sheets; ice shelves
glacio-eustasy, 189–90, 445
glaciofluvial landforms, 232–3
  eskers, 187
  kames and kame terraces, 291
  outwash, 365, 438–9
glaciofluvial streams, 232–3
glaciomarine deposits, 510
glaciotectonism, 233
glacis, 233, 377
Glacken, Clarence, 315
Glen Roy, 371
Glen's law, 233–4, 272
gley soils, 233, 266
gligai, 230
glint line, 234

Global Atmospheric Research Programme (GARP), 12, 223, 332, 440
global environmental change, 234
see also **climatic change**
global ocean circulation, 234–6
Global Sea Level Observing System (GLOSS), 236
global warming, 236
   and carbon cycle, 57
   and deforestation, 129
   impacts, 236
   see also **greenhouse effect**
globes, analemnas on, 20
Gloger's rule, 236
GLOSS, 236
glow curves, 504
gnammas, 236, *237*
gneiss, 236
Goldich weathering series, *69*, 557
goletz terraces, 236
Gondwanaland, 109, **236**
gorges, 236
gouffres, 236
government
   and conservation, 564
   meteorological expenditure, 30
   response to climatic change, 30
graben, 236, 426, *427*
grade
   channel, 236–7
   dynamic equilibrium substituted for, 161
   slope, 237–8
graded bedding, 237
graded time, 238
gradient, hydraulic *see* **hydraulic gradient**
gradient wind, **238**, 562
   see also **geostrophic wind**
gradualism, 523
gradually varied flow, 81, 212, **238–9**
Graf, W.L., 85
grain Reynolds numbers, 425
Grand Coulee, 523, 524
granite, 239
   clitter, 98
   decomposed, 244
   grus derived from, 244
   metamorphosis to gneiss, 236
   origins, 393
   pressure release joints, 403
   springs on, *469*
   tafoni in, 493
   tors on, 244, **511–12**
granules, size of, 373
grasses, fenland, 376
grasslands, 239
   biological productivity, 60
   biomass, 60
   bush encroachment, 72
   distribution, *61*
   herbivores on, 253
   pampas, 475
   soils, 72, 89
   steppes, 475
   see also **savannah**
gravel
   braided rivers on, 71

lag, 297
   in regs, 420
gravimetric method, 239
gravitational slumps, 510
gravity, 239
   anomalies in, 67, 288
   and hydraulic potential, 260
gravity faulting, 239
gravity waves, 38, **239–40**, 330
   lee waves, 253, *304*, 329
grazing
   hydrological effects, 240
   and plant species diversity, 400
grazing food chains, 215, 515
Great Barrier Reef, 45–6
great interglacial, **240**, *378*, 378
Great Lakes, 471
Great Rift Valley, *427*
greenhouse effect, 77, **240–1**
   and carbon cycle, 76
   and deforestation, 129
   gases causing, 240; CFCs, 89; ozone, 3
   human vulnerability to, 29–30
   see also **global warming**
Greenland
   alluvial fans, *17*
   frost creep, 220
   ice sheet, 230, *231*
grey wethers (sarsens), 439
greywacke, 241
grezes litées, 241
Griffith Taylor, T., 97
grikes, 97, 121, **241**, *375*
Grime, J.P., 409
grooves, p-form, 385
gross primary productivity, 59, 60
ground frost, 241
ground ice, 241–2
   ice wedges, 273
   landforms *see* **thermokarst**
   pore ice, 398
   segregated, *242*, **272–3**, 380, *450*
ground moraine *see* **till**
groundwater, 242–4
   in alluvial fill, 18
   bank storage, 43–4
   connate water not involved in circulation of, 108
   and effective rainfall, 170, *171*
   equipotentials, 185
   flow, 244, *366*; Darcy's law, 124–6
   fluctuations: and dry valley formation, 157
   hydraulic gradient, 124–6, 260
   hydrostatic pressure, 268
   mapping, 261
   meteoric, 331
   as percentage of earth's water, 267
   perched, 379
   phreatic divides, 386
   phreatophytes' use of, 386
   piezometric surface, 388
   pumping: cones of depression caused by, 107; and saltwater intrusion, 229–30
   recharge, 418
   sapping by, 439
   storage, *263*, 264

tracer studies, 513
   urban areas, *529*
   water supply from, 560
   zones, 242–3
growan, 244
groynes, 244
grumusol, 244
grus, 244
guano, 244, 385
Günz glaciation, *378*
gullies, 244, 520
   from collapsed pipes, 389
   headward erosion, 250
   types: barrancas, 45; dongas, 153
   see also **arroyos**
gulls, 244
gumbel distributions, 244, 312–13
gumbo, 244
gustiness, 244
Gutenberg Discontinuity, 113
guyots, 245
gypcrete, 245
gypsum, 245, 331, 486
gyres, ocean, 120, 234, 245, 547
gyttja, 245

habitats, 246–8
   biocoenosis, 55–6
   biotopes as, 63–4
   colonization, 103
   diversity: adaptive radiation in response to, 4–5; ecotypes adapted to, 168; *see also* **species: diversity**
   ecological isolation, 287
   forest, 129–30
   and genetics, 224
   niches compared with, 354
   polyclimax vegetation, 396
   refugia, 358, **420**
   stress and disturbance in, 409
   see also **niches**
haboob, 248
hadal zone, 248, *361*
Hadley cells, **248–9**, 517, 518
   and anticyclone origins, 26
   antitrades as upper limits of, 26
   dishpan simulation, 146
Haeckel, Ernst, 567
haematite, 418
haffs, 249, 350
haggs, 249
hail, 249
   suppression, 555
hairpin (parabolic) **dunes**, 371
haldenhangs, 249
half-life, 249
haloclasty, 249, 436
halogenated fluorocarbons, 249
halons, 249
   see also **chlorofluorocarbons**
halophytes, 249, 435
haloseres, 403
hamadas, 249
hanging valleys, 249
hardness, 249
hardpans, 249, 392
hare, snowshoe, 402
harmattan, 250
harmonic analysis, 250

Hawaii
  evolutionary symbiosis, 489
  lava cave systems, 81
  vicariant species on, 537
Hawaiian eruptions, 250, 540,
  541
hazards, 143–4, 348–9
  expansive soils as, 199
  reduction, 27
  see also particular hazards,
    especially floods
haze, 160, 250
  Arctic, 32
  urban areas, 532
head see solifluction
headcuts, 250
headland banks, 437
headward erosion, 250
headwaters, 250
health, meteorology and, 28, 61
heat
  latent, 106, 249, 301–2
  sensible, 249, 451, 531
  storage in soil, 334
  transfer, 250–1; from oceans to
    atmosphere, 330; in
    permafrost, 475; in
    stratosphere, 478
  see also energy
heat budget, 250–2
  satellite data, 440
  urban, 531
Heat Capacity Mapping
  Mission, 527
heat equator, 501
heat island, urban, 531–2
heathlands, 252
heating, degree days used in
  determining, 130
heave, 108, 220, 323
  and ice segregation, 272–3, 350
  nubbins produced by, 356
helical flow, 252
helictites, 252–3
helm wind, 253
hemera, 253
hemicryptophytes, 417, 418
Hennig, W., 91
herbicides, 384
herbivores, 215, 253, 400, 515
heterosphere, 253
heterotrophs, 215, 253
hiatus, 253
hierarchies, systems organized
  as, 491
high energy window, 253
high pressure
  Arctic, 33
  cols separating areas of, 102
  subtropical: horse latitudes,
    255–6
  see also anticyclones
hillslopes, 459
  angles; initial yield, 21; internal
    shearing resistance, 21;
    limiting, 305–6; plane sliding
    friction, 21; repose, 21, 21,
    109, 237; residual shear, 21
  aspect, 35, 248
  colluvium on, 102–3
  denudation rates, 134

erosion: basal sapping, 47;
  latitudinal differences, 149;
  nivation, 355; raindrop
  impact, 414; unequal, 522
evolution, 459; modelling, 357,
  358; by parallel retreat, 42,
  329, 371; by slope
  replacement, 459
flow processes, 253–4, 256, 366,
  414, 433–4, 441; and
  infiltration, 277; lateral, 303,
  507; see also overland flow
graded, 237–8
mass movements see
  avalanches; falls;
  landslides, etc.
measurement, 97
profile: best units analysis, 55;
  models, 68
quasi-equilibrium, 407
retreat: Bubnoff units for, 72
saturation at base, 441
sediment transport:
  congelifluction, 107–8; frost
  creep, 220; modelling, 68
and soils, 97, 462–4, 511; catena
  concept, 80, 462, 463
stability analysis, 472
terracettes on, 499
types: adret, 5; boulder-
  controlled, 67; constant, 109;
  haldenhangs, 249; ice contact,
  271; pediments, 233, 377,
  377, 455–6, 482, 545;
  Richter, 426; strength
  equilibrium, 479; threshold,
  506; transportational, 514;
  ubac, 521; waning, 544–5;
  waxing, 109, 554; weathering-
  limited, 557
unequal, 522
vegetation, 248
see also slope
Himalayas
  formation, 364, 500
  gravity anomaly, 288
  river capture in, 76
histosols, 452
Hjulström curve, 68, 116, 254
hoar frost, 254
hodograph, 254–5
hogbacks, 255
hohlkarren, 292
Holarctic kingdom, 211
Holarctic realm, 204
Holocene, 228, 255
  altithermal (hypsithermal), 19,
    268, 351
  Blytt–Sernander model, 65
  boundary with Pleistocene, 392
  dating: by lichenometry, 305
  Flandrian Transgression, 207–8,
    253
  interpluvial, 281
  lake nutrient status during, 359
  Little Climatic Optimum, 307
  Little Ice Age, 95, 325, 350, 351
  Younger Dryas, 566
holokarst, 255
homeostasis, 303, 491
hominid evolution, 314
Homo erectus, 314

Homo habilis, 314
Homo sapiens, 314
homoclines, 255
homoiothermy, 255
hoodoos, 255, 256
Hooke's law, 337–8
horizons see soil horizons
horns, glacial, 255
horse latitudes, 255–6
horsts, 256
Horton, R.E., 27
  laws of drainage composition,
    16, 257, 343
  overland flow model, 256, 277,
    280, 372, 414, 524
  stream ordering method, 362,
    363
hot spots, 257, 522, 540
hot springs, 257
Hubbard Brook Experimental
  Catchment, 56, 130
Hudson Bay, 190
Huggett, R., 79, 161
human biometeorology, 61
human ecology, 316
human geography
  links with physical geography, 58
human impact, 177–81, 316, 345
  on channel capacity, 83
  climatic: urban areas, 530–3; see
    also global warming;
    nuclear winter
  on colonizing species, 103
  deforestation, 129
  on denudation, 135
  on desertification, 139–40
  on ecosystems, 58, 166, 167;
    coral, 112–13; and energy
    use, 175–6; using fire, 175,
    206, 258, 314; hunter-
    gatherers, 258; islands, 14,
    286, 400; reduced by
    environmental management,
    181–2; resistance and
    resilience to, 167, 470–2;
    species introduction, 24,
    282–3, 286
  on genetics, 63
  geomorphological, 27
  hydrological, 423
  on nutrient status, 191, 359
  on ozone layer, 89, 367–8
  on peatlands, 375
  on salinity, 435
  on soil erosion, 461
  on soil formation, 331
  on succession, 484
  on vegetation, 146, 398, 482
  weather modification, 555–6
  wilderness not subject to, 561
  see also pollution
humans
  comfort: and climatic factors,
    97, 103, 450–1, 563
  diet, 150
  domestication by, 152
  earth's carrying capacity for,
    166–7
  evolution, 314
  meteorological influences on,
    28–31, 61
  period of inhabiting earth, 25

humans (*continued*)
  relationship with environment,
    166–7, **314–17**; and energy,
    175–6; *see also* **hazards**
humate, 257
humicrete, 257
humidity, 257–8
  absolute, **257–8**, 560
  and dewpoint, 140
  and evaporation, 191, 193
  measurement, 193, 268
  and mist persistence, 337
  radiosonde data on, 411
  relative, 52, 97, **257**
  on thermodynamic diagrams, 504
  urban areas, 532
hummock-hollow cycle, 421
hums, 257
humus, 93, 258
  see also **moder; mor; mull**
hunter-gatherers, 152, 175, **258**
Hunza valley, *139*
hurricanes (tropical cyclones),
    516–17
  on Beaufort scale, 52
  modification, 555–6
  storm surge generation, 477
Hutton, James, 79, 122, 393,
    522–3
HYDRA, 193
hydration, 258
  obsidian: dating by, 360
hydraulic autogeometry, 258
hydraulic conductivity, **106–7**,
    124, *125*, 126
  in diffusion equation, 141–2
  and intrinsic permeability, 282
  measurement, 106–7, 381, *383*
  permeability distinguished from,
    381
  and specific storage, 258
hydraulic design, 261
hydraulic diffusivity, 258
hydraulic force, 258–9
hydraulic geometry, 259
  autocorrelation, 258
  and minimum variance theory,
    336
  regime theory similar to, 422
hydraulic gradient, 124, **260**
  and Darcy's law, 125–6
  equipotentials indicating, 185
  and infiltration, 276
  and intermittent springs, 280
  and secondary flows, 446
hydraulic head, 124, *125*, **260**
  aquifer: reduced by draw down,
    155
  in diffusion equation, 141
  and hydraulic diffusivity, 258
  and hydraulic potential, 260
hydraulic jumps, **260**, 261
hydraulic potential, 260
  equipotentials of, 185
  and throughflow, 254
hydraulic radius, 89, **260**, 513
hydraulics, **260–1**
  see also **flow, equations for**
hydrocarbons
  and photochemical smog, 460
hydrodynamic levelling, 261
hydrofracturing, 261

hydrogen ion concentration *see*
  pH
hydrogen/deuterium variations,
    288
hydrogeological maps, 261
hydrogeomorphology, 261
hydrographs, 261–2
  analysis and separation
    techniques, 262, 525
  for channel reach, 347
  channel storage effects, 83
  components: base flow, 47,
    137–8, *243*, 262, *263*, 264;
    delayed flow, 131, 408;
    interflow, *263*, 280;
    quickflow, 131, **408**, 525; *see
    also* **run-off**
  depletion curves, **137**, 262
  flashiness, 208
  peak discharge, 375
  prediction: based on partial area
    models, 277, 372
  recession limb, 137, 262, **418**
  rising limb, 428
  storm, 256, 262, *264*
  unit, 442, **524–5**
hydroisostasy, 262
hydrolaccoliths, 262–3
hydrological cycle, 263–4
  and fire, 206–7
  hydrometeorological studies of,
    266
  and land use, **299–300**, 423;
    forests, 216–17; grazing, 240;
    urban, 301, 528–30; water
    balance studies of, 546
  maps of elements of, 264–5
  phreatophytes' impact on, 386
  theories on, 265
  see also **base flow; discharge;
    evaporation; infiltration;
    rainfall; run-off; storage;
    throughflow; water
    balance**, *etc.*
**hydrological experiments,
    laboratory**, 297
hydrological maps, 264–5, 546
hydrological models, 357
  calibration, 371–2
  stochastic, 477
  storage equations in, 477
hydrological year, 553
hydrology, 265–6
  international programmes, 281
  meteorology applied to, 266
  studied in experimental
    catchments, 199, 423
hydrolysis, 266
hydrometeorology, 266
hydrometry, 266, 267
hydromorphy, 266
hydrophytes, 247, 266
hydroseres, 403, 451, 485
hydrosphere, 266–7
hydrostatic equation, 224, **268**,
    444
hydrostatic pressure, 55, 268
hydrothermal alteration, 268
hyetograph, 268
hygrograph, 268
hygrometer, 268
hypabyssal rock, 268

hypolimnion, 268
hypotheses, multiple working,
    274–5
hypsithermal (altithermal) phase,
    19, 268, 351
hypsographic curves, 268
hypsographic geography, 268
hypsometry, 269
hysteresis, 269
hythergraphs, 97

ITCZ *see* **intertropical
    convergence zone**
ice, 270
  cave systems in, 81
  chattermarks produced by, 88
  cornices, *113*, 114
  deformation and shear stress,
    233–4, **272**, 425–6
  dirt cones, 143
  flow: basal sliding, **47**, **105**, 272,
    487, 505; compressing and
    extending, 105
  flyggbergs shaped by, 213
  pressure melting point, 403, 495
  regelation, 420
  remote sensing, 527
  in rivers: breaking up, 128
  sastrugi in, 439
  seracs, 451
  sublimation onto, 482
  types: anchor, 20; basal, 47;
    frazil, 217; needle, **350**, 356,
    479; pore, *242*, 398; rime,
    428; superimposed, 486;
    temperate, 495; *see also*
    **ground ice; sea ice;
    stagnant ice topography;
    thermokarst**
  see also **freeze-thaw cycles;
    frost; glaciers; hail**
ice ages, 270–1
  Milankovitch hypothesis, 335–6
  snowblitz theory, 460
  see also **glacials**
ice blink, 271
ice caps, 271
  glaciers, 201
  ice fields distinguished from, 271
  ice streams within, 273
  melting: and sea level changes,
    189, 190
  see also **ice sheets**
ice contact slopes, 271
ice cored moraine, 271
ice cores
  acidity profile, 4
  temperature reconstruction
    from, 241
ice crystals, 460
  in clouds: precipitation
    production, 98–9; virgation,
    539
ice dams, 271
  drainage following melting, 209,
    290, 366, 523–4
ice domes, 271, 365
ice edge, 271
ice falls, 271, 361
ice fields, 271
ice floes, 271–2
ice fog, 33, **272**, 532

ice front, 272
ice jams, 272
ice rind, 272
ice segregation, *242*, **272–3**, 380, 450
  frost heaving by, 220, 350
  see also **pingos**
ice sheets, 232, **273**
  distribution: and denudation, 135; during glacials, 230–1
  growth: snowblitz theory, 460
  ice streams within, 273
  isostatic response to, 288
  loess bordering, 308
  proglacial lakes, 404
  see also **ice caps**
ice shelves, 232, **273**
  ablation, 1
  basal debris, 510
  calving from, 75, 273–4
  ice front, 272
ice streams, **273**, 365
ice wedges, *242*, 273
ice-dumped moraine, 341–2
ice-marginal rivers, braided, 71
ice-marginal valleys, talsands in, 494
icebergs, 75, **273–4**
  basal debris, 510
  distribution, *382*
Iceland, *464*
icelandic eruptions, 540, *541*
icings, 274, 348
igneous rocks, 274
  acidity, 4
  basic, 48
  cupolas in, 117
  formed from magma, 312
  hypabyssal, 268
  intrusion *see* **intrusions**
  metamorphosis, 236
  see also **basalt; gabbro; granite**
ignimbrite, *539*
illuviation, 274
imbrication, 274
impeded dunes, 274
impermeability, 31–2, 274
  and spring location, *469*
impervious rocks, 274, 383
impervious surfaces, urban, 528
in and out channels, 274
inceptisols, 452
incised meanders, 327
inconsequent streams, 274
indeterminacy, 274–5
index cycle, 276
Indian Ocean
  circulation, 234, *235*, 236
  southern oscillation over, 312
  water masses, 552
Indian Remote Sensing satellites, 527
Indian subcontinent
  monsoon climate, 340
Indo-Gangetic canals, 422
induration, 276
Indus River, 76
industry
  applied meteorology in, 28
  energy use, 175–6
infiltration, *243*, *263*, **276–8**
  and effective rainfall, *171*

and fire, 206–7
hillslopes, 253–4
improved by drainage, 298
reduced by grazing, 240
and vegetation, 277
infiltration capacity, 256, 276, 433–4
inflorescence, 278
influent streams, 278
infrared emissivity, 251
infrared radiation, 6, *173*, 411
  heat loss by, 251
  measurement, 278, 501–2; from satellites, 526, 527
infrared thermometer, 278
ingrown meanders, 327
inheritance
  Darwinian ideas, 127
  and evolution, 197
  mutations, 347
initial yield, angle of, 21
inliers, 278
insecticides, 384
insects
  coevolution with plants, 101–2
  parasitoid, 401
inselbergs, **278–9**, 434
  dilation joints, 143, *455*
  see also **tors**
insequent streams, 270, 274
inshore zone, *50*
insolation, 279
  variations: Croll-Milankovitch model, 335, 336
insolation weathering, 279
instability
  atmospheric, 37, 536–7
  channels, 528
  population, 402
  see also **stability**
instantaneous unit hydrographs, 524
intact strength, 279
intensity, rainfall *see* **rainfall intensity**
interbasin transfers, 279–80
interception, *243*, **280**
  by trees, 216
  and effective rainfall, 169, *171*
  see also **stem flow**
interflow, *263*, **280**
  see also **lateral flow; throughflow**
interfluves, 280, *463*
interglacials, **280**
  interstadials compared with, 281
  in Penck and Bruckner model, 378
  Pleistocene, 240, 378
  pollen analysis evidence on, 395
  sea level transgressions during, 189, 207–8, 253, 426
  see also **glacials; Holocene**
Intergovernmental Panel on Climatic Change, 240
intermittent springs, 280
intermittent streams, 280–1
internal friction, angle of, 218
internal shearing resistance, angle of, 21

International Association of Scientific Hydrology (IASH), 265
International Decade for Natural Disaster Reduction, 349
International Hydrological Decade, 265, 281, 423
International Hydrological Programme, 265, **281**
International Society of Biometeorology, 61
International Union of Geodesy and Geophysics, 265
interpluvials, 281
interstadials, 281
  Allerød, 15
  Bølling, 65
  pollen analysis evidence on, 395
interstices, 281–2, 398
intertropical convergence zone (ITCZ), 184, **282**, 517
  and Hadley cells, 248
  as heat equator, 501
  poleward limit: and drought, 156
interzonal soil, 282
intrenched meanders, 282
intrinsic permeability, 282, 381
introduced species, 24, **282–3**, 286, 400
  see also **alien species**
intrusions, 283, *284*, 542
  batholiths, 49, 66, 117, *284*
  bosses, 66
  bysmaliths, 73, *284*
  dykes, 161, *284*, 428
  laccoliths, *284*, 297
  lopoliths, *284*, 309
  phacoliths, *284*, 385
  plutons, 393
  ring complex, 428
  sills, *284*, 457
  stocks, 477
  xenoliths in, 565
inversions *see under* **temperature**
inverted barometer effect, 283
inverted relief, 283
involution, 283
ion concentration, 283–4
  and specific conductance, 106
ionic wave technique, 145
ionosphere, **284–5**, 312
iron
  in soils and sediments, 418–19
irrigation
  ditches for: design, 421–2
  and leaching requirement, 304
  and salinity problems, 435
island arcs, 285
islands, 287
  animals and plants: aliens, 14, **286**, 400; endemics, 174, 226, 285; equilibrium theory, 285–6; species diversity, 150; taxon cycles, 494; vicariants, 537
  barrier, 45
  continental, 109
  isthmuses connecting, 289
  oceanic, 109
  tombolos connecting, 511
  see also **coral algal reefs**

isochrones, 287
isoclines, 287
isolation
  biotic, 63
  ecological, 287
isometric growth, 15, 16
isopleths, 287
isostasy, 48, 287–8, 325–6, 445
  decantation caused by, 190
  and raised beach formation, 416
  rate of, 351
isotopes, 288–9
  radioactive decay, 249, 411;
    dating using, 173, 288, 398
isthmuses, 289, 300, 350

Jamaica, cockpit karst in, 101
Japanese Earth Resources
  Satellite, 527
Jeffreys, H., 562
JERS, 527
jet streams, 290
jökulhlaups, 209, 271, 290
Johnson, D.W., 100
joint probability estimates, 290
joints
  pressure release, 403
  and rock mass strength, 431
  and rock quality, 431–2
  sheeting caused by, 454–5
  widened by ice: in vallon de
    gelivation, 535
joules, 164
Jurassic, 228, 362
juvenile water, 190, 290

K-cycle, 291
K-selection, 397, 409–10
Kalahari, mekgacha in, 327
Kali Gandaki gorge, 364
kamenitza, 291
kames and kame terraces, 291
kaolin, 291
Karakoram Mountains, 139
karren, 291–3
  see also clints
karst, 293
  causse synonymous with, 81
  groundwater in, 242–3
  landforms: abîmes, 1; blind
    valleys, 64; bogaz, 65; clints
    and grikes, 97, 121, 292, 375;
    dayas, 127; dolines, 151–2,
    293, 453, 458, 488, 496, 533;
    kamenitza, 291; karren,
    291–3; mogotes, 338;
    pavements, 374, 375; poljes,
    394; shakeholes, 453;
    sinkholes, 458, 533; swallow
    holes, 396–7, 488
  organic action producing, 58
  pseudokarst similar to, 406
  springs, 424, 469
  types: cockpit (kegelkarst), 101,
    151, 293, 338, 513; holokarst,
    255; merokarst, 328;
    phytokarst, 58, 387;
    polygonal, 396; tower karst
    (turmkarst), 513, 520
  see also caves; karren
karstic processes
  blueholes caused by, 65

kata-fronts, 293
katabatic flows, 293
  Antarctic, 23
  bora, 65–6
  chinook, 89, 214
  föhn, 214
  mistral, 337
kavirs, 293, 391
keels, 443
kegelkarst, 101, 151, 293, 338,
  513
Kelvin scale, 495
kelvin waves, 294
kettle holes, 294
khamsin wind, 294
Kilimanjaro, Mt, 75
Kimberley, 186
Kimberlite pipes, 141
kinematic waves, 294
kinematics, 294
kinetic energy, 174, 294
  atmosphere, 225
  gradually varied flow, 238, 239
  see also potential energy; wind
King, L.C., 123
kingdoms, animal and plant,
  294–5
klippes, 295
Kluftkarren see grikes
knickpoints, 295
knock-and-lochan topography,
  295
Köppen, W., 66, 96, 295
koniology, 296
kopjes, 296
krotovinas, 296
krummholz, 296
kumatology, 296
kunkar, 74
kunkur, 74
kurtosis, 296

laboratory experiments, fluvial
  and hydrological, 297
laccoliths, 284, 297
ladybirds, introduced, 282–3
lag gravel, 297
lag time, 297
lagoons, 45, 46, 112, 249
  sapropels in, 439
lahars, 297
lakes
  acid rain effects, 3
  benthic organisms, 54
  bottom-set beds, 67
  deflation: lunettes in lee of, 309
  eutrophication, 14, 191, 359
  evaporation, 190
  former shorelines, 371
  ice in, 271
  ice-dammed, 366; sudden
    drainage, 209, 290, 523–4
  limnology as study of, 306
  orientation, 363
  overflow channels from, 366
  pluvial terraces, 498–9
  polders reclaimed from, 394
  sapropels in, 439
  sediments: nutrient status
    changes indicated by, 359
  shoals in, 455
  siltation, 458

thermocline in, 503
  types: crater, 75; meres, 328;
    oxbow, 367; paternoster, 374;
    pluvial, 190, 393; proglacial,
    404; seasonal (chotts), 90;
    tarns, 494; thaw, 363, 500–1
  varves deposited in, 535
  zones: abyssobenthic, 1;
    abyssopelagic, 1–2; aphotic,
    26; aphytic, 26; epilimnion,
    183; hypolimnion, 268;
    photic, 385; sublittoral, 482
Lamarck, Jean-Baptiste de, 196
Lamarckianism, 127
Lamb, H.H., 90
laminar boundary layer, 68
laminar flow, 212, 297–8, 520
  Darcy's law applicable to, 126
  Reynolds number, 425
laminar sublayer, 68, 69
land breezes, 329, 444
land bridges, 300
land capability, 298
land drainage, 205–6, 298, 560
  see also channelization
land systems, 299
land units, 299
land use
  hydrological effects see under
    hydrological cycle
  and weather, 556
Land Use Capability Maps, 298
landfill, 300
landforms
  allometric growth, 15–16
  and climate, 96–7, 184
  human influence on, 25
  mapping, 27, 342
  morphology: and equifinality,
    184
  morphometry, 342, 343
  and vegetation, 387
  see also specific landforms
Landsat, 525–6, 527
landscape ecology, 301
landscapes
  aesthetic character, 430
  aesthetic degradation, 8
  conservation, 108
  evaluation, 301, 430
  maturity, 325
  protection, 405–6
  and soils, 80, 462–4
landslides, 301, 302, 321, 323,
  324, 459
  backwalls, 42
  lahars, 297
  limiting angles for, 306
  rotational, 323, 324, 432–3
  slab failures, 302, 459
  threshold slopes for, 506
  toppling failures, 302, 511
  translational, 323, 459, 513
  wedge failures, 302, 557
lapiés, 291, 293
Laplace's equation, 141
lapse rates, 301
  dry adiabatic, 301, 536–7
Late Glacial, 65, 301
  interstadials: Allerød, 15;
    Bølling, 65
  Younger Dryas stadial, 566

latent heat, 301–2
  released by condensation, 106
  transfer in Hadley cells, 249
lateral accretion, 302–3, 393–4
lateral flow, 303, 507
  see also **interflow; throughflow**
lateral migration, 303
lateral moraine, 340, 341
laterite, 303, 347, 392
  pisoliths in, 389
latitude
  and solar radiation, 251
  and species diversity, 150
  and thermocline characteristics, 503
latosols, 303
Laurasia, 109, 303
lava, 303, *304*
  caves in, 81
  coulées, 114
  fissure eruptions, 207
  fragments fused into agglomerate, 9
  louderbacks, 309
  necks, 350
  planeze, 389
  plateaux, *541*
  see also **pumice**
le Chatelier principle, 303–4
leaching, 304
  by throughflow, 277
  and soil nutrient status, 358
leaching requirement, 304
lead isotopes, 288
lee depressions, 304
lee dunes, 159
lee eddies, 304, 433
lee waves, 253, **304,** 329
legumes, nitrogen fixation by, 354, 488
lenticular clouds, 304
Leopold, L.B., 259, 430
lessivage, 305
levées, 305, 432
  in backswamps, 42
  crevasses in, 116
lichenometry, 305
Lichty, R.W., 237, 238
Liebig, Justus von, 306
life cycle analysis, 305
life forms, 305, 417–18
ligands, 88–9
light, plant growth and, 246
lightning, r305, *306*
  fulgurite formation, 222
Likens, G.E., 60
limestone
  arches in, *32*
  biological weathering, 387
  caves, 81; speleothems in, 252–3, 468
  coastal platforms: vasques on, 535
  fluviokarst landforms, 213
  gouffres in, 236
  hums, 257
  landforms *see* **karst**
  notches in, 356
  pavements, 241, 374, *375*
  pot-holes in, 399
  shillow on, 455
  soils on, 423, 498

solution, 9, 81, **466;** carbonation in, 77; in permafrost regions, 222; supersaturation, 486; *see also* **karst**
  suffosion over, 485
  tepees, 496
  see also **tufa**
limiting angles, slope, 305–6
limiting factors, 306
limnology, 306
Lindemann, R.L., 165
line squalls, 470
lineaments, 306
linear dunes, *158*
Linnaeus, Carolus, 494
Linton, D.L., 511–12
liquid limits, 39, 306
literature, environmental ideas in, 315
lithic sandstone, 438
lithification, 307
lithology, 307
lithoseres, 403
lithosols, 307
lithosphere, 307
  isostatic equilibrium, 287–8
  plates *see* plate tectonics; plates subduction, 285, 364, 390, 391
  tectonics as study of, 495
  see also **crust; mantle**
lithostratigraphy, 478
litter, 307
Little Climatic Optimum, 307
Little Ice Age, 95, 325, 351
load, stream *see* **bed load; dissolved load; sediment transport; suspended load**
load structures, 307, 365
local climate, 307–8
  see also **microclimate; urban areas**
local winds, 308, 327
  see also **katabatic flows; mountain winds; sea breezes**
lochans, 295
locusts, plagues of, 165–6
lodgement till, 108, *509,* **510**
loess, 5, 130, **308**
  see also **coversand; parna deposits**
logan stones, 308
logistic equation, 355
Lone Tree Creek, *448,* 449
long profiles, 308–9
  knickpoints in, 295
longshore bars, *50*
longshore currents, 118–20, 309
longshore drift, 309
  reduced by groyne construction, 244
  and spit formation, 468
lopoliths, *284,* 309
louderbacks, 309
Lovelock, J.E., 223
low flow analysis, 309
low pressure
  Antarctic, 23
  Arctic, 33
  ITCZ, 156, **184,** 248, **282,** 501, 517
  in lee depressions, 304

  in thermal depressions, 501
  troughs, 518
  see also **cyclones**
lunar tides, 508
lunettes, 159, **309,** 371
  clay, 93
  see also **parna deposits**
Lusitanian flora, 309–10
Lyell, Charles, 79, 196, 523
lynchets, 310
lynx, 402
lysimeters, *192,* 193, **310**
  percolation gauges beneath, 379

MRCFE, 345
maars, 311
Mabbut, J.A., 71–2
MacArthur, R.H., 285–6, 471
McGee, W.J., 227
machair, 311
Mackenzie Delta, 388
macroclimate, 307
macrofossils, 311, 375
macrometeorology, 311–12
macrotidal estuaries, 188–9
macrotidal ranges, *508*
Maddock, T., 259
maelstroms, 312
magma, 312
  sea floor subducted into, 482
  solidification into pluton, 393
  in volcanic eruptions, 540, 542
magnetic equator, 4
magnetic poles
  agonic line through, 9
  changing position, 399
magnetic storms, 312
magnetism, 370, 500
  and continental drift, 109
  reversal, 425; evidence of, 370
magnitude and frequency effects, 312–14, 350
Malaysia, Wallace's line through, 543, *544*
Maldive Islands, *112*
Malham Cove, *375*
mallee, 314
Malthus, Thomas, 316, 397
mammilated surfaces, 314
man *see* **human impact; humans**
management
  coastal, 181
  drainage basins, 428
  environmental, *178,* 180, 181–2, 405
Mandelbrot, B.B., 217
mangrove swamps, *112,* **317,** *318*
  see also **salt marshes**
manned earth resources satellites, 317–19
Manning equation, 319–20, 522
  roughness coefficient estimated from particle size, 480
  use in discharge measurement, 145
Man's role in changing the face of the earth (MRCFE), 345
mantle, 320
  hot spots, 257, 522
  Mohorovičić discontinuity above, 338

mantle (*continued*)
plumes, **320**, 522
rheology: indicated by isostatic
rebound, 288
see also **asthenosphere;
lithosphere**
mapping cameras, 6
maps
area measurement on, 389
contours on, 110
denudation, 135
edge-lines on, 169
form lines on, 217
geomorphological, 27
hydrogeological, 261
hydrological, **264–5**, 546
isopleths on, 287
morphological, 342
maquis, 320
margalitic soil horizons, 320
marginal channels, 320
marginal sea coasts, *101*
Marine Observation Satellite,
526–7
maritime air masses, 11
maritime climates, 95
Mark, D.M., 15, 16
market system, 315
and resource use, 176–7
Markov process, 321
marl, 213
Marsh, George Perkins, 316,
345
marshes, 488
paludal sediments, 370
watten, 553
see also **salt marshes**
marsupials, 204
Maspalomas, *514*
mass
conservation: continuity
equations for, 110
transfer in systems, 491
mass balance, 321
mass movements, 321–5
cone formation, 338
levée formation, 305
likelihood, 472
protalus rampart formation, 404
and soils, *463*
types *see* **avalanches; creep;
falls; landslides,** *etc.*
mass strength, 325
slopes adjusted to, 479
mass wasting
gelifluction, 220, **224**, 322,
392–3
periglacial, 380
solifluction, 19, 322, **464**
sturzstrom, 482
massive rock, 325
Matanuska Glacier, 510
matric potential, 384–5, 485
measurement, 496
mattoral, 325
maturity, 325
Maunder minimum, 325
**mean, accumulated departure
from,** 2
meanders, 326–7
and bank erosion, 43
and channel classification, 87, 88

and cut-off development, 121
and helical flow, 252
intrenched, 282
and minimum variance theory,
336
and navigability, 349–50
oxbow, 367
point bar deposits on, 393–4
and pool–riffle sequence, 397
and river currents, 120
slip-off slopes, 459
underfit streams, 521
valley, 521
mechanical weathering, 556
congelifraction, 108
exfoliation, 140, 198, 362,
454–5, 468
freeze–thaw processes in, 108,
218
by frost, 10, 220, **221–2**, 380
hydrofracturing, 261
index of, 557
by insolation, 279
by pressure release, 143, 403,
455, 525
by salt, 249, 258, 436, 493
mechanics, fluid, 212–13, 260–1
see also **flow**
medial moraine, 340, 341
Mediterranean climate, 327
Mediterranean vegetation
chaparral, 88
garrigue, 223
maquis, 320
mattoral, 325
megashears, 327
megathermal climates, 327
mekgacha, 327
melting
glaciers *see* **ablation; meltwater**
snow, 461
Melton, M.A., 67
meltout till, 509
meltwater, 327–8
flow through moulins, 343
ice-dammed, 271; sudden
drainage, 290
rapid fluctuations, 271, 290
regelation, 47, 420
rock flour in, 430
see also **stagnant ice
topography**
meltwater streams
deposition, 365; eskers, 187
in and out channels, 274
marginal channels, 320
on outwash, 365
sandur deposition, 438–9
sediment loads, 232–3, 327
subglacial: overflow channels
reinterpreted as, 366;
urstromtäler formation, 533
varve deposition, 535
Mendel, Gregor, 127, 197
Mercalli scale, 163, 328
mercury barometer, 44
Mercury satellites, 317
meres, 328
meridional circulation, 328
angular momentum, 328, 339
Atlantic Ocean, *551*
in Ferrel cell, 205

in Hadley cells, 248–9
meridional index, 90
merokarst, 255, 328
mesas, 328, *329*
mesoclimate, 307–8
mesometeorology, 328–30
mesopelagic zone, *361, 378*
mesophytes, 330
mesoscale cellular convection,
330
mesoscale precipitation areas,
219–20
mesosphere, 330
mesothermal climates, 330
mesotidal estuaries, 188
mesotidal ranges, *508*
mesotrophic nutrient status,
358, 359
Mesozoic era, 228, 500
Messinian salinity crisis, 330–1
metamorphic aureole, 39
metamorphism, 236, **331**
metapedogenesis, 331
metasomatism, 331
meteoric water, 331
meteorites
and denudation, 133
impact on earth, 35, 115
meteorological agencies, 30,
554–5
meteorological data
for weather forecasting, 440,
554
meteorological satellites, 332–3,
439–41
cloud data, 351–2
VISSR in, 539
meteorology, 333
air motion central to, 561
applied, 28–31; in agriculture,
9–10, 334; in hydrology, 266
Arctic, 32–3
dynamic, **161**, 386
macroscale, 311–12
mesoscale, 328–30
microscale, 334; *see also*
microclimate
physical, 386–7
polar, 22–4, 32–3
synoptic, 490
tropical, 517–18
urban, 530–3
see also **biometeorology;
climatology**
Meteosat, 352
METROMEX project, 532
mice, island evolution of, 286
micro-atolls, 38, 419
micro-erosion meter, 333–4
micro-organisms
biotechnological use, 62–3
in decomposition, 56, 128, 215,
307; and nitrogen fixation,
354–5
energy-transfer role, 165
in stromatolite formation, 481
see also **algae; plankton**
microbiocoenosis, 56
microclimate, 333
urban, 530–3
microcracks, 333
micrometeorology, 334

microsomia (nanism), 348
microtidal estuaries, 188
microtidal ranges, *508*
microwave radiation
  in SLAR, 456–7, 527
mid-ocean ridges, 334
  see also guyots; islands; sea-
    mounts
migration, 334–5
Milankovitch hypothesis, 335–6
Milne, G., 80
mima mounds, 336
mimicry, plant, 102
Mindel glaciation, *378*
minerals
  accessory, 2
  anisotrophism, 21–2
  biogeochemical cycling, 56–7
  chelation, 88–9
  cleavage, 94
  conchoidal fractures, 106
  crystallization, 69
  evaporites, 194
  polymorphism, 396
minimum acceptable flow, 336
minimum variance theory, 336
Miocene, 74, 228
  Messinian salinity crisis, 330–1
mirages, 336
mires, 336, 358
  see also bogs; peat
misfit (underfit) streams, 76,
  521–2, 534
Mississippi Delta, 89, 345
Missoula, Lake, 523–4
mist, 337
mistral, 337
mixing corrosion, 337
mixing models, 337
mobile belts, 337
models
  atmospheric, 146, 224–5
    prediction poor, 38
  boundary conditions in, 67–8
  catastrophe theory in, 79
  drainage basins, 458, **553**
  hillslope evolution, 357, 358
  hydrological, 357, 371–2, 476–7,
    477, 553
  numerical, **357–8**, 553;
    calibration, 371–2
  simulation, 458
  stochastic, 357, **476**, 553
moder, 337
modulus of elasticity, 337–8
Möbius, K., 56, 568
mogotes, 338
Mohorovičić discontinuity, 338
Mohr–Coulomb equation, 170,
  338, 514
Moh's scale, 249
moisture
  antecedent, 24
  soil *see* soil moisture
moisture index, 338
molards, 338
molasse, 338
mole drainage, 205–6, 298
molecular drive, 197
mollisols, 452
momentum
  angular, 21, 328, 338–9

budgets, 338–9
monadnocks, 339
monera, 294
mongoose, 400
monoclimax theory, **339**, 396,
  398
monoclines, 214, 340
monoglaciation, 340
monophylesis, 340
monsoon forests, 340
monsoons, 311–12, **340**, 518
montane forest, *75*
Monteith, J.L., 195
Monument Valley Tribal Park,
  329
monumented sections, 340
moon
  gravitational attraction *see* **tides**
  perigee, 379
mor, 340
moraine, *232*, **340–2**
  De Geer, 130, 545
  ice-cored, 271
  swales in, 488
  terminal, 497
  see also till
MORECS, 195
morphogenetic regions, 342
morphological mapping, 342
morphology
  denudation chronology based
    on, 136
  and equifinality, 184
morphometry, 342–3
morphotectonics, 343
mortality rates, 397
mosaic-cycle concept, 343
Moscow, hydrological cycle in,
  529
motion, equations of, 183, 562
  and Coriolis force, 113
  in GCMs, 224
  in numerical models, 358
mottled zone, 343
moulins, 343
mountain building *see* orogeny
mountains, 345
  animals, 345, 567
  biomes: distribution, *61*
  climate and meteorology, 343–4
  denudation rates, 134
  depressions in lee of, 304
  endemic species on tops of, 174
  glaciation level, 231
  glaciers, 232
  orographic precipitation, 364
  piedmonts, 388
  pressure differences across, 339
  rain shadow effect, 414
  rivers: step-pool systems in, 475
  rotor streaming over, 433
  sea floor, 245, 445
  summit levels: concordant, 106;
    uniform, 230
  surrounded by ice (nunataks),
    358
  timberline, 511
  vegetation, 32, 247–8, 344, 345
  winds, 344–5; fohn, 214
  see also orogeny
mud, sand deposition on, 307
mud balls, 34

mud banks, slacks between, 459
mud flats *see* salt marshes
mud flows, 322
mud lumps, 345
mud slides, 324
mud volcanoes, 345
mull, 346
multiband cameras, 6
multiple working hypotheses,
  274–5
multispectral scanners, **346–7**,
  501
  on satellites, 319, 525–6, 527
Mungo, Lake, 309
murram, 347
Murray, Sir John, 24
muskeg, 347
Muskingum method, 347
mutation, 347
  inheritance, 197
  by saltation, 436
mutualism, 347, 489

Namib Desert, *278, 455*
nanism, 348
nappes, 283, 348
NASA, 317–19, 525–6, 527
natality rates, 397
natural hazards *see* disasters;
  hazards
natural resources *see* resources
natural selection, 126, 127,
  196–7
  mutations incorporated by, 347
  r- and *K*-types, 183, 397, **409–10**
natural vegetation *see*
  vegetation
navigation, channels affected
  by, 349- 50
Neanderthal man, 314
neap tides, 508
near infrared film, 6
Nearctic realm, 204, 543
nearshore, *45*
nearshore currents, 118–20
nebkhas, 159, 350
neck cut-offs, 121
necks, 350
needle ice, 350
  nubbin formation, 356
  striated soil formation 479
negative feedback, 303, 491
  and quasi-equilibrium, 407
nehrungs, 350
nekton, 350
neo-Darwinism, 350–1
neo-Malthusianism, 315
neocatastrophism, 350
Neogaea, 204
Neogene, 228
neoglacial, 351
neotectonics, 351
Neotropical kingdom, 211
Neotropical realm, 204, 543
nephanalysis, 351–2
nephoscope, 352
neptunism, 352
neritic zone, 352, *361*
ness, 352
net primary productivity, 59,
  60, **352–3**, 515
  and eutrophication, 191

net radiation, 252, **353**
networks, 353–4
 drainage *see* **drainage networks**
neutron probes, 354
nevé, 354
New Zealand, braiding in, *70*
Newton's Law, 561–2
niches, 354
 differentiation: and latitude, 150
 diversification within, 19
 diversity: adaptive radiation in
  response to, 4–5
 population growth in, 397
 and vicariance, 537–8
 see also **biotopes**
nimbostratus clouds, *99*, 100,
 545
nimbus clouds, 100
 see also **cumulonimbus**
Nimbus satellites, 332
nitrates
 loss from disturbed forests, 470
 water pollution by, 321
nitric acids, atmospheric, 3
nitrification, 354
nitrogen
 in atmosphere, 36
 eutrophication by, 191
 fixation by azotobacters, 41, 488
nitrogen cycle, *57*, **354–5**
nitrogen oxides, 360
 air pollution, 3, 460
 ozone destruction, 367
nival ridges, 404
nivation, 355, 380
nivometric coefficient, 355
NOAA satellites, 332
Noah's flood, 143, 523–4
noctilucent clouds, 355
noise pollution, 396
non-linear systems, 355
 catastrophism in, 79
nonconformities, 355, *521*
noosphere, 62
normal cycle *see* **cycle of**
 erosion
normal faults, 203, **356**
 in rift valley, 426, *427*
normal stress, 356, 454
North American Cordillera, 499
Northern Lights, 39
Norwegian Sea, 234, 236
notches, 356
Notogaea, 204
nowcasting, 554
nubbins, 356
nuclear energy, 176, 315
nuclear plants
 decomissioning, 128
 waste, 356–7
nuclear winter, 357
nudation, 483
nuée ardente, 357
numerical models, 357–8
 parameterization, 371–2
nunataks, 34, 358
nutrient cycling, 128, 167
nutrient status, 358–9
 eutrophic, 191, 321, 359
 peat, 359, 375–6
nutrients

 in oceans, 353, 361, 528 and el
  Niño effect, 172
 plant requirements, 352–3
 transfer: by symbiotic fungi,
  488–9; by transpiration,
  513–14

oases, 360
oblique aerial photography, 7
oblique slip faults, 203
obsequent streams, 360
observational networks, design
 of, 275
obsidian hydration dating, 360
occlusions, *201*, 219, **360**
occult deposition, 360–1
 see also **acid rain**
ocean trenches
 island arcs associated with, 285,
  390
oceanography, 361
oceans, 361
 abysses, 1
 atmospheric interaction with,
  11–12; tropics, 312, 517–18
 benthic organisms, 54
 biological productivity, 60, 353,
  361; and downwelling, 153
 biomass, 60
 biosphere in, 62
 chemical composition, 464
 circulation, **234–6**, 245, *551*;
  teleconnections, 495
 commensalism in, 103
 currents, 120, *551*; divergence,
  149, 528; modified by
  Coriolis effects, 234
 deep water: source regions, 234,
  235, 236, *550*; temperature
  and salinity, 552
 deep-sea sediments: climatic
  change evidence in, 94, 335–6
 downwelling, 153
 energy exchange on surface, 11
 gyres, 120, 234, 245, 547
 horse latitudes, 255–6
 islands, 109, 287, *see also*
  islands
 kelvin waves in, 294
 lithosphere under, 307;
  subduction, 364, 390
 mantle plumes below, 320
 mesoscale cellular convection
  over, 330
 microwave sensing, 527
 mid-ocean ridges in, 334
 moho depth below, 338
 nutrients in, 353, 361, 528; and
  el Niño effect, 172
 percentage of earth's water in,
  267
 pollution, 320–1
 sedimentation: and sea level
  changes, 190
 surface temperatures: anomalies,
  12; and tropical cyclone
  formation, 516, 517
 thermocline, 234, *503*, 547
 tides, 507, 508
 trophic levels in, 515
 upwelling, 518, **528**; el Niño
  effect, 172

 water masses, 547–52
 zones, *361*; abyssobenthic, 1;
  abyssopelagic, 1–2, *361*;
  aphotic, 26; aphytic, 26;
  bathypelagic, *361*, 378;
  euphotic, 378; hadal, 248,
  361; hypolimnion, 268;
  mesopelagic, *361*, 378; neritic,
  352, *361*; pelagic, *361*, 378;
  photic, *361*, 385
 see also **Atlantic Ocean;**
  coasts; Pacific Ocean
Odum, E.P., 59, 60
Odum, H.T., 491, 515
ogives, 361
oil pollution, salt marsh, 471
oil-shale, 362
oktas, 362
Old Man of Hoy, *472*
Oligocene, 74, 228
oligotrophic bogs, 376
oligotrophic nutrient status,
 353, 358, 359
ombotrophic bogs, 376, 421
ombotrophic plants, 362
omnivores (diversivores), 150
onion-weathering *see* **exfoliation**
ontogeny, 362
oolite, 362
open systems, 491
optimization, 371–2
orbits
 aphelion, 26
 apogee, 27
 ecliptic, 164, 335, 399
 perihelion, 380
order, stream *see* **stream order**
Ordovician, 228
 glaciations, 270, 271
organic chemicals, synthetic,
 384
organic materials
 decomposition, 56, 128
 fermentation, 62–3
 heterotrophs dependent on, 253
 produced by autotrophic
  organisms, 40, 215
 soil, 257, 258, 331, 337, 416
 streams dammed by, 363
 in suspended sediment, 447
 see also peat; sapropels
organic weathering, 363
organisms *see* **animals; micro-**
  organisms; plants *and*
  specific aspects, e.g. **evolution**
organochlorine pesticides, 384
organophosphate pesticides, 384
Oriental realm, 204, 543
oriented lakes, 363
oroclines, 363
orogenic eustasy, 190
orogeny, 363–4
 alpine, 19
 molasse diagnostic feature of,
  338
 and plate boundaries, 390
 rate, 351
 see also **epeirogeny**
orographic clouds, 100
orographic precipitation, 24,
 364
osage-type streams, 521–2

osmosis, 365
OSTA, 319
outlet glaciers, 365
  ice streams, 273
  iceberg calving from, 274
outliers, 365
outwash, 365, 438–9
outwash fans, *232*
overbank deposits, 210, 365, 394
overdeepening, glacial, 365–6
overflow channels, 366
overland flow, *243, 263,* 366, 414
  and antecedent moisture, 264
  and depression storage, 138
  hillslopes, **253–4**, 256, 277, 366
  Hortonian model, **256**, 277,
    280, *366*
  measurement, 229
  in partial area models, 372
  and saturation deficit, 441
  sheetflow component, 454
  source area, 161, 372, 441,
    466–7
  stream solute concentration
    diluted by, 277
  surface detention preceding, 486
overthrust faults, 366
overtopping, 366–7
overturned folds, *214*
overwash fans, *45*
overwashing, 366, 367
oxbow lakes, 367
oxidation, 367, 419
  and desert varnish formation,
    139
  and weathering rind formation,
    557
oxisols, 452
oxygen
  in atmosphere, 36
  dissolved, 55, 56, **147**
  isotopes, 288
  in photosynthesis, 386
oxyseres, 403
ozone, 3, 37, 89, 249, **367–8**
ozonosphere, 37

pF, 384–5
pH, 385
  acid precipitation, 2, 3
  fenland, 376
  solonchak soils, 464
  solonetz soils, 464
p-forms, 385
Paarl Mountain, *237*
Pacific coasts, 36, 369
Pacific Ocean
  atmospheric circulation over,
    517–18
  circulation, 234, *235,* 236
  el Niño effect, 172
  guyots in, 245
  ITCZ, 282
  island arcs, 285
  island biogeography, 285–6
  southern oscillation over, 312
  tropical cyclones, 516, 517
  volcanoes around, *542*
  water masses, *548–50,* 552
pack ice, 396, 443
padang, 369
paired terraces, 48

palaeobotany, 369
Palaeocene, 74, 228
palaeochannels, 369
palaeoclimatic history, 94–5
palaeoclimatology, **369**, 408
palaeoecology, **369**, 408
  pollen analysis in, 394–5
Palaeogene, 228
palaeogeography, 370
palaeohydrology, 370
palaeolimnology, 408
palaeomagnetism, 370
  continental drift evidence from,
    109
palaeontology, catastrophism
  in, 79
palaeosols, 370
Palaeotropical kingdom, 211
Palaeozoic era, 228, 500
Palearctic realm, 204, 543
Paley, William, 115
pali ridges, 370
pallid zone, 370
palm savannah, *442*
Palmer, L., 85
palsas, 370
paludal sediments, 370
paludification, 370
palynology, 394–5
pampas, 475
panbiogeography *see* vicariance
  biogeography
Panchgani, *160*
panchromatic film, 6
panfans, 371
Pangaea, 74, 109, 371, 374
  see also Laurasia
panoramic cameras, 6
panplains, 371
pans, 371
pantanal savannah, 371
parabolic dunes, *158, 159,* 371
paraconformities, *521*
parallel drainage, *155*
parallel retreat, 42, *329,* 371
parallel roads, 371
parameterization, 371–2
parapatric species, 489
parasites, 372
  carnivorous, 77
  control by fungicides, 384
parasitoids, insect, 401
parna deposits, 372
partial area models, 277, 372,
    524
partial duration series, 372
particle sizes, **373–4,** 378
  alluvial fill, 18
  alluvium, 18
  and bed load transport, 157
  channel perimeter, 557–8
  and channel roughness, 480
  and competence, 104
  distribution: kurtosis, 296;
    skewness, 459; sorting, 466
  fluvial sediments, *446,* 447, 487
  and Reynolds number, 425
  in sandstone, 438
  terminal grades, 496–7
  and threshold velocity, 254
particles

form, 372–3 roundness, 433;
    sphericity, 468
  rollability, 432
  sizes *see* **particle sizes**
Pascal's law, 268
passiflora, mimicry by, 102
passive margins, 320, 374
pastoral zone, 85
patch reefs, 419
paternoster lakes, 374
patterned ground, 374
  cryostatic pressure as cause of,
    117
  earth hummocks, 163
  and needle ice formation, 350
paved beds, 34
pavements
  limestone, 241, 374, *375*
  shell, 455
  stone, 477; gibbers, 230
peak discharge, 375
Pearson type III distribution,
    375
peat, 375–6
  accumulation, 370, 421
  bogs *see* **bogs**
  in fens, 205
  gyttja, 245
  haggs, 249
  macrofossils in, 311, 375
  nutrient status, 359, 375–6
  palsas, 370
  roddons on, 432
pebbles, size of, 373
pedalfers, 377
pediments, 377
  fans, 43
  glacis slope, 233, 377
  shore platforms, 455–6
  suballuvial benches on, 482
  waning slopes, 545
pedocals, 377
pedons, 377–8
peds, 378
pelagic zone, *361,* 378
peléean eruptions, 540, *541*
Penck and Bruckner model, 378
Penck, Walther, 295, 459
peneplains, *109,* 122
  denudation chronology, 136
  endrumpfs, 174
  as final stage in erosion cycle,
    122
  monadnocks on, 339
  see also **planation surfaces**
penitent rocks, 378
Penman formula, 195, 399
Penman–Monteith revision, 195
per cent silt slay, 378, 557–8
perception
  of disasters, 144
  of drought, 155–6
  environmental, 315–16
perched blocks, 379
perched groundwater, 379
percolation, 379
percolation gauges, *192,* 379
percoline, 379
pereletok, 379
perennial streams, 379
periclines, 379
perigee, 379

periglacial areas, 117, **379–80**
landforms: alluvial fans, *17*;
asymmetric valleys, 36;
fragipans, 217; icings, 274;
patterned ground, 117, 163,
350, **374**; thufurs, 507
processes, 117; cambering, **244**,
75, *302*, 534; frost creep, 220;
frost wedging, 220, 221, 380;
gelifluction, 220, **224**, 322,
392- 3; ice segregation, 220,
272–3, 350, 380; landslides,
306; solifluction, 19, 322,
464; weathering, 221–2, 380
see also **freeze–thaw cycles;
meltwater; permafrost**
perihelion, 380
permafrost, 380–1
active layer *see* **active layer**
congelifluction in, 107–8
freezing fronts, 218
frost creep over, 220
gelifluction above, **224**, 322,
392–3
ground ice as proportion of, 241
heat flow in, 475
ice wedges, 273
oriented lakes in, 363
pereletok above, 379
study of (geocryology), 227
taliks in, *381*, 494
thawing: landforms of *see*
thermokarst
and valley bulge formation, 534
see also **tundra**
permeability, 274, **381**, 381, 383
affected by coordination
numbers, 111
and cone of depression shape,
107
hydraulic conductivity
distinguished from, 106
intrinsic, 282
and throughflow, 507
see also **hydraulic conductivity**
permeameter, 106, 381, 383
Permian, 228
**Permo-Carboniferous
glaciations,** 270
persistence, ecosystem, 383
perturbations, 383
Peru, cushion bog in, *66*
perviousness, 381, 383
pesticides, 384
biological magnification, 59
pests
biological control, **58**, 282–3
ecological explosions, 165–6
resistance to pesticides, 384
phacoliths, *284*, 385
phanerophytes, 417–18
phenetic classification, 494
phenological vicariance, 538
phenology, 385
phoresy, 385
phosphate rock, 385
phosphates
in guano, 244
**phosphorus, eutrophication
caused by,** 191
photic zone, *361*, 385
photochemical smog, 460

photogrammetry, 385
photosynthesis, 386
forests, 129
oceans, 62
and productivity, 352
respiration reverse of, 424
role in carbon cycle, 76
solar energy fixed by, 164
phreatic divides, 386
phreatic zone, 242, 243
phreatophytes, 386
phylogenesis, 386
phylogenetic classification, 494
**phylogenetic systematics
(cladistics),** 91–3, 538
phylogenetic trees, 92
physical hydrology, 265
physical meteorology, 386–7
physiography, 387
phytocoenosis, 56
phytogeography, 387
phytogeomorphology, 387
phytokarst, 58, 387
phytomass, 60
phytoplankton
decomposition, 439
productivity, 389, 528
phytosociology, 388
piedmont glaciers, 388
piedmonts, 388
piezometer, 106, 388
piezometric surface, 388, 547
pingos, *242, 262–3,* 380, **388**
types, 275
see also **palsas**
pinnate drainage, *155*
pipes, 388–9
piprakes (needle ice), 350
pisoliths, 389
pitometer, 389
placer deposits, 76
plagioclimax, 146
planar slides, *302*
planation surfaces, 186–7
denudation chronology, 136, 187
silcrete on, 457
types: etchplains, 189;
pediments, 233, **377**, 482,
545; shore platforms, 455–6
plane bed, 54, **389,** 389
plane sliding friction, angle of,
21
planetary albedo, 13
planeze, 389
planimeter, 389
plankton, 378, **389**
decomposition into sapropel,
439
in gyttja, 245
productivity, 352
respiration rates, 59
planktonic algae
dimethylsulphide production,
143
planning
environmental, 181
landscape evaluation for, 301
plant associations, 35, 388
plant communities
disclimax, 146
ecotones between, 168
overlapping within, 114

peatland, 376
phytosociological studies of, 388
polyclimax, 396
priseres, 403
seral, **451**, 482
serclimax, 451
seres, 451
subclimax, 482
subseres, 483
plant cover, 114–15
plantae, 295
plants
adaptation, 246–8; to drought,
247, 565; to snow and ice, 117
allelopathy, 14–15
amensalism, 19
aposematic coloration, 27
areographic studies of, 34
biotic potential, 63
calcium tolerance, 74
circadian rhythm, 90
classification: by life forms, 305,
417–18; into kingdoms,
294–5; phytosociological, 388
and climate, 246–7, 333, 417–18
coevolution with insects, 101–2
cold-hardiness, 247
colonizing, **103**, 300
communities *see* **plant
communities**
competition between, 104–5,
168
conservation, 108
consociation, 108
decomposition, 307;
denitrification by, 354
die-back, 141
diseases and pests, 282–3, 384
dispersal, 146
distribution, 211; endemic, 174;
polytopic, 396; study of, 387;
vicariant, 537–9; zonal, 567
domestication, 152; and energy
use, 175
as dominant organisms, 152–3
dwarf, 348
ecotypes, 168
effective rainfall for, 169
exotic, 198
fermentation, 63
as food for herbivores, 253
fossil, 311, 369, 375
and geomorphology, 58
inflorescence, 278
introduced: anthropochores, 24
on islands, 285–6
life cycle phenology, 385
and light, 246
limestone weathering by, 387
Lusitanian, 309–10
mimicry, 102
mutualism between animals
and, 347
nanism, 348
nitrogen fixation, 354, 488
nutrients, 352–3; and
eutrophication, 191
palaeobotanical studies, 369
peat formation from, 375–6
photosynthesis, 386
phytokarst produced by, 58
pollen: analysis of, 394–5

polymorphic, 396
polytopic, 396
productivity, 59, 60, 352–3
respiration, 164, 165, 424
on serpentine soils, 168–9
species diversity, 467
species–area curves, 467
stem flow down, 475
symbiotic bacteria on, 41, 354,
    488
and temperature, 246–7
tolerance limits, 511
topographic influences on,
    247–8
transpiration, 194, 247, **513–14**,
    see also **evapotranspiration**
types: angiosperms, 20;
    anthropochores, 24; aquatic
    macrophytes, 31; arctic-
    alpine, 32; ephemeral, 183;
    epiphyte, 183; evergreen, 196;
    geophyte, 229; halophyte,
    249, 435; hydrophyte, 266;
    mesophyte, 330; ombotrophic,
    362; parasitic, 372;
    phreatophyte, 386; salt-marsh,
    435; savannah, 441;
    sciophyte, 442; sclerophyllous,
    442
on waste land, 433
water availability for, 247, 352
and wind, 247
zonation, 567
see also **vegetation**
plastic limits, 39, 389
plasticity, 389–90
index, 39
soils: and activity ratio, 4
plate tectonics, 390–1
and coastal classification, 100,
    101
and continental drift, 109
mechanism, 320, 391
and orogeny, 364
and Tethys Ocean, 500
see also **plates; terranes**
plateaux, 391
former: accordant summits as
    evidence of, 2
lava, *541*
plates
boundaries, 390 and hot spots,
    257; seismic activity along,
    450; subduction at, 482;
    volcanoes on, 540
passive margins, 320, 374
playas, 65, 391, 435
**Playfair's law (of accordant
    junctions)**, 2, **391**
Pleistocene, 228, **391–2**
cambering during, 75
climatic fluctuations: loess
    deposits as record of, 308;
    Milankovitch hypothesis,
    335–6
coral destruction during, 124
glacials *see* **glacials**
great interglacial, **240**, 378
ice sheets: distribution, 230–1;
    loess bordering, 308
interpluvial, 281
lakes, 499

Late Glacial period, 15, 65, **301**,
    566
Penck and Bruckner model, 378
pluvials, *392*, 393
sea level changes, 48, 124,
    189–90; and dry valley
    formation, 157
valley bulge formation during,
    534
plinian eruptions, 392, 540, *541*
plinthite, 392
Pliocene, 74, 228
boundary with Pleistocene,
    391–2
plots, erosion, 392
ploughing blocks, 392–3
**ploughing, lynchets produced
    by**, 310
plucking, 393
plutonic rock, 393
plutonism, 393
plutons, 393
see also **intrusions**
pluvial terraces, 498–9
pluvials, 393
insolation during, 336
lakes: evaporation, 190
pluviometric coefficient, 393
pneumatolysis, 268
podzols, 257, 305, 393
POEM, 527
poikilothermy, 393
point bar deposits, 393–4
polar air masses, 11
polar front, 225, **394**
polar meteorology, 22–4, 32–3
**Polar Orbit Earth Observing
    Mission**, 527
polar platforms, 527
polders, 394
pole, cold, 102
pole, magnetic, 9, 399
policy-making
and climatic change, 30
cost-benefit analysis in, 114
**politics, ecological influence
    on**, 167
poljes, 394
pollen analysis, 394–5
deforestation evidence from, 129
pollutants, 395
occult deposition, 360–1
sediment-associated transport,
    447
see also **carbon dioxide; dust;
    nitrous oxides; nuclear
    waste**
pollution, 395–6
air *see* **air pollution**
control and legislation, 396
critical load, 116
marine, 320–1
salt marshes, 471
thermal, 502
urban areas, 486, 529
water *see under* **water**
see also **aesthetic degradation**
polychronism, 396
polyclimax vegetation, 396
polygonal cracks, 273
polygonal karst, 396
polygons, tundra, 380

polymorphism, 396
polynyas, 396
polypedons, 396
polytopy, 396
ponors, 396–7
pool–riffle sequence, 397
and channelization, *84*
and energy conversion, 239
and secondary flows, 446
Popov, L.V., 87
population
abundance: estimation, 1
and carrying capacity, **77–8**,
    166–7, 409
changes, 397–8
density dependence, 77–8, 133,
    397–8
dynamics, 397–8; and
    competition, 105; and
    predation, 400–3
equilibrium, 397; and predation,
    402
explosions, 165–6
fluctuations: and species
    survival, 383
human *see* **human impact;
    humans**
limited by territorial behaviour,
    500
pore ice, *242*, 398
pore water pressure, 398
and soil strength, 170, 454
porosity, 21, 111, **398**
positive feedback, 460, 491
Post Glacial, 65
postclimax, 398
postclisere, 98
pot-holes, 81, 399, *400*
potamology, 398
potassium argon dating, 398
potato blight warnings, 52
potential climax, 398
potential energy, 174, **398–9**
atmospheric, 225, 249, *399*, 517
gradually varied flow, 238, 239
soil water, 384
see also **kinetic energy**
potential evaporation, 191–3,
    399
potential evapotranspiration,
    194, 195, 399
and climatic classification, 96,
    501
measurement, 310
in moisture index, 338
and soil moisture deficits, 461–2
in water balance calculations,
    546
potential temperature, 399, 504
potentiometric surface, 547
potholes, 385
Powell, J.W., 47, 227
prairie soil, 72
prairies, 475
Pratt, J.H., 288
Pre-Boreal, 65
Precambrian
glaciations, 270
see also **cratons; shields**
precession of the equinoxes,
    335, 399

precipitation, *263,* 399
  acid, 2–4
  in altithermal phase, 19
  and altitude, 344, 345
  Antarctic, 24
  availability *see* effective rainfall
  below vegetation, 506
  chemical composition, 465
  and denudation rates, 135
  deserts, 138
  double mass analysis, 153
  drainage basin response to:
    modelling, 458
  efficiency, 482
  flooding caused by, 209
  forecasting, 554
  formation, 98–9, 266
  frontal, 219–20, 545
  hydrometeorological studies,
    266
  on hythergraph, 97
  interception *see* **interception**
  Köppen's classification based
    on, 295
  lack of *see* **drought**
  mapping, 265
  measurement, 268, **412–14, 505**
  modification, 555, 556
  in moisture index, 338
  orographic, 24, 364
  partition into evaporation and
    run-off, 545–7
  in pluvials, 393
  pluviometric coefficient, 393
  and sediment yield, *449,* 450
  and soil moisture deficits, 461
  surface detention, 486
  and temperature, 412
  in tropical cyclones, 516
  tropics, 517
  urban areas, 532–3
  in water balance calculations,
    545–7
  see also **dew; hail; rainfall;**
    snow
preclisere, 98
predation, 400–3
  and coevolution, 101
  and population fluctuations, 383
predators
  carnivores as, 77
  introduced, 282–3, 400
preservation, conservation seen
  as, 108
pressure, 403
  and altitude, 344
  Antarctic, 23
  Arctic, 33
  Bernouilli's theorem, 55
  Buys Ballot's law, 73
  in cyclones, 122, 123
  gradient, 562; and sea breezes,
    444
  high *see* **high pressure**
  hydrostatic, 268
  low *see* **low pressure; troughs**
  measurement, 44–5; by
    radiosonde, 411
  sea level adjustments to changes
    in, 283
  surfaces: thickness, 102, 503,
    505

  on tephigrams, 496, *497*
  on thermodynamic diagrams,
    504
pressure bulb recorder, *473*
pressure melting point, 403, 495
pressure release, 403, 525
  effects, 143, 455
pressure ridges
  in sea ice, 443
prevailing wind, 153, **403**
prey *see* **predation; predators**
prickly pear, 283
primarrumpfs, 403
primary productivity, 59, 60
priseres, 403
probabilistic models, 476
probability
  distributions: Pearson type III,
    375
  joint estimates of, 290
  see also **Markov process**
probable maximum
  precipitation, 403–4
process–response systems *see*
  systems
processes
  equations of motion
    representing, 183
  see also **geomorphological**
    processes
product life analysis, 305
profiles
  convexity, 342
  glacial troughs, 366
  rivers, 308–9; knickpoints in,
    295
  soil *see* **soil profiles**
  superimposed, 486
proglacial environment, 176
proglacial lakes, 404
progradation, 404
protalus ramparts, 404
protection
  conservation seen as, 108
  ecosystems and landscapes,
    405–6
Proterzoic, stromatolites in,
  481
protista, 294
proximal troughs, 406
psammoseres, 403
pseudokarst, 406
pudding stones (sarsens), 439
pumice, 406
punas, 406
punctuated aggradational
  cycles, 406
punctuated equilibrium, 79, 197
push moraines, 341–2
pyroclastic rocks, 406

quadrats, 1, 115
quartz, 438
  in granite, 239
  in greywacke, 241
  in grus, 244
  silt-sized: formation, 308
quasi-biennial oscillation, 312
quasi-equilibrium, 185, 407
Quaternary, 228, 407
  ecology, 407–8

  environmental reconstruction:
    by pollen analysis, 394–5
  glaciation *see* **glacials,** *etc.*
  glacio-eustatic sea level changes,
    189–90
  interglacials, 280
  see also **Holocene; Pleistocene;**
    Pliocene
quick clays, 408
  rotational failure, 433
  sensitivity, 451
quickflow, 131, **408**
  unit response graph, 525
quicksands, 408

r-selection, 183, 397, **409–10**
radar, 456–7
Radarsat, 527
radial drainage, *155*
radiation, 410–11
  and dew formation, 140
  ecotoxicological studies of, 168
  electromagnetic *see*
    electromagnetic radiation
  and fog formation, 100, 214
  heat transfer by, 250–1
  infrared *see* **infrared radiation**
  interception: in canopy, 75
  long-wave, *410,* 411; absorption
    by greenhouse gases, 240
  measurement, 4, 505
  net, 252, *353;* and climatic
    classification, 96; urban areas,
    531
  reflected *see* **albedo**
  solar *see* **solar radiation**
  in stratosphere, 478 ultraviolet
    see **ultraviolet radiation**
radiation balance, 252
radiative forcing, 411
radioactivity
  decay, 249 dating using, 77,
    173, 288
  from nuclear reactors, 128
  nuclear waste, 356–7
radiocarbon dating, 77
radioisotopes, 411
radiosondes, 411
  data plotted on thermodynamic
    diagrams, 496, *497,* 504
  thickness data from, 505
radon gas, 411–12
radwaste, 356–7
rain days, 412
rain factor, 412
rain forest *see* **tropical forest**
rain gauges, 412–14
  networks, 505
rain shadow, 344, 414
rainbows, 414
raindrops
  erosion by, 414
  sizes, 412
  sunlight refraction, 414
rainfall, 414
  deforestation effects on, 129
  depth: and area, 138; and
    duration, 138; probable
    maximum, 403–4
  drainage basin response to,
    263–4
  dust particles in, 64

effective, **169–70**, *171*, 442, 524, 525
frontal, 219–20
intensity, **279**, 412, **414**; and discharge, 110; and infiltration, 277; and overland flow, 256; and storm area, 404
measurement, 268, **412–14**, **505**
Mediterranean climate, 327
modification, 555, 556
monsoon climate, 340
pluviometric coefficient, 393
and sediment yield, *449*, 450
simulation, 415–16
soil moisture preceding, 24
on squall lines, 470
tropics, 517
urban areas, 532–3
see also **precipitation**
rainfall run-off, 414
raised beaches, *232*, 416
raised bogs, 359, **421**
Ramsay, A.C., 80–1
randklufts, 416
random-walk networks, 416
randomness, 476–7
rankers, 416
rapids, 416
Ras al Khaimah, *436*
rated sections, 417
rating curves, 145, **417**
rational formula, 417
Raunkiaer's life forms, 305, 417–18
Ray, John, 115
reaches, channel, 418
inflow and outflow hydrographs, 347
reaction time, 418
recession limbs, hydrograph, 137, 418
recharge, 418
recombinant DNA, 63
reconnaissance cameras, 6
rectangular drainage, *155*
recumbent folds, 214–15, 348, 418
recurrence intervals
discharge, 209–10
floods, 266, 313, 418, 425; design discharge based on, 140
red beds, 418–19
redox potential, 419
reefs, 419
gross primary productivity, 60
see also **atolls; barrier reefs; bioherms; coral algal reefs**
reflective beaches, 419
reflectivity, 12
refraction, wave, 419–20
refugia, 358, **420**
refuse *see* **waste**
regelation, 47, **420**, **509**
regeneration complex, 421
regime, rivers, 145, **421**
regime theory, 261, 421–2
regolith, 422
on graded slopes, 237–8
regosols, 422
regs, 420, 451

regur, 422
rejuvenation, 422
knickpoints indicating, 295
relative age, 422
relative humidity, 52, 257, *257*, 560
on climograph, 97
measurement, 268
and mist persistence, 337
relaxation time, 422
relevés, 388
relief, inverted, 283
religion, environmental ideas in, 315–16
remanié glacier, 422
remote sensing, 422–3
atmospheric conditions, 332–3, 439–41
digital image processing in, **142**, 346
by earth resources satellites, 317- 19, 525–8
electromagnetic radiation wavelengths for, 172, *173*, 346
by multispectral scanners, 319, 346- 7, 501, 525–6, 527
phytogeomorphological use, 387
sideways looking airborne radar (SLAR), **456–7**, 527
by thermal infrared linescanners, 501–2
and weather forecast improvements, 29
see also **aerial photography; meteorological satellites**
rendzina, 423
renewable resources, 181, 182
repose, angle of, 21
constant slopes, 109
graded slopes, 237
representative basins, 423
reproduction rates
and natural selection, 409, 410
resequent streams, **423**
reservoir rocks *see* **groundwater**
reservoirs
compensation flow below, 104
hydrological effects, 424
rivers dammed to create, 124
residence time, 424
residual shear, angle of, 21
residual strength, 424, 454
resilience, ecosystem, 167, 470–1
resistance
channel, 83; *see also* **roughness**
ecosystems, 470, 471
to erosion, 134, 325, 431
resource processes, *178*, 179–80
rationalization, 181–2
resources
conservation, 108, 181–2
development: geomorphology applied to, 27
efficient use, 176–7, 181–2
exploitation, 430
sustainability, 488
respiration, 424
energy associated with, 164, 165
and productivity, 59
response analysis, 424
resultant winds, 424
resurgence, 424–5

retention curves, 425
retention forces, 425
retrogradation, 425
return periods *see* **recurrence intervals**
reversed faults, 203, 425
reversed polarity, 425
reversing dunes, *158*, 425
Reynolds number, 212, **425**
boundary layer, 68
and Darcy's law validity, 126
laminar flow, 298, 425
turbulent flow, 68, 520
rheidity, 425
rheology, 425–6
mantle, 288
rheotrophic nutrient status, 358, 376
rhexistasy, 62
rhizomes, 426
rhizosphere, 426
rhodoliths, 426
Rhone valley, mistral in, 337
Rhossili, *420*
rhourds, 426
rhythmite, 535
rias, 426
ribbed moraines, 545
Richter denudation slopes, 426
Richter scale, 164, 426
ridges, beach *see* **berms** and *under* **beaches**
ridges, high pressure, 518
riedel shears, 426
riegels, 426
riffles *see* **pool–riffle sequence**
rift valleys, 426, *427*
rifted margins, 374
rillenkarren, *291*, 292, 293
rills, 426, 428, 433–4
rime, 270, **428**
ring complex, 428
rinnenkarren, 292
rip currents, 120, **428**
and edge waves, 169, *170*
ripples
adhesion, 5
current, 118, *119*
current bedding, 118
migrating: cross-lamination, 116
sand, *49*, 212
size, 54
rising limb, hydrograph, 428
Riss glaciation, 190, *378*
river basins *see* **drainage basins**
rivers
accordant junctions law (Playfair's law), 2, 391
and acid rain, 3
avulsion, 41
base flow *see* **base flow**
blind valleys, 64
braided *see* **braided rivers**
capture, 76
channel characteristics and flow *see* **channels**
channelization, 54, 83, *84*
classification, **83–8**
dams on *see* **dams; reservoirs**
deposition by *see* **bedforms; deltas; floodplains; meanders**

rivers (*continued*)
discharge *see* **discharge**
dissection by, 146, 205
dissolved oxygen in, 55, 56, **147**
distributaries, 148
divagation, 148
doabs between, 150
dry weather flow, 157
equilibrium in, 184–5
floods *see* **flood frequency analysis; floods**
flow in *see* **channels: flow; discharge**
form ratio, 217
gallery forests along, 223
headwaters, 250
ice in, 128
interbasin transfers between, 279–80
landscapes, 430
load *see* **bed load; dissolved load; suspended load**
long profile, 295, 308–9
maturity, 325
meandering *see* **meanders**
metamorphosis, 369, **428–30**
minimum acceptable flow, 336
navigational use, 349–50
pollution, 502, 529
potamology, 398
rapids on, 416
regimes, 145, **421**
rejuvenation, 295, 422
sediment transport *see* **sediment transport; sediment yield**
slip-off slopes, 459
solute concentration *see* **dissolved load**
thalassostatic terraces, 500
tidal bores, 66
types: affluent, 8; allogenic, 15; effluent, 171; ephemeral, 183, 543; ice-marginal, 71; influent, 278; intermittent, 280–1; misfit (underfit), 76, **521–2**, 534; perennial, 379; periglacial, 327; *see also* **drainage; drainage networks; meltwater**
uniclinal shifting, 522
waterfalls on, 553
weirs on *see* **weirs**
yazoos parallel to, 566
see also **estuaries**
riverscapes, 430
robber economy, 430
roches moutonnées, 430, 510
rock avalanches, 41, *302*
sturzstrom, 482
rock drumlins, 430
rock falls, *302*, 321, 323, *324*
berghlaups, 54
bergsturz, 54
magnitude and frequency, 313
and scree formation, 442
rock flour, 430
rock glaciers, 430–1
rock salt, caves in, 81
rock slides, 323
rock steps (riegels), 426
rock varnish
cation-ratio dating using, 80

rocks
accessory minerals in, 2
aphanitic texture, 26
autochthonous formation, 39
bioclastic, 55
cambering, 75
case hardening, 78
cataclasis, 78
chattermarks on, 88
cleavage, 94
conchoidal fractures, 106
congelifluction, 107–8
and denudation rates, 134
dilation joints in, 143
erodibility, 186
erosion *see* **erosion**
facets, 202
facies, 202
fatigue failure, 203
faulting *see* **faults**
flow through: diffusion equation for, 141–2
folding *see* **folds**
foliation, 215
fracturing, 71, **217**, 261
frozen *see* **permafrost**
fulgurites in, 222
geodes in, 227
gnammas in, 236, *237*
hardness, 249
hogback ridges, 255
hydrothermal alteration, 268
inliers, 278
interstices in, 281–2, 398
lithology, 307
magnetic orientation, 370
mammilated surfaces, 314
mass strength, 325, 431, 479
massive, 325
mesas in, 328, *329*
metasomatism, 331
microcracks in, 333
nonconformities between, 355
penitent, 378
percolation through, 379
permeability, 31–2, 274, 282, 381
perviousness, 274, 383
piezometric head, 388
plasticity, 389–90
porosity, 398
pressure release, 143, 403, 455, 525
quality indices, 431–2
resistance, 134, 325, 431
retention forces, 425
saturation, 441
specific retention, 468
specific yield, 468
stratigraphy, 478
strength, 279
striations on, 479–80
types: acid, 4; argillaceous, 34; basal complex, 46; basement complex, 48; brash, 71; cap-rock, 76; eugeogenous, 189; metamorphic, 39, 236, **331**; plutonic, 393; *see also* **igneous rocks; sedimentary rocks; volcanic rocks**
vughs in, 281, 542
weathering *see* **weathering**

see also **geology**
Rocky Mountains
chinook, 89
frost creep, 220
roddons, 432
Rogen moraines, 341
rollability, 432
roots, adventitious, 5
Ross Sea, 234
Rossby waves, 38, 311, **432**, 495
and Antarctic climate, 22
and cyclone development, 122
dishpan simulation, 146
rotating current meters, 118
rotational failures, 432–3
rotational slides, 323, *324*
rotor streaming, 433
roughness, 53, 83, 433
Manning coefficient, 319–20;
estimated by Strickler equation, 480
sandbed channels, 438
and velocity, 53, 212
roundness, particle, 372, 433
ruderal vegetation, 433
run-off, 414, 433–4
and antecedent moisture, 24
and antecedent precipitation index, 24
contributing areas for, 110–11
contributions to, 263–4
dilution effect, 143
distribution graphs of, 148
and effective rainfall, 169, *171*
and fire, 206–7
Horton model, **256**, 277, 280, 414, 524
and land use, 299, 300, 434;
deforestation, 129–30;
grazing, 240
mapping, 265
monitoring, 392
partial area models, 277, **372**, 524
and precipitation, 145
and rainfall intensity, 279
rational formula for estimation, 417
snowmelt component, 461
urban, 300, 434
as water balance component, 545–7
run-off plots, 392
rundkarren, 292
runnels, solution, 292
ruwares, 434

S-curve, 442
s-**selection**, 409
sabkha, 435
Sahara Desert, temperature in, 251
Sahel, 140, 156
St Helena, alien species on, 286
St Louis, climate of, 532–3
Salar of Uyuni, *392*
salars, 435
salcrete, 435
salinity, 435
Tethys Ocean, 330–1
water masses, 547, 552

salinization, 435
 soils, 464
salt
 caves in, 81
 dissolution: and subsidence, 208
 haze caused by, 250
salt domes, 436, *437*
salt flats, 436
salt lakes, former, 436
salt marshes, *188,* **435**
 on barrier beaches, *45*
 die-back, 141
 oil pollution, 471
 see also **mangrove swamps**
salt weathering (haloclasty), 249,
 436, 493
salt wedging, 222
saltation, 52, 104, **436**
salts *see* **salinity**
sampling
 ergodic hypothesis, 186
San Andreas fault, 327, 390
sand, **437**
 adhesion ripples, 5
 bedforms, 53–4
 deposition on silts or muds, 307
 fulgurites in, 222
 particle size, 373
 proportion in soils, *463*
 quicksands, 408
 rocks composed of, 34
 singing, 458
 talsands, 494
 wind movement, 155, 438; sand
  roses depicting, 437
 see also **beaches; berms;**
  dunes; sandstone
sand banks, 437
sand dunes *see* **dunes**
sand grains
 chattermarks on, 88
 packing: coordination numbers
  as representation of, 111
 saltation, 104, 436
sand lenses, 365, 437
sand ripples, *49,* 212
sand roses, 437
sand volcanoes, 437–8
sandar, 438–9
 see also **outwash**
sandbed channels, 438
 bedforms and flow regimes,
  212, 222
sandstone, 438
 arkose, 34, 438
 chines in, 89
 flysch deposits, 213
 inselbergs in, 279
 molasse, 338
 sarsens, 439
 yardangs in, 566
 see also **quartz**
sandstorms, 438
sanitary landfill, 300
sapping, cliff, 439
saprolite, 439
sapropels, 439
saprophytes, 439
sarns, 439
sarsens, 439
sastrugi, 439
satellites

earth resources: manned,
  317–19; unmanned, 525–8
 meteorological, 13, **332–3,**
  439–41; cloud data, 351–2;
  VISSR in, 539
saturated adiabatic lapse rate,
 537
saturated throughflow, 507
saturated wedges and zones,
 441, 466
saturation deficit, 254, **441**
 in partial area models, 372
saturation overland flow *see*
 overland flow
saucer blowouts, 64
Saussure, H.B. de, 430
savannah, **441,** *442*
 climate, 482
 types: bushveld, 73; cerrado, 82;
  chapada, 88; pantannal, 371
scablands, 441
 catastrophic explanation for, 79,
  523–4
scale
 and allometric growth, 15, 16
scale insects, biological control
 of, 282–3
scars, 442
scavengers, carnivorous, 77
Scheidegger, A.E., *362,* 363
schists, 215
Schmidt hammer test, 249
Schumm, S.A., 48, 105, 237,
  238, 428, 491, 557–8
 channel classification, 85, 88
Schwartz, M.L., 72
sciophytes, 442
Sclater, Philip Lutley, 203,
  543–4
sclerophyllous plants, 442
scoria, 442
Scotland, machair in, 311
screes, 442
 angle of repose, 21
 free face above, 217–18
 stratified (grezes litées), 241
 waning slopes, 544–5
sea
 ice in *see* **sea ice**
 polders reclaimed from, 394
 pollution, 320–1
 shoals in, 455
 surface temperatures: anomalies,
  12; and tropical cyclone
  formation, 516, 517
 temperatures: and coral reef
  formation, 111, 113
 zones: epilimnion, 183; neritic,
  352; sublittoral, 482
sea breezes, 329, **444**
sea caves, blow-holes in, 64
sea cliffs, 442
sea floor
 mountains, 245, 446
 spreading, 39, 390
 subduction, 482
sea ice, 443–4
 Arctic, 33, *382, 443*
 attachment to bottom, 20
 distribution, *382, 443*
 expansion during glacials, 230,
  231

floating, 271–2
 formation, 272
sea level, **444**
 adjustment to pressure changes,
  283
 as base level, 47
 changes, 444–5; atoll sensitivity
  to, 38; and bluehole
  formation, 65; and coastal
  classification, 100; and coral
  reef development, 124;
  drainage basin response, 48;
  eustatic, **189–90,** 262, 416,
  445; evidence of, *445,* 483;
  Flandrian transgression,
  207–8, 253; and hydroisostasy,
  262; and Messinian salinity
  crisis, 330–1; monitoring,
  236; and raised beaches, 49,
  416; rate, 445; and ria
  formation, 426; and shoreline
  recession, 72; and
  thalassostatic terraces, 500
 extreme: probability of extreme
  currents and, 290
 isostatic changes, 325–6, 445
 mean, 325–6
 measurements: hydrodynamic
  levelling, 261
 seiches in, 450
 storm surges, 477
 surface in relation to geoid, 227
sea mounts, 245, 446
Seasat, 527
seawater
 chemical composition, 464
 desalinization, 138
 dimethylsulphide in, 143
 freezing, 217
 intrusion into coastal aquifers,
  229–30
sebkha (sabkha), 435
secondary flows, 252, 446
secondary productivity, 59–60
sedges, 376
sediment budget, 448–9
sediment delivery ratio, 449
sediment flows, 510
sediment loads, 307
 meltwater streams, 232–3, 327
 particle size distribution, *446,*
  447
 see also **bed load; suspended**
  load
sediment transport, **446–9**
 artificial channels, 421–2
 and channel classification, 85,
  87, 88
 and competence, 104
 downstream of reservoirs, 424
 equilibrium, 184
 frictional resistance to, 218
 glaciers, 67, 105
 magnitude and frequency
  effects, 313
 measurement, 53
 modelling, 358
 monitoring, 392
 by overtopping flows, 366
 by overwashing flows, 367
 periglacial areas, 380
 by saltation, 436

sediment transport (*continued*)
sandbed channels, 438
by sheetflow, 454
and stream power, 478
by termites, *498*
threshold flow velocity for, **254**, 505
by tidal currents, 121
and tractive force, 513
see also **bed load; suspended load**
sediment traps, 229
sediment yield, 449–50
denudation rates calculated from, 137
and fire, 207
increased by grazing, 240
and land use, 299–300
suspended load, 448
urban areas, 300
sedimentary landforms
mass balance, 321
sedimentary rocks, 450
bedding planes between, 53
connate water in, 108
diatremes in, 141
dip, 143
feather edges, 205
frost weathering, 221
interstices in, 281, 282
law of superposition applied to, 486
red beds, 418–19
solute concentration of rivers on, 465, 466
types: arenaceous, 34; beach rock, 51; greywacke, 241; oolite, 362; shale, 213, 362, **453**, *469*; *see also* **clay; limestone; sandstone**
unconformities, 521
valley bulges, 534
void ratio, 540
sedimentation
accretion through, 2
bars, 46
cycles, 123
on floodplains, 210
ocean basins: and sea level changes, 190
sediments
allochthonous, 15
armoured mud balls of, 34
autochthonous, 15, 39
bedding, 116, 117–18, 302–3; false, 202; graded, 237
bioturbation, 64
in bottom-set beds, 67
clastic, 93
dating *see* **dating techniques**
deep-sea: climatic evidence in, 94, 335–6, 497–8
deposited by river currents, 120–1
diagenesis, 140
fabric analysis, 202
facies, 202
floodplain, 210
fossils in *see* **fossils**
geest, 224
induration, 276
internal friction, 218

krotovinas in, 296
lake: nutrient status changes indicated by, 359
lateral accretion, 302–3, 393–4
liquid limit, 306
lithification, 307
load structures, 307
mass movements *see* **landslides; mass wasting**, *etc.*
paludal, 370
particle size *see* **particle size**
per cent silt clay, 378
pollen separated from, 395
pore water pressure, 398
Quaternary ecology reconstructed from, 407–8
red beds, 418–19
in submarine canyons, 483
see also **alluvium**
seeding, clouds, 555, 556
seepage slopes, *463*
segregated ice, *242*, **272–3**, 380, 450
see also **pingos**
seiches, 450
seifs, 450
seismicity, 450
see also **earthquakes; plate tectonics**
self-mulching, 450
selva *see* **tropical forests**
semi-arid areas, 450
braided rivers, 71
brousse tigrée, 71–2
calcrete, 74
pans, 371
phreatophytes, 386
pipes, 388–9
playas, 391
pluvial terraces, 499
semi-deserts, 450
sensible heat, 451
transfer in Hadley cells, 249
in urban areas, 531
sensible temperature, 450–1
sensitivity, soil, 451
seracs, 451
seral communities, 61, 451
serclimax, 451
seres, 451, 484
convergence on monoclimax, 339, 396
see also **priseres; subseres**
serir, 451
Sernander, R., 65
serpentine rocks, soils and plants on, 168–9
sestons, 451–2
Seventh Approximation, 452–3
sewage
dry weather flow, 157
treatment, 63
sexual selection, 197
shakeholes *see* **dolines**
shale, 453
flysch deposits, 213
oil, 362
springs on, *469*
Shannon–Weiner function, 149
sharp-crested weirs, 453–4, 534, *558*
Shea, J., 522–3

shear box, 454, 478
shear fractures, 426
shear strength, 424, 454
and coordination numbers, 111
measurement, 454, 514
shear stress, 454, *479*
and avalanche initiation, 41
bed, 259
and competence, 104
in du Boys equation, 157
in ice, 233–4, 272
in laminar flow, 297
normal stress perpendicular to, 356
resistance to *see* **shear strength**
and secondary flows, 446
soils, 338
sheared margins, 374
shearing resistance, internal, angle of, 21
shearing surface
reorientation on, 20–1
sheet lightning, 305
sheetflow, 433, 454
sheeting, 454–5
sheetwash
on stone pavements, 477
shell pavements, 455
Shepard, F.P., 100
shields, 455
basal complex underlying, 46
basement complex, 48
glint line, 234
knock-and-lochan topography, 295
shillow, 455
shoals, 455
shore platforms, 455–6
shores, 455
Shreve, R.L., *362*, 363
shrinkage limit, 39
shrub–coppice dunes (nebkhas), 159
sial, 456
sichelwannen, 81, *385*, 456
sideways looking airborne radar (SLAR), 456–7
sieve deposits, 457
sieving, particle size measurement by, 373–4
silcrete, 160, 343, 457
silica, 89, 458
sills, *284*, 457
silt
drapes, 365, 458
particle size, 373, 458
as proportion of sediment, 378, 557–8
proportion in soils, *463*
sand deposition on, 307
siltation, 458
Silurian, 228
silver iodide, 555
Silver Springs ecosystem, 59, 491, 515
sima, 458
simulation, 458
drainage networks, 416
rainfall, 415–16
successional processes, 485
singing sands, 458
sinkholes *see* **dolines**

sinter, 458, 519
sinuosity, channel, 458–9
  classification based on, *86,* 87–8
  and cut-off development, 121
siphons, 459
sirocco, 294
Skempton, A.W., 4
skewness, 459
Skokholm, mice on, 286
Skye, *376*
Skylab, 319
slab failures, *302,* **459**
  see also **toppling failures**
slacks, 459
slaking, 459
SLAR, 456–7
slates, 215
slickenside, 459
slides *see* **landslides**
slip face, 459
slip off slopes, 459
slips, 459
  see also **landslides**
Slobodkin, L.B., 165
slope, 459
  clinographic curves representing,
    97
  consequent streams in direction
    of, 108
  constant, 109
  edge-lines representing break in,
    169
  ice sheets, 273
  measurement, 97
  as morphometric parameter, 342
  and rollability, 432
  and sediment movement, 218
  threshold, 506
  see also **hillslopes**
slope area method, 145, 267
slope failure *see* **mass
  movements,** *etc.*
slope winds *see* **anabatic flows;
  katabatic flows**
slump, *302*
smog, 395, **460**
snouts, glacial, 460
snow, 460
  Antarctic, 24
  Arctic, 33
  avalanches, *40,* 41, 322; remanié
    glaciers fed by, 422
  blizzards, 64
  compaction into glacier ice, 270
  cornices, *113,* 114
  forecasting, 38
  plants' adaptation to, 117
  sastrugi in, 439
  see also **firn; glaciers; snowfall**
snow gauges, 412
snow line, 460
snow patches, 460
  debris slides over, 404
  glacierets developed from, 232
  nivation below, 355
snowblitz theory, 460
snowfall
  measurement, 412
  modification, 555
  nivometric coefficient, 355
snowflakes, 460
snowmelt, 461

flooding caused by, 209
in forests, 217
see also **meltwater**
snowshoe hare, 402
sodium carbonate
  in solonetz soils, 464
soil erosion, 461
  and grazing, 240
  human influence on, 331
  and land use, 299–300
  monitoring, 392
  by overland flow, 366
  pipes as evidence of, 388
  by raindrop impact, 414
  and scabland formation, 441
  thresholds in, 505
soil horizons
  absent from lithosols, 307
  lateral flow along, 303
  margalitic, 320
  in podzols, 393
  rendzina, 423
  solum, 464
soil loss equation, universal,
  392
soil moisture
  antecedent, 24
  and antecedent precipitation
    index, 24
  and atterberg limits, 39
  capillary forces binding to soil
    particles, 75–6
  deficits, 461–2
  drainage from, 138
  and evapotranspiration, 194–5,
    461–2
  in expansive soils, 198–9
  field capacity, 205, 461
  measurement: electrical
    resistance block, 172;
    gravimetric method, 239;
    neutron probe, 354;
    piezometer, 388
  pore water pressure, 398
  recharge, 418
  retention curves, 425
  retention forces, 425
  storage, *263,* 264
  suction, 384–5, 485;
    measurement, 496
soil orders, 452–3
**Soil Survey of England and
  Wales,** 298
soil water
  acid precipitation effects on, 3
  groundwater distinguished from,
    242
  storage: and infiltration, 276–7
soils, 461
  accommodation, 2
  acidic: plants suited to, 74
  activity ratio, 4, 9
  advantages of trees to, 9
  aeration zone, 5–6
  aggregation ratio, 9
  agricultural capability, 298
  agrometeorological studies, 10
  on alluvial fill, 18
  atterberg limits, 39
  bearing capacity, 51–2
  bioturbation, 64

capillary potential, 384–5, 485;
    measurement, 496
carapace, 76
catena concept, 80, 462
cation exchange, 47
cation exchange capacity, 48
cheluviation, 89
chronosequences, 90
classification, 34, **452–3**
claypans, 93
clinosequences, 97
cohesion, 102, 338; and
    atterberg limits, 39;
    measurement, 454; and slope
    stability, 472
compaction, 104
congelifluction, 107–8
conservation, 300, 461
consolidation, 104, **108,** 108,
    451
creep, 116, 323, 325; *see also*
    solifluction
crusts, 117
cryostatic pressure in, 117
cutans in, 121
deserts, 493
drainage, 205–6, 298
duricrusts, 160
eluviation in, 173
energy fluxes in, 251
epipedons, 183
erodibility, 187
erosion *see* **soil erosion**
erosivity, 187
evapotranspiration from, 310
falls, *324*
ferricrete in, 205
field capacity, 205, 461
and fire, 206–7
floodplain, 210
formation: in biostatic periods,
    62; deposition concomitant
    with, 117; human influence,
    331
from former environment
    (palaeosols), 370
fragipans at base of, 217
frost heaving, 220
frozen *see* **permafrost**
gelifluction, 220, **224,** 322,
    392–3
hardpans, 249, 392
heat storage, 334
hillslope, 462, *463*; flow
    processes on, 253–4
horizons *see* **soil horizons**
humate in, 257
humus in, 258
hydraulic conductivity, 106–7;
    measurement, 381, *383*
hydromorphy, 266
ice in *see* **ground ice;
  permafrost**
illuviation within, 274
impervious, 274
infiltration into *see* **infiltration**
instability: and terracette
    formation, 499
in interglacials, 280
interstices in, 398
K-cycle, 291
krotovinas in, 296

soils (*continued*)
leaching, 304
lessivage, 305
mottled zone, 343
needle ice in, 350
nutrient status, 358–9
organic material, 257, 258, 331;
moder, 337; mor, 340; mull,
346
oxidation, 367
pallid zone, 370
patterned ground features, 374
pedons, 377–8
peds, 378
percolation through, 379
permeability, 274, 381; intrinsic,
282
perviousness, 383
piezometric head, 388
pipes in, 388–9
and plant growth, 248
plasticity, 4
plinthite on, 392
podzolization, 305
polypedons in, 396
pore water pressure, 398
porosity, 398, 434
profiles, 462; cumulative, 117;
in laterite, 303; stone lines in,
477; *see also* **soil horizons**
residual strength, 424, 454
rhizosphere, 426
salinity problems, 435
salinization, 464
saturation, 254, **441**; and
overland flow, 264
self-mulching, 450
sensitivity, 451
on serpentine rocks, 168
shear strength, 454, 514
shear stress, 338
sheetflow over, 454
and slope processes, 462–4
specific retention, 468
specific yield, 468
stability, 472
strength, 451; Mohr–Coulomb
equation, 338; and pore water
pressure, 170; and threshold
slope, 506
striated, 479
structure, 462; brodel, 71;
crumb, 117; and infiltration,
277
texture, 462, *463*; and
infiltration, 277
thermal inertia, 251
throughflow *see* **throughflow**
toposequences, 511
trafficability, 513
translational slides, 513
types, 34, **452–3**; alfisols, 14,
452; aridisols, 34; azonal, 41;
brown forest, 71; brunizem,
72; chernozem, 89; expansive,
198–9; gley, 233, 266;
grumusol, 244; impermeable,
274; interzonal, 282; laterites,
303; pedalfers, 377; pedocals,
377; podzols, 305, 393;
rankers, 416; red beds,
418–19; regosols, 422; regur,

422; solonchaks, 464;
solonetz, 464; terra rossa,
498; volcanic, 20; zonal, 567
water in *see* **soil moisture; soil
water**
wetting front in, 276, 560
solar constant, 464
solar radiation, 464
absorption, 184, 333, 334, *410*
and altitude, 344
atmospheric effects on, *410*
backscattering by dust, 160,
357, *410*
cyclic variations: and ice ages,
271
deserts, 251
and evaporation, 191
fixation through photosynthesis,
164
intensity, 279
measurement, 4
photosynthesis into food, 164,
386
and plant growth, 246
reflected, 13, 184, *see also*
albedo
and species diversity, 467
tropics, 517
solar tides, 508
solfataras, 464
solifluction, 19, 322, **464**
see also **gelifluction**
solonchaks, 464
solonetz soils, 464
solstices, 464
solum, 464
solutes, 464–6
in urban areas, 529
solution
caves formed by, 81
as corrosive process, 114, 337
limestone *see under* **limestone**
sorting, 466
source area, 466–7
dynamic, 161; in partial area
models, 372; and saturated
wedges, 441
South Platte River, *429*
Southern Oscillation, 312, 495,
518
Space Shuttle, 319
Spacelab, 319
speciation *see* **Darwinism;
evolution**
species
abundance, 1
adaptation: convergent evolution
resulting from, 197–8; to
habitat differences, 168
adaptive radiation, 4–5
alien *see* **alien species**
allopatric, 16, 537, 538
amensal, 19
biotic isolation, 63
biotic potential, 63, 397
clines, 97
coevolution, 101–2
colonizing, **103**, 300; taxon
cycles, 494
commensal, 103

competition between, 104–5;
and carrying capacity, 78; and
predation, 400
consociation, 108
creationist view of origins,
115–16
disjunct distribution, 146
diversity *see* **diversity**
ecotypes, 168
endangered, 173–4
endemic, 174; and genetic drift,
226; on islands, 285; in
refugia, 420
ephemeral, 183
evolution *see* **evolution**
exotic, 198
extinction *see* **extinction**
introduced, 282–3, 286, 400
island, 285–6, 400
natural selection *see* **natural
selection**
polytopic, 396
population explosions, 165–6
protection, 405
resemblances between: cladistic
view of, 91–3
survival in refugia, 420
symbiotic *see* **symbiosis**
sympatric, 489
vicariant, 537–9
species composition
community concept applied to,
104
succession *see* **succession**
species-area curves, 467
specific conductance, 106, 466
specific heat, 250
specific (intrinsic) permeability,
282
specific retention, 468
specific tractive force, 350
specific yields, 468
speleology, 468
speleothems, 468
helictites, 252–3
supersaturation important for
growth, 486
Sphagnum, 376
lake basins invaded by, 359
macrofossils, 311
in muskeg, 347
in raised bogs, 421
sphenochasms, 468
sphenopiezms, 468
sphericity, particle, 372, 468
spheroidal weathering, 468
spits, 350, 468, 511
barrier, 46
spitzkarren, 292
spodosols, 452
Spokane flood, 523–4
sporopellinin, 394–5
SPOT satellites, 526
spring sapping, 439
springs, 468–70
artesian, 35
geysers, 229
hot, 257, 458
intermittent, 280
karst, 469
mapping, 261
resurgence, 424–5

water supplies from, 560
spurs, truncated, 518
squall lines, 470
stability
atmosphere, 504, 518, **536–7**
channels, 85, *87*, 88, 421–2
ecosystems, 167, 383, **470–2;**
and diversity, 150, 471
population: and predation,
402–3
slopes, 472
see also **instability**
stable equilibrium, 303–4, 472
carrying capacity as, 78
stacks, 472
stadials, 472
see also **interstadials**
staff gauges, 473–4
stage recorders, 267, 417, 473,
474
float recorders, 208–9
stilling wells, 475–6
stagnant ice topography, 474
disintegration moraines, *341,*
342
eskers, 187
kames, 291
kettle holes, 294
see also **till**
stalactites and stalagmites, 468
star dunes, *158,* 474
state, equation of
in GCMs, 224
static pressure, 55
stationary waves, tropospheric,
64
steady flow, 474
uniform, 89, 212, 218–19, **522**
steady state, 185, 475
and graded time, 238
limitations of assumption, 506
steam fog, 214, 475
see also **frost smoke**
Stefan–Boltzmann law, 411
Stefan's method, 475
stem flow, *243,* 475
step–pool systems, 475
steppes, 475
stick slip, 475
stilling wells, 475–6
stillstand, 476
stochastic models, 357, **476, 553**
drainage network simulation,
416
stochastic processes, 476–7
stocks, 477
Stoke's law, 497
stone lines, 477
stone pavements, 130, 230, **477**
storage, *263,* 264, 477
bank, 43–4
in depressions, 138
equation for, 110, 477
and infiltration, 276–7
and run-off, 263–4
surface, *263,* 264, **486–7**
and water balance calculation,
546
storativity, 258
storm hydrographs, 262, *264*
and overland flow, 256
peak discharge, 375

storm run-off, 363, 433
exhaustion effects, 198
storm surges, 477
storms, 52
dust, 160, 248
magnetic, 312
maximum rainfall from, 404
return period: and overtopping,
366
see also **thunderstorms;**
tornadoes
stormwater drainage systems,
528
Strahler, A.N., 226, *362,* 363,
491
strain, 478, 479
and elasticity, 337–8
strandflats, 478
stratified screes (grezes litées),
241
stratiform clouds, 98
stratigraphy, 478
hiatuses in, 253
Quaternary, 407–8
strato-volcanoes, 478
stratocumulus clouds, 99
stratosphere, 478
boundary with troposphere, 518
ozone layer, 367–8; depletion,
89, 249
quasi-biennial oscillation in, 312
radiation interactions in, *410*
stratus clouds, 98, *99,* 100
seeding, 555
stream hydrographs *see*
hydrographs
stream networks *see* **drainage**
networks
stream numbers, 257
stream order, *362,* **363,** 363
and bifurcation ratio, 55
drainage basin morphometry
based on, 343
in Horton's laws, 257
stream power, 478
and minimum variance theory,
336
stream sinking
karst landforms dependent on,
293
see also **dolines**
streamlines, 169, **478**
streams *see* **channels; drainage**
networks; rivers, *etc.*
strength
mass: slope angle adjusted to,
479
rock, 325, 431
soils, 451; Mohr–Coulomb
equation, 338
strength equilibrium slopes, 479
stress, 478, 479
effective, 170; and slab failure,
459
and elasticity, 337–8
normal, 356
and plasticity, 389–90
shear *see* **shear stress**
soil cohesion independent of,
102
stress ecology, 479
stress tolerant species, 409

striated soil, 479
striations, 479–80
Strickler equation, 480
strike, 480
strike-slip faults, 203
megashear, 327
string bogs, 480
strip cameras, 6
stromatolites, *480,* 481–2
Strombolian eruptions, 540, *541*
sturzstrom, 482
Sub-Atlantic zone, 65
Sub-Boreal zone, 65
subaerial processes, 482
suballuvial benches, 482
subantarctic trough, 23
subclimax community, 482
subduction, sea floor, 482
subglacial environment, 176,
482
lakes: sudden drainage, 290
landforms: crag-and-tail, 115,
341; drumlins, 156, *341*
moraine, 340–1
streams: cavitation erosion, 81;
esker formation, 187; tunnel
valley formation, 520;
urstromtäler formation, 533
till, 509
subhumid province, 482
sublimation, 270, 482
sublimation till, 510
sublittoral zone, 482
submarine canyons, 482
submerged forests, 483
subsequent streams, *76,* 360,
483
subseres, 403, 483
subsidence, 323, 483
flashes produced by, 208
subsurface flow *see* **throughflow**
succession, 167, **483–5**
allogenic, 15, 40
autogenic, 40
climax vegetation as final stage
of, 97
and environmental protection,
405
evidence for, 485
primary, 484
ruderal vegetation, 433
salt marshes, 435
secondary, 484
see also **climax vegetation;**
priseres; seral
communities; serclimax;
seres; subseres
suction, soil moisture, 384–5,
485
measurement, 496
Suess effect, 485
suffosion, 379, **485**
sugar cane, productivity of, 353
Sukachev, V.N., 57
sulphate aerosols, 143
sulphation, 485–6
sulphur
as fungicide, 384
sulphur dioxide, 3, 156, 360
oxidation, 485–6
sulphuric acid, atmospheric,
2–3

summit planes, 187
summits
   accordant, 2
   uniform, 230
sun
   declination shown by
      analemnas, 20
   orbit round *see* **earth**
   see also **solar radiation**
sunspots, 486
   Maunder minimum, 325
supercooling, 486
superimposed drainage, 154,
   486
superimposed ice, 185, 486
superimposed profiles, 486
superposition, law of, 486
supersaturation, 486
supervised classification, 142
supraglacial environment, 176,
   486
supraglacial till, *509*, 510
surf, 486
surf zone, *50*, 486
   on dissipative beaches, 147
   longshore currents in, 309
surface detention, 486
surface run-off *see* **run-off**
surface storage, 138, *263*, 264,
   486–7
surface tension, 75–6, **487**
surges, storm, 477
surging glaciers, 487, 505
   push moraines, 342
   seracs on, 451
survival of the fittest, 197
suspended load, 307, **487**
   and channel classification, 85,
      87
   chemical composition, 447
   concentration: exhaustion
      effects, 198
   in density currents, 133
   and discharge, 269
   levée formation, 305
   organic material in, 447
   particle sizes, *446*, 447, 487
   rating curve, 417
   relative importance, 448
   seston in, 451
   urban areas, 529
   see also **sediment yield**
sustainability, 108, **488, 564**
suturing, 488
swales, 488
swallets, 488
swallow holes, 151–2, 396–7
   avens, 41
   see also **shakeholes**
swamps, 488
   mangrove, *112*, *317*, *318*
swash, 42
   and longshore drift, 309
   overtopping, 366
   overwashing, 367
swash zone, *50*
   berms, 54–5
Swaziland, badlands in, *43*
symbiosis, 488–9
   coevolution leading to, 102
   commensalism a weak form of,
      103

   coral ecosystem, 112, 113
   mutualism, 347
   parasitism a form of, 372
symmetrical folds, *214*
sympatry, 489
   see also **isolation**
synclines, 214, 489
   phacoliths below, 385
synecology, 489
synforms (antiforms), 26
synoptic climatology, 489–90
synoptic meteorology, 490
   see also **cyclones**
**synthetic aperture radar
   (SAR),** 457, 527
synthetic organic compounds
   non-biodegradability, 56
synthetic unit hydrographs, 524
system albedo, 13
systematics (taxonomy), 494
systems, 490–2
   analysis, 357; problems in, 492
   boundaries imprecise, 492
   continuity equations relating to,
      110
   equilibrium, 184–5, 505–6;
      minimal change, 336; stable,
      303–4, 472
   fluxes through, 213, 491
   Gaia concept in terms of, 223
   general system theory of, **226**,
      491
   non-linear, 79, 355
   numerical modelling, 357
   quasi-equilibrium, 407
   reaction time, 418
   relaxation time, 422
   steady state, 475, 506
   thermodynamic equilibrium:
      Van't Hoff's rule, 535
syzygy, 492

tafoni, 493
taiga, 66, 493
takyrs, 493–4
taliks, *381*, 494
talsands, 494
talus, 494
   constant slope, 109
   waning slope, 544–5
tanks, 494
Tansley, A.G., 167
tarns, 494
taxon cycle, 494
taxonomy, 494
tear faults, 494
tectonic activity
   alluvial fans associated with, 17
   Cainozoic, 351
   cave formation rarely associated
      with, 81
   cheiorographic coasts, 88
   and denudation rates, 134
   in mobile belts, 337
   rates, 351
   stillstands, 476
   see also **faults; folds; plate
      tectonics**
tectonic geomorphology, 343
tectonics, 495
Tees, River, *416*
teleconnections, 495

temperate ice, 495
temperature, 495
   accumulated, 2, *see also* **degree
      days**
   and adiabatic processes, 5
   air parcels, 504
   in altithermal phase, 19
   and altitude, 344, 345, 536–7
   ambient, 19
   and anabatic winds, 20
   Antarctic, 22, 23
   in antipleions, 26
   Arctic, 33
   and atmospheric stability, 536–7
   and biological productivity, 59
   on climogram, 97
   cold pole, 102
   comfort zone, 103
   continental climates, 109
   of crystallization, 69
   deforestation effects on, 129
   degree days, 130
   and denudation rates, 135
   deserts, 251
   dewpoint, 140
   discomfort zone, 495–6
   dry bulb, 52, 495–6
   effective, 103, 495–6
   in energy balance models, 224
   equilibrium, 184
   in GCMs, 225
   and heat flow, 250
   and humidity, 257–8, 560
   on hythergraph, 97
   and ice fog formation, 272
   increase *see* global warming;
      greenhouse effect
   inversions, **283**, 537; Antarctic,
      23; Arctic, 33
   Köppen's classification based
      on, 295
   lapse rates, 301, 536–7
   measurement, 495, 503, 504;
      continuous, 504
   Mediterranean climate, 327
   megathermal climates, 327
   melting point, 403
   mesothermal climates, 330
   nuclear winter, 357
   in occlusion, 360
   ozone layer's role in controlling,
      367
   periglacial climate, 380
   permafrost, 381
   and plant growth, 246–7
   potential, 399, 504
   and potential evapotranspiration,
      195
   and precipitation, 412
   radiosonde data on, 411
   ranges: near surface, 333
   reconstruction: by
      dendroecology, 132; from ice
      cores, 241
   sea: and coral reef formation,
      111, 113
   and sea level changes, 190
   sea surface: anomalies, 12; and
      tropical cyclone formation,
      516, 517
   sensible, 450–1
   stratosphere, 478

on tephigrams, 496, *497*
in thermal efficiency index, 501
thermal equator, 501
in thermocline, 503
on thermodynamic diagrams, 504
and thickness, 505
and tree growth, 511
tropics, 517
urban areas, 531–2
Van't Hoff's rule on, 535
and vapour pressure, 535
warm sector, 545
water: and dissolved oxygen concentration, 147
water masses, 547, 552
wet-bulb *see* **wet-bulb temperature**
wind chill, 563
see also **freeze–thaw cycles**
temperature humidity index, 495–6
tensile stress, 479
tensiometers, 496
tepees, 496
tephigrams, 496, *497*, 504
tephra, 496
tephrochronology, 496
terloughs, 496
terminal grades, 496–7
terminal moraines, 497
terminal velocity, 497
terminations, 497–8
termites and termite mounds, 498
terra rossa, 498
terraces
in alluvial fill, 18
goletz, 236
kame, 291
lynchets, 310
outwash, 365
paired, 48
pluvial, 498–9
thalassostatic, 500
terracettes, 499
terranes, 499–500
terrestrial magnetism *see* **magnetism**
territories, animal, 500
Tertiary, 228, 500
Tethys Ocean, 330–1, 500
thalassostatic terraces, 500
thalweg, 500
currents along, 120
thaw lakes, 500–1
orientation, 363
thematic mapper, 526
thermal depressions, 501
thermal efficiency, 501
thermal equator, 501
thermal inertia, soil, 251
thermal infrared linescanners, 501–2, 526, 527
thermal pollution, 502
thermal remanent magnetism, 370
thermal sensors
on unmanned satellites, 527
thermal wind, 502–3
thermistors, 503

thermoclasty (insolation weathering), 279
thermocline, 503
circulation below, 234, 236, *551*
water masses above, 547, *551*
thermocouples, 504, 505
thermodynamic diagrams, 504
atmospheric stability assessed by, 536–7
tephigrams, 496, *497*
thermodynamic equations, 504
in GCMs, 224
thermodynamics
atmospheric, 386–7
laws of, 515, 561–2
thermographs, 504
thermokarst, 380, 504
alases, 12
baydzharakhs, 49
beaded drainage, 51
palsas, 370
pingos, *242*, 262–3, 275, 380, **388**
thaw lakes, 363, 500–1
thermoluminescence, 504
thermopile, 505
thermosphere, 37
thickness, 503, **505**
and cold pole location, 102
Thiessen polygons, 505
tholoids, 505
Thomas, W.L., 345
Thornthwaite, C.W.
climatic classification, 96, 138–9, 482, 501, 546
thresholds, 505–6
complex responses to, 105
shear stress, 104
slopes, 506
velocity, 104, *254*, 505
see also **Hjulstrom curve**
throughfall, *243*, **506**
throughflow, *243*, *254*, *263*, **507**
hydraulic gradient, 260
leaching by, 277
in partial area models, 372
and saturation, 441, 466
see also **interflow; lateral flow**
throw, 507
thrust faults, 203, 348, 366, 507
thufurs, 507
thunderstorms, 507
circulation in, 330
floods caused by, 404
lightning associated with, 305
tidal palaeomorphs, 507
tidal prism, 507
tidal waves (tsunamis), 518–19
tides, 507–9
anamolistic cycle, 20
antipodal bulge effect, 26
bores, 66
currents, *121*, 312, 507
diurnal, 148
energy: and delta morphology, *132*
harmonic analysis, 250
power generation, 508–9
range, 507, *508*; and estuary classification, 188–9
response analysis, 424
till, 509–10

consolidation, 108
diluvialists' explanation for, 143
drumlins in, 156, *341*
fabric analysis, 202, 510–11
terminal grades in, 496–7
see also **moraine**
tillite, 511
tilting siphon rain guge, 412, *413*, 414
timberline (treeline), 511, 514
time
geological, 227, 228
graded, 238
lag in, 297
Tiros-I satellite, 332
tjaele, 511
TOGA, 12
tolerance, 513
tombolos, 511
topography
and local winds, 308
and mesoscale circulation, 329–30
topology, network, 354
toposequences, 511
toppling failures, *302*, **511**
see also **slab failures**
topset beds, 511
toreva blocks, 512
tornadoes, 512
torrents, 513
tors, 98, 184, 244, **511–12**
total dissolved solids, 147–8
tower karst (turmkarst), 513, 520
toxins
biological magnification, 58–9
tracers, 513
dilution, 143, **145**, 267
tractive force, 258–9, **513**, *513*
du Boys equation based on, 157
trade winds, 513
antitrades above, 26
convergence at ITCZ, 282
easterly waves in, 163
and Hadley cells, 248
trafficability, 513
trailing-edge coasts, *101*
transform boundaries, 390
translational slides, 323, 459, **513**
transmissivity, 258, 278
transpiration, 194, 247, 352, 399, **513–14**
forests, 216–17
phreatophytes, 386
sclerophyllous plants, 442
and soil moisture, 461
see also **evapotranspiration**
transport *see* **sediment transport**
transportational slopes, 463, 514
transverse dunes, *158*, 514
transverse ribs, 365, 514
travertine, 519
tree-ring studies, 131–2
treeline (timberline), 511, 514
trees
acid precipitation effects on, *2*, *3*
in agroforestry, 9
altitudinal and latitudinal limits, 511, 514

trees (*continued*)
banding, 71–2
decline, 216
die-back, 141
wind effects on, 296, 564
see also **forests**
trellis drainage, *155*
trenches, arroyo, 35
Triassic, 228
triaxial apparatus, 478, 514
Troll, C., 301
Trombe's curves, 515
trophic levels, 515
biological magnification, 58–9
carnivores within, 77
energy flow through, 165, 215,
515; by photosynthesis, 386
stability, 471
tropical air masses, 11
tropical cyclones, 348, **516–17**
modification, 555–6
storm surge generation, 477
tropical forests, 183–4, 517
biomass, 60
conservation, 286
deforestation, 129
distribution, *61*
litter in, 307
monsoon forests, 340
productivity, 60, 184, 353
respiration rates, 59
see also **mangrove swamps**
tropical meteorology, 517–18
Tropical Ocean Global
   Atmosphere Programme,
   12
tropicoid palaeo-earth, 96
tropopause, 518
troposphere, 518
ageostrophic motion, 8–9
antitrades in, 26
Arctic, 33
blocking in, 64
boundary with stratosphere, 518
clouds in *see* **clouds**
cold pole, 102
jet streams in, 290
lapse rates, 301
ozone concentration, 3
radiation interactions in, *410*
stationary waves in, 64
subsidence through: and
anticyclone origins, 26
temperature, 478; inversion, 23
vertical motion, 536
water vapour in, 37
trottoirs, 114, 518
trough blowouts, 64–5
troughs, 518
equatorial, 184; *see also*
   intertropical convergence
   zone
subantarctic, 23
see also **cyclones**
truncated spurs, 518
tsunamis, 518–19
tufa, *392*, **519**
tuff, 519
tundra, 519
biomass, 60
distribution, *61*
net primary productivity, 353

see also **periglacial areas;**
   permafrost
tundra polygons, 380
tunnel valleys, 520
tunnelling, 520
turbidity currents, 483, **520**
turbulence, 520
atmospheric, 334, 520; urban
   areas, 530
boundary layer, 68–9, 520
in rapids, 416
see also **eddies**
turbulent flow, 212, 520
Reynolds number, 425, 520
velocity: boundary conditions,
   67–8
Turkey, beach rock in, *51*
turmkarst (tower karst), 513,
   520
typhoons, 477, **516–17**

USSR
channel classification, 87
meteorological satellites, 332
weather modification, 555
ubac, 521
ultisols, 452
ultrasonic gauging, 145, 267
ultraviolet radiation, 164
absorption by ozone layer, 367,
   368
and plant growth, 246
unconformities, 521
angular, 21, *521*
disconformities, 146, *521*
under-drainage, 205
underfit (misfit) streams, 76,
   521–2, 534
underplating, 522
undertow *see* **rip currents**
undertow springs, 470
unequal slopes, law of, 522
uniclinal shifting, 522
uniclinal strata, 522
uniform slopes, law of, 522
uniform steady flow, 212, **522**
Chézy equation's assumption of,
   89, 522
and frictional resistance, 218–19
tractive force in, 513
uniformitarianism, 314, **522–4**
catastrophism compared with,
   79, 523
in dry valley explanations, 157
inferences on Quaternary
   ecology based on, 407
and neocatastrophism, 350
uniqueness, landscape, 430
unit hydrograph, 524, 525
extended by S-curve, 442
unit response graph, 525
United Kingdom
drought, 156
minimum acceptable flow
   requirements, 336
pollution control, 396
weather forecasting, 490
United States
applied geomorphology, 27
applied meteorology, 29, 30
Department of Agriculture: soil
   classification, 452–3

environmental impact
   assessment, 180
environmental management, 182
expansive soils: damage caused
   by, 199
meteorological satellites, 332
pluvial terraces, 499
soil erosion, 461
tornadoes, 512
Weather Bureau: satellite
   nephanalysis, 351–2
weather modification, 555–6
universal soil loss equation, 392
unloading, 525
unmanned earth resources
   satellites, 525–8
unstable channels, 528
unstable equilibrium, 528
unsteady flow, 528
Upper Teesdale, 420
upslope fog, 214
upwelling, 528
and atmospheric circulation,
   518
el Niño effect, 172
and net primary productivity,
   353
ural-type glaciers, 528
uranium
dating using, 207, 528
radon gas from, 411–12
urban areas
boundary layer (UBL), 530,
   531, 532
dry deposition, 3
flooding, 300, 529, *530*
hydrological effects, 300, 434,
   528–30
meteorology, 530–3
occult deposition, 361
pollution, 486, 529
sediment yield, 300
wind shadows, 563
urstromtaler, 533
uvalas, 533

V-notch (sharp-crested) weirs,
   453–4, 534, *558*
vadose zone, 242–3
percolation through, 379
valley fills, 18
valley sand (talsand), 494
valleys
adret side, 5
asymmetry, 36; explanations for,
   274
blind, 64
bogs in, 376
bulges in, 534
cross-profiles, 116–17
drowned: fiards, 205; fiords,
   206; rias, 426; valloni, 535
dry, 156–7, 327, 439
glacial: overdeepening, 365–6;
   paternoster lakes in, 374;
   riegels in, 426
hanging, 249
interfluves between, 280
meandering, 327, 507, 521, **534**
rift, 426, *427*
tidal palaeomorphic, 507
tunnel, 520

U-shaped: processes forming,
184
ubac side, 521
winds, 329–30, 344–5, **534–5**
see also **gorges**
vallon de gelivation, 535
valloni, 535
values, landscape, 301
**Vancouver, urban boundary
layer in,** *531*
Van't Hoff's rule, 535
vapour pressure, 535
and potential evapotranspiration,
195
and relative humidity, 257
variance, analysis of, 275
varves, 535
vasques, 535
vauclusian springs, 469
veering winds, 561
vegetation
albedo, 13
associations, 35
biomass, 60
in biostatic periods, 62
changes: reconstruction from
pollen analysis, 394–5
and climatic classification, 96,
295
climax *see* **climax vegetation**
contributing areas determined
by amount of, 110–11
decomposition, 307
and denudation rates, 135
dune, 100, *159*, 371; and
blowouts, 64
equilibrium with environment
see **climax vegetation**
estuarine, *188*
evapotranspiration *see*
evapotranspiration
High Arctic islands, 33
human impact on, 129–30, 398
and infiltration, 277
interception by, 169, *171*, 216,
243, **280**
interglacial, 280
and landforms, 387
limiting angles for, 305–6
mountain, 344, 345
natural, 349
precipitation below, 506
regeneration complex, 421
removal: by fire, 206;
desertification caused by, 140
ruderal, 433
and run-off, 434
and sediment yield, 299–300
on slopes, 248, 305–6
stem flow down, 475
succession *see* **succession**
tundra, 519
and water balance, 546
see also **plants** *and particular
types, e.g.* **forests**
vegetation banding, 71–2
vegetative cycling
and soil nutrient status, 358
velocity
airflow, 562 and angular
momentum, 21
and Bernouilli effect, 55

boundary layer, 68–9
channel flow: and cavitation, 81;
and competence, 104; and
drainage density, 154; and
hydraulic jump, 260; and
pool-riffle formation, 397
in Darcy's law, 125
in discharge measurement, 145
earth, 114
entrainment, 116, 254, 505
and Froude number, 222
geostrophic wind, 229
glaciers, 105–6, 487
ice shelves, 273
kinematic waves, 294
laminar flow, 297
mass movements, 322–3
measurement, 267, **536**; by
pitometer, 389; using current
meters, 118
navigable channels, 349, 350
overland flow, 366
profile, 535–6
rapids, 416
rills, 428
rip currents, 428
and roughness, 53, 212
sturzstrom, 482
terminal, 497
threshold, **254**, 505
tidal currents, 121
turbulent flow, 520; boundary
conditions, 67–8
uniform steady flow, **212**, 522
and vorticity, 540–1
see also **hydraulic conductivity**
velocity area method, 144–5,
417, **535**
ventifacts, 536
see also **dreikanters**
vertical aerial photography, 7
vertical motion *see under*
atmosphere
vertical stability *see under*
atmosphere
vertisols, 198–9, 452
vesicular structure, 537
vesuvian eruptions, 540, *541*
vicariance biogeography, 537–9
see also **cladistics; endemism**
vidicom cameras, 526
Vigil Network, 340, 423
virgation, 539
viruses, 384
viscosity, 425, **539**
in laminar flow, 298
and terminal velocity, 497
visibility reduction *see* **fog;
haze; mist; smog**
**visible infrared spin scan
radiometer (VISSR),** 539
void ratio, 540
angle of internal shearing
resistance dependent on, 21
reduced by compaction, 104
volcanic eruptions, 187
ash from: dating based on, 496
ice core evidence on, 4
magnitude: dust veil index, 161
nuée ardente from, 357
pyroclastic rocks from, 406
tephra ejected by, 496

types, 540, *541*; fissure, 207;
Hawaiian, 250, 540; plinian,
392, 540, *541*
volcanic rocks, 539
andosols on, 20
extrusive, 201
fission track dating, 207
pumice, 406
scoria, 442
solute concentration of rivers
on, 465
volcanic sandstone, 438
volcanoes, *539, **540–1**, 542*
cones: adventive, 5; lahars on,
297; tholoids, 505
craters, 115; lakes in, 75, 311;
maars, 311
dormant, 153
eruption *see* **volcanic eruptions**
fumaroles, 222
island arcs associated with, 285
lava from *see* **lava**
mud, 345
pali ridges on, 370
solfataras, 464
strato-volcanoes, 478
see also **vulcanism**
volumetric gauging, 144
Von Bertalanffy, L., 226, 491
vortex streets, 329
vorticity, 540–2
Rossby waves, 432
tropical cyclones, 516
vughs, 281, 542
vulcanian eruptions, 540, *541*
vulcanism, 542
diatremes caused by, 141
mid-ocean ridges, 334
tsunamis generated by, 518
see also **earthquakes; volcanic
eruptions; volcanoes**

wackes, 438
wadis, 543
wakes, circulations in, 328
Waldsterben and *Waldschaden,*
216
Walker circulation, 312, 517–18
Wallace, Alfred Russel
on island biogeography, 285
on natural selection, 127, 196–7
zoogeographical regions, 203–4,
543–4, 567–8
Wallacea, 543
Wallace's line, 543
wandering rivers, 70
waning slopes, 544–5
warm fronts, 201, 219, 360, **545**
warm ice *see* **temperate ice**
warm occlusions, 360
warm sector, 545
warping, 545
wash load, 487
washboard (De Geer) **moraines,**
130, 545
washover fans, 367
waste
disposal in landfill sites, 300
nuclear, 356–7
pollution by, 395
ruderal vegetation on, 433
waste water, treatment of, 63

water
aggresivity, 9
artesian, 35
brackish, 69
caballing, 74
capillary forces, 75–6
carbon dioxide in, 77
chemical composition, 147,
  283–4, 424, 464–6, 465–6;
  urban areas, 529; *see also*
  dissolved load
connate, 108
eutrophication, 191, 359
flow *see under* **channels**
hydraulic force (tractive force),
  258–9, 513
in hydrosphere, 266–7
juvenile, 190, 290
meteoric, 331
in mists, 337
nuclear waste disposal in, 356
oxygen in, 55, 56, 146
passage through rocks *see*
  impermeability; permeability
and plant growth, 247, 352
pollution, 171, **395**; critical load,
  116; marine, 320–1;
  measurement, 55; thermal,
  502; urban, 529; *see also*
  eutrophication
pressure melting point, 403
salinity problems, 435
siltation in, 458
specific conductance, **106**, 466
storage *see* **storage**
temperature, 502; and dissolved
  oxygen concentration, 147
waste: treatment, 63
see also **ice; lakes; oceans;
  rivers,** *etc.*
water balance, 545–7
in antecedent moisture
  estimation, 24
climatic classification based on,
  96
in discharge estimation, 48, 49
and evaporation measurement,
  193
and interbasin transfers, 280
lysimeters based on, 310
in Muskingum method, 347
urban areas, 528–9
water level, measurement of,
  208–9, 473–4, 475–6
water masses, 547–52
density differences, 234
water resources
international research, 281
mapping, 261
planning and management, 261,
  428, 546–7
water supply, 261
and catchment control, 80
interbasin transfers, 279–80
from wells, 560
water table, 547
in bogs, 376
cone of depression, 107
draw down, 155
hydraulic gradient as, 260
and intermittent streams, 280–1

lowering: and cave formation,
  81; by land drainage, 298
mapping, 185, 261
mobility, 243
perched, 379
and phreatic divides, 386
and pore water pressure, 398
and specific yield, 468
and springs, 280, 469
wetlands, 560
water vapour, 36, 37, 263, 267
and altitude, 344
condensation *see* **condensation**
and ice fog formation, 272
plants' loss *see* **transpiration**
pressure, 535
sublimation, 270, 482
see also **evaporation;
  evapotranspiration;
  humidity**
water year, 546, 553
waterfalls, 553
waterlogging
and peat formation, 375
watersheds, 553
underground (phreatic divides),
  386
waterspouts, 553
watten, 553
wave-cut platforms *see* **shore
  platforms**
wave-normal currents, 118
waves, 553
amplitude: and notches, 356
associated with tidal currents,
  121
atmospheric *see under*
  atmosphere
backwash, 42
and beach profiles, 49, *50*
and beach zones, 49
coastal bars at breakpoint of, 44
coastal classification based on,
  100
constructive, 419
and delta morphology, *132*
on dissipative beaches, 147
edge, 169, *170*; and
  overwashing, 367; rip currents
  associated with, 120
fetch, 205
gravity, 239–40
high energy window, 253
kelvin, 294
kinematic, 294
and longshore drift, 309
and nearshore currents, 118
reflection, 419
refraction, 419–20
and shore platform formation,
  455
surfzone, 486
tidal: bores, 66
see also **tsunamis**
waxing slopes, *109*, 554
wealth, transfer of, 315
weather, 554
air motion central to, 561
extreme: storm surges caused
  by, 477
forecasting, 29, 37–8, 220; and
  air velocity, 563; computer-

based, 440, 490, 554;
  improvements, 31; satellite
  data for, 440; synoptic charts
  for, 490; temperature
  humidity index in, 496;
  tephigrams for, 496
human activities influenced by,
  28- 31, 61–2; agriculture,
  9–10, 28, 29
modification, 555–6
perturbations, 383
stationary anomalies, 64
types: cyclonic, 122, 123, 200;
  synoptic approach to
  classifying, 489–90, 556
see also **climate** *and particular
  weather elements*
weather services, 490, 554–5
cost-effectiveness, 30
weathered rock
corestones in, 113
woolsack in, 564
weathering, 556–7
biological: chelation caused by,
  88–9
chemical *see* **chemical
  weathering**
deep, 128
eluvium produced by, 173
and ferrallitization, 205
Goldich series, *69*
hillslopes, *463*; form controlled
  by, 557
in situ, 71
le Chatelier principle applied to,
  303–4
measurement: by micro-erosion
  meter, 333–4
mechanical *see* **mechanical
  weathering**
in microcracks, 333
organic, 363
and pediment formation, 377
rates, 556–7
and soil nutrient status, 358
tafoni formed by, 493
and tor formation, 511
weathering front, 557
weathering index, 557
weathering rinds, 557
Weddell Sea, 234, *235*
wedge failures, *302,* 557
weeds, 384
Wegener, A., 109
**weighted mean percentage silt
  clay,** 557–8
weirs, 558–60
discharge measurement at, 111,
  144, 267, **453–4**
types: broad-crested, 71, *559*;
  sharp-crested, 453–4, 534, *558*
Weismann, August, 127
wells, 560
cones of depression around, 107
piezometric surface, 388
pumping from: draw down
  caused by, 155
Werner, A.G., 352
westerlies, 560
long waves in, 22–3
wet-bulb temperatures, 560
and absolute humidity, 258, 560

on climograph, 97
and comfort zone, 103
in temperature humidity index, 495–6
and vapour pressure, 535
wetlands, 560
see also **bogs; fens; marshes**
wetted perimeter, 560
wetting front, 276, 560
whirlwinds, 560–1
white alkali soils, 464
White River, *326*
Whittaker, R.H., 60, 97
width–depth ratio, 561
Wien displacement law, 411
wilderness, 561
Williams, G.P., 82
Wilson, E.O., 285–6
wind, 225, **561–3**
in Antarctic, 23–4
in Arctic, 33
deflation by, 130, 160, 371
direction: backing and veering, 561; and Coriolis force, 113–14, 562–3; dominant, 153; on hodograph, 254- 5; prevailing, 403; resultant, 424; on wind rose, 563
distribution in boundary layer, 171–2
in doldrums, 151
drift potential, 155
and dry deposition, 156
duststorm generation, 248
forecasting, 563
hodographs, 254–5
landforms associated with: pans, 371; sastrugi, 439; stone pavements, 477; ventifacts, 536; *see also* **dunes**
on leeward slopes, 328
local, **308,** 327
loess deposition, 308
measurement, 20, 75, 561
microclimatic effects, 333
ocean currents driven by, 120

sand movement, 155, 157, 437, 438; *see also* **dunes**
soil erosion, 461
speed, 561; Beaufort scale, 52; calm, 75; and deforestation, 129; and evaporation, 191, 193, 195; gustiness, 244; hodograph of, 254–5; on wind rose, 563
on squall lines, 470
stress, 11
in tornadoes, 512
in tropics, 516, 517–18
turbulence, 520, 530
types: anabatic, **20,** 344, 534–5; anticyclonic, 25; antitrade, 26; antitriptic, 562, 563; auge, 39; bise, 64; bora, **65–6,** 293; chinook, **89,** 293; cyclonic, 122; cyclostrophic, 123; etesian, 189; Eulerian, 562, 563; föhn, 54, **214,** 293; geostrophic, 171, **229,** 502, 562; glacier, 293; gradient, **238,** 562; harmattan, 250; helm, 253; khamsin, 294; mistral, 293, **337;** monsoon, 340, 518; mountain, 344- 5; slope *see* **anabatic flows; katabatic flows; thermal,** 502–3; **trade, 26,** 163, 248, 282, 513; tropospheric, 518; valley, 329–30, 344–5, **534–5;** westerlies, 22–3, 560; whirlwinds, 560–1
in urban areas, 530–1
and vegetation, 247, 296, 564
waves produced by, 553
wind chill, 563
wind gaps, *76*
wind rose, 563
wind shadow, 563
windthrow, 564
winter talus ridges, 404
WOCE, 12
Wolman, M.G., 299

Wood, A., 109
woodland
caatinga, 74
canopies, 75
savannah, 441, *442*
see also **forests; trees**
woolsack, 564
World Climate Applications Programme, 30
World conservation strategy, 564
World Meteorological Assocation, 265
World Ocean Circulation Experiment, 12
wrench faults, 564
Würm glaciation, 190, *378*

xenoliths, 565
xerophytes, 247, 565
xeroseres, 403, 451, 565
xerosphere, 565
see also **arid zone; deserts**
xerothermic index, 565

yardangs, 566
yazoos, 566
yields sustainable, 488
Younger Dryas, 566
Young's modulus of elasticity, 337–8

zeugen, 567
zibars, 567
zonal circulation, 567
zonal index, 90, 276
zonal soils, 567
zonation, 345, 567
zoocoenosis, 56
zoogeographical regions, 203–4, 543–4, 567–8
zoogeography, 567–8
zooplankton, 389, 439